ATLAS FOR
COMPUTING
MATHEMATICAL
FUNCTIONS

ATLAS FOR COMPUTING MATHEMATICAL FUNCTIONS

An Illustrated Guide for Practitioners

With Programs in C and Mathematica

WILLIAM J. THOMPSON

A Wiley-Interscience Publication
JOHN WILEY & SONS, INC.
New York • Chichester • Weinheim • Brisbane • Singapore • Toronto

This text is printed on acid-free paper.

Library of Congress Cataloging in Publication Data:

Thompson, William J. (William Jackson), 1939–
    Atlas for computing mathematical functions : an illustrated
guidebook for practitioners, with programs in C and Mathematica /
William J. Thompson.
      p. cm.
    "A Wiley-Interscience publication."
    Includes bibliographical references and index.
    ISBN 0-471-00260-7 (cloth : alk. paper)
    1. Functions—Computer programs.   2. Science—Mathematics—
Computer programs.   3. C (Computer program language)
4. Mathematica (Computer program language)   I. Title.
QA331.T385   1997
511.3'3—dc20   96-32557
                                                CIP

Printed in the United States of America

10 9 8 7 6 5 4 3 2 1

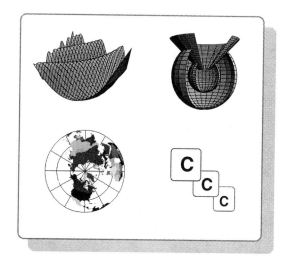

CONTENTS

PART II. THE COMPUTER INTERFACE

21 The C Driver Programs 797

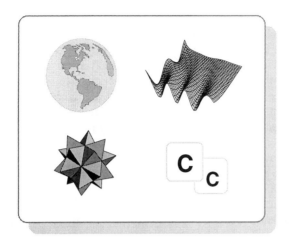

PREFACE

When you plan to travel, rather than merely reading about places you would like to visit, I'm sure that you also consult maps and atlases to learn more about these places. When you use mathematical functions to solve practical or research problems, you probably often wished for a pictorial guide and computer programs, in addition to the usual text, formulas, and tables.

Four themes underlie my approach to the *Atlas for Computing Mathematical Functions*. The *atlas* theme comprises more than 700 pictures to guide your thinking about the functions. The *computing* aspect consists of programs provided in the C language and in the *Mathematica* system for doing mathematics by computer. The *mathematical* viewpoint summarizes and references key formulas for each function, especially formulas needed for computing it accurately and efficiently. The *functions* delimit the field—specifically to numerical functions from mathematics that are used frequently by practitioners in areas such as applied mathematics, statistics, physics and chemistry, computer science, and engineering. More than 150 of these so-called special functions are included in the *Atlas*.

The rationale for such a reference work is that printed tables of functions, as in Abramowitz and Stegun's *Handbook of Mathematical Functions* (100 graphics in 1000 pages), nowadays serve a very limited purpose—preliminary exploration and checking the correctness of computer programs for calculating the functions. The need to visualize what is to be computed is increasingly important, because the variety of mathematical functions needed in applications is constantly increasing. As a beginning college student I was much influenced in my development by the classic book of Jahnke and Emde, *Tables of Functions*, which has 200 graphics in 400 pages. The carefully hand-drawn three-dimensional graphics inspired me to understand the mathematics. On the other hand, only when I began research would I venture into the forest of formulas in *Higher Transcendental Functions* edited by Erdélyi and coworkers. In the *Atlas* there are more than 700 graphics, most prepared using *Mathematica*, with nearly one third of the material being pictorial.

Few reference books of mathematics have addressed at a satisfactory level the need for visualization. Those who prepare mathematical aids to computation still emphasize either analytical aspects (formulas and theorems) or computational aspects (algorithms and source programs). Part I of the *Atlas* summarizes and references for each function the analysis background, displays the function in several 3D and 2D figures, summarizes numerical techniques needed for stable, efficient, and accurate computation (typically to 10 digits), and provides the C function code, which is on the CD-ROM provided. All the C function computations have been checked against those from *Mathematica* and from the Fortran-90 version of the *Atlas*, which is in preparation.

I designed the *Atlas* to be used interactively, so that you can explore the functions by running the *Mathematica* notebooks that produce the function views. Annotated listings of the notebooks are in Part II, as are annotated driver programs for checking the C functions. Extensive cross-referencing and a complete index of functions are provided, both in the printed book and on the CD-ROM, which has complete source programs.

Like Atlas of classical mythology, the author of a work such as this bears the major burden, but several people have helped with preparation of this book. Clarick Talmadge and George Rawitscher provided the major scientific reviews. Editorial staff at Wiley-Interscience—John Falcone, Greg Franklin, and Lisa Van Horn —have been most helpful. Finally, the staff at Zanana Press prepared the camera-ready copy expeditiously.

WILLIAM J. THOMPSON

Chapel Hill, North Carolina
November 1996

INTRODUCTION

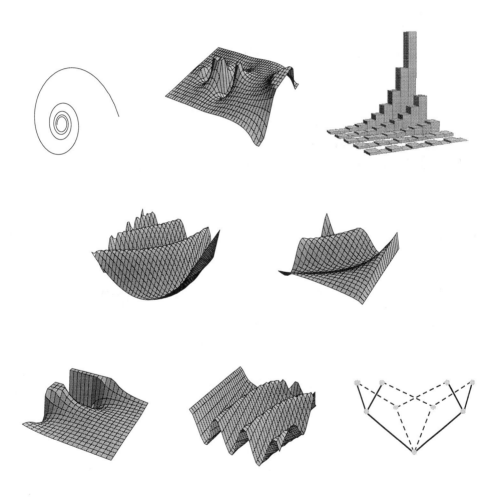

INTRODUCTION

Motifs on the previous page:

The Sici spiral parametrized by $Si(t)$ and $Ci(t)$. (Exponential Integrals and Related Functions; Figure 5.2.4)

Gamma function in the complex plane; x in [−4, 4] along the front, y in [−2, 2] on the side. (Gamma and Beta Functions; below Figure 6.1.1)

Fibonacci polynomials, $F_n(x)$, as a 3D bar chart drawn for $x = 0$ (0.25) 1.0 on the side and $n = 0$ (1) 10 at the front. (Combinatorial Functions; Figure 7.4.1)

Chebyshev polynomial, $T_n(x)$, surface with x in [−1, 1] at front and n in [2, 8] at side. (Orthogonal Polynomials; Figure 11.2.1)

Surface of the Legendre function $P_n^i(x)$ for x in [−1, 1] along the front and n in [1, 5] on the side. (Legendre Functions; Figure 12.2.2)

Irregular Kelvin function $ker_n x$ as a 3D surface; x in [0.5, 4] at front, n in [0, 5] at side. (Bessel Functions; Figure 14.3.6)

Surface of the Jacobi function $cn(u \mid m)$; u in [0, 20] at front, m in [0, 0.99] at side. (Elliptic Integrals and Elliptic Functions; Figure 17.3.5)

Tree diagram for the 9-j angular momentum recoupling coefficient. (Miscellaneous Functions for Science and Engineering; Figure 19.7.3)

INTRODUCTION

The Introduction describes first how you can use the *Atlas* in its print version to look at graphics and to read descriptions of functions that you want to compute. The Computer Interface then describes how you can use the notebooks for the *Mathematica* system and the source programs for the C functions, all of which are on the CD-ROM, to produce your own graphics and numerical values of the functions.

THE ATLAS OF FUNCTIONS

What This Atlas Contains

The *Atlas* comprises descriptions from analytical, visual, and computational viewpoints of more than 150 of the special functions that are useful in applied mathematics, the natural sciences, engineering, and statistics.

Analytical materials emphasize properties of the functions that are most useful for computing them by the methods developed for the C functions in the *Atlas*. Analytical results are referenced to standard mathematics books and to specialized monographs and articles.

Visual materials comprise more than 700 graphics of the functions from many viewpoints, chosen to help you understand general properties of the functions and to clarify the algorithms used when computing them. By using a computer to run the indicated *Mathematica* notebooks, you can impose your own point of view on these graphics, and you can explore particular regions of the functions in more detail.

Computational materials include complete C source programs for every function that is described, as well as parts of the notebooks for generating test values for the functions. Most of the functions are computed in double precision C to an accuracy of at least 10 decimal digits. Tables of sample test values are provided for each function.

How to Use the Atlas

One way to use the *Atlas* is like a geographical or astronomical atlas—to look at the pictures in order to learn about the countries of the world or the stars in the sky. Here the mathematical functions are the countries or stars. You can, however, explore the functions in more detail on your own by running the *Mathematica* notebooks. To indicate this, we put in the margin the marker *M* in a shadow box, as shown at left, except in figure captions, where the explicit notebook and cell to use are given.

If you want to compute numerically a few values of the function, the last section of a *Mathematica* notebook cell will usually generate numerical values of the function in that cell. If you need many values of the function, such as within another C program, it is much more efficient to use the C version of the function. There are C driver programs for every function so that you can test your implementation of the function. Computer materials—all of which are on the CD-ROM that comes with the *Atlas*—are described in more detail in the next section, The Computer Interface.

Here is a chart summarizing how to use the *Atlas*:

	Analysis	Visualization	Computing
Function descriptions: Part I, Chapters 4–19	✔	✔	✔
Mathematica notebooks: Chapter 20 and CD-ROM *M*	✔	✔	✔
C functions: Part I and CD-ROM			✔
C drivers: Chapter 21 and CD-ROM C			✔

There is one more indicator to help you use the *Atlas*. This is the warning sign shown in the margin. Wherever you see this, there is a rocky road ahead—inconsistent definitions of functions, tricky mathematics (branch cuts and such), and perhaps difficult algorithms or coding.

About the Production of the Atlas

The *Atlas for Computing Mathematical Functions* was prepared completely by computer, but with considerable human intervention. The text and layout were prepared using Word by Microsoft, the equation processor was MathType by Design Science, the function graphics used *Mathematica* by Wolfram Research. All figures were composed using ClarisDraw by Claris, and the C programs were developed using C++ by Symantec. The total of these components needs about 500 Mbyte of disk storage, with the *Atlas* occupying about 200 Mbyte of this.

THE COMPUTER INTERFACE

In this section of the Introduction we describe how the *Atlas for Computing Mathematical Functions* interfaces to computers. The interface occurs at three levels; the on-line guide to locate a function in the text or in the machine-readable version, the *Mathematica* notebooks to visualize the function, and the C functions and their driver programs for computing the function numerically. Finally, we summarize for Fortran and Pascal programmers how programming constructs in these languages relate to those in C as used in the *Atlas*.

What the CD-ROM Contains

The CD-ROM that comes with the *Atlas* contains machine-readable forms of all programs in the *Atlas*. These programs are of two kinds. First, there are the *Mathematica* notebooks for the visualizations, for the algebraic generation of some functions, and for calculating numerical test values; details about the structure and use of the notebooks is given below in the section Exploring Functions with *Mathematica* and in the first three sections of Chapter 20. Second, there are the source programs for the functions in C, as described below in The C Functions: No Assembly Required. The C driver programs—including annotated listings—are described in detail in Chapter 21.

The ReadMe file on the CD-ROM gives you up-to-date instructions on how to install the *Atlas* programs for several computer systems.

How to Locate a Function

There are several ways to locate a function in the *Atlas*. The first way is to use the table of contents, in which the functions are organized by the major families in which they occur. For example, the Legendre polynomials belong to the class of orthogonal polynomials, so will be found in Chapter 12. The Legendre functions of the second kind, Q_n^m in standard notation, are neither polynomials nor are they orthogonal. Chapter 12 is therefore the appropriate place to find them, and there they are in Section 12.2.3. In the same chapter you will find other interesting Legendre functions; the familiar spherical Legendre function P_n^m, the toroidal Legendre functions, and those appropriate to conical coordinates. Nine of the functions are not naturally part of any major family, so there is the catch-all Chapter 19 for Miscellaneous Functions.

Another way to locate a function is through the Index of Function Notations, which matches the symbols to the names, such as $P_n(x)$ to the Legendre polynomial. Each index entry indicates the section of the *Atlas* in which the function is defined.

The third way to locate a function is through its name in the regular Index of Subjects and Authors. The most common name is in the main entry, with less common names indexed to the common name. For example, the Basset function is another name for the regular hyperbolic Bessel function in Section 14.2.3.

Exploring Functions with *Mathematica*

Almost all of the graphics in the *Atlas* have been produced by using *Mathematica*. You can make these graphics yourself by running the notebooks whose annotated listings are given in Chapter 20. We give two examples—first the gamma function (Section 6.1.1), then the Fibonacci polynomials and numbers.

```
(* GammaFunction cell *)

(* Gamma Function in 3D *)
GammaPlot[ReOrIm_] := Plot3D[
 ReOrIm[Gamma[x+I*y]],{x,-4,4},
 {y,0.0001,1},PlotPoints->{80,20},
 Ticks->{Automatic,None,None},
 AxesEdge->{{-1,-1},{1,-1},{-1,-1}},
 PlotRange->{-6,6},
 ViewPoint->{0.0,-3.0,0.5},
 DisplayFunction->Identity]

RealPlot = Show[GammaPlot[Re],
 DisplayFunction->$DisplayFunction];

ImaginaryPlot = Show[GammaPlot[Im],
 DisplayFunction->$DisplayFunction];
```

```
(* Gamma Function contours; log scale *)
GammaAbsContour= ContourPlot[
 Log[Abs[Gamma[x+I*y]]],
 {x,-4,4},{y,-2,2},PlotRange->{0,3},
 PlotPoints->{120,60},
 AspectRatio->Automatic,
 Contours->15,ContourShading->False,
 ContourSmoothing->Automatic];
```

```
(* Gamma Function in 2D *)
(* Real-axis plot; poles at integers <0 *)
GammaXXPlotRe = Plot[Re[Gamma[x]],
 {x,-4,4},PlotPoints->80,
 Ticks->{Automatic,None},
 AxesLabel->{"x",""}];
```

```
(* Real-axis plot; reciprocal of gamma *)
RecGammaXXPlotRe = Plot[1/Re[Gamma[x]],
 {x,-4,4},PlotPoints->80,
 Ticks->{Automatic,None},
 AxesLabel->{"x",""},
 PlotStyle->Dashing[{0.02,0.02}]];
```

```
GammaYYPlotRe = Plot[Re[Gamma[I*y]],
 {y,-2,2},PlotPoints->60,Ticks->Automatic,
 AxesLabel->{"y",""}];
```

```
GammaYYPlotIm = Plot[Im[Gamma[I*y]],
 {y,-2,2},PlotPoints->60,Ticks->Automatic,
 AxesLabel->{"y",""}];
```

```
(* Gamma Function: Numerical *)
GammaTable1 = Table[{x,N[Gamma[x],10]},
 {x,0.5,2,0.5}]
GammaTable2 = Table[{x,N[Gamma[x],10]},
 {x,5,20,5}]
```

$\{\{0.5, 1.772453851\}, \{1., 1.\},$
$\{1.5, 0.8862269255\}, \{2., 1.\}\}$
$\{\{5, 24.\}, \{10, 362880.\},$
$\{15, 8.7178291210^{10}\},$
$\{20, 1.21645100410^{17}\}\}$

We often use transections (slices) through a surface to get a more detailed view of a function, as you see above for the curves of $\Gamma(x)$ and $1/\Gamma(x)$ below the contour plot and for the real and imaginary parts of $\Gamma(iy)$ below these curves. The annotated version of this cell is in Section 20.6.

As our second example of exploring with *Mathematica*, consider the Fibonacci polynomials and numbers in Section 7.4.1. Compute them by running cell Fibonacci of notebook Combinatorial, which is listed and annotated in Section 20.7 and is available in machine-readable form off the CD-ROM.

```
(* Fibonacci cell *)

<<Graphics`Graphics`
<<Graphics`Graphics3D`

(* Fibonacci Numbers *)

(* Definition by recursion *)
FibNum[n_] := FibNum[n-1]+FibNum[n-2]
FibNum[1] = 1;    FibNum[2] = 1;
FibNumbers = Table[{n,N[FibNum[n]]},
 {n,1,10}]
```

$\{\{1, 1.\}, \{2, 1.\}, \{3, 2.\}, \{4, 3.\}, \{5, 5.\}, \{6, 8.\},$
$\{7, 13.\}, \{8, 21.\}, \{9, 34.\}, \{10, 55.\}\}$

```
(* Fibonacci Numbers: Log-base-10 Plot *)
nmax = 9;
LogFibNum = Table[Log[10,N[FibNum[n]]],
  {n,1,nmax}];
FibNumChart = BarChart[LogFibNum,
  BarStyle->{GrayLevel[0.8]},
  BarEdges->False,
  AxesLabel->{"n","log10[F(n)]"}];
```

log10[F(n)]

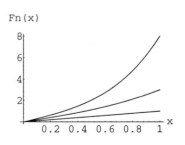

```
(* Fibonacci Polynomials *)
(* Definition by recursion *)
FibPoly[n_,x_] :=
  x*FibPoly[n-1,x]+FibPoly[n-2,x]
FibPoly[1,x_] = 1;  FibPoly[2,x_] = x;

(*  Fibonacci Polynomials: Algebraic *)
nmax = 6
Clear[x];
FibPolyAlg =
Table[Expand[FibPoly[n,x]],{n,1,nmax}]

(*  Fibonacci Polynomials: Numerical *)
FibTable[x_,nmax_] := Table[
  N[FibPoly[n,x],10],{n,1,nmax}]

nmax = 10;
FibTable000 = FibTable[0.00,nmax]
FibTable025 = FibTable[0.25,nmax]
FibTable050 = FibTable[0.50,nmax]
FibTable075 = FibTable[0.75,nmax]
FibTable100 = FibTable[1.00,nmax]
```

6

$$\{1, x, 1 + x^2, 2x + x^3, 1 + 3x^2 + x^4,$$
$$3x + 4x^3 + x^5\}$$

$\{1., 0., 1., 0., 1., 0., 1., 0., 1., 0.\}$

$\{1., 0.25, 1.0625, 0.515625, 1.19140625,$
$0.8134765625, 1.394775391, 1.16217041,$
$1.685317993, 1.583499908\}$

$\{1., 0.5, 1.25, 1.125, 1.8125, 2.03125, 2.828125,$
$3.4453125, 4.55078125, 5.720703125\}$

$\{1., 0.75, 1.5625, 1.921875, 3.00390625,$
$4.174804688, 6.135009766, 8.776062012,$
$12.71705627, 18.31385422\}$

$\{1., 1., 2., 3., 5., 8., 13., 21., 34., 55.\}$

```
(*  Fibonacci Polynomials: 3D Bar Charts *)

BarChart3D[{FibTable100,FibTable075,FibTable050,
     FibTable025,FibTable000},
     XSpacing->0.5,Boxed->False, Axes->False,
     SolidBarEdgeStyle->GrayLevel[0.5],
     ViewPoint->{2.94,-1.66,0.245}];
```

```
(*  Fibonacci Polynomials: 2D Curves *)

FibPoly2D[n_] := Plot[FibPoly[n,x],{x,0,1},
  AxesLabel->{" x","Fn(x)"},PlotRange->All,
  DisplayFunction->Identity];

Show[FibPoly2D[2],FibPoly2D[4],FibPoly2D[6],
  DisplayFunction->$DisplayFunction];
```

Fn(x)

Figures in the *Atlas* often have more labeling than in the *Mathematica* graphics. If so, the figure caption reads "Adapted from the output of cell Fibonacci ...". If a visualization comes directly from executing a cell then "Adapted from the output of" is omitted.

How do you know which notebook and cell to execute to visualize a given function? Just look in the text for each function at the second entry "C Function and *Mathematica* Notebook".

Each cell in a notebook usually runs independently of the other cells, rare exceptions being carefully noted in the first cell of the notebook. After you have loaded a notebook, to run a cell just click in it by using the mouse. Some packages may be loaded more than once in a given notebook. Details about using the notebooks are given in Chapter 20.

The C Functions: No Assembly Required

Each C function in the *Atlas* comes in two forms—printed and machine-readable on the CD-ROM. The printed version is given under Algorithm and Program in the section describing a function, and it gives only the function in that section. To have a program that runs, you need a main driver program for input and output, sometimes you need functions called by the function of interest, and you will always need the standard input-output library. So, you are thinking that you are given a jigsaw puzzle of function fragments:

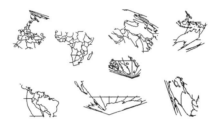

In many computer libraries, function fragments have to be assembled by the user to make a program that may be usable.

However, for the C programs on the CD-ROM no assembly is required:

In the *Atlas* a function with its driver is already a working program.

It is best to use the programs in their full source form off the CD-ROM. An example of such a complete program is that for the Fibonacci function (Section 7.4.1), which is the C analogue of the *Mathematica* program shown above. The program is listed on the next page.

```
#include <stdio.h>
#include <math.h>

main()
{
/* Function Name: Fibonacci */

int n;
double x,Fibonacci(int n, double x);

printf("Fibonacci polynomial Tests");
n = 1;
while ( n >= 0 )
  {
  printf("\n\nInput n, x ( n<0 to end): ");
  scanf("%i%lf",&n,&x);
  if ( n < 0 )
    {
    printf("\nEnd  Fibonacci polynomial Tests");
    exit(0);
    }
  else
    {
    printf("Fibonacci(%i,%g) = %g",
           n,x,Fibonacci(n,x) );
  } } }

double Fibonacci(int n, double x)
/* Fibonacci polynomial */
{
if ( n <= 0 )  return 0;
if ( n == 1 )  return 1;
return  x*Fibonacci(n-1,x)+Fibonacci(n-2,x);
}
```

In this example of a complete program, the header files for the standard input-output library (stdio.h) and the standard mathematics functions in C (math.h) are included. Here, no other functions are used by Fibonacci. If they had been needed, they would be included below Fibonacci if they were not listed elsewhere in the *Atlas*. In Part I of the *Atlas* only function listings are given; the annotated driver programs are in Chapter 21.

Hints for Fortran and Pascal Programmers

The functions in the text and on the CD-ROM are coded in C conforming to the ANSI/ISO standard, as described in the books by Schildt [Sch90], by Deitel [Dei94], and by Plauger and Brodie [Pla96]. Our C functions are easily translated to Fortran or Pascal, with translation to Pascal line-by-line being straightforward. Fortran programmers should be careful with arrays to make them start at zero rather than at unity. With Fortran 77 (but not Fortran 90) you also have to reprogram parts of the few function programs that use recursion. We discuss in Section 3.2 under what circumstances recursive evaluation is appropriate. A Fortran-90 version of the *Atlas* is in preparation.

To help Fortran and Pascal programmers, an outline of the main differences between C, Fortran, and Pascal follows. It will help you to translate the C programs in the *Atlas* to

Fortran or Pascal; it does not substitute for learning thoroughly the language of your choice. To become a competent programmer in C, the book by Müldner and Steele [Mül88] is suitable if you are experienced in other computer languages. Kerrigan's book [Ker91] is specifically aimed to help the transition from Fortran to C, and the book by Shammas [Sha88] introduces C to Pascal programmers. The texts by Thompson [Tho92] and by Glassey [Gla93] provide introductions to numerical programming in C.

The following summarizes the aspects of C as used in the *Atlas*, together with workable counterparts in Fortran and Pascal. Not indicated are correspondences between the three languages for topics that are highly dependent on the computing environment, such as file declarations, especially since these are not necessary for the *Atlas*. In the table, items in italics are generic names that are substituted by actual names.

C	Fortran	Pascal
Overall Structure and Procedure Declarations		
main ()	program *name*	program *name* ;
.	.	.
function ()	*function* ()	*function* ();
.	.	.
void *function* ()	subroutine *name* ()	procedure ();
.	.	.
.	.	main program
Data Type Declarations		
double *name* ;	real *name*	var *name* : real;
int *name* ;	integer *name*	var *name* : integer;
double array [*SIZE*];	real array (*SIZE*)	var array [1..*SIZE*] of real;
int array [*SIZE*];	int array (*SIZE*)	var array [1..*SIZE*] of integer;
char *name* ;	character *name*	var *name* : char;
double rmin,dr;	real rmin,dr	var rmin,dr: real;
int nterms,kmax;	integer nterms,kmax	var nterms,kmax: integer;
Input and Output		
Console:		
scanf(*input list*);	read(*,*format*) *input list*	read(*input list*);
printf(*format,* *output list*);	write(*,*format*) *output list*	write(*output list*);
printf("Input x1 ");	write(*,*)'Input x1'	write('Input x1');
scanf("%lf",x1);	read(*,*) x1	read(x1);

C	Fortran	Pascal

Control Structures

```
while

while (condition )
  {action }

while (x1 != 0 )
  {
printf("Input x1:");
  }
```

```
while  (condition )
do
while (x1 <> 0 ) do
begin
writeln('Input x1');
end;
```

```
if

if (condition )

if ( x1 == 0 )
  { printf ("End");
  }
```

```
if (condition ) then

if ( x1.EQ.0 ) then
   write (*,*) 'End'
```

```
if (condition ) then

if ( x1 = 0 )
begin
   writeln ("End");
end;
```

```
if .. else

if (condition)
  {action 1}
else
  {action 2}
```

```
if (condition) then
   action 1
else
   action 2
```

```
if (condition) then
begin
   action 1;
end;
else
begin
   action 2;
end;
```

```
if ( den == 0 )
  { x1d2 = 0;
  }
else
  { x1d2 =
  (x1*x2+y1*y2)/den;
  }
```

```
if ( den.EQ.0) then
   x1d2 = 0;
else
   x1d2 =
   (x1*x2+y1*y2)/den
```

```
if ( den = 0 )
begin
  x1d2 := 0;
end;
else
begin
  x1d2 :=
  (x1*x2+y1*y2)/den;
end;
```

```
for

for (loop condition)
  { action }

for(k=1;k<=kmax;k++)
  {
  term = r*term;
  sum = sum+term;
  }
```

```
(Sometimes DO loop)

do 2 k=1,kmax,1
   term = r*term
   sum = sum+term
2 continue
```

```
for loop condition
do begin action end;

for k:=1 to kmax do
begin
   term := r*term;
   sum := sum+term;
end;
```

C	Fortran	Pascal
switch		

switch (*expression*)
  {
  case *constant* 1:
    {*action* 1}; break;
  case *constant* 2:
    {*action* 2}; break;
  .
  default: {*action*};
  }
	goto(*st* 1, *st* 2, .),
	      expression
	st 1 *action* 1
	st 2 *action* 2
	  .
	C *st* 1 & *st* 2 are
	     statement numbers
		case *condition* of
		  1: *action* 1;
		  2: *action* 2;
		  .
		end;

switch (choice)
  {
  case 1:
    series=PSexp;
    break;
  case 2:
    series=PScos;
    break;
  }
	goto (1,2),choice
	1 series=PSexp
	2 series=PScos
		case choice of
		  1: series :=PSexp;
		  2: series :=PScos;

Program Operators

Arithmetic Power

pow (*x,y*)	$x **y$	(Requires function)

Assignment

variable=expression;	*variable=expression*	*variable*:=*expression* ;

Increment & Decrement

+ +	increment by 1
- -	decrement by 1

Arithmetic Compare

Equal to;	= =	.EQ.	=
Not equal to;	!=	.NE.	<>
Less than;	<	.LT.	<
Less than or equal;	<=	.LE.	<=
Greater than;	>	.GT.	>
Greater than or equal;	>=	.GE.	>=

Logical Operators

And;	&&	.AND.	AND
Or;	||	.OR.	OR
Not;	!	.NOT.	NOT

File Names for PC-Based Systems

Some computer operating systems restrict the length of the components of file names to at most 8 characters; for example, `BesselAiry.m` (the name of a *Mathematica* file in the *Atlas*) is not allowed. Files with abbreviated names are therefore also provided on the CD-ROM; thus, `BslAry.m` instead of `BesselAiry.m`.

The Appendix gives a complete list of these abbreviated file names.

Reliability of Programs: Disclaimer

The programs in this book and on the CD-ROM have been written carefully. Use of these programs is, however, at your own risk. The author and publisher disclaim all liability for direct or consequential damages resulting from use of the programs.

References on the Computer Interface

[Dei94]   Deitel, H. M., *C; How to Program*, Prentice Hall, Englewood Cliffs, New Jersey, 1994.

[Gla93]   Glassey, R., *Numerical Computation Using C*, Academic Press, San Diego, 1993.

[Ker91]   Kerrigan, J. F., *From Fortran to C*, Windcrest Books, Blue Ridge Summit, Pennsylvania, 1991.

[Ker93]   Kerrigan, J. F., *Migrating to Fortran 90*, O'Reilly and Associates, Sebastapol, California, 1993.

[Mül88]   Müldner, T., and P. W. Steele, *C as a Second Language*, Addison-Wesley, Reading, Massachusetts, 1988.

[Pla96]   Plauger P. J., and J. Brodie, *Standard C; A Reference*, Prentice Hall, Upper Saddle River, New Jersey, 1996.

[Sch90]   Schildt, H., *The Annotated ANSI C Standard*, Osborne McGraw-Hill, Berkeley, California, 1990.

[Sha88]   Shammas, N., *Introducing C to Pascal Programmers*, Wiley, New York, 1988.

[Tho92]   Thompson, W. J., *Computing for Scientists and Engineers*, Wiley, New York, 1992.

PART I
THE FUNCTIONS

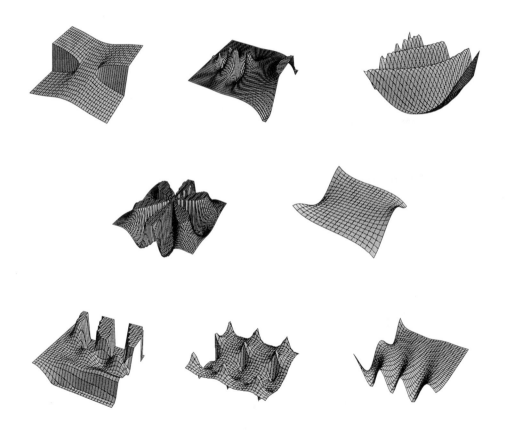

PART I
THE FUNCTIONS

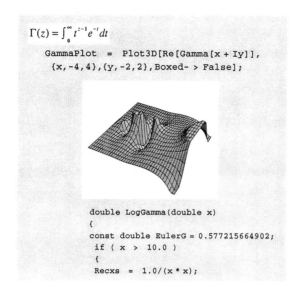

$$\Gamma(z) = \int_0^\infty t^{z-1} e^{-t} dt$$

```
GammaPlot  =  Plot3D[Re[Gamma[x + Iy]],
   {x,-4,4},{y,-2,2},Boxed- > False];
```

```
double LogGamma(double x)
{
const double EulerG = 0.577215664902;
if ( x > 10.0 )
{
Recxs  =  1.0/(x * x);
```

Chapter 1

INTRODUCTION TO THE FUNCTIONS

More than 150 special functions are described in the *Atlas,* similar in coverage and number to those tabulated in the *Handbook of Mathematical Functions* edited by Abramowitz and Stegun. What you will find to be quite different here is that the extensive tabulations characterizing earlier reference books of applied mathematics have been supplanted by extensive graphics; thus, the emphasis in computing the functions has shifted from numerical computing to visual computing.

The paradigm for describing and computing each function is illustrated for the gamma function, $\Gamma(z)$ in Section 6.1.1, shown in the motif above. First, there is mathematical analysis such as key formulas for the function, then there is a *Mathematica* cell in a notebook to make graphical and numerical objects, such as GammaPlot for the 3D surface of $\Gamma(z)$ shown above. (The program is somewhat longer than indicated here.) Then there is a C function for computing the function accurately and efficiently.

The *Atlas* contains more than 700 surfaces and curves to illustrate properties of the functions described, and these are only a starting point for your more detailed explorations using the *Mathematica* programs provided on the CD-ROM that comes with the *Atlas.* Even if you are not experienced in using this system for doing mathematics by computer, you can still use the notebooks effectively because in Chapter 20 they are all listed with annotations that key each part of a program to formulas, figures, and tables in the *Atlas.*

The C functions that are provided for each mathematical function described in the *Atlas* also mark a departure from most previous works. Each C function has a main driver program (listed and annotated in Chapter 21), so you can check your computer's implementation of the program read off the CD-ROM. Tables that give enough values of the function to check each route through the C function are given in the subsection Test Values at the end of each function description. Note that the complete source program needed to run each function is provided on a CD-ROM file, although the listed function shows only a window into this file with the new function and also any auxiliary functions needed for computing it but not appearing elsewhere in the *Atlas.*

How the Function Descriptions Are Organized

All the functions in Part I are described in a uniform scheme, as we now summarize. Please follow along with the example of the gamma function in Section 6.1.1.

Keywords. These are used to characterize terms and alternate nomenclature for the function. For example, one might distinguish the usual complete gamma function from the incomplete gamma function in Section 6.3.1.

C Function and Mathematica Notebook. The name of the C function—such as Log-Gamma—and of the appropriate *Mathematica* cell and notebook—such as GammaFunction and GmBt—are given here. Note that in the *Atlas* programming variables are always printed in Courier typeface.

Background. References to sources that are useful for understanding the mathematical analysis of the function and for applications of the function are given here. References for a given chapter are listed at the end of that chapter. For example, many texts and most reference handbooks have good coverage of the gamma function, and we point you to appropriate sections of eight books.

Function Definition. We give the definition of the function that we are using; usually this coincides with that used in Abramowitz and Stegun. More than one definition is given if more than one is used for computing the function. Our emphasis in formulas is on computational practicality, rather than analytical tractability or interest. For $\Gamma(z)$ we remind you that when $z = n$, a positive integer, the gamma function is a factorial.

Visualization. Many views of the function are given, usually starting with 3D surfaces then simplifying to curves of the function. The figure captions mention the *Mathematica* programs that are needed to generate the graphics; labels and captions have usually been added to these graphics to produce the figures. Again, we tend towards views that clarify computational aspects. For example, our C function for $\ln \Gamma(z)$ is for real and positive arguments z, since the graphics make it clear that logarithms are often inappropriate when z is complex or nonpositive real.

Use the *Mathematica* programs if you want to investigate the function in a different region of parameter space. For example, the behavior of $\Gamma(z)$ near its singularities at the nonpositive integers depends on how you vary x and y, as Figure 6.1.3 indicates.

Algorithm and Program. Here we describe the algorithms used for the C program. For $\Gamma(x)$ with $x > 0$ we use recursion and an asymptotic expansion, but the algorithm is often much longer than this. We often give suggestions and references for other algorithms.

Programming Notes. If a lot of code is to be explained we describe it after the program. For simple code—as for $\Gamma(x)$—we merge this with the previous section.

Test Values. Brief tables of test values that exercise all routes through the C program are given and the accuracy of the values is specified; it is usually 10 digits. The check values are produced by our *Mathematica* programs, usually as the last part of the matching cell; if you need more test values, run this part of the cell. The numerical output was copied to the text electronically, then each entry was checked against the value obtained by using the C function in its driver program and against the value from corresponding programs in the Fortran-90 version of the *Atlas*, which is in preparation. Checks were also made against published tables, but many of these are not sufficiently accurate.

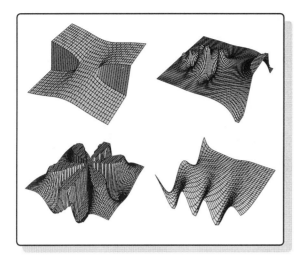

Chapter 2

A VISUAL TOUR OF THE ATLAS

Fly over this chapter to get an overview of the topics in the *Atlas* and to see how its main themes—analysis, visualization, algorithms, and programs—are interrelated. Each chapter description and graphic summarizes a topic that occurs frequently in the chapters of Part I. For example, the location of branch cuts—depicted in the graphic for Chapter 4—are a constant source of ambiguity when computing functions, singularities of the gamma function (Chapter 6) make computing these functions quite challenging, and multiple roots of functions (Chapter 13) hinder the computation of eigenvalues.

This visual tour is also intended to point out graphical features of the *Atlas* that are accessible by using the *Mathematica* notebooks, such as parametric curves generated by functions—the Sici spiral in Chapter 5 and the Cornu spiral in Chapter 10. Coordinate systems that are useful in science and engineering are also described; toroidal coordinates (Chapter 12) are shown as examples of the coordinate systems included in the *Atlas*—the spherical polar, toroidal, and conical systems (Chapter 12), prolate and oblate spheroidal coordinates (Chapter 13), and parabolic cylinder coordinates (Chapter 18).

Because we emphasize efficient computing of more than 150 functions, usually to 10-digit accuracy, each chapter describes appropriate algorithms for its C functions; for example, use of the Landen transform for reducing elliptic integrals to integrals over the circular functions. A view of the Landen transform is shown here for Chapter 17.

Part II of the *Atlas* is The Computer Interface. This has the very long Chapter 20, with all the *Mathematica* notebooks annotated to explain what each cell is computing. Chapter 21 has the driver programs for testing the C functions in Part I; the tables of test values and the output from the driver program that you run should agree to the indicated number of digits (usually ten) if you input the function arguments given in the test tables. Note that—as discussed under The C Functions; No Assembly Required, in the Introduction—the C driver, the C function, and any other C functions required (except ANSI C library functions) are on the CD-ROM as complete programs that are ready to load into your computer, compile, link, and run.

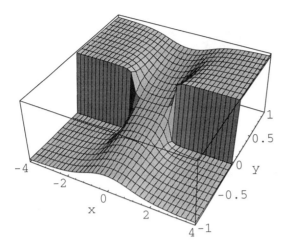

CHAPTER 4 Computing elementary transcendental functions is not so elementary, as indicated by this surface of the imaginary part of arctanh z which has branch cuts along $|x| \geq 1$. Our C functions compute complex principal values.
(Figure 4.3.3)

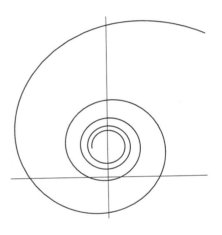

CHAPTER 5 The cosine and sine integrals—Ci(t) and Si(t)—generate the x and y coordinates of the Sici spiral. Here the spiral is shown with t in the range 2 (inside) to 30 (outside). Chapter 5 has *Mathematica* and C functions for exponential integrals of the first and second kind, logarithmic integrals, as well as the cosine and sine integrals.
(Figure 5.2.4)

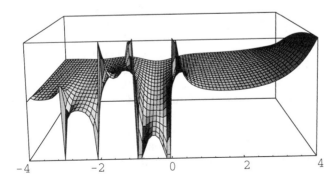

CHAPTER 6 The real part of the gamma function, $\Gamma(z)$, in the complex plane $z = x + iy$. Here x is in [−4, 4] along the front and y ranges from 0 to 1 front to back. Singularities at the nonpositive integers are problematic for computers.
(Figure 6.1.1)

CHAPTER 7 Among the ten combinatorial functions described and coded in Chapter 7 are the binomial coefficients $(n\ m)$, given by $n!/[m!(n-m)!]$, shown here as a bar chart in three dimensions, with $n = 2, 4, 6$ front to back and $m = 0\ldots6$ along the front. You can compute $(n\ m)$ by using *Mathematica* cell `Binomial` in notebook `Combinatorial` or by running the C function `Binomial`. The notebook is annotated in Section 20.7 and the C driver program is listed in Section 21.7; all the programs are on the CD-ROM that comes with the *Atlas*.

(Figure 7.2.1)

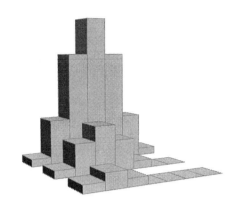

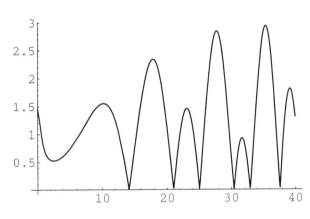

CHAPTER 8 Riemann's zeta function, ς in Section 8.3, is one of several functions for computing the sums of reciprocal powers. Here we show $|\varsigma(1/2+iy)|$ for y in $[0, 40]$, with the zeros along this trajectory. In addition to C functions for $\varsigma(s)$ with real $s > 1$, Chapter 8 includes several other functions of interest for number theory.

(Figure 8.3.2)

CHAPTER 9 The chi-square probability density as a function of the number of degrees of freedom, nu = v, is one of 32 functions useful in statistics that are included in the *Atlas* as both *Mathematica* and C functions. Most of the common discrete distributions are included, as well as density and cumulative distributions for normal and other continuous probability distributions.

(Figure 9.3.6)

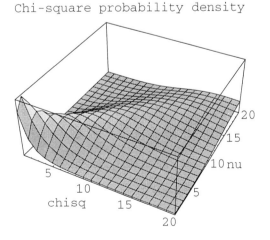

Chi-square probability density

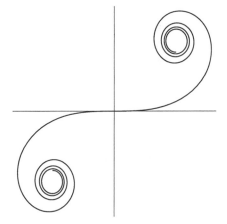

CHAPTER 10 The Cornu spiral is one of the five graphics you see when you run *Mathematica* cell `Fresnel` in the notebook ERFD, which is annotated in Section 20.10 and is on the CD-ROM that comes with the *Atlas*. Use this chapter to understand and compute efficiently the error function, the Fresnel integrals $C(t)$ and $S(t)$ that form the x and y coordinates for the spiral, and the Dawson integral.
(Figure 10.2.3)

CHAPTER 11 The seven classical orthogonal polynomials—such as the Chebyshev polynomial of the second kind at right—are described by 28 graphics, prepared by running the *Mathematica* cell for each polynomial. There are seven C functions to compute the polynomials efficiently in double precision. Chapter 11 starts with an overview of the properties of orthogonal polynomials.
(Figure 11.2.2)

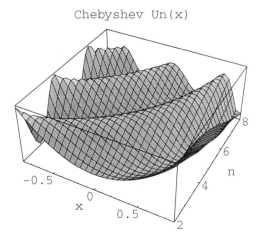

Chebyshev Un(x)

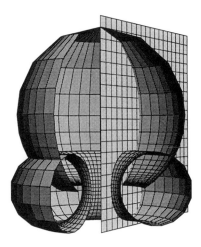

CHAPTER 12 Legendre functions are of several varieties—spherical Legendre functions P_n^m and Q_n^m as well as toroidal and conical kinds. Toroidal functions arise when solving the Laplace equation in toroidal coordinates, shown at left. Chapter 12 depicts the coordinate system appropriate for spherical, toroidal, and conical coordinates when describing each of the five Legendre functions. Driver programs for testing the C functions are in Section 21.12 and on the CD-ROM.
(Figure 12.3.1)

CHAPTER 13   Spheroidal wave functions are challenging to compute. For example, the roots of $U(L)$ shown at right for order $n = 4$ are different from the simple $n(n+1)$ rule for their spherical analogs, the Legendre functions P_n^m. We provide 10-digit accuracy in *Mathematica* and C functions for the eigenvalues, the angular, and the radial functions.

(Figure 13.1.3)

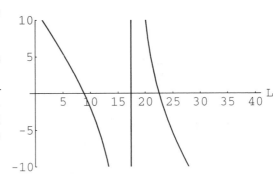

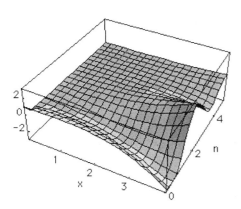

CHAPTER 14   The Kelvin function $\mathrm{ber}_n x$ at left is one of the fourteen Bessel functions depicted through 120 graphics. There are *Mathematica* notebooks for functions of integer order, the Kelvin functions, for functions of half-integer order, and another notebook for Airy functions and their derivatives. The C functions run efficiently and have driver programs in Section 21.14 so you can test the source programs on the CD-ROM.

(Figure 14.3.1)

CHAPTER 15   Struve, Anger, and Weber functions have applications in science and engineering. For example, the Struve function $\mathbf{H}_n(x)$ whose real part and imaginary parts are shown at right is used in designing microphone diaphragms. This chapter gives background to these functions, as well as sources on their analysis and computation. The 36 graphics from the *Mathematica* notebook and the four C programs enable you to comprehend these functions and to compute them efficiently to 10-digit accuracy. As in the other chapters of the *Atlas*, there are extensive references to the mathematical sources used.

(Figure 15.1.2)

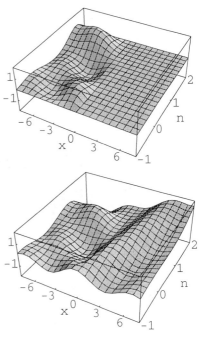

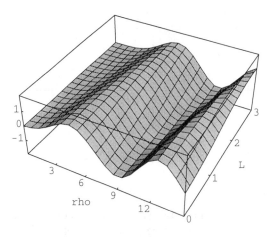

CHAPTER 16 Coulomb wave functions—such as $F_L(\eta,\rho)$ shown at left for $\eta = 1.5$—are often needed in quantum mechanics. This chapter describes the hypergeometric functions $_2F_1$, $_1F_1$, and U, as well as the regular and irregular Coulomb functions and their first derivatives. We have both *Mathematica* and C functions for visualizing and computing these functions.

(Figure 16.3.2)

CHAPTER 17 Elliptic integrals and functions are described, including computing the integrals by the Landen transform; successive transforms indicated at right convert an elliptic integral into a circular one.

(Figure 17.2.3)

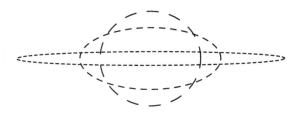

CHAPTER 18 Parabolic cylinder functions $U(a,x)$ and $V(a,x)$ solve the wave equation in parabolic cylinder coordinates. The surface of $e^{-x^2/4}V(a,x)$ is shown at left. This chapter depicts the coordinates and provides *Mathematica* and C functions for computing these functions. Extensive references to the mathematical sources are given.

(Figure 18.2.4)

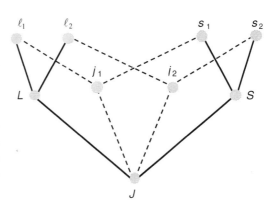

CHAPTER 19 Angular momentum coupling coefficients—the 3-*j*, 6-*j*, and the 9-*j* symbols shown at right—are among the nine miscellaneous functions for science and engineering in Chapter 19. This chapter includes *Mathematica* and C functions for visualizing and computing these functions, all of which come in a complete C version on the CD-ROM accompanying the *Atlas*.
(Figure 19.7.3)

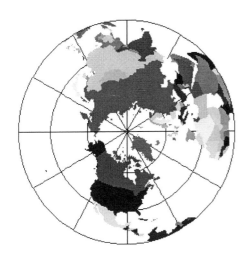

CHAPTER 20 The *Mathematica* notebooks used in Chapters 4 through 19 are listed and annotated in the corresponding Sections 20.4 through 20.19. For each cell there are suggestions for how to use it to explore the properties of the functions it generates, followed by the cell with annotations keying the coding in the cell to figures and tables in the text. This is important reading if you have difficulty using a notebook cell.

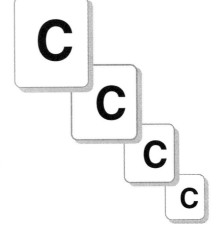

CHAPTER 21 Driver programs for the C functions in Chapters 4 through 19 are listed in the corresponding Sections 21.4 through 21.19. Each driver program is annotated to describe what is computed, how it is used, and how it relates to the tables of test values in the main text. This is required reading for serious users of the *Atlas*.

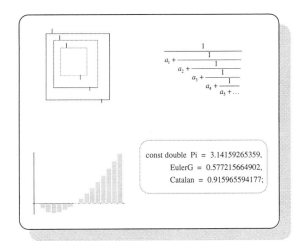

Chapter 3

COMPUTING STRATEGIES

This chapter describes various strategies used consistently throughout the *Atlas* for computing mathematical functions, emphasizing those that may be unfamiliar to practitioners of applied mathematics for the special functions. The topics—some are indicated schematically in the above motif—are; recursion (Section 3.2), continued fractions (Section 3.3), asymptotic expansions (Section 3.4), the Euler-Maclaurin summation formula (Section 3.5), the accuracy and precision of the functions (Section 3.6), and the values of mathematical constants used in the *Atlas* (Section 3.7). We begin by summarizing general strategies for computing the more than 150 functions described.

3.1 GENERAL COMPUTING STRATEGIES

Scope of Special Functions. The range of functions described in the *Atlas* is approximately that in the handbooks edited by Abramowitz and Stegun [Abr64] and by Erdélyi et al [Erd53]. Our coverage is usually narrower than theirs for functions that have more than three parameters and arguments; such functions are difficult to visualize effectively and are also usually difficult to compute reliably and simply over a wide range of their parameters, even though there may exist elegant and powerful mathematical analysis of their properties. The *Atlas* has more material on probability distributions (Chapter 9) and Legendre functions (Chapter 12) than in Abramowitz and Stegun, and more on the spheroidal and Coulomb wave functions (Chapter 13 and Section 16.3).

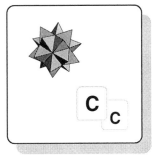

Functions that are nontrivial and often-used special cases of more general functions are presented separately, especially if this clarifies the analysis, visualization, and algorithms used to compute them numerically. When programmed in C, such functions execute more efficiently than a general function.

Our numerical methods usually cover a much broader range of parameter space than most previous published tables, and they are of uniform accuracy—almost always at least 10 decimal digits. Details are given in Section 3.6.

Complex Variables. Many of the functions described are defined for complex variable $z = x + iy$. For most functions, however, it is both difficult and inefficient to write a C program that gives accurate results throughout the complex plane. Only for the elementary transcendental functions (Chapter 4) are our C functions coded for complex variables, because for these functions explicit expressions in terms of real functions of x and y can be obtained. For all other C functions in the *Atlas*, modified functions that are explicitly real occur in our C programs. For example, the Bessel functions of arguments ix and $\sqrt{i}\,x$ are among the fourteen Bessel functions included in Chapter 14.

On the other hand, the *Mathematica* functions—the majority of which were coded by other scientists—often work over a wide area of the complex plane, but at a large penalty in execution speed. We often define our functions in the notebooks in terms of such functions. Thus, when our C and *Mathematica* results in the tables of test values at the end of each section in Part I are found to agree to fractional accuracy of 10^{-10} we can be assured that the C function is correct to this accuracy.

Choice of Algorithms. The algorithms used for the C functions were chosen using four criteria, in the following priority; controllable accuracy, clarity of explanation, simplicity of coding, speed of execution. The accuracy goal for the C functions—which are programmed using double-precision variables—is usually fractional accuracy of 10^{-10}, as discussed in Section 3.6. The *Mathematica* functions are usually accurate to at least machine precision (10^{-16} for the system on which the *Atlas* was developed); exceptions are noted. Numerical values from the notebooks are printed to 10 digits.

Clarity of description is a prime consideration for our algorithms, therefore all the algorithms used are documented within the *Atlas*. If we know of algorithms that are more complicated and are robust under expansion of parameter ranges or accuracy criteria, we reference them. In a very few instances we have "hard-wired" formulas of fixed accuracy, but usually only to bridge the gap between a power series and an asymptotic expansion.

The methods used are (by order of decreasing frequency); power series, asymptotic expansions (Section 3.4), iteration or recursion (Section 3.2), numerical quadrature (integration), continued fractions (Section 3.3), and hard-wired code.

Implementation. Simplicity of coding in C is an important criterion, since the code should be reliable and also feasible to translate to Fortran or Pascal, as discussed in Hints for Fortran and Pascal Programmers in The Computer Interface part of the Introduction. (A Fortran-90 version of the *Atlas* is in preparation.)

The *Mathematica* notebooks in Chapter 20 and on the CD-ROM use only the functions and packages provided with Version 2.2 of this system. No modules or packages were specially written for the *Atlas*. The cells within a notebook can usually be run in any order; the few exceptions are clearly documented in the notebook.

The C functions in the *Atlas* require only standard input-output and mathematical function libraries. There are no other header files. The source functions on the CD-ROM are complete except for the two standard libraries, as discussed under The C Functions: No Assembly Required under The Computer Interface.

3.2   ITERATION AND RECURSION

Throughout the *Atlas* three different repetition patterns in mathematics and computing are distinguished—iteration, recurrence, and recursion. We are careful with the usage of these terms, so an explanation of them is appropriate here, although there is a similar discussion in Section 3.4 of the text on numerical computing [Tho92]. The terms are compared graphically in Figure 3.2.1.

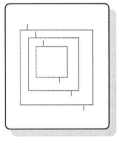

By *iteration* is understood repetition of an analytical or numerical calculation, reusing the same algorithm, parameters, and perhaps memory locations, repeating the calculation until a criterion, such as a convergence limit, is satisfied. Series are usually summed by iteration over partial sums, and the `for` and `while` statements in C are useful for iteration. In the left diagram in Figure 3.2.1 the elongated rectangle indicates a sequence of program steps. The sequence is entered (top, left arrow), executed, then repeated (broken arrow re-entering at top). Eventually, some criterion is satisfied, so the iteration terminates (bottom arrow).

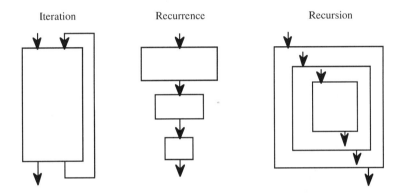

Iteration   Recurrence   Recursion

FIGURE 3.2.1 Three kinds of repetition used for computing mathematical functions.

In *recurrence* a formula is used repeatedly to relate successive values of a function when a parameter changes by uniform steps. For example, power-series terms are usually obtained by recurrence with respect to the number of the term in the series. Recurrence is indicated pictorially in Figure 3.2.1, where each rectangle has a procedure for producing this value of a function from preceding values. For example, the first box may have $k = 1$, the second box $k = 2$, etc. The procedure within each box is often the same, only the value of a control parameter (such as k) changes. Recurrence also requires a stopping criterion. Iteration and recurrence are very similar and are often not distinguished.

In *recursion* a formula refers to itself, as indicated in Figure 3.2.1 by the nesting of the boxes, each describing the same function. When programming recursion, a copy of the function has to be saved for each recursion—each box on the right in Figure 3.2.1. The number of recursions must therefore be finite, and there must be a path out of each function copy, as indicated by arrows at the bottom of each box. Recursion is used in the *Atlas* only when it makes the programming easier to understand and code but does not lessen its efficiency of execution.

Comparing Iteration and Recursion. We compare the relative efficiency of iteration and recursion for the Chebyshev polynomials of the first kind, $T_n(x)$ in Section 11.2.1, by using *Mathematica* programs that compute algebraic or numerical expressions for the polynomials by iteration or by recursion.

Orthogonal polynomials have relatively simple recurrence relations (Table 11.1.3), so the recursive function evaluation capabilities of C (for numerical values) and of *Mathematica* (especially for algebraic values) might be useful. For $T_n(x)$ we have the relation

$$T_0(x) = 1 \qquad T_1(x) = x$$
$$T_n(x) = 2xT_{n-1}(x) - T_{n-2}(x)$$

(3.2.1)

which can be used for algebraic or numerical evaluation. Another formula, suitable for numerical evaluation, is

$$T_n(x) = \cos(n \arccos x) \qquad x \text{ in } [-1,1]$$

(3.2.2)

In Section 11.2.1 we compute $T_n(x)$ numerically by using polynomial formulas for $n < 4$ and (3.2.2) for $n \geq 4$. Our present interest is algorithms using (3.2.1).

For algebraic evaluation our goal is to obtain polynomial expressions for $T_n(x)$. The iterative method in *Mathematica* uses (3.2.1) to relate increasing array elements, `TnIt[i]` in Figure 3.2.2, starting with `TnIt[0]` and `TnIt[1]` given explicitly. One can then program the iteration up to some maximum n value, `maxn` in the program.

```
TnIter[maxn_] :=
 Do[
  (
  TnIt[i]   =
    Simplify[ 2*x*TnIt[i-1] - TnIt[i-2] ]
  ), {i,2,maxn,1} ]

(* Table of polynomial formulas *)
TableTnIt[name_,maxn_] :=
  (
  TnIter[maxn];
  TableForm[Table[{n,Collect[TnIt[n],x]},
    {n,0,maxn}],
  TableHeadings ->
   {None,{"n",""<>ToString[name]<>"(x)"}}]
  )

(* Starting elements *)
TnIt[0] = 1;   TnIt[1] = x;

(* Example: Time iteration up to n = 12 *)
Timing[TableTnIt[Tn,12];]
```

FIGURE 3.2.2 A *Mathematica* program for algebraic Chebyshev polynomials of the first kind, computed by iteration.

The result of executing this program is an array of polynomial formulas for $T_n(x)$, such as in Table 11.2.1, saved as `TableTnIt`; an estimate of the execution time is also obtained.

The recursion method uses (3.2.1) to relate function expressions for increasing n, `TnRec[n,x]` in Figure 3.2.3, with the formulas for `TnRec[0,x]` and `TnRec[1,x]` given explicitly. For each n this program segment executes recursively. If one wants all the formulas up to some maximum n, `maxn`, there is much repeated evaluation of lower-order expressions. The recursion program is as follows.

```
TnRec[n_,x_] :=
 Simplify[ 2*x*TnRec[n-1,x] - TnRec[n-2,x] ]

(* Table of polynomial formulas *)
TableTnRec[name_,maxn_] := TableForm[
 Table[{n,Collect[TnRec[n,x],x]},{n,0,maxn}],
 TableHeadings ->
 {None,{"n",""<>ToString[name]<>"(x)"}}]

(* Starting functions *)
TnRec[0,x_] = 1;
TnRec[1,x_] = x;

(* Example: Time recurrence up to n = 12 *)
Timing[TableTnRec[Tn,12];]
```

FIGURE 3.2.3 *Mathematica* program for algebraic Chebyshev polynomials of the first kind, computed by recursion.

The execution time to produce a table up to order n by iteration is orders of magnitude faster than recursion for $n > 10$. Figure 3.2.4 shows the timing for $n = 5$ to 20 by iteration and for n values from 5 to 15 by recursion.

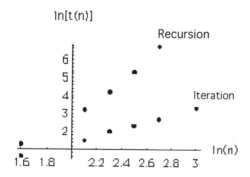

FIGURE 3.2.4 Log-log plot of *Mathematica* program execution times for tables of Chebyshev polynomials of the first kind up to the nth order, using the iterative algorithm (Figure 3.2.2) or the recursive algorithm (Figure 3.2.3).

By making a least-squares fit of these timing data, we find the approximate scaling relations for the dependence of execution times on maximum n value:

$$t_n^{(iter)} \approx c_i n^2 \qquad\qquad t_n^{(recur)} \approx c_r n^5 \qquad\qquad (3.2.3)$$

in which c_i and c_r are constants depending upon the computer system. For iteration one might expect a time proportional to n, the number of times (3.2.1) is applied. However,

Simplify and Collect operations (Figure 3.2.2) also take times that increase roughly linearly with polynomial order. Thus, a table of values up to n produces n^2 timing. The recursive algorithm is very inefficient because of repetitive calculation of lower-order functions and because parts of the code have to be copied many times. Simplify and Collect operations (Figure 3.2.3) also add to the computational burden. The overall time to compute a table of $T_n(x)$ up to $n = 15$ by recursion is fifty times longer than by iteration, and much more storage is needed.

For numerical evaluation, the simplest formula is to use (3.2.2) directly:

```
double ChebyT(int n, double x)
/* Chebyshev Polynomial: first kind, order n */
{ return cos( ( (double) n )*acos(x)); }
```

FIGURE 3.2.5 C function for numerical Chebyshev polynomials of the first kind, computed by direct evaluation using (3.2.2).

Alternatively, one can use recursion in C to produce a function that is not much longer:

```
double ChebyT_Rec(int n, double x)
/* Chebyshev Polynomial: first kind, order n */
/* Evaluation by recursion */
{
if ( n == 0 )  return 1.0;  /* T0 */
if ( n == 1 )  return  x;   /* T1 */
/* For n >= 2 use recursion */
return
  2.0*x*ChebyT_Rec(n-1,x)  -  ChebyT_Rec(n-2,x);
}
```

FIGURE 3.2.6 C function for numerical Chebyshev polynomials of the first kind, computed by recursion using (3.2.1).

The running times for the two numerical programs—direct (Figure 3.2.5) and recursive (Figure 3.2.6)—can be compared easily. As expected, the execution time using the direct algorithm is nearly independent of n. The recursion algorithm becomes slower than the direct one at $n = 4$ and its time increases roughly as n^3. For example, recursion is about 100 times slower than direct computation for $n = 12$.

Thus, recursion is generally not efficient for applications in the *Atlas*. We use it occasionally, as for the gamma function in Section 6.1.1 and the psi function in Section 6.2.1.

3.3   CONTINUED FRACTIONS AND RATIONAL APPROXIMATIONS

It is sometimes efficient as well as accurate to evaluate a special function by using an expression for it as a continued fraction, indicated schematically by the motif at right. If the terms a_k and b_k in the fraction are polynomials in variable x, and if the continued fraction is truncated after a finite number of terms, then one has a

$$a_0 + \cfrac{b_1}{a_1 + \cfrac{b_2}{a_2 + \cfrac{b_3}{a_3 + \cfrac{b_4}{a_4 + \cfrac{b_5}{a_5 + \ldots}}}}}$$

rational-function approximation to the continued fraction and the special function.

Techniques for evaluating continued fractions are described clearly in Section 5.2 of the numerical recipes book by Press et al [Pre92]. A more formal presentation is in Jones and Thron [Jon80]; Chapters 5 and 6 describe how to represent analytic functions by continued fractions and how to make rational approximations by using Padé approximants.

Continued fractions and rational approximations are used occasionally in the *Atlas*. For example, the C program for the incomplete gamma function in Section 6.3.1 is based on a continued fraction, as are the *Mathematica* and C functions for spheroidal wave function eigenvalues in Section 13.1.3. *Mathematica* has several standard packages that allow construction of rational approximations to functions, for example, `Calculus`Pade`` and `NumericalMath`Approximations``. The latter is used in Section 14.5 to devise accurate and efficient rational approximations used in our C programs for the Airy functions and their derivatives.

3.4   USING ASYMPTOTIC EXPANSIONS

An expansion of a function in inverse powers of its argument (or of one of its parameters) is called an asymptotic expansion if the series, although formally divergent, gives a sufficiently close representation of the function if the number of terms is fixed but the magnitude of the argument tends to infinity. A formal definition is given in Chapter 8 of Whittaker and Watson [Whi27]. In Olver [Olv74] there are extensive applications to special functions, while the monograph by Dingle [Din73] emphasizes applications to theoretical physics.

```
k = 0;
term = 1.0;   termold = 1.0;
tkM1 = -1.0; tnMtkP1 = 2*n+1;
sum = 1.0;    ratio = 10.0;
while ( ratio > eps )
{ k++;
  tkM1 += 2.0; tnMtkP1 -= 2.0;
  term *=  tkM1*tnMtkP1*MRxs;
  sum += term;
  ratio = fabs(term/sum);
  termnew = fabs(term);
  if( termnew > termold  )
  { printf
    ("Diverging after %i
      terms",k);
  ratio = 0.0; }
  else { termold = termnew; } }
```

Asymptotic expansions are used frequently in the *Atlas* in order to produce accurate expressions for functions when their arguments are large in magnitude, but other parameters are relatively small. To indicate that an expansion is asymptotic—rather than a convergent power series—we use the "tilde" symbol ~.

As an example of how the convergence of an asymptotic expansion is monitored in our C functions, consider the modified Struve function $\mathbf{L}_n(x)$ in Section 15.1. It has an asymptotic expansion given by (15.1.12), in which the terms decrease with summation index k for only a finite number of terms (depending on n and x), then they diverge. The motif above gives a fragment of the C code for function `StruveL`, showing how one checks for convergence to sufficient accuracy, the control parameter here being variable `eps`, the desired fractional accuracy in the series.

Unlike convergent power series, whose accuracy is (in principle) limited only by roundoff errors, asymptotic series cannot be made to be of arbitrary accuracy for a given argument value. This is illustrated for the ψ function in Section 6.2.1, which has an asymptotic expansion given by (6.2.7). We plot in Figure 3.4.1 the smallest x value at which a given accuracy is attained against the number of terms used.

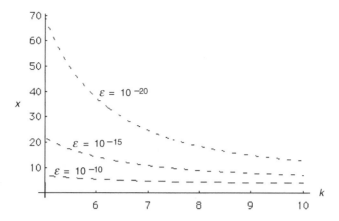

FIGURE 3.4.1 To use the psi function asymptotic series (6.2.7) for a given accuracy ε, the lowest x value at which the series can be used depends upon the number of terms in the asymptotic expansion, k, as shown.

The smallest argument value of any function in the *Atlas* for which an asymptotic expansion is used is always such that fractional accuracy of 10^{-10} or better is obtained. If less accuracy is sufficient for your computing problem, you can usually improve the efficiency of the C function by switching to the asymptotic expansion at a smaller x value.

3.5   EULER-MACLAURIN SUMMATION FORMULA

We use the Euler-Maclaurin summation formula in several sections of the *Atlas*, especially to estimate the contributions of high-order terms in power series. For example, it is used in Section 8.3 to obtain efficiently 10-digit relative accuracy in $\zeta(s) - 1$, where $\zeta(s)$ is the Riemann zeta function. It is used similarly for other sums of reciprocal powers in Section 8.4.

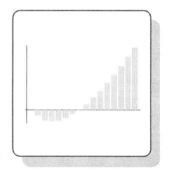

For applications in the *Atlas*, the summation formula for a function $f(x)$ can be written as

$$\sum_{k=M}^{N} f(k) \sim \int_{M}^{N} f(x)dx + [f(M) + f(N)]/2$$
$$+ \sum_{k=2}^{\infty} \frac{(-1)^k B_k f^{(k-1)}(x)\big|_{M}^{N}}{k!} \tag{3.5.1}$$

in which the B_k are the Bernoulli numbers described in Section 8.1. A clear derivation of this formula is given in Section 6.2 of Edwards' monograph on the Riemann zeta function [Edw74], and applications of the formula are also discussed there. Ignoring the summation terms on the right side of (3.5.1) produces the trapezoid rule for integration. For our uses, the series is summed explicitly through $k = M - 1$ terms and $N \to \infty$. Therefore we have an estimate of the remainder sum, provided that we can integrate the remainder function and can evaluate enough of the derivative terms in the sum on the right side before this asymptotic expansion begins to diverge.

The even-k Bernoulli numbers grow explosively with k for $k > 12$, as indicated in the motif above for the logarithm of the magnitude of B_{2k} against k, and shown in more detail in Figure 8.1.2; the odd-k numbers are zero for $k > 1$. In the Euler-Maclaurin formula (3.5.1), however, this growth is damped by the factorials in the denominator, producing coefficients that decrease rapidly in magnitude as k increases, as shown in Figure 3.5.1.

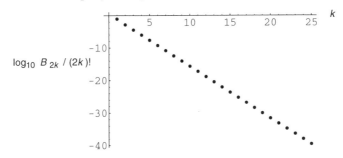

FIGURE 3.5.1 The coefficients of the correction terms in the Euler-Maclaurin summation formula (3.5.1) decrease with k, as shown by the dots.

For the simple power functions that we sum in Sections 8.3 and 8.4, use of two terms in the Bernoulli-number series is enough to give 10-digit relative accuracy.

3.6 ACCURACY AND PRECISION OF THE FUNCTIONS

Our goal throughout the *Atlas* is to compute functions numerically to at least 10 decimal digits of relative accuracy. Often, of course, one can do much better; for example, the evaluation of orthogonal polynomials (Chapter 11) is affected only by the precision of the computer arithmetic. With our goal of 10-digit accuracy, the precision of the computer should not affect the results, provided that floating-point arithmetic is done in double-precision mode.

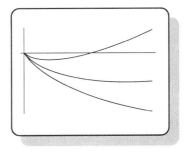

When recurrence relations are used to produce functions of various orders, the direction of the iteration must be such that the function values generally increase, otherwise many significant figures can be lost by subtractive cancellation. This is a particular problem for many of the Bessel functions (Chapter 14), so you will find that the algorithms are designed carefully to minimize subtractive cancellation.

When asymptotic expansions (Section 3.5) are used to approximate a function, accuracy can be comprised if the expansion is used for too small a value of the argument. For example, the Dawson integral $F(x)$ described in Section 10.3 has an asymptotic expansion whose jth term is $(2j+1)!!/(2x^2)^j$ relative to its first term. For fixed x, the terms decrease with j for only a finite number of terms, then they diverge. This is shown schematically in the motif above. Here the magnitude of the terms are plotted as a function of j for $x = 2, 4,$ 6 from top to bottom; more detail is given in Figure 10.3.2. To attain the accuracy goal of 10 digits, special methods are required, as described in detail in Section 10.3.

3.7   MATHEMATICAL CONSTANTS USED IN THE ATLAS

const double Pi = 3.14159265359,
EulerG = 0.577215664902,
Catalan = 0.915965594177;

Throughout the *Atlas* the mathematical constants used in the C functions are named and assigned consistentl, usually to 12 significant digits—as shown in the adjacent motif for π, for Euler's constant (γ), and for Catalan's constant. With our accuracy goal of 10 digits for function values (as discussed in Section 3.6) this is sufficient in almost all cases. Rare exceptions are noted, and the values of the constants are then given to more digits.

Mathematica users can produce mathematical constants quickly by executing lines such as N[Pi,12], N[EulerGamma,12], or N[Catalan,12]. Moreover, the constants have numerical values that can be found to any precision.

REFERENCES ON COMPUTING STRATEGIES

[Abr64]   Abramowitz, M., and I. A. Stegun, *Handbook of Mathematical Functions,* Dover, New York, 1964.

[Din73]   Dingle, R. B., *Asymptotic Expansions: Their Derivation and Interpretation,* Academic Press, London, 1973.

[Edw74]   Edwards, H. M., *Riemann's Zeta Function,* Academic Press, New York, 1974.

[Erd53]   Erdélyi, A. (ed.) *Higher Transcendental Functions,* McGraw-Hill, New York, 1953; reprint edition, Krieger, Malabar, Florida, 1981.

[Jon80]   Jones, W. B., and W. J. Thron, *Continued Fractions,* Addison-Wesley, Reading, Massachusetts, 1980.

[Olv74]   Olver, F. W. J., *Asymptotics and Special Functions,* Academic Press, New York, 1974.

[Pre92]   Press, W. H., S. A. Teukolsky, W. T. Vetterling, and B. P. Flannery, *Numerical Recipes in C,* Cambridge University Press, New York, second edition, 1992.

[Tho92]   Thompson, W. J., *Computing for Scientists and Engineers,* Wiley, New York, 1992.

[Whi27]   Whittaker, E. T., and G. N. Watson, *A Course of Modern Analysis,* Cambridge University Press, Cambridge, England, fourth edition, 1927.

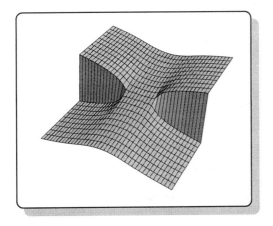

Chapter 4

ELEMENTARY
TRANSCENDENTAL FUNCTIONS

The elementary transcendental functions—the exponential, circular, and hyperbolic functions, plus their inverses—are considered here, just as in Chapter 4 of the handbook of mathematical functions edited by Abramowitz and Stegun [Abr64]. From the computational viewpoint, these functions are straightforward to compute in C for real arguments because their basic components—the exponential, logarithm, cosine, sine, tangent, and some of their inverses—are available as C standard library functions. These component functions are both accurate and efficient to compute. If the functions are required for complex arguments, z, the mathematical analysis required is straightforward and the resulting formulas are widely available in text and reference books. Discussions of the algorithms, C function code, and test values are therefore brief.

A major emphasis in this chapter of the *Atlas* is on visualizations; see Chapter 2. By understanding how to look at and interpret the surfaces and curves of these elementary functions, you will be well prepared for the higher transcendental functions in the following chapters. In particular, many functions are best viewed as functions of complex variables or as multiparameter functions. We illustrate how 3D surfaces of such functions can be visualized, and we also usually provide transections of these surfaces as curves in 2D.

You can generate the graphics and numerical test values by using *Mathematica*. For each function, the appropriate cell and notebook that you should run are given. Annotated listings of all notebooks are in Chapter 20, The *Mathematica* Notebooks. Driver programs for the C functions are listed in Chapter 21.

4.1 EXPONENTIAL AND LOGARITHMIC FUNCTIONS

The exponential and logarithmic functions provide a good starting point for learning how to visualize mathematical functions. In this section functions that are familiar for real arguments are used to establish the pattern for the visual explanations of function properties that is used throughout the *Atlas*.

4.1.1 Exponentials

Keywords
exponential

C Function and *Mathematica* Notebook

ExpZ (C); cell Exp&Log in notebook ElemTF (*Mathematica*)

Background

The exponential function, e^z with z a complex variable, is discussed extensively from the analytical viewpoint in Section 4.2 of Abramowitz and Stegun [Abr64], in Chapters 26 and 27 of Spanier and Oldham [Spa87], and in Section 1.2 of Gradshteyn and Ryzhik [Gra94].

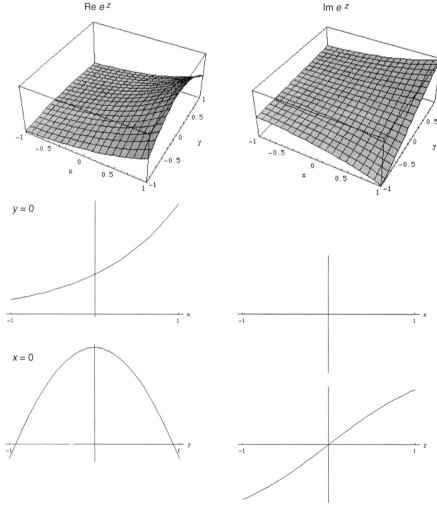

FIGURE 4.1.1 Exponential function of a complex variable; real parts (left) and imaginary parts (right). (Adapted from cell Exp&Log in *Mathematica* notebook ElemTF.)

Function Definition

For the purpose of interpreting the graphics in Figure 4.1.1 and devising an algorithm for the C function, one has the elementary formula

$$e^z \equiv e^{x+iy} = e^x \cos y + ie^x \sin y \qquad (4.1.1)$$

Visualization

The real and imaginary parts of the exponential function e^z with complex argument $z = x + iy$ are shown as the surfaces and curves in Figure 4.1.1. Proceeding down the left column you see the real part of the surface of e^z, then the real part of the curve when $y = 0$, namely e^x, followed by the real part of the curve for $x = 0$, namely, $\cos y$. Similarly, descending the right side of the column, start with the imaginary part of the surface e^z, then see the imaginary-part curve for $y = 0$, which is zero, then the imaginary-part curve for $x = 0$, which is just $\sin y$ according to (4.1.1).

Curves such as those in Figure 4.1.1 are useful for understanding in more detail the surfaces shown throughout the *Atlas*. Moreover, special values of the parameters, such as complex variable z being replaced by the real variable x, often simplify the expression for the general function, as we see in the example of the exponential function.

Algorithm and Program

The C program for the exponential can be computed directly from C standard library functions—exp, cos, and sin—by using (4.1.1), as follows.

```
void ExpZ(double x, double y,
          double *ExpZRe, double *ExpZIm)
/* Exponential of complex argument z = x + i y  */
{
double expx;

expx = exp(x);
*ExpZRe = expx*cos(y);
*ExpZIm = expx*sin(y);
}
```

Test Values

Test values can be generated easily by using a pocket calculator or by running the cell Exp&Log in *Mathematica* notebook ElemTF. Samples are given in Table 4.1.1.

M

TABLE 4.1.1 Test values to 10 digits of the complex exponential defined by (4.1.1). The notation in the table is $(a, b) = a + i b$.

$z = (x, y)$	$(-1, -1)$	$(-1, 1)$	$(1, -1)$	$(1, 1)$
e^z	(0.1987661103, −0.3095598757)	(0.1987661103, 0.3095598757)	(1.468693940, −2.287355287)	(1.468693940, 2.287355287)

4.1.2 Logarithms

Keywords
logarithm

C Function and *Mathematica* Notebook
LogZ (C); cell Exp&Log in notebook ElemTF (*Mathematica*)

Background
The logarithm function—the inverse of the exponential in Section 4.1.1—is discussed from the analytical viewpoint in Section 4.1 of Abramowitz and Stegun [Abr64], in Chapter 25 of Spanier and Oldham [Spa87], and in Section 1.5 of Gradshteyn and Ryzhik [Gra94].

Function Definition
The logarithm of a complex variable, z, is usually not uniquely defined. Most authors distinguish between the *principal branch* of the logarithm, denoted $\ln z$, and the general logarithm, denoted $\text{Ln} z$. The principal value is given by

$$\ln z = \ln r + i\theta \qquad z = re^{i\theta} \qquad -\pi < \theta \leq \pi \qquad (4.1.2)$$

By restricting θ to lie between $-\pi$ and π, with $\theta = \pi$ defining the angle of the branch cut, we can write for the general logarithm

$$\text{Ln} z = \ln z + ik\pi \qquad (4.1.3)$$

in which k is any integer. The branch cut is evident on the right side of Figure 4.1.2.

Visualization
The real and imaginary parts of the principal branch of the logarithm, $\ln z$, for complex argument $z = x + iy$ are shown as the surfaces and curves in Figure 4.1.2. Moving down the left column we see first the real part of the surface of $\ln z$ with a singularity at $z = 0$. This singularity is also apparent in the real part of the curves for either $y = 0$ or $x = 0$.

Going down the right side of Figure 4.1.2, we see the imaginary part of the surface with its branch cut along the negative x axis ($\theta = \pi$), then the imaginary-part plots for $y = 0$ or $x = 0$, again showing the branch-cut discontinuities at $x = 0$ and $y = 0$, respectively.

Algorithm and Program
The principal value of the logarithm of z can be computed from C library functions by using (4.1.2). By noting that the C standard function atan2 returns θ in the range $-\pi$ to π, we have the correct branch cut for the logarithm in function LogZ.

The LogZ function does not check for the singularity at $z = 0$, but our driver program in Section 21.4 makes a check.

```
void LogZ(double x, double y,
          double *LogZRe, double *LogZIm)
/* Principal value of logarithm
   of complex argument  z = x + i y  */
{
*LogZRe = 0.5*log(x*x+y*y);
*LogZIm = atan2(y,x);
}
```

Test Values

Check values of the complex logarithm can be generated easily by using a pocket calculator or by running the cell Exp&Log in *Mathematica* notebook ElemTF. Samples are given in Table 4.1.2.

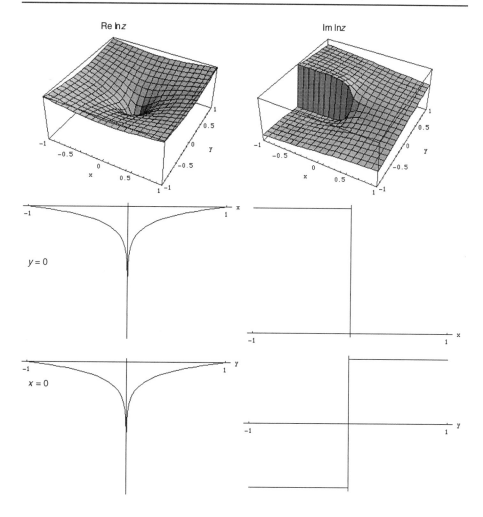

**TABLE 4.1.2 Test values to 10 digits of the principal value of the complex logarithm (4.1.2).
The notation is** $(a, b) = a + i\,b$**.**

$z = (x, y)$	$(-1, -1)$	$(-1, 1)$	$(1, -1)$	$(1, 1)$
ln z	(0.3465735903, −2.356194490)	(0.3465735903, 2.356194490)	(0.3465735903 −0.7853981634)	(0.3465735903, 0.7853981634)

FIGURE 4.1.2 Principal value logarithm function of a complex variable; real parts (left) and imaginary parts (right). Note the branch cut in the imaginary part along the negative x axis. (Adapted from cell Exp&Log in *Mathematica* notebook ElemTF.)

4.2   CIRCULAR AND INVERSE CIRCULAR FUNCTIONS

The circular and inverse circular functions are generally straightforward to compute, so that their main purpose in the *Atlas* is to show how function properties are to be envisioned.

4.2.1   Circular Functions

The cosine, sine, and tangent functions are considered together. Their reciprocals—the secant, cosecant, and cotangent, respectively—are considered primarily from graphical viewpoints. The inverse circular functions—the arccos, arcsine, and arctan—are covered in Section 4.2.2. Analytical properties of circular functions are derived in many texts.

Keywords
circular function, cosecant, cosine, cotangent, secant, sine, tangent

C Functions and *Mathematica* Notebook
CosZ, SinZ, TanZ (C); Cell Circular in notebook ElemTF (*Mathematica*)

Background
Many of the properties of the circular functions are tabulated in Section 4.3 of Abramowitz and Stegun [Abr64], Chapters 32–34 of Spanier and Oldham [Spa87], and Sections 1.3, 1.4 of Gradshteyn and Ryzhik [Gra94].

Function Definitions
We relate the cosine, sine, and tangent of a complex variable $z = x + iy$ to functions of the real variables x and y by [Abr64, 4.3.56]

$$\cos z = \cos x \cosh y - i \sin x \sinh y \qquad (4.2.1)$$

for the cosine, by [Abr64, 4.3.55]

$$\sin z = \sin x \cosh y + i \cos x \sinh y \qquad (4.2.2)$$

for the sine, and by [Abr64, 4.3.57]

$$\tan z = (\sin 2x + i \sinh 2y)/(\cos 2x + \cosh 2y) \qquad (4.2.3)$$

for the tangent. These relations are useful for interpreting the function surfaces in the following visualizations. The secant, cosecant, and cotangent functions are the reciprocals of the above three functions, respectively. They are shown by the curves in Figure 4.2.2.

Visualization
The real and imaginary parts of the cosine, sine, and tangent functions for complex argument $z = x + iy$ are shown as the surfaces in Figure 4.2.1. Note that the real parts (Re) of each surface resemble approximately the corresponding real functions because we have allowed y to vary over a comparatively small range (–1 to 1) compared with x (–4 to 4). The factor $\cosh y$ in (4.2.1) and (4.2.2) varies by only 54% for y in $[-1, 1]$. On the other hand, the imaginary parts (Im) of the surfaces are much more convoluted because the $\sin x$ and $\cos x$ terms in (4.2.1) and (4.2.2) are modulated by $\sinh y$, whose relative variation for y in $[-1, 1]$ is much greater.

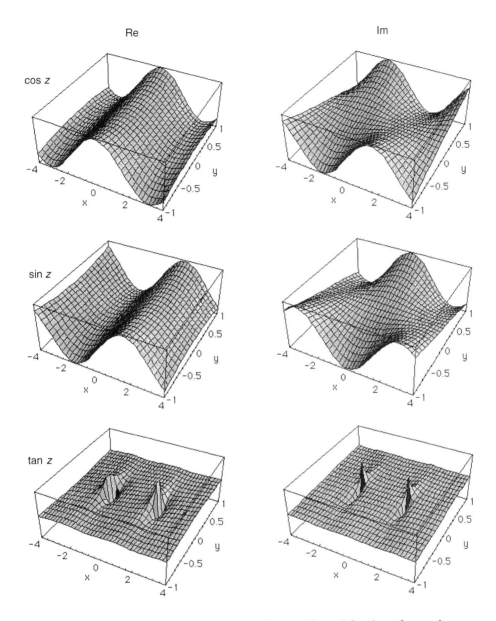

FIGURE 4.2.1 Cosine (top), sine (middle), and tangent (bottom) functions of a complex variable; real parts (left) and imaginary parts (right). (Adapted from cell `Circular` in *Mathematica* notebook `ElemTF`.)

The surfaces of the real and imaginary parts of the tangent function in Figure 4.2.1 are dominated by the poles at $y = 0$ for $x = \pm\pi/2$. The details of the surfaces are therefore much suppresssed for other regions of the section of the complex plane over which the function tan$z$ is mapped.

Curves of the circular functions and of their reciprocals as functions of the real variable x are also interesting to visualize. Look at the curves in Figure 4.2.2. Since the reciprocal functions—secant, cosecant, and cotangent—as well as the tangent have singularities, we plot the function values only in the range $[-2, 2]$.

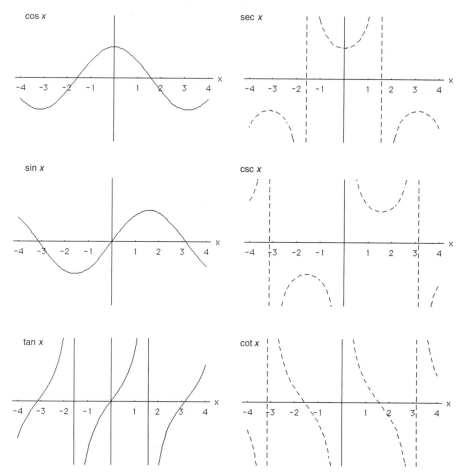

FIGURE 4.2.2 Left: Curves of the circular functions—cosine (top), sine (middle), and tangent (bottom) functions of a real variable. Right: Curves of the reciprocals of the circular functions—secant (top), cosecant (middle), and cotangent (bottom). (Adapted from cell `Circular` in *Mathematica* notebook `ElemTF`.)

Algorithms and Programs

For real arguments, the circular functions shown in Figure 4.2.2 can be computed accurately and efficiently from the C standard library functions as `cos()`, `sin()`, and `tan()`. Test values can be generated easily by using a pocket calculator. The corresponding reciprocal functions—sec, csc, and cot of real argument—are obtained in C as reciprocals of the circular functions, except near the zeros or poles of the latter. In *Mathematica* the corresponding functions are `Cos[]`, `Sin[]`, `Tan[]`, `Sec[]`, `Csc[]`, and `Cot[]`, as used in the `Circular` cell of notebook `ElemTF`.

With complex arguments, the circular functions depicted in Figure 4.2.1 can be computed in terms of circular and hyperbolic functions with real arguments, as follows. Formulas (4.2.1)–(4.2.3) are used directly, as coded in the C functions `CosZ`, `SinZ`, and `TanZ` that follow.

```
void CosZ(double x, double y,
          double *CosZRe, double *CosZIm)
/* Cosine of complex argument z = x + i y */

{
*CosZRe =  cos(x)*cosh(y);
*CosZIm = -sin(x)*sinh(y);
}
```

```
void SinZ(double x, double y,
          double *SinZRe, double *SinZIm)
/* Sine of complex argument z = x + i y  */

{
*SinZRe = sin(x)*cosh(y);
*SinZIm = cos(x)*sinh(y);
}
```

The singularities of the tangent (Figure 4.2.1, bottom) are located at the zeros of $\cos z$, that is, at $z = (2n+1)\pi/2$, n an integer. We do not check for the singularities within the function, but the driver program quits if $z = 0$. The C function `TanZ` for a complex argument is as follows:

```
void TanZ(double x, double y,
          double *TanZRe, double *TanZIm)
/* Tangent of complex argument z = x + i y */
{
double twox,twoy,recden;

twox = 2.0*x;
twoy = 2.0*y;
recden = 1.0/(cos(twox)+cosh(twoy));

*TanZRe = sin(twox)*recden;
*TanZIm = sinh(twoy)*recden;
}
```

Test Values

Check values of the cosine, sine, and tangent for complez arguments z can be generated from formulas (4.2.1)–(4.2.3) or by running the cell `Circular` in *Mathematica* notebook `ElemTF`, the numerical output from which is given in Table 4.2.1.

TABLE 4.2.1 Test values to 10 digits of the circular functions for complex arguments. The notation is $(a, b) = a + i b$.

$z = (x, y)$	$(-1, -1)$	$(-1, 1)$	$(1, -1)$	$(1, 1)$
$\cos z$	(0.8337300251, −0.9888977058)	(0.8337300251, 0.9888977058)	(0.8337300251, 0.9888977058)	(0.8337300251, −0.9888977058)
$\sin z$	(−1.298457581, −0.6349639148)	(−1.298457581, 0.6349639148)	(1.298457581, −0.6349639148)	(1.298457581, 0.6349639148)
$\tan z$	(−0.2717525853, −1.083923327)	(−0.2717525853, 1.083923327)	(0.2717525853, −1.083923327)	(0.2717525853, 1.083923327)

4.2.2   Inverse Circular Functions

The inverse circular functions of complex arguments—arccosine, arcsine, and arctangent—are considered together. Their analytical properties are derived in many texts and are listed in many handbooks, such as Abramowitz and Stegun [Abr64, Section 4.4].

Keywords

arccosine, arcsine, arctangent, inverse circular function

C Functions and *Mathematica* Notebook

`InvCosZ`, `InvSinZ`, `InvTanZ` (C); cell `InverseCircular` in the notebook `ElemTF` (*Mathematica*)

Background

Many properties of the inverse circular functions are tabulated in Section 4.4 of [Abr64] and in Chapter 35 of Spanier and Oldham [Spa87].

Function Definitions

The inverse circular functions are multiple-valued, even for real arguments. We present formulas and C functions for the principal values only. The inverse cosine, sine, and tangent of complex variable $z = x + iy$ are related to functions of the real variables x and y by formulas consistent with [Abr64, 4.4.38]

$$\arccos z = \pm\{\arccos \beta - i\,\mathrm{sgn}(y)\ln[\alpha + (\alpha^2 - 1)^{1/2}]\} \tag{4.2.4}$$

for the inverse cosine, by [Abr64, 4.4.37]

$$\arcsin z = \pm\{\arcsin \beta + i\,\mathrm{sgn}(y)\ln[\alpha + (\alpha^2 - 1)^{1/2}]\} \tag{4.2.5}$$

for the inverse sine, in both of which α and β are given by

$$\left.\begin{array}{c}\alpha \\ \beta\end{array}\right| = \left[\sqrt{(x+1)^2 + y^2} \pm \sqrt{(x-1)^2 + y^2}\right]\Big/2 \tag{4.2.6}$$

The sign of y, $\mathrm{sgn}(y)$, produces consistency with the *Mathematica* results in Figure 4.2.3.

For the inverse tangent [Abr64, 4.4.39] gives

$$\arctan z = \frac{1}{2}\arctan\left(\frac{2x}{1-x^2-y^2}\right) + \frac{i}{4}\ln\left[\frac{x^2+(y+1)^2}{x^2+(y-1)^2}\right] \qquad z^2 \ne -1 \qquad (4.2.7)$$

Visualization

Real and imaginary parts of the principal values of the arccosine, arcsine, and arctangent functions for complex argument $z = x + iy$ are shown as the surfaces in Figure 4.2.3. The branch cuts for each function are specified in Table 4.2.2.

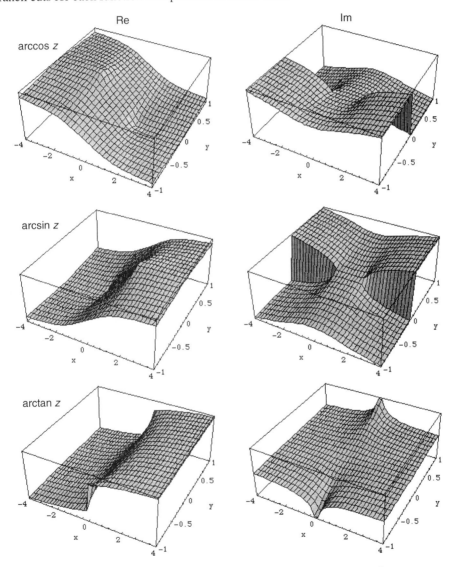

FIGURE 4.2.3 Principal values of the arcosine (top), arcsine (middle), and arctangent (bottom) functions of a complex variable; real parts (left) and imaginary parts (right). (Adapted from cell `InverseCircular` in *Mathematica* notebook `ElemTF`.)

The branch cuts for the inverse circular functions, depicted in Figure 4.2.3, are specified in Table 4.2.2. They agree with those in Abramowitz and Stegun [Abr64] and in *Mathematica*. Curves of principal values of inverse circular functions are shown in Figure 4.2.4. These curves are sections of the curves in Figure 4.2.2 rotated by $\pi/2$.

Table 4.2.2 Branch-cut discontinuities of inverse circular functions.

Inverse circular function	Branch-cut discontinuities
$\arccos z$	$(-\infty,-1)$ and $(1,\infty)$
$\arcsin z$	$(-\infty,-1)$ and $(1,\infty)$
$\arctan z$	$(-i\infty,-i]$ and $[i,i\infty)$

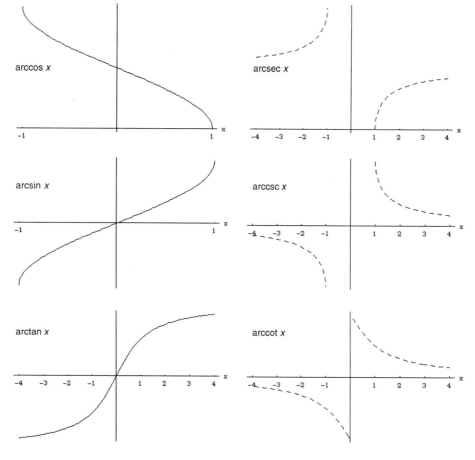

FIGURE 4.2.4 Curves of the inverse circular functions—arccosine (top left), arcsine (middle left), and arctangent (bottom left), and of arcsecant (top right), arccosecant (middle right), and arccotangent (bottom right) as functions of real variable x. (Adapted from cell `InverseCircular` in *Mathematica* notebook `ElemTF`.)

Algorithms, Programs, and Programming Notes

Our algorithms for the principal values of the inverse circular functions of complex variable z—the arccos, arcsin, and arctangent—use formulas (4.2.4)–(4.2.7) directly.

By choosing the plus signs in (4.2.4) and (4.2.5) one obtains functions whose branch cuts agree with those in Table 4.2.2 and with the conventions used in *Mathematica*. For the arcsine, however, when z is real ($y = 0$) our C function InvSinZ may give the supplement of the angle that you want. You must determine from the context in which you are using these inverse functions which of the signs in (4.2.4) and (4.2.5) are appropriate and whether you want just the principal values as the C functions compute them.

The C function for the arccosine is InvCosZ, given as follows.

```
void InvCosZ(double x, double y,
         double *InvCosZRe, double *InvCosZIm)
/* Inverse cosine of complex argument z = x + i y */
{
double ysqd,rtpos,rtneg,alpha,beta,sign;

ysqd = y*y;
rtpos = sqrt(pow(x+1,2)+ysqd);
rtneg = sqrt(pow(x-1,2)+ysqd);
alpha = 0.5*(rtpos + rtneg);
beta = 0.5*(rtpos - rtneg);
sign = ( y > 0 ) ? 1: -1;

*InvCosZRe = acos(beta);

*InvCosZIm = - sign*log(alpha+sqrt(alpha*alpha-1));
}
```

Our C function for the principal value of the arcsine is InvSinZ, given as follows. Note the warning given above for the choice of sign from (4.2.5).

```
void InvSinZ(double x, double y,
         double *InvSinZRe, double *InvSinZIm)
/* Inverse sine of complex argument z = x + i y   */
{
double ysqd,rtpos,rtneg,alpha,beta,sign;

ysqd = y*y;
rtpos = sqrt(pow(x+1,2)+ysqd);
rtneg = sqrt(pow(x-1,2)+ysqd);
alpha = 0.5*(rtpos + rtneg);
beta = 0.5*(rtpos - rtneg);
sign = ( y > 0 ) ? 1: -1;

*InvSinZRe = asin(beta);

*InvSinZIm = sign*log(alpha+sqrt(alpha*alpha-1));
}
```

Programming the principal value of the inverse tangent of complex variable z requires care for a different reason than for the inverse cosine and sine described above. The C standard library function `atan2` is required with *two* arguments, rather than the simpler `tan` function. With this precaution, `InvTanZ` is a direct transcription of (4.2.7).

The driver program for the inverse circular functions checks for the condition $z^2 \neq -1$ ($x = 0$, $y = \pm 1$) and uses it to terminate execution. You might choose to move this check into the C function.

```
void InvTanZ(double x, double y,
             double *InvTanZRe, double *InvTanZIm)
/* Inverse tangent of complex argument z = x + i y */
{
double xsqd,ysqd;

xsqd = x*x;
ysqd = y*y;

*InvTanZRe = 0.5*atan2(2.0*x,1.0 - xsqd - ysqd);
*InvTanZIm = 0.25*log(
   (xsqd+pow(y+1.0,2.0))/(xsqd+pow(y-1.0,2.0)));
}
```

The inverse functions that are related to the reciprocals of the circular functions—the arcsecant, arccosecant, and arccotangent on the right side of Figure 4.2.4—can be obtained from the above functions by using the following identities for complex variable z:

$$\operatorname{arcsec} z = \arccos 1/z \quad \operatorname{arccsc} z = \arcsin 1/z \quad \operatorname{arccot} z = \arctan 1/z \qquad (4.2.8)$$

Test Values

Check values of the inverse cosine, sine, and tangent principal values for complez arguments z can be generated from formulas (4.2.4)–(4.2.7) or by running the cell `Inverse-Circular` in *Mathematica* notebook `ElemTF`, the numerical output from which is given in Table 4.2.3.

TABLE 4.2.3 **Test values to 10 digits of inverse circular functions for complex arguments. The notation is** $(a, b) = a + i\ b$.

$z = (x, y)$	$(-4, -1)$	$(-4, 1)$	$(4, -1)$	$(4, 1)$
$\arccos z$	(2.889413245, 2.096596457)	(2.889413245, −2.096596457)	(0.2521794087, 2.096596457)	(0.2521794087, −2.096596457)
$\arcsin z$	(−1.318616918, −2.096596457)	(−1.318616918, 2.096596457)	(1.318616918, −2.096596457)	(1.318616918, 2.096596457)
$\arctan z$	(−1.338972522, −0.0557858878)	(−1.338972522, 0.0557858878)	(1.338972522, −0.0557858878)	(1.338972522, 0.0557858878)

4.3 HYPERBOLIC AND INVERSE HYPERBOLIC FUNCTIONS

The hyperbolic and inverse hyperbolic functions are straightforward to compute. One of their purposes in the *Atlas* is to show how function properties can be displayed. Many such properties are given in Sections 4.5 and 4.6 of Abramowitz and Stegun [Abr64]. Note also that—within factors of $\pm i$—they are related to the surfaces of the circular and the inverse-circular functions that are displayed in Figures 4.2.1 and 4.2.3.

4.3.1 Hyperbolic Functions

The hyperbolic cosine (cosh), sine (sinh), and tangent (tanh) functions are considered together. Their reciprocals—the sech, csch, and coth, respectively—are considered mainly from graphical viewpoints. Inverse hyperbolic functions—the arccosh, arcsinh, and arctanh—are covered in Section 4.3.2. Analytical properties of the hyperbolic functions are derived in many texts.

Keywords

hyperbolic cosine, hyperbolic function, hyperbolic sine, hyperbolic tangent

C Functions and *Mathematica* Notebook

CoshZ, SinhZ, TanhZ (C); Cell Hyperbolic in notebook ElemTF (*Mathematica*)   *M*

Background

Properties of the hyperbolic functions are tabulated in Section 4.5 of Abramowitz and Stegun [Abr64] Chapters 28–30 of Spanier and Oldham [Spa87], and Sections 1.3, 1.4 of Gradshteyn and Ryzhik [Gra94].

Function Definitions

We relate hyperbolic functions of a complex variable $z = x + iy$ to functions of real variables x and y by [Abr64, 4.5.50]

$$\cosh z = \cosh x \cos y + i \sinh x \sin y \tag{4.3.1}$$

for the hyperbolic cosine, by [Abr64, 4.5.49]

$$\sinh z = \sinh x \cos y + i \cosh x \sin y \tag{4.3.2}$$

for the hyperbolic sine, and by [Abr64, 4.5.51]

$$\tanh z = (\sinh 2x + i \sin 2y)/(\cosh 2x + \cos 2y) \tag{4.3.3}$$

for the hyperbolic tangent. These relations will help you to interpret the function surfaces in the following visualizations. The hyperbolic secant, cosecant, and cotangent functions are the reciprocals of the above three functions, respectively, and are shown by the curves in Figure 4.3.2.

Visualization

The real and imaginary parts of the hyperbolic cosine, sine, and tangent functions for complex argument $z = x + iy$ are shown as the surfaces in Figure 4.3.1. The real parts (Re) of each surface resemble approximately the corresponding hyperbolic functions for real arguments because we allow y to vary over a comparatively small range (-1 to 1) compared with x (-4 to 4).

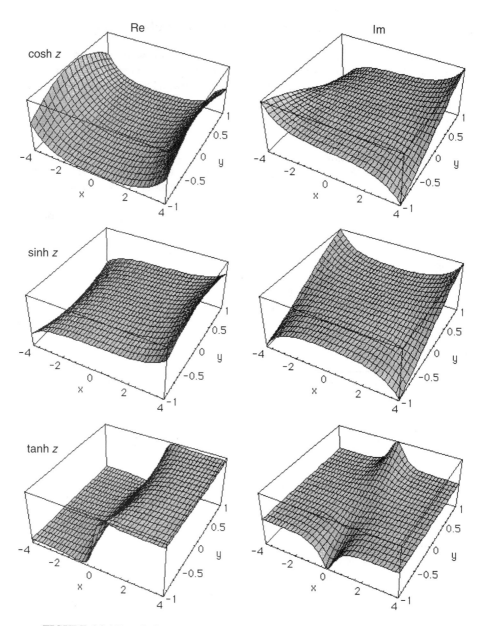

FIGURE 4.3.1 Hyperbolic cosine (top), sine (middle), and tangent (bottom) functions of a complex variable; real parts (left) and imaginary parts (right). (Adapted from cell `Hyperbolic` in *Mathematica* notebook `ElemTF`.)

Curves of the hyperbolic functions and of their reciprocals as functions of the real variable x are also interesting to visualize. Look at the curves in Figure 4.3.2. The functions vary rapidly as x increases, therefore we plot the function values only for x in $[-2, 2]$. The reciprocal functions—sech, csch, and coth—have singularities, so they are shown only over a limited range of function values.

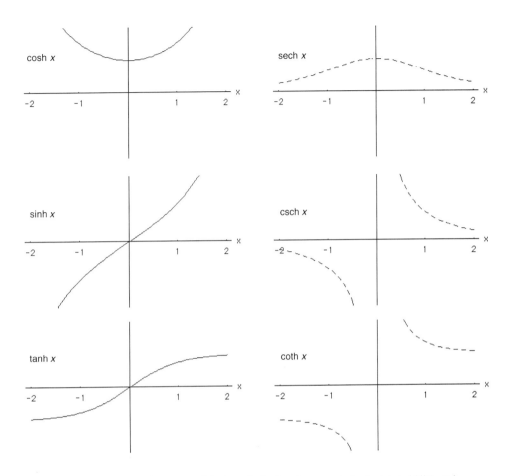

FIGURE 4.3.2 Left: Curves of the hyperbolic functions—cosh (top), sinh (middle), and tanh (bottom) functions of a real variable. Right: Curves of the reciprocals of the hyperbolic functions—sech (top), csch (middle), and coth (bottom). (Adapted from cell `Hyperbolic` in *Mathematica* notebook `ElemTF`.)

Algorithms and Programs

For real arguments, the hyperbolic functions shown in Figure 4.3.2 can be computed accurately and efficiently from the C standard library functions as `cosh()`, `sinh()`, and `tanh()`. Test values can be generated easily by using a pocket calculator. The corresponding reciprocal functions—sech, csch, and coth of real argument—are obtained in C as reciprocals of the hyperbolic functions. In *Mathematica* the corresponding functions are `Cosh[]`, `Sinh[]`, `Tanh[]`, `Sech[]`, `Csch[]`, and `Coth[]`, as used in the `Hyperbolic` cell of notebook `ElemTF`.

M

With complex arguments, the hyperbolic functions depicted in Figures 4.3.1 and 4.3.2 can be computed in terms of circular and hyperbolic functions with real arguments. Formulas (4.3.1)–(4.3.3) are used directly, as coded in the C functions `CoshZ`, `SinhZ`, and `TanhZ` that follow.

```
void CoshZ(double x, double y,
          double *CoshZRe, double *CoshZIm)
/* Hyperbolic cosine of
   complex argument z = x + i y  */
{
*CoshZRe = cosh(x)*cos(y);
*CoshZIm = sinh(x)*sin(y);
}
```

```
void SinhZ(double x, double y,
          double *SinhZRe, double *SinhZIm)
/* Hyperbolic sine of
   complex argument z = x + i y  */
{
*SinhZRe = sinh(x)*cos(y);
*SinhZIm = cosh(x)*sin(y);
}
```

The C function `TanhZ` for a complex argument is:

```
void TanhZ(double x, double y,
          double *TanhZRe, double *TanhZIm)
/* Hyperbolic tangent of
   complex argument z = x + i y  */
{
double twox,twoy,recden;

twox = 2.0*x;
twoy = 2.0*y;
recden = 1.0/(cosh(twox)+cos(twoy));

*TanhZRe = sinh(twox)*recden;
*TanhZIm = sin(twoy)*recden;
}
```

There is a singularity of $\tanh z$ at $z = 0$, so the driver program quits if $z = 0$.

Test Values

Check values of the hyperbolic cosine, sine, and tangent for complex arguments z can be generated from formulas (4.3.1)–(4.3.3) or by running the cell `Hyperbolic` in *Mathematica* notebook `ElemTF`, the numerical output from which is given in Table 4.3.1.

TABLE 4.3.1 Test values to 10 digits of hyperbolic functions for complex arguments. The notation is $(a, b) = a + i\, b$.

$z = (x, y)$	$(-1, -1)$	$(-1, 1)$	$(1, -1)$	$(1, 1)$
$\cosh z$	(0.8337300251, 0.9888977058)	(0.8337300251, −0.9888977058)	(0.8337300251, −0.9888977058)	(0.8337300251, 0.9888977058)
$\sinh z$	(−0.6349639148, −1.298457581)	(−0.6349639148, 1.298457581)	(0.6349639148, −1.298457581)	(0.6349639148, 1.298457581)
$\tanh z$	(−1.083923327, −0.2717525853)	(−1.083923327, 0.2717525853)	(1.083923327, −0.2717525853)	(1.083923327, 0.2717525853)

4.3.2 Inverse Hyperbolic Functions

The inverse hyperbolic functions of complex arguments—arccosh, arcsinh, and arctanh—are considered together. Their analytical properties are derived in many texts and are listed in many handbooks, such as Abramowitz and Stegun [Abr64, Section 4.6].

Keywords
arccosh, arcsinh, arctanh, inverse hyperbolic function

C Functions and *Mathematica* Notebook
InvCoshZ, InvSinhZ, InvTanhZ (C); cell InverseHyperbolic in the notebook ElemTF (*Mathematica*)

$\boxed{M}$

Background
Many properties of the inverse hyperbolic functions are tabulated in Section 4.6 of [Abr64] and in Chapter 31 of Spanier and Oldham [Spa87].

Function Definitions
For complex arguments, inverse hyperbolic functions are multiple-valued. We give formulas and C functions for the principal values. Our definitions produce compatibility between results obtained using C standard library functions and *Mathematica*. See the discusson under Programming Notes below. The inverse hyperbolic cosine, sine, and tangent functions of complex variable $z = x + iy$ can be related to circular functions of z, which can then be related to functions of the real variables x and y.

For the inverse hyperbolic cosine we combine [Abr64, 4.6.15] and our (4.2.4) to give

$$\operatorname{arccosh} z = \ln[\alpha + (\alpha^2 - 1)^{1/2}] + i\operatorname{sgn}(y)\arccos\beta \qquad (4.3.4)$$

Here sgn(y) is the sign of y and

$$\beta\Big| = \left[\sqrt{(x+1)^2 + y^2} \pm \sqrt{(x-1)^2 + y^2}\right]/2 \qquad (4.3.5)$$

For the inverse hyperbolic sine, by using [Abr64, 4.6.14] and our (4.2.5), we have

$$\operatorname{arcsinh} z = \operatorname{sgn}(x)\ln[\alpha' + (\alpha'^2 - 1)^{1/2}] - i\arcsin\beta' \qquad (4.3.6)$$

in terms of the sign of x, with α' and β' given by

$$\beta' \left| = \left[\sqrt{(y-1)^2 + x^2} \pm \sqrt{(y+1)^2 + x^2} \right] \middle/ 2 \right. \tag{4.3.7}$$

For the inverse hyperbolic tangent, [Abr64, 4.6.16] with our (4.2.7) gives

$$\operatorname{arctanh} z = \frac{1}{4} \ln \left[\frac{y^2 + (x+1)^2}{y^2 + (x-1)^2} \right] + \frac{i}{2} \arctan \left(\frac{2y}{1 - x^2 - y^2} \right) \qquad z^2 \neq 1 \tag{4.3.8}$$

Visualization

Real and imaginary parts of the principal values of the arccosh, arcsinh, and arctanh functions for complex argument $z = x + iy$ are shown by surfaces in Figure 4.3.3. The branch cuts for each function are specified in Table 4.3.2.

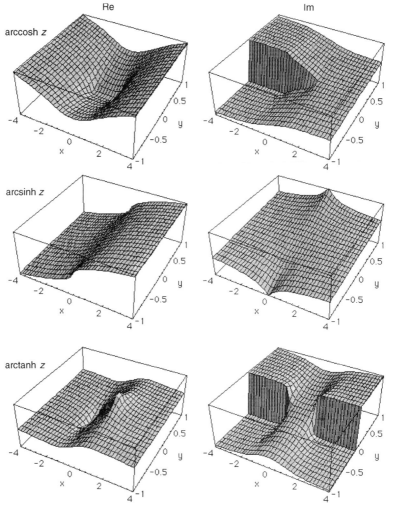

FIGURE 4.3.3 Principal values of the arcosh (top), arcsinh (middle), and arctanh (bottom) functions of a complex variable; real parts (left) and imaginary parts (right). (Adapted from cell `InverseHyperbolic` in *Mathematica* notebook `ElemTF`.)

Branch cuts for the inverse hyperbolic functions, depicted in Figure 4.3.3, are specified in Table 4.3.2. They agree with those in Abramowitz and Stegun [Abr64] and in *Mathematica*. Curves of principal values of inverse circular functions are shown in Figure 4.3.4. These curves are sections of the curves in Figure 4.3.2 rotated by $\pi/2$.

M

Table 4.3.2 Branch-cut discontinuities of inverse hyperbolic functions.

Inverse hyperbolic function	Branch-cut discontinuities
arccosh$z$	$(-\infty, 1)$
arcsinh$z$	$(-i\infty, -i)$ and $(i, i\infty)$
arctanh$z$	$(-\infty, -1]$ and $[1, \infty)$

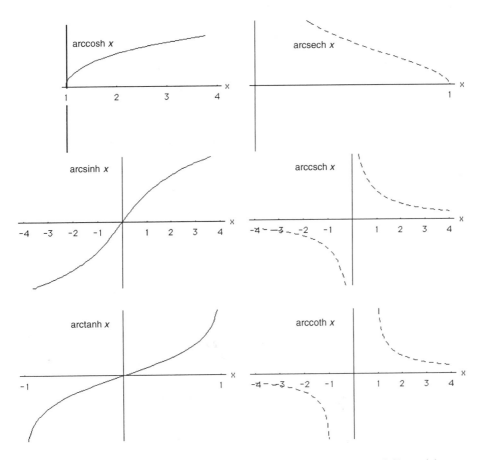

FIGURE 4.3.4 Curves of the inverse hyperbolic functions—arccosh (top left), arcsinh (middle left), and arctanh (bottom left), and of arcsech (top right), arccsch (middle right), and arccoth (bottom right) as functions of real variable x. (Adapted from the cell In-verseHyperbolic in *Mathematica* notebook ElemTF.)

Algorithms and Programs

Our algorithms for the principal values of the inverse hyperbolic functions of complex variable z—the arccosh, arcsinh, and arctanh—use formulas (4.3.4)–(4.3.8). The C function for the arccosh is `InvCoshZ`, given as follows.

```c
void InvCoshZ(double x, double y,
    double *InvCoshZRe, double *InvCoshZIm)
/* Inverse hyperbolic cosine
   of complex argument z = x + i y */
{
double ysign,ysqd,rtpos,rtneg,alpha,beta;

ysign = ( y > 0 ) ? +1.0: -1.0;
ysqd = y*y;
rtpos = sqrt(pow(x+1,2)+ysqd);
rtneg = sqrt(pow(x-1,2)+ysqd);
alpha = 0.5*(rtpos + rtneg);
beta = 0.5*(rtpos - rtneg);

*InvCoshZRe = log(alpha+sqrt(alpha*alpha-1));
*InvCoshZIm = ysign*acos(beta);
}
```

The C function for the principal value of the arcsinh is `InvSinhZ`, as follows.

```c
void InvSinhZ(double x, double y,
    double *InvSinhZRe, double *InvSinhZIm)
/* Inverse hyperbolic sine
   of complex argument z = x + i y */
{
double xsign,xsqd,rtpos,rtneg,alphap,betap;

xsign = ( x > 0 ) ? +1.0: -1.0;
xsqd = x*x;
rtpos = sqrt(pow(y-1,2)+xsqd);
rtneg = sqrt(pow(y+1,2)+xsqd);
alphap = 0.5*(rtpos + rtneg);
betap = 0.5*(rtpos - rtneg);

*InvSinhZRe =
  xsign*log(alphap+sqrt(alphap*alphap-1));
*InvSinhZIm = - asin(betap);
}
```

For the inverse hyperbolic tangent we code (4.3.8) directly. The driver program for the inverse hyperbolic functions checks the condition $z^2 \neq 1$ ($x = \pm 1, y = 0$) and uses it to terminate execution. You might choose to move this check into C function `InvTanhZ`.

```
void InvTanhZ(double x, double y,
        double *InvTanhZRe, double *InvTanhZIm)
/*  Inverse hyperbolic tangent
        of complex argument z = x + i y  */
{
double xsqd,ysqd;

xsqd = x*x;
ysqd = y*y;

*InvTanhZRe =
    0.25*log((ysqd+pow(x+1.0,2.0))/
    (ysqd+pow(x-1.0,2.0))));
*InvTanhZIm = 0.5*atan2(2.0*y,1.0 - xsqd - ysqd);
}
```

The inverse functions that are related to the reciprocals of the hyperbolic functions—the arcsech, arccsch, and arccoth on the right side of Figure 4.3.4—can be obtained from the above functions by using the following identities for complex variable z:

$$\text{arcsech } z = \text{arccosh } 1/z \quad \text{arccsch } z = \text{arcsinh } 1/z \quad \text{arccoth } z = \text{arctanh } 1/z \quad (4.3.9)$$

Programming Notes

By including the signs in (4.3.4) and (4.3.6) with the C standard library functions `arccos` and `arcsin`, we obtain inverse hyperbolic functions whose branch cuts agree with those in Figure 4.3.3, Table 4.3.2, *Mathematica*, and with Abramowitz and Stegun [Abr64, Section 4.6].

Test Values

Check values of the inverse cosh, sinh, and tanh principal values for complez arguments z can be generated by running the cell `InverseHyperbolic` in *Mathematica* notebook `ElemTF`, the numerical output from which is given in Table 4.3.3.

TABLE 4.3.3 Test values to 10 digits of inverse hyperbolic functions for complex arguments. The notation is $(a, b) = a + i b$.

$z = (x, y)$	$(-4, -1)$	$(-4, 1)$	$(4, -1)$	$(4, 1)$
arccosh$z$	(2.096596457, −2.889413245)	(2.096596457, 2.889413245)	(2.096596457, −0.2521794087)	(2.096596457, 0.2521794087)
arcsinh$z$	(−2.122550124, −0.2383174618)	(−2.122550124, 0.2383174618)	(2.122550124, −0.2383174618)	(2.122550124, 0.2383174618)
arctanh$z$	(−0.2388778613, −1.508618830)	(−0.2388778613, 1.508618830)	(0.2388778613, −1.508618830)	(0.2388778613, 1.508618830)

REFERENCES ON ELEMENTARY TRANSCENDENTAL FUNCTIONS

[Abr64]   Abramowitz, M., and I. A. Stegun, *Handbook of Mathematical Functions,* Dover, New York, 1964.

[Gra94]   Gradshteyn, I. S., and I. M. Ryzhik, *Table of Integrals, Series, and Products*, Academic Press, Orlando, fifth edition, 1994.

[Spa87]   Spanier, J., and K. B. Oldham, *An Atlas of Functions*, Hemisphere Publishing, Washington, D.C., 1987.

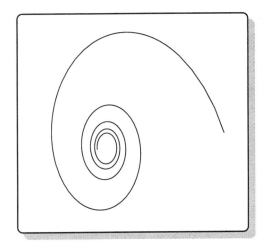

Chapter 5

EXPONENTIAL INTEGRALS
AND RELATED FUNCTIONS

Exponential and logarithmic integrals are related to the sine and cosine integrals, therefore they are considered together here, as in Chapter 5 of the handbook of mathematical functions edited by Abramowitz and Stegun [Abr64].

5.1   EXPONENTIAL AND LOGARITHMIC INTEGRALS

There is more than one kind of exponential integral. For the first—the exponential integral of the first kind (Section 5.1.1)—the integrand has no singularities. For the second—the exponential integral of the second kind (Section 5.1.2)—the integrand has a singularity on the real axis and the principal value of a contour integral is used to define this exponential integral. For the third kind of exponential integral—the logarithmic integral in Section 5.1.3—one uses as argument of the integral of the second kind the logarithm of x rather than x. We explore each integral in turn.

5.1.1   Exponential Integral of the First Kind

Keywords
exponential integral of first kind

C Function and *Mathematica* Notebook
ExpIntFirst in the driver ExpInts (C); cell ExpIntFirst in the notebook ExpInt (*Mathematica*) $\boxed{M}$

Background

The exponential integral of the first kind, $E_n(z)$ with z a complex variable, provides the starting point for computing other special functions, such as the cosine and sine integrals in Section 5.2. The analysis of $E_n(z)$ is discussed in Chapter I of the handbook of functions by Jahnke and Emde [Jah45], in Section 9.7 of Erdélyi et al [Erd53b] as a special case of the incomplete gamma function, in Section 5.1 of Abramowitz and Stegun [Abr64], and in Section 8.2 of Gradshteyn and Ryzhik [Gra94].

Numerical recipes for computing $E_n(z)$ are discussed in Section 6.3 of Press et al [Pre92]. Methods of higher accuracy have been given by Amos [Amo80] and by Chiccoli et al [Chi90].

Function Definition

We define the exponential integral of the first kind in terms of complex z, in order to clarify the connections to other functions and to the visualizations. The integral is defined by

$$E_n(z) \equiv \int_1^\infty \frac{e^{-zt}}{t^n}\, dt \qquad \mathrm{Re}\, z > 0 \qquad n = 0, 1, 2, \ldots \qquad (5.1.1)$$

By a simple change of variable, this may also be expressed as

$$E_n(z) = z^{n-1} \int_z^\infty \frac{e^{-t}}{t^n}\, dt \qquad \mathrm{Re}\, z > 0 \qquad n = 0, 1, 2, \ldots \qquad (5.1.2)$$

It is convenient for visualizations to define a modified exponential integral of the first kind, $E_n'(x)$ for positive real argument x, as

$$E_n'(x) \equiv E_n(x) / x^{n-1} = \int_x^\infty \frac{e^{-t}}{t^n}\, dt \qquad n = 0, 1, 2, \ldots \qquad (5.1.3)$$

Visualization

The real and imaginary parts of the exponential integral $E_n(z)$ for complex argument $z = x + iy$ are shown as the surfaces in Figures 5.1.1 ($n = 1$) and 5.1.2 ($n = 4$).

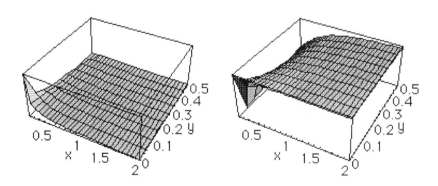

FIGURE 5.1.1 Exponential integral of the first kind (5.1.1) with $n = 1$; real part (left) and imaginary part (right). (Cell ExpIntFirst in the *Mathematica* notebook ExpInt.)

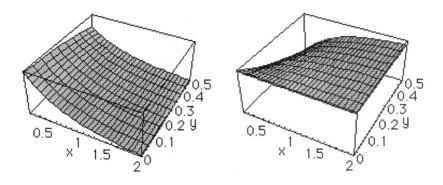

FIGURE 5.1.2 Exponential integral of the first kind (5.1.1) with $n = 4$; real part (left) and imaginary part (right). (Cell `ExpIntFirst` in the *Mathematica* notebook `ExpInt.`)

In the left panel of each figure the curve on the front surface at $y = 0$ is just the exponential integral of the first kind for a real argument x. Figure 5.1.3 shows the modified exponential integrals, $E'_n(x)$ in (5.1.3), for $n = 0$–3. These functions change more uniformly with n than does $E_n(x)$, allowing the same scales to be used for their display. However—except for $n = 0$, for which $E'_0(x) = e^{-x}$—the modified functions are singular near the origin, so we start the displays at $x = 0.002$.

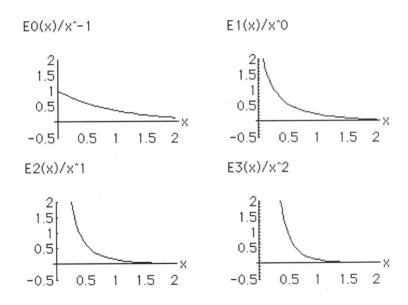

FIGURE 5.1.3 Modified exponential integrals given by (5.1.3) for real argument x and $n = 0$–3. (Cell `ExpIntFirst` in *Mathematica* notebook `ExpInt.`)

Algorithm and Program

This section provides a function to compute $E_n(x)$ with good accuracy. If the exponential integral is required for an arbitrary z, modify and run the cell ExpIntFirst in *Mathematica* notebook ExpInt. Our algorithm for $E_n(x)$ uses three methods: the explicit expression for $n = 0$, a series expansion for $x \le 1$, and a continued-fraction expansion (Section 3.3) for $x > 1$. The corresponding formulas are:

$$E_0(x) = e^{-x}/x \qquad x > 0 \tag{5.1.4}$$

$$E_n(x) = \frac{(-x)^{n-1}}{(n-1)!}[-\ln x + \psi(n)] - \sum_{\substack{m=0 \\ m \ne n-1}}^{\infty} \frac{(-x)^m}{[m-(n-1)]m!} \tag{5.1.5}$$

$$0 < x \le 1 \qquad n > 0$$

where the digamma function for integer argument n is given in terms of Euler's γ constant by

$$\psi(1) = -\gamma \qquad \psi(n) = -\gamma + \sum_{m=1}^{n-1}\frac{1}{m} \qquad n > 1 \tag{5.1.6}$$

Finally, for $x > 1$ there is the rapidly convergent even-form continued fraction

$$E_n(x) = e^{-x}\left[\frac{1}{x+n-}\ \frac{1.n}{x+n+2-}\ \frac{2(n+1)}{x+n+4-}\cdots\right] \qquad n > 0 \tag{5.1.7}$$

This is evaluated by using the Lenz algorithm described in references in Section 3.3.

The resulting C function, ExpIntFirst, is as follows. (The function Psi is listed two pages ahead.)

```
/* nearly smallest floating point number */
#define MINFP 1.0e-20

double ExpIntFirst(int n, double x,
    double eps)
/* Exponential Integral of First Kind,
    order  n >= 0, argument  x > 0,
    convergence criterion  eps   */


{
double Mx,sum,Psi(int n),powX,top,mFact,
    term,ratio,b,c,d,f,a;
int nM1,i,m;

if ( n == 0 )      return  exp(-x)/x;

nM1 = n - 1;
```

```
if ( x <= 1.0 )  /* use series expansion */
  {
  /* Initialize power series */
  Mx = - x;
  sum = -log(x) + Psi(n);
  powX = 1.0;
  if ( nM1 > 0 )
    {
    for ( i = 1; i <= nM1; i++ )
      powX *= Mx/i;
    }
  sum *= powX;
  /* Include power series */
  m = 0;
  top = 1.0;
  mFact = 1.0;
  term =  1.0;
  ratio = 10.0;
  while ( ratio > eps )
    {
    if ( m != nM1 )
      {  /* add in term of series */
      term = top/((m - nM1)*mFact);
      sum -= term;
      ratio = fabs(term/sum);
      }  /* then update pieces */
    m += 1;
    top *= Mx;
    mFact *= m;
    }
  return sum;
  }

/* else use Lentz algorithm
   for continued fraction */
m = 1;
b = x+n;
c = 1.0/MINFP;
d = 1.0/b;
f = d;
ratio = 10.0;
while ( ratio > eps )
  {
  a = -m*(nM1+m);
  b += 2.0;
  d = 1.0/(a*d+b);
  c = b+a/c;
  mFact = c*d;
  f *= mFact;
  m += 1;
  ratio = fabs(mFact-1.0);
  }
return  exp(-x)*f;
}
```

```
double Psi(int n)
/* Psi function for positive integer  n  */

{
double  PsiSum;
const double EulerG = 0.5772156649;
int  m;

PsiSum = - EulerG;
if ( n == 1 )    return  PsiSum;
for ( m = 1; m <= n-1; m++ )
  { PsiSum += 1.0/m;    }
return  PsiSum;
}
```

Programming Notes

The variable MINFP defined at the top of the program is used to control for underflow in the continued-fraction expansion (5.1.7). It should be about the size of—but not smaller than—the smallest floating-point number that can be represented in the computer in which function ExpIntFirst is used.

The input variable eps is used to control accuracy in both the series expansion (5.1.5) and the continued-fraction expansion (5.1.7). A value of about 10^{-10} will give about 10-digit fractional accuracy in $E_n(x)$. A more complete discussion of truncation errors is given by Amos [Amo80].

Test Values

Check values of the exponential integral of the first kind, $E_n(x)$, can be generated by running the ExpIntFirst cell in *Mathematica* notebook ExpInt. Examples are given in Table 5.1.1.

TABLE 5.1.1 Test values to 10 digits of the exponential integral of the first kind, (5.1.1), for positive real argument x.

n/x	0.5	1.0	1.5	2.0
1	0.5597735948	0.2193839344	0.1000195824	0.0489005107
3	0.2216043643	0.1096919672	0.0567394902	0.0301333798

5.1.2   Exponential Integral of the Second Kind

Keywords

exponential integral of second kind

C Function and *Mathematica* Notebook

ExpIntSecond in driver ExpInts (C); cell ExpIntSecond in the notebook ExpInt (*Mathematica*)

Background

The exponential integral of the second kind, $Ei(x)$ with x a positive real variable, is discussed in Chapter I of the handbook of functions by Jahnke and Emde [Jah45], in Section 9.7 of Erdélyi et al [Erd53b] as a special case of the incomplete gamma function, and in Section 5.1 of Abramowitz and Stegun [Abr64]. Recipes for computing $Ei(x)$ are presented in Section 6.3 of Press et al [Pre92].

Function Definition

The exponential integral of the second kind is defined by

$$Ei(x) \equiv -\;\;\Xsum_{-x}^{\infty} \frac{e^{-t}}{t}\,dt \qquad\qquad x>0 \qquad\qquad (5.1.8)$$

in which the Cauchy principal value is to be taken because the integrand is singular at $t=0$. This definition may also be expressed more clearly as

$$Ei(x) = \;\;\Xsum_{-\infty}^{x} \frac{e^{t}}{t}\,dt \qquad\qquad x>0 \qquad\qquad (5.1.9)$$

in which the necessity of going into the complex plane to bypass the singularity at $t=0$ is also evident, since $x>0$. If x were allowed to be negative we would have just $E_1(x)$ of Section 5.1.1 without a singularity in the integrand.

To clarify these relations, Figure 5.1.4 shows contours near the origin in the complex plane of the absolute value of e^v / v, where $v = t + iu$.

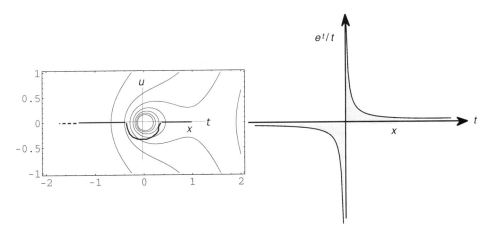

FIGURE 5.1.4 Contours of the absolute value of the integrand e^v/v in the complex plane (left) and this integrand along the real line $v = t$ (right).

The contour lines in Figure 5.1.4 can be obtained by running the cell `ExpIntSecond` in *Mathematica* notebook `ExpInt`. The solid line in the complex-plane plot is the required integration path, in the limit that the semicircle below the origin shrinks to zero radius. As is clear from the right-hand view, the integrand is highly singular near $t=0$. The required integral is the shaded area up to $t=x$, after the effect of the singularity has been surgically adjusted.

Visualization

The exponential integral of the second kind, $\mathrm{Ei}(x)$, is shown in Figure 5.1.5.

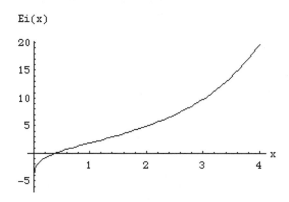

FIGURE 5.1.5 Exponential integral of the second kind, (5.1.9), as a function of $x > 0$. The plot starts at $x = 0.001$, close to the singularity of the integrand at $t = 0$. (Cell ExpIntSecond in *Mathematica* notebook ExpInt.)

The integral changes sign near $x = 0.3725$ because of cancellation between the pole term and the integral along the real axis.

Algorithm and Program

This section provides a function to compute $\mathrm{Ei}(x)$ with good accuracy. If this integral is required with very high accuracy, modify and run the cell ExpIntSecond in *Mathematica* notebook ExpInt. Our algorithm for the error integral of the second kind uses a power series expansion for small x and an asymptotic series for large x. The power series formula is [Abr64, 5.1.10]

$$\mathrm{Ei}(x) = \gamma + \ln x + \sum_{m=1}^{\infty} \frac{x^m}{m!\, m} \qquad x > 0 \qquad (5.1.10)$$

where γ is Euler's gamma constant. For the asymptotic series (Section 3.4) we have

$$\mathrm{Ei}(x) \sim \frac{e^x}{x}\left(1 + \sum_{m=1}^{\infty} \frac{m!}{x^m}\right) \qquad x \gg 0 \qquad (5.1.11)$$

obtained from the asymptotic series for the integral of the first kind—given as 5.1.51 in [Abr64] or (7) in Section 9.7 of [Erd53b]—through the relation for $x > 0$ $\mathrm{Ei}(x) = -E_1(-x)$.

To see that care is needed when estimating the switchover x value between the two series, consider the terms of the power series for $x = 10$ and $x = 15$ and for the asymptotic series for $x = 15$ and $x = 20$, shown in Figure 5.1.6.

As a compromise between the power series (5.1.10) and the asymptotic series (5.1.11), we switch series at $x = -\ln \varepsilon$. The resulting C function, ExpIntSecond, is given on the next two pages.

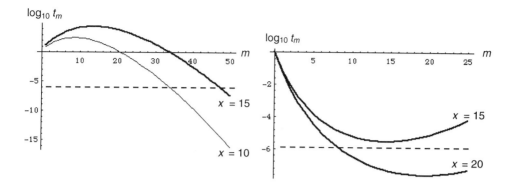

FIGURE 5.1.6 Terms of the power series (5.1.10) in left panel and of the asymptotic series (5.1.11) in right panel for indicated x values. The dashed lines show an accuracy value, $\varepsilon = 10^{-6}$. (Adapted from output of cell ExpIntSecond in *Mathematica* notebook ExpInt.)

```
double ExpIntSecond(double x, double eps)
/* Exponential Integral of Second Kind,
   argument  x > 0, convergence criterion eps */

{
double xSwitch,sum,term,ratio,sterm,termold;
const double EulerG = 0.577215664902;
int m;

xSwitch = - log(eps);

if ( x <= xSwitch )   /* use series expansion */
  {
  /* Initialize power series */
  sum = EulerG+log(x);
  /* Include power series */
  m = 0;
  term =   1.0;
  ratio = 10.0;

  while ( ratio > eps )
    {
    m += 1;
    term *= x/m;
    sterm = term/m;
    sum += sterm;
    ratio = fabs(sterm/sum);
    }

  return sum;
  }
```

```
/* Else use asymptotic series */
m = 1;
term = 1.0;
sum = 1.0;
ratio = 10.0;

while ( ratio > eps )
  {
  termold = term;
  term *= m/x;
  if ( termold/term < 1.0 )
    {
  printf
("\n!!ExpIntSecond not converged: try smaller eps");
    return  exp(x)*sum/x;
    }
  sum += term;
  ratio = fabs(term/sum);
  m += 1;
  }

return  exp(x)*sum/x;
}
```

Programming Notes

The input variable `eps` is used to control accuracy in both the series expansion (5.1.10) and the asymptotic expansion (5.1.11). A value of about 10^{-10} will give about 10-digit fractional accuracy in $\mathrm{Ei}(x)$ for $x < 10$.

Test Values

Check values of the exponential integral of the second kind, $\mathrm{Ei}(x)$, can be generated by running the `ExpIntSecond` cell in *Mathematica* notebook `ExpInt`. Table 5.1.2 gives examples for both small and large x. Note that for large x, (5.1.11) shows that the behavior of the integral is dominated by the exponential.

TABLE 5.1.2 Test values to 10 digits of the exponential integral of the second kind for real argument x.

x	0.1	0.2	0.3	0.4	0.5
$\mathrm{Ei}(x)$	−1.622812814	−0.8217605879	−0.3026685393	0.1047652186	0.4542199049

x	1	2	3	4	5
$\mathrm{Ei}(x)$	1.89511786	4.954234356	9.933832571	19.63087447	40.18527536

5.1.3 Logarithmic Integral

Keywords
logarithmic integral

C Functions and *Mathematica* Notebook
LogIntegral (C); cell LogIntegral in *Mathematica* notebook ExpInt

M

Background
The logarithmic integral, $\mathrm{li}(x)$, occurs in many contexts. For example, it is related to the distribution of prime numbers. An extensive analytical treatment is given by Koosis [Koo88] and formulas are presented in Section 5.1 of Abramowitz and Stegun [Abr64].

Function Definition
The logarithmic integral, $\mathrm{li}(x)$, is defined by

$$\mathrm{li}(x) \equiv \int_0^x \frac{dt}{\ln t} \qquad x > 1 \qquad (5.1.12)$$

As for $\mathrm{Ei}(x)$, the principal value of the integral is to be taken, since the integrand has a singularity on the real axis at $t = 1$. By making the transformation $t = e^{-u}$, we see that

$$\mathrm{li}(x) = -E_1(-\ln x) = \mathrm{Ei}(\ln x) \qquad x > 1 \qquad (5.1.13)$$

with the relation to Ei being of practical use. The asymptotic behavior of $\mathrm{li}(x)$ follows from that of $\mathrm{Ei}(x)$, given by (5.1.11), thus

$$\mathrm{li}(x) \xrightarrow[x \to \infty]{} \frac{x}{\ln x} \qquad (5.1.14)$$

Visualization
To visualize the logarithmic integral, it is interesting to plot the integrand in (5.1.12).

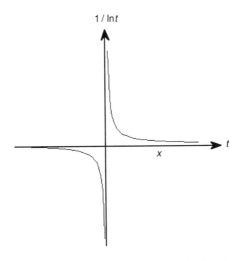

FIGURE 5.1.7 Integrand of the logarithmic integral (5.1.12), showing the singular region near $t = 1$. The shaded area shows the integrand for $x = 1.6$.

Note that in Figure 5.1.7 the integrand is exactly antisymmetric about $t = 1$. The singularity structure of the integrand is similar to that shown for $\text{Ei}(x)$ in Figure 5.1.4.

The logarithmic integral is readily envisioned by running the `LogIntegral` cell in the *Mathematica* notebook `ExpInt`, thus producing Figure 5.1.8.

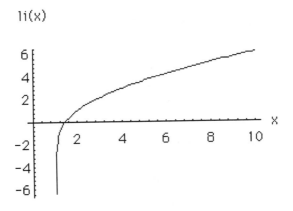

FIGURE 5.1.8   Logarithmic integral (5.1.12) for x ranging from 1.001 to 10.   (Cell `LogInt` in *Mathematica* notebook `ExpInt`.)

The zero of $\text{li}(x)$ that is seen in Figure 5.1.8 occurs near $x = 1.45137$.

Algorithm, Program, and Programming Notes

Computing the logarithmic integral requires only substitution of $\ln x$ for x in the exponential integral of the second kind, $\text{Ei}(x)$, of Section 5.1.2. The same considerations about series convergence apply, except that the x range that can be included is much larger. However, since most of us become nervous whenever instructions for hardware or software read "some assembly required," we provide a new function, in accordance with our precept discussed in The Computer Interface chapter.

```
double LogIntegral(double x, double eps)
/* Logarithmic Integral, argument  x > 1,
   convergence criterion  eps  */


{
double xSwitch,xLog,sum,term,ratio,sterm,
   termold;
const double EulerG = 0.577215664902;
int m;

xSwitch = 1.0/eps;
xLog = log(x);
```

```
if ( x <= xSwitch )   /* use series expansion */
  {
  /* Initialize power series */
  sum = EulerG+log(xLog);
  /* Include power series */
  m = 0;
  term =  1.0;
  ratio = 10.0;
  while ( ratio > eps )
    {
    m += 1;
    term *= xLog/m;
    sterm = term/m;
    sum += sterm;
    ratio = fabs(sterm/sum);
    }
  return sum;
  }

/* Else use asymptotic series */
m = 1;
term = 1.0;
sum = 1.0;
ratio = 10.0;
while ( ratio > eps )
  {
  termold = term;
  term *= m/xLog;
  if ( termold/term < 1.0 )
    {
    printf
("\n!!LogIntegral not converged: try smaller eps");
    return  x*sum/xLog;
    }
  sum += term;
  ratio = fabs(term/sum);
  m += 1;
  }
return  x*sum/xLog;
}
```

Test Values

Test values of logarithmic integral li(x) can be obtained by running cell `LogIntegral` in *Mathematica* notebook `ExpInt`. Samples are given in Table 5.1.3.

$\boxed{M}$

TABLE 5.1.3 Test values to 10 digits of the logarithmic integral (5.1.12).

x	2.0	4.0	6.0	8.0	10.0
li(x)	1.045163780	2.967585095	4.222222391	5.253718300	6.165599505

5.2   COSINE AND SINE INTEGRALS

Keywords
cosine integral, sine integral

C Functions and *Mathematica* Notebook

CosIntegral and SinIntegral (C); cell Cos&SinIntegral in *Mathematica* notebook ExpInt

Background
Cosine and sine integrals arise, for example, in the analysis of diffraction of waves. Analysis of these integrals is presented in Chapter I of Jahnke and Emde's handbook of functions [Jah45], in Section 5.2 of [Erd53b], in Chapter 38 of Spanier and Oldham [Spa87] and in Section 5.2 of Abramowitz and Stegun [Abr64].

Beware that there are several definitions of the cosine and sine integrals in various sources, but that the notation used for them usually indicates the differences.

Function Definition
We use the definition of the cosine integral, $\mathrm{Ci}(z)$, and of the sine integral, $\mathrm{Si}(z)$, in terms of complex variable, z, as in Abramowitz and Stegun, 5.2.2 and 5.2.1, respectively. These definitions are

$$\mathrm{Ci}(z) \equiv \gamma + \ln z + \int_0^z \frac{\cos t - 1}{t}\, dt \tag{5.2.1}$$

in which γ is Euler's gamma constant, and

$$\mathrm{Si}(z) \equiv \int_0^z \frac{\sin t}{t}\, dt \tag{5.2.2}$$

These integrals have the symmetry relations

$$\mathrm{Ci}(-z) = \mathrm{Ci}(z) - i\pi \qquad \mathrm{Si}(-z) = -\mathrm{Si}(z) \tag{5.2.3}$$

Therefore, when z is real, one needs only to compute the functions for positive x then use these symmetries if x is negative. The functions provided here compute the cosine and sine integrals for positive, real arguments.

Visualization

The cosine and sine integrals—$\mathrm{Ci}(z)$ and $\mathrm{Si}(z)$ in (5.2.1) and (5.2.2)—can be envisioned by running the Cos&SinIntegral cell in *Mathematica* notebook ExpInt. We note from (5.2.1) that the cosine integral has a logarithmic singularity as $z \to 0$, since the integral part of its definition gives a well-behaved function, with the −1 term canceling out the $1/t$ part of the integrand that would give rise to a logarithmic singularity. Therefore, in Figures 5.2.1 and 5.2.3 we start x at 0.1 rather than at zero.

As a function of the complex variable $z = x + iy$ the cosine and sine integrals vary relatively slowly with y. We therefore show their variation with y only from $y = 0$ to $y = 0.5$, with the $y = 0$ functions—which are purely real—providing the usual integrals.

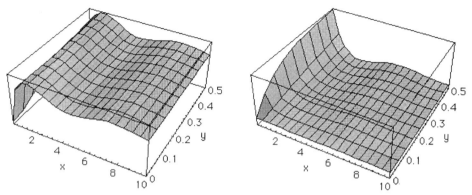

FIGURE 5.2.1 Cosine integrals in the complex plane, showing real part (left) and imaginary part (right) for $x = 0.1$ to 10. (Cell `Cos&SinIntegral` in *Mathematica* notebook `ExpInt.`)

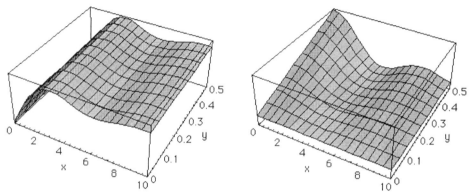

FIGURE 5.2.2 Sine integrals in the complex plane, showing real part (left) and imaginary part (right) for $x = 0$ to 10. (Cell `Cos&SinIntegral` in *Mathematica* notebook `ExpInt.`)

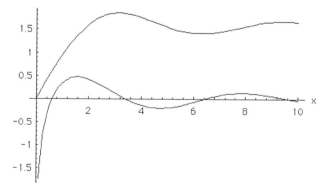

FIGURE 5.2.3 Cosine and sine integrals on the real axis, cosine integral (lower, x from 0.1 to 10) and sine integral (upper, $x = 0$ to 10). (Cell `Cos&SinIntegral` in *Mathematica* notebook `ExpInt.`)

From these figures we can be convinced that $\mathrm{Ci}(x)$ asymptotes to zero and that $\mathrm{Si}(x)$ asymptotes to $\pi/2$. This can also be seen if we define Cartesian coordinates by the parametric equations

$$x = \mathrm{Ci}(t) \qquad\qquad y = \mathrm{Si}(t) \tag{5.2.4}$$

to generate the spiral shown in Figure 5.2.4.

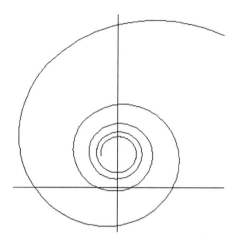

FIGURE 5.2.4 The Sici spiral, obtained from the cosine and sine integrals by (5.2.4) with t from 2 to 30. (Cell Cos&SinIntegral in *Mathematica* notebook ExpInt.)

The Sici spiral—perhaps named after an obscure Italian mathematician—can be computed and displayed by running the last part of the Cos&SinIntegral cell in the *Mathematica* notebook ExpInt. In Figure 5.2.4 the point of convergence of the spiral is $(x,y) = (0, \pi/2)$, the asymptotic values of $\mathrm{Ci}(x)$ and $\mathrm{Si}(x)$. Remarkably, the radius of curvature of the Sici spiral is $R = e^{-s}$, where s is the length of the arc measured from the origin. This property was sometimes used to design French curves in the ancient art of graphics using mechanical devices. The Sici spiral is similar to the Cornu spiral discussed in Section 10.2.

Algorithm and Programs
Most uses of the cosine and sine integrals restrict their arguments to z being real, that is, to Figure 5.2.3 rather than Figures 5.2.1 and 5.2.2. Further, if $x < 0$ then $\mathrm{Ci}(x)$ has an imaginary part according to (5.2.3), since it has the logarithmic term in its definition (5.2.1). We therefore provide functions to compute to moderate accuracy $\mathrm{Ci}(x)$ and $\mathrm{Si}(x)$ for positive values of x. If either function is required for some arbitrary z, just modify and run the Cos&SinIntegral cell in the *Mathematica* notebook ExpInt.

Our algorithms for $\mathrm{Ci}(x)$ and $\mathrm{Si}(x)$ use a combination of power series and rational-fraction approximations for the related auxiliary functions $f(x)$ and $g(x)$ similar to those in Section 5.2 of Abramowitz and Stegun [Abr64]. The power series used for $x \leq 15$ are

$$\mathrm{Ci}(x) = \gamma + \ln x + \sum_{n=1}^{\infty} \frac{(-x^2)^n}{2n\,(2n)!} \tag{5.2.5}$$

for the cosine integral, and

$$Si(x) = x \sum_{n=0}^{\infty} \frac{(-x^2)^n}{(2n+1)(2n+1)!} \qquad (5.2.6)$$

for the sine integral. These are formulas 5.2.16 and 5.2.15 in [Abr64]. Fewer than 30 terms are needed for 10-digit accuracy. For $x > 15$ we use for the cosine integral

$$Ci(x) = f(x)\sin x - g(x)\cos x \qquad (5.2.7)$$

and for the sine integral

$$Si(x) = \pi/2 - f(x)\cos x - g(x)\sin x \qquad (5.2.8)$$

which are from [Abr64] 5.2.9 and 5.2.8, respectively. The auxiliary functions f and g are approximated by rational fractions, as follows:

$$f(x) \approx \frac{1 + \sum_{n=1}^{5} a_{fn} t^n}{x \left(1 + \sum_{n=1}^{5} b_{fn} t^n \right)} \qquad t = 1/\sqrt{x} \qquad (5.2.9)$$

and

$$g(x) \approx \frac{1 + \sum_{n=1}^{5} a_{gn} t^n}{x^2 \left(1 + \sum_{n=1}^{5} b_{gn} t^n \right)} \qquad t = 1/\sqrt{x} \qquad (5.2.10)$$

which result in Ci and Si accurate to at least 10 digits. The a and b coefficients in these expansions are given as elements of constant arrays at the start of each C function below.

The C programs are listed separately for $Ci(x)$ and $Si(x)$, although there is considerable repetition. If you always require both integrals, you should re-assemble the functions carefully for greater efficiency. Function `CosIntegral` that computes for x positive real is given on this and the following page.

```
double CosIntegral(double x, double eps)
/* Cosine Integral of positive argument x */
{
double t,sum,term,Tn,ratio,
  fNum,fDen,fRat,gNum,gDen,gRat;
const double EulerG = 0.577215664902;
const double af[] =
{0.0,1.5174925627,2.41568720382,
-13.5963712331,54.2431202954,-48.5270735746};
const double bf[] =
{0.0,1.51749287668,2.41565271558,
-13.5948378168,56.2062565896,-44.9587260941};
 const double ag[] =
{0.0,4.84019146854,8.91175071572,
-49.899430686,174.744490325,-156.250555088};
const double bg[] =
{0.0,4.840194966833,8.91136645122,
-49.8823576188,180.33451762,-121.288632506};
```

```
if ( x <= 15.0 )
/* use power series to relative accuracy eps*/
 {
 t = x*x;
 sum = EulerG+log(x);
 term = 1.0;
 Tn = 2.0;
 ratio = 10.0;
 while ( ratio > eps )
   { term *= -t/(Tn*(Tn-1.0));
     sum += term/Tn;
     ratio = fabs(term/sum);
     Tn += 2.0; }
 return    sum;
 }
/* Else use rational fraction */
t = 1.0/sqrt(x);
fNum = t*(t*(t*(t*(
 af[5]*t+af[4])+af[3])+af[2])+af[1])+1.0;
fDen = x*(t*(t*(t*(t*(
 bf[5]*t+bf[4])+bf[3])+bf[2])+bf[1])+1.0);
fRat = fNum/fDen;
gNum = t*(t*(t*(t*(
 ag[5]*t+ag[4])+ag[3])+ag[2])+ag[1])+1.0;
gDen = x*x*(t*(t*(t*(t*(
 bg[5]*t+bg[4])+bg[3])+bg[2])+bg[1])+1.0);
gRat = gNum/gDen;
return   fRat*sin(x) - gRat*cos(x);
 }
```

The C program for `SinIntegral` to compute $Si(x)$ in (5.2.2) for x positive real follows. Although it is repetitious of `CosIntegral`, it avoids the awkwardness in C when one wants to return more than one value.

```
double SinIntegral(double x, double eps)
/* Sine Integral of positive argument x*/
{
double t,sum,term,TnP1,ratio,
 fNum,fDen,fRat,gNum,gDen,gRat;
const double EulerG = 0.577215664902;
const double PiD2 = 1.57079632679;
const double af[] =
{0.0,1.5174925627,2.41568720382,
-13.5963712331,54.2431202954,-48.5270735746};
const double bf[] =
{0.0,1.51749287668,2.41565271558,
-13.5948378168,56.2062565896,-44.9587260941};
const double ag[] =
{0.0,4.84019146854,8.91175071572,
-49.899430686,174.744490325,-156.250555088};
const double bg[] =
{0.0,4.840194966833,8.91136645122,
-49.8823576188,180.33451762,-121.288632506};
```

```
if ( x <= 15.0 )
/* use power series to relative accuracy eps*/
  {
  t = x*x;
  sum = 1.0;
  term = 1.0;
  TnP1 = 3.0;
  ratio = 10.0;
  while ( ratio > eps )
    {
    term *= -t/(TnP1*(TnP1-1.0));
    sum += term/TnP1;
    ratio = fabs(term/sum);
    TnP1 += 2.0;
    }
  return    x*sum;
  }

/* Else use rational fraction */
t = 1.0/sqrt(x);
fNum = t*(t*(t*(t*(
 af[5]*t+af[4])+af[3])+af[2])+af[1])+1.0;
fDen = x*(t*(t*(t*(t*(
 bf[5]*t+bf[4])+bf[3])+bf[2])+bf[1])+1.0);
fRat = fNum/fDen;
gNum = t*(t*(t*(t*(
 ag[5]*t+ag[4])+ag[3])+ag[2])+ag[1])+1.0;
gDen = x*x*(t*(t*(t*(t*(
 bg[5]*t+bg[4])+bg[3])+bg[2])+bg[1])+1.0);
gRat = gNum/gDen;
return  PiD2 - fRat*cos(x) - gRat*sin(x);
}
```

If you usually compute the pair $Ci(x)$ and $Si(x)$ for the same argument, combine the two functions for greater efficiency.

Programming Notes

Functions `CosIntegral` and `SinIntegral` compute the polynomials in (5.2.9) and (5.2.10) by using the Horner polynomial algorithm, which accumulates the terms in the series iteratively, starting from the highest term. This is efficient, since it avoids direct computation of powers of the variable t.

The input parameter `eps` is used to check the convergence of the power-series expansions (5.2.5) and (5.2.6) to relative accuracy `eps`. An input value of 10^{-10} will give about this accuracy in the functions. Such accuracy will also match that obtained in the rational-fraction approximations (5.2.9) and (5.2.10).

Test Values

Check values of $Ci(x)$ and $Si(x)$ can be obtained by running the `Cos&SinIntegral` cell in *Mathematica* notebook `ExpInt`. Examples are given in Tables 5.2.1 and 5.2.2, following.

M

TABLE 5.2.1 Test values to 10 digits of the cosine integral for positive real argument x.

x	0.5	1.0	1.5	2.0
$\text{Ci}(x)$	-0.1777840788	0.3374039229	0.4703563172	0.4229808288
x	5	10	15	20
$\text{Ci}(x)$	-0.1900297497	−0.045456433	0.0462786777	0.0444198208

TABLE 5.2.2 Test values to 10 digits of the sine integral for positive real argument x.

x	0.5	1.0	1.5	2.0
$\text{Si}(x)$	0.4931074180	0.9460830704	1.324683531	1.605412977
x	5	10	15	20
$\text{Si}(x)$	1.549931245	1.658347594	1.618194444	1.548241701

REFERENCES ON EXPONENTIAL INTEGRALS AND RELATED FUNCTIONS

[Abr64]   Abramowitz, M., and I. A. Stegun, *Handbook of Mathematical Functions*, Dover, New York, 1964.

[Amo80]   Amos, D. E., ACM Transactions on Mathematical Software, **6**, 365 (1980); **6**, 420 (1980).

[Chi90]   Chiccoli, C., S. Lorenzutta, and G. Maino, Computing, **45**, 269 (1990).

[Erd53b]   Erdélyi, A., et al, *Higher Transcendental Functions*, Vol. 2, McGraw-Hill, New York, 1953; reprint edition, Krieger, Malabar, Florida, 1981.

[Gra94]   Gradshteyn, I. S., and I. M. Ryzhik, *Table of Integrals, Series, and Products*, Academic Press, Orlando, fifth edition, 1994.

[Jah45]   Jahnke, E., and F. Emde, *Tables of Functions*, Dover, New York, fourth edition, 1945.

[Koo88]   Koosis, P., *The Logarithmic Integral*, Cambridge University Press, Cambridge, 1988.

[Pre92]   Press, W. H., S. A. Teukolsky, W. T. Vetterling, and B. P. Flannery, *Numerical Recipes in C*, Cambridge University Press, New York, second edition, 1992.

[Spa87]   Spanier, J., and K. B. Oldham, *An Atlas of Functions*, Hemisphere Publishing, Washington, D.C., 1987.

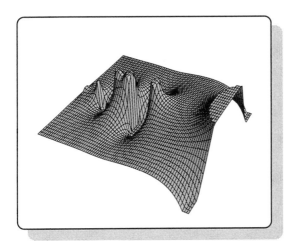

Chapter 6

GAMMA AND BETA FUNCTIONS

The gamma function, its derivatives (the digamma and other polygamma functions), and the beta function (products and quotients of gamma functions) are interrelated, as are the incomplete gamma and beta functions. They are therefore considered together, as in Chapter 6 of the mathematical-functions handbook edited by Abramowitz and Stegun [Abr64].

6.1   GAMMA FUNCTION AND BETA FUNCTION

The gamma function provides all that is needed for computing the beta function, so it is explored first. Note that this section covers the complete gamma and beta functions, with the incomplete functions being described in Section 6.3.

6.1.1 Gamma Function

Keywords
gamma function, complete gamma function

C Function and *Mathematica* Notebook
LogGamma (C); cell GammaFunction in notebook GmBt (*Mathematica*)

M

Background
The gamma function, $\Gamma(z)$ with z a complex variable, provides the starting point for computing many other special functions used from number theory to mathematical physics. The analysis and properties of $\Gamma(z)$ are discussed in Chapter 12 of Whittaker and Watson [Whi27], in Chapter II of the handbook of functions by Jahnke and Emde [Jah45], in Sections 1.1–1.6 of Erdélyi et al [Erd53a], in Chapter 1 of Magnus and Oberhettinger [Mag49], in Section 6.1 of Abramowitz and Stegun [Abr64], in Chapter 11 of Luke [Luk69], in Chapter 43 of Spanier and Oldham [Spa87], and in Sections 8.31–8.33 of Gradshteyn and Ryzhik [Gra94].

The function $\Pi(z) \equiv \Gamma(z+1)$ is called the factorial function (Section 7.1), since it coincides with the factorial $z!$ when z is a nonnegative integer.

Function Definition

The gamma function for complex z, $\Gamma(z)$, is defined by

$$\Gamma(z) \equiv \int_0^{\infty} t^{z-1}e^{-t}\,dt = \int_0^1 \left(\ln 1/u\right)^{z-1}du \qquad \mathrm{Re}\,z > 0 \qquad (6.1.1)$$

From this, it is easy to verify that $\Gamma(n+1) = n!$ when n is a nonnegative integer, suggesting that the gamma function increases rapidly in magnitude as the magnitude of z increases. This surmise is verified in the following visualizations.

Visualization

The gamma function is shown from several viewpoints. Figure 6.1.1 shows $\Gamma(z)$ in the complex plane $z = x + iy$.

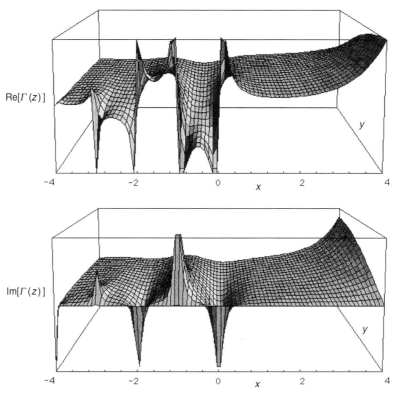

FIGURE 6.1.1 Gamma function in the complex plane, truncated for values exceeding 6 in magnitude; real part (upper) and imaginary part (lower). Shown for x in [–4, 4] and y in (0, 1]. (Adapted from cell `GammaFunction` in *Mathematica* notebook `GmBt`.)

Another view of the surface $\mathrm{Re}\,\Gamma(z)$, with x in [–4, 4] and y in [–2, 2], provides the motif at the front of this chapter. The poles of $\Gamma(z)$ at zero and the negative integers, seen in Figure 6.1.1, are emphasized by contour plots and $\Gamma(z)$ along the axes, as shown for $|\Gamma(z)|$ in Figures 6.1.2 and 6.1.3.

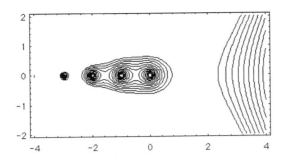

FIGURE 6.1.2 Contours of the logarithm of the absolute value of $\Gamma(z)$ in the complex plane for x in $[-4, 4]$ and y in $[-2, 2]$. Four simple poles are shown at $x = 0, -1, -2, -3$, with that at $x = -4$ not shown. (Cell GammaFunction in *Mathematica* notebook GmBt.)

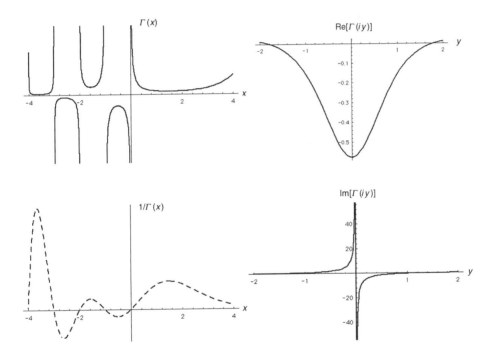

FIGURE 6.1.3 Left side: Gamma function along the real (x) axis in upper panel (truncated for values exceeding 6 in magnitude). Lower panel shows its reciprocal. Right side: Gamma function along the imaginary (y) axis, real part (upper) and imaginary part (lower), for y in $[-2, 2]$. (Adapted from cell GammaFunction in *Mathematica* notebook GmBt.)

Algorithm, Program, and Programming Notes

For computing $\Gamma(x)$ for $x > 0$ to accuracy better than 10 digits we use recurrence on x so that $\ln\Gamma(x')$ is estimated by an asymptotic series where $x' > 10$. For $x < 10$ we use

$$\Gamma(z) = \Gamma(z+1)/z \qquad (6.1.2)$$

applied to the logarithm of the gamma function to move its argument to be greater than 10, then a single recursion of the function produces $\ln\Gamma(x)$. The asymptotic expansion is given by [Abr64, 61.40]

$$\ln\Gamma(x) \sim (x-1/2)\ln x - x + \ln\sqrt{2\pi} + \sum_{k=1}^{\infty} \frac{B_{2k}}{2k(2k-1)x^{2k-1}} \tag{6.1.3}$$

in which B_{2k} is an even-order (nonzero) Bernoulli number (Section 8.1). The coefficients in this series diverge for $k > 4$, following the rapid divergence of the Bernoulli numbers (Figures 8.1.1, 8.1.2). The series is therefore divergent for any x, and must be truncated appropriately. For example, when $x > 10$ the series is convergent through $k = 5$ and produces better than 10-digit accuracy in $\ln\Gamma(x)$. We therefore choose to sum it through this order. If higher acuracy is required the lowest x value that is used must be increased.

The function is coded as follows:

```
double LogGamma(double x)
{
/* Log of gamma function by asymptotic series */

double Recxs,sum,term,x10;

const double EulerG = 0.577215664902,
  hlfLn2Pi = 0.918938533205;
const double b[] =
  {0.0, 0.0833333333333, -0.00277777777778,
   0.000793650793651, -0.000595238095238,
   0.000841750841751};
int xInt,k,x11;

if ( x > 10.0 ) /* use asymptotic series */
  {
  Recxs = 1.0/(x*x);
  term = x;
  sum = (x-0.5)*log(x)-x+hlfLn2Pi;
  for ( k = 1; k <= 5; k++ )
    {
    term *= Recxs;
    sum += b[k]*term;
    }
  return  sum;
  }

if ( x > 0 ) /* recurrence to x>10 */
  {
  x -= 1.0;
  x11 = 11.0 - x;
  x10 = x + (double) x11;
  xInt = x;
  sum = 0.0;
  for ( k = 1; k <= x11-1; k++ )
    {
    sum -= log(x10 - (double) k );
    }
  return sum+LogGamma(x10);
  }
```

```
if ( x == 0.0 )
 {
printf("\n!!x=0 in LogGamma; zero returned");
return 0.0;
 }


printf("\n!!x < 0 in LogGamma; zero returned");
printf("\nUse reflection formula");
return  0.0;
 }
```

Note that our algorithm and program are optimal for $x > 10$. If most of your computing requires $0 < x < 10$, then downshift x by using the recurrence (6.1.2) in the opposite direction until the argument is less than unity. A rapidly-convergent power series, such as [Abr64, 6.1.33], can then be used.

For negative x values the reflection formula [Abr64, 6.1.17] can be used as

$$\Gamma(x) = \frac{\pi}{-x\,\Gamma(-x)\sin(\pi x)} \qquad x \neq 0, -1, -2, \ldots \qquad (6.1.4)$$

Therefore—except at the poles of the Γ function at zero and the negative integers (Figures 6.1.1, 6.1.3)—the C function `LogGamma` can be used to compute $\ln\Gamma(-x)$, this can be exponentiated and inserted in (6.1.4). Note that it is foolhardy to attempt to code this for $\ln\Gamma(x)$, since $\Gamma(x)$ is negative half of the time when $x < 0$ (Figure 6.1.3).

Extensive performance evaluations of various programs for $\Gamma(x)$ and $\ln\Gamma(x)$, implemented in Fortran on several computers, are given by Cody [Cod91]. Bohman and Fröberg [Boh92] provide power series for computing accurately $\Gamma(x)$ when x is close to a small integer. They also provide an algorithm for finding the z values at which either the real or the imaginary part of $\Gamma(z)$ vanishes.

Test Values

Numerical values of $\Gamma(x)$, or of $\Gamma(z)$ for complex arguments, can be obtained readily by running cell `GammaFunction` in *Mathematica* notebook `GmBt`. Test values for our C function, `LogGammaFunc`, are given in Table 6.1.1.

TABLE 6.1.1   Test values to 10 digits of the gamma function for positive real argument x.

x	0.5	1.0	1.5	2.0
$\Gamma(x)$	1.772453851	1.000000000	0.8862269255	1.000000000
$\ln\Gamma(x)$	0.5723649429	0	-0.1207822376	0

x	5	10	15	20
$\Gamma(x)$	24.00000000	362880.0000	$8.717829120 \times 10^{10}$	$1.216451004 \times 10^{17}$
$\ln\Gamma(x)$	3.178053830	12.80182748	25.19122118	39.33988419

6.1.2 Beta Function

Keywords
beta function, complete beta function

C Function and *Mathematica* Notebook
BetaFunc (C); cell BetaFunction in notebook GmBt (*Mathematica*)

Background
The beta function, $B(z,w)$ with z and w complex variables, occurs when computing many other special functions, as in probability theory and mathematical physics. Properties of $B(z,w)$ are given in Section 1.5 of Erdélyi et al [Erd53a], in Chapter 1 of Magnus and Oberhettinger [Mag49], in Section 6.2 of Abramowitz and Stegun [Abr64], and in Section 8.38 of Gradshteyn and Rhyzik [Gra94].

Function Definition
The beta function for complex z and w, $B(z,w)$, is defined by

$$B(z,w) \equiv \int_0^1 t^{z-1}(1-t)^{w-1}\, dt \qquad \operatorname{Re} z > 0, \quad \operatorname{Re} w > 0 \qquad (6.1.5)$$

Alternative formulas are

$$B(z,w) \equiv 2\int_0^{\pi/2}(\sin\theta)^{2z-1}(\cos\theta)^{2w-1}\, d\theta$$

$$= \frac{\Gamma(z)\,\Gamma(w)}{\Gamma(z+w)} \qquad (6.1.6)$$

The second of these expressions is the most practical one for computing $B(z,w)$. It also shows the symmetry

$$B(z,w) = B(w,z). \qquad (6.1.7)$$

which is illustrated in Figure 6.1.5 that follows.

Section 1.5 of Erdélyi et al [Erd53a] provides an extensive list of properties of the beta function.

Visualization
If one tries to show the beta function in the full complexity used for the gamma function in Section 6.1.1, one has a formidable range of parameter space to visualize, since there are now two complex variables to consider.

As the relation to the gamma functions in (6.1.4) shows, $B(z,w)$ generally has poles at the poles of $\Gamma(z)$ and of $\Gamma(w)$, plus zeros at the poles of $\Gamma(z+w)$. The beta-function surfaces analogous to those shown in Figure 6.1.1 are therefore enormously complicated, and would not likely improve our understanding of $B(z,w)$.

On the other hand, if z and w are constrained to real positive values, the surfaces of the beta function are remarkably simple, as shown in Figure 6.1.4.

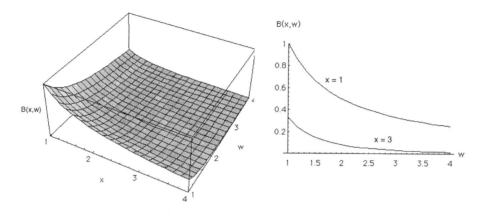

FIGURE 6.1.4 Left: Beta function surface in the real x and real w variables, both in the range [1, 4]. Right: Slices of the beta function along the $x = 1$ and $x = 3$ directions. (Adapted from cell `BetaFunction` in *Mathematica* notebook `GmBt`.)

Notice that for $x \geq 1$ and $w \geq 1$ the beta function changes gently with x and w, even though the gamma function diverges rapidly as its argument increases. To emphasize this, Figure 6.1.5 shows contours of $B(x,w)$ in the same range of arguments as in Figure 6.1.4. Note the symmetry of the contours about the $x = w$ axis because of the symmetry (6.1.5).

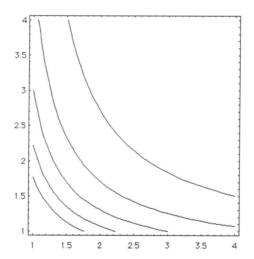

FIGURE 6.1.5 Contours of the beta function in the real x and real w variables. (Cell `BetaFunction` in *Mathematica* notebook `GmBt`.)

Algorithm, Program, and Programming Notes

The algorithm for computing $B(x,w)$, with x and w real and positive, is very simple. It requires only use of the relation in (6.1.4) in terms of the gamma functions, as follows.

```
double BetaFunc(double x, double w)
{
/* Beta function for real & positive arguments
   from log gamma functions */

double LogGamma(double x);

return  exp(LogGamma(x) + LogGamma(w)
            - LogGamma(x+w));
}
```

Notice that very large values of x and w can be accommodated in this method of computation, since only at the last step—computation of the exponential—is there any chance of overflow. If the beta function is required with real but negative arguments, the code for LogGamma should be modified as described in the text below the listing of this C function in Section 6.1.1.

Test Values

Numerical values of $B(x,w)$ can be obtained readily by running cell BetaFunction in *Mathematica* notebook GmBt. Test values for our C function are given in Table 6.1.2.

TABLE 6.1.2 **Test values to 10 digits of the beta function for positive real argument** x **and** w.

x, w	0.5, 2.0	2.0, 5.0	2.5, 7.5	3.0, 11.0
$B(x,w)$	1.333333333	0.0333333333	0.0068549766	0.0011655011

6.2   PSI (DIGAMMA) AND POLYGAMMA FUNCTIONS

The psi function, $\psi(z)$, also called the digamma function, along with its derivatives as functions of z, called the polygamma functions, all arise from the gamma function, $\Gamma(z)$, explored in Section 6.1.1. Nomenclature and relations are summarized in Table 6.2.1.

TABLE 6.2.1 **Nomenclature of gamma and polygamma functions.**

n	Function	Name
—	$\Gamma(z)$	gamma
—	$\psi(z) = d\ln\Gamma(z)/dz$	digamma (psi)
1	$\psi^{(1)}(z) = \psi'(z) = d\psi(z)/dz$	trigamma
2	$\psi^{(2)}(z) = d\psi'(z)/dz$	tetragamma
$\vdots$	$\vdots$	$\vdots$
n	$\psi^{(n)}(z)$	$(n+2)$ - gamma

We now explore each of these functions that derive from the gamma function.

6.2.1   Psi Function

Keywords
psi function, digamma function

C Function and *Mathematica* Notebook
PsiFunc (C); cell PsiFunction in notebook GmBt (*Mathematica*)

Background
The psi function often occurs in conjunction with the gamma function, $\Gamma(z)$ with z a complex variable, as the starting point for computing many other special functions. Analysis and properties of $\psi(z)$ are discussed in Chapter II of the handbook of functions by Jahnke and Emde [Jah45], but with a different definition than the one we use. The psi function is also discussed in Section 1.7 of Erdélyi et al [Erd53a], in Chapter 1 of Magnus and Oberhettinger [Mag49], in Section 6.3 of Abramowitz and Stegun [Abr64], in Chapter 44 of Spanier and Oldham [Spa87], and in Section 8.36 of Gradshteyn and Ryzhik [Gra94]. Analytical results are derived in Section 4.5 of the text by González [Gon92].

Function Definition
The psi function for complex z, $\psi(z)$, is defined by

$$\psi(z) \equiv \frac{d \ln \Gamma(z)}{dz} = \frac{1}{\Gamma(z)} \frac{d\Gamma(z)}{dz} \tag{6.2.1}$$

This is well-defined, since $\Gamma(z)$ has no zeros. Psi has zeros at the poles of $\Gamma(z)$, as is clear in the following visualizations.

Visualization
Just as with the gamma function in Section 6.1.1, the psi function is shown from several viewpoints. Figure 6.2.1 shows $\psi(z)$ in the complex plane.

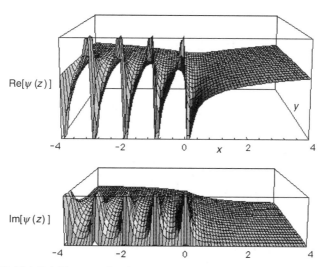

FIGURE 6.2.1 Psi (digamma) function in the complex plane; real part (upper, range –6 to 6) and imaginary part (lower, range 0 to 6). Shown for x in [–4, 4] and y in (0, 1]. (Adapted from cell PsiFunction in *Mathematica* notebook GmBt.)

Contours of the digamma function in the complex plane—complete with the second-order poles at zero and the negative integers—are shown in Figure 6.2.2.

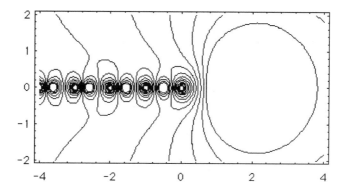

FIGURE 6.2.2 Contours of the logarithm of the absolute value of $\psi(z)$ in the complex plane, shown for x in [–4, 4] and y in [–2, 2]. Second-order poles of $\psi(z)$ occur along the real axis at zero and the negative integers. (Cell `PsiFunction` in *Mathematica* notebook `GmBt`.)

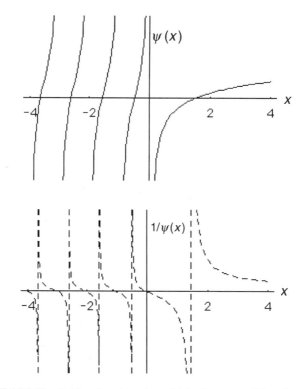

FIGURE 6.2.3 Top: Psi function along the real (x) axis, truncated for values exceeding 6 in magnitude) for x in [–4, 4]. Bottom: Reciprocal of the psi function for x in [–4, 4] with the same truncation. (Adapted from cell `PsiFunction` in *Mathematica* notebook `GmBt`.)

Figure 6.2.3 shows ψ along the real axis, similarly to the left side of Figure 6.1.3 for its parent function Γ. Since $\psi(x)$ has zeros as well as poles (for $x \leq 0$), its reciprocal (bottom of Figure 6.2.3) has poles matching the zeros of $\psi(x)$.

Algorithm and Program

Our algorithm for computing $\psi(x)$, with x real and positive, uses different methods depending upon the x value, as summarized in Table 6.2.2. For completeness, we give the expressions for complex argument z.

TABLE 6.2.2 Formulas for calculating psi in various ranges of $z = x$, a positive real variable.

z range	Equation in [Abr64]	Formula	
$(0, 0.5)$	6.3.7	$\psi(z) = \psi(1-z) - \pi / \tan(\pi z)$	(6.2.2)
$[0.5, 1.5]$	6.3.15	$\psi(1+z) = 1 - \gamma + 1/(2z) - \pi / [2\tan(\pi z)]$ $-1/(1-z^2) - \sum_{n=1}^{\infty} [\zeta(2n+1)-1] z^{2n}$	(6.2.3)
$[0.999, 1.001]$	6.3.14 (adapted)	$\psi(1+z) = -\gamma + z/(1+z) + [\zeta(2)-1]z$ $-[\zeta(3)-1]z^2 + [\zeta(4)-1]z^3 \dots$	(6.2.4)
$(1.5, 2.5)$	6.3.5	$\psi(z) = 1/(z-1) + \psi(z-1)$	(6.2.5)
$[2.5, 10]$	6.3.5, 6.3.6	$\psi(n+z) = \psi(1+z) + \sum_{k=0}^{n-2} 1/(k+1+z)$	(6.2.6)
$(10, \infty)$	6.3.18	$\psi(z) \sim \ln z - 1/(2z) - 1/(12z^2) + 1/(120z^4)$ $-1/(252z^6) + 1/(240z^8) - 1/(132z^{10})$	(6.2.7)

In the table, γ is the Euler gamma function and ζ is the Riemann zeta function (Section 8.1). Although we code it with $z = x$, function `PsiFunc` is easily adapted to a complex variable, except for the poles at the negative integers.

```
double PsiFunc(double x)
{
/* Psi function for real positive arguments
   from various series */

double xM1,t,sum,tPower,xF,s;
const double MEulerG = -0.577215664902,
  OneMEulerG = 0.422784335098,
  Pi = 3.14159265359, PiD2 = 1.57079632679,
  r[10] = {0.0,0.202056903160,0.036927755143,
  0.008349277382,0.002008392826,0.000494188604,
  0.000122713348,0.000030588236,0.000007637198,
  0.000001908213};
int n,k;
```

```
if ( x < 0.5 )
 return  PsiFunc(1.0 - x) - Pi/tan(Pi*x);

if ( x <= 1.5 )
 {
 xM1 = x - 1.0;

 if ( fabs(xM1) <= 0.001 )
  return  MEulerG+xM1/x+xM1*(0.644934066848 -
   xM1*(0.202056903160 - 0.082323233711*xM1));

 t = xM1*xM1;
 sum = OneMEulerG + 0.5/xM1 - PiD2/tan(Pi*xM1)
      - 1.0/(1.0-t);
 tPower = t;
 for ( n = 1; n <= 9; n++ )
  {
  sum -= r[n]*tPower;
  tPower *= t;
  }
 return  sum;
 }

if (  x < 2.5 )
 return  1.0/(x - 1.0) + PsiFunc(x-1.0);

if ( x <= 10.0 )
 {
 n = x;  n = n - 1;
 xF = x - (double) n;
 sum = 0.0;
 for ( k = 0; k < n; k++ )
  sum += 1.0/(k+xF);
 return  sum + PsiFunc(xF);
 }

/* Else   x > 10; use asymptotic series  */
s = 1.0/x;
t = s*s;
return
log(x) - 0.5*s-0.083333333333*t*(1.0-t*(0.1-
t*(0.047619047619 - t*(0.05 - 0.090909090909*t)))));
 }
```

Programming Notes

Function `PsiFunc` makes extensive use of recursion as discussed in Section 3.2, in spite of warnings given there about its overuse. It is, however, both convenient and practical to use in this case, since the recursion is at most one level deep and it avoids repetitive coding. For x near unity, expression (6.2.4) is preferred over (6.2.3) because of subtractive cancellation between the leading terms in the latter when z is near zero. It is adapted from [Abr64, 6.3.14] in that the coefficients of the powers of z have unity subtracted to improve convergence. As given, the accuracy obtained is better than 12 digits.

The asymptotic expansion (6.2.7) carried out to the terms shown already has coefficients that are increasing in magnitude, as always happens eventually with such expansions (Section 3.4). For $x > 10$, convergence of this expansion to about 12 digits through the number of terms given is assured by the negative powers of x. If such an asymptotic expansion is attempted to be used with user-variable accuracy then the x value at which the switch to the expansion is made has to depend upon the accuracy, as described in the detailed example in Section 3.4. There is no way that arbitrary accuracy can be obtained by using an asymptotic expansion, even if the computer does the arithmetic exactly!

Performance evaluations of various programs for $\psi(x)$, implemented in Fortran on several computers, are given by Cody [Cod91]. Other algorithms and test results are given by Bowman [Bow84].

Test Values
Values for checking function `PsiFunc` for $\psi(x)$ can be obtained readily by running cell `PsiFunction` in *Mathematica* notebook GmBt. Test values are given in Table 6.2.3.

TABLE 6.2.3 Test values to 10 digits of the psi function for positive real argument x.

x	0.5	1.5	2.0	3.0	15.0
$\psi(x)$	−1.963510026	0.0364899740	0.4227843351	0.9227843351	2.674346662

Extensive tabulations of the digamma function for real arguments are in Tables 6.1 and 6.3 in Abramowitz and Stegun [Abr64] and in Table 6.8 therein for complex arguments.

6.2.2 Polygamma Functions

Keywords
polygamma functions

C Function and *Mathematica* Notebook
`PolyGamma1` and `Polygamma2` (C); cell `PolygammaFunction` in the *Mathematica* notebook GmBt.

Background
The polygamma functions—$\psi^{(n)}(z)$ with n a positive integer and z complex—occur in conjunction with the gamma function, $\Gamma(z)$ in Section 6.1.1 and its logarithmic derivative, the psi (digamma) function in Section 6.2.1.

One application of polygamma functions is for summing any convergent infinite series whose general term is a rational function of the index. Such a series can always be reduced to a finite series of psi and polygamma functions. An example using the digamma function (Section 6.2.1) is given in Section 1.7.4 of Erdélyi et al [Erd53a]. Another example, requiring also the tri- and tetra-gamma functions ($n = 1$ and 2, in the notation of Table 6.2.1), is in Section 6.8 of Abramowitz and Stegun [Abr64]. Suppose that the denominators of the terms of such series to be summed over k contain products of the form

$$d_m(k) = \prod_{p=1}^{s} (k + \gamma_p)^m \tag{6.2.8}$$

in which the roots γ_p are distinct and in which m is the maximum number of repeated roots. Then the series sum involves at most $\psi^{(m-1)}(z)$. For example, if roots repeat no more than three times, then functions only up to the tetragamma function ($n = 2$) are required. If all the roots are distinct ($m = 1$) then the digamma (psi) function in Section 6.2.1 is involved.

The analysis and properties of $\psi^{(1)}(z) = \psi'(z)$ are discussed in Section II.5 of Jahnke and Emde [Jah45]. The polygamma functions for $n = 1, 2, \ldots$ are covered extensively in Section 6.4 of Abramowitz and Stegun [Abr64].

Function Definition

The polygamma function of order n and complex z, $\psi^{(n)}(z)$, can be defined in terms of the psi function (Section 6.2.1) by

$$\psi^{(n)}(z) \equiv \frac{d^n \psi(z)}{dz^n} \qquad n = 1, 2, \ldots \tag{6.2.9}$$

The nomenclature for the gamma and polygamma functions is given in Table 6.2.1.

An alternative expression for $\psi^{(n)}(z)$—useful for algebraic computations and sometimes useful in numerical work—is the series

$$\psi^{(n)}(z) = (-1)^{n+1} n! \sum_{k=0}^{\infty} \frac{1}{(z+k)^{n+1}} \qquad z \neq -1, -2, \ldots \tag{6.2.10}$$

This expansion is the basis for applying polygamma functions in summing series, as discussed above under Background. Further, when $z = x$, a real variable, the terms are everywhere positive, therefore $\psi^{(n)}(x)$ is positive or negative according as n is odd or even.

A further property deriving from (6.2.10) is the relation to the Riemann zeta function, $\zeta(n)$, explored in Section 8.3. Namely

$$\psi^{(n)}(1) = (-1)^{n+1} n! \zeta(n+1) \tag{6.2.11}$$

which suggests the possibility of expanding $\psi^{(n)}(z)$ in terms of zeta functions, as we do in the computing algorithm below.

Visualization

Because the polygamma function depends on two variables, n (integer) and z (complex), it is not practical to show it from the many viewpoints (Figures 6.2.1–6.2.3) used for the digamma function. Rather, Figure 6.2.4 on the following page shows $|\psi^{(n)}(x)|$ over ranges of n and x that are often of interest for computing the functions.

Since—according to (6.2.10)— $\psi^{(n)}(x)$ has a pole of order $n+1$ on the real axis at the negative integers and at zero, $\psi^{(n)}(x)$ begins to increase rapidly as $x \to 0$ and this increase is more rapid as n increases, as the right panels in Figure 6.2.4 show.

Algorithm and Program

For computing $\psi^{(n)}(x)$ with x real and positive, we use different methods depending upon the x value. The formulas are extensions of those for $\psi(x)$, summarized in Table 6.2.1. For completeness, we give the expressions for complex argument z.

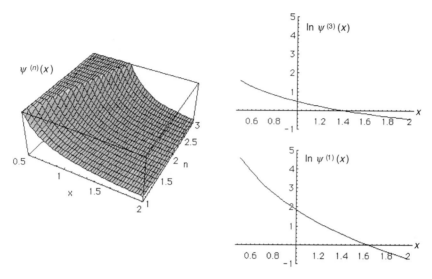

FIGURE 6.2.4 Left: Surface of absolute value of the polygamma functions along the real axis, x, as a function of n (treated as a continuous variable), truncated for values exceeding 5. Right: Logarithms of absolute values of polygamma functions for x in [0.5, 2], with $n = 1$ (trigamma) bottom and $n = 3$ (pentagamma) top. (Adapted from cell `Poly-gammaFunction` in *Mathematica* notebook GmBt.

TABLE 6.2.4 Formulas for calculating polygamma functions in various ranges of x, which is a positive real variable.

z range	Equation in [Abr64]	Formula	
(0, 0.5)	6.4.7	$\psi^{(n)}(z) = (-1)^n \psi^{(n)}(1-z) - \pi \dfrac{d^n \cot(\pi z)}{dz^n}$	(6.2.12)
[0.5, 1.5]	6.4.1	$\psi^{(n)}(z) = \dfrac{d^n \psi(z)}{dz^n}$ with (6.2.3) for $\psi(z)$	(6.2.13)
[0.999, 1.001]		$\psi^{(n)}(1+z) = \dfrac{d^n \psi(1+z)}{dz^n}$ with $\psi(1+z)$ from (6.2.4)	(6.2.14)
(1.5, 2.5)	6.4.6	$\psi^{(n)}(z) = (-1)^n n!/(z-1)^{n+1} + \psi^{(n)}(z-1)$	(6.2.15)
[2.5, 10]	6.4.6	$\psi^{(n)}(z) = \psi^{(n)}(z-m) + (-1)^n n! \displaystyle\sum_{k=1}^{m} \dfrac{1}{(z-k)^{n+1}}$	(6.2.16)
$(10, \infty)$	6.4.11	$\psi^{(n)}(z) \sim (-1)^{n-1} \left[\dfrac{(n-1)!}{z^n} + \dfrac{n!}{2z^{n+1}} \right.$ $\left. + \displaystyle\sum_{k=1}^{\infty} B_{2k} \dfrac{(2k+n-1)!}{(2k)!\, z^{2k+n}} \right]$	(6.2.17)

In (6.2.16) m is the largest integer such that, with $z = x$ real, $|x - m| < 1$. Equation (6.2.16) is used extensively in this algorithm, since for all x between 0 and 10, the range is modified by using (6.2.12), (6.2.15), or (6.2.16) until either this formula—or (6.2.14) for x near unity—is applied.

The derivatives in (6.2.13) are carried out analytically to produce formulas for various n. We provide the resulting code for $n = 1$ (trigamma) and $n = 2$ (tetragamma). For $n = 1$,

$$\psi^{(1)}(1+z) = \frac{-1}{2z^2} - \frac{2z}{(1-z^2)^2} + \frac{\pi^2/2}{\sin^2(\pi z)} - \frac{2}{z}\sum_{k=1}^{\infty} r_k^{(1)} z^{2k} \qquad (6.2.18)$$

with expansion coefficients are given in terms of Riemann zeta functions of odd order by

$$r_k^{(1)} = k\,[\zeta(2k+1) - 1] \qquad (6.2.19)$$

For $n = 2$ the expansion used is

$$\psi^{(2)}(1+z) = \frac{1}{z^3} - \frac{2(1+3z^2)}{(1-z^2)^3} - \frac{\pi^3 \cos(\pi z)}{\sin^3(\pi z)} - \frac{2}{z^2}\sum_{k=1}^{\infty} r_k^{(2)} z^{2k} \qquad (6.2.20)$$

with expansion coefficients

$$r_k^{(2)} = k(2k-1)[\zeta(2k+1) - 1] \qquad (6.2.21)$$

The series (6.2.18) and (6.2.19) converge much more rapidly than the series 6.4.9 in Abramowitz and Stegun [Abr64], since the coefficients decrease rapidly with k, as can be seen in the r arrays in the function below. Near $z = 1$ the leading terms of expansions (6.2.18) and (6.2.20) are sensitive to subtractive cancellation and roundoff errors. Hence our use of approximation (6.2.14).

If corresponding expansions are required for $n > 2$, the cell `PolygammaFunction` in *Mathematica* notebook GmBt can be used to perform algebraic differentiation of (6.2.3). Before coding the result or applying the `CForm[]` function to produce efficient C code, some simplification of rational fractions and of trigonometric expressions is necessary.

In the asymptotic approximation (6.2.17) the B_{2k} are Bernoulli numbers (Section 8.1.1). We write out the expansions explicitly for $n = 1$ and $n = 2$ as

$$\psi^{(1)}(z) = s\left[1 + (1/2)s + (1/6)t - (1/30)t^2 + (1/42)t^3 - (1/30)t^4 + (5/66)t^5\right] \qquad (6.2.22)$$

$$\psi^{(2)}(z) = -t\left[1 + s + (1/2)t - (1/6)t^2 + (1/6)t^3 - (3/10)t^4 + (5/6)t^5\right] \qquad (6.2.23)$$

in both of which $s = 1/z$ and $t = s^2$. To the number of terms given, at least 10-digit accuracy is achieved for $z = x > 10$. For fixed x, the convergence worsens as n increases because of the divergence of the Bernoulli numbers (Figure 8.1.2).

The warning (Section 6.2.1, Programming Notes, and Section 3.4) for the psi function about excessive reliance on asymptotic expansions, should be especially heeded for the higher-order polygamma functions. The smallest x value at which (6.2.17) is used should depend upon both n and the accuracy desired.

Finally, we have the programs for the polygamma functions. For $n = 1$ (trigamma) there is `PolyGamma1`, then `PolyGamma2` for $n = 2$ (tetragamma).

```
double PolyGamma1(double x)
{
/* Polygamma function for real positive arguments
   for  n = 1  */
double xM1,t,sum,tPower,xF,s;
const double Pi = 3.14159265359,PiD2 = 1.57079632679,
r1[10] =
{0.,0.202056903160,0.073855510287,0.025047832146,
 0.008033571304,0.00247094302,0.00073628008,
 0.00021411765,0.000061097581,0.000017173914};
int k,m;

if ( x < 0.5 )
 return   - PolyGamma1(1.0 - x) +
            pow(Pi/sin(Pi*x),2.0);

if ( x <= 1.5 )
 {
 xM1 = x - 1.0;
 if ( fabs(xM1) <= 0.001 )
  return   1.0/(x*x) + 0.644934066848
   - xM1*(0.404113806319 - xM1*(0.246969701133
   - xM1*0.147711020573)));

 t = xM1*xM1;
 sum = -0.5/t - 2.0*xM1*(pow(1.0-t,-2.0)) +
         0.5*pow(Pi/sin(Pi*xM1),2.0);
 tPower = 2.0*xM1;
 for ( k = 1; k <= 9; k++ )
  {
  sum -= r1[k]*tPower;
  tPower *= t;
  }
 return  sum;
 }

if (  x < 2.5 )
 return  -pow(x-1.0, -2.0) + PolyGamma1(x-1.0);

if ( x <= 10.0 )
 {
 m = x;  m = m - 1;
 xF = x - (double) m;
 sum = 0.0;
 for ( k = 1; k <= m; k++ )
  sum -= pow(x-(double) k,-2.0);
 return  sum + PolyGamma1(xF);
 }

/* Else  x > 10; use asymptotic series  */
s = 1.0/x;
t = s*s;
return  s*(1.0+0.5*s
 +0.166666666667*t*(1.0-t*(0.2-t*(0.142857142857
 -t*(0.2-0.454545454545*t)))));
}
```

The program for the tetragamma function ($n = 2$) is as follows.

```c
double PolyGamma2(double x)
{
/* Polygamma function for real positive arguments
   for  n = 2  */
double xM1,t,sum,tPower,xF,s;
const double Pi = 3.14159265359,PiD2 = 1.57079632679,
r2[11] =
{0.0,0.20205690316,0.22156653086,0.125239160729,
 0.056234999130,0.022238487185,0.008099080940,
 0.002783529504,0.000916463717,0.000291956546,
 0.000090617267};
int k,m;

if ( x < 0.5 )
  return   PolyGamma2(1.0 - x) -
           2.0*cos(Pi*x)*pow(Pi/sin(Pi*x),3.0);

if ( x <= 1.5 )
  {
  xM1 = x - 1.0;
  if ( fabs(xM1) <= 0.001 )
   return   -2.0/pow(x,3) - 0.404113806319
    +xM1*(0.493939402267 - xM1*(0.44313306172
    - xM1*0.346861239689)));

  t = xM1*xM1;
  sum = pow(xM1,-3.0)-2.0*(1.0+3.0*t)*pow(1.0-t,-3.0)
        -cos(Pi*xM1)*pow(Pi/sin(Pi*xM1),3.0);
  tPower = 2.0;
  for ( k = 1; k <= 10; k++ )
   {
   sum -= r2[k]*tPower;
   tPower *= t;
   }
  return  sum;
  }

if (  x < 2.5 )
  return  2.0*pow(x-1.0,-3.0) + PolyGamma2(x-1.0);

if ( x <= 10.0 )
  {
  m = x;  m = m - 1;
  xF = x - (double) m;
  sum = 0.0;
  for ( k = 1; k <= m; k++ )
   sum += pow(x-(double) k,-3.0);
  return  2.0*sum + PolyGamma2(xF);
  }

/* Else   x > 10; use asymptotic series  */
s = 1.0/x;
t = s*s;
return  -t*(1.0+s
+0.5*t*(1.0-t*(0.3333333333*t*(1.0-0.9*t)))));
}
```

Programming Notes

`PolyGamma1` and `PolyGamma2` make extensive use of recursion (Section 3.2) to reduce the range of x in (0, 10] to the range [0.5, 1.5], thus allowing (6.2.18)–(6.2.20) to be applied with the variable in the series not exceeding 0.5 in magnitude. Recursion is both convenient and practical to use in this case, since the recursion is at most one level deep and it avoids much repetitive coding. We also use asymptotic expansions (Section 3.4) for $x > 10$, as discussed in the programming notes in Section 6.2.1 for the psi function.

The driver program for this pair of functions generates both $\psi^{(1)}(x)$ and $\psi^{(2)}(x)$. If you need a higher-order function ($n > 2$), the methods summarized in Table 6.2.4 may be extended judiciously, except for the asymptotic series (6.2.17), which needs a higher starting value than $x = 10$ if $n > 3$. A possible remedy is to use the series (6.2.10) directly, but its convergence to produce high *relative* accuracy is poor unless n is very large.

Program test examples for alternative algorithms that include higher-order polygamma functions are given by Bowman [Bow84].

Test Values

Check values for functions `PolyGamma1` ($n = 1$, trigamma) and `PolyGamma2` ($n = 2$, tetragamma) can be obtained by running cell `PolygammaFunction` in *Mathematica* notebook `GmBt`. Test values are given in Table 6.2.5.

TABLE 6.2.5 **Test values to 10 digits of the polygamma functions for positive real argument x.**

x	0.5	1.5	2.0	15.0
$\psi^{(1)}(x)$	4.934802201	0.9348022005	0.6449340668	0.06893822785
$\psi^{(2)}(x)$	–16.82879664	–0.8287966442	–0.4041138063	–0.00475060272

Tabulations of the tri-, tetra-, and pentagamma function for real arguments are in Tables 6.1–6.3 of Abramowitz and Stegun [Abr64].

6.3   INCOMPLETE GAMMA AND BETA FUNCTIONS

The incomplete gamma and beta functions arise in many situations, such as in statistics where they are used to describe χ^2 and variance-ratio distributions.

6.3.1   Incomplete Gamma Function

Keywords

incomplete gamma function, regularized incomplete gamma function, chi-squared distribution

C Function and *Mathematica* Notebook

`IncGamma` (C); cell `IncompleteGammaFunction` in *Mathematica* notebook `GmBt`.

Background

The incomplete gamma function occurs in conjunction with the gamma function, $\Gamma(z)$ in Section 6.1.1 when the upper limit on the defining integral (6.1.1) is finite rather than infinite. The incomplete gamma function is used in statistics, where it describes the cumulative probability distribution for χ^2, as presented in Section 9.3.3, in Chapter 17 of Johnson et al [Joh94a], in Section 26.4 of Abramowitz and Stegun [Abr64], in Chapter 45 of Spanier and Oldham [Spa87], and in Section 8.35 of Gradshteyn and Ryzhik [Gra94].

An extensive discussion of analytical properties of incomplete gamma functions is given in Chapter 9 of Erdélyi et al [Erd53b]. The variety of definitions is explained carefully therein. We use the definitions and notations of Abramowitz and Stegun.

Function Definition
The incomplete gamma function, $\gamma(a,x)$, is defined by

$$\gamma(a,x) \equiv \int_0^x e^{-t} t^{a-1} dt \qquad \mathrm{Re}\, a > 0 \qquad (6.3.1)$$

Thus, as $x \to \infty$ $\gamma(a,x) \to \Gamma(a)$, the conventional gamma function in Section 6.1.1. Alternatively—and usually more relevant for statistics—one can compute the ratio of $\gamma(a,x)$ to $\Gamma(a)$. Thereby, one defines the regularized incomplete gamma function $P(a,x)$ by

$$P(a,x) \equiv \gamma(a,x)/\Gamma(a) \qquad (6.3.2)$$

which is well-defined, since $\Gamma(a)$ has no zeros. The function $\Gamma(a,x) \equiv \Gamma(a) - \gamma(a,x)$ is also used in the literature.

In the following we show both forms of the function, although we give the C function for $P(a,x)$ only. The form (6.3.1) can then be obtained by multiplying by $\Gamma(a)$. Note that to proceed in the opposite order would give numerically unstable results if values near the poles of a (at the nonpositive integers) were to be computed.

Visualization
The incomplete gamma functions depend on two variables, a (complex) and x (real). Since the functions are most often used for real values of a, in the following figures we show them as functions of two real variables, a and x. Figure 6.3.1 shows the surface of $\gamma(a,x)$ over the a range –4 to 4. As with the complete gamma function (Section 6.1.1), there are poles at zero and the negative integer values of a as long as x is nonzero. As x—the upper limit to the integral in (6.3.1)—increases, $\gamma(a,x)$ approaches the regular gamma function.

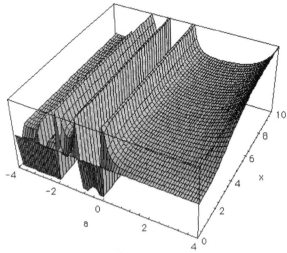

FIGURE 6.3.1 Incomplete gamma function (6.3.1) as a function of a and x. (*Mathematica* cell IncompleteGammaFunction in the notebook GmBt.)

If a slice of the surface in Figure 6.3.1 is taken with x constant, then $\gamma(a,x)$ in two dimensions shows its similarity to $\Gamma(a)$. Compare the left sides of Figures 6.3.2 and 6.1.3, in both of which the range of the function shown is –6 to 6. For example, when $x = 5$ and $a = 4$ the incomplete function value is 73% that of the complete function.

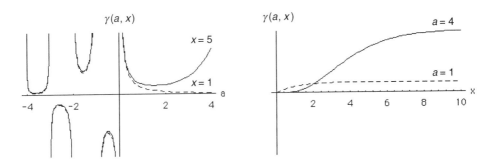

FIGURE 6.3.2 Incomplete gamma functions, showing (left) slices at constant x values of $x = 1$ and 5, and (right) slices at constant a values of $a = 1$ and 4. (Adapted from *Mathematica* cell `IncompleteGammaFunction` in notebook `GmBt`.)

If Figure 6.3.1 is sliced so that x is varying and a is constant then the rapid approach of $\gamma(a,x)$ to $\Gamma(a)$ becomes even clearer, as seen on the right side of Figure 6.3.2.

The regularized incomplete gamma function, $P(a,x)$ in (6.3.2), has a much smoother dependence on a than does $\gamma(a,x)$, since as $x \to \infty$ the regularized function tends to unity for any a. This behavior is shown in Figure 6.3.3.

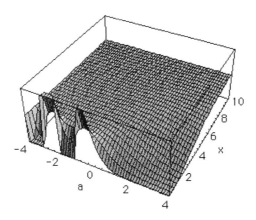

FIGURE 6.3.3 Regularized incomplete gamma function (6.3.2) as a function of a and x. (*Mathematica* cell `IncompleteGammaFunction` in the notebook `GmBt`.)

Slices of the $P(a,x)$ surface along constant x, as shown on the left side of Figure 6.3.4, illustrate that the regularized function has no poles.   Indeed, for any x value $\gamma(-n,x) = 1$ for n a nonnegative integer.   The graphic on the right side of this figure shows how rapidly $P(a,x)$ approaches unity as x increases, especially when a is small.   On the other hand, since the maximum of the integrand for the gamma function in (6.3.1) is at $t = t_m = a - 1$, if the upper limit of the integral (x) is less than t_m then a significant part of the integral is lost, as the graphic on the right makes clear.

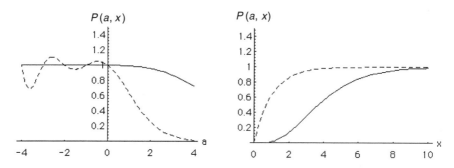

FIGURE 6.3.4 Regularized incomplete gamma function, showing (left) slices with constant x values of $x = 1$ (dashed) and 5 (solid), and (right) slices at constant a values of $a = 1$ (dashed) and 4 (solid). (Adapted from *Mathematica* cell `IncompleteGammaFunction` in notebook `GmBt`.)

Algorithm and Program

To compute $P(a, x)$ with a real and positive, depending upon the x value we use two different methods. If $x < a + 1$ we use a power series that is derived by combining equations 6.5.4 and 6.5.29 in [Abr64] to obtain

$$P(a, x) = \frac{e^{-x} x^a}{\Gamma(a)} \sum_{k=0}^{\infty} t_k x^k \qquad (6.3.3)$$

in which the coefficients satisfy

$$t_{k+1} / t_k = 1/(a+1+k) \qquad t_0 = 1/a \qquad (6.3.4)$$

For $x \geq a + 1$ we use the recurrence formula 6.5.21 in [Abr64] applied n times

$$P(a, x) = P(a+n, x) + \frac{e^{-x} x^a}{\Gamma(a+1)} \sum_{k=0}^{n-1} t_k \qquad t_0 = 1 \quad t_k / t_{k-1} = x/(a+k) \quad k > 0 \qquad (6.3.5)$$

in which $n = \mathrm{ceil}(x - a)$ is the integer difference between x and a, rounded upwards. The function value $P(a+n, x)$ can therefore be computed by (6.3.3). We thus have the following program for the regularized incomplete gamma function.

```
double IncGamma(double a, double x, double eps)
{
/* Regularized Incomplete Gamma function */

double LogGamma(double x),gln,ap,sum,del,ratio,
  term,denterm;
int n,k;
```

```
  gln = LogGamma(a);
  if ( x < a+1.0 )    /* use series */
   {
   ap = a;
   sum = 1.0/a;
   del = sum;
   ratio = 10.0;
   while ( ratio > eps )
    {
     ap++;
     del *= x/ap;
     sum += del;
     ratio = fabs(del/sum);
    }
   return  sum*exp(-x+a*log(x) - gln);
   }

  /* Else use a power series plus above series */
  n = ceil(x - a);
  k = 1;
  sum = 1.0;
  term = 1.0;
  denterm = a;
  while ( k <= n-1 )
   {
   denterm += 1.0;
   term *= x/denterm;
   sum += term;
   k += 1;
   }
  return  IncGamma( a + (double) n, x, eps)
   + (pow(x,a)/exp(x+LogGamma(a+1.0)))*sum;
  }
```

Programming Notes

Function `IncGamma` uses `LogGamma` (Section 6.1.1) to compute the complete gamma function for regularizing the incomplete function, as in (6.3.2). Input parameter `eps` determines the convergence of the series in (6.3.3) and gives the fractional accuracy for the result.

Test Values

Check values for the regularized incomplete gamma function `IncGamma`, encoding $P(a,x)$ in (6.3.2), can be obtained by running the cell `IncompleteGammaFunction` in *Mathematica* notebook `GmBt`. The test values given in Table 6.3.1 include cases that use both expansion methods, (6.3.3) and (6.3.5).

TABLE 6.3.1 Test values to 10 digits of the regularized incomplete gamma functions for positive real arguments a and x.

a, x	1, 1	3, 1	2, 5	5, 5
$P(a,x)$	0.6321205588	0.08030139707	0.9595723180	0.5595067149

6.3.2   Incomplete Beta Function

Keywords
incomplete beta function, regularized incomplete beta function, F distribution, variance-ratio distribution

C Function and *Mathematica* Notebook
IncBeta (C); cell IncompleteBetaFunction in the *Mathematica* notebook GmBt

Background
The incomplete beta function occurs in conjunction with the beta function, $B(z,w)$ in Section 6.1.2 when the upper limit on the defining integral (6.1.3) is less than unity. This function is used in statistics, for example, where it describes the cumulative probability distribution for the ratio of two variances, as presented in Section 9.3.4, in Chapter 27 of Johnson et al [Joh94b], in Sections 6.6 and 26.5 of Abramowitz and Stegun [Abr64], and in Chapter 58 of Spanier and Oldham [Spa87].

   Be aware that the notations and definitions of incomplete beta functions are somewhat variable. As usual, we follow Abramowitz and Stegun.

Function Definition
The incomplete beta function, $B_x(a,b)$, is defined by

$$B_x(a,b) \equiv \int_0^x t^{a-1}(1-t)^{b-1}dt \tag{6.3.6}$$

Thus, as $x \to 1$ we have $B_x(a,b) \to B(a,b)$, the conventional beta function in Section 6.1.2. Note the awkward notation—which we tolerate only for conformity with Abramowitz and Stegun—that in (6.3.6) the variable a corresponds to z in (6.1.3), while b corresponds to w therein.

   An alternative function that is usually more relevant for use in statistics is the ratio of $B_x(a,b)$ to $B(a,b)$. We define the regularized incomplete beta function $I_x(a,b)$ by

$$I_x(a,b) \equiv B_x(a,b) / B(a,b) \tag{6.3.7}$$

This ratio is well-defined except when $a+b$ is a negative integer or zero. In the following, we show both forms of the function, but we give the C function only for $I_x(a,b)$.

   The equivalence of the complete beta functions under interchange of a and b, (6.1.5), does not hold generally for the incomplete functions. Rather, for $I_x(a,b)$ there holds

$$I_x(a,b) = 1 - I_{1-x}(b,a) \tag{6.3.8}$$

Therefore, necessarily $I_{0.5}(a,a) = 0.5$, a result that is seen in the visualizations following and that may be useful for checking program correctness.

Visualization
The incomplete beta functions depend on two complex variables, a and b, as well as on the real variable, x. Since the functions are most often used for real values of a and b, in the following graphics we show them as functions of three real variables. We use $B(x,a,b)$ and $I(x,a,b)$ to label $B_x(a,b)$ and $I_x(a,b)$, respectively.

   Figure 6.3.5 on the next page shows surfaces of $B_x(a,b)$ for $x = 0.5$ and $x = 0.8$. Slices through these figures for constant a and x values are also shown.

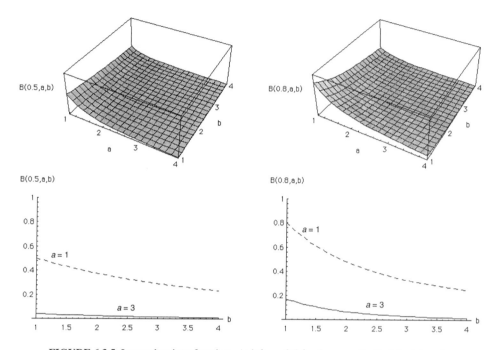

FIGURE 6.3.5 Incomplete beta function. At left $x = 0.5$ for the upper limit of the integral in (6.3.6); as a function of a and b (top), and slices with constant values of $a = 1$ and 3. At right $x = 0.8$; as a function of a and b (top), and slices with constant values of $a = 1$ and 3. (Adapted from *Mathematica* cell `IncompleteGammaFunction` in notebook `GmBt`.)

The regularized incomplete beta function, $I_x(a,b)$ shown in Figure 6.3.6 on the following page, is even more monotonous than $B_x(a,b)$. The regularized function approaches unity if x approaches unity or if $b \gg a$, with both behaviors being evident in the figure.

Algorithm and Program

To compute the regularized function, $I_x(a,b)$ with a and b real and positive, we use two different formulas, depending upon the x value. If $x < (a+1)/(a+b+2)$ we use the continued-fraction expansion that is 26.5.8 in [Abr64], namely,

$$I_x(a,b) = \frac{x^a(1-x)^{b-1}}{aB(a,b)} \left(\frac{e_1}{1+} \frac{e_2}{1+} \frac{e_3}{1+} \cdots \right)$$

$$e_1 = 1 \qquad e_{2m} = -\frac{(a+m-1)(b-m)x}{(a+2m-2)(a+2m-1)(1-x)} \qquad (6.3.9)$$

$$e_{2m+1} = \frac{m(a+b-1+m)x}{(a+2m-1)(a+2m)(1-x)}$$

If x exceeds the above value, we compute with $1-x$ and use the symmetry (6.3.8).

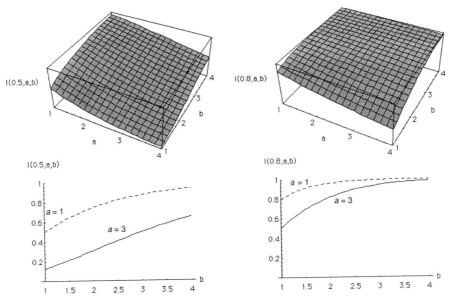

FIGURE 6.3.6 Regularized incomplete beta function. At left $x = 0.5$ for upper limit in (6.3.6); as a function of a and b (top), and slices with constant values of $a = 1$ and 3. At right $x = 0.8$; as a function of a and b (top), and slices with constant values of $a = 1$ and 3. (Adapted from *Mathematica* cell `IncompleteGammaFunction` in the notebook GmBt.)

The program for the regularized incomplete beta function is as follows.

```
double IncBeta
  (double x, double aIn, double bIn, double eps)
{
/* Regularized Incomplete Beta function */

double LogGamma(double x),a,b,
  offset,bmult,z,error,floatm,
twicefm,dc,aPb,aM1,aP1,factor,h,dm,cf,delta,ratio;
static double tiny = 1.0e-20;
int m;

if ( x == 1.0 || x == 0.0 ) return  x;

bmult = exp(LogGamma(aIn+bIn)-LogGamma(aIn) -
         LogGamma(bIn)+aIn*log(x)+bIn*log(1.0 - x));

if ( x < (aIn+1.0)/(aIn+bIn+2.0) )
   { /*  set up for continued fraction of  x  */
   a = aIn;
   b = bIn;
   z = x;
   offset = 0.0;
   bmult /= aIn;
   }
```

```
    else
    { /*  set up for continued fraction of  1 - x   */
    a = bIn;
    b = aIn;
    z = 1.0 - x;
    bmult = -bmult/bIn;
    offset = 1.0;
    }

    aPb = a + b;
    aM1 = a - 1.0;
    aP1 = a + 1.0;
    dm = 0.0;
    h = tiny;
    cf = h;
    ratio = 10.0;
    m = 0;
    while ( ratio > eps )
    {
    floatm = m;
    twicefm = 2*m;
    if ( m )
      dc = floatm*(b-floatm)*z/
          ((aM1+twicefm)*(a+twicefm));
    else
      dc = 1.0;
      dm = 1.0 + dm*dc;
      cf = 1.0 + dc/cf;
      if ( dm == 0.0 )          dm = tiny;
      if ( cf == 0.0 )          cf = tiny;
      dm = 1.0/dm;
      delta = dm*cf;
      h *= delta;
      dc = -(a+floatm)*(aPb+floatm)*z/
          ((a+twicefm)*(aP1+twicefm));
      dm = 1.0 + dm*dc;
      cf = 1.0 + dc/cf;
      if ( dm == 0.0 )       dm = tiny;
      if ( cf == 0.0 )    cf = tiny;
      dm = 1.0/dm;
      delta = dm*cf;
      h *= delta;
      ratio = fabs(delta - 1.0);
      m++;
      }
    return   h*bmult+offset;
    }
```

Programming Notes

Our function `IncBeta` is adapted from the program `incbeta(aa,bb,x)` on pages 580 and 581 of Baker's handbook [Bak92]. It is slightly more efficient and it has controllable accuracy through the input parameter `eps`, which terminates the expansion in continued fractions as soon as terms contribute less than `eps` in magnitude. Brown and Levy [Bro94] have also described a method with high accuracy.

Function `IncBeta` calls `LogGamma` (Section 6.1.1) to compute the complete beta function and thereby to regularize the incomplete function, as in (6.3.7). The complete machine-readable program on the CD-ROM includes the gamma function code.

Test Values

Check values for the regularized incomplete beta function, `IncBeta`, encoding $I_x(a,b)$ in (6.3.7), can be obtained by running the cell `IncompleteBetaFunction` in *Mathematica* notebook `GmBt`. Test values given in Table 6.3.2 include cases using both ranges of x.

Note that when both a and b are positive integers the defining integral (6.3.6) can be performed directly and the complete beta function can be obtained in terms of factorials. Thus, one can produce algebraic expressions for $I_x(a,b)$ that can be used for calculating and testing.

TABLE 6.3.2 Test values to 10 digits of the regularized incomplete beta functions for positive real arguments a, b, and x.

x, a, b	0.5, 1, 1	0.5, 3, 1	0.8, 3, 1	0.8, 3, 5
$I_x(a,b)$	0.5000000000	0.1250000000	0.5120000000	0.9953280000

REFERENCES ON GAMMA AND BETA FUNCTIONS

[Abr64]   Abramowitz, M., and I. A. Stegun, *Handbook of Mathematical Functions*, Dover, New York, 1964.

[Bak92]   Baker, L., *C Mathematical Function Handbook*, McGraw-Hill, New York, 1992.

[Boh92]   Bohman, J., and C.-E. Fröberg, Mathematics of Computation, **58**, 315, 1992.

[Bow84]   Bowman, K. O., Communications in Statistics: Simulation and Computation, **13**, 409 (1984).

[Bro94]   Brown, B. W., and L. B. Levy, ACM Transactions on Mathematical Software, **20**, 393 (1994).

[Cod91]   Cody, W. J., ACM Transactions on Mathematical Software, **17**, 46 (1991).

[Erd53a]   Erdélyi, A., et al, *Higher Transcendental Functions*, Vol. 1, McGraw-Hill, New York, 1953; reprint edition, Krieger, Malabar, Florida, 1981.

[Erd53b]   Erdélyi, A., et al, *Higher Transcendental Functions*, Vol. 2, McGraw-Hill, New York, 1953; reprint edition, Krieger, Malabar, Florida, 1981.

[Gon92]   González, M. O., *Complex Analysis*, Marcel Dekker, New York, 1992.

[Gra94]   Gradshteyn, I. S., and I. M. Ryzhik, *Table of Integrals, Series, and Products*, Academic Press, Orlando, fifth edition, 1994.

[Jah45]   Jahnke, E., and F. Emde, *Tables of Functions*, Dover, New York, fourth edition, 1945.

[Joh94a]   Johnson, N. L., S. Kotz, and N. Balakrishnan, *Distributions in Statistics: Continuous Univariate Distributions-1*, Wiley, New York, second edition, 1994.

[Joh94b]   Johnson, N. L., S. Kotz, and N. Balakrishnan, *Distributions in Statistics: Continuous Univariate Distributions-2*, Wiley, New York, second edition, 1994.

[Luk69]   Luke, Y. L., *The Special Functions and Their Approximations*, Vol. 1, Academic Press, New York, 1969.

[Mag49]   Magnus, W., and F. Oberhettinger, *Formulas and Theorems for the Special Functions of Mathematical Physics*, Chelsea Publishing, New York, 1949.

[Spa87]   Spanier, J., and K. B. Oldham, *An Atlas of Functions*, Hemisphere Publishing, Washington, D.C., 1987.

[Whi27]   Whittaker, E. T., and G. N. Watson, *A Course of Modern Analysis*, Cambridge University Press, Cambridge, England, fourth edition, 1927.

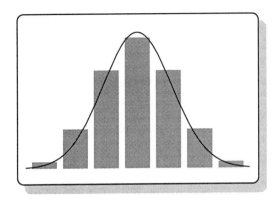

Chapter 7

COMBINATORIAL FUNCTIONS

Combinatorial functions occur frequently when computing mathematical functions, particularly those used for number theory (Chapter 8) and probability distributions (Chapter 9). The functions described in this chapter gradually increase in complexity and range of parameters, so that the later functions usually make use of the earlier ones, particularly factorials (Section 7.1) and binomial coefficients (Section 7.2.1). A general source for combinatorial identities is the monograph by Riordan [Rio79]. The combinatorial functions most often used in number theory—Bernoulli and Euler functions—are described in Chapter 8.

Most of the coefficients in functions described here are either integers or rational fractions, so that the algorithms we provide usually result in exact computations, limited only by the computer word size used. Whenever possible we use integer arithmetic, which—although it limits the range of values that can be stored in a given word size—allows exact computation. Most of the programs can be adapted easily to use floating-point arithmetic, thus extending the range of values computable, but at the expense of decreased accuracy.

7.1 FACTORIALS AND RISING FACTORIALS

Factorials and rising factorials are the basic functions for calculating combinatorials. It is therefore essential to compute them efficiently. Factorials are gamma functions (Chapter 6) having positive arguments

$$n! \equiv \Gamma(n+1) \tag{7.1.1}$$

Factorials therefore share many of the analytical properties of gamma functions. Formula (7.1.1) is seldom efficient for computing exact factorials, but it is useful for deriving approximations to them.

Rising factorials are special cases of Pochhammer's symbol (Chapters 6, 16), but are most easily related to factorials, therefore they are included in this section.

7.1.1   Factorial Function

Keywords
combinatorial, factorial, gamma function, integer programming

C Function and *Mathematica* Notebook

M   Factorial (C); cell Factorial in notebook Combinatorial (*Mathematica*)

Background
Factorials are ubiquitous in applied numerical mathematics, usually occurring whenever terms are combined, as in Taylor series expansions, combinatorials, and permutations.

Function Definition
The factorial function is defined for nonnegative integers, n, by

$$n! \equiv n(n-1)(n-2)....3.2 \qquad n > 1$$
$$1! = 0! \equiv 1 \tag{7.1.2}$$

The so-called *double factorial* function can be defined by

$$(2n)!! \equiv 2n(2n-2)(2n-4)....4.2 = 2^n n! \qquad n > 1$$
$$2!! = 2 \qquad 0!! \equiv 1$$

$$(2n+1)!! \equiv (2n+1)(2n-1)....3.1 = \frac{(2n+1)!}{2^n n!} \qquad n > 0 \tag{7.1.3}$$

$$1!! = 1$$

Since it can—but with some loss in efficiency—be computed directly from the factorial, we do not consider it further.

An approximation to $n!$ that has a relative error of less than 10^{-10} for $n > 10$ is obtained from the first few terms of Stirling's asymptotic expansion for the gamma function (Section 6.1), namely,

$$\ln(n-1)! = \ln \Gamma(n) = (n-1/2)\ln n - n + \ln\sqrt{2\pi}$$
$$+ \frac{1}{12n}\left(1 - \frac{1}{30n^2}\left(1 - \frac{2}{7n^2}\left(1 - \frac{3}{4n^2}\right)\right)\right) + ... \tag{7.1.4}$$

which is written in Horner polynomial form for convenient computation. The first term of this approximation, also neglecting 1 and 1/2 compared with n, produces $n! \approx \sqrt{2\pi}\,(n/e)^n$, an expression that is often used in statistical mechanics.

Visualization
The factorial increases so quickly with n that it is appropriate to show it on a log scale, as in Figure 7.1.1, where we show the factorial and two approximations to it for n between 1 and 10. The first approximation—often used in statistical mechanics, where n is typically of order 10^{20}—is quite close even for $n = 10$. The second approximation, Stirling's formula (7.1.4), is quite accurate for small n, as shown here.

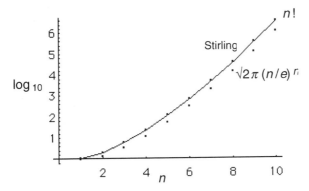

FIGURE 7.1.1 Two approximations for the factorial function are compared with exact values shown by the solid line. (Adapted from cell `Factorial` in *Mathematica* notebook `Combinatorial`.)

Algorithm and Program

The algorithm for the factorial uses the definition (7.1.1) to provide the recurrence relation

$$n! \equiv n(n-1)! \qquad n > 1 \qquad 1! = 0! \equiv 1 \qquad (7.1.5)$$

The corresponding program is as follows.

```
long int Factorial(int n)
/* Factorial function */
{
long int term,i;

if ( n < 2 ) return 1;
else
  {
  term = 1;
  for ( i = 2; i <= n; i++ )
    {
    term *= i;
    }
  return term;
  }
}
```

Programming Notes

Because C allows recursive calls to functions, it would seem appropriate to avoid the iteration loop over **i** and to invoke `Factorial` recursively. However, as discussed in general in Section 3.2, using recursion produces a function that executes about 40% slower than the one given above.

Since $n!$ increases very rapidly with n, it is very easy to overflow the word size used to store the factorial. Plotting $\log_2 n!$ against n gives an *upper* bound on the number of bits in the word that can be used for given maximum n, since at least one bit in the word will be used as a sign bit. Figure 7.1.2 shows the relationship between number of bits and argument n.

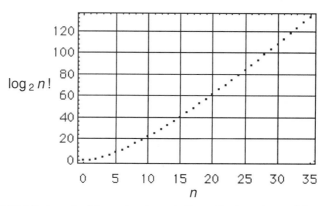

FIGURE 7.1.2 A graph of $\log_2 n!$ against n indicates the dependence of the number of bits in a word needed to store $n!$ and the largest value of n whose factorial can be stored in such a word. For example, a 32-bit signed integer word overflows at 13!. (Adapted from cell `Factorial` in *Mathematica* notebook `Combinatorial`.)

The function program `Factorial` does not have built-in checking for word overflow, since this would reduce the efficiency of the calculation and it is better done at a higher program level, such as the C driver program in Section 21.7. When used in binomial and multinomial coefficient calculations (Section 7.2), the range of factorial-function arguments can be greater than when a single factorial is calculated.

If approximate decimal factorials for $n > 10$ are sufficient, rather than the exact integer factorial produced by `Factorial`, then use a Stirling expansion such as (7.1.4) which then gives at least 10 decimal digits of accuracy.

Test Values

Table 7.1.1 gives test values for spot checking the function `Factorial`, which can be obtained by using the '!' operator in *Mathematica*. Another check is to verify the recurrence relation (7.1.2).

TABLE 7.1.1 Exact test values for the factorial function (7.1.2).

n	$n!$
0	1
1	1
10	3 628 800
20	2 432 902 008 176 640 000
30	265 252 859 812 191 058 636 308 480 000 000

7.1.2   Rising Factorial Function

The rising factorial function has two nonnegative integer parameters, n and m. It is related to factorials (Section 7.1.1) and to binomial coefficients (Section 7.2.1).

Keywords
combinatorial, factorial, gamma function, integer programming, Pochhammer symbol, rising factorial

C Function and *Mathematica* Notebook
RisingFactorial (C); cell RisingFactorial in the notebook Combinatorial
(*Mathematica*)

Background
The most common occurrence of rising factorials is in computing gamma and beta functions (Chapter 6) or hypergeometric functions with integer parameters (Chapter 16).

Function Definition
The rising factorial $(n)_m$ is defined by

$$(n)_m \equiv n(n+1)...(n+m-1) \qquad m>0$$
$$(n)_0 = 1 \tag{7.1.6}$$

The relation to the factorial function

$$(n)_m = \frac{(n+m-1)!}{(n-1)!} \qquad n>0 \tag{7.1.7}$$

is useful for analytical work but does *not* provide an efficient way of computing the rising factorial. (Consider, for example, $n = 100$, $m = 1$.) Similarly, the relation to gamma functions with integer arguments

$$(n)_m = \frac{\Gamma(n+m)}{\Gamma(n)} \qquad n>0 \tag{7.1.8}$$

is more appropriate for deriving analytical approximations to $(n)_m$ in terms of approximations for the gamma function than for numerical calculations.

Tables of rising factorials are prepared efficiently as follows. If n is fixed and m is increasing by unit steps then the recurrence relation

$$(n)_{m+1} = (n+m)(n)_m \tag{7.1.9}$$

can be used. If n is increasing by unit steps and m is fixed, an appropriate recurrence is

$$(n+1)_m = \frac{n+m}{n}(n)_m \qquad n>0$$
$$(1)_m = m! \tag{7.1.10}$$

Visualization

As n increases from zero, the rising factorial increases very rapidly with m. This is illustrated in Figure 7.1.3 for small values of n and m.

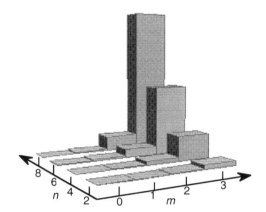

FIGURE 7.1.3 Three-dimensional bar-chart view of rising factorials, shown on the vertical axis for $n = 2, 4, 6, 8$ and $m = 0, 1, 2, 3$. Numerical values are given in Table 7.1.2. (Adapted from cell `RisingFactorial` in *Mathematica* notebook `Combinatorial`.)

In many applications, the parameter m in $(n)_m$ is the proportional to the power of a term in a series expansion, typically $(n)_m x^m$ or $x^m/(n)_m$. Such series are therefore very likely to have poor numerical convergence if the rising factorial occurs in numerators.

Algorithm, Program, and Programming Notes

Program `RisingFactorial` uses definition (7.1.6) directly for computation of a single rising factorial. As discussed under Visualization, the rapid increase of $(n)_m$ with both n and m can produce overflow of the rising factorial for quite modest values of the function arguments. For example $(10)_8$ and $(20)_7$ both overflow a 32-bit word that has one sign bit.

Test Values

Test values are readily obtained from the factorial relation (7.1.7). The recurrence relations in n and m—(7.1.9) and (7.1.10), respectively—may also be used to check the correctness of `RisingFactorial`.

TABLE 7.1.2 Exact test values for the rising factorial function $(n)_m$.

$n \setminus m$	0	1	2	3
2	1	2	6	24
4	1	4	20	120
6	1	6	42	336
8	1	8	72	720

```
long int RisingFactorial(int n, int m)
/* Rising Factorial function */
{
int i;
long int product,term;

if ( m == 0 )  return 1;
else
  {
  product = 1;
  term = n;
  for ( i = 0; i <= m-1; i++ )
    {
    product *= term;
    term += 1;
    }
  return  product;
  }
}
```

7.2 BINOMIAL AND MULTINOMIAL COEFFICIENTS

Binomial coefficients arise in combinatorial analysis when one computes the number of ways of partitioning n things into two bins, with m in one bin and $(n-m)$ in the other. The generalization of the binomial coefficient to the multinomial coefficient (Section 7.2.2) gives the number of ways of partitioning n things into m bins, with n_1 in the first bin, n_2 in the second bin, and n_m in the mth bin.

7.2.1 Binomial Coefficients

Keywords
binomial, combinatorial, distribution, factorial

C Function and *Mathematica* Notebook
Binomial (C); cell Binomial in notebook Combinatorial (*Mathematica*)

$\boxed{M}$

Background
In addition to their use in combinatorial analysis, in the theory of discrete probability distributions (Section 9.2) binomial coefficients give the weights associated with probabilities of events for which there are only two possible outcomes. A comprehensive discussion of binomial coefficients in the context of probability theory is given in Section VI.2 of Feller's text [Fel68].

Section 24.1.1 and Table 24.1 in Abramowitz and Stegun [Abr64] give formulas and numerical values, respectively. In Section 2.5.3 of Erdélyi et al [Erd53a] relations involving binomial coefficients and incomplete beta functions (our Section 6.3.2) are given.

Function Definition

For nonnegative integers n and m the binomial coefficient is defined in terms of the factorial functions in Section 7.1.1 by

$$\binom{n}{m} \equiv \left| \begin{array}{ll} \dfrac{n!}{m!\,(n-m)!} & 0 \le m \le n \\[2mm] 0 & m < 0,\, m > n \end{array} \right. \tag{7.2.1}$$

By setting the binomial coefficient to zero for m outside the range $[0, n]$, use of the coefficients in situations in which n varies is often simplified.

Visualization

We show two ways of visualizing binomial coefficients. The first, Table 7.2.1, is the Pascal triangle, which illustrates the recurrence relation

$$\binom{n}{m} \equiv \binom{n-1}{m} + \binom{n-1}{m-1} \qquad n > 0 \tag{7.2.2}$$

which is shown by the arrows in Table 7.2.1. Because of the restrictions in (7.2.1), this relation is valid for all m.

TABLE 7.2.1 **Pascal triangle for binomial coefficients, showing (7.2.2) for $n = 4$ and $m = 2$.**

$n \setminus m$	0	1	2	3	4	5
0	1					
1	1	1				
2	1	2	1			
3	1	3 →	3	1		
4	1	4	6	4	1	
5	1	5	10	10	5	1
⋮	⋮	⋮	⋮	⋮	⋮	⋮

The Pascal triangle also illustrates the symmetry relation

$$\binom{n}{m} \equiv \binom{n}{n-m} \qquad n > 0 \tag{7.2.3}$$

This formula is used in efficient algorithms for the binomial coefficients, as discussed below under Algorithm and Program.

Our second visualization of the binomial coefficients is a three-dimensional view, as shown in Figure 7.2.1. Here n and m form two of the axes and the coefficient value is on the vertical axis.

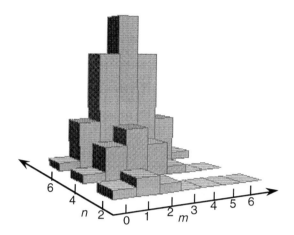

FIGURE 7.2.1 Three-dimensional bar-chart view of binomial coefficients, shown on the vertical axis for $n = 2, 4, 6$. (Adapted from cell `Binomial` in *Mathematica* notebook `Combinatorial`.)

The distribution of binomial coefficients, when plotted against m for fixed n, rapidly approaches a Gauss distribution centered on $m = n/2$ and having variance $\sigma^2 = n/2$. This is shown for $n = 6$ in the graphic at the head of this chapter. Specifically, as can be derived from the discussion in Section 6.1 of [Joh92], one has that

$$\binom{n}{m} \approx \frac{2^{n-1/2}}{\sqrt{2\pi}} e^{-(m-n/2)^2/\sqrt{n}} \tag{7.2.4}$$

Binomial coefficients, as they appear in the binomial distribution (Section 9.2.1), are also related in a similar way to Poisson distributions (Section 9.2.6).

Algorithm and Program

For efficient computation of the binomial coefficient—that also reduces the chance of word overflow—the algorithm minimizes computing cost by finding which of the two arguments in the denominator of (7.2.1) is the larger, L, and which is the smaller, S. Then we factor $L!$ from the numerator, leaving $n - L$ terms, so that, expressed as a rising factorial (Section 7.1.2), $n!/L! = (L+1)_{n-L}$. Thus

$$\binom{n}{m} = \frac{(L+1)_{n-L}}{S!} \tag{7.2.5}$$

The largest number that can occur when computing the binomial coefficient is then $n!/(n/2)!$.

The program for the binomial coefficients is as follows.

```
long int Binomial(int n, int m)
/* Binomial function */
{
int nMm,Large,Small;
if ( m > n ) return 0;
else
  {
  nMm = n-m;
  if ( m > nMm )
    {
    Large = m; Small = nMm;
    }
  else
    {
    Large = nMm; Small = m;
    }
  return
    RisingFactorial
    (Large+1,n-Large)/Factorial(Small);
  }
}
```

Programming Notes
The factorization method described under Algorithm delays word overflow, but cannot eliminate it. For a 32-bit word with one sign bit, the binomial coefficient with $n = 17$ and $m = 9$, is computed correctly as 24310, but for $n = 18$ and $m = n/2 = 9$ (the worst case for this n) the calculation fails. The program does not check for word overflow.

Test Values
Table 7.2.1 gives test values of the binomial coefficients for small n and m. The Pascal triangle relation (7.2.2) and the symmetry (7.2.3) can be used for any n and m as a further check.

7.2.2   Multinomial Coefficients

Keywords
combinatorial, distribution, factorial, multinomial

C Function and *Mathematica* Notebook

Multinomial (C); cell Multinomial in notebook Combinatorial(*Mathematica*)

Background
Multinomial coefficients are a generalization of binomial coefficients. They give the number of possible ways of distributing n objects among m partitions, with n_1 in the first partition, n_2 in the second, and so on. Applications of multinomial coefficients to probability

theory are given in Section VI.9 of Feller's text [Fel68], while extensive formulas for multinomial coefficients are given in Section 24.1.2 of Abramowitz and Stegun [Abr64].

Function Definition
Multinomial coefficients of order $m \geq 2$ are defined by

$$(n; n_1, n_2, \ldots, n_m) \equiv \frac{n!}{n_1! n_2! \ldots n_m!} \qquad \sum_{i=1}^{m} n_i = n \tag{7.2.6}$$

Binomial coefficients (Section 7.2.1) are thus the simplest of multinomial coefficients, having $m = 2$ in (7.2.6);

$$\binom{n}{m} = (n; m, n - m) \tag{7.2.7}$$

Visualization
Since the multinomial $(n; n_1, n_2, \ldots, n_m)$ has m distinct arguments, the summation condition in (7.2.6) removing say n_m, its display is quite complicated. Figure 7.2.2 shows the simplest example, the trinomial ($m = 3$) excluding the case $n_2 = 0$ which produces just the binomial shown in the figure. For $n = 2, 4, 6$ there are 6 three-dimensional bar charts.

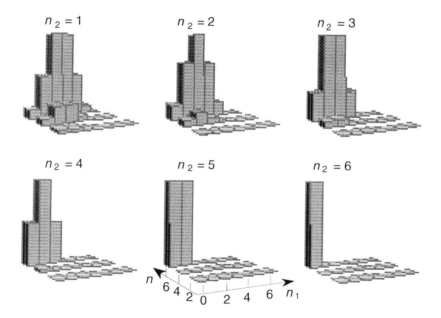

FIGURE 7.2.2 Three-dimensional bar charts of the trinomial coefficients, shown for $n = 0, 2, 4, 6$, and the corresponding range of n_1, [0, 6], for $n_2 = 1$ to 6. Vertical scales are chosen so that the maximum coefficient for given n_2 is the same height in each chart. Actual values are given in Table 7.2.2. (Adapted from cell Binomial in *Mathematica* notebook Combinatorial.)

Algorithm and Program

For efficiency and for reducing the chance of word overflow, we use the same strategy as for the binomial coefficient (Section 7.2.1) by finding which argument in the denominator of (7.2.6) is the largest, n_ℓ. Then we factor n_ℓ from the numerator, leaving $n - n_\ell$ terms, so that, as a rising factorial (Section 7.1.2), $n!/n_\ell! = (n_\ell + 1)_{n-n_\ell}$. Thus

$$(n; n_1, n_2, \ldots, n_m) \equiv (n_\ell + 1)_{n-n_\ell} \left/ \prod_{i=1, i \neq \ell}^{m} n_i! \right. \tag{7.2.8}$$

The largest number that can occur when computing the multinomial coefficient is then $n!/(n/2)!$. The resulting program for function `Multinomial` is given on the following page.

Programming Notes

The factorization method delays word overflow, but cannot eliminate it, just as for the binomial coefficient (Section 7.2.1). The program does not check for word overflow. Also, the summation condition in (7.2.6) should be checked by the calling program. One way to ensure this condition is to construct the last bin size, n_m, as n minus the sum of all the other bin sizes. Then, after checking that this result is not negative, the function is ready to be used. The driver program `Multinomial.C` in Section 21.7.2 shows how this can be done. Note that the function `Multinomial` assumes that m is at least 2. For $m = 2$, however, it is more efficient to use the function `Binomial` (Section 7.2.1).

The functions `RisingFactorial` and `Factorial` that are required are in the machine-readable file.

Test Values

Table 7.2.2 gives test values of the trinomial coefficient $(4; n_1, n_2, n_3)$, with the constraint that $n_3 = 4 - n_1 - n_2$ else the coefficient is set to zero. A multinomial coefficient is invariant under permutations among the arguments n_i, $i = 1, \ldots, m$, as seen in Table 7.2.2 for interchange of n_1 and n_2. Also, a trinomial with either n_1 or n_2 zero is just a binomial coefficient, as the table also shows.

TABLE 7.2.2 Test values of trinomial coefficients ($m = 3$) for $n = 4$.

$n_1 \setminus n_2$	0	1	2	3	4
0	1	4	6	4	1
1	4	12	12	4	0
2	6	12	6	0	0
3	4	4	0	0	0
4	1	0	0	0	0

```
long int Multinomial(int n, int m, int bin)
/* Multinomial function */
{
int nMm,Large,i,iLarge;
long int ProdFact;

/* Find largest bin */
Large = bin[1];
for ( i = 2; i <= m; i++ )
  {
  if ( bin[i] > Large )
    {
    Large = bin[i];
    iLarge = i;
    }
  }

/* Make product of other bin factorials */
ProdFact = 1;
for ( i = 1; i <= m; i++ )
  {
  if ( i != iLarge )
    ProdFact *= Factorial(bin[i]);
  }

return
  RisingFactorial
  (bin[iLarge]+1,n-bin[iLarge])/ProdFact;
}
```

7.3   STIRLING NUMBERS OF FIRST AND SECOND KINDS

Stirling numbers occur in two main contexts; the first is combinatorial problems in discrete mathematics and the second is in the finite-difference calculus of numerical analysis.

A wide variety of notations is used for the Stirling numbers, as described in the introduction to Chapter 24 of Abramowitz and Stegun [Abr64].

7.3.1   Stirling Numbers of the First Kind

Keywords
combinatorials, numerical analysis, Stirling numbers

C Function and *Mathematica* Notebook
Stirl_1 (C); cell StirlingNumbers in notebook Combinatorial (*Mathematica*)

M

Background

Stirling numbers of the first kind, $S_{1,n}^{(m)}$, occur most frequently in the context of enumerating the number of permutations of n symbols that have exactly m cycles. This is given by $(-1)^{n-m} S_{1,n}^{(m)}$, a positive integer.

In finite-difference calculus in numerical analysis, let $\Delta^n f(x)$ denote the nth finite difference of the well-behaved function $f(x)$, where

$$\Delta f(x) \equiv f(x+1) - f(x) \qquad \Delta^{n+1} f(x) \equiv \Delta\left(\Delta^n f(x)\right) \tag{7.3.1}$$

are the successive finite differences. Then the nth derivative can be obtained in terms of the $S_{1,n}^{(m)}$ from

$$\frac{d^m f(x)}{dx^m} = m! \sum_{n=m}^{\infty} \frac{S_{1,n}^m}{n!} \Delta^n f(x) \tag{7.3.2}$$

assuming convergence of the series. Formulas for Stirling numbers of the first kind are given in Section 24.1.3 of Abramowitz and Stegun [Abr64].

Function Definition

Stirling numbers of the first kind, $S_{1,n}^{(m)}$, can be defined through their generating function

$$x(x-1)...(x-n+1) = \sum_{m=0}^{n} S_{1,n}^m \frac{x^n}{n!} \tag{7.3.3}$$

in which the correspondence of our notation to that in Section 24.1.3 of [Abr64] is that $S_{1,n}^{(m)} = S_n^{(m)}$. If either n or m is negative, or if $m > n$, then $S_{1,n}^{(m)} = 0$.

Visualization

Since Stirling numbers involve counting combinations, they show dependencies on n and m that are similar to those of binomial and multinomial coefficients (Section 7.2). Figure 7.3.1 shows the magnitudes of the numbers of the first kind.

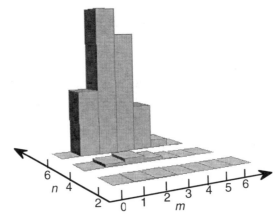

FIGURE 7.3.1 Stirling numbers of the first kind. The number of the first kind, $S_{1,n}^{(m)}$, is displayed with a factor $(-1)^{n-m}$ that makes a nonnegative value. (Adapted from the cell StirlingNumbers in *Mathematica* notebook Combinatorial.)

Algorithm and Program

For efficient numerical calculation of $S_{1,n}^{(m)}$ we use the recurrence relation

$$S_{1,n}^{(m)} = S_{1,n-1}^{(m-1)} - (n-1)S_{1,n-1}^{(m)} \tag{7.3.4}$$

in conjunction with the special values

$$S_{1,m}^{(m)} = 1 \qquad S_{1,n}^{(1)} = (-1)^{n-1}(n-1)! \tag{7.3.5}$$

The C program is very simple:

```
long int Stirl_1(int n, int m)
/* Stirling numbers of the first kind */
/* Needs Factorial */
{
if ( m < 0 || n < 0 || m > n ) return 0;
if( m == n ) return 1;
if( !m ) return 0;
if( m == 1 )
    return ((n-1)%2?-1:1)*Factorial(n-1);
return Stirl_1(n-1,m-1)-(n-1)*Stirl_1(n-1,m);
}
```

Programming Notes

The program for function `Stirl_1` uses recursion, which is allowed in C. In principle, this could overflow the computer memory if n is large enough, but in practice the word size is likely to be overflowed if many recursions are needed to generate $S_{1,n}^{(m)}$.

The simplicity gained by using recursion outweighs its possible inefficiencies of execution-time and storage, of the kind discussed in Section 3.2. Some efficiency might be obtained by using a fairly complex iterative scheme based on (7.3.4). By comparison, when programming the function `Factorial` (Section 7.1.1) it is better to use iteration.

Test Values

Table 7.3.1 gives test values of Stirling numbers of the first kind. A more extensive table is Table 24.3 of [Abr64]. The recurrence relation (7.3.4), although complicated, can be used as a general check.

TABLE 7.3.1 Test values of Stirling numbers of the first kind.

$n \backslash m$	0	1	2	3	4
0	1	0	0	0	0
1	0	1	0	0	0
2	0	-1	1	0	0
3	0	2	-3	1	0
4	0	-6	11	-6	1

7.3.2   Stirling Numbers of the Second Kind

Keywords
combinatorials, numerical analysis, Stirling numbers

C Function and *Mathematica* Notebook
`Stirl_1`; (C); cell `StirlingNumbers` in notebook `Combinatorial` (*Mathematica*)

Background
Stirling numbers of the second kind, $S_{2,n}^{(m)}$, give the number of ways of partitioning n symbols into m nonempty sets. In the finite-difference calculus, with $\Delta^n f(x)$ the nth finite difference of the function $f(x)$, we have the relation inverse to (7.3.2), namely

$$\Delta^m f(x) = m! \sum_{n=m}^{\infty} \frac{S_{2,n}^m}{n!} \frac{d^n f(x)}{dx^n} \qquad (7.3.6)$$

assuming convergence of the series.

The $S_{2,n}^{(m)}$ also occur in summing powers of integers, in relations such as

$$\sum_{k=0}^{n} k^m = \sum_{k=0}^{m} S_{2,m}^{(k)} k! \binom{n+1}{k+1} \qquad (7.3.7)$$

Formulas for Stirling numbers of the second kind are given in Section 24.1.4 of Abramowitz and Stegun [Abr64].

Function Definition
Stirling numbers of the second kind, $S_{2,n}^{(m)}$, can be defined through the generating function

$$x^n = \sum_{m=0}^{n} S_{2,n}^m \, x(x-1)...(x-m+1) \qquad (7.3.8)$$

The correspondence of our notation to that in Section 24.1.4 of [Abr64] is $S_{2,n}^{(m)} = S_n^{(m)}$. If either n or m is negative, or if $m > n$, then $S_{2,n}^{(m)} = 0$.

Visualization
The Stirling numbers of the second kind, $S_{2,n}^{(m)}$, are shown for small values of n and m in the right-hand panel of Figure 7.3.2 on the next page.

Algorithm and Program
Numerical calculation of $S_{2,n}^{(m)}$ is done efficiently by using the recurrence relation

$$S_{2,n}^{(m)} = S_{2,n-1}^{(m-1)} + n S_{2,n-1}^{(m)} \qquad (7.3.9)$$

and the special values

$$S_{2,m}^{(m)} = 1 \qquad S_{2,1}^{(1)} = 1 \qquad S_{2,n}^{(2)} = 2^{n-1} - 1 \qquad (7.3.10)$$

The C program is then simply:

```
long int Stirl_2(int n, int m)
/* Stirling numbers of the second kind */
/* Needs IntPow */
{
  if ( m < 0 || n < 0 || m > n ) return 0;
  if( m == n )  return 1;
  if( !m ) return 0;
  if( m == 1 )  return 1;
  if( m == 2 )  return IntPow(2,n-1)-1;
  return m*Stirl_2(n-1,m)+ Stirl_2(n-1,m-1);
}
```

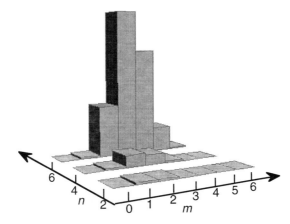

FIGURE 7.3.2 Stirling numbers of the second kind plotted as functions of m and n. (Adapted from cell StirlingNumbers in the *Mathematica* notebook Combinatorial.)

Programming Notes

Function Stirl_2 uses recursion (Section 3.2), as for Stirl_1. This could overflow the computer memory if n is large enough, but in practice the word size is likely to overflow first if many recursions are needed to generate $S_{2,n}^{(m)}$. The simplicity gained by using recursion outweighs the efficiency that might be gained by implementing a complex iterative scheme based on (7.3.9).

The function IntPow(i, n) returns integer i to the integer n power. It is provided with the machine-readable function and the driver program described in Section 21.7.

Test Values

Table 7.3.2 gives test values of Stirling numbers of the second kind. A more extensive compilation is Table 24.4 of [Abr64]. Recurrence relation (7.3.9) can be used as a general check.

TABLE 7.3.2 Test values of Stirling numbers of the second kind.

$n \setminus m$	0	1	2	3	4
0	1	0	0	0	0
1	0	1	0	0	0
2	0	1	1	0	0
3	0	1	3	1	0
4	0	1	7	6	1

7.4 FIBONACCI AND LUCAS POLYNOMIALS

Fibonacci and Lucas polynomials are used frequently in a wide variety of contexts, such as number theory, cryptography, and enumerating configurations of organic molecules. See, for example, the conference proceedings edited by Philippou et al [Phi84, Phi86] and the journal Fibonacci Quarterly [Fib63]. Fibonacci numbers and Lucas numbers are sequences obtained by evaluating the polynomials with their argument $x = 1$ and incrementing the order of the polynomials by unity to generate successive terms in the sequence.

7.4.1 Fibonacci Polynomials and Fibonacci Numbers

Keywords
combinatorials, Fibonacci numbers, Fibonacci polynomials

C Function and *Mathematica* Notebook
Fibonacci (C); cell Fibonacci in notebook Combinatorial (*Mathematica*)

Background
Applications of Fibonacci polynomials in many contexts are discussed by Bicknell [Bic70], Hoggatt and Bicknell [Hog73], Philippou et al [Phi84, Phi86], and in articles appearing in the Fibonacci Quarterly [Fib63].

Function Definition
We define the Fibonacci polynomials, $F_n(x)$, by the recurrence relation

$$F_{n+1}(x) = xF_n(x) + F_{n-1}(x) \qquad n > 2 \tag{7.4.1}$$

with the initial values

$$F_1(x) = 1 \qquad F_2(x) = x \tag{7.4.2}$$

The Fibonacci *numbers*, usually denoted by F_n, are thus Fibonacci polynomials with $x = 1$. The order of $F_n(x)$ is $n-1$. The definition of $F_n(x)$ is sometimes extended to include $n \leq 0$, but we do not do this here.

If (7.4.1) is considered as a finite-difference approximation for the derivative of F_n with respect to n, then we find that $\ln F_n \approx (n-1)/2$ for large n, a relation consistent with the bar chart of F_n in the lower left part of Figure 7.4.1.

Examples of formulas for Fibonacci polynomials obtained from (7.4.1) are:

$$F_2(x) = x \qquad F_3(x) = 1 + x^2 \qquad F_4(x) = 2x + x^3$$
$$F_5(x) = 1 + 3x^2 + x^4 \qquad F_6(x) = 3x + 4x^3 + x^5 \tag{7.4.3}$$

Note that definitions (7.4.1) and (7.4.2) are consistent with definitions used in *Mathematica* for Fibonacci numbers, as well as that used in *Maple* [Cha91, Section 4.1.8] for Fibonacci numbers and polynomials. In *Maple* $n \leq 0$ is also allowed.

Visualization

The Fibonacci polynomials, $F_n(x)$, have the nonnegative integers, n, as the range of the first argument, and a continuous variable, x, as the range of the second argument. It is therefore convenient to display a histogram (bar chart) for each x value, as in Figure 7.4.1.

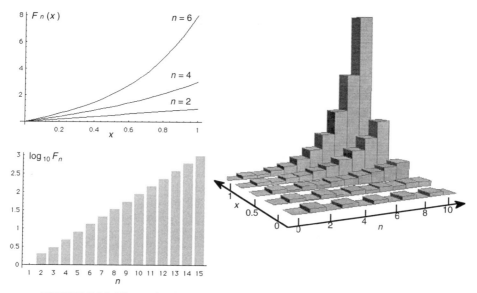

FIGURE 7.4.1 Fibonacci polynomials and numbers. Left: Polynomials as functions of x (top) and the numbers as functions of n. Right: Polynomials as functions of n and x for $n = 0,...,10$ and $x = 0, 0.25, 0.50, 0.75,$ and 1.0. (Adapted from the output of cell `Fibonacci` in *Mathematica* notebook `Combinatorial`.)

Algorithm and Program

The algorithm for numerical computation of $F_n(x)$ uses the recurrence relation (7.4.1) with the starting values for $n = 0$ and $n = 1$ given by (7.4.2). Recursion is used in the following C program, without much loss in efficiency of execution. The same algorithm is used for symbolic computation to obtain formulas for $F_n(x)$—such as (7.4.3)—by running cell `Fibonacci` in the *Mathematica* notebook `Combinatorial`. The C program is on the next page.

Programming Notes

Function `Fibonacci` assumes that $n \geq 0$, and it will fail if negative values of n are used. For $n > 0$ the result will overflow the word size only if large values of n and x are used.

```
double Fibonacci(int n, double x)
/* Fibonacci polynomial */
{
if ( n <= 0 )    return 0;
if ( n == 1 )    return 1;
return  x*Fibonacci(n-1,x)+Fibonacci(n-2,x);
}
```

Test Values

Test values of $F_n(x)$ are given in Table 7.4.1 for small values of x and n. For $x = 0.5$ the values have been rounded to four decimal digits. The other values are exact. Fibonacci numbers are obtained when $x = 1$.

TABLE 7.4.1 **Test values to 10 digits of Fibonacci polynomials and numbers.**

$x \backslash n$	2	3	4	5	6
0	0	1	0	1	0
0.50	0.50000000	1.25000000	1.12500000	1.81250000	2.03125000
1.00	1	2	3	5	8

A test that can be made for $x = 1$ is that the Fibonacci numbers, F_n, satisfy

$$F_n(1) = F_n = \left[\frac{1}{\sqrt{5}}\left(\frac{1+\sqrt{5}}{2}\right)^n\right] \tag{7.4.4}$$

in which $[...]$ means "the nearest integer to." For example,

$$\frac{1}{\sqrt{5}}\left(\frac{1+\sqrt{5}}{2}\right)^6 = 8.0249 \tag{7.4.5}$$

so that $F_6(1) = F_6 = 8$, in agreement with Table 7.4.1. The "golden-ratio" formula (7.4.4) also provides a way to generate Fibonacci numbers that is especially practical for large n.

7.4.2   Lucas Polynomials

Keywords

combinatorials, encryption, Fibonacci numbers, Lucas numbers, Lucas polynomials

C Function and *Mathematica* Notebook

Lucas (C); cell Lucas in notebook Combinatorial (*Mathematica*)

Background

Fibonacci and Lucas polynomials—which satisfy the same recurrence relations but differ in their initial values—are examples of Fibonacci *sequences*, as discussed by Bicknell [Bic70] and by Hoggatt and Bicknell [Hog73].

Function Definition

Lucas polynomials, $L_n(x)$, can be defined by the same recurrence relation as for Fibonacci polynomials, namely

$$L_{n+1}(x) = xL_n(x) + L_{n-1}(x) \tag{7.4.6}$$

but with a different lower limit for n and different initial values, namely

$$L_0(x) = 0 \qquad L_1(x) = x \tag{7.4.7}$$

The Lucas *numbers*, usually denoted by L_n, are just Lucas polynomials having $x = 1$.

The function $L_n(x)$ is an nth-order polynomial. Examples of formulas for $L_n(x)$ are:

M

$$L_2(x) = 2 + x^2 \qquad L_3(x) = 3x + x^3 \qquad L_4(x) = 2 + 4x^2 + x^4$$
$$L_5(x) = 5x + 5x^3 + x^5 \qquad L_6(x) = 2 + 9x^2 + 6x^4 + x^6 \tag{7.4.8}$$

Beware that the relation between n and the order of the Lucas polynomial varies among authors.

Visualization

Lucas polynomials, $L_n(x)$, have the nonnegative integers, n, as the range of their first argument, and a continuous variable, x, as the range of their second argument. It is therefore convenient to display a histogram for each x value, as shown in Figure 7.4.2.

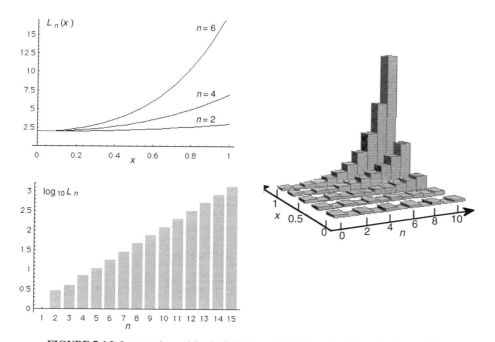

FIGURE 7.4.2 Lucas polynomials. Left: Polynomials (top) as functions of x for $n = 2, 4,$ 6, and numbers (bottom) for n 1 to 15. Right: Bar charts of the polynomials for $n = 0,\ldots,10$ and $x = 0, 0.25, 0.50, 0.75,$ and 1.0. (Adapted from the output of cell Lucas in *Mathematica* notebook Combinatorial.)

Since Lucas and Fibonacci polynomials differ only because of their different initial values, they look very similar in their graphics—compare Figures 7.4.1 and 7.4.2—especially as x increases.

Algorithm and Program
The algorithm for computing $L_n(x)$ numerically uses recurrence relation (7.4.6) with the starting values for $n = 0$ and $n = 1$ given in (7.4.7). Recursion is used in the C program, without significant loss in efficiency of execution. The same algorithm is used for symbolic computation using cell `Lucas` in *Mathematica* notebook `Combinatorial` to produce formulas such as in (7.4.8).

```
double Lucas(int n, double x)
/* Lucas polynomial */
{
if ( n <= 0 ) return 2;
if ( n == 1 ) return x;
return  x*Lucas(n-1,x)+Lucas(n-2,x);
}
```

Programming Notes

The programming of function `Lucas` assumes that $n \geq 0$, and it will fail if negative n is used. For $n > 0$ the result will overflow the word size only for large n and x.

Test Values

Test values of $L(n, x)$ are given in Table 7.4.2 for small values of x and n. For $x = 0.5$ the values are accurate to ten decimal digits. The Lucas numbers are obtained when $x = 1$.

TABLE 7.4.2 Test values to 10 digits of Lucas polynomials and numbers.

$x \setminus n$	2	3	4	5	6
0	2	0	2	0	2
0.50	2.25000000	1.62500000	3.06250000	3.15625000	4.64062500
1.00	3	4	7	11	18

REFERENCES ON COMBINATORIAL FUNCTIONS

[Abr64]   Abramowitz, M., and I. A. Stegun, *Handbook of Mathematical Functions*, Dover, New York, 1964.

[Bic70]   Bicknell, M., Fibonacci Quarterly, **8**, 407 (1970).

[Cha91]   Char, B. W. et al, *Maple V Library Reference Manual*, Springer, New York, 1991.

[Erd53a]   Erdélyi, A., et al, *Higher Transcendental Functions*, Vol. 1, McGraw-Hill, New York, 1953; reprint edition, Krieger, Malabar, Florida, 1981.

[Fel68]   Feller, W., *Introduction to Probability Theory and Its Applications*, Wiley, New York, third edition, 1968.

[Fib63]   Fibonacci Quarterly, (1963–).

[Hog73]   Hoggatt, V. E., and M. Bicknell, Fibonacci Quarterly, **11**, 271 (1973).

[Joh92]   Johnson, N. L., S. Kotz, and A. W. Kemp, *Univariate Discrete Distributions*, Wiley, New York, second edition, 1992.

[Phi84]   Philippou, A. N., G. E. Bergum, and A. F. Horadam (eds.), *Fibonacci Numbers and Their Applications*, D. Reidel, Dordrecht, 1984.

[Phi86]   Philippou, A. N., A. F. Horadam, and G. E. Bergum, (eds.), *Applications of Fibonacci Numbers*, D. Reidel, Dordrecht, 1984.

[Rio79]   Riordan, J., *Combinatorial Identities*, Krieger Publishing, Huntington, New York, 1979.

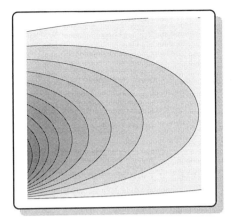

Chapter 8

NUMBER THEORY FUNCTIONS

This chapter emphasizes functions used for number theory, corresponding closely to the functions considered in Chapter 23 of Abramowitz and Stegun [Abr64], except that the polylogarithms (Section 8.5), which are not considered generally in [Abr64], are included here. The functions in this chapters can be computed accurately and efficiently in C by using floating-point arithmetic. By comparison, most of the C programs for combinatorial functions in Chapter 7 are programmed in long-integer arithmetic, although it is easy to adapt the programs to floating-point arithmetic. For number theory functions requiring exact calculations on integers, and thereby arbitrarily long computer words, we recommend the procedures described in Section 4.4 of the *Maple* library reference manual [Char91] and in the *Mathematica* guide to standard packages [Boy92].

M

The chapter begins with Bernoulli and Euler numbers and the corresponding polynomials (Sections 8.1 and 8.2), which are sometimes considered as combinatorial functions (Chapter 7). Then follow the Riemann zeta function (Section 8.3) and other sums of reciprocal powers (Section 8.4). Polylogarithms are described from graphical and computational viewpoints in Section 8.5.

8.1   BERNOULLI NUMBERS AND BERNOULLI POLYNOMIALS

The Bernoulli numbers and their extensions, the Bernoulli polynomials, are among the most interesting and important number sequences. A 150-page bibliography of the Bernoulli numbers covering the years 1713 to 1990 is in Dilcher et al [Dil91], while an extensive treatment of the zeros of Bernoulli polynomials has been given by Dilcher [Dil88]. Many formulas are described in Sections 9.6 and 9.71 of Gradshteyn and Ryzhik [Gra94].

We describe first the formulas and function for Bernoulli numbers (Section 8.1.1), which are required for computing Bernoulli polynomials (Section 8.1.2).

8.1.1   Bernoulli Numbers

Keywords
Bernoulli numbers, combinatorials, number theory functions

C Function and *Mathematica* Notebook

BernoulliNum (C); cell Bernoulli in notebook NumberTheory (*Mathematica*)

Background

Extensive background to Bernoulli numbers is given in the references cited in [Dil91], while Section 1.13 of Erdélyi et al [Erd53a] and Section 23.1 of [Abr64] have summaries of their mathematical properties. Asymptotic expansions (Section 3.4) often require Bernoulli numbers, as do Euler-Maclaurin summation formulas (Section 3.5).

Function Definition

Bernoulli numbers B_n and Bernoulli polynomials $B_n(x)$ can be related through [Erd53a, Section 1.13 (3)]

$$B_n(x) = \sum_{r=0}^{n} \binom{n}{r} B_r \, x^{n-r} \tag{8.1.1}$$

Given the definition

$$B_r = B_r(1) \tag{8.1.2}$$

we have that

$$\sum_{r=0}^{n-1} \binom{n}{r} B_r = 0 \tag{8.1.3}$$

from which, by letting $n \to n+1$, we deduce the recurrence relation

$$B_n = -\frac{1}{n+1} \sum_{r=0}^{n-1} \binom{n+1}{r} B_r \qquad n > 0 \qquad B_0 = 1 \tag{8.1.4}$$

that can be used to generate the B_n iteratively. Also

$$B_1 = -1/2 \qquad B_{2n+1} = 0 \qquad n > 0 \tag{8.1.5}$$

Since the initial value in the iteration is an integer and the binomial coefficients in (8.1.4) are integers, the quantity $(n+1)! B_n$ is an integer and thereby B_n is a rational fraction. Nonzero Bernoulli numbers alternate in sign, as (8.1.4) suggests.

A useful approximation to B_{2n} can be obtained from its bounds [Abr64, 23.1.15] as

$$B_{2n} \approx (-1)^{n+1} \, 2 \, (2n)! \, / (2\pi)^{2n} \tag{8.1.6}$$

which has a fractional accuracy of better than 10^{-10} for $n > 17$. The numbers $B_{2n}/(2n)!$ that are required for using the Euler-Maclaurin summation formula (Sections 3.5 and 8.3) decrease rapidly with n, as this formula shows.

Visualization

Bernoulli numbers, B_n, for small n are shown in the upper panel of Figure 8.1.1. They increase in magnitude monotonically and very rapidly as n increases beyond 6, as implied by (8.1.6) and shown in Figure 8.1.2. Since B_n is computed in decimal approximation, word-size overflow is less of a problem than for the combinatorial functions where we use exact integer computations.

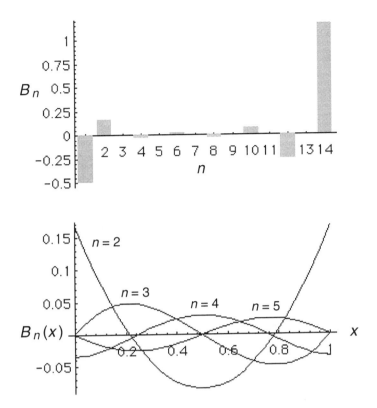

FIGURE 8.1.1 Bernoulli numbers (upper panel) for $n = 1$ to 14, Bernoulli polynomials (lower panel) for $n = 2$ to 5 and x in [0,1]. (Adapted from the output of cell `Bernoulli` in *Mathematica* notebook `NumberTheory`.)

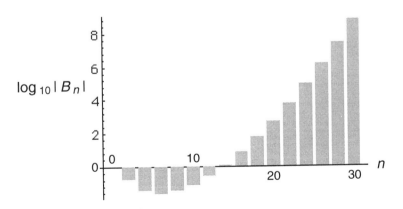

FIGURE 8.1.2 Rapid variation of the magnitude of the Bernoulli numbers for even n as n increases, shown on a base-10 logarithmic scale. (Adapted from the output of cell `Bernoulli` in the *Mathematica* notebook `NumberTheory`.)

Algorithm and Program

Formula (8.1.4) can be rewritten to give an efficient algorithm for generating the Bernoulli numbers B_i iteratively for $i = 0$ to $i = n$, namely,

$$B_1 = -1/2 \qquad B_i = 0 \quad \text{if } i \text{ odd} \quad i > 1 \qquad B_0 = 1$$

$$B_i = \left[(i-1)/2 - \sum_{r=1}^{[i/2]-1} \binom{i+1}{2r} B_{2r} \right] / (i+1) \qquad \text{if } i \text{ even} \qquad i \geq 2 \qquad (8.1.7)$$

Thus we obtain the program $\texttt{BernoulliNum}$ for an array B of the first $n+1$ Bernoulli numbers:

```
void BernoulliNum(int n, double B[])
/* Bernoulli numbers from B(0) to B(n) */
{
double sum;
long int Binomial(int n, int m);
int i,r,i2;

for ( i = 0; i <= n; i++ )
  {
  if ( i == 1 ) B[1] = -0.5;
  else
    { /* Other odds are zero */
    if ( i%2 == 1) B[i] = 0;
    else
      {/* Evens are nonzero */
      if ( i == 0 ) B[0] = 1;
      else
        {
        i2 = i/2;
        sum = (1-i)/2.0;
        /* Sum over previous values */
        for ( r = 1; r <= i2-1;r++ )
          {
          sum += Binomial(i+1,2*r)*B[2*r];
          }
        B[i] = -sum/(i+1);
        }
      }
    }
  }
}
```

Alternative algorithms, especially suitable for large n, are given by Fillebrown [Fil92].

Programming Notes

Function $\texttt{BernoulliNum}$ uses the integer function $\texttt{Binomial}$ from Section 7.2.1. Therefore, if a 32-bit word is used to store the binomial it will overflow if $n \geq 18$. If this creates a problem for your application, you could increase the range of n by programming a floating-point version of the binomial-coefficient function.

Test Values
Test values of decimal approximations to B_n obtained by running the cell `Bernoulli` in *Mathematica* notebook `NumberTheory` are given in Table 8.1.1. Rational-fraction and decimal values up to $n = 60$ are given in Table 23.2 of [Abr64].

$\boxed{M}$

TABLE 8.1.1 Test values to 10 digits of Bernoulli numbers.

n	2	4	6	8
B_n	0.1666666667	−0.0333333333	0.0238095238	−0.0333333333

8.1.2 Bernoulli Polynomials

Keywords
Bernoulli numbers, Bernoulli polynomials, binomial coefficients

C Function and *Mathematica* Notebook
`BernoulliPoly` (C); cell`Bernoulli` in notebook `NumberTheory`(*Mathematica*)

$\boxed{M}$

Background
Background to the Bernoulli polynomials is given in references cited in [Dil91], while Section 1.13 of Erdélyi et al [Erd53a] and Section 23.1 of Abramowitz and Stegun [Abr64] summarize their analytical properties.

Function Definition
Bernoulli polynomials, $B_n(x)$, can be related to the Bernoulli numbers of Section 8.1.1, B_n, by (8.1.1). Some explicit expressions—which can be obtained by running the cell `Bernoulli` in *Mathematica* notebook `NumberTheory`—are:

$\boxed{M}$

$$B_0(x) = 1 \qquad B_1(x) = -\frac{1}{2} + x \qquad B_2(x) = \frac{1}{6} - x + x^2$$

$$B_3(x) = \frac{x}{2} - \frac{3x^2}{2} + x^3 \qquad B_4(x) = -\frac{1}{30} + x^2 - 2x^3 + x^4 \qquad (8.1.8)$$

Visualization
The Bernoulli polynomials, $B_n(x)$, for small values of n and x are shown in the lower panel of Figure 8.1.1. Note that for larger values of n or x $B_n(x)$ will increase in magnitude very rapidly.

$\boxed{!}$

Algorithm and Program
The algorithm for $B_n(x)$ is a rewriting of (8.1.7) for efficiency so that odd-order Bernoulli numbers are not used, namely,

$$B_0(x) = 1 \qquad B_1(x) = -\frac{1}{2} + x$$

$$B_n(x) = x^{n-1}(1 - n/2) + \sum_{i=1}^{[n/2]} \binom{n}{2i} B_{2i} \, x^{n-2i} \qquad n \geq 2 \qquad (8.1.9)$$

Thus, the program for the function `BernoulliPoly` is as follows:

```
double BernoulliPoly
 (int n, double x, double B[])
/* Bernoulli polynomial B(n,x) */
{
double sum,Power(double xx, int j);
long int Binomial(int n, int m);
int i;

if ( n == 0 )   return 1;
if ( n == 1 )   return -0.5 + x;
else
  {
  sum = Power(x,n-1)*(x-n/2.0);
  /* Accumulate sum over even terms */
  for ( i = 1; i <= n/2; i++ )
    {
    sum += Binomial(n,2*i)*B[2*i]*
      Power(x,n-2*i);
    }
  return sum;
  }
}
```

Programming Notes

Before `BernoulliPoly` is called with arguments n and x to calculate $B_n(x)$, the array `B[]` must be available and contain the Bernoulli numbers B_0 through B_n in elements `B[0]` through `B[n]`. This can be done by calling function `BernoulliNum` (Section 8.1.1).

Function `Power` must be used in this function because the C function `pow` returns 0^0 as negative infinity, instead of unity. The machine-readable program with its driver (Section 21.8) is pre-assembled with the needed functions.

Test Values

Test values of the Bernoulli polynomials can be obtained from Table 8.1.2. The symmetry $B_n(1-x) = (-1)^n B_n(x)$ and the relation to the Bernoulli numbers, $B_n = B_n(0)$, from Section 8.1.1 can also be used to check the polynomial values.

TABLE 8.1.2 Test values to 10 digits of Bernoulli polynomials.

$x \setminus n$	2	3	4	5
0	0.1666666667	0	−0.0333333333	0
0.50	−0.0833333333	0	0.0291666667	0
1.00	0.1666666667	0	−0.0333333333	0

The explicit formulas for the low-order polynomials that are given in (8.1.8) can also be evaluated for given x to generate test values. Table 23.1 in [Abr64] gives the coefficients of $B_n(x)$ for $n \leq 15$. Note that the symmetry of the Bernoulli polynomials (Figure 8.1.1) requires that they vanish at the x values in Table 8.1.2 if n is odd; when x is nonzero this is a good test of roundoff errors in their computation.

8.2   EULER NUMBERS AND EULER POLYNOMIALS

Euler numbers and Euler polynomials, along with the Bernoulli functions (Section 8.1), are among the most interesting number sequences. An extensive treatment of the zeros of the Euler polynomials is given by Dilcher [Dil88]. Sections 9.63 and 9.72 of Gradshteyn and Ryzhik [Gra94] have many formulas. We describe first the formulas and function for Euler numbers, which are needed for computing Euler polynomials in Section 8.2.2.

8.2.1   Euler Numbers

Keywords
Euler numbers, combinatorials

C Function and *Mathematica* Notebook
EulerNum (C); cell Euler in notebook Combinatorial (*Mathematica*) $\boxed{M}$

Background
Extensive background to Euler numbers is given in references cited in [Dil91], Section 23.1 of Abramowitz and Stegun [Abr64] and Section 1.14 of Erdélyi et al [Erd53a] have summaries of mathematical properties.

Function Definition
For computational purposes, the Euler polynomials, $E_n(x)$, and Euler numbers, E_r, can be related through [Erd53a, 1.14 (4)]

$$E_n(x) = \sum_{r=0}^{n} \binom{n}{r} 2^{-r} E_r \left(x - \frac{1}{2} \right)^{n-r} \tag{8.2.1}$$

so that, formally,

$$E_n = 2^n E_n(1/2) \tag{8.2.2}$$

It is practical to consider (8.2.1) as defining the polynomial in terms of the numbers E_r. The latter can be obtained by using the identity [Erd53a, 1.14 (17)]

$$\sum_{r=0}^{n} \binom{2n}{2r} E_{2r} = 0 \tag{8.2.3}$$

to deduce that

$$E_{2n} = -\sum_{r=0}^{n-1} \binom{2n}{2r} E_{2r} \tag{8.2.4}$$

Given that

$$E_0 = 1 \qquad E_{2n+1} = 0 \tag{8.2.5}$$

the Euler numbers can be generated iteratively. Since the initial value in the iteration is an integer and the binomial coefficients occurring in (8.2.4) are integers, the Euler numbers are integers. Also, they alternate in sign, as (8.2.4) suggests.

Visualization
Views of the Euler numbers, E_n, for n and x are shown in the upper panel of Figure 8.2.1.

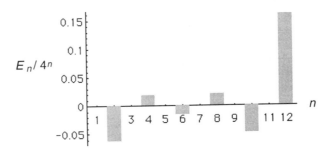

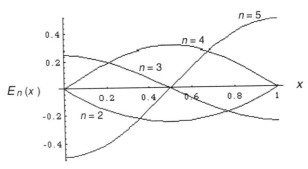

FIGURE 8.2.1 Euler numbers (upper panel) for $n = 1$ to 12, with a scale factor 4^n divided out. Euler polynomials (lower panel) for $n = 2$ to 5 and x in [0,1]. Note that $E_n = 2^n E_n(1/2)$. (Adapted from output of cell Euler in *Mathematica* notebook Number–Theory.)

For larger values of n than shown in Figure 8.2.1, E_n increases monotonically in magnitude very rapidly, as shown in Figure 8.2.2. For example, $E_{30} \approx 10^{25}$.

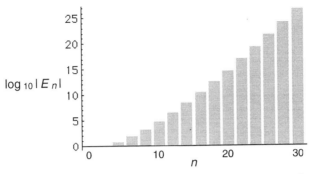

FIGURE 8.2.2 Rapid variation of the magnitude of the Euler numbers as n increases by steps of 2, plotted on a base-10 logarithmic scale. (Adapted from output of cell `Euler` in *Mathematica* notebook `Combinatorial`.)

A useful approximation to E_n is obtained from the inequalities in [Abr64, 23.1.15], giving

$$E_{2n} \approx (-1)^n \, 4^{n+1} \, (2n)! / \pi^{2n+1} \tag{8.2.6}$$

which has a relative accuracy of better than 10^{-10} for $n \geq 10$.

Algorithm and Program
The result (8.2.4) can be written to give an efficient algorithm for generating the Euler numbers E_i iteratively for $i = 0$ to $i = n$, namely

$$E_i = 0 \quad \text{if } i \text{ odd} \qquad E_0 = 1$$

$$E_i = - \sum_{r=0}^{i/2-1} \binom{i}{2r} E_{2r} \quad \text{if } i \text{ even} \tag{8.2.7}$$

Thus we obtain the program `EulerNum` for an array E of the first $n+1$ Euler numbers, that is given on the next page with the Programming Notes.

Test Values
Table 8.2.1 gives test values of E_n. Note that for odd n the Euler numbers are zero.

TABLE 8.2.1 **Exact test values for the Euler numbers.**

n	E_n
0	1
1	0
2	−1
3	0
4	5
14	−199 360 981
16	19 391 512 145

```
void EulerNum(int n, long int E[])
/* Euler numbers from E(0) to E(n) */
{
long int sum,Binomial(int n, int m);
int i,r,i2;

for ( i = 0; i <= n; i++ )
  {
  if ( i%2 == 1)E[i] = 0;/* odds are zero */
  else
    {
    if ( i == 0 ) E[0] = 1;
    else
      {
      i2 = i/2;
      sum = 0; /* sum over previous values */
      for ( r = 0; r <= i2-1; r++ )
        {
        sum += Binomial(i,2*r)*E[2*r];
        }
      E[i] = -sum;
      }
    }
  }
}
```

Programming Notes

Function `EulerNum` requires function `Binomial` (Section 8.1.1). For a 32-bit long-integer word size E_n will overflow the word size for $n \geq 16$. If decimal approximations suffice, reprogram `EulerNum` with double-precision variables or use (8.2.6).

8.2.2   Euler Polynomials

Keywords
Euler polynomials

C Function and *Mathematica* Notebook
`EulerPoly` (C); cell `Euler` in notebook `Combinatorial` (*Mathematica*)

Background
The background to the Euler polynomials $E_n(x)$ is discussed in Section 8.2.1 with the Euler numbers E_n.

Function Definition
The general definition of the Euler polynomial, (8.2.1), can be simplified so that only the nonzero Euler numbers, E_{2i}, with i a nonnegative integer, are involved in the summation. Thus, we have the formula

$$E_n(x) = \sum_{i=0}^{[n/2]} \binom{n}{2i} E_{2i}\left(x - \frac{1}{2}\right)^{n-2i} \Big/ 2^{2i} \tag{8.2.8}$$

in which $[m]$ denotes the integer part of m. The symmetry property under reflection about $x = 1/2$ is

$$E_n(1-x) = (-1)^n E_n(x) \tag{8.2.9}$$

Explicit expressions for the lowest-order polynomials are obtained by running the cell   M
Euler in *Mathematica* notebook NumberTheory. For example,

$$E_0(x) = 1 \qquad E_1(x) = -\frac{1}{2} + x \qquad E_2(x) = -x + x^2$$

$$E_3(x) = \frac{1}{4} - \frac{3x^2}{2} + x^3 \qquad E_4(x) = x - 2x^3 + x^4 \tag{8.2.10}$$

Visualization
Euler polynomials, $E_n(x)$, for small values of n and x are shown in the lower panel of Figure 8.2.1. Note the symmetry property (8.2.9) with respect to reflection about $x = 1/2$.

Algorithm and Program
The algorithm for the Euler polynomial is a direct implementation of formula (8.2.8), so the corresponding program is as follows:

```
double EulerPoly(int n, double x, long int E[])
/* Euler polynomial E(n,x) */
{
double sum,Power(double xx, int j);
long int Binomial(int n, int m);
int i;

sum = 0;  /* Accumulate sum over even terms */
for ( i = 0; i <= n/2; i++ )
  {
  sum += Binomial(n,2*i)*E[2*i]*
     Power(x-0.5,n-2*i)/Power(2.0,2*i);
  }
return sum;
}
```

Programming Notes
Before EulerPoly is called with arguments n and x to calculate $E_n(x)$, the array E[] must be available and contain the Euler numbers E_0 through E_n in elements E[0] through E[n]. This can be done by calling the function EulerNum (Section 8.2.1).

The function Power must be used in this function because the C function pow returns 0^0 as negative infinity, instead of unity. Both EulerNum, Power, and a driver program (Section 21.8) to test function EulerPoly are provided with the prefabricated machine-readable program.

Test Values

Test values of the Euler polynomials can be obtained from Table 8.2.2. Note that since $E_n(1/2) = E_n/2^n$, values of the Euler numbers, E_n, from Section 8.2.1 can also be used to check the polynomial values.

TABLE 8.2.2 Exact test values of Euler polynomials.

$x \setminus n$	2	3	4	5
0	0	−0.25	0	−0.5
0.50	−0.25	0	0.3125	0
1.00	0	−0.25	0	0.5

Explicit formulas for the low-order polynomials given in (8.2.10) can also be evaluated for given x to generate test values. Symmetry (8.2.9) exemplified in Table 8.2.2 can also be used.

8.3   RIEMANN ZETA FUNCTION

Keywords

number theory functions, Riemann zeta function, zeta function

C Function and *Mathematica* Notebook

RiemannZeta (C); cell RiemannZeta in notebook NumberTheory (*Mathematica*)

Background

The Riemann zeta function, $\zeta(z)$, has a central role in number theory and the function also occurs when evaluating integrals and summing series. Introductory-level treatments of the zeta function are given in the texts by Edwards [Edw74] and by González [Gon92, Section 4.7, and more advanced treatments are in Chapter 13 of Whittaker and Watson [Whi27] and in the monograph by Karatsuba and Voronin [Kar92]. The use of calculations with many significant digits to investigate Riemann's hypothesis that all the zeros of $\zeta(z)$ have $\mathrm{Re}\, z = 1/2$ is described in Varga's book [Var90] and illustrated in Figure 8.3.2 below. Formulas are summarized in Section 23.2 of Abramowitz and Stegun [Abr64] and in Section 9.5 of Gradshteyn and Rhyzik [Gra94].

Function Definition

Riemann's zeta function for complex argument z is defined by

$$\zeta(z) = \sum_{k=1}^{\infty} \frac{1}{k^z} \qquad \mathrm{Re}\, z > 1 \tag{8.3.1}$$

Although $\zeta(z)$ is analytic throughout the complex z plane except for a simple pole at $z = 1$, our computational interest is for real $z = s$, for which it is convenient to write (8.3.1) as

$$\zeta^*(s) \equiv \zeta(s) - 1 = \frac{1}{2^s} + \sum_{k=3}^{\infty} \frac{1}{k^s} \qquad s > 1 \tag{8.3.2}$$

Thus, the fractional accuracy of computing the left side of this formula is the ratio of the sum starting at $k = 3$ to the first term on the right side.

Visualization

Figure 8.3.1 shows surfaces of $\zeta(z)$ as functions of $z = x + iy$ depicted as real and imaginary parts. Figure 8.3.2 gives the magnitude and contours of constant phase of $\zeta(z)$ and its behavior along the line $z = 1/2 + it$ where Riemann's hypothesis places the zeros.

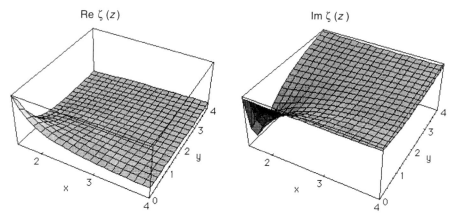

FIGURE 8.3.1 Riemann zeta function surfaces as functions of complex variable $z = x + iy$, real part (left) and imaginary part (right). (Adapted from output of cell `RiemannZeta` in *Mathematica* notebook `NumberTheory`.)

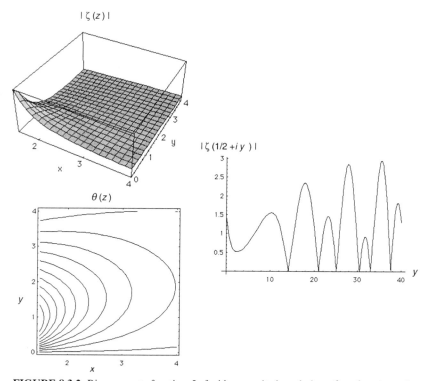

FIGURE 8.3.2 Riemann zeta function: Left side; magnitude and phase θ as functions of complex variable $z = x + iy$. Right side; Magnitude of ζ along the critical line. (Adapted from the output of cell `RiemannZeta` in *Mathematica* notebook `NumberTheory`.)

Algorithm and Program

We approximate the sum in (8.3.2) by using the Euler-Maclaurin summation formula (Section 3.5). By following the development in Section 6.4 of [Edw74], we can show that if fractional accuracy of ε is required for $\zeta^*(s)$ then the series should be summed directly up to $k = N-1$ where N is given by

$$N = \left[\left(\frac{s(s+1)(s+2)(s+3)(s+4)2^s}{30240\varepsilon} \right)^{1/(s+5)} \right] \tag{8.3.3}$$

in which $[x]$ denotes the integer part of x. The Euler-Maclaurin formula then produces

$$\zeta^*(s) = \sum_{k=2}^{N-1} \frac{1}{k^s} + \frac{N^{1-s}}{s-1} + \frac{N^{-s}}{2} + \frac{sN^{-(1+s)}}{12} - \frac{s(s+1)(s+2)N^{-(3+s)}}{720}$$

$$+ \frac{s(s+1)(s+2)(s+3)(s+4)N^{-(5+s)}}{30240\varepsilon} \tag{8.3.4}$$

If the estimated N in (8.3.3) is less than 10, we set $N = 10$.

Thus we obtain the efficient program `RiemannZeta` for $\zeta^*(s) = \zeta(s) - 1$ with real $s > 1$ and specified fractional accuracy `eps` as:

```
double RiemannZeta(double s, double eps)
/*  Riemann zeta - 1  for  s > 1  */
{
double NsTerm,sum,fN,negS;
int N,k;

/* Estimate N for accuracy  eps  */

NsTerm =
s*(s+1.0)*(s+2.0)*(s+3.0)*(s+4.0)/30240.0;
N = pow(NsTerm*pow(2.0,s)/eps,1.0/(s+5.0));
if ( N < 10 )  N = 10;

fN = N;
negS = -s;
sum = 0.0; /* Direct sum */

for ( k =2; k <= N-1; k++ )
  {
  sum += pow( (double) k, negS);
  }

/* Add Euler-Maclaurin correction terms */
sum += pow(fN,negS)*(0.5 + fN/(s-1.0)
 +s*(1.0-
(s+1.0)*(s+2.0)/(60.0*fN*fN))/(12.0*fN))
 +NsTerm/pow(fN,s+5.0);
return  sum;
}
```

Test Values

Decimal values of $\zeta(n)$ to 20 decimals for the integers from $n = 2$ to 42 are given in Table 23.3 of [Abr64]. For $n > 30$ these are not sufficiently accurate for checking 10-digit relative accuracy in $\zeta(s) - 1$ when s is an integer.

The entries in Table 8.3.1 were checked against the output of cell `RiemannZeta` in *Mathematica* notebook `NumberTheory`. About $12 + s \log_{10} 2$ decimal digits have to be carried when computing $\zeta(s)$ to ensure 10-digit relative accuracy in $\zeta^*(s) = \zeta(s) - 1$.

TABLE 8.3.1 Test values for $\zeta(s) - 1$ to 10 digits relative accuracy.

s	2.5	3.0	3.5
$\zeta^*(s)$	0.3414872573	0.2020569032	0.1267338673

s	10.0	20.0	30.0
$\zeta^*(s)$	$0.9945751278 \times 10^{-3}$	$0.9539620339 \times 10^{-7}$	$0.9313274324 \times 10^{-10}$

8.4   OTHER SUMS OF RECIPROCAL POWERS

Keywords

number theory functions, sums of reciprocal powers

C Functions and *Mathematica* Notebook

`EtaNum`, `LambdaNum`, `BetaNum` (C); cell `SumRecPower` in notebook `NumberTheory` (*Mathematica*)

Background and Function Definitions

There are three commonly used sums of reciprocal powers of integers in addition to the Riemann zeta function (Section 8.3) $\zeta(s)$. These functions are defined as follows:

$$\eta(s) = \sum_{k=1}^{\infty} \frac{(-1)^{k-1}}{k^s} \qquad s \geq 1 \tag{8.4.1}$$

$$\lambda(s) = \sum_{k=0}^{\infty} \frac{1}{(2k+1)^s} \qquad s > 1 \tag{8.4.2}$$

and

$$\beta(s) = \sum_{k=0}^{\infty} \frac{(-1)^{k-1}}{(2k+1)^s} \qquad s \geq 1 \tag{8.4.3}$$

The first two series are related to $\zeta(s)$ by

$$\eta(s) = (1 - 2^{1-s})\zeta(s) \qquad s > 1 \qquad \eta(1) = \ln 2$$
$$\lambda(s) = (1 - 2^{-s})\zeta(s) \qquad s > 1 \tag{8.4.4}$$

Other formulas are given in Section 23.2 of Abramowitz and Stegun [Abr64] and in Sections 0.23, 0.24 of Gradshteyn and Ryzhik [Gra94].

Special values of the sums are useful for analytical work or for checking the computer programs that follow. Table 8.4.1 gives examples.

TABLE 8.4.1 Special values of sums of reciprocal powers. The value $\beta(2)$ is called Catalan's constant. Numerical values are given to 11 digits.

n	$\zeta(n)$	$\eta(n)$	$\lambda(n)$	$\beta(n)$
1	diverges	$\ln 2$ 0.69314718056	diverges	$\pi/4$ 0.78539816340
2	$\pi^2/6$ 1.6449340668	$\pi^2/12$ 0.82246703342	$\pi^2/8$ 1.2337005501	0.91596559418
3	 1.2020569032	0.90154267737	1.0517997903	$\pi^3/32$ 0.96894614626
4	$\pi^4/90$ 1.0823232337	$7\pi^4/720$ 0.94703282950	$\pi^4/96$ 1.0146780316	0.98894451741

Visualization

To anticipate the computational scheme described in Algorithms and Programs below, it is most useful to visualize the deviations of these functions from unity on a logarithmic scale, as in Figure 8.4.1. As s increases, the leading s-dependent term in each series predominates—hence the approximate power-law behavior of each function.

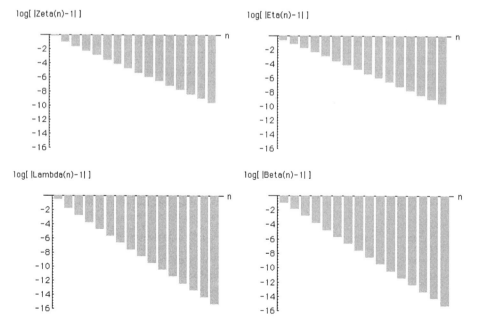

FIGURE 8.4.1 Absolute values of deviations of sums of reciprocal powers from unity, shown on log-10 scales for the functions ζ, η, λ, and β, with $s = n = 2, 3, \ldots$. (Adapted from output of cell SumRecPower in *Mathematica* notebook NumberTheory.)

Algorithms and Programs

The algorithms and C programs for $\eta(s)$ and $\lambda(s)$ are described first. The formulas in (8.4.4) can be used directly, given program `RiemannZeta` in Section 8.3 to compute $\zeta^*(s) \equiv \zeta(s) - 1$. Thus, we use:

$$\eta^*(s) \equiv \eta(s) - 1 = (1 - 2^{1-s})\zeta^*(s) - 2^{1-s} \qquad s > 1$$

$$\eta^*(1) = \ln 2 - 1 = -0.30685281944 \qquad (8.4.4)$$

$$\lambda^*(s) \equiv \lambda(s) - 1 = (1 - 2^{-s})\zeta^*(s) - 2^{-s} \qquad s > 1 \qquad (8.4.5)$$

As explained in relation to computing the zeta function in Section 8.3, the relative accuracy of these functions is best expressed by subtracting unity from them.

The function `EtaNum` for $\eta^*(s)$ is as follows:

```
double EtaNum(double s, double eps)
/*   Eta reciprocal sum - 1   for  s >= 1   */
{
double Spow,RiemannZeta(double s, double eps);

if ( s == 1 )   return  -0.30685281944;/* ln 2 - 1 */
Spow = pow(2.0,1.0-s);
return  (1.0 - Spow)*RiemannZeta(s,eps) - Spow;
}
```

Function `LambdaNum` for $\lambda^*(s)$ is simply:

```
double LambdaNum(double s, double eps)
/*   Lambda reciprocal sum - 1   for  s >= 1   */
{
double Spow,RiemannZeta(double s, double eps);

if ( s == 1 )
  {
printf("\n\n!!Lambda diverges for s=1; return zero");
return  0.0;
  }
Spow = pow(2.0,-s);
return  (1.0 - Spow)*RiemannZeta(s,eps) - Spow;
}
```

Test values for the above two functions are given after the description of `BetaNum`.

The function $\beta(s)$ defined by (8.4.3), which is sometimes called the Catalan beta function, can be separated into two series of positive terms that converge rapidly. Again, we program the function with unity subtracted, to obtain

$$\beta^*(s) \equiv \beta(s) - 1 = \beta_{+1}(s) - \beta_{-1}(s) \qquad s \geq 1 \qquad (8.4.6)$$

where

$$\beta_p(s) = \sum_{k=1}^{\infty} \frac{1}{(4k+p)^s} \qquad s \geq 1 \qquad (8.4.7)$$

We approximate the sum in (8.4.7) by using the Euler-Maclaurin summation formula (Section 3.5). If fractional accuracy of ε is required for $\beta^*(s)$ then the series should be summed directly up to $k = N - 1$ where N is given by

$$N = \left[\left(\frac{32 s(s+1)(s+2)(s+3)(s+4) 2^s}{945 \varepsilon} \right)^{1/(s+5)} \right] \bigg/ 4 + 1 \qquad (8.4.8)$$

in which $[x]$ denotes the integer part of x. The Euler-Maclaurin formula then produces

$$\beta_p(s) = \sum_{k=1}^{N-1} \frac{1}{(4k+p)^s}$$

$$+ f^{-s} \left[\frac{1}{2} + \frac{f}{4(s-1)} + \frac{s}{3f} - \frac{4s(s+1)(s+2)}{45f^3} + \frac{32s(s+1)(s+2)(s+3)(s+4)}{945 f^5} \right] \qquad (8.4.9)$$

where $f = 4N + p$. If the estimated N in (8.3.3) is less than 5, we set $N = 5$.

Thus, one obtains the efficient program `BetaNum` for $\beta^*(s)$ with real $s \geq 1$ and specified fractional accuracy `eps` as:

```
double BetaNum(double s, double eps)
/*  Beta reciprocal sum - 1  for  s >= 1  */
{
double BetaSum(double p, double s, double eps);

if ( s == 1 )   return -0.214601836603;/* Pi/4 - 1 */

return  BetaSum(1.0,s,eps)  - BetaSum(-1.0,s,eps);
}

double BetaSum(double p, double s, double eps)
/* Beta reciprocal sum for even or odd powers */
{
double Term,sum,fN,fk,negS,f4Np;
int N,k;

/* Estimate N for accuracy  eps  */
Term = 32.0*s*(s+1.0)*(s+2.0)*(s+3.0)*(s+4.0)/945.0;
N = pow(Term/eps,1.0/(s+5.0));
if ( N < 5 )  N = 5;

fN = N;
fk = 0.0;
negS = -s;
sum = 0.0; /* Direct sum */
for ( k = 1; k <= N-1; k++ )
  {
  fk += 1.0;
  sum += pow( 4.0*fk+p, negS);
  }
```

```
/* Add Euler-Maclaurin correction terms */
f4Np = 4.0*fN+p;
sum += pow(f4Np,negS)*(0.5 + 0.25*f4Np/(s-1.0)
  +s*(1.0
  -4.0*(s+1.0)*(s+2.0)/(15.0*f4Np*f4Np))/(3.0*f4Np))
  +Term/pow(f4Np,s+5.0);
return sum;
}
```

Programming Notes

The numerical accuracy of the functions can be controlled by input parameter eps, and the accuracy is otherwise limited only by the accuracy of the floating-point calculations.

Test Values

Decimal values of $\eta(s)$, $\lambda(s)$, and $\beta(s)$ to 20 decimals for the integers from $n = 2$ to 38 are given in Table 23.3 of [Abr64]. For $n > 18$ these are not sufficiently accurate for all the functions to check 10-digit relative accuracy of the functions minus unity—$\eta^*(s)$, $\lambda^*(s)$, and $\beta^*(s)$—when s is an integer.

The entries in Table 8.4.2 were checked against the output of cell SumRecPower in *Mathematica* notebook NumberTheory. About $12 + s \log_{10} 4$ decimal digits have to be carried when computing the functions to ensure 10-digit relative accuracy.

TABLE 8.4.2 Test values to 10 digits relative accuracy for sums of reciprocal powers minus unity, as given in (8.4.1) to (8.4.3).

s	2.5	3.0	3.5
$\eta^*(s)$	−0.1328001110	−0.09845732263	−0.07244642223
$\lambda^*(s)$	0.1043435731	0.05179979026	0.02714372255
$\beta^*(s)$	−0.05137782596	−0.03105385374	−0.01859748873

s	10.0	15.0	20.0
$\eta^*(s)$	$-0.9604924017 \times 10^{-3}$	$-0.3044878690 \times 10^{-4}$	$-0.9533884185 \times 10^{-6}$
$\lambda^*(s)$	$0.1704136304 \times 10^{-4}$	$0.6972470313 \times 10^{-7}$	$0.2868076975 \times 10^{-9}$
$\beta^*(s)$	$-0.1683597380 \times 10^{-4}$	$-0.6965915738 \times 10^{-7}$	$-0.2867867258 \times 10^{-9}$

8.5   POLYLOGARITHMS

Keywords

Clausen integral, dilogarithm, Jonquière function, number theory functions, polylogarithms, Riemann zeta function, Spence integral

C Function and *Mathematica* Notebook

PolyLog; cell PolyLogarithm in notebook NumberTheory (*Mathematica*)

Background

Extensive background to polylogarithms and associated functions is given in the references cited in the monograph by Lewin [Lew81]. Their relations to other functions in the *Atlas* are summarized in Table 8.5.1 below. Polylogarithms are sometimes called Jonquière functions. They occur in many physical problems, such as computing Feynman diagrams in quantum mechanics, resonances in cavities, and statistical mechanics. An extensive reference list is given in Section 1.12 of [Lew81].

Function Definition

The polylogarithm of order n, $\mathrm{Li}_n(z)$, is defined by

$$\mathrm{Li}_n(z) \equiv \sum_{k=1}^{\infty} \frac{z^k}{k^n} \qquad n = 2,3,4,\ldots \qquad |z| \le 1 \tag{8.5.1}$$

This definition may be extended to include $n = 0$ and 1, as in Table 8.5.1 where relations to other functions are summarized. It is usual to extend the definition of $\mathrm{Li}_n(z)$ to the whole of the complex plane by writing

$$\mathrm{Li}_n(z) \equiv \int_0^z \frac{\mathrm{Li}_{n-1}(t)}{t} dt \tag{8.5.2}$$

As Table 8.5.1 below shows, $\mathrm{Li}_2(z)$ then involves integrating over the logarithm of a complex variable, thus introducing branch cuts into this and higher-order polylogarithms, as is evident in Figure 8.5.1 following. We assume the extended definition (8.5.2) when writing down some of the relations given under Algorithm and Program.

Polylogarithms are related to several other functions, most of which are described in the *Atlas*, as indicated in Table 8.5.1.

Table 8.5.1 Relations between polylogarithms and other functions in the *Atlas*.

n	Polylogarithm relations	Related function [section]
0	$\mathrm{Li}_0(z) = z/(1-z) \qquad |z| < 1$	geometric series
1	$\mathrm{Li}_1(z) = -\ln(1-z)$	logarithm [4.1.2]
2	$\mathrm{Li}_2(x) = S(1-x) \qquad x \ge 0$ $\mathrm{Li}_2(e^{i\theta}) = \pi^2/6 - \theta(2\pi-\theta)/4 + iC(\theta)$ $0 \le \theta \le 2\pi$	Spence integral [19.4] Clausen integral [19.5]
n	$\mathrm{Li}_n(1) = \zeta(n) \qquad \mathrm{Li}_n(-1) = -(1-2^{1-n})\zeta(n)$	Riemann zeta [8.3]

The most common polylogarithm with $n > 1$ is the dilogarithm, $n = 2$, discussed and computed from different viewpoints in Sections 19.4 and 19.5. The trilogarithm ($n = 3$) is discussed in Chapter 6 of Lewin's monograph [Lew81], and higher-order functions are summarized in Chapter 7 therein. Dilogarithms of complex argument—shown in Figure 8.5.1—are presented in Chapter 5 of [Lew81].

Visualization

Polylogarithms are shown for both complex arguments (Figure 8.5.1) and for real arguments (Figure 8.5.2). For complex arguments, the function must have branch cuts, just as for the conventional logarithm ($n = 1$) shown in Figure 4.1.2. (For $n = 1$ there is also a singularity in the real part of the logarithm, which adds to its complexity.) The *Mathematica* function PolyLog places the branch cut along the real axis in $(1, \infty)$.

M

For real arguments, polylogarithms are smoothly varying functions of both x and n, as shown by the surfaces and curves in Figure 8.5.2. As n increases, $\mathrm{Li}_n(x)$ varies more slowly with x, since the higher powers of x in the series (8.5.1) are suppressed by the larger denominators.

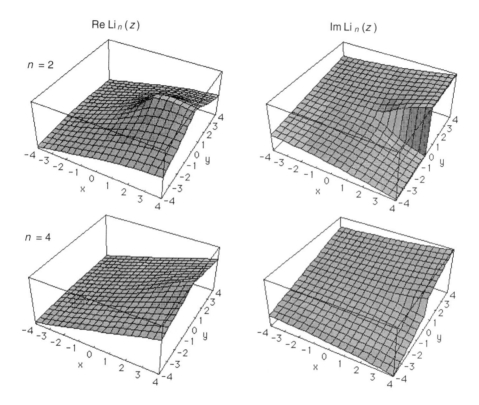

FIGURE 8.5.1 Polylogarithm surfaces for complex arguments as a function of order n for $n = 2$ (dilogarithm, top) and $n = 4$ (tetralogarithm, bottom). Notice the the branch cut along the real axis in $(1, \infty)$. (Adapted from the output of cell PolyLogarithm in *Mathematica* notebook NumberTheory.)

Polylogarithms for real arguments, $z = x$, are considered for our C functions. Figure 8.5.2 therefore shows the surfaces of $\mathrm{Li}_n(x)$ as a function of x and n, and also transection curves of these surfaces as functions of x in $[-1, 1]$ for n fixed. Note that as the order of the polylogarithm increases, the function varies more slowly with x, as is consistent with the integral definition (8.5.2).

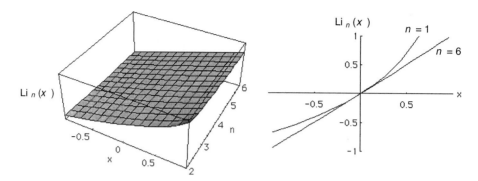

FIGURE 8.5.2 Polylogarithms for real arguments. Left: Surface as a function of x and n for $n = 2$ to $n = 6$. Right: Curves as a function of x for n fixed; $n = 1$ (related to conventional logarithm) and $n = 5$ (pentalogarithm). (Adapted from output of cell `PolyLoga-rithm` in *Mathematica* notebook `NumberTheory`.)

Algorithm and Program

Algorithms and programs for computing $\mathrm{Li}_n(x)$ in the x range $[-1, 1]$ are provided. The functions are complex for $x > 1$, while for $x < -1$ the following inversion equation given by Lewin [Lew81, (7.20)] can be used:

$$\mathrm{Li}_n(x) = (-1)^{n-1} \mathrm{Li}_n(1/x) - \frac{[\ln(-x)]^n}{n!}$$

$$-\sum_{r=1}^{[n/2]} \frac{(2^{2r-2} - 2)\, \pi^{2r} \,|\, B_{2r}\,|}{(2r)!\,(n-2r)!} [\ln(-x)]^{n-2r} \qquad (8.5.3)$$

in which $[n/2]$ denotes the integer part of $n/2$. (Note that Lewin's notation for Bernoulli numbers, B_{2r}, differs from what we use in Section 8.1.) Function `PolyLog` below does not use this formula.

For x in $[-1, 0)$ we use the reflection formula [Lew81, A.2.7 (5)]

$$\mathrm{Li}_n(x) = 2^{1-n}\, \mathrm{Li}_n(x^2) - \mathrm{Li}_n(-x) \qquad (8.5.4)$$

which therefore requires only that $\mathrm{Li}_n(x)$ with x in $(0, 1]$ be computed, since $\mathrm{Li}_n(0) = 0$. If the series (8.4.1) is then computed directly, its terms are positive and monotonically decreasing. For $n = 2$, the dilogarithm, there is a symmetry formula [Lew81, (1.11)]

$$\mathrm{Li}_2(x) = \pi^2/6 - \ln x \ln(1-x) - \mathrm{Li}_2(1-x) \qquad (8.5.5)$$

which is used to reduce the x range to $(0, 1/2]$. Corresponding formulas for $n > 2$ are not known.

```
double PolyLog(int n, double x, double eps)
/* Polylogarithm of order n >= 2, |x| <= 1.0,
   to accuracy  eps   */
{
double PolyLogSum(int n, double x, double eps);
double Pi2D6 = 1.64493406685; /* Pi^2/6 */
```

```
if ( x < -1.0 )
  {
printf("\n\n!!PolyLog x<-1; use inversion equation");
printf("\nZero returned");  return 0.0;
  }
if ( x > 1.0 )
  {
 printf("\n\n!!PolyLog x>1;\nZero returned");
 return 0.0;
 }
/* x in [-1,1] */
if ( x < 0.0 )
 { /* use reflection formula */
 return
  PolyLog(n,x*x,eps)*pow(2.0, 1.0 - (double)n)
  - PolyLog(n,-x,eps);
 }
if ( x == 0.0 )  return  0.0;
/* For x in (0,1] */
if ( n == 2 )
 {
 if ( x <= 0.5 )
  { /* use series */
  return  PolyLogSum(2,x,eps);
  }
 if ( x == 1.0 )  return  Pi2D6;
 /* x in (0.5,1) use symmetry formula */
 return  Pi2D6-log(x)*log(1.0-x)
  - PolyLog(2,1.0-x,eps);
 }
/* For n > 2 use series brute force */
return  PolyLogSum(n,x,eps);
}

double PolyLogSum(int n, double x, double eps)
/* Brute-force sum for polylogarithm series */
{
double Mfn,fk,xterm,sum,ratio,eps10,term;

Mfn = -n;
fk = 1.0;
xterm = 1.0;
sum = 1.0;
ratio = 10.0;
eps10 = 0.1*eps; /* realistic convergence */
while ( ratio > eps )
 { /* k sum started at 2 */
 fk += 1.0;
 xterm *= x;
 term = xterm*pow(fk,Mfn);
 sum += term;
 ratio = fabs(term/sum);
 }
return  x*sum;
}
```

Programming Notes

The programming of function `PolyLog` for the polylogarithm $\text{Li}_n(x)$ with $n > 1$ and x in $[-1, 1]$ uses recursion to implement formulas (8.5.4) and (8.5.5), which reduces the x range to (0, 1], or to (0, 1/2] for $n = 2$. When $|x| \leq 0.5$ the series (8.5.1) converges to accuracy $\varepsilon = 10^{-10}$ within about 20 terms for all n.

When $|x| > 0.5$ the series converges much more slowly, especially for $n = 3$ (the trilogarithm). For example, accuracy $\varepsilon = 10^{-10}$ in $\text{Li}_3(x)$ requires more than 100 terms for $x > 0.9$ and the values are sensitive to roundoff errors. (At $x = 1$ the identity with the Riemann zeta function may be used.) For $n \geq 5$ and for all x in $[-1, 1]$ the series converges to accuracy $\varepsilon = 10^{-10}$ within about 100 terms.

Test Values

Test values of the polylogarithm $\text{Li}_n(x)$—obtained by running cell `PolyLogarithm` in *Mathematica* notebook `NumberTheory`—are given in Table 8.5.2. A brief tabulation of the dilogarithm in the guise of the Spence integral $\text{Li}_2(x)$ is Table 27.7 in Abramowitz and Stegun [Abr64].

TABLE 8.5.2 Test values to 10 digits of polylogarithms

x / n	2	3	4	5
−0.9	−0.7521631792	−0.8186382015	−0.8564782888	−0.8771870234
−0.3	−0.2800743338	−0.2896400341	−0.2946800958	−0.2972913961
0.3	0.3261295101	0.3124001779	0.3059945353	0.3029324080
0.9	1.299714723	1.049658950	0.9640053712	0.9292671964

REFERENCES ON NUMBER THEORY FUNCTIONS

[Abr64]   Abramowitz, M., and I. A. Stegun, *Handbook of Mathematical Functions*, Dover, New York, 1964.

[Boy92]   Boyland, P. et al, *Guide to Standard Mathematica Packages*, Wolfram Research, Champaign, Illinois, second edition, 1992.

[Cha91]   Char, B. W. et al, *Maple V Library Reference Manual*, Springer, New York, 1991.

[Dil88]   Dilcher, K., *Zeros of Bernoulli, Generalized Bernoulli and Euler Polynomials*, American Mathematical Society, Providence, Rhode Island, 1988.

[Dil91]   Dilcher, K., L. Skula, and I. Sh. Slavutskiĭ, *Bernoulli Numbers Bibliography (1713–1990)*, Queen's University, Kingston, Ontario, 1991.

[Edw74]   Edwards, H. M., *Riemann's Zeta Function*, Academic Press, New York, 1974.

[Erd53a] Erdélyi, A., et al, *Higher Transcendental Functions*, Vol. 1, McGraw-Hill, New York, 1953; reprint edition, Krieger, Malabar, Florida, 1981.

[Fil92]   Fillebrown, S., Journal of Algorithms, **13**, 431 (1992).

[Gon92]   González, M. O., *Complex Analysis*, Marcel Dekker, New York, 1992.

[Gra92]   Gradshteyn, I. S., and I. M. Ryzhik, *Table of Integrals, Series, and Products*, Academic Press, Orlando, fifth edition, 1994.

[Kar92]   Karatsuba, A. A., and S. M. Voronin, *The Riemann Zeta-Function*, Walter de Gruyter, Berlin, 1992.

[Lew81]   Lewin, L., *Polylogarithms and Associated Functions*, Elsevier North Holland, New York, 1981.

[Var90]   Varga, R. S., *Scientific Computation on Mathematical Problems and Conjectures*, Society for Industrial and Applied Mathematics, Philadelphia, 1990.

[Whi27]   Whittaker, E. T., and G. N. Watson, *A Course of Modern Analysis*, Cambridge University Press, Cambridge, England, fourth edition, 1927.

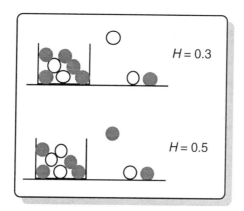

H = 0.3

H = 0.5

Chapter 9

PROBABILITY DISTRIBUTIONS

This chapter is probably one of the most important in the *Atlas*. Section 9.1 summarizes general properties of probability distribution functions (PDFs), then the following three sections describe discrete PDFs (Section 9.2), probability distributions related to the normal or Gauss probability distribution (Section 9.3), and other continuous probability distributions (Section 9.4).

The analytical materials for this chapter rely extensively on the compendia on probability distributions written by Johnson et al, namely [Joh70, Joh92, Joh94a, Joh94b]. These reference books and "Kendall's" compendium [Stu94] provide indispensable background to the origins, uses, and interpretations of the PDFs for which visualization and functions are provided in the *Atlas*. The handbook of mathematical functions by Abramowitz and Stegun [Abr64] also has an extensive tabulation (Chapter 26) of formulas and numerical tables for probability functions.

Mathematica notebooks for quick calculations and for visualizing the probability distributions are provided as usual. They are referenced primarily in the figure captions, since they are used mainly to help with visualizing PDFs. Because there are many functions described in this chapter, a separate notebook is provided for each of Sections 9.2, 9.3, and 9.4. Cells within each notebook correspond to each subsection and specific PDF.

Visualizations of probability distribution functions are very straightforward to generate and are usually easy to interpret. The probability-density surfaces (for 3D views) and curves (for 2D views) are smooth, unimodal, and positive. Similarly, cumulative-density surfaces and curves are usually smooth and monotonic.

Computing PDFs is also generally straightforward, for the following reasons. First, they are positive-valued functions defined for real variables, x, without singularities on the real axis. Second, they are normalizable, so they tend to zero sufficiently fast as $|x| \to \infty$. Third, they can be computed readily with accuracy of 10 digits, corresponding to a sample size N about 10^{20} for $\sqrt{N}$ (Poisson) statistics, which is much larger than ever encountered.

9.1   OVERVIEW OF PROBABILITY DISTRIBUTION FUNCTIONS

In this section we summarize terminology and formulas used to characterize a PDF. Derivations are in Johnson et al [Joh70, Joh92, Joh94a, Joh94b] or are referred to there, in Abramowitz and Stegun [Abr64, Chapter 26], or in Chapter 3 of Stuart and Ord [Stu94].

A probability distribution function is often characterized by its moments and by quantities derived from them, with the most common of these being the mean, standard deviation, skewness, and excess. The *mean value*, m, is defined for the distribution function with density $\rho(x)$ by

$$m \equiv \int_{-\infty}^{\infty} x\rho(x)\, dx \qquad (9.1.1)$$

The nth central moment, μ_n, is then defined in terms of m by

$$\mu_n \equiv \int_{-\infty}^{\infty} (x-m)^n \rho(x)\, dx \qquad (9.1.2)$$

so that $\mu_0 = 1$ because of normalization of the total probability to unity, while $\mu_1 = 0$ because of definition (9.1.1) for m. For discrete distributions whose arguments are the integers, the definitions (9.1.1) and (9.1.2) may be used if the distributions $\rho(x)$ are discretized by multiplying them by Dirac delta functions (delta distributions) located at the integers. Thus, the integrations are replaced by summations.

The *standard deviation*, σ, is defined by

$$\sigma \equiv \sqrt{\mu_2} \qquad (9.1.3)$$

Normalized moments for $n = 3$ and $n = 4$ are of particular interest. Coefficients of *skewness* and of *excess* are defined by

$$\gamma_1 \equiv \frac{\mu_3}{\sigma^3} \qquad \text{skewness} \qquad (9.1.4)$$

$$\gamma_2 \equiv \frac{\mu_4}{\sigma^4} - 3 \qquad \text{excess} \qquad (9.1.5)$$

For a distribution that is symmetric about its mean value, the moments with n odd are zero, so its skewness is zero. If $\gamma_1 > 0$ the distribution is skewed toward the right side of the mean. The excess, γ_2, measures in relative terms how the fourth moment differs from that of the Gauss (normal) distribution, for which $\mu_4 = 3\sigma^4$. If $\gamma_2 > 0$ then the wings of the distribution are generally longer than those of a Gauss distribution.

Properties of discrete probability distribution functions are summarized in Tables 9.2.1 and 9.2.2, while those of continuous PDFs are given in Tables 9.3.1 and 9.3.2.

9.2   DISCRETE PROBABILITY DISTRIBUTIONS

Discrete probability distributions are characterized by being defined at discrete (usually integer) values of the parameter that is summed over to obtain cumulative probability distributions and moments of the PDF, as in (9.1.1). Therefore, a graph of the PDF is a histo-gram (bar chart), rather than a smooth curve. For distribution P we use the notation $P(k)$ to indicate the probability that value k is obtained.

This section summarizes the properties of the most commonly used discrete PDFs, pro-vides visualizations of them based on the *Mathematica* notebook DPDF, and also gives convenient C functions for computing them efficiently. Relevant background for this sec-tion is Chapter 7 (combinatorial functions), from which some of the functions needed are also drawn.

Tables 9.2.1 and 9.2.2 summarize relevant properties of the discrete distributions con-sidered.

TABLE 9.2.1 Properties of discrete probability distribution functions; the binomial, the negative binomial, and the geometric distributions.

Distribution function	Domain	Point probability	Parameter limits	Moments: $m, \sigma^2, \gamma_1, \gamma_2$
binomial	$k = 0,1,\dots$	$\binom{N}{k}p^k$ $\times (1-p)^{N-k}$	$0 < p < 1$	$Np, Np(1-p),$ $\dfrac{1-2p}{\sqrt{Np(1-p)}},$ $\dfrac{1-6p(1-p)}{Np(1-p)}$
negative binomial (Pascal)	$k = 0,1,\dots$	$\binom{N+k-1}{k}$ $\times p^N(1-p)^k$	$0 < p < 1$	$\dfrac{N(1-p)}{p}, \dfrac{N(1-p)}{p^2},$ $\dfrac{2-p}{\sqrt{N(1-p)}},$ $\dfrac{1}{N}\left(6+\dfrac{p^2}{1-p}\right)$
geometric	$k = 0,1,\dots$	$p(1-p)^k$	$0 < p < 1$	$\dfrac{1-p}{p}, \dfrac{1-p}{p^2},$ $\dfrac{2-p}{\sqrt{1-p}}, 6+\dfrac{p^2}{1-p}$

TABLE 9.2.2 Properties of discrete probability distribution functions; the hypergeometric, logarithmic series, and Poisson distributions.

Distribution function	Domain	Point probability	Parameter limits	Moments: m, σ^2, γ_1, γ_2
hypergeo-metric	$[\max(0, n-N_2),$ $\min(n, N_1)]$	$\dfrac{\dbinom{N_1}{k}\dbinom{N_2}{n-k}}{\dbinom{N}{n}}$	N_1 and N_2 integers, $N = N_1 + N_2$	$\dfrac{nN_1}{N}$, $\dfrac{nN_1N_2(N-n)}{N^2(N-1)}$, $\dfrac{(N_2-N_1)(N-2n)}{N-2}$ $\times\sqrt{\dfrac{N-1}{nN_1N_2(N-n)}}$ γ_2 is complicated
logarithmic series	$k = 1, 2, \ldots$	$\dfrac{\alpha\theta^k}{k}$ $\alpha \equiv \dfrac{-1}{\ln(1-\theta)}$	$0 < \theta < 1$	$\dfrac{\alpha\theta}{1-\theta}$, $\dfrac{\alpha\theta(1-\alpha\theta)}{(1-\theta)^2}$ γ_1 and γ_2 are complicated
Poisson	$k = 0, 1, \ldots$	$\dfrac{e^{-m}m^k}{k!}$	$0 < m < \infty$	m, m, $\dfrac{1}{\sqrt{m}}$, $\dfrac{1}{m}$

9.2.1   Binomial Distribution

Keywords
binomial, distribution

C Function and *Mathematica* Notebook
BinomDist (C); cell BinomialPDF in *Mathematica* notebook DPDF

Background
If a total of N independent trials is made in a statistical test and if in each trial there is a probability p that outcome E will occur, then the number of trials in which E occurs is represented by random variable x which satisfies a binomial distribution with parameters N and p. Extensive discussions of the binomial distribution are given in Feller [Fel68, Section 6.2] and in Johnson et al [Joh92, Chapter 3], including analytical approximations to the distribution that are useful in statistics and probability theory.

Function Definition

The binomial distribution function, B, is defined in terms of the binomial coefficients (Section 7.2.1) by

$$B_{N,p}(k) \equiv \binom{N}{k} p^k (1-p)^{N-k} \qquad k = 0,1,2,\ldots,N \qquad 0 \le p \le 1 \qquad (9.2.1)$$

with N a positive integer. Note that the binomial distribution is a discrete distribution in terms of N and k but it is continuous in p.

From the binomial theorem we have the sum rule for any allowed p

$$\sum_{k=0}^{N} B_{N,p}(k) = 1 \qquad (9.2.2)$$

This may be useful to check the function values, as described below in Programming Notes.

Visualization

The binomial distribution, $B_{N,p}(k)$, depends upon three variables, two of which are integers (N and k) and one of which is continuous (p). Since N, the number of trials, is often held fixed during a statistical test, we consider this to be the controlling parameter. Then, for a given choice of probability per trial, p, with $0 < p < 1$, the binomial probability, B, depends upon k, with $0 \le k \le N$. One can therefore make three-dimensional bar charts (histograms) showing B as a function of N, p, and k. Two such charts are shown in Figure 9.2.1.

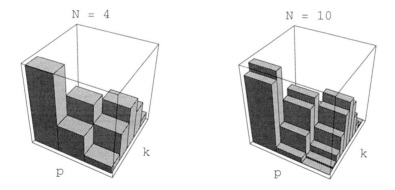

FIGURE 9.2.1 Binomial distributions for $N = 4$ and $N = 10$. In each 3D bar chart p has the values 0.1, 0.3, 0.5 from left to right, while k varies from 0 to N by integer steps. (Cell `BinomialPDF` in *Mathematica* notebook `DPDF`.)

Because of the symmetry that follows immediately from (9.2.1), namely,

$$B_{N,p}(k) = B_{N,1-p}(N-k) \qquad (9.2.3)$$

it is not necessary to plot the binomial distributions for $p > 0.5$, since they are just those for $p < 0.5$ but plotted with k running in the reversed direction. Note also that if $p = 0.5$, then the binomial distribution is just proportional to the binomial coefficient of Section 7.2.1 that is displayed in Figure 7.2.1.

The maximum of the binomial distribution function as a function of p is obtained for the integer k that is closest to pN, as one would expect from the simple interpretation of p (probability per trial) and N (total number of trials). This behavior is also evident in Figure 9.2.1.

Algorithm and Program
The program for function `BinomDist` is a direct transcription of formula (9.2.1), with the assumption that k lies in $[0, N]$.

```
double BinomDist(int N, double p, int k)
/* Binomial distribution function */
{
long int Binomial(int N, int k);
if ( p <= 0 )     return 0;
if ( p >= 1 )     return 0;
return
  Binomial(N,k)*pow(p,k)*pow(1-p,N-k);
}
```

Programming Notes
The function `Binomial`, which is needed by `BinomDist`, is included with the machine-readable program.

Test Values
Check values of the binomial distribution function $B_{N,p}(k)$ are readily generated, by hand calculation or by running cell `BinomialPDF` in *Mathematica* notebook `DPDF`. A sample of exact values is given in Table 9.2.1. Note that the sum rule on k, (9.2.2), is satisfied.

TABLE 9.2.1 Exact values for the binomial distribution function (9.2.1) with $N = 4$.

$p \backslash k$	0	1	2	3	4
0.1	0.6561	0.2916	0.0486	0.0036	0.0001
0.3	0.2401	0.4116	0.2646	0.0756	0.0081

9.2.2   Negative Binomial (Pascal) Distribution

Keywords
distribution, negative binomial, Pólya distribution

C Function and *Mathematica* Notebook

`NegBinomDist` (C); cell `NegativeBinomialPDF` in notebook `DPDF` (*Mathematica*)

Background
In a statistical test the negative binomial distribution gives the distribution of the number of failures that occur in a sequence of trails before N successes have occurred, given a success

probability per trail of p. Section 6.8 of Feller [Fel68] and Chapter 5 of Johnson et al [Joh92] have extensive discussions of the negative binomial distribution and its applications.

Function Definition

The negative binomial distribution function $N_{N,p}(k)$ is defined in terms of the binomial coefficients (Section 7.2.1) by

$$N_{N,p}(k) \equiv \binom{N+k-1}{k} p^N (1-p)^k$$

$$k = 0, 1, 2, \ldots \qquad 0 \le p \le 1 \qquad (9.2.4)$$

If N is an integer then this is sometimes called the Pascal distribution. The negative binomial distribution is then a discrete in terms of N and k but continuous in p. Alternative names for this distribution are binomial waiting-time distribution and Pólya distribution. The geometric distribution (Section 9.2.3) has $N = 1$.

The maximum of the negative binomial distribution function as a function of p is obtained for the integer k that is closest to satisfying $pk = (1-p)N$, which just balances failures against successes. The mean value, m, of the distribution (see Table 9.2.1) satisfies this equation with $k = m$.

Visualization

The function $N_{N,p}(k)$ depends upon three variables; two are usually integers (N and k) and one is continuous (p). Since N is often held fixed during a statistical test, we consider this as the controlling parameter. Then, for a given choice of probability per trial, p, with $0 < p < 1$, the negative binomial probability depends upon k, with $0 \le k \le N$. We can therefore make three-dimensional bar charts showing the distribution as a function of N, p, and k, as shown in Figure 9.2.2.

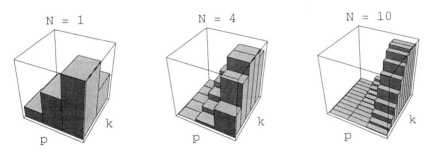

FIGURE 9.2.2 Negative binomial distributions for $N = 1$, 4, and $N = 10$. In each bar chart p has the values 0.1, 0.3, 0.5 from left to right, while k varies from 0 to N by integer steps. The leftmost distribution with $N = 1$ is also that of the geometric distribution (Section 9.2.3) (Cell NegativeBinomialPDF in *Mathematica* notebook DPDF.)

Algorithm and Program

The program for function NegBinomDist comes directly from (9.2.4). It is given on the following page.

Programming Notes

Function `Binomial` (Section 7.2.1), which is needed by `NegBinomDist,` is provided in the machine-readable program.

```
double NegBinomDist(int N, double p, int k)
/* Negative binomial distribution function */
{
if ( p <= 0 )        return 0;
if ( p >= 1 )        return 0;
return
  Binomial(N+k,k)*pow(p,N)*pow(1-p,k);
}
```

Test Values

Test values of the negative binomial distribution function $N_{N,p}(k)$ are readily generated, by hand calculation or by running cell `NegativeBinomialPDF` in *Mathematica* notebook DPDF.

TABLE 9.2.2 **Exact check values of the negative binomial distribution function (9.2.4) with $p = 0.3$.**

$N \backslash k$	0	1	2	3
1	0.30000	0.21000	0.14700	0.10290
2	0.09000	0.12600	0.13230	0.12348
3	0.02700	0.05670	0.07938	0.09261

9.2.3   Geometric Distribution

Keywords

distribution, geometric

C Function and *Mathematica* Notebook

`GeometricDist` (C); cell`GeometricPDF` in notebook DPDF (*Mathematica*)

Background

The geometric distribution is the negative binomial distribution (Section 9.2.2) with $N = 1$. Johnson et al [Joh92, Section 5.2] discuss the geometric distribution and its relation to the continuous exponential distribution.

Function Definition

The geometric distribution function $G_p(k)$ is defined by (9.2.4) with $N = 1$, thus

$$G_p(k) \equiv N_{1,p}(k) = p(1-p)^k$$

$$k = 0,1,2,... \qquad 0 < p < 1$$

(9.2.5)

Since—for p fixed—successive terms of this distribution are in the constant ratio $1-p$. the function $G_p(k)$ describes a geometric series, which can be summed to give, for any allowed p,

$$\sum_{k=0}^{\infty} G_p(k) = 1 \qquad (9.2.6)$$

Differentiation of the geometric series with respect to p readily leads to the formulas for the moments of the geometric distribution that are given in Table 9.2.1. Because of the connection to the negative binomial distribution, one can also explore the properties of $G_p(k)$ by setting NN = 1 in the NegativeBinomialPDF cell of *Mathematica* notebook DPDF.

As $p \rightarrow 1$ the envelope of the PDF of the geometric distribution tends to an exponential distribution, as suggested in Figure 9.2.3 below. This behavior and its ramifications are discussed in Section 5.2 of [Joh92].

Visualization

Since the function $G_p(k)$ depends upon just two variables, with k a nonnegative integer and p continuous, it is relatively easy to visualize. For a given choice of probability per trial, p, with $0 < p < 1$, the geometric-distribution probability depends upon k, with $0 \le k \le \infty$. We can therefore make a three-dimensional bar chart showing the distribution as a function of p and k, as shown in Figure 9.2.3. The geometric-series behavior of the PDF as a function of k for given p is quite evident, as is the increasing tendency of the distribution to resemble an exponential distribution as p increases towards 1.

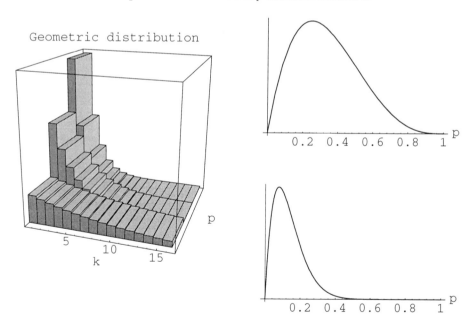

FIGURE 9.2.3 Geometric distributions. Left view; $p = 0.5, 0.3, 0.1$ from front to rear, with k varing from 0 to 15 by integer steps. Right views; $k = 3$ (top) and $k = 12$ (bottom) (Cell GeometricPDF in *Mathematica* notebook DPDF.)

When considered as a function of p with k fixed (right views in Figure 9.2.3), $G_p(k)$ shows a maximum at $p = 1/(1+k)$, as is easily derived from (9.2.5).

Algorithm and Program

The program for function GeometricDist comes directly from (9.2.5).

```
double GeometricDist(double p, int k)
/* Geometric distribution function */
{
if ( p <= 0 )      return 0;
if ( p >= 1 )      return 0;
return   p*pow(1-p,k);
}
```

Programming Notes

Function GeometricDist is self-contained.

Test Values

Test values of the geometric probability distribution function $G_p(k)$ are readily generated by hand calculation. The sum rule (9.2.6) can be used to check that the sum over k converges to 1 uniformly from below as k increases indefinitely. The convergence is more rapid as $p \rightarrow 1$. For example, with $k \leq 15$ we get a sum of 0.8146979811 for $p = 0.1$, a sum of 0.9966767069 for $p = 0.3$, and a sum of 0.9999847412 for $p = 0.5$. This convergence behavior and numerical values of the PDF can be explored by running the Geometric-PDF cell in *Mathematica* notebook DPDF.

TABLE 9.2.3 Test values to 10 digits of the geometric distribution function (9.2.5).

$p \setminus k$	0	5	10	15
0.1	0.1000000000	0.0590490000	0.0348678440	0.0205891132
0.3	0.3000000000	0.0504210000	0.0084742575	0.0014242685
0.5	0.5000000000	0.0156250000	0.0004882812	0.0000152588

9.2.4   Hypergeometric Distribution

Keywords

distribution, hypergeometric

C Function and *Mathematica* Notebook

HyprGmtrcDist (C); cell HypergeometricPDF in *Mathematica* notebook DPDF

Background

In a statistical test the hypergeometric distribution gives the distribution of the number of successes in picking at random and without replacement a given kind of object from a total of N objects. In Section 2.6 of Feller [Fel68] and in Chapter 6 of Johnson et al [Joh92] there are extensive discussions of the hypergeometric distribution and its applications.

Function Definition

The hypergeometric distribution function $H_{N,N_1,n}(k)$ is defined in terms of the binomial coefficients (Section 7.2.1) by

$$H_{N,N_1,n}(k) \equiv \frac{\binom{N_1}{k}\binom{N-N_1}{n-k}}{\binom{N}{n}} \tag{9.2.7}$$

in which N_1, n, and k are to be chosen relative to N so that none of the binomial coefficients is zero.

An explanatory example of using the hypergeometric distribution is illustrated in the motif at the beginning of this chapter. Here we have a total of $N = 10$ balls, originally all in a container. Of these, $N_1 = 4$ are white and $N - N_1 = 6$ are black. Suppose that we draw out of the container $n = 3$ balls without replacing them. If $k = 0, 1, 2$, or 3 balls are drawn, the probability of them being white is $H_{10,4,3}(k)$. For example, $H_{10,4,3}(1) = 0.5$ and $H_{10,4,3}(2) = 0.3$, as indicated in the sketch.

The function $H_{N,N_1,n}(k)$ depends upon four nonnegative-integer variables. Since N is usually fixed during a statistical test, we consider this as the controlling parameter. Then N_1 and n can vary up to N. The range of k is limited by the requirement that one doesn't run out of white balls, $k \leq N_1$, and that one doesn't run out of black balls before the number of trails is exhausted, $n - k \leq N - N_1$.

From this discussion, or by direct inspection of (9.2.7), we have the symmetry relation

$$H_{N,N_1,n}(k) = H_{N,N-N_1,n}(n-k) \tag{9.2.8}$$

in which the right-hand side corresponds to keeping count of black balls and the left-hand side corresponds to keeping count of white balls.

The sum rule for the hypergeometric PDF follows from an identity for binomial coefficients, given that all balls of a given color must eventually be drawn. Thus

$$\sum_{k=0} H_{N,N_1,n}(k) = 1 \tag{9.2.9}$$

in which the k summation upper limit is given by the vanishing of $H_{N,N_1,n}(k)$.

Visualization

Because the hypergeometric distribution is a function of four variables, it is quite difficult to visualize its behavior as these variables change. By using the `HypergeometricPDF` cell in *Mathematica* notebook DPDF, we can make a 3D bar chart for a given choice of N and N_1, as shown in Figure 9.2.4.

N1 = 4

N1 = 8

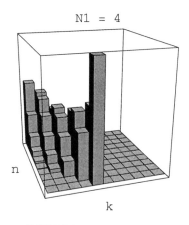
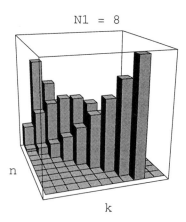

FIGURE 9.2.4 Hypergeometric distributions for $N = 10$, with $N_1 = 4$ and 8. In each bar chart n has the values 1 to 10 from left to right, while k varies from 0 to 10 by integer steps. (Cell `HypergeometricPDF` in *Mathematica* notebook `DPDF`.)

Notice in the figure that, since you can't get more balls of a given color than their are trials, the constraints $k \leq n$ and $n - k \leq N - N_1$ force the probability function values to zero over much of the range of k. In particular, for $n = N_1$ (rightmost n value in the two 3D bar charts) there is a unique $k = N_1$. The sum rule (9.2.9) requires that as the number of non-zero values increases, the height of each decreases, since they are each positive—as also seen in Figure 9.2.4.

Algorithm and Program
The program for function `HyprGmtrcDist` comes directly from (9.2.7), as follows:

```
double HyprGmtrcDist
  (int N, int N1, int n, int k)
/* Hypergeometric distribution function */
{
double num;

if ( k > N1  ||  k > n )            return 0;
if ( n-k > N-N1  ||  n > N )return 0;
num = Binomial(N1,k)*Binomial(N-N1,n-k);
return  num/Binomial(N,n);
}
```

Programming Notes
Function `Binomial`, as used by `HyprGmtrcDist`, is obtained from the machine-readable complete program version. Since $H_{N,N_1,n}(k)$ is the quotient of combinatorials, it is a rational fraction. If such values are needed—rather than the decimal approximations that we give— then the program can be modified easily to return the numerator and denominator in (9.2.7) separately.

Test Values
Check values of the hypergeometric distribution function $H_{N,N_1,n}(k)$ are readily generated, by hand calculation or by running cell `HypergeometricPDF` in *Mathematica* notebook DPDF.

M

TABLE 9.2.4 **Test values to 10 digits of the hypergeometric probability distribution function (9.2.7) with $N = 10$ and $N_1 = 4$.**

$n \backslash k$	0	1	2	3	4
1	0.6000000000	0.4000000000	0	0	0
2	0.3333333333	0.5333333333	0.1333333333	0	0
3	0.1666666666	0.5000000000	0.3000000000	0.0333333333	0
4	0.0714285714	0.3809523810	0.4285714286	0.1142857143	0.0047619048

9.2.5 Logarithmic Series Distribution

Keywords
distribution, logarithmic series

C Function and *Mathematica* Notebook
`LogSeriesDist` (C); cell `LogSeries` in notebook DPDF (*Mathematica*)

M

Background
The logarithmic series distribution is described in Chapter 7 of Johnson et al [Joh92]. It is so called because its PDF has values in the same ratio as in the series expansion of $\ln(1-\theta)$. Among its many uses are those in ecology, especially studies of population growth.

Function Definition
The logarithmic series distribution function $L_\theta(k)$ with parameter θ is defined by

$$L_\theta(k) \equiv -\frac{1}{\ln(1-\theta)} \frac{\theta^k}{k} \qquad 0 < \theta < 1 \tag{9.2.10}$$

Successive terms of the k-dependent part of this function are the same as in the series expansion of $-\ln(1-\theta)$, so the sum rule over k for any allowed θ is

$$\sum_{k=1}^{\infty} L_\theta(k) = 1 \tag{9.2.11}$$

Differentiation of the logarithmic series with respect to θ leads readily to the formulas for the first two moments of the logarithmic series distribution given in Table 9.2.2. The higher moments are quite complicated, but the third and fourth central moments are given in (7.15) in Section 7.1 of [Joh92].

Visualization

Since the function $L_\theta(k)$ depends upon just two variables, with k a positive integer and θ continuous, it is easy to visualize. Since—for a given choice of θ, with $0 < \theta < 1$—the logarithmic series distribution probability depends upon k, we can make a 3D bar chart showing the distribution as a function of θ and k, as shown in Figure 9.2.5.

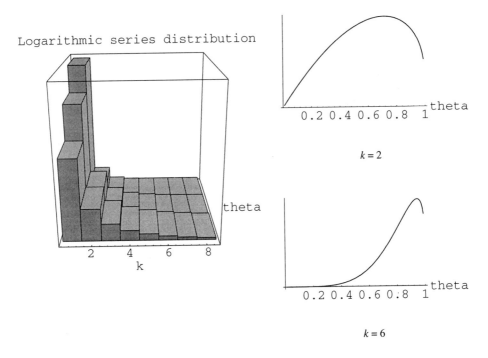

FIGURE 9.2.5 Logarithmic-series. Left: Distributions for $\theta = 0.2$, 0.5, 0.8 from front to back, with k varying from 1 to 8 by integer steps. Right panels: θ varying from 0.01 to 0.99 with $k = 2$ (top) and $k = 6$ (bottom). (Cell `LogSeriesPDF` in *Mathematica* notebook `DPDF`.)

Algorithm, Program, and Programming Notes

The program for function `LogSeriesDist` comes directly from (9.2.10). Function `LogSeriesDist` is self-contained.

```
double LogSeriesDist(double theta, int k)
/* Logarithmic series distribution function */

{
if ( theta <= 0 )              return 0;
if ( theta >= 1 )              return 0;
return - pow(theta,k)/(k*log(1-theta));
}
```

Test Values

Test values of the logarithmic series probability distribution function $L_\theta(k)$ are readily generated by using a calculator or by running cell `LogSeriesPDF` in *Mathematica* notebook `DPDF`. The sum rule (9.2.11) can be used to check that the sum over k converges to 1 uniformly from below as k increases indefinitely. Convergence is more rapid as $\theta \to 0$. More numerical values of the PDF can be explored by modifying and running the `LogSeriesPDF` cell.

TABLE 9.2.5 Test values to 10 digits of logarithmic series distribution function (9.2.10).

θ / k	1	2	3
0.2	0.8962840235	0.0896284024	0.0119504536
0.5	0.7213475204	0.1803368801	0.0601122934
0.8	0.4970679476	0.1988271791	0.1060411622

9.2.6 Poisson Distribution

Keywords
distribution, Poisson

C Function and *Mathematica* Notebook
`PoissonDist` (C); cell `PoissonPDF` in notebook `DPDF` (*Mathematica*)

Background

The well-known Poisson distribution is described extensively in Section 6.5 of Feller [Fel68] and in Chapter 4 of Johnson et al [Joh92]. This distribution is the limit of the binomial distribution (Section 9.2.1) in which N tends to infinity and p tends to zero such that the product $Np = m$. Such relationships are summarized in Figure 9.3.1.

Uses of the Poisson distribution range from studies of death rates by mule kicks in the Prussian Army Corps to time variations in the number of alpha particles emitted by radioactive materials and of photon decays from excited states of atoms. Haight's monograph [Hai67] describes the Poisson distribution and many of its applications.

Function Definition

The Poisson PDF with parameter m, $P_m(k)$, is defined by

$$P_m(k) \equiv \frac{e^{-m}m^k}{k!} \qquad m > 0 \qquad (9.2.12)$$

This PDF gives the probability of observing k random and independent events if their mean number is m.

Since the k-dependent parts in (9.2.12) are just those in the series expansion of e^m, the sum rule over k for any allowed m is

$$\sum_{k=0}^{\infty} P_m(k) = 1 \qquad (9.2.13)$$

Formulas for the moments of the Poisson PDF—as given in Table 9.2.2—are easily derived from the series expansion of e^m. In particular, both the mean value and the variance of the Poisson PDF are given by m. Hence one has the well-known and often-abused standard deviation for this PDF given by $\sqrt{m}$.

As m increases, the Poisson PDF tends to the Gauss (normal) distribution discussed in Section 9.3.1. This property is related to the corresponding limit for the binomial coefficients (Section 7.2.1) and is evident in the bar chart Figure 9.2.6 for $m = 18$.

Visualization

Since $P_m(k)$ depends upon just two variables, with k a nonnegative integer and m positive, it is easy to visualize. For given m, the Poisson distribution depends on k, so we can make a 3D bar chart of the distribution as a function of m and k, as in Figure 9.2.6.

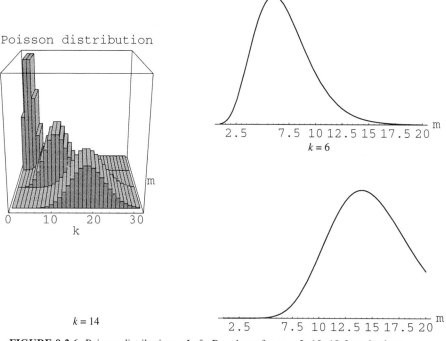

FIGURE 9.2.6 Poisson distributions. Left: Bar charts for $m = 2$, 10, 18 from back to front, with k varying from 0 to 30 by integer steps. Right: (top) $k = 6$ and m varying from 1 to 10; (bottom) $k = 14$ and m varying from 1 to 10. (Cell `PoissonPDF` in *Mathematica* notebook `DPDF`.)

As a function of m (side panels in Figure 9.2.6), $P_m(k)$ has its maximum at $m = k$, as is easily verified from the definition (9.2.12). With m fixed and k varying (bar charts in Figure 9.2.6), the maximum of $P_m(k)$ is at $k = [m]$, the integer part of m.

Algorithm, Program, and Programming Notes

The program for function `PoissonDist` comes directly from (9.2.12). Function `PoissonDist` requires `Factorial` (Section 7.1.1), which is provided with the machine-readable function.

```
double PoissonDist(double m, int k)
/* Poisson distribution function */
{
long int Factorial(int k);

if ( m <= 0 )      return 0;
return exp(-m)*pow(m,k)/Factorial(k);
}
```

Test Values

Test values of the Poisson PDF $P_m(k)$ are readily generated by using a hand-held calculator. The sum rule (9.2.13) can be used to check that the sum over k converges to 1 uniformly from below as k increases. The convergence and numerical values of the PDF can be explored by running the cell `PoissonPDF` in *Mathematica* notebook `DPDF`.

TABLE 9.2.6 Test values to 10 digits of the Poisson distribution function (9.2.12).

m / k	0	10	20
2	0.1353352832	$0.381898505 \times 10^{-4}$	$0.5832924198 \times 10^{-13}$
10	$0.4539992976 \times 10^{-4}$	0.1251100357	$0.1866081314 \times 10^{-2}$

9.3 NORMAL PROBABILITY DISTRIBUTIONS

Continuous probability distributions—particularly those related to normal distributions—are characterized by continuous variation of the parameters determining their probability densities, $\rho(x)$ in the notation of (9.1.1). Thus, moments and cumulative distribution functions of continuous PDFs are obtained by integration, as compared with summations for the discrete PDFs in Section 9.2. Also, the functions considered in this section are *densities* rather than *probabilities*. Therefore, $\rho(x)dx$ in the limit of very small dx gives the probability that a value between x and $x + dx$ is obtained. If the variable in the density is changed from x to y then the density is changed from $\rho(x)$ to $\rho(y)(dy/dx)$.

In Table 9.3.1 we summarize properties of the continuous distributions related to the normal distributions considered in this section. Much more detail is given in Chapter 7 of Feller's text [Fel68] and in Chapter 16 of the treatise by Stuart and Ord [Stu94]. We use central distributions when convenient and we often choose the scale for the variable x to minimize the number of parameters. Especially in the C functions, we often specialize to standardized distributions.

Many of the distributions discussed in this section are limiting cases of the discrete distributions considered in Section 9.2. Further, the normal-related distributions are interconnected through various limiting conditions on their parameters. We summarize these relations in Figure 9.3.1.

TABLE 9.3.1 Properties of continuous probability distributions related to the normal distribution.

Function	Domain	Probability density, $\rho(x)$	Parameter limits	Moments: $m,\ \sigma^2,\ \gamma_1,\ \gamma_2$
Gauss (normal)	$(-\infty, \infty)$	$\dfrac{e^{-(x-m)^2/2\sigma^2}}{\sigma\sqrt{2\pi}}$	$\sigma > 0$	$m,\ \sigma^2, 0, 0$
chi-square	$[0, \infty)$	$\dfrac{x^{v/2-1}e^{-x/2}}{2^{v/2}\ \Gamma(v/2)}$	$v > 0$	$v,\ 2v,\ \dfrac{2^{3/2}}{\sqrt{v}},\ \dfrac{12}{v}$
F-distribution (variance-ratio)	$[0, \infty)$	$\dfrac{v_1^{v_1/2}\ v_2^{v_2/2}}{B(v_1/2, v_2/2)}$ $\times f^{v_1/2-1}$ $\times(v_2+v_1 f)^{-(v_1+v_2)/2}$	$v_1 > 0$ $v_2 > 0$	$v_2/(v_2-2)\quad v_2 > 2,$ $\dfrac{2 v_2^2(v_1+v_2-2)}{v_1(v_2-2)^2(v_2-4)}$ $v_2 > 4,$ γ_1 and γ_2 complicated
t-distribution	$(-\infty, \infty)$	$\dfrac{\left(1+x^2/2\right)^{-(v+1)/2}}{\sqrt{v}\ B(1/2, v/2)}$	$v > 0$	$0,\ \dfrac{v}{v-2}\quad v > 2,$ $0,\ \dfrac{6}{v-4}\quad v > 4$
lognormal	$(0, \infty)$	$\dfrac{e^{-(\ln x-\varsigma)^2/2\sigma^2}}{\sqrt{2\pi}\ \sigma x}$ $\omega = e^{\sigma^2}$	$\sigma > 0$	$e^{\varsigma}\omega,\ \omega(\omega-1)e^{2\varsigma},$ $(\omega+2)\sqrt{\omega-1},$ $(\omega-1)$ $\times(\omega^3+3\omega^2+6\omega+6)$

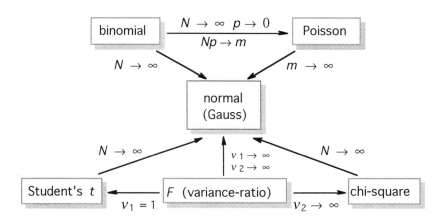

FIGURE 9.3.1 Interrelations between discrete probability distributions (Section 9.2), the normal distribution (Section 9.3.1), and three continuous distributions related to the normal distribution.

To visualize the continuous probability densities, use either *Mathematica* notebook NPDF (for distributions related to the normal distribution) or notebook ocPDF (for other continuous distributions). For computing continuous distributions, we give convenient functions programmed in C. In many instances, both the density distribution and the cumulative distribution are of interest, so we provide functions for both.

9.3.1 Gauss (Normal) Probability Function

Keywords
distribution, Gauss

C Function and *Mathematica* Notebook
GaussDist (C); cell NormalPDF in notebook NPDF (*Mathematica*)

Background
The very well known Gauss (normal) distribution is described extensively in many books, for example, in Chapter 7 of Feller's text [Fel68], Chapter 13 of Johnson et al [Joh94a], and Section 26.2 of Abramowitz and Stegun [Abr64]. This distribution is a limit of the binomial distribution (Section 9.2.1) and of the Poisson distribution described in Section 9.2.6.

Uses and abuses of the Gauss distribution are very extensive, since its probability density function is very convenient to use and much analysis has been done on it and related distributions summarized in Table 9.3.1 and Figure 9.3.1. Whether a given random variable describing a probabilistic situation approximates a Gaussian distribution is often not investigated by those who use it.

The *central limit theorem* provides a major theoretical argument for using the Gauss distribution. In Lyapunov's version of the theorem [Joh94, Section 13.2] one considers a set $\{X_1, X_2, ..., X_n\}$ of n independent, identically distributed random variables with finite mean and standard deviation. The distribution of their standardized sum has the property that

$$\left(\sum_{j=1}^{n} X_j - n\,E[X]\right)\Bigg/\sqrt{n\,\mathrm{var}(X)} \xrightarrow[n\to\infty]{} G(x) \tag{9.3.1}$$

where $G(x)$ is the standardized Gauss distribution discussed below. In (9.3.1) $E[X]$ denotes the expected value of x and $\mathrm{var}(X)$ is the variance over the set.

Function Definition
The Gauss PDF with mean value m and standard deviation σ, $G_{m,\sigma}(x)$, is usually defined by its density distribution

$$G_{m,\sigma}(x) \equiv \frac{1}{\sqrt{2\pi}\,\sigma}\, e^{\left[\frac{-1}{2}\left(\frac{x-m}{\sigma}\right)^2\right]} \tag{9.3.2}$$

A linear transformation to move the x origin to zero and to measure x in units of the standard deviation σ produces the *standardized form* of the Gauss distribution as the univariate (x) probability density with no parameters. Thus

$$G(x) \equiv \frac{1}{\sqrt{2\pi}} e^{(-x^2/2)}$$

(9.3.3)

This PDF gives the probability density for a normally-distributed random variable x having mean value zero and standard deviation unity. Note that the cumulative PDF of the Gauss distribution is easily obtained in terms of the error function (Section 10.1).

The standardization (normalization) of $G(x)$ is

$$\int_{-\infty}^{\infty} G(x)\,dx = 1$$

(9.3.4)

Moments of the Gauss PDF—as given in Table 9.3.1—are easily derived. The skewness, γ_1 in (9.1.4), is zero because the distribution is symmetric about its mean value. The excess, γ_2 in (9.1.5), is zero, by design.

Visualization

We display the Gauss distribution as a function of σ and x, recognizing that the linear transformation that converts $G_{m,\sigma}(x)$ into the standardized form, $G(x)$, can always be made, as discussed below (9.3.2).

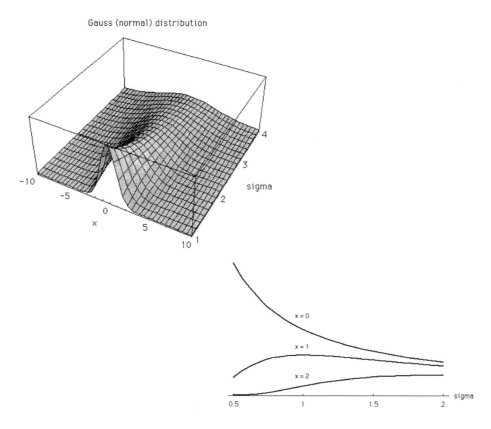

FIGURE 9.3.2 Gauss (normal) distributions. Left: Surface with σ and x varying. Right: Curves with σ varying, for $x = 0$, 1, 2. (Adapted from output of cell `NormalPDF` in the *Mathematica* notebook NPDF.)

Algorithm, Program, and Programming Notes

The program for function `GaussDist` comes directly from (9.3.3).

```
double GaussDist(double x)
/* Gauss distribution function;
 in standardized form */
{
const double PiFactor = 0.398942280401;

return PiFactor*exp(-x*x/2);
}
```

Test Values

Test values of the standardized Gaussian PDF $G(x)$ are readily generated by using a hand-held calculator or by running `NormalPDF` cell in *Mathematica* notebook `NPDF`.

$\boxed{M}$

TABLE 9.3.2 Test values to 10 digits of the standardized Gauss (normal) distribution function (9.3.2).

x	0	± 2	± 4
$G(x)$	0.3989422804	$5.399096651 \times 10^{-2}$	$1.338302258 \times 10^{-4}$

9.3.2 Bivariate Normal Probability Function

Keywords
distribution, bivariate normal

C Function and *Mathematica* Notebook
`BivNormDist` (C); cell `BivariatePDF` in notebook `NPDF` (*Mathematica*)

$\boxed{M}$

Background
The bivariate normal PDF is a generalization of the Gauss (normal) distribution to two random but *correlated* variables, x and y. It is described, for example, in Section 26.3 of Abramowitz and Stegun [Abr64]. The density distribution is such that in the absence of correlation it is the product of two standardized Gauss distributions as given by (9.3.3).

Function Definition
The standardized bivariate normal PDF (mean value of x any y both zero and standard deviations both unity), $B_\rho(x,y)$, is defined by

$$B_\rho(x,y) \equiv \frac{1}{2\pi\sqrt{1-\rho^2}} e^{\left[-\frac{1}{2}\left(\frac{x^2-2\rho xy+y^2}{1-\rho^2}\right)\right]} \tag{9.3.5}$$

in which ρ is the correlation coefficient between x and y variables. It lies in the range $(-1, 1)$. If $\rho \to +1$ then a large value of x is most probably associated with a large value of y. On the other hand, as $\rho \to -1$ then a large value of x is most probably associated with a small value of y and vice versa. This is illustrated in the visualizations below.

A glance at Figure 9.3.3 suggests that some of the correlation between x and y can, in some sense, be removed by a rotation of coordinates. Indeed, a rotation by $\pi/4$ that produces

$$x' = (x - y)/\sqrt{2} \qquad y' = (x + y)/\sqrt{2} \qquad (9.3.6)$$

leads to a new distribution, $B'_\rho(x', y')$, given by

$$B'_\rho(x', y') \equiv \frac{1}{\pi\sqrt{1 - \rho^2}} e^{\left[-\frac{1}{2}\left(\frac{x'^2}{a_-^2} + \frac{y'^2}{a_+^2}\right)\right]} \qquad a_\pm = \sqrt{1 \pm \rho} \qquad (9.3.7)$$

in which x' and y' are not correlated, except that their joint distribution function has variances $a_\pm$ that depend on each other through the correlation coefficient ρ. Note that the surfaces of constant B'_ρ are ellipses in the (x', y') plane, having axes a_- and a_+.

Visualization

We display first in Figure 9.3.3 surfaces of the bivariate PDF density (9.3.5) as a function of x and y for a given correlation coefficient ρ.

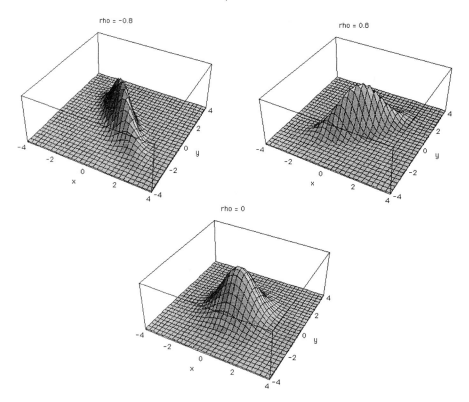

FIGURE 9.3.3 Bivariate normal probability density surfaces for correlation coefficient $\rho = -0.8, 0, 0.8$, with both x and y varying from -4 to 4. (Cell `BivariatePDF` in *Mathematica* notebook `NPDF`.)

After the rotation that leads to (9.3.7), we have distributions as shown in Figure 9.3.4 for $\rho = \pm 0.8$. Clearly, the new variables are uncorrelated but, as expected, their joint probability density depends on ρ.

Rotated bivariate distributions

rho = -0.8

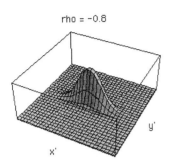

rho = 0.8

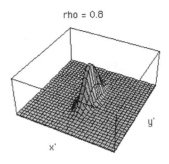

FIGURE 9.3.4 Rotated bivariate normal probability density surfaces for correlation coefficients $\rho = -0.8$ and $\rho = 0.8$, with both x' and y' varying from -4 to 4. (Cell `BivariatePDF` in *Mathematica* notebook `NPDF`.)

To convince yourself that the surfaces of constant B'_ρ are ellipses, look in Figure 9.3.5 at contour plots of the density functions shown in Figure 9.3.4. Clearly, (9.3.7) is fulfilled.

Contours of logs of rotated bivariate PDFs

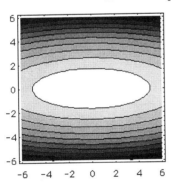
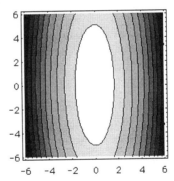

FIGURE 9.3.5 Contours (on log scales) of rotated bivariate normal probability density surfaces for correlation coefficient $\rho = -0.8$ (left) and $\rho = 0.8$ (right), with both x' and y' varying from -6 to 6. (Cell `BivariatePDF` in *Mathematica* notebook `NPDF`.)

Algorithm, Program, and Programming Notes

The program for function `BivNormDist` comes directly from (9.3.5). It is given on the next page.

```
double BivNormDist(double x, double y,
   double r)
/* Bivariate Normal distribution function */
{
const double TwoPi = 6.28318530718;
double del;

del = 1-r*r;
return
   exp(-(x*x-2*r*x*y+y*y)/(2*del))
   /(TwoPi*sqrt(del));
}
```

Test Values

Test values of the bivariate normal PDF $B_\rho(x,y)$ are readily generated by using a hand-held calculator or by running the cell `BivariatePDF` in *Mathematica* notebook NPDF. Table 9.3.3 gives convenient test values.

TABLE 9.3.3 Test values to 10 digits of the bivariate normal PDF (9.3.5) for $\rho = 0.8$.

(x,y)	$(0,-2)$	$(-1,0)$	$(1,2)$
$B_\rho(x,y)$	0.0010254672	0.0661427277	0.0217737221

9.3.3   Chi-Square Probability Functions

Keywords
distribution, chi-square

C Functions and Mathematica Notebook
ChiSqDnsty, ChiSqCum (C); cel ChiSquarePDF in notebook NPDF (*Mathematica*)

Background
The chi-square probability distribution is the distribution followed by the sums of squares of independent unit normal variables, with each following the Gauss (normal) distribution. It is described extensively, for example, in Chapter 18 of Johnson et al [Joh94a], while Section 26.4 of Abramowitz and Stegun [Abr64] has an extensive list of relevant formulas.

Applications of the chi-square distribution in statistics are very numerous. In scientific applications, estimates of confidence limits for estimated model parameters are described in Section 15.6 of the book of numerical recipes by Press et al [Pre92]. Computing the cumulative chi-square distribution as a special case of the incomplete gamma function (our Section 6.4.1) is discussed in Section 6.2 of Press et al.

It is important to note a difference between the chi-square functions described here compared with those in the remainder of this chapter, as follows. We often give only the probability *density* function (ChiSqDnsty here). For chi-square, however, it is more usual to quote the *cumulative* probability function (ChiSqCum here), which is the probability that an observed chi-square is less than a given chi-square. We specify this in more detail in the Function Definition section and the Algorithm and Program section.

Function Definition

The chi-square probability *density* function, $C(\chi^2, v)$, is defined by

$$C(\chi^2, v) \equiv \frac{(\chi^2)^{v/2-1} e^{-\chi^2/2}}{2^{v/2} \Gamma(v/2)} \tag{9.3.8}$$

in which χ^2 is the sum of the squares of v identically and normally distributed random variables each having unit standard deviation. The positive integer v is called the number of degrees of freedom. The normalization of C is such that it has unit integral if integrated from $\chi^2 = 0$ to $\chi^2 \to \infty$, as discussed below for the cumulative distribution. Moments of this distribution are given in Table 9.3.1. Since v is an integer, Γ is straightforward to evaluate, as described below under Algorithms, Programs, and Programming Notes.

The chi-square *cumulative* probability function, $P(\chi^2, v)$, is defined by

$$P(\chi^2, v) \equiv \int_0^{\chi^2} C(t, v)\, dt = \frac{1}{2^{v/2} \Gamma(v/2)} \int_0^{\chi^2} t^{v/2-1} e^{-t/2}\, dt \tag{9.3.9}$$

This function approaches unity monotonically as χ^2 increases towards infinity. It gives the probability that the chi-square for a correct model with v degrees of freedom should be less than χ^2.

Also in common use is the complement of P, namely

$$Q(\chi^2, v) \equiv 1 - P(\chi^2, v) \tag{9.3.10}$$

which gives the probability that by chance the observed chi-square is greater than χ^2 even if the model is correct for the number of degrees of freedom, v. As χ^2 increases towards infinity, Q tends to zero monotonically, whereas it begins at zero for $\chi^2 = 0$.

Visualization

We show several graphics of the chi-square distributions defined by (9.3.8) and (9.3.9). First, in Figure 9.3.6, surfaces of the chi-square probability density (9.3.8) as a function of χ^2 and v are shown. Notice that the density maximizes roughly at $\chi^2 = v$, which gives rise to the rule of thumb that a fit between model and data is acceptable when the "chi-square per degree of freedom" is about unity. Differentiation of (9.3.9) with respect to χ^2 shows that the maximum is actually at $\chi^2 = v - 2$ for $v > 2$.

Figure 9.3.7 relates to the cumulative χ^2 distributions. The left-hand display is just $P(\chi^2, v)$ in (9.3.9) as a function of χ^2 and v. The monotonic increase toward unity as χ^2 increases is evident. The right-hand side of Figure 9.3.7 shows the cumulative-distribution plot in terms of the variables χ^2/v and v. Recalling the discussion of Figure 9.3.6, we see that in this representation the cumulative distribution is a much more uniform function of v than in the χ^2 and v representation.

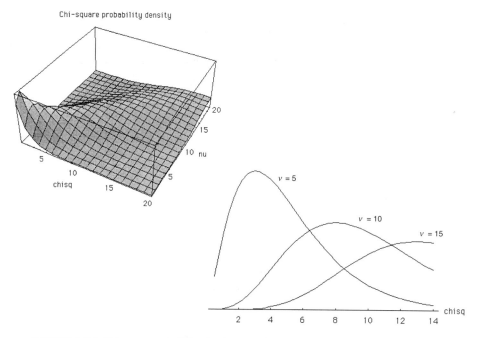

FIGURE 9.3.6 Chi-square probability densities. Left: Surfaces corresponding to (9.3.8).
Right: (Cell `ChiSquarePDF` in *Mathematica* notebook `NPDF`.)

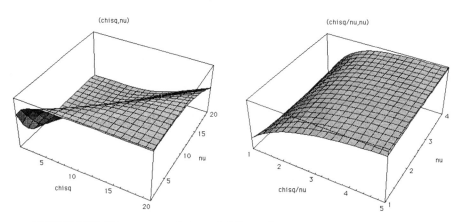

FIGURE 9.3.7 Chi-square cumulative probability surfaces corresponding to (9.3.9) in the
two representations discussed under Visualization. (Cell `ChiSquarePDF` in *Mathematica*
notebook `NPDF`.)

Algorithms, Programs, and Programming Notes

We describe first the function `ChiSqDnsty`, which comes directly from (9.3.8). The only
complexity involves the algorithm for the Γ function of half-integer argument $v/2$ therein.
By using the recurrence properties of this function, plus the special values $\Gamma(2/2) = 1$ and
$\Gamma(1/2) = \sqrt{\pi}$, one obtains

$$\Gamma(v/2) = \begin{vmatrix} \dfrac{(v-2)(v-4)\ldots 2}{2^{v/2-1}} & v \text{ even} & v > 2 \\[3ex] \dfrac{(v-2)(v-4)\ldots 1\sqrt{\pi}}{2^{(v-1)/2}} & v \text{ odd} & v > 1 \end{vmatrix} \tag{9.3.11}$$

This is coded as the variable `gamfact` in `ChiSqDnsty`. The special cases of (9.3.8) for $v = 1$ and $v = 2$ are computed separately.

```
double ChiSqDnsty(double chisq,int nu)
/* Chi-square Probability Density function */
{
const double SqrtPi = 1.77245385091;
double gamfact,nuD2;
int factor;

nuD2 = nu/2.0;

if ( nu == 1 )
return exp(-chisq/2)/(sqrt(2*chisq)*SqrtPi);

if ( nu == 2 )
return exp(-chisq/2)/2;

/* Compute rest of function */
if ( nu % 2 == 1 )
  /*  nu  is odd */
 gamfact = SqrtPi/pow(2.0,nuD2-0.5);
else
   /* nu is even */
 gamfact = 2.0/pow(2.0,nuD2-1.0);

factor = nu - 2;
while ( factor > 2 )
  {
 gamfact *= factor;
 factor -= 2;
  }

return
 pow(chisq,nuD2-1.0)*exp(-chisq/2)/
   (pow(2.0,nuD2)*gamfact);
}
```

The second function is `ChiSqCum`, the cumulative distribution (9.3.9). Although we could specialize the incomplete gamma function (Section 6.4.1), this is not efficient. Figures 9.3.6 and 9.3.7 convince us that values of χ^2/v of order unity are the most appropriate, so a power series in which the variable is of this order is likely to converge quickly. We use formula 26.4.7 in [Abr64], namely

$$P(\chi^2, v) = \frac{(\chi^2/2)^{v/2}}{\Gamma(v/2)} \sum_{n=0}^{\infty} \frac{(-\chi^2/2)^n}{n!(v/2+n)} \tag{9.3.12}$$

in which the ratio of successive terms decreases faster than χ^2/v. Programming of (9.3.12) is done efficiently by writing it as

$$P(\chi^2, v) = \frac{(\chi^2/2)^{v/2}}{\Gamma(v/2)} \sum_{n=0}^{\infty} T_n$$

$$T_0 = 2/v \qquad T_n = S_n/d_n$$

$$S_0 = 1 \qquad S_n = -\frac{(\chi^2/2)}{n} S_{n-1} \tag{9.3.13}$$

$$d_0 = v/2 \qquad d_n = d_{n-1} + 1$$

The summation is terminated as soon as the magnitude of the ratio of a term to the partial sum is less than the input value eps.

```
double ChiSqCum(double chisq,int nu,double eps)
/* Chi-square Cumulative Distribution function:
    Convergence to fractional accuracy  eps  */
{
const double SqrtPi = 1.77245385091;
double gamfact,nuD2,chD2,Term,S,D,sum;
int factor,n;


nuD2 = nu/2.0;
chD2 = chisq/2.0;


/* Compute remaining factor */
if ( nu == 1 )   gamfact = SqrtPi;
else
if ( nu == 2 )   gamfact = 1.0;
else
  {
  if ( nu % 2 == 1 )
    /*  nu  is odd */
    gamfact = SqrtPi/pow(2.0,nuD2-0.5);
  else
    /* nu is even */
    gamfact = 2.0/pow(2.0,nuD2-1.0);

  factor = nu - 2;
  while ( factor > 2 )
    {
    gamfact *= factor;
    factor -= 2;
    }
  }
```

```
/* Compute the series */
Term = 1/nuD2;
S = 1;
D = nuD2;
sum = Term;
n = 1;

while ( fabs(Term/sum) >= eps )
  {
  S *= -chD2/n;
  D += 1;
  Term = S/D;
  sum += Term;
  n += 1;
  }

return
  pow(chD2,nuD2)*sum/gamfact;
}
```

Test Values

Test values of the chi-square density and cumulative functions can be generated by running `ChiSquarePDF` cell in *Mathematica* notebook NPDF. Table 9.3.4 gives appropriate test values for the density function (9.3.8).

TABLE 9.3.4 Test values to 10 digits of the chi-square density (*C*) function.

χ^2 / v	1	2	5
2.0	0.1037768744	0.1839397206	0.1383691658
4.0	0.0269954832	0.0676676416	0.1439759107
6.0	0.00810876956	0.0248935342	0.0973043467

Table 9.3.5 shows test values for the chi-square cumulative function (9.3.9).

TABLE 9.3.5 Test values to 10 digits of the cumulative distribution (*P*) function.

χ^2 / v	1	2	5
2.0	0.8427007929	0.6321205588	0.1508549639
4.0	0.9544997361	0.8646647168	0.4505840486
6.0	0.9856941216	0.9502129316	0.6937810816

9.3.4   *F*- (Variance-Ratio) Distribution Functions

Keywords
distribution, *F* test, variance-ratio test

C Functions and *Mathematica* Notebook
FDnsty, FQCum (C); cell F-DistributionPDF in notebook NPDF (*Mathematica*)

Background
The *F* probability distribution—which is also called the variance-ratio distribution— is used in statistics to estimate the probability that two normally-distributed samples of independent random variables have the same variance. It is described in Chapter 27 of Johnson et al [Joh94b], while Section 26.6 of Abramowitz and Stegun [Abr64] gives relevant formulas. Applications of the *F* distribution are described in Sections 6.4 and 14.2 of the numerical recipes book of Press et al [Pre92]. Computing the cumulative *F* distribution as a special case of the incomplete beta function (our Section 6.4.2) is discussed in Section 6.4 of Press et al.

A difference between the *F*-distribution functions described here compared with others in this chapter, is as follows. We often provide only the probability *density* function (FDnsty here). For the *F* distribution, as for chi-square (Section 9.3.3), it is more usual to quote unity minus the *cumulative* probability function (FQCum here), which is the probability that an observed *F* ratio is *more* than a given *F* ratio. This is specified in more detail in the Function Definition and the Algorithms and Programs sections.

Function Definition
The *F* (variance-ratio) probability *density* function, $F_{v_1 v_2}(f)$, is defined by

$$F_{v_1 v_2}(f) \equiv \frac{v_1^{v_1/2} v_2^{v_2/2}}{B(v_1/2, v_2/2)} \, f^{v_1/2-1} (v_2 + v_1 f)^{-(v_1+v_2)/2} \qquad f \geq 0 \qquad (9.3.14)$$

in which v_1 and v_2 are the number of degrees of freedom of the first and second distributions, respectively, and *B* is the beta function (Section 6.1.2). The ratio of variances whose probability distribution is considered is σ_1/σ_2. The normalization of *F* is such that it has unit integral if integrated from $f = 0$ to $f \to \infty$, as discussed below for its cumulative distribution. Moments of *F* are summarized in Table 9.3.1. The density function is straightforward to evaluate, as described under Algorithms and Programs.

The *F*-distribution *cumulative* probability function that we discuss, $Q(F_{v_1 v_2}, f)$, is defined by

$$Q(F_{v_1 v_2}, f) \equiv \int_f^\infty F_{v_1 v_2}(t)\, dt = 1 - P(F_{v_1 v_2}, f) \qquad (9.3.15)$$

in which *P* is the increasing function of *f* that we usually discuss. The function *Q* approaches zero monotonically as *f* increases towards infinity, and it is unity at *f* = 0. The significance level at which the hypothesis that sample 1 has smaller variance than sample 2 can be rejected is *Q*. Thus, as *Q* *decreases* in value one's confidence that sample 1 has a variance greater than that of sample 2 *increases*.

Visualization

We show views of the F distributions defined by (9.3.14) and (9.3.15). Figure 9.3.8 shows the variance-ratio probability density as a function of f and v_2 with v_1 fixed. Since the density varies smoothly with v_1, we show only one surface. If both numbers of degrees of freedom are the same and are equal to v, the density (9.3.14) simplifies to a function having its maximum at $f = (v-2)/(v+2)$, as seen on the right side of the figure.

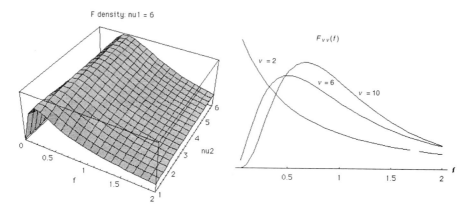

FIGURE 9.3.8 F (variance-ratio) probability densities. Left: Surfaces corresponding to (9.3.14) for $v_1 = 6$. Right: Curves of f for $v_1 = v_2 = v$. (Adapted from cell `F-Distribu-tionPDF` in *Mathematica* notebook `NPDF`.)

Figure 9.3.9 illustrates the cumulative F distribution, Q in (9.3.15). The monotonic decrease from unity to zero as f increases is evident.

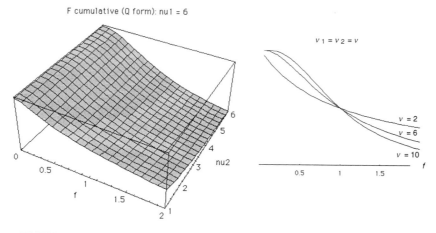

FIGURE 9.3.9 F (variance-ratio) cumulative distributions, Q corresponding to (9.3.15). Left: Surfaces for $v_1 = 6$. Right: Curves of Q for $v_1 = v_2 = v$. (Adapted from cell `F-DistributionPDF` in *Mathematica* notebook `NPDF`.)

Algorithms and Programs

We list the program for the density function, FDnsty (first box), whose algorithm comes directly from (9.3.14). Then we list the program for the cumulative distribution function FQCum (second box) from (9.3.15).

```
double FDnsty(int nu1, int nu2, double f)
/* F (Variance-Ratio) Density */
{
double LogGam(double x),floatn1,floatn2,
  halfn1,halfn2,halfn1n2,betaFact;

floatn1 = nu1;
halfn1 = 0.5*floatn1;
floatn2 = nu2;
halfn2 = 0.5*floatn2;
halfn1n2 = halfn1+halfn2;
betaFact = exp(LogGam(halfn1n2) -
  LogGam(halfn1) - LogGam(halfn2));

returnbetaFact*pow(floatn1,halfn1)*
  pow(floatn2,halfn2)*pow(f,halfn1-1.0)*
  pow(floatn2+floatn1*f,-halfn1n2);
}
```

```
double FQCum
   (int nu1, int nu2, double f, double eps)
/* F (Variance-Ratio) Cumulative */
{
double floatn1,floatn2,halfn1,halfn2,x,
  IncBeta
  (double a, double b, double x, double eps);

floatn1 = nu1;
halfn1 = 0.5*floatn1;
floatn2 = nu2;
halfn2 = 0.5*floatn2;
x = floatn2/(floatn2+floatn1*f);

return IncBeta(halfn2,halfn1,x,eps);
}
```

The function FQCum computes the F-distribution cumulative distribution (9.3.15), which decreases as its argument f increases, as shown in Figure 9.3.9. The algorithm is obtained by comparing the form of the density function (9.3.14) with the definition of the incomplete beta function (Section 6.4.2), whereby we see that

$$Q(F_{v_1 v_2}, f) = I_x(v_2/2, v_1/2) \qquad x = \frac{v_2}{v_2 + v_1 f} \qquad (9.3.16)$$

Therefore, we need only to borrow the function `IncBeta` from Section 6.4.2 to obtain the cumulative distribution. Clearly, if one wants the more usual P distribution, just compute $1 - Q$.

Programming Notes

For the function `FDnsty`, only the (complete) beta function in the denominator requires particular care. We use the expression (6.1.4) giving B in terms of gamma functions (Section 6.1.1). Since the numbers of degrees of freedom—v_1 and v_2—may be large, we combine logarithms of gamma functions then compute the exponential of the result to get B, thus greatly extending the range over which the function can be used.

For function `FQCum`, an additional argument—beyond v_1, v_2, and f—is needed to control the numerical accuracy of the continued-fraction expansion that is used when computing the incomplete beta function I_x in (9.3.16). The argument `eps` that is input by the function user is passed to `IncBeta`. A fractional accuracy of the result which is at least `eps` will be obtained. Thus, for an accuracy of about 10 significant digits use an `eps` of about 10^{-10}.

Test Values

Test values of the F-distribution density and cumulative (Q) functions can be generated by running the `F-DistributionPDF` cell in *Mathematica* notebook `NPDF` to provide the   M
numerical values. Table 9.3.6 gives appropriate values for the density function (9.3.14).

TABLE 9.3.6 **Test values to 10 digits of the F-distribution density function. Note that $F(0) = 0$.**

$F_{v_1 v_2}(f) / f$	0.5	1.0	1.5
$v_1 = 5, v_2 = 3$	0.5715855636	0.3611744789	0.2235922237
$v_1 = 10, v_2 = 3$	0.5928437662	0.4041228115	0.2467396298

Table 9.3.7 gives test values for the cumulative function (9.3.15).

TABLE 9.3.7 **Test values to 10 digits of the F-distribution cumulative function Q. Note that $Q(0) = 1$.**

$Q(F_{v_1 v_2}, f) / f$	0.5	1.0	1.5
$v_1 = 5, v_2 = 3$	0.7673760820	0.5351452100	0.3922763792
$v_1 = 10, v_2 = 3$	0.8219925926	0.5676627970	0.4085743713

9.3.5   Student's *t*-Distribution Functions

Keywords

distribution, *t* test, means test

C Functions and *Mathematica* Notebook

[M] StudTDnsty, StudTACum (C); cell StudentT-PDF in the notebook NPDF (*Mathematica*)

Background

Student's *t* probability distribution is often used in statistics to estimate the probability that two normally-distributed samples of independent random variables have the same mean. (The term "Student" is the pseudonym of a statistician, not a reference to student ownership of this distribution.) The *t* distribution is described in Chapter 28 of Johnson et al [Joh94b], while Section 26.7 of Abramowitz and Stegun [Abr64] has relevant formulas and emphasizes the cumulative distributions. Applications of the *t* distribution are described in Sections 6.4 and 14.2 of the numerical methods book by Press et al [Pre92]. How to compute the cumulative *t* distribution as a special case of the incomplete beta function (our Section 6.4.2) is discussed in Section 6.4 of Press et al.

The cumulative *t*-distribution function described here, as well as the *F* distribution in Section 9.3.4, differ from others in this chapter as follows. We typically provide only the probability *density* function (StudTDnsty here). For the *t* distribution it is standard to quote the *cumulative* probability function for the *absolute value* of *t* (StudTACum here), which is the probability that an observed *t* value is less in magnitude than a given *t*. This cumulative distribution is specified in more detail in the Function Definition section and in the Algorithms and Programs section.

Function Definition

Student's *t*-test probability *density* function, $S(t,v)$, is defined by

$$S(t,v) \equiv \frac{1}{\sqrt{v}\, B(1/2, v/2)(1+t^2/v)^{(v+1)/2}} \tag{9.3.17}$$

in which v is the number of degrees of freedom and B is the beta function (Section 6.1.2). The normalization of S is such that it has unit integral if integrated from $t \to -\infty$ to $t \to \infty$, as discussed below for its cumulative distribution. Notice that negative values of t are considered, with $S(t,v)$ being symmetric under reflection of t in the origin. This density function is straightforward to evaluate, as described under Algorithms and Programs.

The *t*-distribution *cumulative* probability function that we discuss, $A(t,v)$, is defined by

$$A(t,v) \equiv \int_{-t}^{t} S(x,v)\, dx \tag{9.3.18}$$

Thus, *A* tests whether the *absolute value* of *t* is likely to be significant, with a *small* value of *A increasing* one's confidence that two means are different.

Visualization

We show graphics of the Student's *t* distributions defined by (9.3.17) and (9.3.18). Figure 9.3.10 is the surfaces of $S(t,v)$ as a function of t (for nonnegative t) and of v.

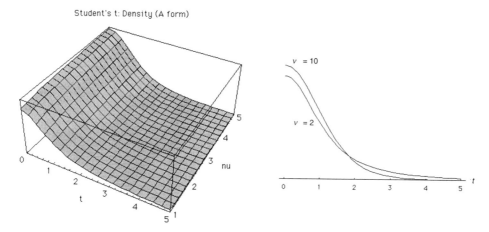

Student's t: Density (A form)

FIGURE 9.3.10 Student's t density surface corresponding to (9.3.17). Left: Surface of the density function S. Right: Curves of S at $v = 2$ and $v = 10$. (Adapted from output of cell StudentT-PDF in *Mathematica* notebook NPDF.)

As v increases, the t distribution approaches rapidly that of the unit normal (Gauss) distribution (Section 9.3.1). Two examples are shown in Figure 9.3.11.

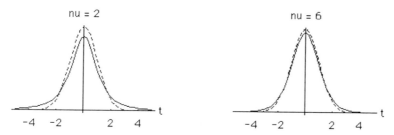

FIGURE 9.3.11 Comparison of Student's-t probability densities. Gauss (normal) shown dashed and Student's t shown by full lines, corresponding to (9.3.17) for $v = 2$ and 6. (Cell StudentT-PDF in *Mathematica* notebook NPDF.)

Figure 9.3.12 on the following page illustrates the cumulative t distribution, A in (9.3.18). The monotonic increase from zero to unity as t increases is evident.

Algorithms and Programs

In the first box we give the program for the density function, StudTDnsty, whose algorithm comes directly from (9.3.17). The algorithm for the cumulative distribution, StudTACum (second box), is derived by comparing (9.3.17) with the definition of the incomplete beta function (Section 6.4.2), which shows that

$$A(t,v) = 1 - I_x(v/2, 1/2, t) \tag{9.3.19}$$

Therefore, we use function IncBeta from Section 6.4.2 to obtain the cumulative distribution.

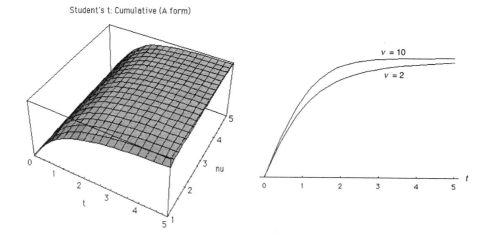

Student's t: Cumulative (A form)

FIGURE 9.3.12 Student's t cumulative probabilities corresponding to (9.3.18) and (9.3.19). Left: Surface as a function of ν and t. Right: Curves as a function of t for $\nu = 2$ and $\nu = 10$. (Adapted from cell `StudentT-PDF` in *Mathematica* notebook `NPDF`.)

```
double StudTDnsty(int nu, double t)
/*  Student's t: Density Distribution */
{
double LogGamma(double x),floatn,halfn,betaFact;
static double LogSqrtPi = 0.572364942925;

floatn = nu;
halfn = 0.5*floatn;
betaFact = exp(LogGamma(halfn+0.5) -
           LogGamma(halfn) - LogSqrtPi);

returnbetaFact/(sqrt(floatn)*
       pow(1+t*t/floatn,halfn+0.5)));
}
```

```
double StudTACum(int nu, double t, double eps)
/*  Student's t: Cumulative Distribution */
{
double floatn,x,IncBeta
  (double a, double b, double x, double eps);

floatn = nu;
x = floatn/(floatn+t*t);

return  1.0 - IncBeta(0.5*floatn,0.5,x,eps);
}
```

Programming Notes

For the function `StudTDnsty`, the complete beta function in the denominator requires care. We use expression (6.1.4), giving B in terms of gamma functions (Section 6.1.1). Since the number of degrees of freedom, v, may be large, we combine logarithms of gamma functions then compute the exponential of the result to get B, thus extending the parameter range over which `StudTDnsty` can be used.

For `StudTACum` an additional argument is needed to control the numerical accuracy of the continued-fraction expansion used for computing the incomplete beta function I_x in (9.3.19). The argument `eps` input by the function user is passed to `IncBeta`, giving a fractional accuracy of the result which is at least `eps`. For example, an accuracy of about 6 significant digits is obtained with an input `eps` of about 10^{-6}.

Test Values

Test values of the t-distribution density (S) and cumulative (A) functions can be generated by running the `StudentT-PDF` cell in *Mathematica* notebook NPDF. Table 9.3.8 gives appropriate values for the density function (9.3.17). Note that the dependence on v is quite weak, consistent with the approach of Student's t distribution to the Gauss distribution as v increases, as shown in Figure 9.3.11.

TABLE 9.3.8 **Test values to 10 digits of the t-distribution density function, S.**

$S(t,v)$ / t	0	2	4
$v = 1$	0.3183098862	0.0636619772	0.0187241111
$v = 2$	0.3535533906	0.0680413817	0.0130945700
$v = 5$	0.3796066898	0.0650903103	0.0051237271

Table 9.3.9 gives test values for the cumulative distribution function (9.3.18) or (9.3.19).

TABLE 9.3.9 **Test values to 10 digits of the t-distribution cumulative function, A. Note that $A(0, v) = 0$ and that $A(\infty, v) = 1$.**

$A(t,v)$ / t	2	4	6
$v = 1$	0.7048327647	0.8440417392	0.8948630866
$v = 2$	0.8164965809	0.9428090416	0.9733285268
$v = 5$	0.8980605212	0.9896765845	0.9981538617

9.3.6   Lognormal Distribution

Keywords
distribution, Gauss, lognormal

C Functions and *Mathematica* Notebook
`LogNDnsty`, `LogNCum` (C); cell `LognormalPDF` in notebook `NPDF` (*Mathematica*)

Background
The lognormal distribution is described in several books, for example, Chapter 14 of Johnson et al [Joh94a]. It is a variation on the normal (Gauss) distribution that is obtained by transforming the random variable X, of the kind we consider in Section 9.3.1, to U given by

$$U = (\log X - \zeta)/\sigma \qquad (9.3.20)$$

in which ζ and σ are parameters. Because the central limit theorem provides a justification for using the Gauss distribution for a *sum* of random variables, it therefore justifies use of the lognormal distribution for a *product* of random variables, since the logarithm of such a product is the sum of logarithms. Such a development of the lognormal distribution is discussed in Section 14.2 of [Joh94a].

Function Definition
The two-parameter lognormal PDF $L_{\zeta,\sigma}(x)$ is defined by the density distribution

$$L_{\zeta,\sigma}(x) \equiv \frac{1}{\sqrt{2\pi}\,\sigma x}\, e^{\left[-\frac{1}{2}\left(\frac{\ln x - \zeta}{\sigma}\right)^2\right]} \qquad x > 0 \qquad (9.3.21)$$

in which the $1/x$ factor ensures that the probability content over small intervals is the same in the lognormal and normal distributions. The standardization (normalization) of $L_{\zeta,\sigma}(x)$ is therefore unity. Moments of the lognormal distribution—as given in Table 9.3.1—are easily derived.

In the Gauss distribution (Section 9.3.1) it is possible to make the distribution symmetric by displacing the x axis by the mean value. A similar transformation can also be made for the lognormal distribution by replacing x in (9.3.21) by $xe^{-\zeta}$. This can be done only if one knows ζ, however, just as one has to know m to centralize the Gauss distribution. In practical use, the parameter ζ or m is often one of those to be estimated. The maximum (mode) of the density distribution (9.3.21) is at x_m given by

$$x_m = e^{\zeta - \sigma^2} \qquad (9.3.22)$$

which is evident in the visualizations that follow.

The cumulative distribution of $L_{\zeta,\sigma}(x)$ is obtained as

$$P_{\zeta,\sigma}(x) \equiv \int_0^x L_{\zeta,\sigma}(t)\,dt = \frac{1}{2}\,\mathrm{erfc}\left(\frac{\zeta - \ln x}{\sqrt{2}\,\sigma}\right) \qquad x > 0 \qquad (9.3.23)$$

in terms of the complementary error function, erfc (Section 10.1).

Visualization

We display the lognormal distributions as functions of σ and x for ζ fixed, both for the density distribution in Figure 9.3.13 and for the cumulative distribution in Figure 9.3.14. Notice how similar the density distribution graphic is to the Gauss distribution density, Figure 9.3.2, but that the variables have quite different values. The chosen values $\zeta = -0.7$ and $\zeta = 0.9$ make the maxima of the density distribution (9.3.21) occur near $x = 0.5$ and $x = 2.5$, respectively, for $\sigma = 0$, according to (9.3.22).

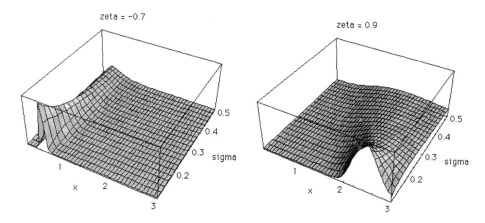

FIGURE 9.3.13 Lognormal density distributions (9.3.22) for $\zeta = -0.7$ and 0.9, with σ varying from 0.1 to 0.5 from front to back and x varying from 0 to 3. (Cell Lognormal-PDF in *Mathematica* notebook NPDF.)

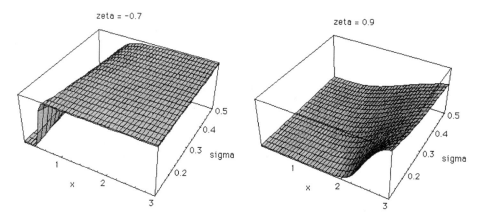

FIGURE 9.3.14 Lognormal cumulative distributions (9.3.23) for $\zeta = -0.7$ and 0.9, with σ varying from 0.1 to 0.5 from front to back and x varying from 0 to 3. (Cell Lognormal-PDF in *Mathematica* notebook NPDF.)

Algorithms, Programs, and Programming Notes

The program for the density function `LogNDist` comes directly from (9.3.21). The constant `PiFactor` is just $1/\sqrt{2\pi}$.

```
double LogNDist
  (double zeta, double sigma, double x)
/* Lognormal distribution: density */
{
const double PiFactor = 0.398942280401;

return PiFactor*
  exp(-0.5*pow((log(x)-
  zeta)/sigma,2.0))/(sigma*x);
}
```

For the cumulative distribution function `LogNCDF` we code the expression (9.3.23), using the error-function routine `Erf` from Section 10.1.

```
double LogNCDF
  (double zeta, double sigma, double x)
/* Lognormal distribution: cumulative */
{
double Erf(double u);
static double sqrt2 = 1.41421356237;

return 0.5*(1 - Erf((zeta-log(x))/
  (sqrt2*sigma)));
}
```

Test Values

Check values of the lognormal density distribution $L_{\zeta,\sigma}(x)$ and the lognormal cumulative distribution $P_{\zeta,\sigma}(x)$ are readily generated by running the `LognormalPDF` cell in *Mathematica* notebook NPDF.

TABLE 9.3.10 Test values to 10 digits of the lognormal density distribution (9.3.21) for $\zeta = 0.5$, $\sigma = 0.3$.

x	1.0	1.5	2.0
$L_{\zeta,\sigma}(x)$	0.3315904626	0.8435973428	0.5404422251

TABLE 9.3.11 Test values to 10 digits of the lognormal cumulative distribution (9.3.22) for $\zeta = 0.5$, $\sigma = 0.3$.

x	1.0	1.5	2.0
$P_{\zeta,\sigma}(x)$	0.0477903523	0.3763366741	0.7401551945

9.4   OTHER CONTINUOUS PROBABILITY DISTRIBUTIONS

In this section we include continuous probability distributions that are not directly related to the normal (Gauss) distributions. We emphasize central distributions in standardized forms, in order to minimize the number of parameters when visualizing and computing the distributions, but this is easily changed if necessary. The *Mathematica* notebook for the other continuous probability distribution functions is ocPDF.

M

Definitions and lowest moments for these distributions are summarized in Table 9.4.1.

TABLE 9.4.1 Properties of other continuous probability distribution functions.

Function	Domain	Probability density, $\rho(x)$	Parameter limits	Moments: m, σ^2, γ_1, γ_2
Cauchy (Lorentz)	$(-\infty, \infty)$	$\dfrac{1}{\pi(1+x^2)}$		0, higher moments undefined
exponential	$[0, \infty)$	e^{-x}		1, 1, 2, 6
Pareto	$[k, \infty)$	$\dfrac{ak^a}{x^{a+1}}$	$a > 0,$ $k > 0$	$\dfrac{ak}{a-1}$ $a > 1,$ $\dfrac{ak^2}{(a-1)^2(a-2)}$ $a > 2,$ $\dfrac{2(a+1)}{a-3}\sqrt{\dfrac{a-2}{a}}$ $a > 3,$ $\dfrac{6[a(a^2+a-6)-2]}{a(a-3)(a-4)}$ $a > 4$
Weibull	$(0, \infty)$	$cx^{c-1}e^{-x^c}$	$c > 0$	See Section 9.4.4.
logistic	$(-\infty, \infty)$	$\dfrac{e^{-x}}{(1+e^{-x})^2}$		$0, \dfrac{\pi^2}{3}, 0, \dfrac{6}{5}$
Laplace	$(-\infty, \infty)$	$\dfrac{1}{2}e^{-|x|}$		0, 2, 0, 6
Kolmogorov-Smirnov	$[0, \infty)$	$8x\displaystyle\sum_{j=1}^{\infty}(-1)^{j-1}j^2$ $\times e^{-2j^2x^2}$		0.868731, 0.067773, 0.860425, 0.881601
beta	$[0,1]$	$\dfrac{x^{p-1}(1-x)^{q-1}}{B(p,q)}$	$p > 0$ $q > 0$	See Section 9.4.8.

9.4.1 Cauchy (Lorentz) Distribution

Keywords
distribution, Cauchy, Lorentz

C Functions and *Mathematica* Notebook

M

`CauchyDnsty`, `CauchyCum` (C); cell `CauchyPDF` in notebook `ocPDF` (*Mathematica*)

Background
The Cauchy distribution—known to physicists as the Lorentz distribution—is described in several sources, for example, Chapter 16 of Johnson et al [Joh94a]. The geometry for one application of the Cauchy distribution is shown in Figure 9.4.1. Consider the distribution of the point of intersection, P, of a fixed line with the line, CP. If angle OCP has a uniform random distribution between $-\pi/2$ and $\pi/2$ then length OP has a Cauchy density distribution with mean value zero.

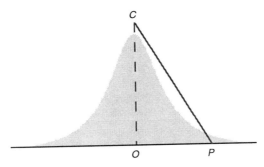

FIGURE 9.4.1 Geometry for generating a Cauchy distribution of the length OP if the angle OCP has a uniform random distribution. The curve shown is the Cauchy density distribution.

Suppose that C is a point source of particles or rays that are emitted at random angles. They will then impact along the line OP with a Cauchy density distribution.

Function Definition
The two-parameter Cauchy PDF $C_{m\lambda}(x)$ is defined by the density distribution

$$C_{m\lambda}(x) \equiv \frac{1}{\pi\lambda} \frac{1}{1+[(x-m)/\lambda]^2} \tag{9.4.1}$$

Here m is the mean value and mode of the distribution, while λ is its width parameter. In science and engineering it is usual to discuss the Full Width at Half Maximum (FWHM) of the distribution, which is the distance between the two points where the density values are one half of the maximum value. Thus, for the Cauchy distribution the FWHM is just 2λ. Moments of the Cauchy distribution—except for the first moment (m)—are undefined because the necessary integrals (9.1.2) are divergent.

By comparison with the Gauss (normal) distribution (Section 9.3.1), the Cauchy (Lorentz) distribution has much longer tails, as illustrated in Figure 19.6.1 when discussing the convolution of these two functions to form the Voigt (plasma dispersion) function.

For the Cauchy distribution (9.4.1) it is possible to make the distribution symmetric by displacing the x axis by the mean value, and a standard form can be obtained by setting $\lambda = 1$, so that x is measured in units of λ. The standard Cauchy density distribution is then given by

$$C(x) \equiv \frac{1}{\pi} \frac{1}{1+x^2}$$
(9.4.2)

The cumulative distribution of $C(x)$ is obtained as

$$P(x) \equiv \int_0^x C(t)\,dt = \frac{1}{2} + \frac{1}{\pi} \arctan x$$
(9.4.3)

in which the arctangent lies between $-\pi/2$ and $\pi/2$.

Visualization
We display the Cauchy distributions as functions of λ and x, both for the density distribution in Figure 9.4.2 and for its cumulative distribution in Figure 9.4.3. Notice how similar the density-distribution graphic is to that for the Gauss distribution density, Figure 9.3.2, but that the variables have quite different values.

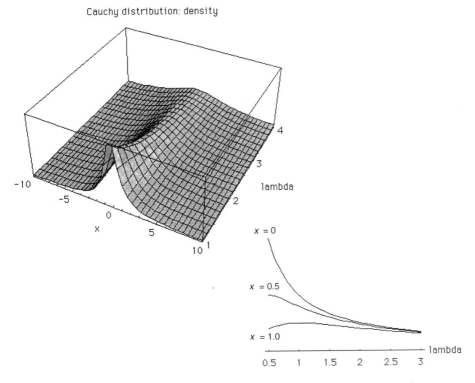

FIGURE 9.4.2 Cauchy density distributions (9.4.1). Left: Surfaces for $m = 0$, with λ varying from 1 to 4 from front to back and x varying from –10 to 10. Right: Curves for λ varying and x fixed. (Adapted from cell CauchyPDF in *Mathematica* notebook ocPDF.)

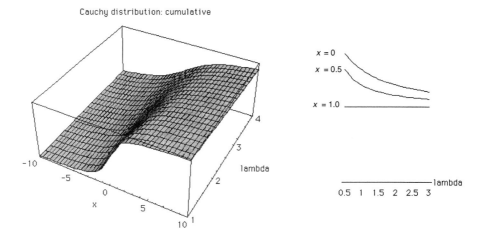

FIGURE 9.4.3 Cauchy cumulative distributions for $m = 0$. Left: Surfaces with λ varying from 1 to 4 from front to back and x varying from -10 to 10. Right: Curves for λ varying and x fixed. (Adapted from cell CauchyPDF in *Mathematica* notebook ocPDF.)

Algorithms, Programs, and Programming Notes

Our programs for the standard Cauchy distributions come directly from the definitions. CauchyDnsty for the density is from (9.4.2), and CauchyCum for the cumulative distribution is from (9.4.3).

```
double CauchyDnsty(double x)
/* Gauss distribution density;
   in standard form */
{
const double Pi = 3.14159265359;

return 1/(Pi*(1+x*x));
}
```

To constrain the arctangent in (9.4.3) to the specified range, we use the library function atan2 rather than the function atan.

```
double CauchyCum(double x)
/* Gauss distribution cumulative;
   in standard form */
{
const double Pi = 3.14159265359;

return 0.5 + atan2(x,1)/Pi;
}
```

Test Values

Check values for the standard Cauchy density distribution $C(x)$ and its cumulative distribution $P(x)$ are readily generated by running the cell `CauchyPDF` in the *Mathematica* notebook `ocPDF`. They are given in Table 9.4.2.

TABLE 9.4.2 Test values to 10 digits of the standard Cauchy density distribution (9.4.2) and of the standard Cauchy cumulative distribution (9.4.3).

x	-2	0	2
$C(x)$	0.0636619772	0.3183098862	0.0636619772
$P(x)$	0.1475836177	0.5000000000	0.8524163823

9.4.2 Exponential Distribution

Keywords
distribution, exponential

C Functions and *Mathematica* Notebook
`ExpDnsty`, `ExpCum` (C); cell `ExponentialPDF` in notebook `ocPDF` (*Mathematica*)

Background
The exponential distribution is described in several sources, for example, Chapter 19 of Johnson et al [Joh94a]. It is applied in many situations in which a statistical distribution without memory is expected. For example, the radioactivity decay of atomic nuclei follows very accurately an exponential distribution with time as the variable.

Function Definition
The one-parameter exponential PDF $E_\lambda(x)$ is defined by its density distribution

$$E_\lambda(x) = \lambda e^{-\lambda x} \qquad x > 0 \qquad (9.4.4)$$

where λ is the "decay" parameter. This distribution can be standardized by setting $\lambda = 1$, so that x is measured in units of $1/\lambda$, giving

$$E(x) = e^{-x} \qquad x > 0 \qquad (9.4.5)$$

The cumulative exponential distribution of $C_\lambda(x)$ is obtained as

$$C_\lambda(x) \equiv \int_0^x E_\lambda(t)\,dt = 1 - e^{-\lambda x} \qquad x > 0 \qquad (9.4.6)$$

For the standardized distribution (9.4.5) the cumulative distribution becomes

$$C(x) = 1 - e^{-x} \qquad x > 0 \qquad (9.4.7)$$

The first four moments of the standardized exponential distribution are simply 1, 1, 2, 6 (Table 9.4.1). Generally, the *m*th moment for $m \geq 0$ is $m!$.

Visualization

We show the exponential distributions as functions of λ and x, for the density distribution in Figure 9.4.4 and for the cumulative distribution in Figure 9.4.5. The corresponding standardized distributions have $\lambda = 1$, the front plane in each graphic.

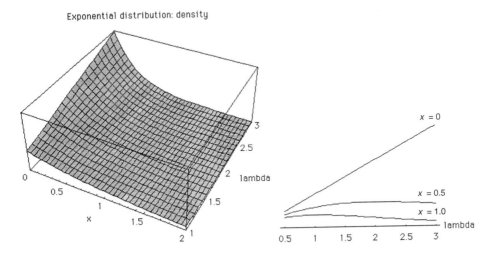

FIGURE 9.4.4 Exponential density distributions (9.4.4). Left: Surfaces with λ varying from 1 to 3 from front to back and x varying from 0 to 2. Right: Curves with x fixed and λ varying. (Adapted from cell `ExponentialPDF` in *Mathematica* notebook `ocPDF`.)

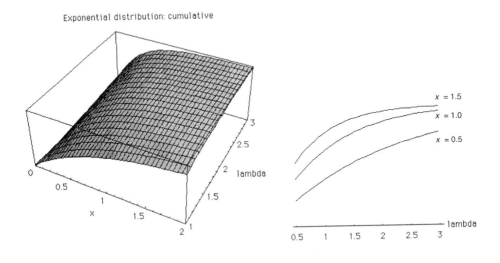

FIGURE 9.4.5 Exponential cumulative distributions (9.4.5). Left: Surfaces with λ varying from 1 to 3 from front to back and x varying from 0 to 2. Right: Curves with x fixed and λ varying. (Adapted from cell `ExponentialPDF` in *Mathematica* notebook `ocPDF`.)

Algorithms, Programs, and Programming Notes

Our programs for the standard exponential distributions come directly from the definitions, (9.4.5) for `ExpDnsty` and (9.4.7) for `ExpCum`. They are the simplest in the *Atlas*.

```
double ExpDnsty(double x)
/*  Exponential distribution density;
    in standard form */
{
return exp(-x);
}
```

```
double ExpCum(double x)
/*  Exponential distribution cumulative;
    in standard form */
{
return 1.0 - exp(-x);
}
```

Test Values

Test values for the standard exponential density distribution $E(x)$ and its cumulative distribution $C(x)$ are readily generated by running the cell `ExponentialPDF` in *Mathematica* notebook `ocPDF`, or by using a pocket calculator. They are given in Table 9.4.3.

TABLE 9.4.3 Test values to 10 digits of the standard exponential density distribution (9.4.5) and of the exponential cumulative distribution (9.4.7). The two values sum to unity.

x	0	1	2	3
$E(x)$	1.000000000	0.3678794412	0.1353352832	0.04978706837
$C(x)$	0.000000000	0.6321205588	0.8646647168	0.95021293163

9.4.3 Pareto Distribution

Keywords

distribution, Pareto, logexponential

C Functions and *Mathematica* Notebook

`ParetoDnsty`, `ParetoCum` (C); cell `ParetoPDF` in notebook `ocPDF` (*Mathematica*)

Background

The Pareto distribution—named after a Swiss professor of economics—is described in several sources, for example, in Chapter 20 of Johnson et al [Joh94a]. Because it is a simple power-law distribution, it is widely used, especially whenever a long right-side tail of the density distribution is expected. For example, in econometrics the distribution of the wealth of individuals is sometimes described by the Pareto distribution.

Function Definition

We define the standardized two-parameter Pareto PDF $P_{a,k}(x)$ by its density distribution

$$P_{a,k}(x) \equiv \begin{vmatrix} 0 & x < k \\ \dfrac{ak^a}{x^{a+1}} & x \geq k \qquad a > 0 \qquad k > 0 \end{vmatrix} \qquad (9.4.8)$$

in which a and k are parameters. The original definition replaces the function for $x \geq k$ by $(k/x)^a$, but this is less convenient for analysis and applications. The cumulative Pareto distribution $P_{a,k}(x)$ is then obtained as $F_{a,k}(x)$ given by

$$F_{a,k}(x) = \begin{vmatrix} 0 & x < k \\ 1 - \left(\dfrac{k}{x}\right)^a & x \geq k \qquad a > 0 \qquad k > 0 \end{vmatrix} \qquad (9.4.9)$$

If the variable x in (9.4.8) is transformed to $\ln x$ and if x is still used as the density variable, then (9.4.8) is transformed into

$$P'_{a,k}(x) \equiv \begin{vmatrix} 0 & x < k \\ ak^a e^{-a \ln x} & x \geq k \end{vmatrix} \qquad (9.4.10)$$

This is an exponential distribution (Section 9.4.2) in terms of the variable $\ln x$, which justifies calling it the lognormal distribution. Thus, the Pareto distribution is related to the exponential distribution just as the lognormal distribution in Section 9.3.6 is related to the normal distribution in Section 9.3.1.

Visualization
We show Pareto distributions as functions of k and x for a fixed a, for the density distribution in Figure 9.4.6 and for the cumulative distribution in Figure 9.4.7.

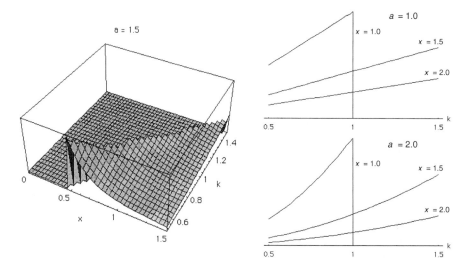

FIGURE 9.4.6 Pareto density distributions (9.4.8). Left: Surfaces with x and k varying for $a = 1.5$. Right: Curves for $a = 1.0$ (top) and for $a = 2.0$ (bottom). (Adapted from cell ParetoPDF in *Mathematica* notebook ocPDF.)

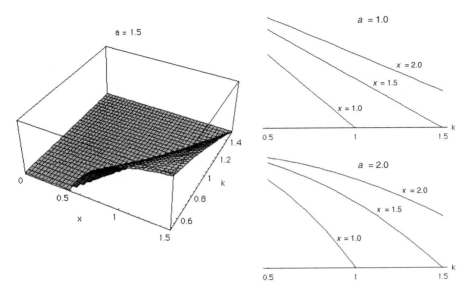

FIGURE 9.4.7 Pareto cumulative distributions (9.4.9). Left: Surfaces with x and k varying for $a = 1.5$. Right: Curves for $a = 1.0$ (top) and for $a = 2.0$ (bottom). (Adapted from cell `ParetoPDF` in *Mathematica* notebook `ocPDF`.)

Note in both figures the diagonal discontinuity in the Pareto distributions at threshold value $x = k$. Apart from this, they are like most continuous probability distributions, being generally featureless.

Algorithms, Programs, and Programming Notes
The programs for the Pareto distributions come directly from the definitions, (9.4.8) for `ParDnsty` and (9.4.9) for `ParCum`.

```
double ParDnsty(double a, double k, double x)
/*  Pareto distribution density */
{
if ( x < k ) return 0;
return a*pow(k/x,a)/x;
}
```

```
double ParCum(double a, double k, double x)
/*  Pareto distribution cumulative */
{
if ( x < k ) return 0;
return 1 - pow(k/x,a);
}
```

Test Values

Test values for the Pareto density distribution $P_{a,k}(x)$ and its cumulative distribution $F_{a,k}(x)$ can be generated easily by running the cell `ParetoPDF` in *Mathematica* notebook `ocPDF`, or by using your pocket calculator. They are given in Table 9.4.4.

**TABLE 9.4.4   Test values to 10 digits of the Pareto density distribution (9.4.8)
and of the Pareto cumulative distribution (9.4.9) for $a = 1.5$ and $k = 1$.**

x	1.0	1.5	2.0
$P_{a,k}(x)$	1.500000000	0.5443310540	0.2651650429
$F_{a,k}(x)$	0	0.4556689460	0.6464466094

9.4.4   Weibull Distribution

Keywords
distribution, Weibull

C Functions and *Mathematica* Notebook

`WeibullDnsty`, `WeibullCum` (C); cell `WeibullPDF` in notebook `ocPDF` (*Mathematica*)

Background

The Weibull distribution—named after a twentieth-century Swedish physicist, and also known as the Weibull-Gnedenko distribution—is a generalized distribution similar to the Gauss (normal) distribution of Section 9.3.1 and including the exponential distribution (Section 9.4.2). It is described, for example, in Chapter 21 of Johnson et al [Joh94a]. The Weibull distribution is often used in the early stages of a statistical analysis, since it allows flexibility in the choice of PDF.

Function Definition

The complete Weibull PDF $W_{c,a,\beta}(x)$ is sometimes defined by the density distribution

$$W_{c,a,\beta}(x) \equiv c\, e^{\left[-\left(\frac{x-\beta}{\alpha}\right)^c\right]} \Bigg/ \left(\frac{x-\beta}{\alpha}\right)^{c-1} \tag{9.4.11}$$

which has three parameters. It is—as for the normal and exponential distributions—more convenient to consider the centralized and standardized one-parameter Weibull distribution $W_c(x)$ with density distribution given by

$$W_c(x) \equiv c\, x^{c-1} e^{-x^c} \qquad x > 0 \tag{9.4.12}$$

The cumulative Weibull distribution $C_c(x)$ is then given by

$$C_c(x) = 1 - e^{-x^c} \qquad x > 0 \tag{9.4.13}$$

If we choose $c = 1$ in the standardized Weibull distribution, we obtain the standardized exponential distribution in Section 9.4.2. Moments of the Weibull distribution are complicated when c is arbitrary. They are discussed in Section 21.3 of [Joh94a].

Visualization

We show standardized Weibull distributions as functions of c and x, both for the density distribution in Figure 9.4.8 and for the cumulative distribution in Figure 9.4.9. For $c < 1$ the density peaks at $x = 0$, while for $c \geq 1$ it peaks at $x = (1 - 1/c)^{1/c}$.

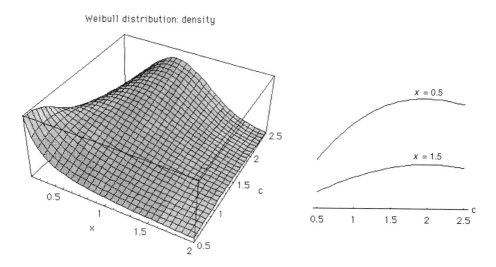

FIGURE 9.4.8 Weibull density distributions (9.4.12). Left: Surface with x and c varying. Note that x starts at 0.1. Right: Curves with c varying and x fixed. (Adapted from cell WeibullPDF in *Mathematica* notebook ocPDF.)

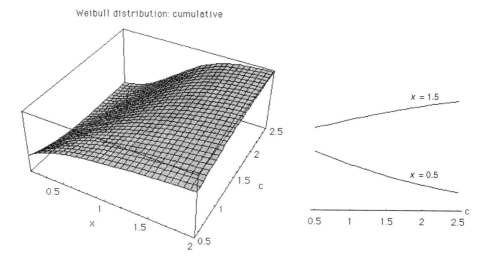

FIGURE 9.4.9 Weibull cumulative distributions (9.4.13). Left: Surface with x and c varying. Note that x starts at 0.1. Right: Curves with c varying and x fixed. (Adapted from cell WeibullPDF in *Mathematica* notebook ocPDF.)

In Figures 9.4.8 and 9.4.9, x starts at 0.1 because for $c < 1$ the values of the distribution functions are large and rapidly varying when x is small.

Algorithms, Programs, and Programming Notes

The programs for the Weibull distribution functions are coded directly from the definitions, (9.4.12) for WeibullDnsty and (9.4.13) for WeibullCum.

```
double WeibullDnsty(double c, double x)
/*  Weibull distribution density, x > 0 */
{
return c*pow(x,c-1)*exp(-pow(x,c));
}
```

```
double WeibullCum(double c, double x)
/*  Weibull distribution cumulative, x > 0 */
{
 1 - exp(-pow(x,c));
}
```

Test Values

Check values for the Weibull density distribution $W_c(x)$ and its cumulative distribution $C_c(x)$ can be generated easily by running the cell WeibullPDF in the *Mathematica* notebook ocPDF, or by using your pocket calculator. For $c = 1$ the distributions coincide with those for the exponential distribution in Section 9.4.2. Numerical values are given in Tables 9.4.5 and 9.4.6. Note that $C_c(x) = 1 - 1/e$ for any c.

TABLE 9.4.5 Test values to 10 digits of the Weibull density distribution (9.4.12).

x	0.5	1.0	1.5
$c = 1$	0.6065306597	0.3678794412	0.2231301601
$c = 2$	0.7788007831	0.7357588823	0.3161976737

TABLE 9.4.6 Test values to 10 digits of the Weibull cumulative distribution (9.4.13).

x	0.5	1.0	1.5
$c = 1$	0.3934693403	0.6321205588	0.7768698399
$c = 2$	0.2211992169	0.6321205588	0.8946007754

9.4.5 Logistic Distribution

Keywords
distribution, logistic

C Functions and *Mathematica* Notebook
`LgstcDnsty`, `LgstcCum` (C); cell `LogisticPDF` in notebook `ocPDF` (*Mathematica*)   `M`

Background
The logistic distribution is similar to the well-known Gauss (normal) distribution in Section 9.3.1, but it is sometimes more convenient to use. It is described in Chapter 23 of Johnson et al [Joh94b], in a monograph on the logistic distribution by Hosmer and Lemeshaw [Hos89], and in the book edited by Balakrishnan [Bal92]. Differential equations relevant to the logistic distribution are discussed in Thompson's text on numerical methods and computing [Tho92].

Function Definition
The complete logistic PDF *cumulative* distribution $C_{a,\beta}(x)$ is sometimes defined as the two-parameter function

$$C_{a,\beta}(x) \equiv \frac{1}{1 + e^{-[(x-\alpha)/\beta]}} \qquad (9.4.14)$$

For many applications, especially for computing, it is more convenient to standardize the distribution by setting $\alpha = 0$, $\beta = 1$, so that x is centered and is measured in units of $1/\beta$. The standardized logistic cumulative distribution is then given by

$$C(x) \equiv \frac{1}{1 + e^{-x}} = \frac{1}{2}[1 + \tanh(x/2)] \qquad (9.4.15)$$

The corresponding standardized logistic density distribution $L(x)$ is then given by

$$L(x) = \frac{e^{-x}}{(1 + e^{-x})^2} = \frac{1}{4} \operatorname{sech}^2(x/2) \qquad (9.4.16)$$

The second equality gives rise to an alternative name among many—the *sech-square* distribution. The low-order moments of the standardized logistic distribution are quite simple and are given in Table 9.4.1.

The differential equation satisfied by the cumulative distribution, $C(x)$, is

$$\frac{dC(x)}{dx} = C(x) - C^2(x) \qquad (9.4.17)$$

sometimes called Bernoulli's equation. If x is interpreted as time, then this equation can be interpreted as a growth equation in which the first term on the right-hand side would give rise to exponential increase, but the second term gives rise to damping that produces long-term equilibrium as $C \rightarrow 1$. In this context, the logistic cumulative distribution is called the sigmoid curve of growth or Verhulst's curve. This equation becomes clear in the Visualization, Figure 9.4.11.

Another connection is that between the centralized Gauss (normal) distribution (Section 9.3.1) and the standardized logistic distribution $L(x)$. Both the density and cumulative functions of these two distributions are very similar in shape. For example, suppose that we match a normal distribution so that it has the same Full Width at Half Maximum

(FWHM) as $L(x)$. It is easy to show that the FWHM of $L(x)$ is $2\ln(3+\sqrt{8}) = 3.525494$. If we equate this to the FWHM of a normal density distribution, which is $\sqrt{2\ln 2}\,\sigma$, then we find that $\sigma = 1.497140$. The two distributions are compared in Figure 9.4.10. A similar analysis for the cumulative distributions is given in Section 22.4 of [Joh94b].

Visualization
We show standardized logistic distributions as a function of x for the density distribution $L(x)$ and its comparison with the normal distribution discussed in the preceding paragraph (Figure 9.4.10), and for the cumulative logistic distribution $C(x)$ (Figure 9.4.11).

Logistic distribution: density

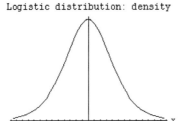 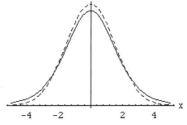

FIGURE 9.4.10   Left: Standardized logistic density distribution (9.4.16). Right: Comparison of a standardized logistic density distribution (9.4.16), shown as the solid curve, with a Gauss (normal) distribution—the dashed curve—having the same FWHM. (Cell `LogisticPDF` in *Mathematica* notebook `ocPDF`.)

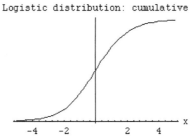

FIGURE 9.4.11   Standardized logistic cumulative distributions (9.4.15). See also the discussion below the di.fferential equation (9.4.17) of the growth equation interpretation of this curve. (Cell `LogisticPDF` in *Mathematica* notebook `ocPDF`.)

Algorithms and Programs
The programs for the logistic distribution functions are coded directly from the definitions, namely (9.4.16) for `LgstcDnsty` and (9.4.15) for `LgstcCum`.

```
double LgstcDnsty(double x)
/*  Logistic distribution density */
{
double expMx;
expMx = exp(-x);
return   expMx*pow((1+expMx),-2.0);
}
```

```
double LgstcCum(double x)
/* Logistic distribution cumulative */
{
return  1.0/(1+exp(-x));
}
```

Test Values

Test values for the logistic density distribution $L(x)$ and the cumulative distribution $C(x)$ can be generated easily by running the cell `LogisticPDF` in *Mathematica* notebook `ocPDF`, or by using a pocket calculator. Numerical values are given in Table 9.4.7.

M

TABLE 9.4.7 Test values to 10 digits of the logistic distributions (9.4.16) and (9.4.15).

x	−2.0	0	2.0
$L(x)$	0.1049935854	0.2500000000	0.1049935854
$C(x)$	0.1192029220	0.5000000000	0.8807970780

9.4.6 Laplace Distribution

Keywords

distribution, Laplace

C Functions and *Mathematica* Notebook

`LaplaceDnsty`, `LaplaceCum` (C); cell `LaplacePDF` in notebook `ocPDF` (*Mathematica*)

M

Background

The Laplace distribution is similar to the exponential distribution in Section 9.4.2, having the same shape for $x > 0$, but it is defined also for $x < 0$. This distribution is described in Chapter 24 of Johnson et al [Joh94b].

If two independent random variables have identical exponential distributions, then their difference has a Laplace distribution. Related distributions are discussed in Section 24.6 of [Joh94b]. For example, the Cauchy density distribution (Section 9.4.1) and the Laplace density distribution are Fourier integral transforms of each other.

Function Definition

We discuss only the standardized Laplace distribution, since in many applications—such as computing—it is most convenient to work with this form. The standardized Laplace density distribution is given by

$$L(x) \equiv \frac{1}{2} e^{-|x|} \tag{9.4.18}$$

Note that the density distribution is discontinuous at $x = 0$, but this usually does not give any problems in statistical applications. The standardized cumulative distribution, $C(x)$, is given by

$$
C(x) = \begin{vmatrix} \dfrac{1}{2} e^x & x \le 0 \\ 1 - \dfrac{1}{2} e^{-x} & x > 0 \end{vmatrix}
\tag{9.4.19}
$$

The moments of the standardized Laplace distribution are given simply by

$$
\mu_n = n!
\tag{9.4.20}
$$

Therefore, the moments considered in Table 9.4.1 are especially simple, $m = 0$, $\sigma^2 = 2$, $\gamma_1 = 0$, $\gamma_2 = 6$.

Visualization

We show standardized Laplace distributions as a function of x for the density distribution $L(x)$ in Figure 9.4.12 and for the cumulative distribution $C(x)$ in Figure 9.4.13.

Laplace distribution: density

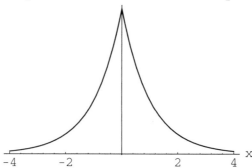

FIGURE 9.4.12 Standardized Laplace density distribution (9.4.18). (Cell `LaplacePDF` in *Mathematica* notebook `ocPDF`.)

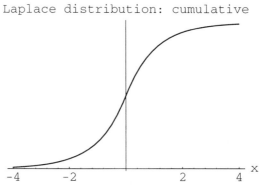

Laplace distribution: cumulative

FIGURE 9.4.13 Standardized Laplace cumulative distribution (9.4.19). (Cell `Laplace-PDF` in *Mathematica* notebook `ocPDF`.)

Algorithms and Programs

The programs for the Laplace distribution functions are coded directly from the definitions, (9.4.16) for `LaplaceDnsty` and (9.4.15) for `LaplaceCum`.

```
double LaplaceDnsty(double x)
/*  Laplace distribution density */
{
return    0.5*exp(-fabs(x));
}
```

```
double LaplaceCum(double x)
/*  Laplace distribution cumulative */
{
if ( x <= 0 )      return  0.5*exp(x);
return    1.0 - 0.5*exp(-x);
}
```

Test Values

Test values for the Laplace density distribution $L(x)$ and the cumulative distribution $C(x)$ can be generated by running the cell `LaplacePDF` in *Mathematica* notebook `ocPDF`, or by using a pocket calculator. The numerical values given in Table 9.4.8 indicate that $C(x) = L(x)$ for $x \le 0$ and $C(x) = 1 - L(x)$ for $x > 0$.

M

TABLE 9.4.8 **Test values to 10 digits of the Laplace distributions (9.4.18) and (9.4.19).**

x	−2.0	0	2.0
$L(x)$	0.06766764162	0.5000000000	0.06766764162
$C(x)$	0.06766764162	0.5000000000	0.9323323584

9.4.7 Kolmogorov-Smirnov Distribution

Keywords

Kolmogorov-Smirnov distribution, distribution, Kolmogorov-Smirnov

C Functions and *Mathematica* Notebook

`KSDnsty`, `KSQCum` (C); cell `KolmogorovSmirnovPDF` in notebook `ocPDF` (*Mathematica*)

M

Background

The Kolmogorov-Smirnov distribution—unlike other distributions in this chapter—does not depend upon assumptions about the probability distributions to which it is applied. The analysis of this distribution is described in Section 33.2 of Johnson and Kotz [Joh70] and extensive references are given in Section 33.1 of Johnson et al [Joh94b]. Applications are emphasized in Section 14.3 of Press et al [Pre92], where its use to estimate the probability that two distributions have different cumulative probability distributions is explained.

Function Definition

The Kolmogorov-Smirnov density distribution $K(x)$ is given by

$$K(x) = 8x \sum_{j=1}^{\infty} (-1)^{j-1} j^2 e^{-2j^2 x^2} \tag{9.4.21}$$

The Q-cumulative distribution $Q_{KS}(x)$ is then given by

$$Q_{KS}(x) = 2 \sum_{j=1}^{\infty} (-1)^{j-1} e^{-2j^2 x^2} \tag{9.4.22}$$

$$Q_{KS}(0) = 1 \qquad Q_{KS}(\infty) = 0$$

Moments of the Kolmogorov-Smirnov distribution about the origin, μ_n', can be obtained in terms of the Riemann zeta function in Section 8.1, $\zeta(n)$, as

$$\mu_1' = \sqrt{\frac{\pi}{2}} \ln 2$$
$$\mu_n' = \frac{\Gamma(n/2+1)\left(1-2^{1-n}\right)\zeta(n)}{2^{n/2-1}} \qquad n > 1 \tag{9.4.23}$$

By using these values to calculate central moments, we obtain the numerical values of the moment quantities given in Table 9.4.1.

Visualization

We show Kolmogorov-Smirnov distributions as a function of x for the density distribution $K(x)$ and the cumulative distribution $Q_{KS}(x)$ in Figure 9.4.14. The curves start at $x = 0.1$, since the convergence of the series in (9.4.21) and (9.4.22) is slow for x less than this. See the discussion in Programming Notes below.

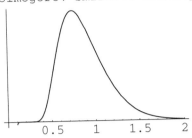
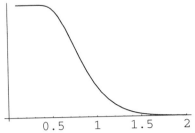

FIGURE 9.4.15 Kolmogorov-Smirnov density distribution (9.4.21) and Q cumulative distribution (9.4.22). The curves start at $x = 0.1$. (Cell `KolmogorovSmirnovPDF` in *Mathematica* notebook `ocPDF`.)

Algorithms and Programs

The programs for the Kolmogorov-Smirnov distribution functions—coded from the definitions (9.4.21) for `KSDnsty` and (9.4.22) for `KSQCum`—are on the following page.

```
double KSDnsty(double x, double eps)
/*  Kolmogorov-Smirnov distribution density;
    fractional accuracy  eps  */
{
double mtxSqd,sum,phase,term,fj,fjSqd;

/* If x is too small for series to
   converge, just quit */
if ( fabs(x) < eps )  return 0.0;

/* Else sum the series */
mtxSqd = - 2.0*x*x;
fj = 1.0;
sum = exp(mtxSqd);
term = sum;
phase = 1.0;
while ( fabs(term/sum) > eps )
  {
  fj += 1.0;
  fjSqd = fj*fj;
  phase = - phase;
  term = phase*fjSqd*exp(mtxSqd*fjSqd);
  sum += term;
  }
return 8*x*sum;
}
```

```
double KSQCum(double x, double eps)
/*  Kolmogorov-Smirnov Q distribution cumulative;
    fractional accuracy  eps  */
{
double mtxSqd,sum,phase,term,fj;

/* If   x  is too small for series to
   converge, just quit */
if ( fabs(x) < eps )  return 1.0;

/* Else sum the series */
mtxSqd = - 2.0*x*x;
fj = 1.0;
sum = exp(mtxSqd);
term = sum;
phase = 1.0;
while ( fabs(term/sum) > eps )
  {
  fj += 1.0;
  phase = - phase;
  term = phase*exp(mtxSqd*fj*fj);
  sum += term;
  }
return 2*sum;
}
```

Programming Notes

Both series require checks for convergence. We use the input variable `eps` to specify the relative accuracy of the last term that was included in the sum to the partial sum. For most x values that are relevant for applications (x about unity), one does somewhat better than accuracy `eps`, since the series is then rapidly convergent. In either series, for $x > 1$ the fractional accuracy is better than 1% (`eps` = 0.01), even if only the first term is used.

Care needs to be taken in programming (9.4.21) and (9.4.22) to take account of the slow convergence of the series when x is very small. For example, we use the floating-point representation of j rather than the obvious integer representation, in order to extend the largest j value that can be used in the series. Nevertheless, since the series is alternating it is difficult to ensure high relative accuracy. Therefore, if $x <$ `eps` the functions return the values for $x = 0$, namely zero or unity.

Test Values

Check values for the Kolmogorov-Smirnov density distribution $K(x)$ and its cumulative distribution $Q_{KS}(x)$ can be generated by running the cell `KolmogorovSmirnovPDF` in *Mathematica* notebook `ocPDF`. Numerical values are given in Table 9.4.9.

TABLE 9.4.9 Test values to 10 digits of the Kolmogorov-Smirnov distributions (9.4.21) and (9.4.22).

x	0.5	1.0	1.5
$K(x)$	0.6395828509	1.0719548558	0.1333072274
$Q_{KS}(x)$	0.9639452437	0.2699996717	0.0222179626

9.4.8   Beta Distribution

Keywords

distribution, beta

C Functions and *Mathematica* Notebook

`BetaDnsty`, `BetaCum` (C); cell `BetaPDF` in notebook `ocPDF` (*Mathematica*)

Background

The beta distribution is described from its probability and statistics aspects in Chapter 25 of Johnson et al [Joh94b], whereas formulas are emphasized in Section 26.5 of Abramowitz and Stegun [Abr64]. The F distribution (Section 9.3.4) uses essentially the same function, but its parameters and arguments are sufficiently different from those of the beta function that we describe the latter separately here. See also Section 6.3.2 for the regularized incomplete beta function.

Function Definition

The beta density distribution $B_{p,q}(x)$ is defined by

$$B_{p,q}(x) \equiv \frac{x^{p-1}(1-x)^{q-1}}{B(p,q)} \qquad p > 0 \quad q > 0 \qquad 0 \le x \le 1 \qquad (9.4.24)$$

where $B(p,q)$ is the beta function in Section 6.1.2. The cumulative distribution $C_{p,q}(x)$ is given by

$$C_{p,q}(x) = \int_0^x \frac{t^{p-1}(1-t)^{q-1}}{B(p,q)}\, dt = I_x(p,q) \tag{9.4.25}$$

in which $I_x(p,q)$ is the incomplete beta function (Section 6.3.2).

Moments of the beta distribution about the origin, μ_n', can be obtained in terms of the gamma function in Section 6.1.1 as

$$\mu_n' = \frac{\Gamma(p+n)\,\Gamma(p+q)}{\Gamma(p)\,\Gamma(p+q+n)} \tag{9.4.26}$$

By using these values to calculate central moments, we can obtain the following expressions for the mean, variance, skewness, and excess, which are all defined in Section 9.1:

$$m = \frac{p}{p+q} \qquad \sigma^2 = \frac{pq}{(p+q)^2(p+q+1)} \qquad \gamma_1 = \frac{2(p-q)}{p+q+2}$$

$$\gamma_2 = \sqrt{\frac{p+q+1}{pq}}\left\{\frac{3(p+q+1)\left[2(p+q)^2 + pq(p+q-6)\right]}{pq(p+q+2)(p+q+3)} - 3\right\} \tag{9.4.27}$$

For example, when $q = p$ one obtains

$$m = \frac{1}{2} \qquad \sigma^2 = \frac{1}{4(2p+1)}$$

$$\gamma_1 = 0 \qquad \gamma_2 = \frac{-6\sqrt{2p+1}}{p(2p+3)} \tag{9.4.28}$$

Therefore, the mean is at $x = 0.5$ for any $q = p$, the width of the distribution decreases as p increases, and it is symmetric about the mean. Its excess is always negative and decreases with increasing p.

Visualization

We show beta distributions as function of q and x for fixed p, for the density distribution $B_{p,q}(x)$ in Figure 9.4.15 and for the cumulative distribution $C_{p,q}(x)$ in Figure 9.4.16.

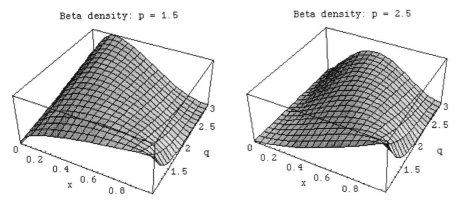

FIGURE 9.4.15 Beta density distribution (9.4.24) for $p = 1.5$ (left panel) and $p = 2.5$ (right panel). (Cell BetaPDF in *Mathematica* notebook ocPDF.)

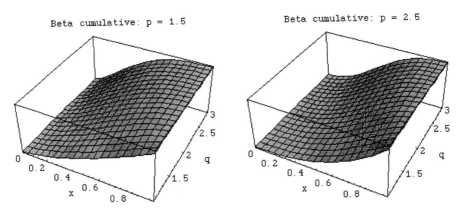

FIGURE 9.4.16 Beta cumulative distribution (9.4.25) for $p = 1.5$ (left panel) and $p = 2.5$ (right panel). (Cell BetaPDF in *Mathematica* notebook ocPDF.)

Algorithms, Programs, and Programming Notes

The programs for the beta distribution functions are coded from the definitions, (9.4.24) for BetaDnsty and (9.4.25) for BetaCum. The argument eps input by the function user is passed to the incomplete beta function, IncBeta, and it determines the fractional accuracy of the result. For about 6 significant digits of accuracy, use eps about 10^{-6}.

```
double BetaDnsty
  (double p, double q, double x)
/* Beta Density Distribution */
{
double LogGamma(double x),betaFact;

betaFact = exp(LogGamma(p+q) -
       LogGamma(p) - LogGamma(q));

returnbetaFact*pow(x,p - 1.0)*
       pow(1.0 - x,q - 1.0);
}
```

```
double BetaCum
  (double p, double q, double x, double eps)
/* Beta Cumulative Distribution */
{
double IncBeta
  (double a, double b, double x, double eps);

return IncBeta(p,q,x,eps);
}
```

Test Values

Check values for the beta density distribution $B(x)$ and its cumulative distribution $C(x)$ can be generated by running the cell `BetaPDF` in *Mathematica* notebook `ocPDF`. Numerical values are given in Tables 9.4.10 and 9.4.11.

M

TABLE 9.4.10 Test values to 10 digits of beta density distribution (9.4.24) for $q = 2$.

x	0.25	0.50	0.75
$p = 1.5$	1.406250000	1.325825215	0.8118988160
$p = 2.5$	0.8203125000	1.546796084	1.420822928

TABLE 9.4.11 Test values to 10 digits of beta cumulative distribution (9.4.25) for $q = 2$.

x	0.25	0.50	0.75
$p = 1.5$	0.2656250000	0.6187184335	0.8930886977
$p = 2.5$	0.0898437500	0.3977475644	0.7916013456

REFERENCES ON PROBABILITY DISTRIBUTION FUNCTIONS

[Abr64]   Abramowitz, M., and I. A. Stegun, *Handbook of Mathematical Functions,* Dover, New York, 1964.

[Bal92]   Balakrishnan, N., (ed.) *Handbook of the Logistic Distribution*, Dekker, New York, 1992.

[Fel68]   Feller, W., *An Introduction to Probability Theory and Its Applications*, Wiley, New York, third edition, 1968.

[Hai67]   Haight, F. A., *Handbook of the Poisson Distribution*, Wiley, New York, 1967.

[Hos89]   Hosmer, D. W., and S. Lemeshaw, *Applied Logistic Regression*, Wiley, New York, 1989.

[Joh70]   Johnson, N. L., and S. Kotz, *Distributions in Statistics: Continuous Univariate Distributions-2*, Houghton Mifflin, Boston, 1970.

[Joh92]   Johnson, N. L., S. Kotz, and A. W. Kemp, *Univariate Discrete Distributions*, Wiley, New York, second edition, 1992.

[Joh94a]   Johnson, N. L., S. Kotz, and N. Balakrishnan, *Distributions in Statistics: Continuous Univariate Distributions-1*, Wiley, New York, second edition, 1994.

[Joh94b]   Johnson, N. L., S. Kotz, and N. Balakrishnan, *Distributions in Statistics: Continuous Univariate Distributions-2*, Wiley, New York, second edition, 1994.

[Pre92]   Press, W. H., S. A. Teukolsky, W. T. Vetterling, and B. P. Flannery, *Numerical Recipes*, Cambridge University Press, New York, second edition, 1992.

[Stu94]   Stuart, A., and J. K. Ord,, *Kendall's Advanced Theory of Statistics*, Vol. 1, Edward Arnold, London, sixth edition, 1994.

[Tho92]   Thompson, W. J., *Computing for Scientists and Engineers*, Wiley, New York, 1992.

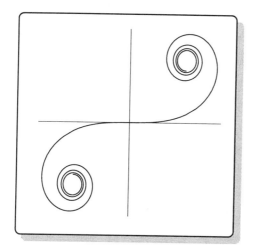

Chapter 10

ERROR FUNCTION,
FRESNEL AND DAWSON INTEGRALS

The error function, Fresnel's integrals, and Dawson's integral are interrelated, since the latter functions can be computed in terms of error functions with complex arguments. They are considered together here, just as in Chapter 7 of the handbook of mathematical functions edited by Abramowitz and Stegun [Abr64].

10.1   ERROR FUNCTION

Keywords
error function, normal distribution, cumulative distribution

C Function and *Mathematica* Notebook
ErfReal (C); cell Erf in notebook ErfFD (*Mathematica*)

Background
The error function, $\mathrm{erf}(z)$, in which z is a complex variable, provides the starting point for computing—analytically and numerically—many other special functions, such as the Fresnel and Dawson integrals in Sections 10.2 and 10.3. Its name arises because it is related simply to the cumulative distribution function for the Gauss (normal) distribution discussed in Section 9.3.1. The relation is given by (10.1.3) below.

The analysis of $\mathrm{erf}(z)$ is discussed in many texts on probability theory, in Chapter 3 of the handbook of functions by Jahnke and Emde [Jah45], in Section 9.9 of Erdélyi et al [Erd53b], and in Section 7.1 of Abramowitz and Stegun [Abr64].

Function Definition

We give the definition of the error function, $\text{erf}(z)$, in terms of a complex variable, z, in order to clarify the connections to other functions and to the visualizations. The error function is defined by

$$\text{erf}(z) \equiv \frac{2}{\sqrt{\pi}} \int_0^z e^{-t^2} dt \qquad (10.1.1)$$

Notice immediately the symmetry relation

$$\text{erf}(-z) = -\text{erf}(z) \qquad (10.1.2)$$

For the standardized Gauss probability density distribution, $G(x)$ given by (9.3.3), one has a cumulative distribution, $\Phi(x)$, given by

$$\Phi(x) \equiv \int_{-\infty}^x G(x)\,dx = \frac{1}{\sqrt{2\pi}} \int_0^z e^{-t^2/2} dt = \frac{1}{2} + \frac{1}{2}\text{erf}(x/\sqrt{2}) \qquad (10.1.3)$$

Since for a normalized distribution $\Phi(x)$ must approach unity for large x, this convinces us that $\text{erf}(x)$ has the same limit, as does $\text{erf}(z)$. Similarly, if (real) x in $z = x + iy$ is negative but large in magnitude we have $\text{erf}(z) \rightarrow -1$.

Successive derivatives of the error function of real argument are related to the Hermite polynomials, $H_n(x)$ in Section 11.5, by

$$\frac{d^{n+1}\text{erf}(x)}{dx^{n+1}} = \frac{(-1)^n 2H_n(x)e^{-x^2}}{\sqrt{\pi}} \qquad (10.1.4)$$

Visualization

The real and imaginary parts of $\text{erf}(z)$ for argument $z = x + iy$ are shown as the surfaces in Figure 10.1.1.

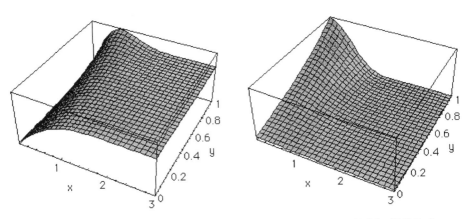

FIGURE 10.1.1 Error function, with real part (left) and imaginary part (right). (Cell `Erf` in *Mathematica* notebook `ErfFD`.)

Note in the left panel that the curve on the front surface at $y = 0$ is just the error function as used in probability—apart from the rescaling of x that is indicated in (10.1.3). Figure 10.1.2 shows $\text{erf}(x)$ in more detail.

erf(x)

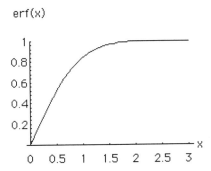

FIGURE 10.1.2 The error function for real argument x. (Cell Erf in *Mathematica* notebook ErfFD.)

Algorithm, Program, and Programming Notes

Most applications of the error function restrict its argument to special trajectories in the complex plane then they develop appropriate formulas and algorithms. For example, the Fresnel and Dawson integrals (Sections 10.2 and 10.3) are treated in this way. Therefore, this section provides a program for computing erf(x) to moderate accuracy. If erf(z) is needed for some arbitrary z, modify and run the Erf cell in the *Mathematica* notebook   **M**
ErfFD.

As algorithm for erf(x), we use the relation to the regularized incomplete gamma function $P(a,x)$ in Section 6.3, namely,

$$\text{erf}(x) = \begin{vmatrix} P(1/2, x^2) & x > 0 \\ -P(1/2, x^2) & x < 0 \end{vmatrix} \qquad (10.1.5)$$

Note that relation 6.5.16 in Abramowitz and Stegun [Abr64] is incomplete, since it ignores   **!**
the sign change for negative x that is required by (10.1.2).

```
double ErfReal(double x, double eps)
{
/* Error function of real argument  x
   to accuracy  eps  */

double IncGamma(double a, double x, double eps);

/* Erf is a regularized incomplete gamma function */

if ( x > 0 )  return  IncGamma(0.5, x*x, eps);
return  - IncGamma(0.5, x*x, eps);
}
```

The fractional accuracy of ErfReal is controlled by input parameter eps. Its use is illustrated by the driver program for the error function in Section 21.10.

Test Values

Check values of erf(x) can be generated by running the Erf cell in *Mathematica* notebook   **M**
ErfFD. Examples are given in Table 10.1.1.

TABLE 10.1.1 Test values to 10 digits of the error function for real argument x.

x	0.5	1.0	1.5
erf(x)	0.5204998778	0.8427007929	0.9661051465

10.2   FRESNEL INTEGRALS

Keywords
Fresnel cosine integral, Fresnel integrals, Fresnel sine integral

C Functions and *Mathematica* Notebook

 FresnelCS (C); cells Fresnel and FresnelApprox in the *Mathematica* notebook ErfFD

Background
The Fresnel integrals often arise when analyzing the of diffraction of waves. Because Fresnel integrals can be defined formally in terms of error functions with complex arguments, they are included in this chapter.

   The analysis of Fresnel integrals is presented in Section III.4 of Jahnke and Emde's handbook of functions [Jah45], in Section 9.10 of [Erd53b], in Chapter 39 of Spanier and Oldham [Spa87], in Section 7.1 of Abramowitz and Stegun [Abr64], and in Section 8.25 in Gradshteyn and Ryzhik [Gra94]. Beware that definitions of the Fresnel integrals are not consistent between various sources, as summarized in Table 10.2.1 following.

Function Definition
We use the definition of the Fresnel cosine integral, $C(z)$, and of the Fresnel sine integral, $S(z)$, in terms of complex variable, z, as in Abramowitz and Stegun, 7.3.1 and 7.3.2, respectively:

$$C(z) \equiv \int_0^z \cos(\pi t^2 / 2)\, dt \qquad S(z) \equiv \int_0^z \sin(\pi t^2 / 2)\, dt \qquad (10.2.1)$$

These functions have the symmetry relations

$$C(-z) = -C(z) \qquad S(-z) = -S(z) \qquad (10.2.2)$$

The following two pairs of functions are also called Fresnel integrals:

$$C_1(x) \equiv \sqrt{\frac{2}{\pi}} \int_0^x \cos(t^2)\, dt = C\left(\sqrt{\frac{2}{\pi}}\, x\right)$$
$$S_1(x) \equiv \sqrt{\frac{2}{\pi}} \int_0^x \sin(t^2)\, dt = S\left(\sqrt{\frac{2}{\pi}}\, x\right) \qquad (10.2.3)$$

$$C_2(x) \equiv \frac{1}{\sqrt{2\pi}} \int_0^x \frac{\cos(t)}{\sqrt{t}}\, dt = C\left(\sqrt{\frac{2x}{\pi}}\right)$$
$$S_2(x) \equiv \frac{1}{\sqrt{2\pi}} \int_0^x \frac{\sin(t)}{\sqrt{t}}\, dt = S\left(\sqrt{\frac{2x}{\pi}}\right) \qquad (10.2.4)$$

   Unfortunately, the symbols C and S are sometimes also used for these functions or for quantities proportional to them. Table 10.2.1 may be used to clarify the interrelations.

TABLE 10.2.1 Relations between notations for Fresnel integrals.

Source	Fresnel integrals defined as	Relation of source notation (left) to present notation (right)
Abramowitz and Stegun [Abr64]	$C(z)$, $S(z)$ as in (10.2.1)	same
Erdélyi et al [Erd53b], Section 9.10	$C(x)$, $S(x)$	$C(x) = C_2(x)$ $S(x) = S_2(x)$
Jahnke and Emde [Jah45], Section III.4	$C(z)$, $S(z)$	$C(z) = C_2(z)$ $S(z) = S_2(z)$
Maple [Cha92], Sections 2.1.129, 2.1.130	$C(x)$, $S(x)$	same
Mathematica [Wol91], Section 3.2	$C(z)$, $S(z)$	same
Spanier and Oldham [Spa87], Section 39.2	$C(x)$, $S(x)$	$C(x) = C_2(x)$ $S(x) = S_2(x)$

Visualization

For purposes of visualization, it is most interesting to relate the Fresnel cosine and sine integrals—$C(z)$ and $S(z)$ in (10.2.1)—to the error function (10.1.1). The relation is

$$C(z) + iS(z) = \frac{1+i}{2}\, \mathrm{erf}\left[\frac{\sqrt{\pi}\,(1-i)z}{2}\right] \tag{10.2.5}$$

This enables us to use in the `Fresnel` cell of *Mathematica* notebook `ErfFD` the function `Erf[]` with a complex argument appropriate to (10.2.5). When $z = x$, purely real, the real part of the resulting object will be the cosine integral and the imaginary part will be the sine integral, as shown in Figures 10.2.1 and 10.2.2.

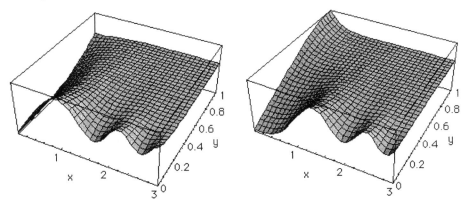

FIGURE 10.2.1 Fresnel integrals in the complex plane, showing real part (left) and imaginary part (right). The front curves are the cosine integral (left) and sine integral (right). (Cell `Fresnel` in *Mathematica* notebook `ErfFD`.)

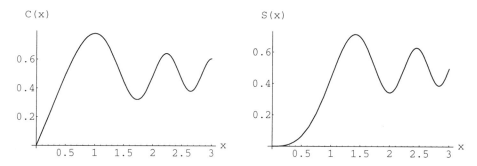

FIGURE 10.2.2 Fresnel integrals on the real axis, cosine integral (left) and sine integral (right). (Cell `Fresnel` in *Mathematica* notebook `ErfFD`.)

From these figures we see that both $C(x)$ and $S(x)$ asymptote to the value 1/2. This behavior becomes even more pronounced if we use the Fresnel integrals to define Cartesian coordinates parametrically by

$$x = C(t) \qquad y = S(t) \tag{10.2.6}$$

The resulting Cornu spiral—named after the French physicist Cornu (1841–1902) who studied optical diffraction—is generated from this pair of parametric equations. It can be computed and displayed by running the last part of the `Fresnel` cell in *Mathematica* notebook `ErfFD`. In Figure 10.2.3 the range of *t* is [−4, 4]. The two points of convergence of the spiral are $(x, y) = (\pm 1/2, \pm 1/2)$. This spiral can be compared to the Sici spiral in Section 5.2.

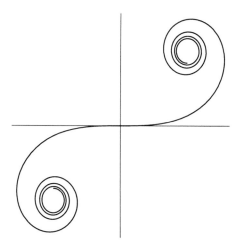

FIGURE 10.2.3 The Cornu spiral, obtained from Fresnel integrals by (10.2.6). (Cell `Fresnel` in *Mathematica* notebook `ErfFD`.)

Algorithm and Program

Most uses of the Fresnel integrals restrict their arguments to *z* being real, that is, to Figure 10.2.2 rather than Figure 10.2.1. Our function computes $C(x)$ and $S(x)$ for positive *x* with accuracy of at least 10 digits. For $x < 0$ the signs of the integrals are reversed, according to (10.2.2). If either integral is required for some arbitrary z, modify and run the `Fresnel` cell in *Mathematica* notebook `ErfFD`.

We use as algorithm for computing $C(x)$ and $S(x)$ three methods, depending upon the range of x. For x in $(0, 1)$ we use the power series 7.3.11 and 7.3.13 in [Abr64], written as

$$C(x) = x \sum_{n=0}^{\infty} \frac{t^n}{(2n)!\,(4n+1)} \qquad t = -\left(\frac{\pi x^2}{2}\right)^2 \qquad (10.2.7)$$

for the Fresnel cosine integral and as

$$S(x) = \frac{\pi x^3}{2} \sum_{n=0}^{\infty} \frac{t^n}{(2n+1)!\,(4n+3)} \qquad t = -\left(\frac{\pi x^2}{2}\right)^2 \qquad (10.2.8)$$

for the Fresnel sine integral. For $x < 1$, only about 10 terms are needed for fractional accuracy ε of about 10^{-10}.

As x increases, both $C(x)$ and $S(x)$ oscillate about 0.5 with decreasing amplitudes, as shown in Figure 10.2.2. It is therefore useful to define auxiliary functions for the Fresnel integrals, f and g, from which the oscillatory behavior has been removed. These functions are defined as in [Abr64, 7.3.5, 7.3.6] by

$$f(x) = [1/2 - S(x)]\cos(\pi x^2/2) - [1/2 - C(x)]\sin(\pi x^2/2) \qquad (10.2.9)$$

$$g(x) = [1/2 - C(x)]\cos(\pi x^2/2) + [1/2 - S(x)]\sin(\pi x^2/2) \qquad (10.2.10)$$

From these relations $C(x)$ and $S(x)$ can be obtained as

$$\begin{aligned} C(x) &= 1/2 + f(x)\sin(\pi x^2/2) - g(x)]\cos(\pi x^2/2) \\ S(x) &= 1/2 - f(x)\cos(\pi x^2/2) - g(x)]\sin(\pi x^2/2) \end{aligned} \qquad (10.2.11)$$

The auxiliary functions are shown in Figure 10.2.4. Note the approximately hyperbolic behavior of $f(x)$ and the rapid decrease of $g(x)$—approximately as $1/x^3$ according to (10.2.13) below—as x increases.

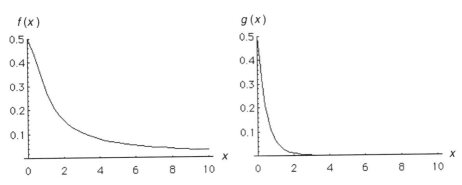

FIGURE 10.2.4 Auxiliary functions for Fresnel integrals, f (left) and g (right) as functions of x. (Adapted from cell `FresnelApprox` in *Mathematica* notebook `ErfFD`.)

For x in $[1, 6)$, we use the smooth x dependence of f and g to develop rational-fraction approximations to the auxiliary functions in this interval. Quotients of polynomials whose numerators are 10th order in x and whose denominators are of 11th order are used. The details of this are carried out by using the *Mathematica* function `RationalInterpolation`, implemented in cell `FresnelApprox` in notebook `ErFD`. The rational-fraction approximations—with coefficients given in function `FresnelCS` by arrays `fn[]` and `fd[]` for f and by `gn[]` and `gd[]` for g—are accurate to fractional accuracy of at least 15 digits. The Fresnel integrals are then computed from (10.2.11).

For $x \geq 6$, the asymptotic expansions of f and g as given by [Abr64, 7.3.27, 7.3.28] are used. Namely

$$\pi x f(x) \sim 1 + \sum_{m=1}^{\infty} 1.3. \ldots (4m-1) t^m \qquad t = -(\pi x^2)^{-2} \qquad (10.2.12)$$

for auxiliary function f, and

$$\pi^2 x^3 g(x) \sim 1 + \sum_{m=1}^{\infty} 1.3. \ldots (4m+1) t^m \qquad t = -(\pi x^2)^{-2} \qquad (10.2.13)$$

for g. These expansions converge to fractional accuracy better than 10^{-12} after about 6 terms for x in the indicated range.

The C code for function `FresnelCS` that returns $C(x)$ and $S(x)$ as arguments for x real and positive is as follows.

```
void FresnelCS(double x,double eps,
  double *FresnelC, double *FresnelS)
{
/* Fresnel Cosine & Sine Integrals of real argument
   by power series (to accuracy  eps ), rational
   approximation, and asymptotic expansion */

double t,twofn,fact,denterm,numterm,term,sum,ratio,
  sumn,sumd,f,g,U,SinU,CosU,eps10,oldterm,absterm;

const double Pi = 3.14159265359,PiD2 = 1.57079632679,

fn[] = {0.49999988085884732562,1.3511177791210715095,
  1.3175407836168659241,1.1861149300293854992,
  0.7709627298888346769,0.4173874338787963957,
  0.19044202705272903923,0.066559988966627697537,
  0.022789258616785717418,0.0040116689358507943804,
  0.0012192036851249883877},

fd[] = {1.0,2.7022305772400260215,
  4.2059268151438492767,4.5221882840107715516,
  3.7240352281630359588,2.4589286254678152943 ,
  1.3125491629443702962,0.5997685720120932908,
  0.2090768075037 8849485,0.0715962163465 7901433,
  0.012602969513793714191,0.0038302423512931250065},

gn[] =
{0.50000014392706344801,0.032346434925349128728,
  0.17619325157863254363,0.038606273170706486252,
  0.023693692309257725361,0.0070920185168450 33662,
  0.0012492123212412087428,0.00044023040894778468486,
  -8.0266827476172521e-6,-1.4033554916580018648e-8,
  2.3509221782155474353e-10},
```

```
  gd[] =
  {1.0,2.0646987497019598937,2.9109311766948031235,
   2.6561936751333032911,2.0195563983177268073,
   1.1167891129189363902,0.5726787747559731727 15,
   0.1940848116959307 0798,0.0763480834143 1248904,
   0.0115732474072078659 77,0.004409927369306731 1209,
   -0.0000907095841042999331 4};

  int k;

  if ( x < 1.0 )
   {
   t = - pow(PiD2*x*x,2.0);
   /* Cosine integral series */
   twofn = 0.0;
   fact = 1.0;
   denterm = 1.0;
   numterm = 1.0;
   sum = 1.0;
   ratio = 10.0;

  while ( ratio > eps )
    {
    twofn += 2.0;
    fact *= twofn*(twofn-1.0);
    denterm += 4.0;
    numterm *= t;
    term = numterm/(fact*denterm);
    sum += term;
    ratio = fabs(term/sum);
    }
  *FresnelC =  x*sum;

    /* Sine integral series */
    twofn = 1.0;
    fact = 1.0;
    denterm = 3.0;
    numterm = 1.0;
    sum = 1.0/3.0;
    ratio = 10.0;

  while ( ratio > eps )
    {
    twofn += 2.0;
    fact *= twofn*(twofn-1.0);
    denterm += 4.0;
    numterm *= t;
    term = numterm/(fact*denterm);
    sum += term;
    ratio = fabs(term/sum);
    }
  *FresnelS =  PiD2*x*x*x*sum;
   }
```

```
else
  {
  if ( x < 6.0 )
  { /* Rational approximation for  f    */
  sumn =  0.0;
  sumd =  fd[11];
  for ( k = 10; k >= 0; k-- )
    {
    sumn = fn[k]+x*sumn;
    sumd = fd[k]+x*sumd;
    }
  f = sumn/sumd;
  /* Rational approximation for  g    */
  sumn =  0.0;
  sumd =  gd[11];
  for ( k = 10; k >= 0; k-- )
    {
    sumn = gn[k]+x*sumn;
    sumd = gd[k]+x*sumd;
    }
  g = sumn/sumd;
  U = PiD2*x*x;
  SinU = sin(U);
  CosU = cos(U);
  *FresnelC = 0.5+f*SinU-g*CosU;
  *FresnelS = 0.5-f*CosU-g*SinU;
  }
  else
    {
    /* x >= 6; asymptotic expansions for  f  and  g */
    t = - pow(Pi*x*x,-2.0);
    /* Expansion for  f    */
    numterm = -1.0;
    term = 1.0;
    sum = 1.0;
    oldterm = 1.0;
    ratio = 10.0;
    eps10 = 0.1*eps;
    while ( ratio > eps10 )
      {
      numterm += 4.0;
      term *= numterm*(numterm-2.0)*t;
      sum += term;
      absterm = fabs(term);
      ratio = fabs(term/sum);
      if ( oldterm < absterm )
        {
        printf
          ("\n\n!!In FresnelCS f not converged to eps");
        ratio = eps10;
        }
      oldterm = absterm;
      }
    f = sum/(Pi*x);
```

```
     /* Expansion for  g   */
    numterm = -1.0;
     term = 1.0;
     sum = 1.0;
     oldterm = 1.0;
     ratio = 10.0;
     eps10 = 0.1*eps;

     while ( ratio > eps10 )
       {
        numterm += 4.0;
        term *= numterm*(numterm+2.0)*t;
        sum += term;
        absterm = fabs(term);
        ratio = fabs(term/sum);
        if ( oldterm < absterm )
          {
          printf
           ("\n\n!!In FresnelCS g not converged to eps");
          ratio = eps10;
          }
         oldterm = absterm;
       }

     g = sum/(pow(Pi*x,2.0)*x);
     U = PiD2*x*x;
     SinU = sin(U);
     CosU = cos(U);
     *FresnelC = 0.5+f*SinU-g*CosU;
     *FresnelS = 0.5-f*CosU-g*SinU;
     }
    }
  }
```

Programming Notes

Since one often needs to compute both $C(x)$ and $S(x)$ for the same argument, their computation is combined in `FresnelCS` and their values are passed to the calling routine as `FresnelS` and `FresnelC`. Look at the driver program in Section 21.10 to see how this is implemented.

In the rational-fraction expansion region of x, $[1, 6)$, we use the Horner polynomial algorithm to accumulate the terms of the numerator and denominator polynomials iteratively, starting from the highest term—in these cases $k = 11$. This algorithm is efficient, since it avoids direct computation of powers of x. Coding of the Horner algorithm is explicit in the `for` loops.

Algorithms for computing the Fresnel integrals to about 16 significant figures have been implemented by Van Snyder [Van93].

Test Values

Check values of $C(x)$ and of $S(x)$ for the three ranges of x over which different methods are used can be obtained by running `Fresnel` cell in *Mathematica* notebook `ErfFD`. The results are given in Table 10.2.1.

M

TABLE 10.2.1 Test values to 10 digits of the Fresnel cosine and sine integrals in (10.2.1) for positive real argument x.

x	0.2	0.4	0.6	0.8
$C(x)$	0.199210576	0.3974807592	0.5810954470	0.7228441719
$S(x)$	0.004187609	0.0333594327	0.1105402074	0.2493413931

x	2.0	3.0	4.0	5.0
$C(x)$	0.4882534061	0.6057207893	0.4984260330	0.5636311887
$S(x)$	0.3434156784	0.4963129990	0.4205157542	0.4991913819

x	6.0	10.0	14.0	18.0
$C(x)$	0.4995314679	0.4998986942	0.4996307680	0.4999826269
$S(x)$	0.4469607612	0.4681699786	0.4772637594	0.4823161686

10.3   DAWSON INTEGRAL

Keywords
Dawson integral, error function

C Function and *Mathematica* Notebook
FDawson (C); cell Dawson in *Mathematica* notebook ErfFD

Background
Dawson's integral arises in diverse areas of science and engineering, such as when computing the Voigt (plasma dispersion) function in Section 19.6. Since it can formally be defined in terms of an error function with complex argument, it is included in this chapter.

Analysis of the Dawson integral is presented in Chapter 42 of Spanier and Oldham [Spa87]. It is mentioned only in passing in Section 7.1 of Abramowitz and Stegun [Abr64], but an extensive table of numerical values (Table 7.5) and references are provided. The Dawson integral is discussed at an introductory level in Sections 10.3 and 10.4 of Thompson's text on computing [Tho92] and a method for computing it is given in Section 6.10 of Press et al [Pres92].

Function Definition
The Dawson integral, $F(x)$, is defined by

$$F(x) \equiv e^{-x^2} \int_0^x e^{t^2}\, dt = \frac{i\sqrt{\pi}\,\mathrm{erf}(-ix)}{2\,e^{x^2}} \qquad (10.3.1)$$

By inspection, the integral has the symmetry relation

$$F(-x) = -F(x) \qquad\qquad (10.3.2)$$

so that it is sufficient to devise computing methods for $x > 0$ only. Limiting behaviors of $F(x)$ are readily derived—for example, following Exercise 10.40 in [Tho92]—in terms of the power series and the asymptotic expansion given as 42:6:1 and 42:6:6, respectively, in Spanier and Oldham [Spa87]:

$$F(x) = x \sum_{j=0}^{\infty} \frac{t^j}{(2j+1)!!} \qquad t = -2x^2 \qquad\qquad (10.3.3)$$

for small x, and the asymptotic expansion

$$F(x) \sim \frac{1}{2x}\left[1 + \sum_{j=1}^{\infty} \frac{(2j+1)!!}{t^j} \right] \qquad t = 2x^2 \qquad\qquad (10.3.4)$$

for large x. In (10.3.5) the tilde ($\sim$) indicates that the series does not converge. (Asymptotic expansions are discussed in Section 3.4.)

Visualization

To compute the Dawson integral for visualization, it is easiest to use its relation to the error function, given in (10.3.1). This enables the `Dawson` cell of *Mathematica* notebook `ErfFD` to use the function `Erf[]` with purely imaginary argument to produce Figure 10.3.1.

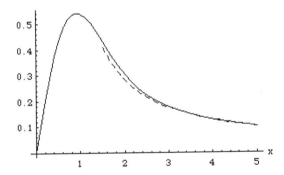

FIGURE 10.3.1 Dawson's integral (10.3.1) shown by the solid curve and the first two terms of the asymptotic approximation (10.3.5) shown by the dashed curve. (Cell `Dawson` in *Mathematica* notebook `ErfFD`.)

Algorithm, Program, and Programming Notes

For computing the Dawson integral, $F(x)$, in a C function, we use power series (10.3.3) when x is less than unity in magnitude. The asymptotic series is unreliable for achieving high accuracy for small values of x, as seen by examining the terms in the asymptotic series by running the `Dawson` cell in *Mathematica* notebook `ErfFD` for various values of x, as shown in Figure 10.3.2 following. For example, use of (10.3.5) with $x < 4$ is not appropriate even if accuracy only of order $\varepsilon = 10^{-6}$ is required.

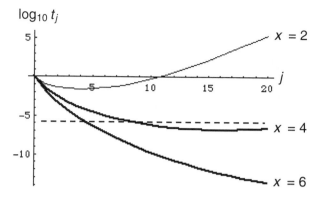

FIGURE 10.3.2 Terms of the asymptotic series (10.3.4) for the Dawson integral for $x = 2$, 4, and 6 on a logarithmic scale. The dashed line shows accuracy value, $\varepsilon = 10^{-6}$. (Adapted from output of cell Dawson in *Mathematica* notebook ErfFD.)

To compute $F(x)$ accurately for $|x| > 1$ we adapt a method described in Sections 6.10 and 13.11 of *Numerical Recipes* [Pre92], which uses an approximation due to Rybicki [Ryb89] that is based on Fourier expansions. Rybicki's result—in the form recommended by Press et al—is

$$F(x) \approx \frac{1}{\sqrt{\pi}} \sum_{n'=-N}^{N} \frac{e^{-(x'-n'h)^2}}{n'+n_0} \tag{10.3.5}$$

$$n_0 = [x/h] \quad x_0 = n_0 h \quad x' = x - x_0 \quad n' = n - n_0$$

which becomes more accurate as h approaches zero. The approximate centering of the summation interval improves the accuracy of the algorithm. If one requires fractional accuracy of order ε, then one can show in a straightforward way that N and h should be related by

$$Nh \approx \sqrt{x^2 - \ln \varepsilon} \tag{10.3.6}$$

In [Pre92] N is defined as 6 and h as 0.4, which gives low accuracy. By choosing N as 38 and using (10.3.6) to compute h, our C function FDawson gives accuracy of about 12 digits for x up to 10. For larger x values the asymptotic expansion (10.3.4) converges rapidly.

The program for the Dawson integral is thus:

```
double FDawson(double x, double eps)
{
/* Dawson Integral of real argument  x
   to fractional accuracy  eps  */

double OneDSqrtPi = 0.564189583548;
  /* is reciprocal of square root of Pi */
double  u,t,tjP1,term,ratio,eps10,
  H,twoH,d1,d2,e1,e2,sum,x2,xp;
int i,nzero;
```

```
/* Number of sum points in Rybicki series;
   this value good until  eps = 10^(-12)  */

const int  NMAX = 38;


if ( x < 0 )    return   - FDawson(-x,eps);


if ( x <= 1.0 ) /* use power series */
  {
  t = - 2.0*x*x;
  tjP1 = 3.0;
  term = 1.0;
  sum = 1.0;
  ratio = 10.0;
  eps10 = 0.1*eps;
  while ( ratio > eps10 )
    {
    term *= t/tjP1;
    sum += term;
    ratio = fabs(term/sum);
    tjP1 += 2.0;
    }
  return  x*sum;
  }


/* x > 1; use Rybicki expansion */
H = pow(x*x - log(eps),0.5)/( (double) NMAX );
twoH = 2.0*H;
term = - H;
/* Series is centered near zero exponent */
nzero = 2*(int)(0.5*x/H+0.5);
xp = x - nzero*H;
e1 = exp(2.0*xp*H);
e2 = e1*e1;
d1 = nzero+1;
d2 = d1 - 2.0;
sum = 0.0;
for ( i = 1; i <= NMAX; i++ )
  {
  term += twoH;
  sum += exp(-term*term)*(e1/d1+1.0/(d2*e1));
  d1 += 2.0;
  d2 -= 2.0;
  e1 *= e2;
  }
return  OneDSqrtPi*exp(-xp*xp)*sum;
}
```

Test Values

Check values of the Dawson integral $F(x)$ can be obtained by running the cell `Dawson` in $\boxed{M}$
Mathematica notebook `ErfFD`. Examples are given in Table 10.3.1.

TABLE 10.3.1 Test values to 10 digits of the Dawson integral. The C function FDawson switches methods at |x| = 1.

x	0.2	0.4	0.6
$F(x)$	0.1947510334	0.3599434819	0.4747632037
x	2.0	4.0	6.0
$F(x)$	0.3013403889	0.1293480012	0.0845426890

REFERENCES ON ERROR FUNCTION, FRESNEL AND DAWSON INTEGRALS

[Abr64]   Abramowitz, M., and I. A. Stegun, *Handbook of Mathematical Functions*, Dover, New York, 1964.

[Erd53b] Erdélyi, A., et al, *Higher Transcendental Functions*, Vol. 2, McGraw-Hill, New York, 1953; reprint edition, Krieger, Malabar, Florida, 1981.

[Gra94]   Gradshteyn, I. S., and I. M. Ryzhik, *Table of Integrals, Series, and Products*, Academic Press, Orlando, fifth edition, 1994.

[Jah45]   Jahnke, E., and F. Emde, *Tables of Functions*, Dover, New York, fourth edition, 1945.

[Pre92]   Press, W. H., S. A. Teukolsky, W. T. Vetterling, and B. P. Flannery, *Numerical Recipes*, Cambridge University Press, New York, second edition, 1992.

[Ryb89]   Rybicki, G. B., Computers in Physics, **3**, 85 (1989).

[Spa87]   Spanier, J., and K. B. Oldham, *An Atlas of Functions*, Hemisphere Publishing, Washington, D.C., 1987.

[Tho92]   Thompson, W. J., *Computing for Scientists and Engineers*, Wiley, New York, 1992.

[Van93]   Van Snyder, W., ACM Transactions on Mathematical Software, **19**, 452 (1993).

[Wol91]   Wolfram, S., *Mathematica: A System for Doing Mathematics by Computer*, Addison-Wesley, Redwood City, California, second edition, 1991.

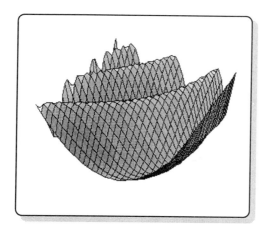

Chapter 11

ORTHOGONAL POLYNOMIALS

Orthogonal polynomials are of very general applicability when computing mathematical functions, especially in science and engineering. This chapter begins with an overview of these polynomials, then specializes in Sections 11.2–11.7 to the standard orthogonal polynomials, both to envision them and to construct C programs for them.

There is an extensive literature on the properties of orthogonal polynomials and on interrelations between them. Emphasizing mathematical analysis are the books by Sansone [San59], Chapter 8 of Luke [Luk69], and Szegö [Sze75], as well as Chapter 10 of the treatise on higher transcendental functions edited by Erdélyi et al [Erd53b]. More practical emphasis is in the monograph on orthogonal polynomials by Beckmann [Bec73] and in Chapter 22 of the handbook of functions edited by Abramowitz and Stegun [Abr64].

The orthogonal polynomials considered are the following: Chebyshev (of first and second kinds, Section 11.2), Gegenbauer (also called ultraspherical, Section 11.3), Hermite (Section 11.4), Laguerre (in generalized form, Section 11.5), Legendre (excluding associated Legendre functions, which are seldom polynomials and are usually not orthogonal, Section 11.6), and Jacobi (Section 11.7). Generalizations are discussed in, for example, [Erd53b, Sections 10.22–10.25].

11.1 OVERVIEW OF ORTHOGONAL POLYNOMIALS

The intrinsic simplicity of orthogonal polynomials has led to very complete mathematical analysis of their properties. This section summarizes in tabular form general properties that are useful for computing them. Numerically, they are also very simple and can be computed to an accuracy limited only by roundoff errors.

Differential Equations
Orthogonal polynomials satisfy linear, second-order differential equations, of the kind shown as (11.1.1) in Table 11.1.1. In general, there will be two linearly independent solutions of these equations. Except for the Chebyshev polynomials, however, only one of

these solutions is a polynomial. For example, the version of (11.1.1) having the Legendre polynomial, $P_n(x)$, as one solution also has a second solution, $Q_n(x)$, containing logarithmic terms, as discussed in Chapter 12.

TABLE 11.1.1 **Differential equations for orthogonal polynomials.**

General differential equation:

$$g_2(x)w'' + g_1(x)w' - g_0(x)w(x) = 0 \qquad (11.1.1)$$

Polynomial	$w(x)$	$g_2(x)$	$g_1(x)$	$g_0(x)$
Chebyshev of first kind	$T_n(x)$	$1-x^2$	$-x$	n^2
Chebyshev of second kind	$U_n(x)$	$1-x^2$	$-3x$	$n(n+2)$
Gegenbauer	$C_n^{(\alpha)}(x)$	$1-x^2$	$-(2\alpha+1)x$	$n(n+2\alpha)$
Hermite	$H_n(x)$	1	$-2x$	$2n$
Laguerre (generalized)	$L_n^{(\alpha)}(x)$	x	$\alpha+1-x$	n
Legendre	$P_n(x)$	$1-x^2$	$-2x$	$n(n+1)$
Jacobi	$P_n^{(\alpha,\beta)}(x)$	$1-x^2$	$\beta-\alpha-(\alpha+\beta+2)x$	$n(n+\alpha+\beta+1)$

The linearity of the differential equations for orthogonal polynomials indicates that a choice of normalization (standardization) is necessary for them. This can be specified by stating their values for some particular x, typically $x=1$, as given in Table 22.4 of [Abr64]. Alternatively, if the value of the orthogonality integral is specified (as in Table 11.1.5 below) then the normalization can be inferred within a sign. For the orthogonal polynomials, simplicity of expression is preferred over simplicity of orthogonality integrals, so these polynomials are not orthonormal.

To compute an orthogonal polynomial with a parameter such as the order n fixed but requiring many closely spaced values of x, it is efficient and it is sometimes accurate to solve the differential equation (11.1.1) numerically from a pair of starting values obtained by other means. In the algorithms and programs that we describe in Sections 11.3–11.8, however, it is assumed that x is fixed and that n may be varying.

Integral Representations

It is often convenient to have available the integral representation of functions. We summarize those for orthogonal polynomials in Table 11.1.2. Integral representations are also discussed in Sections 10.8–10.13 of [Erd53b].

TABLE 11.1.2 Integral representations of orthogonal polynomials.

General integral form for a closed contour C taken around $z = a$ in the positive sense:

$$f_n(x) = \frac{g_0(x)}{2\pi i} \oint [g_1(z,x)]^n g_2(z,x)\, dz \qquad (11.1.2)$$

Polynomial	$g_0(x)$	$g_1(x)$	$g_2(x)$	a
$T_n(x)$	$1/2$	$1/z$	$\dfrac{1-z^2}{z(1-2xz+z^2)}$	0 *
$U_n(x)$	1	$1/z$	$\dfrac{1}{z(1-2xz+z^2)}$	0 *
$C_n^{(\alpha)}(x) \quad \alpha > 0$	1	$1/z$	$\dfrac{1}{z(1-2xz+z^2)^\alpha}$	0 *
$H_n(x)$	$n!$	$1/z$	e^{2xz-z^2}/z	0
$L_n^{(\alpha)}(x)$	$e^x x^{-\alpha}$	$\dfrac{z}{z-x}$	$\dfrac{z^\alpha}{(z-x)e^z}$	x †
$P_n(x)$	$\dfrac{1}{2^n}$	$\dfrac{z^2-1}{z-x}$	$\dfrac{1}{z-x}$	x
$P_n^{(\alpha,\beta)}(x)$	$(1-x)^{-\alpha}$ $\times(1+x)^{-\beta}$	$\dfrac{z^2-1}{2(z-x)}$	$\dfrac{(1-z)^\alpha(1+z)^\beta}{z-x}$	x ††

* Both zeros of $1-2xz+z^2$ outside C.

† Zero outside C.

†† ± 1 outside C.

Recurrence Relations

An orthogonal polynomial labelled by n is a polynomial of order n. A recurrence relation between the polynomials at n, $n-1$, and $n-2$ therefore provides a way to build up the polynomials step by step. Table 11.1.3 following gives the coefficients for the recurrence relations.

Note that the coefficients given in the general relation (11.1.3) differ from those in the corresponding Table 22.7 in Abramowitz and Stegun [Abr64] because we write (11.1.3) as relating the functions at n, $n-1$, and $n-2$, rather than at $n+1$, n, and $n-1$. The former practice is convenient for programming.

TABLE 11.1.3 Recurrence relations for orthogonal polynomials.

General recurrence relation:

$$b_{1n}f_n(x) - (b_{2n} + b_{3n}x)f_{n-1}(x) + b_{4n}f_{n-2}(x) = 0 \qquad (11.1.3)$$

$f_n(x)$	b_{1n}	b_{2n}	b_{3n}	b_{4n}
$T_n(x)$	1	0	2	1
$U_n(x)$	1	0	2	1
$C_n^{(\alpha)}(x)$	n	0	$2(n+\alpha-1)$	$n+2\alpha-2$
$H_n(x)$	1	0	2	$2n-2$
$L_n^{(\alpha)}(x)$	n	$2n+\alpha-1$	-1	$n+\alpha-1$
$P_n(x)$	n	0	$2n-1$	$n-1$
$P_n^{(\alpha,\beta)}(x)$	$2n(n+\alpha+\beta)$ $\times(2n+\alpha+\beta-2)$	$(2n+\alpha+\beta-1)$ $\times(\alpha^2-\beta^2)$	$(2n+\alpha+\beta-2)_3$	$2(n+\alpha-1)(n+\beta-1)$ $\times(2n+\alpha+\beta)$

Explicit formulas for all these orthogonal polynomials can be generated by running the *Mathematica* notebook `OrthoPIter`, as described in Section 20.11.

Relations to Hypergeometric and Parabolic Cylinder Functions

Orthogonal polynomials can be related to hypergeometric functions and parabolic cylinder functions in many ways. Table 11.1.4 gives one such set of relations.

If care is taken that the argument and parameter ranges are appropriate, such relations may be useful for numerical computation of orthogonal polynomials. A power series expansion of the hypergeometric or parabolic-cylinder functions will terminate. Generally, however, such a computation will be less efficient than the methods described separately for each function in Sections 11.2–11.7.

There are two exceptions to the expression of orthogonal polynomials as regular hypergeometric functions. Namely, in terms of the parabolic cylinder function, the Hermite polynomials are given by

$$H_n(x) = 2^n U\left(\frac{1}{2} - \frac{n}{2}, \frac{3}{2}, x^2\right) \qquad (11.1.4)$$

In terms of the *confluent* hypergeometric function, the Laguerre polynomials are

$$L_n^{(\alpha)}(x) = \binom{n+\alpha}{n} M(-n, \alpha+1, x) \qquad (11.1.5)$$

TABLE 11.1.4 Orthogonal polynomials in terms of the hypergeometric functions.

General relation:

$$f_n(x) = d\, F\!\left(-n, b; c; \frac{1-x}{2}\right) \qquad (11.1.6)$$

$f_n(x)$	d	b	c
$T_n(x)$	1	n	$\dfrac{1}{2}$
$U_n(x)$	$n+1$	$n+2$	$\dfrac{3}{2}$
$C_n^{(\alpha)}(x)$	$\dfrac{\Gamma(n+2\alpha)}{n!\,\Gamma(2\alpha)}$	$n+2\alpha$	$\alpha + \dfrac{1}{2}$
$P_n(x)$	1	$n+1$	1
$P_n^{(\alpha,\beta)}(x)$	$\dbinom{n+\alpha}{n}$	$n+\alpha+\beta+1$	$\alpha+1$

Orthogonality Relations

The orthogonality properties of the polynomials that we study in this chapter are of great practical importance for their applications, as discussed by Beckmann [Bec73] and by Sansone [San59], for example, as well as in a myriad of texts in science, engineering, and applied mathematics.

At the level of computing orthogonal polynomials, their orthogonality relations summarized in Table 11.1.5 on the next page can be very useful for checking the average accuracy of their numerical evaluation over their domain of definition. For such checking to be informative, however, the numerical-integration procedure must be much more accurate than the functions being integrated. Note that orthogonal functions must oscillate in sign, as shown in the graphic motif for the Chebyshev polynomial of the first kind that is the motif at the start of this chapter.

Note also that the orthogonal polynomials are not ortho*normal*, as discussed above in Differential Equations.

Iteration or Recursion for Orthogonal Polynomials

Because orthogonality relations satisfy recurrence relations (Table 11.1.3), whether it is more efficient for their numerical computation to use iteration or to use recurrence can be investigated. A comparison of the two methods is carried out in detail in Section 3.2 for the Chebyshev polynomials of the first kind, which are discussed in detail in the next section.

TABLE 11.1.5 Orthogonality relations of orthogonal polynomials.

Orthogonality integral:

$$\int_a^b w(x) f_m(x) f_n(x)\,dx = \delta_{m,n}\, h_n \tag{11.1.7}$$

$f_n(x)$	a	b	$w(x)$	h_n
$T_n(x)$	-1	1	$\dfrac{1}{\sqrt{1-x^2}}$	$\dfrac{\pi}{2}\left(1+\delta_{n,0}\right)$
$U_n(x)$	-1	1	$\sqrt{1-x^2}$	$\dfrac{\pi}{2}$
$C_n^{(\alpha)}(x)$	-1	1	$\left(1-x^2\right)^{\alpha-1/2}$	$\dfrac{2^{1-2\alpha}\,\pi\,\Gamma(n+2\alpha)}{n!\,(n+\alpha)\left[\Gamma(\alpha)\right]^2}$ $\alpha > -\dfrac{1}{2}\quad \alpha \ne 0$
$H_n(x)$	$-\infty$	∞	e^{-x^2}	$\sqrt{\pi}\,2^n\,n!$
$L_n^{(\alpha)}(x)$	0	∞	$e^{-x}x^\alpha$	$\dfrac{\Gamma(\alpha+n+1)}{n!}$
$P_n(x)$	-1	1	1	$\dfrac{2}{2n+1}$
$P_n^{(\alpha,\beta)}(x)$	-1	1	$(1-x)^\alpha(1+x)^\beta$	$\dfrac{2^{\alpha+\beta+1}}{2n+\alpha+\beta+1}$ $\times \dfrac{\Gamma(\alpha+n+1)\Gamma(\beta+n+1)}{n!\,\Gamma(\alpha+\beta+n+1)}$ $\alpha>-1\quad\beta>-1$

11.2   CHEBYSHEV POLYNOMIALS

The Chebyshev polynomials are of two kinds, $T_n(x)$ and $U_n(x)$. They are the solutions—regular at the origin—of differential equations obtained from (11.1.1) by the substitutions indicated in Table 11.1.1, namely,

$$(1-x^2)T_n'' - x T_n' - n^2 T_n = 0 \tag{11.2.1}$$

for the functions of the first kind, and the equation

$$(1-x^2)U_n'' - 3x U_n' - n(n+2)U_n = 0 \tag{11.2.2}$$

for the functions of the second kind. In both equations the variable with respect to which derivatives are taken is x. The two standardized solutions of these equations are, for the polynomial of the first kind,

$$T_n(x) = \cos(n \arccos x) \qquad x \text{ in } [-1,1] \qquad (11.2.3)$$

and, for the polynomial of the second kind,

$$U_n(x) = \frac{\sin[(n+1)\arccos x]}{\sin(\arccos x)} \qquad x \text{ in } [-1,1] \qquad (11.2.4)$$

Although it is not immediately obvious from these expressions, $T_n(x)$ and $U_n(x)$ are both polynomials of order n, assuming that n is a nonnegative integer.

Searching the mathematical literature on Chebyshev polynomials is quite tedious because transliteration of the Cyrillic spelling of the name of 19th-century Russian mathematician Chebyshev into various European languages has been confusingly variable—Chebyshev, Tchebichiev, Tschebyscheff, Tchebychef, and Čebyšev—but the first spelling has gradually become predominant.

By comparison with other orthogonal polynomials—especially the Legendre polynomials, who use they can often displace effectively—the Chebyshev polynomials have been relatively little used in applied mathematics. There is a monograph on their mathematical analysis by Rivlin [Riv74], Fox and Parker [Fox68] have given an extensive treatment of their uses in numerical analysis, and Sections 8.5, 8.6 of Luke [Luk69] discuss expansions into Chebyshev polynomials. An introduction to their uses in physics contexts is given by Thompson [Tho94a]. We point out several applications in the next two subsections.

11.2.1 Chebyshev Polynomials of the First Kind

Keywords
orthogonal polynomials, Chebyshev polynomial of first kind

C Function and *Mathematica* Notebook
ChebyT (C); cell ChebyT in notebook OrthoP (*Mathematica*)

M

Background
The Chebyshev polynomial $T_n(x)$ is a fine gem of mathematics, revealing different facets when illuminated from different angles. For example, from the viewpoint of analysis—as discussed in Rivlin's book [Riv74]—there are expansions in Chebyshev polynomials analogous to Fourier series, as well as interesting semigroup and number theory properties of $T_n(x)$.

From the viewpoint of numerical applications—as emphasized in Fox and Parker's monograph [Fox68]—the minimax property of $T_n(x)$ makes it ideal for approximating functions and data, as well as for economizing series. Further, as discussed in [Fox68] and by authors such as Quarteroni [Qua91], the Chebyshev polynomials are useful for numerical solution of differential and integral equations.

Function Definition
The Chebyshev polynomial of the first kind, $T_n(x)$, can be defined in several ways, for example in terms of the differential equation (11.2.1) or its explicit expression (11.2.3). Alternatively, one can use the relation (11.2.1) to build up the $T_n(x)$, as discussed in

detail in Section 11.2. Other analytical properties of $T_n(x)$ are given in the tables in the overview, Section 11.1: Table 11.1.2 for the integral representation, Table 11.1.4 for its expression in terms of hypergeometric functions, and Table 11.1.5 for orthogonality relations.

Formulas for Chebyshev polynomials—from $n = 0$ up to a chosen maximum value $n = \text{maxn}$—can be generated iteratively by using the appropriate cell in *Mathematica* notebook OrthoPIter. For example, with maxn = 7, we obtain the polynomials given in Table 11.2.1.

TABLE 11.2.1 Lowest-order Chebyshev polynomials of the first kind.

n	$T_n(x)$	n	$T_n(x)$
0	1	4	$1 - 8x^2 + 8x^4$
1	x	5	$5x - 20x^3 + 16x^5$
2	$-1 + 2x^2$	6	$-1 + 18x^2 - 48x^4 + 32x^6$
3	$-3x + 4x^3$	7	$-7x + 56x^3 - 112x^5 + 64x^7$

Visualization

When plotted as functions of n and x, Chebyshev polynomials of the first kind reveal a very interesting surface, showing their relation to cosines, (11.2.3). Since for $n = 0$ and 1 $T_n(x)$ is uninteresting, we start the surface in Figure 11.2.1 at $n = 2$.

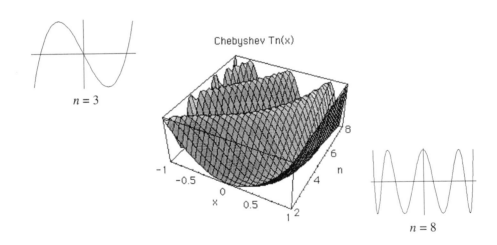

FIGURE 11.2.1 Chebyshev polynomials of the first kind for $n = 2–8$. (Cell ChebyT in *Mathematica* notebook OrthoP.)

Note that $T_n(x)$ is bounded by ± 1 and has the reflection symmetry evident in the figure, $T_n(-x) = (-1)^n T_n(x)$.

Algorithm, Program, and Programming Notes

For computing $T_n(x)$ numerically, the explicit polynomial expressions from Table 11.2.1 for $n < 4$, and the trigonometric form (11.2.3) for larger n are used.

```
double ChebyT(int n, double x)
/* Chebyshev Polynomial: first kind, order n */
{

if ( n < 1 )    return 1.0; /* T0 */
if ( n < 2 )    return x;
if ( n < 3 )    return -1.0+2.0*x*x;   /* T2 */
if ( n < 4 )    return x*(-3.0+4.0*x*x);

/* For n >= 4 use trig formulas */
return  cos( ( (double) n)*acos(x));
}
```

Note that the optimum switchover n value between the two methods depends upon the relative efficiency of computing polynomial, trigonometric, and inverse trigonometric functions in your computer system.

Test Values

Check values of $T_n(x)$ can be generated by running the ChebyT cell in *Mathematica* notebook OrthoP. Examples are given in Table 11.2.2.

$\boxed{M}$

TABLE 11.2.2 **Exact values of the Chebyshev polynomials of the first kind.**

$x \setminus n$	1	2	3	4
−1.0	−1.0	1.0	−1.0	1.0
−0.5	−0.5	−0.5	1.0	−0.5
0.0	0	−1.0	0	1.0
0.5	0.5	−0.5	−1.0	−0.5
1.0	1.0	1.0	1.0	1.0

11.2.2 Chebyshev Polynomials of the Second Kind

Keywords

orthogonal polynomials, Chebyshev polynomial of second kind

C Function and *Mathematica* Notebook

ChebyU (C); cell ChebyU in notebook OrthoP (*Mathematica*)

$\boxed{M}$

Background

By comparison with the Chebyshev polynomial of the first kind, $T_n(x)$ in Section 11.2.1, the polynomial of the second kind, $U_n(x)$, is relatively little used. One reason for this is that near the endpoints $x = \pm 1$ the polynomial changes very rapidly, as seen in Visualization below.

Function Definition

The Chebyshev polynomial of the second kind can be defined in terms of the differential equation (11.2.2) or its explicit expression (11.2.4). Alternatively, one can use the recurrence relation from Table 11.1.3 to build up the $U_n(x)$, as discussed in detail for $T_n(x)$ in Section 11.2. Other analytical properties of $U_n(x)$ are given in the tables in the overview, Section 11.1: Table 11.1.2 for the integral representation, Table 11.1.4 for its expression in terms of hypergeometric functions, and Table 11.1.5 for orthogonality relations.

Formulas for Chebyshev polynomials of the second kind—from $n = 0$ up to a chosen maximum value $n = $ maxn—can be generated iteratively (Section 3.2) by using the cell ChebyU in *Mathematica* notebook OrthoPIter. For example, with maxn $= 7$, one obtains the polynomials in Table 11.2.3.

TABLE 11.2.3 Lowest-order Chebyshev polynomials of the second kind.

n	$U_n(x)$	n	$U_n(x)$
0	1	4	$1 - 12x^2 + 16x^4$
1	$2x$	5	$6x - 32x^3 + 32x^5$
2	$-1 + 4x^2$	6	$-1 + 24x^2 - 80x^4 + 64x^6$
3	$-4x + 8x^3$	7	$-8x + 80x^3 - 192x^5 + 128x^7$

Visualization

A plot of the Chebyshev polynomials of the second kind as functions of n and x constructs a very interesting surface, arising from their relation to sine and cosine functions, as given by (11.2.4). As for $T_n(x)$ in Section 11.2.1, we start the surface of $U_n(x)$ in Figure 11.2.2 at $n = 2$.

Unlike $T_n(x)$, the bounds on $U_n(x)$ increase with n as $U_n(\pm 1) = (\pm 1)^n(n + 1)$. The range of x over which $U_n(x)$ is drawn is therefore decreased, in order that values for smaller n are not dwarfed by those for larger n, and so that values near $x = 0$ (which are zero or ± 1) are similarly not suppressed in the surface plot.

Note the reflection symmetry of $U_n(x)$ that is evident in the figure and is easily derived from (11.2.4), namely, $U_n(-x) = (-1)^n U_n(x)$.

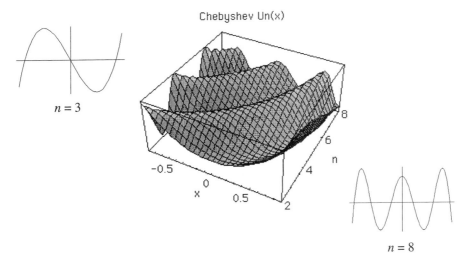

FIGURE 11.2.2 Chebyshev polynomials of the second kind for $n = 2$–8 and $|x|$ bounded by 0.8. (Cell ChebyU in *Mathematica* notebook OrthoP.)

Algorithm and Program

For computing numerical values of $U_n(x)$ we use the explicit polynomial expressions from Table 11.2.3 for $n < 4$, and for most x values the trigonometric form (11.2.4) for larger n. For x near ± 1 this form becomes indeterminate, therefore a suitable approximation—which requires the accuracy parameter eps—is needed, as described in the Programming Notes. The function ChebyU is given on the next page.

Programming Notes

Two considerations affect the programming of function ChebyU for $U_n(x)$. First, the optimum switchover n value between the two methods depends upon the relative efficiency of computing polynomial, trigonometric, and inverse trigonometric functions in your computer system.

The second programming consideration is the indeterminate form of (11.2.4) as x approaches ± 1. To handle this danger zone—for x near 1, say—we expand the numerator and denominator in (11.2.4) in Taylor series about their $x = 1$ values, keeping terms through the squares of the angles. In the numerator, if the first neglected term is to contribute a fractional accuracy ε, then x and n must be related by

$$1 - x^2 < \frac{\sqrt{120\,\varepsilon}}{(n+1)^2} \tag{11.2.5}$$

Within this range of x we use the approximation

$$U_n(x) \approx (\pm 1)^n (n+1)\left[1 - \frac{n(n+2)}{6}(1-x^2)\right] \qquad x \approx \pm 1 \tag{11.2.6}$$

The function ChebyU therefore requires the additional input parameter eps to control the accuracy of this approximation near the endpoints.

```
double ChebyU(int n, double x, double eps)
/* Chebyshev Polynomial: second kind, order n */
{
double xNearOne,nPone,nTerm,sign,theta;

if ( n < 1 ) return    1.0; /* U0 */
if ( n < 2 ) return    2.0*x;
if ( n < 3 ) return    -1.0+4.0*x*x;   /* U2 */
if ( n < 4 ) return    4.0*x*(-1.0+2.0*x*x);

/* For n >= 4 use trig formulas */
/* First check if |x| near 1 */
xNearOne = 1 - x*x;
nPone = n+1;
nTerm = n*(n+2);

if ( xNearOne < sqrt(120.0*eps)/(nPone*nPone) )
  {
  /* Within danger zone for accuracy eps */
  /* Sign negative if n is odd & x is negative:
     else sign is positive */
  sign = ( (n%2==1) && (x<0.0) ) ?  -1.0: 1.0;
  return   sign*nPone*(1.0 - nTerm*xNearOne/6.0);
  }

/* Outside danger zone for accuracy eps */
theta = acos(x);
  return   sin(nPone*theta)/sin(theta);
}
```

Test Values

Check values of $U_n(x)$ can be generated by running the ChebyU cell in *Mathematica* notebook OrthoP, except that the *Mathematica* function is indeterminate at the endpoints $x = \pm 1$, so the limiting values from (11.2.6) have to be used. Examples are given in Table 11.2.4.

TABLE 11.2.4 **Exact values of the Chebyshev polynomials of the second kind.**

$x \setminus n$	1	2	3	4
−1.0	−2.0	3.0	−4.0	5.0
−0.5	−1.0	0	1.0	−1.0
0.0	0	−1.0	0	1.0
0.5	1.0	0	−1.0	−1.0
1.0	2.0	3.0	4.0	5.0

11.3 GEGENBAUER (ULTRASPHERICAL) POLYNOMIALS

Keywords
orthogonal polynomials, Gegenbauer polynomial, ultraspherical polynomial

C Function and *Mathematica* Notebook
Gegenbauer (C); cell Gegenbauer in notebook OrthoP (*Mathematica*)

Background
The Gegenbauer polynomials are the first of the generalized orthogonal polynomials, in the sense that they have an extra parameter (here indicated by α but elsewhere by λ or ν), thus including other orthogonal polynomials as special cases. Gegenbauer polynomials are discussed in detail in Erdélyi et al [Erd53a, Section 3.15, Erd53b, Section 10.9].

The nomenclature *ultraspherical* for these polynomials comes from extending the generating function relation for Legendre polynomials $P_n(x)$

$$(1-2xz+z^2)^{-1/2} = \sum_{n=0}^{\infty} P_n(x)z^n \qquad -1 < x < 1 \qquad |z| < 1 \qquad (11.3.1)$$

For the Gegenbauer polynomials $C_n^{(\alpha)}(x)$ the corresponding relation is

$$(1-2xz+z^2)^{-\alpha} = \sum_{n=0}^{\infty} C_n^{(\alpha)}(x)z^n \qquad -1 < x < 1 \qquad |z| < 1 \qquad (11.3.2)$$

Thus, $C_n^{(1/2)}(x)$ is just a Legendre polynomial. The generalization to associated Legendre functions is given in (11.3.5) below.

Function Definition
The Gegenbauer polynomial can be defined either in terms of its differential equation (Table 11.1.1), its integral representation in Table 11.1.2, its recurrence relation in Table 11.1.3, or its relation to the hypergeometric function (Table 11.1.4).

Explicit Gegenbauer polynomials—from $n = 0$ up to a maximum value $n = \text{maxn}$—can be generated iteratively by inserting the b coefficients into (11.1.3) to obtain the following formula, valid for $\alpha > -1/2, \alpha \neq 0$,

$$C_0^{(\alpha)}(x) = 1 \qquad C_1^{(\alpha)}(x) = 2\alpha x$$
$$C_n^{(\alpha)}(x) =$$
$$\left[2(n+\alpha-1)xC_{n-1}^{(\alpha)}(x) - (n+2\alpha-2)C_{n-2}^{(\alpha)}(x)\right]/n \qquad n > 1 \qquad (11.3.3)$$

For $\alpha = 0$ special formulas are required. These are given by

$$C_0^{(0)}(x) = 1 \qquad C_n^{(0)}(x) \equiv \lim_{\alpha \to 0}\left[\frac{C_n^{(\alpha)}(x)}{\alpha}\right] = \frac{2}{n}T_n(x) \qquad n > 0 \qquad (11.3.4)$$

in which $T_n(x)$ is the Chebyshev polynomial of the first kind (Section 11.2.1). The iterated polynomials can be obtained by running the cell Gegenbauer in *Mathematica* notebook OrthoPIter, as described in Section 20.11. For example, with maxn = 3, we obtain the polynomials in Table 11.3.1.

TABLE 11.3.1 The lowest-order Gegenbauer polynomials.

n	$C_n^{(0)}(x)$	$C_n^{(\alpha)}(x)$ $\quad \alpha \neq 0$
0	1	1
1	$2x$	$2\alpha x$
2	$-1 + 2x^2$	$-\alpha + 2\alpha(1+\alpha)x^2$
3	$-2x + \dfrac{4}{3}x^3$	$2\alpha\left[-(1+\alpha) + \dfrac{4 + 6\alpha + 2\alpha^2}{3}x^2\right]$

The connection to $T_n(x)$ for $\alpha \to 0$ given by (11.3.4) is one of several relations to other special functions. For example,

$$C_n^{(1/2)}(x) = P_n(x) \qquad C_n^{(1)}(x) = U_n(x)$$

$$C_{n-m}^{(m+1/2)}(x) = \frac{(-1)^n (1-x^2)^{-m/2}}{(2m-1)!!} P_n^{(m)}(x) \tag{11.3.5}$$

in which the $P_n^{(m)}(x)$ are associated Legendre functions (Section 12.1).

Visualization

Plotting the Gegenbauer polynomials as functions of n, α, and x produces several very interesting surfaces. It is clearest to show the polynomials of a given order n as a surface mapped out by varying x and α, as in Figure 11.3.1.

In Figure 11.3.1 a scale factor $1/\alpha^2$ has been introduced so that the values of $C_n^{(\alpha)}(x)$ as α increases—which generally grow rapidly with α, as seen in Table 11.3.2 below—do not dwarf those for small α. Note also that as n increases, the polynomials generally oscillate more rapidly as functions of x, since they are nth-order polynomials in x. This is especially noticeable for small values of α, since as α tends to zero $C_n^{(\alpha)}(x)/\alpha$ becomes proportional to the Chebyshev polynomial $T_n(x)$, all of whose zeros lie in $[-1, 1]$.

Algorithm, Program, and Programming Notes

For computing $C_n^{(\alpha)}(x)$ numerically, we use the special formulas in (11.3.4) for $\alpha = 0$. For $\alpha \neq 0$ we use the explicit polynomial expressions from Table 11.3.1 for $n < 4$, and iteration using (11.3.3) for larger n, starting the iteration with the expressions for $n = 2$ and $n = 3$.

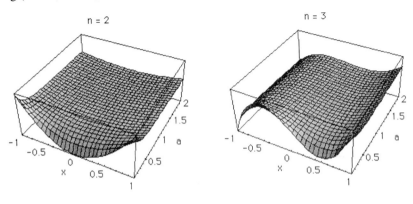

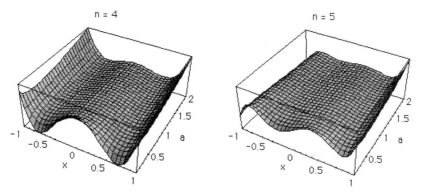

FIGURE 11.3.1 Gegenbauer polynomials divided by α^2 for x in the range -1 to 1 and α in the range 0.1 to 2, for $n = 2, 3$, (foot of previous page) and $n = 4, 5$ (above). In the figure a denotes α. (Cell Gegenbauer in *Mathematica* notebook OrthoP.)

```
double Gegenbauer(double a, int n, double x)
/* Gegenbauer Polynomial, */
/*   parameter a, order n */
{
double floatn,CiM2,CiM1,floati,Ci;
int i;

if ( a <= -0.5 )    return  0.0;
if ( a ==  0.0 )
 {if ( n == 0 )   return  1.0;
 if ( n == 1 )        return  2.0*x;
 floatn = n;      /*   2*Tn(x)/n   */
 return  2.0*cos(floatn*acos(x))/floatn;
 }

/* General case: explicit for  n < 4 */
if ( n < 1 )        return     1.0;  /* C(a)0 */
if ( n < 2 )        return     2.0*a*x;
if ( n < 3 )        return
 -a+2.0*a*(1+a)*x*x;       /* C(a)2 */
if ( n < 4 )        return
 2.0*a*x*(-(1.0+a)+(4.0+2.0*a*(3.0+a))/3.0*x*x);

/* Use iteration for  n >= 4   */
/* C(a)2  then  C(a)3   */
CiM2 = -a+2.0*a*(1+a)*x*x;
CiM1 = 2.0*a*x*(-(1.0+a)+
     (4.0+2.0*a*(3.0+a))/3.0*x*x);
for ( i = 4; i <= n; i++ )
 {floati = i;
 Ci = (2.0*(floati+a-1.0)*x*CiM1 -
     (floati+2.0*a - 2.0)*CiM2)/floati;
 CiM2 = CiM1;
 CiM1 = Ci;
 }
 returnCi;
}
```

Test Values

Test values of $C_n^{(\alpha)}(x)$ can be generated by running the Gegenbauer cell in *Mathematica* notebook OrthoP. Examples are given in Table 11.3.2.

TABLE 11.3.2 **Test values of the Gegenbauer polynomials for** $\alpha = 0.5$ **and** $\alpha = 1.5$. **The values given are exact.**

$x \setminus n$	2	3	4	5
$\alpha = 0.5$				
0	–0.5000000	0	0.3750000	0
0.5	–0.1250000	–0.4375000	–0.2890625	0.089843750
1.0	1.0000000	1.0000000	1.0000000	1.000000000
$\alpha = 1.5$				
0	–1.5000000	0	1.8750000	0
0.5	0.3750000	–1.5625000	–2.2265625	–0.57421875
1.0	6.0000000	10.0000000	15.0000000	21.00000000

11.4   HERMITE POLYNOMIALS

Keywords
orthogonal polynomials, Hermite polynomial

C Function and *Mathematica* Notebook

Hermite (C); cell Hermite in notebook OrthoP (*Mathematica*)

Background
Hermite polynomials $H_n(x)$ are orthogonal polynomials associated with x in the interval $(-\infty, \infty)$ and an exponential weight function. If the weight function in the orthogonality integral (Table 11.1.5) is $w(x) = \exp(-x^2)$ then we have $H_n(x)$. If—as common in statistics applications—the weight function is chosen as $\exp(-x^2/2)$ then one has the function $He_n(x) = H_n(x/\sqrt{2})$, which is also called the Hermite polynomial. We discuss here only $H_n(x)$, whose properties are summarized in Section 10.13 of Erdélyi et al [Erd53b], and (jumbled with the other orthogonal polynomials) in Chapter 22 of Abramowitz and Stegun [Abr64].

Hermite polynomials have many applications. For example, their use in eigenfunction expansions is discussed in Chapter 6 of Morse and Feshbach's treatise [Mor53], and the example of the quantum harmonic oscillator is presented in a myriad of texts on quantum mechanics. Recipes for numerical integration by using Gauss-Hermite quadrature are given in Section 4.5 of Press et al [Pre92]. The statistics connection is established through the weight function $\exp(-x^2/2)$, which is proportional to the standardized normal probability density (Section 9.3.1). In this context, Hermite and other orthogonal polynomials can be used to construct probability densities for pairs of correlated random variables, as discussed in Chapter 6 of Beckmann [Bec73].

Function Definition

The Hermite polynomials can be defined in several ways, for example in terms of the differential equation (11.1.1) in Table 11.1.1, the integral representation in Table 11.1.2, the recurrence relations in Table 11.1.3, or in terms of hypergeometric functions in Table 11.1.4.

Formulas for the Hermite polynomials—from $n = 0$ up to a chosen maximum value $n = $ maxn—can be generated iteratively by using the Hermite cell in *Mathematica* notebook OrthoPIter. For example, with maxn = 7, we obtain the polynomials given in Table 11.4.1.

M

TABLE 11.4.1 **Lowest-order Hermite polynomials.**

n	$H_n(x)$	n	$H_n(x)$
0	1	4	$12 - 48x^2 + 16x^4$
1	$2x$	5	$120x - 160x^3 + 32x^5$
2	$-2 + 4x^2$	6	$-120 + 720x^2 - 480x^4 + 64x^6$
3	$-12x + 8x^3$	7	$-1680x + 3360x^3 - 1344x^5 + 128x^7$

Visualization

If plotted directly as functions of n and x, the surface mapped out by the Hermite polynomials looks featureless, since $H_n(x)$ has such a rapid variation with both n and x that oscillations are generally not visible. By changing from $H_n(x)$ to an appropriately scaled function we can reveal a very interesting surface.

Consider the orthogonality integral for $H_n(x)$ that is given in Table 11.1.5. If we define the function $H'_n(x)$ by

$$H'_n(x) \equiv \frac{H_n(x)}{\sqrt{n! \, 2^n e^{x^2}}} \qquad (11.4.1)$$

then $H'_n(x)$ has the orthogonality integral

$$\int_{-\infty}^{\infty} [H'_n(x)]^2 \, dx = \sqrt{\pi} \qquad (11.4.2)$$

This modified function (no longer a polynomial!) has a scale factor independent of n, a uniform weight in the integral as a function of x, and the same x dependence as the wave function in quantum mechanics. The appropriate function to envision is thus $H'_n(x)$, shown in Figure 11.4.1. Since for $n = 1$ the function is uninteresting, we start the surface at $n = 2$.

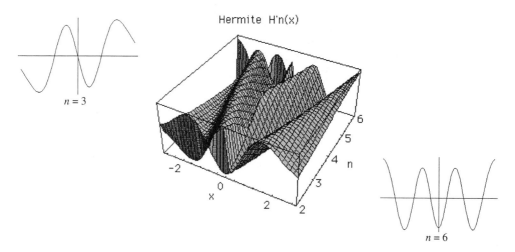

FIGURE 11.4.1 Modified Hermite polynomials, scaled as in (11.4.1), for $n = 2$ to 6. (Cell `Hermite` in *Mathematica* notebook `OrthoP`.)

Algorithm, Program, and Programming Notes

To compute $H_n(x)$ numerically we use explicit polynomial expressions from Table 11.4.1 for $n < 4$, and iteration for larger n.

```
double Hermite(int n, double x)
/* Hermite Polynomial of order n */

{
double HiM2,HiM1,Hi;
int i;

/* Explicit for  n < 4 */
if ( n < 1 ) return   1.0;  /* H0 */
if ( n < 2 ) return   2.0*x;
if ( n < 3 ) return   -2.0+4.0*x*x; /* H2 */
if ( n < 4 ) return   4.0*x*(-3.0+2.0*x*x);

/* Use iteration for  n >= 4   */
/* H2  then  H3  */
HiM2 = -2.0+4.0*x*x;
HiM1 =  4.0*x*(-3.0+2.0*x*x);
for ( i = 4; i <= n; i++ )
  {
  Hi = 2.0*(x*HiM1 - ((double)i-1.0)*HiM2);
  HiM2 = HiM1;
  HiM1 = Hi;
  }

  return  Hi;
}
```

The recurrence formula is obtained from Table 11.1.3 as

$$H_n(x) = 2\left[xH_{n-1}(x) - (n-1)H_{n-2}(x)\right] \qquad (11.4.3)$$

Test Values
Check values of $H_n(x)$ can be generated by running the `Hermite` cell in *Mathematica* notebook `OrthoP`. Examples are given in Table 11.4.2. Notice the very rapid growth of $H_n(x)$ with both n and x, as discussed above Figure 11.4.1.

TABLE 11.4.2 Exact test values of the Hermite polynomials.

$x \setminus n$	1	2	3	4
0.5	1.0	−1.0	−5.0	1.0
1.0	2.0	2.0	−4.0	−20.0
1.5	3.0	7.0	9.0	−15.0
2.0	4.0	14.0	40.0	76.0

11.5   LAGUERRE POLYNOMIALS

Keywords
orthogonal polynomials, Laguerre polynomials

C Function and *Mathematica* Notebook
Laguerre (C); cell Laguerre in notebook OrthoP (*Mathematica*)

Background
The generalized Laguerre polynomials $L_n^{(\alpha)}(x)$ are the second of the generalized orthogonal polynomials after the Gegenbauer polynomials (Section 11.3), since they have an extra parameter—here indicated by α but elsewhere by λ. If $\alpha = 0$ one has the conventional Laguerre polynomials $L_n(x)$. The generalized polynomials are discussed in detail in Erdélyi et al [Erd53b, Section 10.12], where they are called simply Laguerre polynomials, a nomenclature that we follow.

Laguerre polynomials occur as eigenfunctions of the hydrogen atom, as presented in many texts on quantum mechanics. Recipes for numerical integration by using Gauss-Laguerre quadrature are given in Section 4.5 of Press et al [Pre92]. In statistics, Laguerre polynomials can be used to construct probability densities for pairs of correlated random variables, as discussed in Section 6.3 of [Bec73].

Function Definition
Laguerre polynomials can be defined via the differential equation in Table 11.1.1, the integral representation in Table 11.1.2, the recurrence relation in Table 11.1.3, or the relation to the confluent hypergeometric function in Table 11.1.4. Explicit formulas for the

Laguerre polynomials—from $n = 0$ up to a chosen maximum value $n = \texttt{maxn}$—can be generated iteratively by putting the b coefficients into (11.1.3):

$$L_0^{(\alpha)}(x) = 1 \qquad\qquad L_1^{(\alpha)}(x) = 1 + \alpha - x$$

$$L_n^{(\alpha)}(x) =$$

$$\left[(2n + \alpha - 1 - x)L_{n-1}^{(\alpha)}(x) - (n + \alpha - 1)L_{n-2}^{(\alpha)}(x)\right]\Big/ n \qquad n > 1$$

(11.5.1)

M

The iterated polynomials $L_n^{(\alpha)}(x)$ can be obtained by running cell $\texttt{Laguerre}$ in *Mathematica* notebook $\texttt{OrthoPIter}$. For example, with $\texttt{maxn} = 4$, we obtain the polynomials in Table 11.5.1.

TABLE 11.5.1 Lowest-order (generalized) Laguerre polynomials.

n	$L_n^{(\alpha)}(x)$
0	1
1	$1 + \alpha - x$
2	$1 + \dfrac{\alpha(3 + \alpha)}{2} - (2 + \alpha)x + \dfrac{x^2}{2}$
3	$1 + \dfrac{\alpha(11 + 6\alpha^2 + \alpha^3)}{6} - \left[3 + \dfrac{\alpha(5 + \alpha)}{2}\right]x$ $+ \dfrac{3 + \alpha}{2}x^2 - \dfrac{x^3}{6}$

Visualization

We plot in Figure 11.5.1 the $L_n^{(\alpha)}(x)$ as functions of n and x for two values of α, with $\alpha = 0$ giving the conventional Laguerre polynomials.

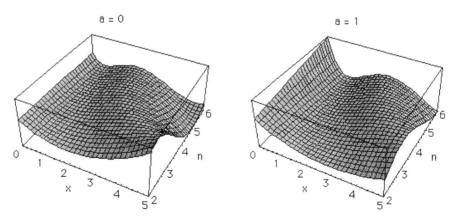

FIGURE 11.5.1 Laguerre polynomials for $n = 2$ to 6 with $\alpha = 0$ and $\alpha = 1$. In the figure, a denotes paameter α. (Cell $\texttt{Laguerre}$ in *Mathematica* notebook $\texttt{OrthoP}$.)

Algorithm, Program, and Programming Notes

For computing $L_n^{(\alpha)}(x)$ explicit polynomial expressions from Table 11.5.1 for $n < 3$ and the recurrence relation (11.5.1) for larger n are used, starting the iteration with the expressions for $n = 1$ and $n = 2$.

```
double Laguerre(double a, int n, double x)
/* Laguerre Polynomial, parameter a, order n */
{
double LiM2,LiM1,floati,Li;
int i;

/* Explicit for  n < 3 */
if ( n < 1 )   return   1.0;  /* L(a)0 */
if ( n < 2 )   return   1.0+a-x;
if ( n < 3 )   return
  1.0+0.5*a*(3.0+a)-x*(2.0+a-0.5*x); /* L(a)2 */

/* Use iteration for  n >= 3  */
/* L(a)1  then  L(a)2  */
LiM2 = 1.0+a-x;
LiM1 = 1.0+0.5*a*(3.0+a)-x*(2.0+a-0.5*x);
for ( i = 3; i <= n; i++ )
  {
  floati = i;
  Li = ((2.0*floati+a-1.0-x)*LiM1 -
      (floati+a - 1.0)*LiM2)/floati;
  LiM2 = LiM1;
  LiM1 = Li;
  }
returnLi;
}
```

Test Values

Test values of the generalized Laguerre polynomial $L_n^{(\alpha)}(x)$ can be generated by running the cell Laguerre in the *Mathematica* notebook OrthoP. Examples are given in Tables 11.5.2 and 11.5.3.

M

TABLE 11.5.2 Test values to 10 digits of the generalized Laguerre polynomials for $\alpha = 0$.

$x \setminus n$	1	2	3	4
0	1.0000000000	1.0000000000	1.0000000000	1.0000000000
1.0	0	−0.5000000000	−0.6666666667	−0.6250000000
2.0	−1.0000000000	−1.0000000000	−0.3333333333	0.3333333333

TABLE 11.5.3 Test values to 10 digits of the generalized Laguerre polynomials for $\alpha = 1.5$.

$x \setminus n$	1	2	3	4
0	2.5000000000	4.3750000000	6.5625000000	9.0234437500
1.0	1.5000000000	1.3750000000	0.7708333333	–0.1015625000
2.0	0.5000000000	–0.6250000000	–1.5208333333	–1.768229167

11.6   LEGENDRE POLYNOMIALS

Keywords
orthogonal polynomials, Legendre polynomial

C Function and *Mathematica* Notebook
Legendre (C); cell Legendre in notebook OrthoP (*Mathematica*)

Background
The Legendre polynomial $P_n(x)$ is probably the best known of the special functions, especially in science and engineering. In these fields it occurs when the Laplacian, ∇^2, is separated in spherical polar coordinates, then $x = \cos\theta$ in terms of polar angle θ.

Note that, although the Legendre polynomial is a special case of the Legendre functions $P_n^m(x)$, the latter are generally not polynomials and are seldom members of a set of orthogonal functions. Legendre functions and their relatives—including P_n^m and Q_n^m—are therefore discussed separately in Chapter 12. Legendre polynomials are special cases of the Gegenbauer polynomials, as shown by (11.3.5). This connection is seldom useful for understanding $P_n(x)$ and it is not efficient for computing $P_n(x)$ numerically.

Analytical properties of $P_n(x)$ are presented in Section 10.10 of Erdélyi et al [Erd53b], and several examples of applications are given in Chapter 5 of Beckmann's monograph on orthogonal polynomials [Bec73]. Legendre polynomials are also used extensively in the theory of rotational symmetry and angular momentum, as illustrated in detail in Thompson's text [Tho94b].

Function Definition
The Legendre polynomial, $P_n(x)$, can be defined in several ways; for example, in terms of the differential equation (11.3.1) or its explicit expression (11.3.3). Alternatively, one can use the recurrence relation (11.2.1) to build up the $P_n(x)$. Other analytical properties of $P_n(x)$ are given in the tables in the overview, Section 11.1: Table 11.1.2 for the integral representation, Table 11.1.4 for its expression in terms of hypergeometric functions, and Table 11.1.5 for orthogonality relations.

Formulas for Legendre polynomials—from $n = 0$ up to a chosen maximum value $n = \text{maxn}$—can be generated iteratively by using cell Legendre in *Mathematica* notebook OrthoPIter. For example, with $\text{maxn} = 7$, we obtain the polynomials in Table 11.6.1.

TABLE 11.6.1 The lowest-order Legendre polynomials for $n = 0 - 7$.

n	$P_n(x)$	n	$P_n(x)$
0	1	4	$(3 - 30x^2 + 35x^4)/8$
1	x	5	$(15x - 70x^3 + 63x^5)/8$
2	$(-1 + 3x^2)/2$	6	$(-5 + 105x^2 - 315x^4 + 231x^6)/16$
3	$(-3x + 5x^3)/2$	7	$(-35x + 315x^3 - 693x^5 + 429x^7)/16$

Visualization

When plotted as functions of n and x, the Legendre polynomials reveal a convoluted surface. For $n = 0$ and 1 the functions are uninteresting, therefore the surface in Figure 11.6.1 is started at $n = 2$.

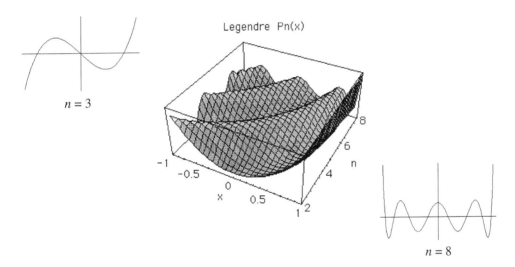

FIGURE 11.6.1 The Legendre polynomials for $n = 2$–8. (Cell `Legendre` in *Mathematica* notebook `OrthoP`.)

Note that $P_n(x)$ is bounded by ± 1 and has the reflection symmetry evident in the figure, $P_n(-x) = (-1)^n P_n(x)$.

Algorithm, Program, and Programming Notes

To compute $P_n(x)$ numerically we use explicit polynomial expressions from Table 11.6.1 for $n < 4$, and iteration with the coefficients from Table 11.1.3 for the $P_n(x)$ with larger n. A recurrence formula that becomes numerically unstable for large n is

$$P_n(x) = \left[(2n - 1)xP_{n-1}(x) - (n - 1)P_{n-2}(x)\right]/n \qquad n > 1 \qquad (11.6.1)$$

The switchover *n* value between the two methods is a compomise between the relative efficiency of computing the polynomial terms and the recurrence formula in your computer system.

```
double Legendre(int n, double x)
/* Legendre Polynomial of order n */
{
double LiM2,LiM1,floati,Li;
int i;

/* Explicit for  n < 4 */
if ( n < 1 )   return   1.0;   /* L0 */
if ( n < 2 )   return   x;
if ( n < 3 )   return   -0.5+1.5*x*x;  /* L2 */
if ( n < 4 )   return   x*(-1.5+2.5*x*x);

/* Use iteration for  n >= 4   */
/* L2   then   L3   */
LiM2 = -0.5+1.5*x*x;
LiM1 = x*(-1.5+2.5*x*x);
for ( i = 4; i <= n; i++ )
  {
  floati = i;
  Li = ((2.0*floati-1.0)*x*LiM1 -
      (floati-1.0)*LiM2)/floati;
  LiM2 = LiM1;
  LiM1 = Li;
  }
  return   Li;
}
```

Test Values

Check values of $P_n(x)$ can be generated by running the `Legendre` cell in *Mathematica* notebook `OrthoP`. The results are given in Table 11.6.2.

TABLE 11.6.2 **Exact test values of the Legendre polynomials.**

$x \setminus n$	1	2	3	4
−1.0	−1.0000000	1.0000000	−1.0000000	1.0000000
−0.5	−0.5000000	−0.1250000	0.4375000	−0.2890625
0.0	0	−0.5000000	0	0.3750000
0.5	0.5000000	−0.1250000	−0.4375000	−0.2890625
1.0	1.0000000	1.0000000	1.0000000	1.0000000

11.7 JACOBI POLYNOMIALS

Keywords
orthogonal polynomials, Jacobi polynomials

C Function and *Mathematica* Notebook
`Jacobi` (C); cell `Jacobi` in notebook `OrthoP` (*Mathematica*) M

Background
Jacobi polynomials, $P_n^{(\alpha,\beta)}(x)$ with x in $[-1, 1]$, are the most general orthogonal polynomials, having two parameters α and β. They are discussed in detail in Erdélyi et al [Erd53b, Section 10.8]. Their use for expanding functions in series of Jacobi polynomials is described in Sections 8.3 and 8.4 of Luke [Luk69]. Among applications are their use in the theory of rotational symmetry (angular momentum), where they are used to define reduced rotation matrix elements, as discussed in Section 6.2.4 of Thompson's text [Tho94b].

Because of their generality, Jacobi polynomials include as special cases all the other orthogonal polynomials with x in the same range. This is discussed in Section 3.13 of [Bec73], and detailed interrelation formulas are given in Section 22.5 of Abramowitz and Stegun [Abr64]. The connections are summarized in Figure 11.7.1.

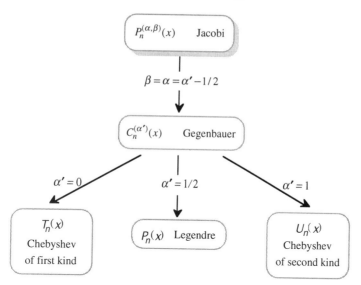

FIGURE 11.7.1 Connection between the Jacobi polynomial and the other polynomials that are orthogonal over x in -1 to 1.

Note that there are factors normalizing these functions that depend on n and α. Further, the weight functions in the orthogonality integrals also vary, as indicated in Table 11.1.5.

Function Definition
Jacobi polynomials, $P_n^{(\alpha,\beta)}(x)$, can be defined through the differential equation in Table 11.1.1, their integral representation in Table 11.1.2, the recurrence relation in Table 11.1.3, or their relation to the confluent hypergeometric function in Table 11.1.4. It

is usual—but not always necessary—to assume that $\alpha > -1$ and $\beta > -1$ so that the weight factor in the orthogonality integral (Table 11.1.5), which is given by $(1-x)^\alpha (1+x)^\beta$, does not contribute a singularity to the integral.

Explicit formulas for Jacobi polynomials—from $n = 0$ up to some maximum value $n = \mathtt{maxn}$—can be generated iteratively by putting the b coefficients into (11.1.3). One obtains thereby

$$P_0^{(\alpha,\beta)}(x) = 1 \qquad\qquad P_1^{(\alpha,\beta)}(x) = \left[\alpha - \beta + (2+\alpha+\beta)x\right]/2$$

$$P_n^{(\alpha,\beta)}(x) = \left[(b_{2n} + b_{3n}x)P_{n-1}^{(\alpha,\beta)}(x) - b_{4n}P_{n-2}^{(\alpha,\beta)}(x)\right]/b_{1n} \qquad n > 1$$

$$b_{1n} = 2n(n+\alpha+\beta)(2n+\alpha+\beta-2)$$

$$b_{2n} = (2n+\alpha+\beta-1)(\alpha^2 - \beta^2) \tag{11.7.1}$$

$$b_{3n} = (2n+\alpha+\beta-2)(2n+\alpha+\beta-1)(2n+\alpha+\beta)$$

$$b_{4n} = 2(n+\alpha-1)(n+\beta-1)(2n+\alpha+\beta)$$

$\boxed{M}$ The $P_n^{(\alpha,\beta)}(x)$ can be obtained iteratively by running cell `Jacobi` in *Mathematica* notebook `OrthoPIter`. For example, with `maxn = 2`, we obtain the polynomials given in Table 11.7.1.

TABLE 11.7.1 The lowest-order Jacobi polynomials.

n	$P_n^{(\alpha,\beta)}(x)$
0	1
1	$\dfrac{\alpha - \beta}{2} + \dfrac{2+\alpha+\beta}{2}x$
2	$\dfrac{-4-(\alpha+\beta)+(\alpha-\beta)^2}{8} + \dfrac{(\alpha-\beta)(3+\alpha+\beta)}{4}x$ $+ \dfrac{12+7(\alpha+\beta)+(\alpha+\beta)^2}{8}x^2$

As can be seen from the iteration formula (11.7.1), the complexity of the coefficients of the Jacobi polynomials increases very rapidly with n.

Visualization

Jacobi polynomials are shown in Figure 11.7.2. Below each surface plot are indicated the polynomials to which this $P_n^{(\alpha,\beta)}(x)$ is proportional, omitting n-dependent factors. The connections are established through Figure 11.7.1 and relations of the $P_n^{(\alpha,\beta)}(x)$ to the hypergeometric function (F) given in Section 10.8 of [Erd53b].

Note that in Figure 11.7.2 the surfaces on the diagonal of the graphics array are not exactly the same as Figures 11.3.1 (T_n), 11.3.2 (P_n), and 11.6.1 (U_n) because their relation to the Jacobi polynomials may depend on n.

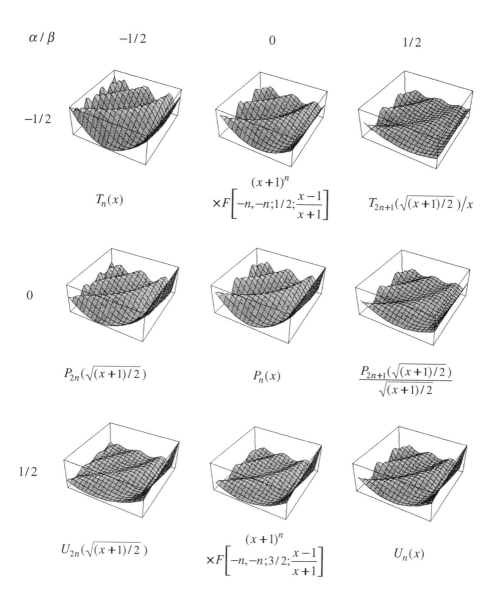

α / β $-1/2$ 0 $1/2$

$-1/2$

$T_n(x)$ $(x+1)^n \times F\left[-n,-n;1/2;\dfrac{x-1}{x+1}\right]$ $T_{2n+1}(\sqrt{(x+1)/2})/x$

0

$P_{2n}(\sqrt{(x+1)/2})$ $P_n(x)$ $\dfrac{P_{2n+1}(\sqrt{(x+1)/2})}{\sqrt{(x+1)/2}}$

$1/2$

$U_{2n}(\sqrt{(x+1)/2})$ $(x+1)^n \times F\left[-n,-n;3/2;\dfrac{x-1}{x+1}\right]$ $U_n(x)$

FIGURE 11.7.2 Jacobi polynomials depend on α, β, n, and x. For each surface, $n = 2$ to 8 from front to back, while x ranges from -1 to 1 from left to right. (Cell `Jacobi` in *Mathematica* notebook `OrthoP`.)

Algorithm, Program, and Programming Notes

To compute $P_n^{(\alpha,\beta)}(x)$ numerically, we use the explicit polynomial expressions from Table 11.7.1 for $n < 3$, and the recurrence relation (11.7.1) for larger n, starting the iteration with the expressions for $n = 1$ and $n = 2$.

```
double Jacobi
  (double a, double b, int n, double x)
/* Jacobi Polynomial;
     parameters a & b, order n */

{
double aPb,aMb,PiM2,PiM1,ab2,floati,
  factor1,factor2,b1i,b2i,b3i,b4i,Pi;
int i;

/* Explicit for  n < 3 */
if ( n < 1 ) return    1.0;  /* P(a,b)0 */

aPb = a+b;
aMb = a-b;
PiM2 = 0.5*(aMb+(2+aPb)*x);

if ( n < 2 )        return    PiM2;

PiM1 = 0.125*(-4.0-aPb+aMb*aMb +
  x*(2.0*aMb*(3.0+aPb) +
  x*(12.0+aPb*(7.0+aPb)))));

if ( n < 3 ) return    PiM1;  /* P(a,b)2 */


ab2 = a*a - b*b;

/* Use iteration for  n >= 3  */
for ( i = 3; i <= n; i++ )
  {
  floati = i;
  factor1 = floati+aPb;
  factor2 = factor1+floati;
  b1i = 2.0*floati*factor1*(factor2-2.0);
  b2i = (factor2-1.0)*ab2;
  b3i = (factor2-2.0)*(factor2-1.0)*factor2;
  b4i = 2.0*(floati+a-1.0)*(floati+b
  -1.0)*factor2;
  Pi = ((b2i+b3i*x)*PiM1 - b4i*PiM2)/b1i;
  PiM2 = PiM1;
  PiM1 = Pi;
  }

  returnPi;
}
```

Test Values

Test values of the Jacobi polynomial $P_n^{(\alpha,\beta)}(x)$ can be generated by running cell `Jacobi` in *Mathematica* notebook `OrthoP`. Examples are given in Tables 11.7.2 and 11.7.3. The reflection symmetry about $x = 0$, $P_n^{(\alpha,\alpha)}(-x) = (-1)^n P_n^{(\alpha,\alpha)}(x)$, allows values for $x < 0$ to be checked readily if $\beta = \alpha$.

TABLE 11.7.2 **Exact test values of the Jacobi polynomials for** $\alpha = \beta = -1/2$.

$x \setminus n$	1	2	3
0	0	−0.3750000000	0
0.5	0.2500000000	−0.1875000000	−0.3125000000
1.0	0.5000000000	0.3750000000	0.3125000000

TABLE 11.7.3 **Exact test values of the Jacobi polynomials for** $\alpha = 1/2$, $\beta = 0$.

$x \setminus n$	1	2	3
0	0.2500000000	−0.5312500000	−0.1796875000
0.5	0.8750000000	0.1796875000	−0.4462890625
1.0	1.5000000000	1.8750000000	2.1875000000

REFERENCES ON ORTHOGONAL POLYNOMIALS

[Abr64]   Abramowitz, M., and I. A. Stegun, *Handbook of Mathematical Functions*, Dover, New York, 1964.

[Bec73]   Beckmann, P., *Orthogonal Polynomials for Engineers and Physicists*, Golem, Golden, Colorado, 1973.

[Erd53a] Erdélyi, A., et al, *Higher Transcendental Functions*, Vol. 1, McGraw-Hill, New York, 1953; reprint edition, Krieger, Malabar, Florida, 1981.

[Erd53b] Erdélyi, A., et al, *Higher Transcendental Functions*, Vol. 2, McGraw-Hill, New York, 1953; reprint edition, Krieger, Malabar, Florida, 1981.

[Fox68]   Fox, L., and I. B. Parker, *Chebyshev Polynomials in Numerical Analysis*, Oxford University Press, London, 1968.

[Luk69]   Luke, Y. L., *The Special Functions and Their Approximations*, Vol. 1, Academic Press, New York, 1969.

[Mor53]   Morse, P. M., and H. Feshbach, *Methods of Theoretical Physics*, McGraw-Hill, New York, 1953.

[Pre92]   Press, W. H., S. A. Teukolsky, W. T. Vetterling, and B. P. Flannery, *Numerical Recipes in C*, Cambridge University Press, New York, second edition, 1992.

[Qua91]   Quarteroni, A., in *Advances in Numerical Analysis*, Vol. 1, pp. 96–146, edited by W. Light, Clarendon Press, Oxford, 1991.

[Riv74]   Rivlin, T. J., *The Chebyshev Polynomials*, Wiley, New York, second edition, 1974.

[San59]   Sansone, G., *Orthogonal Functions*, Dover, New York, 1959.

[Sze75]   Szegö, G., *Orthogonal Polynomials*, American Mathematical Society Colloquium Publications, **23**, fourth edition, 1975.

[Tho94a] Thompson, W. J., Computers in Physics, **8**, 161 (1994).

[Tho94b] Thompson, W. J., *Angular Momentum*, Wiley, New York, 1994.

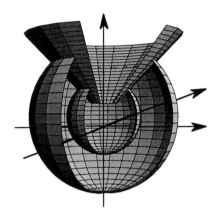

Chapter 12

LEGENDRE FUNCTIONS

The Legendre functions are often used when computing mathematical functions, especially in science and engineering. The chapter begins with an overview of these functions, then specializes in Sections 12.2–12.4 to Legendre functions as they appear when solving the Laplace equation in spherical, toroidal, and conical 3D coordinates, respectively. As in other chapters of the *Atlas*, relevant analytical properties of the functions are summarized, graphics that help when using them are shown, and C programs to calculate them efficiently are constructed.

12.1 OVERVIEW OF LEGENDRE FUNCTIONS

There is an extensive literature on the properties of Legendre functions and on relations between them. Emphasizing mathematical analysis are the texts by Whittaker and Watson [Whi27], by Magnus and Oberhettinger [Mag49, Chapter 4], and the compendium by Erdélyi et al [Erd53a, Chapter 3]. Extensive formulas are given in Jahnke and Emde [Jah45, Chapter 7], Gradshteyn and Ryzhik [Gra94, Sections 8.7 and 8.8] and in Abramowitz and Stegun [Abr64, Chapter 8]. Applications of Legendre functions to the physical sciences and engineering are stressed, for example, in Lebedev [Leb65, Chapters 7 and 8], in Morse and Feshbach [Mor53, Section 10.3], and in Moon and Spencer [Moo61].

For the Legendre functions the notation of Abramowitz and Stegun [Abr64] is followed, although there are many other definitions of these functions, usually differing only by phase factors and normalizations from those presented here. In particular, *Mathematica* uses a different phase than in [Abr64]. Legendre functions are conventionally defined in terms of solutions of Legendre's differential equation [Abr64, Section 8.1]

$$(1-z^2)\frac{d^2w}{dz^2} - 2z\frac{dw}{dz} + \left[\nu(\nu+1) - \frac{\mu^2}{1-z^2}\right]w = 0 \qquad (12.1.1)$$

in which $z = x + iy$ is a complex variable, as are the degree, v, and the order, μ. The solutions of this equation have singularities at $z = \pm 1$ and ∞ as ordinary branch points. Since (12.1.1) is of second order, it has two linearly-independent solutions for each z, v, and μ. The solution that is regular at $z = \pm 1$ is called the Legendre function of the first kind, $P_v^\mu(z)$, while the solution that is irregular at these points is the Legendre function of the second kind, $Q_v^\mu(z)$. Note that the Legendre polynomials in Section 11.6 are special cases of $P_v^\mu(z)$ having $\mu = 0$, v a nonnegative integer, and z a real variable in the range [–1, 1].

Because (12.1.1) is linear in w, its solutions are arbitrary to within normalization factors that are independent of z but may depend upon v and μ, which is the major source of confusion in their definitions, as warned about above. The normalization can be established by relating $P_v^\mu(z)$ and $Q_v^\mu(z)$ to hypergeometric functions, as in Section 3.2 of [Erd53a] and Section 8.1 of [Abr64], or by using integral representations of the Legendre functions, as in [Erd53a, Section 3.7] and [Abr64, Section 8.8].

Numerical computation of the Legendre functions $P_v^\mu(z)$ and $Q_v^\mu(z)$ is done most efficiently and accurately by developing algorithms for special choices of the variables v, μ, and z. Insight into how to devise such algorithms and use the Legendre functions is best obtained by visualizing the functions from many different viewpoints and with several choices of variables. This is done now, first for $P_v^\mu(z)$ (regular at $z = \pm 1$), then for $Q_v^\mu(z)$ (irregular at $z = \pm 1$).

12.1.1 Visualizing Legendre Functions of the First Kind

Legendre functions of the first kind depend upon three variables, all of which may be complex. It is therefore a challenge to map makers to show their general topography. They are shown first by surfaces in the complex plane $z = x + iy$ for $v = n$, a positive integer, ($n = 3$ in Figure 12.1.1) with $\mu = m$, an integer between 0 and n. Then $z = x$ is fixed and surfaces as functions of real and nonnegative variables for n and m (Figure 12.1.2) are shown, then surfaces at points on the imaginary axis $z = iy$ as n and m vary (Figure 12.1.3) are shown. Finally, for Legendre functions of the first kind with v not an integer, the branch-cut structure of the function in the complex z plane is shown. In Sections 12.2 to 12.4, other choices of variables that are more specialized and appropriate to common usages are made.

The following surfaces and curves of Legendre functions of the first kind can be generated readily by running the cell `LegendreP` in *Mathematica* notebook `Legendre`. (For the much simpler Legendre polynomial in Section 11.6 the cell name is `Legendre` but the notebook is `OrthoP`.) Particular care is needed to identify the appropriate Legendre function to display, since in many applications $z = x$ where x is in the range (–1, 1), which is part of the branch cut. As very clearly discussed in Section 3.4 of Erdélyi et al [Erd53a], one may choose as solutions of (12.1.1) linear combinations of $P_v^\mu(z)$ infinitesimally above and below the cut to define a new function, $\mathrm{P}_v^\mu(x)$, by

$$\mathrm{P}_v^\mu(x) \equiv \left[e^{i\mu\pi/2} P_v^\mu(x + i0) + e^{-i\mu\pi/2} P_v^\mu(x - i0) \right] / 2 \qquad (12.1.2)$$

It is very important to note that the function on the left, denoted by a Roman P, is usually distinct from the $P_v^\mu(x \pm i0)$ on the right, which are denoted by Italic P. (Every Roman is an Italian, but not all Italians are Romans.) The distinction of notations is not made in most references, such as [Abr64, Section 8.3], [Leb65, Section 7.12], and the *Mathematica* book [Wol91, p. 578]. In the *Mathematica* function `LegendreP`, the variable `LegendreType` has a default value of `Real`, which results in values from the two sides of the branch cut of $P_v^\mu(z)$ being combined to give functions with branch cuts from $-\infty$ to 1 and from 1 to ∞. In the following visualizations we display this type of Legendre function. The symmetric branch cuts are clear in Figure 12.1.4 but they do not occur in Figure 12.1.1 since $|x| \leq 1$.

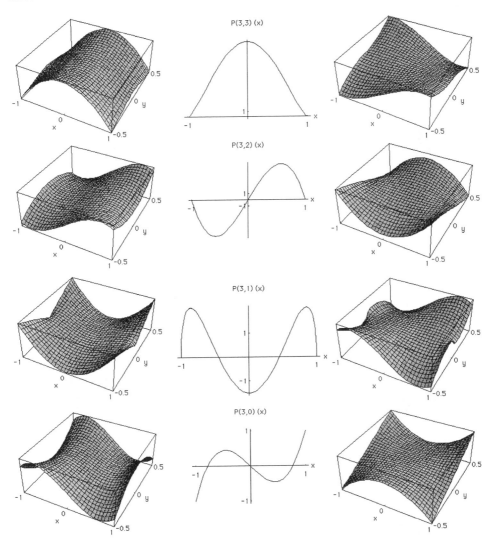

FIGURE 12.1.1 Legendre functions of the first kind in the complex plane for $n = 3$, $m = 0, 1, 2, 3$. Left side; surfaces of real parts. Center; curves of function for real variable x. Note the changes of scale between graphs. Right side; surfaces of imaginary parts. (Cell LegendreP in *Mathematica* notebook Legendre.)

For modified Legendre functions of the first kind, defined as used by default in *Mathematica* and which we denote by $P_\nu^\mu(z)$, the symmetry relation for integer $\nu = n$ and $\mu = m$ is

$$P_n^m(-z) = (-1)^{n-m} P_n^m(z) \qquad (12.1.3)$$

You can see the symmetry (12.1.3) in Figure 12.1.1, which shows the general behavior of $P_\nu^\mu(z)$ in the complex z plane, with the central column of curves illustrating the reflection symmetry for x in the range $[-1, 1]$. Outside this range, the surfaces show branch cuts, as can be verified by modifying the x range in cell `LegendreP` of *Mathematica* notebook `Legendre` then executing this cell.

The modified Legendre function of the first kind, $P_\nu^\mu(x)$, is shown in Figure 12.1.2 for three values of real argument x in $(-1, 1)$ with ν and μ varying to define the surfaces. This visualization clarifies design of the algorithm for computing the conventional Legendre function of the first kind.

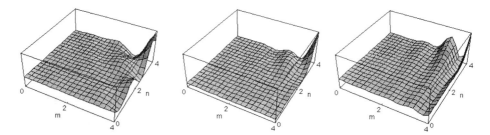

FIGURE 12.1.2 Legendre functions of the first kind as surfaces for real argument x fixed with $\mu = m$ and $\nu = n$ varying. From left to right $x = -0.5, 0, 0.5$. Note that the function is generally increasing with ν for fixed μ and x, thus determining the direction of stable iteration in the algorithm. (Cell `LegendreP` in *Mathematica* notebook `Legendre`.)

In Figure 12.1.3 we show surfaces of $P_\nu^\mu(x)$ with y fixed while ν and μ vary.

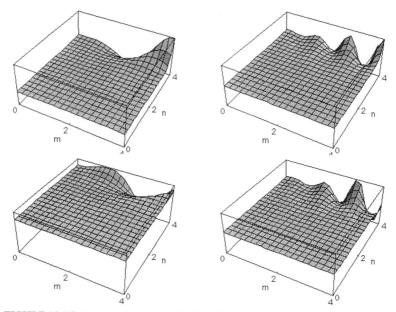

FIGURE 12.1.3 Legendre functions of the first kind as surfaces for imaginary argument y fixed with $\mu = m$ and $\nu = n$ varying. Left; $y = -2$. Right; $y = 2$. Top row; real part of function. Bottom row; imaginary part of function. (Cell `LegendreP` in *Mathematica* notebook `Legendre`.)

The last visualizations in our overview of Legendre functions of the first kind are shown in Figure 12.1.4. Noninteger orders of $P_\nu^\mu(z)$ are shown as surfaces for $\nu = 1.5$ and two values of μ. Because we have extended the x range beyond $x = \pm 1$, the branch cuts are evident.

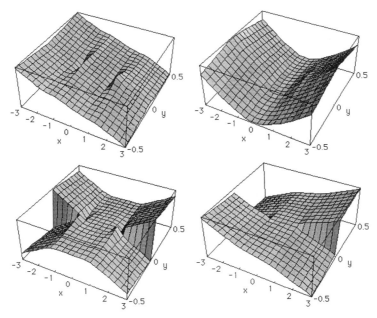

FIGURE 12.1.4 Legendre functions of the first kind and noninteger order (here $\nu = 1.5$) are shown as surfaces for μ fixed. Left; $\mu = 1.5$. Right; $\mu = 0.5$. Top row; real part of function. Bottom row; imaginary part of function. The phase convention of *Mathematica* is used in these views. (Cell LegendreP in *Mathematica* notebook Legendre.)

12.1.2 Visualizing Legendre Functions of the Second Kind

Just as for Legendre functions of the first kind, it is challenging to show the general topography of Legendre functions of the second kind. They are shown first by surfaces in the complex plane $z = x + iy$ for $\nu = n$, a positive integer, ($n = 3$ in Figure 12.1.5) with $\mu = m$, an integer between 0 and n. To elucidate the singularities near $x = \pm 1$, Figure 12.1.6 is developed. We then fix $z = x$ to show the surfaces as functions of real and nonnegative variables for ν and μ (Figure 12.1.7), then surfaces at points on the imaginary axis $z = iy$ as ν and μ vary are shown as Figure 12.1.8. Finally, the branch-cut structure of the function in the complex z plane is shown as Figure 12.1.9. Sections 12.2 to 12.4 show the functions for other choices of variables that are more specialized and appropriate to common usage.

Surfaces and curves of Legendre functions of the second kind can be generated by running the cell LegendreQ in *Mathematica* notebook Legendre. Care is needed to identify the appropriate Legendre function to display, since in many applications $z = x$ where x is in the range $(-1, 1)$, which is part of the branch cut. As discussed clearly in Section 3.4 of Erdélyi et al [Erd53a], one may choose as irregular solutions of (12.1.1) linear combinations of $Q_\nu^\mu(z)$ infinitesimally above and below the cut to define a new function, $Q_\nu^\mu(x)$, by

$$Q_v^\mu(x) \equiv e^{-i\mu\pi}\left[e^{-i\mu\pi/2}Q_v^\mu(x+i0) + e^{i\mu\pi/2}Q_v^\mu(x-i0)\right]\Big/2 \qquad (12.1.4)$$

The function on the left, denoted by Roman Q, is usually distinct from the $Q_v^\mu(x \pm i0)$ on the right, denoted by Italic Q. This notational distinction is not made in most references, such as [Abr64, Section 8.3], [Leb65, Section 7.12], and the *Mathematica* book [Wol91, p. 578]. In *Mathematica* function `LegendreQ` the variable `LegendreType` has a default value of `Real`, and values from the two sides of the branch cut of $Q_v^\mu(z)$ are combined to give functions with branch cuts from $-\infty$ to 1 and 1 to ∞. The visualizations show this type of Legendre function. The symmetric branch cuts are clear in Figure 12.1.9 but do not appear in Figure 12.1.5 because $|x| \le 0.8$.

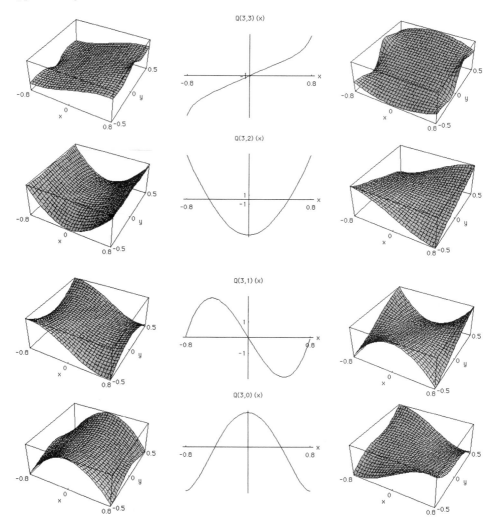

FIGURE 12.1.5 Legendre functions of the second kind in the complex plane for $n = 3$, $m = 0, 1, 2, 3$. Left side; surfaces of real parts. Center; curves of function for real variable x. Note the changes of scale between graphs and that the magnitude of x does not exceed 0.8 because of the singularities at $x = \pm1$. Right side; surfaces of imaginary parts. (Cell `LegendreQ` in *Mathematica* notebook `Legendre`.)

For the modified Legendre function of the second kind, as used by default in *Mathematica* and which we denote by $Q_\nu^\mu(z)$, the symmetry relation for $\nu = n$ and $\mu = m$ is

$$Q_n^m(-z) = (-1)^{n-m} \, Q_n^m(z) \tag{12.1.5}$$

Look at this in Figure 12.1.5, which shows the general behavior of $Q_n^m(z)$ in the complex z plane, with the central column of curves illustrating the reflection symmetry for x in the range $[-0.8, 0.8]$. Outside this range the surfaces show branch cuts, as can be verified by modifying the x range in cell LegendreQ of *Mathematica* notebook Legendre then executing this cell. Note that the *Mathematica* definition of $Q_\nu^\mu(z)$ differs by a phase from that used here and in Chapter 8 of Abramowitz and Stegun [Abr64]. For example, when $\mu = m$, an integer, the phase factor is $(-1)^m$.

The behavior of $Q_\nu^\mu(x)$ near $x = \pm1$ can be investigated analytically and visually by considering the leading term of the function near these singular points. From equations (10), (11), (16), and (17) in Section 3.9.2 of Erdélyi et al [Erd53a], we have that

$$Q_\nu^\mu(x) \propto (1-|x|)^{-|\mu|/2} \qquad \mu > \nu \tag{12.1.6}$$

The functions

$$q_\nu^\mu(x) \equiv (1-|x|)^{|\mu|/2} \, Q_\nu^\mu(x) \qquad \mu > \nu \tag{12.1.7}$$

are therefore regular at the singular points of $Q_\nu^\mu(x)$. Note that the regularizing factors are independent of ν. We see the regularizing behavior in Figure 12.1.6, in which $Q_\nu^\mu(x)$ is starting to diverge even near $x = 0.7$, whereas $q_\nu^\mu(x)$ is well-behaved as $x \to \pm1$.

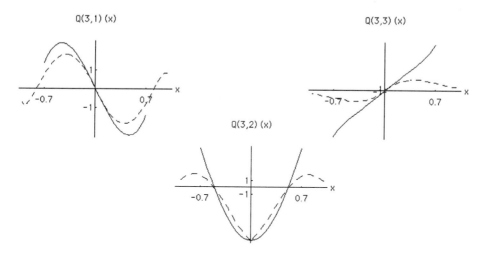

FIGURE 12.1.6 Legendre functions of the second kind, Q, emphasizing the singulariities that develop as x approaches ±1 (solid curves stopped at $x = \pm0.7$). The functions q, related to Q by (12.1.7), do not have such singularities (dashed curves). Curves shown for $n = 3$, $m = 3, 2, 1$. (Cell LegendreQ in *Mathematica* notebook Legendre.)

The behavior of $Q_\nu(x)$ near the singular points $x = \pm1$ is much more complicated than for $\mu \neq 0$. The leading terms are given by equations (12) and (18) in Section 3.9.2 of Erdélyi et al [Erd53a].

The modified Legendre function of the second kind, $Q_\nu^\mu(x)$, is shown in Figure 12.1.7 for three values of real argument x in $(-1, 1)$ with ν and μ varying to define the surfaces. This visualization clarifies devising the algorithm for computing the conventional Legendre function of the second kind.

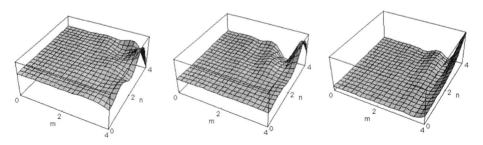

FIGURE 12.1.7 Legendre functions of the second kind as surfaces for real argument x fixed with μ and ν varying. From left to right $x = -0.5, 0, 0.5$. Note that the function is generally increasing with n for fixed m and x, thus determining the direction of stable iteration in the algorithm. (Cell `LegendreQ` in *Mathematica* notebook `Legendre.`)

Figure 12.1.8 shows surfaces of $Q_\nu^\mu(iy)$ with y fixed while ν and μ vary.

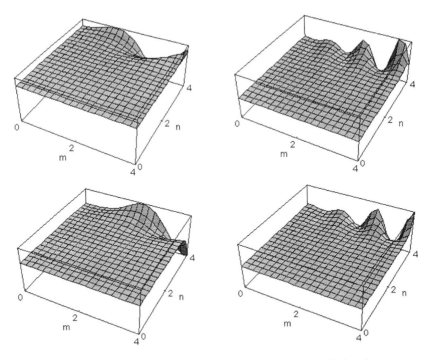

FIGURE 12.1.8 Legendre functions of the second kind as surfaces for imaginary argument y fixed with μ and ν varying. Left; $y = -2$. Right; $y = 2$. Top row; real part of function. Bottom row; imaginary part of function. (Cell `LegendreQ` in *Mathematica* notebook `Legendre.`)

The last visualizations in our overview of Legendre functions of the second kind are shown in Figure 12.1.9. Noninteger orders of $Q_\nu^\mu(z)$ are shown as surfaces for $\nu = 1.5$ and two values of μ. The branch cuts are evident because the x range is extended beyond $x = \pm 1$.

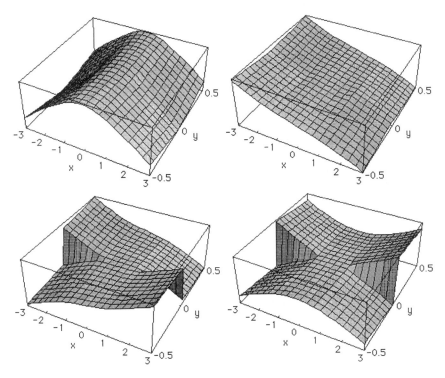

FIGURE 12.1.9 Legendre functions of the second kind and noninteger order (here $\nu = 1.5$) are shown as surfaces for μ fixed. Left; $\mu = 1.5$. Right; $\mu = 0.5$. Top row; real part of function. Bottom row; imaginary part of function. Note the branch cuts along the real axis. The phase convention of *Mathematica* is used in these views. (Cell `LegendreQ` in *Mathematica* notebook `Legendre`.)

12.1.3 Legendre Functions and Coordinate Systems

Legendre functions often appear when solving second-order differential equations in three-dimensional coordinate systems, particularly equations in which the Laplacian operator, ∇^2, appears. General discussions of the separability of the coordinates for the Laplacian are given in several texts, such as Morse and Feshbach [Moo53, Section 5.1] and Moon and Spencer [Moo61, Chapters 3 and 12].

In the following three sections discussion and computation of Legendre functions is specialized to variables of common interest, namely, those involved in solving the Laplace and Poisson equations in spherical, toroidal, and conical coordinates. The solutions then involve the Legendre functions of integer degree ν and order μ, of half-integer degree, and of degree that is $-1/2$ plus imaginary, respectively. Such Legendre functions are just those presented in Chapter 8 of Abramowitz and Stegun [Abr64].

12.2   SPHERICAL LEGENDRE FUNCTIONS

The Legendre functions P_n^m and Q_n^m in which the degree n and the order m are nonnegative integers arise most commonly when solving differential equations involving Laplacians, ∇^2, expressed in spherical polar coordinates. This section therefore begins with an overview of this familiar 3D coordinate system, mainly so that the less familiar toroidal and conical systems in Sections 12.3 and 12.4 can be related to it. We than procede with the usual analytical, graphical, and computational aspects of P_n^m and Q_n^m.

12.2.1 Spherical Polar Coordinates

Spherical polar coordinates (r,θ,ϕ) are related to Cartesian coordinates (x,y,z) by the coordinate transformations

$$x = r\sin\theta\cos\phi \qquad y = r\sin\theta\sin\phi \qquad z = r\cos\theta$$
$$r \geq 0 \qquad 0 \leq \theta < \pi \qquad 0 \leq \phi < 2\pi \tag{12.2.1}$$

To establish analogies for understanding the toroidal and conical coordinate systems in Sections 12.3.1 and 12.4.1, we display surfaces for r constant (spheres centered on the origin), for θ constant (cones with apex at the origin), and for ϕ constant (half planes with the edge along the z axis). These are shown as cutaway views in Figure 12.2.1.

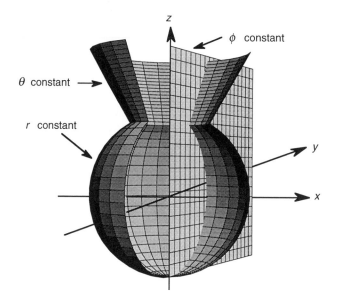

FIGURE 12.2.1 Spherical polar coordinates. Coordinate surfaces are spheres for r constant, right circular cones for θ constant, and half planes for ϕ constant. (Adapted from cell `SphereCoords` in *Mathematica* notebook `LegendreCoords`.)

Consider, for example, a potential function V. In spherical polar coordinates the Laplacian depends on θ as follows

$$r^2 \, \mathbf{V}^2_{\theta,\phi} \, V = \frac{1}{\sin\theta} \frac{\partial}{\partial\theta} \left(\sin\theta \frac{\partial V}{\partial\theta} \right) + \frac{1}{\sin^2\theta} \frac{\partial^2 V}{\partial\phi^2} \qquad (12.2.2)$$

Solutions for V that are of the form

$$V = V(r,\theta,\phi) = R(r) \, w(\cos\theta) \, \Phi_m(\phi) \qquad (12.2.3)$$

have for the θ dependence of their solutions $w = \mathrm{P}_n^m(\cos\theta)$, $w = \mathrm{Q}_n^m(\cos\theta)$, or any linear combination of these. The ϕ dependence is $\Phi_m(\phi) = \cos m\phi$, $\Phi_m(\phi) = \sin m\phi$, or some linear combination of them. If unique solutions to V are required, then m must be an integer. Correspondingly, the degree $\nu = n$ and the order $\mu = m$ of the Legendre functions will be integers satisfying $n \geq m$, as we assume in the rest of this section. The product of the θ– and ϕ-dependent functions in (12.2.3) when suitably normalized are spherical harmonics.

12.2.2 Legendre Functions of the First Kind for Integer m and n

Keywords
associated Legendre functions, Legendre functions of first kind, spherical Legendre function

C Function and *Mathematica* Notebook
PNM (C); cell `LegendreP` in notebook `Legendre` (*Mathematica*)

Background
The introductory material in Section 12.1 and its references provide background on these functions. Applications of the $\mathrm{P}_n^m(\cos\theta)$ in the general study of rotational symmetry (angular momentum), in which $n = \ell$ is identified with the angular momentum and m is its projection onto the z axis, are given in Thompson's text [Tho94], particularly Sections 4.2.1 and 4.2.2. Extensive lists of formulas are given in Chapter 3 of Erdélyi et al [Erd53a] and in Chapter 8 of Abramowitz and Stegun [Abr64].

We distinguish between the Legendre functions on the cut, $\mathrm{P}_n^m(x)$ with $|x| \leq 1$, and the general Legendre function $P_n^m(z)$—note the distinction between Roman P and italic P, as discussed in Section 12.1.1.

Function Definition
Of the many formulas that can be used to define the P_n^m and to relate them to other functions, only those relevant for their computation are cited here. We begin by defining the *modified Legendre function* $S_n^m(x)$ as

$$S_n^m(x) \equiv \sqrt{\frac{(n-m)!}{(n+m)!}} \, \mathrm{P}_n^m(x) \qquad n \geq 0 \qquad n \geq m \qquad (12.2.4)$$

This modified function varies less rapidly with m for given n (and vice versa) than does P_n^m, as is clear from the normalization integral (12.2.8) below. In the following, we write formulas in terms of both S_n^m and P_n^m whenever relevant. A similar renormalization is made in the tabulations by Belousov [Bel62] for Legendre functions up to large integer degree and order.

Symmetry relations reduce the range over which the functions must be computed directly. For n and m integers, one has [Abr64, 8.2.1]

$$\mathrm{P}_{-n}^m(x) = \mathrm{P}_{n-1}^m(x) \qquad (12.2.5)$$

and also [Abr64, 8.2.5]

$$P_n^{-m}(x) = \frac{(n-m)!}{(n+m)!} P_n^m(x) \qquad S_n^{-m}(x) = S_n^m(x) \qquad n \geq 0 \tag{12.2.6}$$

The symmetry with respect to x is

$$P_n^m(-x) = (-1)^{n-m} P_n^m(x) \tag{12.2.7}$$

Therefore, one need consider only $n \geq m \geq 0$ and $x \geq 0$ for direct computation.

The normalization of the Legendre functions on the cut, P_n^m, and of modified Legendre functions, $S_n^m(x)$, is

$$\int_{-1}^{1} \left[P_n^m(x) \right]^2 dx = \frac{2(n+m)!}{(2n+1)(n-m)!} \qquad \int_{-1}^{1} \left[S_n^m(x) \right]^2 dx = \frac{2}{(2n+1)} \tag{12.2.8}$$

This result justifies definition (12.2.4).

Useful recurrence relations on n for fixed m are

$$P_n^m(x) = \left[(2n-1) x P_{n-1}^m(x) - (n+m-1) P_{n-2}^m(x) \right] \big/ (n-m)$$

$$S_n^m(x) = \left[(2n-1) x S_{n-1}^m(x) - \sqrt{(n-1)^2 - m^2}\, S_{n-2}^m(x) \right] \big/ \sqrt{n^2 - m^2} \tag{12.2.9}$$

To initiate the recurrence we use [Abr64, 8.6.17] with (12.2.6) to obtain for $m \geq 0$

$$P_{m-1}^m(x) = 0 \qquad P_m^m(x) = \frac{(2m)! (1-x^2)^{m/2}}{2^m\, m!}$$

$$S_{m-1}^m(x) = 0 \qquad S_m^m(x) = \frac{\sqrt{(2m)!} (1-x^2)^{m/2}}{2^m\, m!} \tag{12.2.10}$$

The relations in (12.2.9) and (12.2.10) are used to generate P_n^m and S_n^m analytically and numerically by running the cell LegendreP in the *Mathematica* notebook Legendre. Algebraic expressions for $n = 0$ to $n = 3$ are given in Tables 12.2.1 and 12.2.2. Note that $P_n^m(x)$ and $S_n^m(x)$ are polynomials in x only if m is even. In spite of this fact, the terminology "associated Legendre polynomials" is often used for these functions. Beware that the *Mathematica* function LegendreP gives a function that differs by the phase factor $(-1)^m$ from that used here and in Abramowitz and Stegun [Abr64].

TABLE 12.2.1 Lowest-order Legendre functions of the first kind for x in $[-1, 1]$. Adapted from output of cell LegendreP in *Mathematica* notebook Legendre.

n/m	0	1	2	3
0	1			
1	x	$\sqrt{1-x^2}$		
2	$-(1-3x^2)/2$	$3x\sqrt{1-x^2}$	$3(1-x^2)$	
3	$-x(3-5x^2)/2$	$-3\sqrt{1-x^2}\,(1-5x^2)/2$	$15x(1-x^2)$	$15\sqrt{1-x^2}\,(1-x^2)$

TABLE 12.2.2 Modified lowest-order Legendre functions of the first kind for x in $[-1, 1]$, as given by (12.2.4). Adapted from output of cell LegendreP in *Mathematica* notebook Legendre.

n/m	0	1	2	3
0	1			
1	x	$\sqrt{(1-x^2)/2}$		
2	$-(1-3x^2)/2$	$x\sqrt{3(1-x^2)/2}$	$\sqrt{3/8}\,(1-x^2)$	
3	$-x(3-5x^2)/2$	$-\sqrt{3(1-x^2)(1-5x^2)/4}$	$\sqrt{15/2}\,x(1-x^2)/2$	$\sqrt{5(1-x^2)}\,(1-x^2)/4$

Visualization

Many views of Legendre functions of the first kind are given in Section 12.1. The present visualizations emphasize aspects that are relevant for computing $P_n^m(x)$ by using iteration formula (12.2.9), in which m and x are fixed and n increases by unit steps, starting at m. Figure 12.2.2 shows relevant graphics.

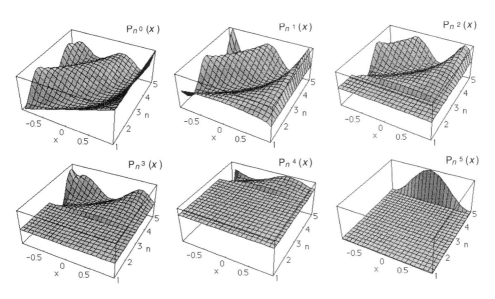

FIGURE 12.2.2 Surfaces of Legendre functions of the first kind as functions of x from -1 to 1 for n from 1 to 5. Each surface has a fixed m value given by the superscript. Note that the function is zero if $n < m$. The curves are shown as continuous functions of n, although for spherical Legendre functions only integer n are relevant. (Adapted from the cell LegendreP in *Mathematica* notebook Legendre.)

Note that in Figure 12.2.2 we use the phase convention of Abramowitz and Stegun [Abr64, Chapter 8]; in *Mathematica* there is a phase factor of $(-1)^m$. For any nonzero $P_n^m(x)$, one having $n \geq m$, the figure shows that the function generally increases in magni-

M

tude as n increases. The iteration should therefore be taken in the direction of increasing n in order to minimize relative errors from roundoff.

The modified Legendre function of the first kind—$S_n^m(x)$ defined by (12.2.4)—removes much of this n dependence, as shown in Figure 12.2.3. For example, the surface of $S_n^2(x)$ is flatter (relative to its extreme values that determine the vertical scale of the surface) than is $P_n^2(x)$. This consideration becomes more important as n increases for fixed m.

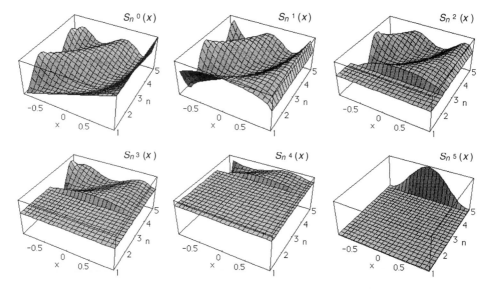

FIGURE 12.2.3 Surfaces of modified Legendre functions of the first kind as functions of x from −1 to 1 for n from 1 to 5. Each surface has a fixed m value given by the superscript. The function is zero if $n < m$, and the curves are shown as continuous functions of n, although for spherical Legendre functions only integer n are relevant. Note that for $m > 0$ the modified function generally varies more slowly with n than does the associated function (Figure 12.2.2). (Adapted from the cell `LegendreP` in *Mathematica* notebook `Legendre`.)

Algorithm, Program, and Programming Notes

To compute $P_n^m(x)$ numerically for $n \geq 0$ and $n \geq m \geq 0$, we use the explicit expressions in Table 12.2.1 for $n < 2$ and the recurrence relation on n for $S_n^m(x)$ given by (12.2.9) for larger n.

```
double PNM(int n, int m, double x)
/* Legendre Function of First Kind
   degree n >= 0, order m >= 0, argument x  */
{
const double ln2 = 0.693147180560;
double  LogGamma(double x),xsq,oneMxsq,xsqrt,
  fm,SnM2m,SnM1m,tiM1,M1Root,Root,Snm;
int  nMm,m2,i;

if ( m > n )  return  0.0;

if ( n == 0 )  return  1.0;
```

```
nMm = n - m;
if ( x < 0 )
  { x = -x; /* then check phase */
    return  ( nMm > 2*(nMm/2) ) ? -PNM(n,m,x):
PNM(n,m,x);
  }

/* Now  n >= m > 0  and  x >= 0  */
xsq = x*x;
oneMxsq = 1.0 - xsq;
xsqrt = sqrt(oneMxsq);
if ( n == 1 )
  {        /* Check  m = 0  or  m = 1  */
   return   ( m == 0 ) ?  x:  xsqrt;
  }

if ( n == 2 )
  {        /* Check m = 0,1, or 2  */
  return ( m == 0 ) ? 1.5*xsq-0.5:
          ( m == 1 ) ? 3.0*x*xsqrt: 3.0*oneMxsq;
  }

/* Now  n > 2, so use iteration */
m2 = m*m;
fm = m;
if ( n == m )
  return exp(LogGamma(2.0*fm+1.0)-fm*ln2-
    LogGamma(fm+1.0))*pow(xsqrt,fm);
SnM2m = 0.0;
SnM1m =
  exp(0.5*LogGamma(2.0*fm+1.0)-fm*ln2-
    LogGamma(fm+1.0))*pow(xsqrt,fm);
tiM1 = 2*m+1;
M1Root = 0.0;
Root = sqrt(tiM1);

for ( i = m+1; i <= n; i++ )
  {
  Snm = (tiM1*x*SnM1m - M1Root*SnM2m)/Root;
  tiM1 += 2.0;
  M1Root = Root;
  Root = sqrt( (double) ((i+1)*(i+1)-m2));
  SnM2m = SnM1m;
  SnM1m = Snm;
  }

return exp(0.5*(LogGamma((double)(n+m+1))
              -LogGamma((double)(n-m+1))))*Snm;
}
```

Starting values for the iteration are obtained from (12.2.10). All factorials are calculated in terms of the logarithms of corresponding gamma functions by using the function LogGamma. Factorials in the numerators and denominators of (12.2.10) and (12.2.4) are combined so as to minimize the likelihood of overflow.

Test Values

Check values of $P_n^m(x)$ can be generated by running the `LegendreP` cell in *Mathematica* notebook `Legendre`. Examples are given in Table 12.2.3.

TABLE 12.2.3 **Test values to 10 digits of the Legendre functions of the first kind of integer degree *n* and order *m*.**

$x \setminus m$	0	1	2	3
n= 1				
0.0	0	1.0000000000		
0.5	0.5000000000	0.8660254038		
1.0	1.0000000000	0		
n= 2				
0.0	−0.5000000000	0	3.0000000000	
0.5	−0.1250000000	1.299038106	2.2500000000	
1.0	1.0000000000	0	0	
n= 4				
0.0	0.3750000000	0	−7.5000000000	0
0.5	−0.2890625000	−1.3531646930	4.2187500000	34.09975027
1.0	1.0000000000	0	0	0

12.2.3 Legendre Functions of the Second Kind for Integer *m* and *n*

Keywords
associated Legendre functions, Legendre functions of second kind, spherical Legendre function

C Function and *Mathematica* Notebook

QN and QNM (C); cell `LegendreQ` in notebook `Legendre` (*Mathematica*)

Background
The introductory material in Section 12.1 (particularly Section 12.1.2) and its references provide background on these functions. Extensive lists of formulas are given in Chapter 3 of Erdélyi et al [Erd53a] and in Chapter 8 of Abramowitz and Stegun [Abr64]. We distin-

guish the Legendre functions on the cut, $Q_n^m(x)$ with $|x| \le 1$, from the general Legendre function $Q_n^m(z)$; note the distinction between Roman Q and italic Q, as discussed in Section 12.1.2.

The major difficulties with the Legendre functions of the second kind is that they have singularities at $x = \pm 1$. The nature of the singularities is discussed in relation to Figure 12.1.6 and equations (12.1.6) and (12.1.7). Computationally, little work has been done on these functions, except for $Q_n(z) = Q_n^0(z)$, probably because of analytical and numerical difficulties of handling the functions near the singularities. We provide C functions for $z = x$, real, and $m = 0, 1, 2$.

Function Definition

Formulas relevant for the computation of the Legendre functions of the second kind are cited, rather than the many formulas that can be used to discuss their analytical properties, as in Chapter 3 of Erdélyi et al [Erd53a]. We must also distinguish between computing the functions on the cut—x in $(-1, 1)$—which are the Q_n^m, and the function defined throughout the complex plane, $Q_n^m(z)$.

Our basic computational definition is that for $m = 0$ and on the cut, equations (27) and (28) of [Erd53a, Section 3.6.2] give

$$Q_n(x) = \frac{1}{2} P_n(x) \ln\left(\frac{1+x}{1-x}\right) - W_{n-1}(x) \qquad |x| < 1 \qquad (12.2.11)$$

where the polynomial of degree $n-1$ is

$$W_{n-1}(z) \equiv \sum_{k=0}^{[(n-1)/2]} \frac{(2n-4k-1)}{(n-k)(2k+1)} P_{n-2k-1}(z) \qquad (12.2.12)$$

with $[x]$ denoting the integer part of x. Given the Legendre polynomials (Section 11.6), we can therefore compute the function of the second kind analytically or numerically. A similar formula holds for the rest of the complex plane, namely,

$$Q_n(z) = \frac{1}{2} P_n(z) \ln\left(\frac{z+1}{z-1}\right) - W_{n-1}(z) \qquad z \text{ not in } [-1, 1] \qquad (12.2.13)$$

For $m > 0$ the Rodrigues' formulas from [Erd53a, Section 3.6.1, (5) and (7)] are used, namely

$$Q_n^m(x) = (-1)^m (1-x^2)^{m/2} \frac{d^m Q_n(x)}{dx^m} \qquad |x| < 1 \qquad (12.2.14)$$

and, elsewhere in the complex plane,

$$Q_n^m(z) = (z^2 - 1)^{m/2} \frac{d^m Q_n(x)}{dx^m} \qquad z \text{ not in } [-1, 1] \qquad (12.2.15)$$

These formulas are used to obtain the analytical expressions in Table 12.2.3 below.

The symmetries of the Legendre functions of the second kind with respect to $z \rightarrow -z$ follow from (12.2.11) to (12.2.15) as

$$Q_n^m(-x) = -(-1)^{n-m} P_n^m(x) \qquad |x| < 1 \qquad (12.2.16)$$

and

$$Q_n^m(-z) = (-1)^{n+1} Q_n^m(z) \qquad z \text{ not in } [-1, 1] \qquad (12.2.17)$$

Recurrence relations on n for fixed m are

$$Q_n^m(z) = \left[(2n-1)z Q_{n-1}^m(z) - (n+m-1)Q_{n-2}^m(z)\right]/(n-m) \qquad (12.2.18)$$

which also holds for the function on the cut. Values from Table 12.2.3 for x on the cut are used to initiate the recurrence.

TABLE 12.2.3 Lowest-order Legendre functions of the second kind for x in $[-1, 1]$. Adapted from the output of cell LegendreQ in *Mathematica* notebook Legendre.

m	$Q_0^m(x)$	$Q_1^m(x)$	$Q_2^m(x)$
0	$\frac{1}{2}\ln\left(\frac{1+x}{1-x}\right)$	$\frac{x}{2}\ln\left(\frac{1+x}{1-x}\right)-1$	$\frac{(3x^2-1)}{4}\ln\left(\frac{1+x}{1-x}\right)-\frac{3x}{2}$
1	$\frac{x}{\sqrt{1-x^2}}$	$-\frac{\sqrt{1-x^2}}{2}\ln\left(\frac{1+x}{1-x}\right)-\frac{x}{\sqrt{1-x^2}}$	$-\frac{3x\sqrt{1-x^2}}{2}\ln\left(\frac{1+x}{1-x}\right)-\frac{3x^2-2}{\sqrt{1-x^2}}$
2	$-\frac{(1+x^2)}{1-x^2}$	$\frac{2}{1-x^2}$	$\frac{3(1-x^2)}{2}\ln\left(\frac{1+x}{1-x}\right)+\frac{x(5-3x^2)}{1-x^2}$

Notice that $Q_n^m(x)$ on the cut exists for $n < m$, but relation (12.2.18) cannot then be used to produce the function for $n > m$. In the case of $P_n^m(x)$, the analogue of (12.2.14) shows that this function is zero if $n < m$, since $P_n(x)$ is a polynomial of degree n and therefore has vanishing derivatives for $m > n$. The behavior of the function near the singularity points follows (12.1.6).

Although there is a branching of the function $Q_n^m(z)$ along the rest of the real axis other than the cut, as seen in Figure 12.1.9, it is common to define $Q_n^m(x)$ by using relations (12.2.13) to 12.2.15) with $z = x$. One thereby obtains the Legendre functions of the second kind for $|x| > 1$ that are given in Table 12.2.4. These functions are tabulated to 6-digit accuracy in [Low45] and are computed by our C function QNM that is described below.

TABLE 12.2.4 Lowest-order Legendre functions of the second kind for $|x| > 1$. Adapted from the output of cell LegendreQ in *Mathematica* notebook Legendre.

m	$Q_0^m(x)$	$Q_1^m(x)$	$Q_2^m(x)$
0	$\frac{1}{2}\ln\left(\frac{x+1}{x-1}\right)$	$\frac{x}{2}\ln\left(\frac{x+1}{x-1}\right)-1$	$\frac{(3x^2-1)}{4}\ln\left(\frac{x+1}{x-1}\right)-\frac{3x}{2}$
1	$\frac{-1}{\sqrt{x^2-1}}$	$\frac{\sqrt{x^2-1}}{2}\ln\left(\frac{x+1}{x-1}\right)-\frac{x}{\sqrt{x^2-1}}$	$\frac{3x\sqrt{x^2-1}}{2}\ln\left(\frac{x+1}{x-1}\right)-\frac{3x^2-2}{\sqrt{x^2-1}}$
2	$\frac{2x}{x^2-1}$	$\frac{2}{x^2-1}$	$\frac{3(x^2-1)}{2}\ln\left(\frac{x+1}{x-1}\right)+\frac{x(5-3x^2)}{x^2-1}$

Visualization

Section 12.1.2 provides many views of Legendre functions of the second kind. The present visualizations emphasize aspects relevant for computing $Q_n^m(x)$, taking into account the singularities at $x = \pm1$. The nature of such singularities is shown in Figure 12.1.6 and in Figure 12.2.4.

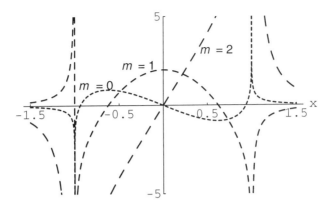

FIGURE 12.2.4 Curves of spherical Legendre functions of the second kind as functions of x from -1.5 to 1.5 for $n = 2$ and $m = 0$, 1, 2. Note the singularities at $x = \pm 1$ and the reflection symmetries given by (12.2.16) and (12.2.17). (Adapted from the cell Legendre Q in *Mathematica* notebook Legendre.)

Since we use iteration formula (12.2.18), in which m and x are fixed and n increases by unit steps, starting at m, it is interesting to display relevant graphics as in Figure 12.2.5.

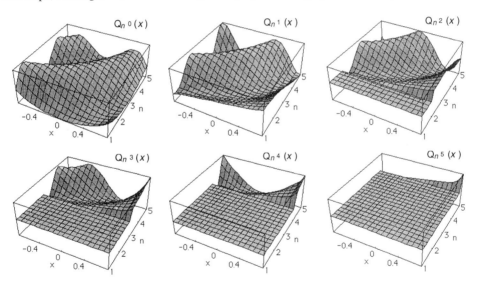

FIGURE 12.2.5 Surfaces of Legendre functions of the second kind as functions of x from -0.8 to 0.8 for n from 1 to 5. Each surface has a fixed m value given by the superscript. Note that the function is taken to be zero if $n < m$. The curves are shown as continuous functions of n, although for spherical Legendre functions only integer n are relevant. (Adapted from the cell LegendreQ in *Mathematica* notebook Legendre.)

In Figure 12.2.5 the phase convention of Abramowitz and Stegun [Abr64, Chapter 8] is used; in *Mathematica* there is a phase factor of $(-1)^m$. The figure shows that the function $Q_n^m(x)$ generally increases in magnitude as n increases for fixed m and x. The iteration

(12.2.18) should therefore be taken in the direction of increasing n in order to minimize relative errors from roundoff.

Algorithm, Program, and Programming Notes

We provide two C functions, QN and QNM. Function QN is for $m = 0$ and $|x| < 1$ only; it uses (12.2.11) to compute $Q_n(x)$ in terms of Legendre polynomials (Section 11.6). It is therefore useful if you need $Q_n(x)$ and $P_n(x)$ together. If you need several P's and Q's for the same x value, then it is more efficient to make an array indexed by n when you first compute the $P_n(x)$, then to use these in (12.2.12) and (12.2.13) for the $Q_n(x)$. Here is the program for QN.

```
double QN(int n, double x)
/* Legendre Function of Second Kind
   degree n >= 0, order m = 0, argument x in (-1,1) */
{
double  Legendre(int n, double x),Lnx,floatn,
  sum,snum,snMk,sden;
int   kmax,k;

Lnx = log((1.0+x)/(1.0-x));

if ( n == 0 )  return  0.5*Lnx;

if ( n == 1 )  return  0.5*x*Lnx - 1.0;

if ( n == 2 )  return  (0.75*x*x-0.25)*Lnx-1.5*x;

floatn = n;
sum = 0.5*Legendre(n,x)*Lnx-(2.0-
1.0/floatn)*Legendre(n-1,x);
kmax = (n - 1)/2;
snum = 2*n - 5;
snMk = n - 1;
sden = 3.0;
for ( k = 1; k<=kmax; k++ )
  {
  sum -= (snum/(snMk*sden))*Legendre(n-2*k-1,x);
  snum -= 4.0;
  snMk -= 1.0;
  sden += 2.0;
  }
return  sum;
}
```

Function QNM computes $Q_n^m(x)$ numerically for $n \geq m = 0,1$, or 2, and any x with $|x| \neq 1$. We start with the explicit expressions in Tables 12.2.3 and 12.2.4, then—if needed (for $n > 2$)—the recurrence relation on n for $Q_n^m(x)$, given by (12.2.18), is used. To modify the function for $m > 2$ the necessary derivatives can be computed algebraically. Alternatively, the methods described in [Low45] can be programmed, but they eventually require algebraic computations to develop the starting expressions. Here is our C function for $Q_n^m(x)$.

```
double QNM(int n, int m, double x)
/* Legendre Function of Second Kind
   degree n >= m, order m = 0,1,2, argument x    */
{
double  phase,Lnx,xsq,xsqM1,xsqrt,QnM2m,QnM1m,
 x3pwr,tiM2,tiM1,tiM,Qnm;
int  i;i

if ( fabs(x) == 1.0 )
 {
 printf("\n!!QNM singular at |x|=1; zero returned");
 return  0.0;
 }

if ( !( (m == 0) || (m == 1) || (m == 2) ) )
 {
 printf("\n!!m not 0,1, or 2 in QNM; zero returned");
 return  0.0;
 }

if ( n < m )  return 0.0;

phase = 1.0;
if ( fabs(x) < 1.0 )
 {
 Lnx = 0.5*log((1.0+x)/(1.0-x));
 xsq = x*x;
 xsqM1 = 1.0 - xsq;
 xsqrt = sqrt(xsqM1);

 switch ( m )
  {
  case 0 : if ( n == 0 ) return Lnx;
           QnM2m = x*Lnx - 1.0;
           if ( n == 1 )  return QnM2m;
           QnM1m = (1.5*xsq-0.5)*Lnx-1.5*x;
           break;
  case 1 : QnM2m = -xsqrt*Lnx - x/xsqrt;
           if ( n == 1 )  return QnM2m;
           QnM1m =
            -3.0*x*xsqrt*Lnx-(3.0*xsq-2.0)/xsqrt;
           break;
  case 2 : QnM2m = 2.0/xsqM1;
           QnM1m =
            3.0*xsqM1*Lnx+x*(5.0-3.0*xsq)/xsqM1;
           break;
  }
 if ( n == 2 )  return QnM1m;
 } /* ends |x| < 1 initialization */
else
 { /* |x| > 1 */
 if ( x < -1.0 )
  { /* Phase is  -1 to power n+1  */
  phase =  ( n%2 > 0 ) ? 1.0: -1.0;
  x = -x;
  }
```

```
Lnx = 0.5*log((x+1.0)/(x-1.0));
xsq = x*x;
xsqM1 = xsq - 1.0;
xsqrt = sqrt(xsqM1);

switch ( m )
  {
  case 0 : if ( n == 0 )   return phase*Lnx;
           QnM2m = x*Lnx - 1.0;
           if ( n == 1 )   return phase*QnM2m;
           QnM1m = (1.5*xsq-0.5)*Lnx-1.5*x;
           break;

  case 1 : QnM2m = xsqrt*Lnx - x/xsqrt;
           if ( n == 1 )   return phase*QnM2m;
           QnM1m =
             3.0*x*xsqrt*Lnx-(3.0*xsq-2.0)/xsqrt;
           break;

  case 2 : QnM2m = 2.0/xsqM1;
           QnM1m =
             3.0*xsqM1*Lnx+x*(5.0-3.0*xsq)/xsqM1;
           break;
  }

  if ( n == 2 )   return phase*QnM1m;
  } /* ends |x| > 1 initialization */

/* Iterate on Q(n,m) for any x; start with Q(3,m) */
tiM2 = 2+m;
tiM1 = 5.0;
tiM = 3-m;

for ( i = 3; i <= n; i++ )
  {
  Qnm = (tiM1*x*QnM1m - tiM2*QnM2m)/tiM;
  tiM2 += 1.0;
  tiM1 += 2.0;
  tiM += 1.0;
  QnM2m = QnM1m;
  QnM1m = Qnm;
  }
return  phase*Qnm;
}
```

Test Values

M

Check values of $Q_n^m(x)$ can be generated by running the LegendreQ cell in *Mathematica* notebook Legendre. The builtin *Mathematica* function LegendreQ provides values for $|x| < 1$ that have the same m dependence of the phase as our function. For x between 1 and 10, values to 6 digits are tabulated in Lowan [Low45] for n up to 10, for m up to 4. Table entries in [Low45] differ by a factor $(-1)^m$ from our sample values in Table 12.2.5, which otherwise agree with these two sources to 10 and 6 digits, respectively.

TABLE 12.2.5 Test values to 10 digits of the Legendre functions of the second kind and of integer degree n and order m.

$x \setminus m$	0	1	2
$n = 1$			
0.5	–0.7253469278	–1.053063345	0
1.5	0.2070784343	–0.4419376421	0
$n = 2$			
0.5	–0.818663268	0.7298060598	4.069272158
1.5	0.0635669991	–0.1998650073	0.9176960858
$n = 4$			
0.5	0.44401745260	1.934086611	–11.03678137
1.5	0.007095922339	–0.03656892314	0.2400431644

12.3 TOROIDAL LEGENDRE FUNCTIONS

The toroidal Legendre functions arise most commonly when solving differential equations involving Laplacians, $\mathbf{\nabla}^2$, expressed in toroidal coordinates. This section therefore begins with an overview of this unfamiliar 3D coordinate system, before proceeding in Section 12.3.2 with the analytical, graphical, and computational aspects of toroidal functions.

12.3.1 Toroidal Coordinates

Toroidal coordinates—(c, θ, ϕ) together with the parameter η—are related to Cartesian coordinates (x, y, z) by the transformations

$$x = \frac{c \sinh \eta \cos \phi}{\cosh \eta - \cos \theta} \qquad y = \frac{c \sinh \eta \sin \phi}{\cosh \eta - \cos \theta} \qquad z = \frac{c \sin \theta}{\cosh \eta - \cos \theta} \qquad (12.3.1)$$

$$c \geq 0 \qquad -\pi < \theta \leq \pi \qquad 0 \leq \phi < 2\pi \qquad 0 \leq \eta$$

The coordinate surfaces associated with this coordinate system are

$$x^2 + y^2 + z^2 + c^2 = 2c\sqrt{x^2 + y^2} \, \coth \eta$$

$$x^2 + y^2 + (z - c \cot \theta)^2 = \frac{c^2}{\sin^2 \theta} \qquad (12.3.2)$$

$$y = x \tan \phi$$

For η constant this generates a torus; for θ constant a spherical bowl is formed, and when ϕ is constant a half-plane is formed. These surfaces are shown as cutaway views in Figure 12.3.1.

Our nomenclature for coordinates is also Hobson's choice [Hob31, Section 253]. It is also essentially that used in Moon and Spencer's field theory handbook [Moo71, pp. 112–115], except that we use ϕ where they use ψ. Hobson provides a clear discussion of the geometry involved, while Moon and Spencer present the related vector calculus, as well as the separation of Laplace's equation in toroidal coordinates.

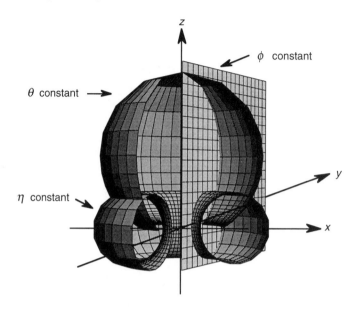

FIGURE 12.3.1 Toroidal coordinates. Coordinate surfaces are spheres for θ constant, toroids for η constant, and half planes for ϕ constant. (Adapted from the cell `Torus-Coords` in *Mathematica* notebook `LegendreCoords`.)

The dependence of the toroids (rings) on the parameter η is both interesting and confusing. If in the first of the equations in (12.3.2) one sets $x = 0 = z$ then solves for y, it is readily found that the circular cross section of the torus has a radius of $c / \sinh\eta$; the center of these circles traces out a circle of radius $c\coth\eta$ in the $x - y$ plane, and this circle is centered on the z axis. Figure 12.3.2 illustrates how the toroids depend on η.

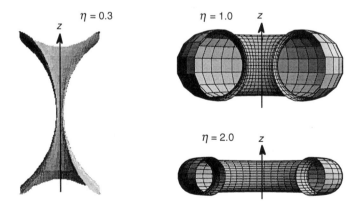

FIGURE 12.3.2 Sections of toroids for different values of parameter η. (Adapted from the cell `TorusCoords` in *Mathematica* notebook `LegendreCoords`.)

As an example of using toroidal coordinates, consider a potential function V for a geometry in which boundary conditions are applied on a toroidal or ring-shaped boundary. Typically, V is an electrostatic potential, as in the worked example in Lebedev's text on the special functions and their applications [Leb65, Section 8.11] and in Hobson's monograph [Hob31, Section 254]. In toroidal coordinates the solutions of Laplace's equation are shown by Hobson to be of the form

$$V(\eta,\theta,\phi) = \sqrt{\cosh\eta - \cos\theta} \; \begin{vmatrix} P^m_{n-1/2}(\cosh\eta) \\ Q^m_{n-1/2}(\cosh\eta) \end{vmatrix} \begin{vmatrix} \cos n\theta \\ \sin n\theta \end{vmatrix} \begin{vmatrix} \cos m\phi \\ \sin m\phi \end{vmatrix} \tag{12.3.3}$$

in which any linear combination of solutions may be taken. If the solutions are periodic in θ and in ϕ, then m and n must be integers. Correspondingly, the degree $v = n - 1/2$ is a half integer and the order $\mu = m$ is an integer. There is no necessary relation between n and m, unlike for the spherical Legendre functions (Section 12.2). Usually, one considers just $n \geq 0$ and $m \geq 0$, since the general symmetries of the Legendre functions can be used for negative degree or order.

The toroidal function of the first kind, $P^m_{n-1/2}(\cosh\eta)$, is regular as $\eta \to 0$, which is exterior to the toroids in Figures 12.3.1 and 12.3.2. The toroidal function of the second kind, $Q^m_{n-1/2}(\cosh\eta)$, is singular as $\eta \to 0$, therefore it is not applicable as part of V in the exterior region. The center of the torus is a circle of radius c in the $x - y$ plane associated with $\eta \to \infty$. In this limit, the function of the first kind is divergent, whereas the function of the second kind converges to zero. Therefore, only $Q^m_{n-1/2}(\cosh\eta)$ is applicable in the interior of the torus.

12.3.2 Toroidal Functions of the First Kind

Keywords
associated Legendre functions, kugel functions, ring functions, toroidal functions

C Function and *Mathematica* Notebook
`ToroidalP` (C); cell `ToroidP` in notebook `Legendre` (*Mathematica*) $\boxed{M}$

Background
The introductory material in Section 12.3.1 and its references provide background on these functions. A few formulas are given in Section 3.13 of Erdélyi et al [Erd53a] and in Section 8.11 of Abramowitz and Stegun [Abr64].

Function Definition
There are many formulas that can be used to define $P^m_{n-1/2}(x)$ and to relate it to other functions, but we cite only those relevant for its computation. In the context of the toroidal functions, the discussion in Section 12.3.1 indicates that—since $x = \cosh\eta \geq 1$—one needs to devise algorithms only for this range of x.

For computing a single value of $P^m_{n-1/2}(x)$ the relation to the hypergeometric function $_2F_1$ that is derived by Snow [Sno42, p. 255] is most useful:

$$P_{n-1/2}^{m}(\cosh\eta) = \tanh^{m}(\eta/2)\cosh^{2n-1}(\eta/2)\frac{\Gamma(m+n+1/2)}{\Gamma(m+1)\,\Gamma(n-m+1/2)}$$

$$\times \,_2F_1\left[1/2-n,1/2-n+m;m+1;\tanh^2(\eta/2)\right]$$

(12.3.4)

in which $\,_2F_1$ can be computed as a power series:

$$\,_2F_1(a,b;c;y) = \sum_{k=0}^{\infty}\frac{(a)_k(b)_k}{k!\,(c)_k}\,y^k$$

(12.3.5)

in terms of the rising factorials (Section 7.1.2), which satisfy

$$(z)_0 = 1 \qquad (z)_k = (z+k-1)(z)_{k-1} \qquad k>0$$

(12.3.6)

For argument x in the toroidal function of the first kind, we need the relations

$$\cosh\eta = x \qquad\qquad t = x+\sqrt{x^2-1}$$

$$\tanh(\eta/2) = \frac{t-1}{t+1} \qquad \cosh(\eta/2) = \frac{\sqrt{t}+1/\sqrt{t}}{2}$$

(12.3.7)

If a series of $P_{n-1/2}^{m}(x)$ values each differing by unity in m or n is required, two starting values of the function can be computed by using (12.3.4), then the recurrence relations of the Legendre functions with respect to μ or to ν can be used. This method was used for the tables of Lowan [Low45].

Visualization

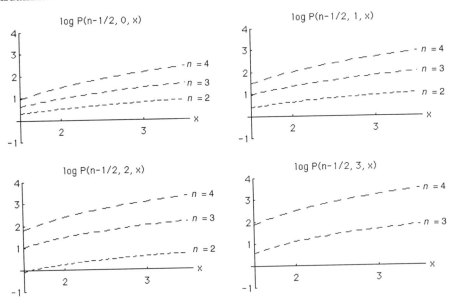

FIGURE 12.3.3 Surfaces of logarithms (base 10) of toroidal functions of the first kind as functions of x from 1.5 to 3.5 for n from 2 to 4. Each surface has a fixed m value given by the second argument in P. For $m=3$ and $n=2$ the function is negative in the x range shown, so its logarithm is not shown. (Adapted from the output of cell `ToroidP` in *Mathematica* notebook `Legendre`.)

Many views of Legendre functions of the first kind are given in Section 12.1.1. The present visualizations emphasize aspects relevant for computing toroidal functions of the first kind, $P_{n-1/2}^m(x)$. To emphasize that this is to be evaluated for half-integer degree $(n-1/2)$ and integer order (m), Figure 12.3.3 shows curves for discrete values of n and m. The increase of $P_{n-1/2}^m$ with increase of n is rapid, therefore base-10 logarithms are displayed, except for $P_{3/2}^3(x)$ which is negative in the x range displayed. This n dependence indicates that a recurrence scheme such as that used for the spherical Legendre functions, (12.2.9), is appropriate to use if a sequence of values with n increasing is needed.

Note that in Figure 12.3.3 we use the phase convention of the Lowan tables [Low45, Table XI]; the Legendre function from *Mathematica* must be multiplied by the factor $(-1)^{m/2}$

Algorithm and Program

To compute a toroidal function of the first kind, $P_{n-1/2}^m(x)$, numerically for $n \geq 0$ and $m \geq 0$, we use the expression (12.3.4).

```
double ToroidalP(int n, int m, double x, double eps)
{
/* Toroidal Function of First Kind by Snow's Method;
   degree n, order m, argument x > 1
   to fractional accuracy  eps  */

double  GamPM(double x),
 Hyper2F1(double a, double b, double c, double y,
 double eps),
  tfact,Tanh,cfact,Cosh,fn,fm,fnPm,fnMm,Factor;

/* Compute  tanh  and  cosh  of half angle */
tfact = x+sqrt(x*x - 1.0);
Tanh = (tfact-1.0)/(tfact+1.0);
cfact = sqrt(tfact);
Cosh = 0.5*(cfact+1.0/cfact);
/* Set up prefactor & arguments of hypergeometric */
fn = n;
fm = m;
fnPm = n+m;
fnMm = n-m;
Factor = pow(Tanh,fm)*pow(Cosh,2.0*fn-1.0)
  *GamPM(fnPm+0.5)/(GamPM(fm+1.0)*GamPM(fnMm+0.5));

return  Factor*Hyper2F1(0.5-fn,0.5-
fnMm,fm+1.0,Tanh*Tanh,eps);
}
```

Programming Notes

Formula (12.3.4) for the toroidal function of the first kind, $P_{n-1/2}^m(x)$, is expressed in terms of the variable $\cosh \eta$, whereas the function argument is x. The relations (12.3.7) are therefore used to make the connection. The factor before the hypergeometric function may require the gamma function (Section 6.1.1) of both positive and negative half-integer arguments. Our usual function for $\ln \Gamma$, LogGamma, provides only for positive arguments;

therefore we use the reflection formula (6.1.4) and this function to produce (in function GamPM) the necessary gamma functions.

The hypergeometric function $_2F_1$ used in (12.3.4) is computed as an infinite series. The input parameter eps for function ToroidalP is passed to Hyper2F1 and the series (12.3.5) is summed until the last term included is smaller in absolute ratio than the partial sum of the series. Thus, for 10-digit accuracy use eps $= 10^{-10}$.

If you need a sequence of $P_{n-1/2}^m(x)$ values each differing by unity in m or n, compute two contiguous values by using (12.3.4), then use the recurrence relations of the Legendre functions with respect to μ or to ν.

Test Values

Check values of $P_{n-1/2}^m(x)$ can be generated by running the ToroidP cell in *Mathematica* notebook Legendre. The x values used in Table 12.3.1 correspond approximately with the η values used in Figure 12.3.2; namely, $\cosh 0.3 = 1.045$, $\cosh 1.0 = 1.543$, $\cosh 2.0 = 3.762$.

TABLE 12.3.1 Test values to 10 digits of the toroidal functions of the first kind of integer degree n and order m.

$x \setminus m$	0	1	2	3
$n= 1$				
1.05	1.018605615	0.1182227606	–0.01150410591	0.003139267015
1.55	1.190914205	0.3823923424	–0.107780622	0.08627174893
3.75	1.773775118	0.8100585134	–0.3506562324	0.4427507852
$n= 2$				
1.05	1.094769063	0.6133032713	0.08254506678	–0.009594658028
1.55	2.147821362	2.708257524	0.9650813277	–0.3130242668
3.75	8.603470038	12.65390795	6.004339176	–2.775390608
$n= 4$				
1.05	1.428509681	2.972307708	3.002775469	1.477032724
1.55	10.72071349	35.47977463	75.97788748	90.08107459
3.75	318.8610903	1108.985367	2720.758516	3956.620685

12.3.3 Toroidal Functions of the Second Kind

Keywords
associated Legendre functions, ring functions, toroidal functions

C Function and *Mathematica* Notebook

ToroidalQ (C); cell ToroidQ in notebook Legendre (*Mathematica*)

Background

The introductory material on toroidal coordinates in Section 12.3.1 and its references provide background on the toroidal functions of the second kind $Q^m_{n-1/2}(x)$. A few formulas are given in Section 3.13 of Erdélyi et al [Erd53a] and in Section 8.11 of Abramowitz and Stegun [Abr64].

As discussed toward the end of Section 12.3.1, $Q^m_{n-1/2}(x)$ is singular as $x = \cosh\eta \to 1$ ($\eta \to 0$). When toroidal coordinates are used to solve Laplace's equation $Q^m_{n-1/2}(\cosh\eta)$ is therefore the solution that is appropriate for the interior of a torus; see also Figure 12.3.2. The nature of the singularities is discussed in Section 12.1.2 in connection with Figure 12.1.6. Since the behavior of an irregular Legendre function near the singularity is independent of the degree, the relation (12.1.6) is appropriate, apart from a complex phase.

Function Definition

Although there are many formulas that relate $Q^m_{n-1/2}(\cosh\eta)$ to other functions, we cite only those relevant for computing it. In the context of toroidal functions, the discussion in Section 12.3.1 indicates that—since $x = \cosh\eta \geq 1$—one needs to devise algorithms only for this range of x.

To compute a single value of $Q^m_{n-1/2}(\cosh\eta)$ the relation to the hypergeometric function $_2F_1$ that is given in Lowan [Low45, (15)] is most useful:

$$Q^m_{n-1/2}(x) = B(n-1/2, m, x) \; _2F_1\left[1/2 + m, 1/2 - m; n+1; -t\right]$$

$$B(n-1/2, m, x) = \frac{(-1)^m \sqrt{\pi/2} \; \Gamma(n+m+1/2)}{\Gamma(n+1)(x^2-1)^{1/4}\left(x+\sqrt{x^2-1}\right)^n} \qquad t = \frac{x - \sqrt{x^2-1}}{2\sqrt{x^2-1}} \qquad (12.3.8)$$

in which $_2F_1$ is computed as a power series:

$$_2F_1(a,b;c;y) = \sum_{k=0}^{\infty} \frac{(a)_k (b)_k}{k!\,(c)_k} \, y^k \qquad (12.3.9)$$

in terms of the rising factorials (Section 7.1.2), which satisfy

$$(z)_0 = 1 \qquad (z)_k = (z+k-1)(z)_{k-1} \qquad k > 0 \qquad (12.3.10)$$

If both the regular and irregular toroidal functions are required, $P^m_{n-1/2}$ and $Q^m_{n-1/2}$, the C functions for $_2F_1$ and for the Γ functions can be shared by the functions `ToroidalP` and `ToroidalQ`.

If a series of $Q^m_{n-1/2}(x)$ values each differing by unity in m or n is required, two starting values of the function can be computed by using (12.3.8), then the recurrence relations of the Legendre functions with respect to μ or to ν can be used. This method was used for the tables prepared by Lowan et al [Low45].

Visualization

Many views of Legendre functions of the second kind are given in Section 12.1.2. Here the visualizations emphasize aspects relevant for computing $Q^m_{n-1/2}(x)$ by using (12.3.8). To emphasize that the toroidal function of the second kind is to be evaluated for a half-integer degree ($n-1/2$) and integer order (m), Figure 12.3.4 shows curves for discrete n

and m. Since the variation of $(-1)^m Q^m_{n-1/2}$ with n is rapid and the function (with indicated phase) is positive in the x range displayed, base-10 logarithm values are displayed.

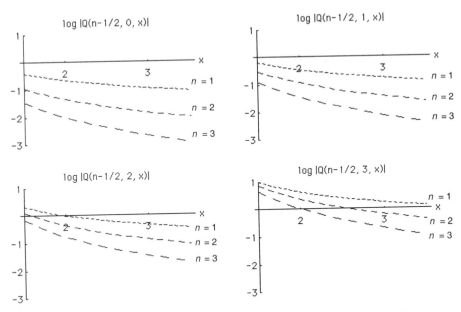

FIGURE 12.3.4 Curves of logarithms (base 10) of toroidal functions of the second kind as functions of x from 1.5 to 3.5 for n from 1 to 3. Each set of curves has a fixed m value, as indicated by the second argument of Q. (Adapted from the output of cell `ToroidQ` in *Mathematica* notebook `Legendre`.)

Note that in Figure 12.3.4 we use the phase convention of the Lowan tables [Low45, Table XIII]; the irregular Legendre function from *Mathematica* must be multiplied by the factor $(-1)^{m/2}$.

Algorithm and Program

To compute a toroidal function of the first kind, $Q^m_{n-1/2}(x)$, numerically for $n \geq 0$ and $m \geq 0$, expression (12.3.8) is used, as follows:

```
double ToroidalQ(int n, int m, double x, double eps)
{
/* Toroidal Function of Second Kind
   degree n, order m, argument x > 1
   to fractional accuracy  eps  */

const double sqrtPiD2 = 1.25331413732;
double  GamPM(double x),
 Hyper2F1(double a, double b, double c, double y,
          double eps),
 xsqM1,xsqrt,t,fn,fm,phase,BigB;
```

```
xsqM1 = x*x - 1.0;
xsqrt = sqrt(xsqM1);
t = 0.5*(x-xsqrt)/xsqrt;
fn = n;
fm = m;
phase = ( m % 2 > 0 ) ? -1.0: 1.0;
BigB = phase*sqrtPiD2*GamPM(fn+fm+0.5)/
  (GamPM(fn+1.0)*sqrt(xsqrt)*pow(x+xsqrt,fn));

return  BigB*Hyper2F1(fm+0.5,0.5-fm,fn+1.0,-t,eps);
}
```

Programming Notes

The hypergeometric function $_2F_1$ used in (12.3.8) for the toroidal function of the second kind, $Q_{n-1/2}^m(x)$, is computed as an infinite series. The input parameter eps for function ToroidalQ is passed to Hyper2F1, and the series (12.3.9) is summed until the last term included is smaller in absolute ratio than the partial sum of the series. Thus, for 10-digit accuracy use eps = 10^{-10}. Function ToroidalQ does not give reliable results for $x < 1.1$ because of the singularity behavior as $x \to 1$, which is given by (12.1.6).

If you need a sequence of $Q_{n-1/2}^m(x)$ values each differing by unity in m or n, compute two contiguous values by using (12.3.8) in function ToroidalQ, then use the recurrence relations of the Legendre functions with respect to μ or to ν. Note from Figure 12.3.4 that $Q_{n-1/2}^m(x)$ generally increases as n decreases, therefore iteration on n should be made in the direction of decreasing n if numerical accuracy is to be maintained.

Test Values

Test values of $Q_{n-1/2}^m(x)$ can be generated by running the ToroidQ cell in *Mathematica* notebook Legendre. The x values used in Tables 12.3.2 and 12.3.3 are similar to the η values used in Figure 12.3.2; $\cosh 0.3 = 1.045$, $\cosh 1.0 = 1.543$, $\cosh 2.0 = 3.762$.

M

TABLE 12.3.2 Test values to 10 digits of the toroidal functions of the second kind of integer degree *n* and order *m*, for *n* = 1 and 2. See also Table 12.3.3.

$x \setminus m$	0	1	2	3
n= 1				
1.10	0.9787602829	−1.947110839	10.08174744	−94.36688829
1.55	0.366464025	−0.5913672219	1.822835727	−8.803853876
3.75	0.07915135627	−0.119839631	0.3080479355	−1.128687736
n= 2				
1.10	0.4817841242	−1.469035460	8.859226479	−87.63344736
1.55	0.1011261719	−0.2656296077	1.074546376	−6.090407374
3.75	0.008067344808	−0.02029440983	0.07236635348	−0.3358563376

TABLE 12.3.3 Test values to 10 digits of the toroidal functions of the second kind of integer degree $n = 4$, and order $m = 0, 1, 2, 3$. See also Table 12.3.2.

$x \setminus m$	0	1	2	3
$n = 4$				
1.10	0.1474438159	−0.7522732684	5.933748782	−67.31716361
1.55	0.009911240131	−0.04596262142	0.2764157227	−2.079100883
3.75	0.0001085364167	−0.0004902463563	0.002726780049	−0.0180578066

12.4 CONICAL LEGENDRE FUNCTIONS

The conical Legendre functions, also just called conical or kegel functions, arise when solving differential equations involving Laplacians, $\mathbf{V}^2$, expressed in spherical coordinates. We therefore begin this section with a summary of this topic, before proceeding in Section 12.4.2 with analytical, graphical, and computational aspects of conical functions.

12.4.1 Laplace Equation on a Cone
Suppose that the Laplace equation, $\mathbf{V}^2 V = 0$, is to be solved using spherical polar coordinates (r, θ, ϕ) in a region interior to a right circular cone that is described by $\theta = \alpha$, a constant angle, as shown in Figure 12.4.1.

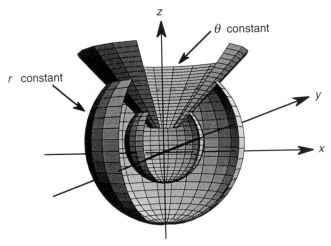

FIGURE 12.4.1 Coordinate surfaces for solving Laplace's equation on a cone that intersects two spheres. (Adapted from cell `ConeCoords` in the *Mathematica* notebook `LegendreCoords`.)

As shown in the monographs by Hobson [Hob31, Section 259] and by Lebedev [Leb65, Section 8.5], if Dirichlet boundary conditions are specified at the inner surface of the cone then an expansion of the potential distribution inside the cone requires the Legendre functions $P_{-1/2+it}(\cos\theta)$ with $\theta \le \alpha$. For example, if the potential is also required to vanish on the surfaces of the spheres that are interior to the cone then t is of the form

$$t = k\pi / \ln(b/a) \tag{12.4.1}$$

where k is an integer of either sign and b/a is the ratio of the radii of the two spheres.

A Legendre function $P_{-1/2+it}(x)$ with t real is called a conical function, which we now explore.

12.4.2 Conical Functions

Keywords
associated Legendre functions, conical functions, kegel functions

C Function and *Mathematica* Notebook
ConicalP (C); cell Conical in notebook Legendre *(Mathematica)*

$\boxed{M}$

Background
The introductory material in Section 12.4.1 and its references provide background on these functions. Some formulas are given in Section 3.14 of Erdélyi et al [Erd53a] and in Section 8.12 of Abramowitz and Stegun [Abr64]. A comprehensive discussion of techniques for computing conical functions is given in the introduction to the tables prepared by Zhurina and Karmazina [Zhu64].

We describe only the function $P_{-1/2+it}(x)$, the function of zeroth order that is regular at $x = 1$ and that is evaluated on the cut in the complex plane from $x = -1$ to 1; Legendre functions on the cut are discussed in relation to (12.1.2) and (12.1.4). There can also be defined conical functions of nonzero order of the regular kind, $P^{\mu}_{-1/2+it}(x)$, and irregular kind, $Q^{\mu}_{-1/2+it}(x)$, as presented in [Erd53a, Section 3.14] and in [Abr64, Section 8.12]. Computationally, however, the only conical function that seems to have been investigated extensively is $P_{-1/2+it}(x)$, both for $|x|<1$ [Zhu64, Part I] and for $|x|>1$ [Zhu64, Part II].

Function Definition
A formula that is relevant for computing the conical function $P_{-1/2+it}(x)$ if $|x|<1$ is given in terms of a hypergeometric function by

$$P_{-1/2+it}(x) = {}_2F_1\left[1/2 - it, 1/2 + it; 1; (1-x)/2\right] \tag{12.4.2}$$

which follows from [Erd53a, 3.4 (6)]. Here the ${}_2F_1$ function (Chapter 16) can be computed as a power series, resulting in

$$P_{-1/2+it}(x) = \sum_{k=0}^{\infty} t_k u^k \qquad u = (1-x)/2 \tag{12.4.3}$$

$$t_0 = 1 \qquad t_k = \frac{(2k-1)^2 + 4t^2}{(2k)^2} t_{k-1} \qquad k > 0$$

This series is convergent if $|u|<1$, albeit slowly when $|t|$ is large.

From (12.4.3) we note that $P_{-1/2+it}(x)$ with t and x real is real-valued, and it is not less than unity. It has the symmetry

$$P_{-1/2-it}(x) = P_{-1/2+it}(x) \qquad (12.4.4)$$

An unfortunate property when computing the conical function is that

$$P_{-1/2+it}(x) \xrightarrow[x \to -1]{} \infty \qquad (12.4.5)$$

Note, however, that in applications to problems such as presented in Section 12.4.1 the cone angle α is likely to lie between 0 and $\pi/2$, therefore $x = \cos\theta$ is likely to lie between 1 and 0, for which the convergence of the series (12.4.3) may be rapid (depending on t), since then u lies between 0 and $1/2$. Convergence properties are illustrated under Test Values below.

An asymptotic expression for the conical function is given by Zhurina and Karmazina [Zhu64], namely,

$$P_{-1/2+it}(\cos\theta) \xrightarrow[t \to \infty]{} \frac{e^{t\theta}}{\sqrt{2\pi t \sin\theta}} \qquad 0 < \theta < \pi \qquad (12.4.6)$$

Visualization

Properties of the conical function discussed under Function Definition, namely $P_{-1/2+it}(x)$ with x in $(-1, 1]$, are shown graphically in Figure 12.4.2 for a range of t values relevant to problems of the kind summarized in Section 12.4.1. The symmetry (12.4.4) makes it unnecessary to show the surface for $t < 0$, but we illustrate this symmetry in the transections for fixed x.

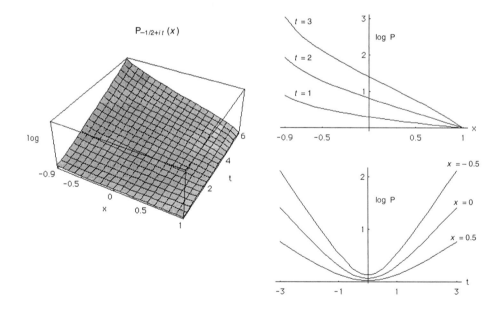

FIGURE 12.4.2 Conical functions on base-10 logarithmic scales. Surfaces in x and t (left side), curves as functions of x for t fixed (right top) and as functions of t for x fixed (right bottom). (Adapted from the output of cell `Conical` in the *Mathematica* notebook `Legendre`.)

Algorithm and Program

To compute the conical function $P_{-1/2+it}(x)$ numerically, expression (12.4.3) is used directly in function `ConicalP`, as follows:

```
double ConicalP(double t, double x, double eps)
{
/* Conical Function of First Kind
   degree -1/2+it, order 0, argument x ( |x| < 1 )
   to fractional accuracy   eps   */

double
u,tterm,sum,ratio,eps10,twokM1,twok,tk,upow,term;
int k;

u = 0.5*(1.0-x);
tterm = 4.0*t*t;
sum = 1.0;
ratio = 10.0;
/* Better convergence when x near -1 */
eps10 = 0.1*eps;
twokM1 = -1.0;
twok = 0.0;
tk = 1.0;
upow = 1.0;
term = 1.0;
k = 1;

while ( ratio > eps )
  {
  twokM1 += 2.0;
  twok += 2.0;
  tk *= (twokM1*twokM1 + tterm)/(twok*twok);
  upow *= u;
  term = tk*upow;
  sum += term;
  ratio = fabs(term/sum);
  k ++;
  } /*  k terms are needed to converge */
return  sum;
}
```

Programming Notes

The series in (12.4.3) needs input parameter `eps` to check convergence. The series is summed until the last term included is smaller in absolute ratio than the partial sum of the series. Thus, for 10-digit accuracy use $eps = 10^{-10}$. The convergence will be very slow if t is large, or if x is near -1, or if high accuracy is required. Representative values for the number of terms needed are given under Test Values. By printing the final value of `k` in `ConicalP`, you will see how many terms are needed.

Test Values

Check values of $P_{-1/2+it}(x)$ can be generated by running the `Conical` cell in *Mathematica* notebook `Legendre`. The x values used in Table 12.4.1 correspond approximately to

$\boxed{M}$

angles of 2.50 radian (143°), 1.57 radian (90°), and 0.64 radian (37°). (In Figure 12.4.1 the angle of the cone is $\alpha = \pi/4 = 45°$.) Suppose that in the situation discussed in Section 12.4.1 the ratio of sphere radii is $b/a = 2$, as shown in Figure 12.4.1, then t_k given by (12.4.1) has the values $t_0 = 0$, $t_1 = 4.53$, and $t_2 = 9.06$—comparable to the t values in Table 12.4.1.

TABLE 12.4.1 Test values to 10 digits of the conical function with parameter t and argument x. The value in parentheses is the number of series terms required for fractional accuracy 10^{-10}.

$t \setminus x$	-0.8		0		0.8	
0	1.641264414	(177)	1.180340599	(31)	1.026512044	(11)
5	5.954332801×10^4	(217)	4.607672600×10^2	(43)	6.011124083	(16)
10	$1.133278339 \times 10^{10}$	(247)	8.376546460×10^5	(53)	1.034347464×10^2	(20)

If accuracy of order 6 digits is sufficient for computing $P_{-1/2+it}(x)$, then fewer than half the number of terms indicated will suffice for $x = -0.8$, with smaller reductions for larger x values.

REFERENCES ON LEGENDRE FUNCTIONS

[Abr64]   Abramowitz, M., and I. A. Stegun, *Handbook of Mathematical Functions*, Dover, New York, 1964.

[Bel62]   Belousov, S. L., *Tables of Normalized Associated Legendre Polynomials*, Pergamon Press, New York, 1962.

[Erd53a]   Erdélyi, A., et al, *Higher Transcendental Functions*, Vol. 1, McGraw-Hill, New York, 1953; reprint edition, Krieger, Malabar, Florida, 1981.

[Gra94]   Gradshteyn, I. S., and I. M. Ryzhik, *Table of Integrals, Series, and Products*, Academic Press, Orlando, fifth edition, 1994.

[Hob31]   Hobson, E. W., *The Theory of Spherical and Ellipsoidal Harmonics*, Cambridge University Press, 1931.

[Jah45]   Jahnke, E., and F. Emde, *Tables of Functions*, Dover, New York, fourth edition, 1945.

[Leb65]   Lebedev, N. N., *Special Functions and Their Applications*, Prentice-Hall, Englewood Cliffs, New Jersey, 1965.

[Low45]   Lowan, A. N., project director, *Tables of Associated Legendre Functions*, Columbia University Press, New York, 1945.

[Mag49]   Magnus, W., and F. Oberhettinger, *Formulas and Theorems for the Special Functions of Mathematical Physics*, Chelsea Publishing, New York, 1949.

[Moo61]   Moon, P., and D. E. Spencer, *Field Theory for Engineers*, Van Nostrand, Princeton, 1961.

[Moo71]   Moon, P., and D. E. Spencer, *Field Theory Handbook*, Springer Verlag, Berlin, second edition, 1971.

[Mor53]   Morse, P. M., and H. Feshbach, *Methods of Theoretical Physics*, McGraw-Hill, New York, 1953.

[Sno42]   Snow, C., *The Hypergeometric and Legendre Functions with Applications to Integral Functions of Potential Theory*, National Bureau of Standards, Washington, D.C., 1942.

[Tho94b]   Thompson, W. J., *Angular Momentum*, Wiley, New York, 1994.

[Whi27]   Whittaker, E. T., and G. N. Watson, *A Course of Modern Analysis*, Cambridge University Press, Cambridge, England, fourth edition, 1927.

[Wol91]   Wolfram, S., *Mathematica: A System for Doing Mathematics by Computer*, Addison-Wesley, Redwood City, California, second edition, 1991.

[Zhu64]   Zhurina, M. I., and L. N. Karmazina, *Tables of the Legendre Functions* $P_{-1/2+i\tau}(x)$, Macmillan, New York, 1964.

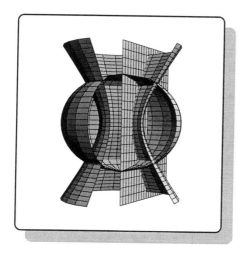

Chapter 13

SPHEROIDAL WAVE FUNCTIONS

Spheroidal wave functions include generalizations of Legendre functions (Chapter 12) and of spherical Bessel functions (Section 14.3.1) that are appropriate for spheroidal coordinates rather than the spherical polar coordinates (Section 12.2.1) in which the latter functions occur. Angular and radial spheroidal functions for prolate or oblate spheroidal coordinates are used primarily in physics and engineering. This chapter begins with an overview of spheroidal wave functions, then specializes in Sections 13.2–13.4 to the prolate and oblate angular functions, as well as to the corresponding radial functions. Visualizations, algorithms, and C programs, are provided, as well as numerical values for checking eigenvalues and wave functions.

The literature on spheroidal wave functions is often in the context of specialized applications. There are, however, two applied mathematics monographs with text and extensive tables that are especially useful; the book by Stratton et al [Stra56], and that by Flammer [Fla57]. Analytical results are derived in the books by Meixner et al [Mei54, Mei80]. Many of these results are summarized in Sections 16.9 – 16.13 of the treatise on higher transcendental functions edited by Erdélyi [Erd53c] and also in Chapter 21 of Abramowitz and Stegun [Abr64]. Since spheroidal wave functions and Mathieu functions have many properties in common, the literature on the latter functions should also be consulted.

Spheroidal wave functions are of significant practical use but are not broadly understood. Our description of them is therefore more complete than for other functions.

13.1 OVERVIEW OF SPHEROIDAL WAVE FUNCTIONS

This section gives an overview of spheroidal wave functions in the context of spheroidal coordinates, of determining the spheroidal eigenvalues, and of solving the scalar (Helmholtz) wave equation in this coordinate system. The spheroidal angular functions are usually expanded in a basis of spherical Legendre functions. Section 12.2 of the *Atlas* thereby provides background for the following introduction.

13.1.1 Spheroidal Coordinates

We define the relation between the spheroidal coordinates—d, η, ξ, and ϕ—and Cartesian coordinates—x, y, and z—as in Flammer [Fla57, Section 2.1]. For *prolate* coordinates:

$$x = \frac{d}{2}\sqrt{(1-\eta^2)(\xi^2-1)}\,\cos\phi \qquad y = \frac{d}{2}\sqrt{(1-\eta^2)(\xi^2-1)}\,\sin\phi$$

$$\zeta = \frac{d\eta\xi}{2} \qquad -1 \le \eta \le 1 \qquad \xi \ge 1 \qquad 0 \le \phi < 2\pi \tag{13.1.1}$$

For *oblate* coordinates:

$$x = \frac{d}{2}\sqrt{(1-\eta^2)(\xi^2+1)}\,\cos\phi \qquad y = \frac{d}{2}\sqrt{(1-\eta^2)(\xi^2+1)}\,\sin\phi$$

$$\zeta = \frac{d\eta\xi}{2} \qquad -1 \le \eta \le 1 \qquad \xi \ge 0 \qquad 0 \le \phi < 2\pi \tag{13.1.2}$$

With these choices of coordinates, the limit $\xi \to \infty$, $d \to 0$, $d\xi/2 = r$ and $\eta = \cos\theta$ produces spherical polar coordinates, as in Section 12.2.1. Many of the spheroidal coordinate systems used by other authors—including Abramowitz and Stegun [Abr64]—do not have this limit property.

Surfaces corresponding to constant parameter values—η, ξ, or ϕ—are illustrated in Figure 13.1.1 for prolate and oblate spheroidal coordinates. The parameter d provides only an overall scale factor, as does r in spherical polar coordinates, and is therefore not interesting to vary in this instance.

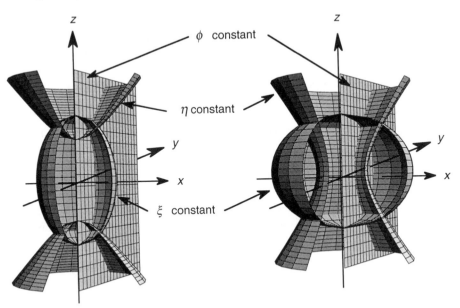

FIGURE 13.1.1 Spheroidal coordinates. Coordinate surfaces are hyperboloids of revolution for η constant, ellipsoids for ξ constant, and half planes for ϕ constant. Left side; prolate coordinates. Right side; oblate coordinates. (Adapted from cell SWFCords in *Mathematica* notebook SWF.)

The sections of the surfaces in Figure 13.1.1 are generated with the following parameter values: $|\eta| = 0.8$ for the hyperboloids, $\xi = 1.3$ for the ellipsoids, and $\phi = \pi/3$ for the half plane. Details of the geometry involved in spheroidal coordinates are given in Section 2.1 of Flammer [Fla57], Sections 21.1–21.3 of Abramowitz and Stegun [Abr64], and Tables 1.06 and 1.07 in Moon and Spencer's field theory handbook [Moo71]. Beware that the choice of coordinate systems and notations is quite variable.

13.1.2 Scalar Wave Equation in Spheroidal Coordinates

To provide a scientific context for spheroidal wave functions, consider solving the scalar wave equation in spheroidal coordinates, as presented in Section 2.1 of Flammer [Fla57] and Section 21.5 of Abramowitz and Stegun [Abr64]. The related, but more complicated, vector wave equation in the context of Maxwell's electromagnetism equations is covered in Chapter 9 of Flammer, in Section 5 of Moon and Spencer [Moo71], and in many technical papers on antenna theory and wave scattering by spheroids.

As shown in the above references, the scalar wave equation for wavenumber k, namely

$$\nabla^2 \psi + k^2 \psi = 0 \qquad (13.1.3)$$

is separable in spheroidal coordinates by writing, for *prolate* coordinates,

$$\psi_{mn} = S_{mn}(c,\eta)\, R_{mn}(c,\xi) \left| \begin{array}{c} \cos m\phi \\ \sin m\phi \end{array} \right. \qquad (13.1.4)$$

in which m is an integer if the ϕ dependence is periodic with period 2π, and n is chosen as an integer. This separability is analogous to that for solving the Laplace equation ($k = 0$) in spherical polar coordinates (Section 12.2.1). In (13.1.4) S_{mn} is called the *prolate spheroidal angular function* when k is real, since in the limit of small nonsphericity η becomes the polar angle θ. The parameter c is given by

$$c \equiv kd/2 = \pi d/\lambda \qquad (13.1.5)$$

where λ is the wavelength corresponding to wavenumber k. Thus, c scales as the ratio of distance to wavelength. (Recall that in the limit of small d and large ξ the combination $d\xi/2$ becomes r, the radial coordinate.)

In (13.1.4) R_{mn} is the *prolate spheroidal radial function*, which becomes a spherical Bessel function (Section 14.3.1) in the limit of zero c. Prolate spheroidal functions satisfy

$$\frac{d}{d\eta}\left[(1-\eta^2)\frac{dS_{mn}(c,\eta)}{d\eta}\right] + \left[\lambda_{mn} - c^2\eta^2 - \frac{m^2}{1-\eta^2}\right]S_{mn}(c,\eta) = 0 \qquad (13.1.6)$$

for the angular function, and

$$\frac{d}{d\xi}\left[(\xi^2-1)\frac{dR_{mn}(c,\xi)}{d\xi}\right] - \left[\lambda_{mn} - c^2\xi^2 + \frac{m^2}{\xi^2-1}\right]R_{mn}(c,\xi) = 0 \qquad (13.1.7)$$

for the radial function. In these two equations the $\lambda_{mn}(c)$ are the prolate spheroidal eigenvalues. As c tends to zero, the spheroidal functions tend to the corresponding spherical functions (at least within normalization factors), as is clear from these equations. One therefore has the requirement on the eigenvalues that

$$\lambda_{mn}(0) = n(n+1) \tag{13.1.8}$$

For *oblate* coordinates the separation of the wave equation is

$$\psi_{mn} = S_{mn}(-ic,\eta)\, R_{mn}(-ic,\xi) \left| \begin{matrix} \cos m\phi \\ \sin m\phi \end{matrix} \right. \tag{13.1.9}$$

where S_{mn} is the *oblate spheroidal angular function* when k is real, and R_{mn} is the *oblate spheroidal radial function*. Oblate spheroidal functions satisfy

$$\frac{d}{d\eta}\left[(1-\eta^2)\frac{dS_{mn}(-ic,\eta)}{d\eta}\right] + \left[\lambda_{mn} + c^2\eta^2 - \frac{m^2}{1-\eta^2}\right] S_{mn}(-ic,\eta) = 0 \tag{13.1.10}$$

for the angular function, and

$$\frac{d}{d\xi}\left[(\xi^2+1)\frac{dR_{mn}(-ic,i\xi)}{d\xi}\right] - \left[\lambda_{mn} - c^2\xi^2 + \frac{m^2}{\xi^2+1}\right] R_{mn}(-ic,i\xi) = 0 \tag{13.1.11}$$

for the radial function. Here $\lambda_{mn}(-ic)$ are the oblate spheroidal eigenvalues. Clearly, the oblate functions collapse to the corresponding spherical functions (Legendre functions and spherical Bessel functions) when c is zero. For purely-imaginary k (damped waves), the meanings of oblate and prolate interchange.

It is usual to present analytical results for spheroidal wave functions in terms of those for the prolate functions. The transition to the oblate functions (angular, radial, or eigenvalues) is made by the rule

$$\text{prolate} \leftrightarrow \text{oblate} \quad \text{by} \quad c \leftrightarrow \pm ic \quad c^2 \leftrightarrow -c^2 \tag{13.1.12}$$

However, all the spheroidal functions are real-valued, in spite of the appearance of this rule. In the following, as in the standard references, we cite results for prolate functions, with the understanding that this rule is used in the analysis (but not in numerical computations!) to obtain results for oblate functions.

Spheroidal wave functions are usually expanded in a basis of corresponding spherical functions (Legendre and Bessel), with the magnitude of c controlling the range of basis functions needed for accurate results. Computing the eigenvalues, λ_{mn}, accurately is thus a major consideration, as discussed in the following section.

13.1.3 Eigenvalues for Spheroidal Equations

Keywords
spheroidal coordinates, spheroidal wave functions

C Function and *Mathematica* Notebook
SWFevalue (C); cell SWFeigenvalue in *Mathematica* notebook SWF

Function Definition
The eigenvalues of the spheroidal wave equations—(13.1.6) and (13.1.7) for the prolate functions, with results for the oblate functions being obtained by the rule $c^2 \to -c^2$ in (13.1.12)—are tricky, tedious, and time-consuming to compute. We present the necessary formulas, with their derivation being given in Flammer's monograph [Fla57, Section 3.1.1]. Here we describe a brute-force, with a modified method being given in

Section 13.1.4. Define two functions that depend on m, n, and c (but not on the eigenvalue λ_{mn}) by

$$\gamma_r^m \equiv (m+r)(m+r+1) + \frac{c^2}{2}\left[1 - \frac{4m^2-1}{(2m+2r-1)(2m+2r+3)}\right] \qquad (13.1.13)$$

$$\beta_r^m \equiv \frac{r(r-1)(2m+r)(2m+r-1)c^4}{(2m+2r-1)^2(2m+2r-3)(2m+2r+1)} \qquad (13.1.14)$$

These functions are combined to define two functions of λ_{mn}, namely

$$U_1(\lambda_{mn}) \equiv \gamma_{n-m}^m - \lambda_{mn} - \cfrac{\beta_{n-m}^m}{\gamma_{n-m-2}^m - \lambda_{mn} - \cfrac{\beta_{n-m-2}^m}{\gamma_{n-m-4}^m - \lambda_{mn} - \cdots}} \qquad (13.1.15)$$

and

$$U_2(\lambda_{mn}) \equiv -\cfrac{\beta_{n-m+2}^m}{\gamma_{n-m+2}^m - \lambda_{mn} - \cfrac{\beta_{n-m+4}^m}{\gamma_{n-m+4}^m - \lambda_{mn} - \cdots}} \qquad (13.1.16)$$

In these two continued fractions, the first terminates with either the term containing γ_0^m or the term with γ_1^m, depending on whether $n-m$ is even or odd, while the second is non-terminating (in principle). As its upper limit is increased, the accuracy with which λ_{mn} can be determined is increased, as discussed below under Algorithm, Program, and Programming Notes.

The eigenvalue λ_{mn}, which depends implicitly upon the nonsphericity parameter c, is the root of the transcendental equation

$$U(\lambda_{mn}) \equiv U_1(\lambda_{mn}) + U_2(\lambda_{mn}) = 0 \qquad (13.1.17)$$

This equation has no closed-form solutions, except if $c = 0$ when $\lambda_{mn} = n(n+1)$ and in the limits of large deformation, namely

$$\lambda_{mn}(c)/c \xrightarrow[c \to \infty]{} 2(n-m)+1 \qquad \text{prolate}$$

$$\lambda_{mn}(c)/c^2 \xrightarrow[c \to \infty]{} -1 \qquad \text{oblate} \qquad (13.1.18)$$

These expressions are from the leading terms of the asymptotic expansions given in Abramowitz and Stegun [Abr64], equations (21.7.6) and (21.8.2), respectively.

Accurate eigenvalues are the essential first step for determining spheroidal wave functions accurately. We therefore provide two ways of computing them, one by using the *Mathematica* notebook SWFevalues, the other by using the C function SWFevalue that is described under Algorithm and Program. The *Mathematica* way uses the root-finding algorithms of this system in order to find solutions of (13.1.17) directly to high accuracy, chosen (as usual in the *Atlas*) to be 10 decimal digits. By rendering the output graphically, we obtain the views of the eigenvalues in Figure 13.1.2.

Approximate eigenvalues can be estimated by expanding them as power series in c or as asymptotic series in c and its inverse powers. The resulting cumbersome formulas—given in Flammer [Fla57] and Abramowitz and Stegun [Abr64, Chapter 21], for example—

M

$!$

are accurate to better than parts per million only for very small or very large c. If the interfocal distance of the spheroidal coordinates $d \approx \lambda$, the wavelength, a condition that is often interesting, then $c \approx 3$, according to (13.1.5). For such c values, for both prolate and oblate coordinates, the series formulas give accuracy of only a few percent for most values of m and n. We therefore use the series expansions in our C function only to provide estimates for the root of (13.1.17).

Visualization

Figure 13.1.2 shows the general dependence of the spheroidal wave function eigenvalues, $\lambda_{mn}(c)$, on m, n, and c, as well as on whether one is using prolate or oblate coordinates (left or right sides in Figure 13.1.1, respectively).

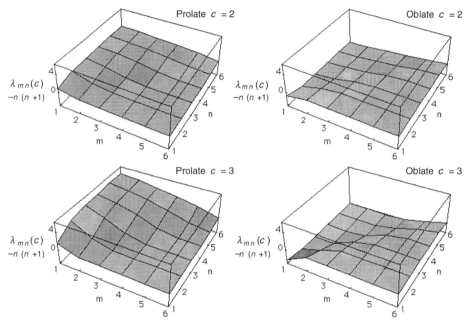

FIGURE 13.1.2 Surfaces of the differences of the eigenvalues of spheroidal wave functions from the spherical value, $n(n+1)$, as a function of parameter c that measures the non-sphericity and of the degree n and order m of the spheroidal wave function. (Adapted from cell `SWFeigenvalue` in *Mathematica* notebook `SWF`.)

Notice that in the range of m, n and c shown, the eigenvalues have a uniform and slow dependence on deformation parameter c, with the $\lambda_{mn}(c)$ for oblate and prolate cases deviating from the spherical value, $n(n+1)$ in opposite directions, consistent with the leading term in a power series expansion in c being quadratic. When c is small, the effects of non-sphericity generally become smaller as n increases.

Although a *Mathematica* calculation is useful if only a few eigenvalues are needed, a compiled C function is much speedier if one needs many values. Such a function is developed in the following.

Algorithm, Program, and Programming Notes

To compute the spherical wave function eigenvalues, $\lambda_{mn}(c)$, numerically, one can start with approximate solutions derived from power series or asymptotic expansions, then refine these solutions by a robust root-finding algorithm. For all positive values of m and n, we begin with a power series expansion in c^2 if $|c|<4.0$. The expansion obtained is of the form

$$\lambda_{mn}(c) \approx \begin{cases} \displaystyle\sum_{k=0}^{4} l_{2k}c^{2k} & \text{prolate} \\[4ex] \displaystyle\sum_{k=0}^{4} (-1)^k l_{2k}c^{2k} & \text{oblate} \end{cases} \qquad (13.1.19)$$

in which $l_0 = n(n+1)$, just as for the spherical functions (associated Legendre functions in Chapter 12), and the higher coefficients are complicated rational functions of m and n that are given in Abramowitz and Stegun [Abr64, 21.7.5] and in Flammer [Fla57, Section 3.1.2]. For $m = n = 0$, for which convergence of the series is slow, Flammer (Table 1) also gives numerical values of the coefficients l_{10}, l_{12}, and l_{14}. The C function LPS crunches out the coefficients and the power series estimate.

For $|c| \geq 4.0$ for all m and n, and for $|c| \geq 3.0$ when $m = n = 0$, asymptotic series are used to estimate $\lambda_{mn}(c)$. For prolate coordinates we use (21.7.6) in [Abr64] and for oblate coordinates we use (21.8.2) therein. The C function LAS computes the asymptotic series estimate.

By using either the power or asymptotic series, we obtain eigenvalue estimates that are usually within 0.2 of the final eigenvalue. We can therefore use a simple and robust root-finding method that starts with the eigenvalue estimate, called Lest in function SWF-evalue. By expanding outwards above and below Lest, the function computes the eigenvalue function U in (13.1.17) until $U = 0$ has been bracketed. The bisectional method is then used to locate the root to within an accuracy of eps, which is an input parameter. Typically, a dozen bisections produce part-per-thousand accuracy in $\lambda_{mn}(c)$ and about 30 bisections will result in 10-digit accuracy, as in Table 13.1.1. The efficiency of the slow-but-steady root-finding algorithm is improved by computing and storing the required coefficients γ_r^m and β_r^m in (13.1.15) and (13.1.16) before the function U is first invoked.

Function SWFevalue requires also a parameter that limits the number of terms in the continued fraction U_2, (13.1.16). This is input as smax; a value of 4 is sufficient for n up to 5; for n about 20, smax should be increased to about 10, depending on the accuracy required for $\lambda_{mn}(c)$. Since both continued fractions used in the algorithm are finite, they are straightforward to compute by starting with the innermost fractions and working outwards, as implemented in C function U.

When c is small in magnitude, the eigenvalue departs steadily from the spherical-coordinates value, $n(n+1)$. One might therefore expect that the roots of the eigenvalue equation (13.1.17) are unique, as has been assumed in some previous investigations, [Stra56, Fla57]. As m and n increase, however, this is no longer necessarily so. For example, for $m = n = 4$, we can see by plotting $U(L)$ against L for oblate deformations of various magnitudes, as in Figure 13.1.3 on the next page, the multiple solutions to $U(L) = 0$. One of these moves steadily away from $n(n+1)$ as the magnitude of c increases and is presumably the appropriate solution. For this reason, we constrain the bracketing of roots from the initial estimate, Lest, to be of small range. The existence of multiple eigenvalues is discussed in [Mei54, Section 3.2] and in [Mei80, Section 3.2].

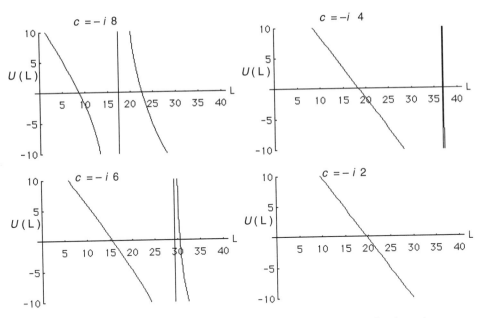

FIGURE 13.1.3 Curves of the function $U(L)$, (13.1.17), whose roots are the eigenvalues $\lambda_{mn} = L$ of the spheroidal wave functions. Shown for $m = n = 4$ as a function of the parameter c that measures the oblateness. As c approaches zero, the root approaches the spherical eigenvalue of $4(4+1) = 20$. Multiple roots appear in the L range $0 - 40$ as c increases in magnitude. (Adapted from cell SWFeigenvalue in *Mathematica* notebook SWF.)

The C function for the spherical wave function eigenvalues is relatively long. The principal function, SWFevalue, first sets up arrays of the coefficients γ_r^m and β_r^m in (13.1.15) and (13.1.16), then estimates (as Lest) the eigenvalue by using either the power series or an asymptotic series coded in functions LPS or LAS, whose formulas are described below (13.1.19). A maximum of NTRY attempts is made to bracket the root. Starting with Lest, the eigenvalue is then bracketed by evaluating U with the bounds on λ_{mn} increasing until $U(L_u)$ and $U(L_d)$ have opposite signs. The root is then refined using bisection until accuracy eps (an input parameter) is obtained or until more than KMAX tries to bisect the root have been made. In the following, the driver program has been omitted; its annotated listing is given in Section 21.13.

```
   /* maximum depth of continued fraction & arrays */
#define MAX 10
   /* maximum number of tries to bracket root */
#define NTRY 6
   /* maximum number of tries to find root */
#define KMAX 40

double SWFevalue(int m, int n, double sign,
                 double c, int smax, double eps)
/* Spheroidal Wave Function Eigenvalue for
   degree n, order m, arguments c & sign;
   For prolate  sign > 0; for oblate  sign < 0.
   Retain  smax  terms in continued fraction  */
```

```
{
double LPS(int m, int n, double csq),
 LAS(int m, int n, double sign, double c),
 U(double B1[], double G1[], double B2[],
 double G2[], int m,int n,int r1Start,int r2Start,
 int smax,double L),
 csq,c2D2,c4,fm2M1,fr,tmPr,mPr,tmPtrM1,
 Lest,Lu,Ld,Uu,Ud,Lstep,Umid,Lrt,dL,Lmid;
double B1[MAX],G1[MAX],B2[MAX],G2[MAX];
int nMm,r1Start,r2Start,r,s,k;
/* Make arrays of terms in continued fractions
   that are independent of eigenvalue */
csq = sign*c*c;
c2D2 = 0.5*csq;
c4 = csq*csq;
nMm = n - m;
fm2M1 = 4.0*m*m - 1.0;
/* Elements for U1 function */
r1Start = ( nMm % 2 == 0 ) ?  0: 1;
r = r1Start;
fr = r;
tmPr = 2*m+r;
mPr = m+r;
tmPtrM1 = 2.0*mPr - 1.0;
for ( s = r1Start; s <=  (nMm+1)/2; s++ )
  {
  B1[s] = (fr+2.0)*(fr+1.0)*(tmPr+2.0)*(tmPr+1.0)*
  c4/((tmPtrM1+4.0)*(tmPtrM1+4.0)*(tmPtrM1+2.0)*
  (tmPtrM1+6.0));
  G1[s] = mPr*(mPr+1.0)+c2D2*(1.0 - fm2M1/
  (tmPtrM1*(tmPtrM1+4.0)));
  fr += 2.0;
  tmPr += 2.0;
  tmPtrM1  += 4.0;
  mPr += 2.0;
  }
/* Elements for U2 function;
   r starts at n-m+2smax for smax terms */
r2Start = nMm+2*smax;
fr = r2Start;
tmPr = 2*m+r2Start;
mPr = m+r2Start;
tmPtrM1 = 2*mPr - 1;
for ( s = 1; s <=  smax; s++ )
  {
  B2[s] = fr*(fr-1.0)*tmPr*(tmPr-1.0)*c4/
  (tmPtrM1*tmPtrM1*(tmPtrM1-2.0)*(tmPtrM1+2.0));
  G2[s] = mPr*(mPr+1.0)+c2D2*(1.0 -
  fm2M1/(tmPtrM1*(tmPtrM1+4.0)));
  fr -= 2.0;
  tmPr -= 2.0;
  tmPtrM1  -= 4.0;
  mPr -= 2.0;
  }
```

```
/* Estimate eigenvalues */
if ( n == 0 )
 { if (fabs(c) < 3.0 )  Lest = LPS(m,n,csq);
 else  Lest = LAS(m,n,sign,c); }

else
 { if ( fabs(c) < 4.0 )  Lest = LPS(m,n,csq);
 else  Lest = LAS(m,n,sign,c); }

/* Bracket the eigenvalue */
Lu = Lest+0.1;
Ld = Lest-0.1;
Uu = U(B1,G1,B2,G2,m,n,r1Start,r2Start,smax,Lu);
Ud = U(B1,G1,B2,G2,m,n,r1Start,r2Start,smax,Ld);
k = 1;

while ( Uu*Ud > 0.0 )
 { Lstep = Lu - Ld;
   if ( fabs(Uu) < fabs(Ud) )
    { Lu += Lstep;
      Uu =
      U(B1,G1,B2,G2,m,n,r1Start,r2Start,smax,Lu); }
   else
    { Ld -= Lstep;
      Ud =
      U(B1,G1,B2,G2,m,n,r1Start,r2Start,smax,Ld); }
 if ( k > NTRY )
 { printf
   ("\n!! SWFevalue bracket > %i tries; 0 returned",
    NTRY);
   return 0.0; }
 k ++;
 }

/* Refine the root to fractional accuracy  eps  by
   using bisection method; start at Lu and Ld */
Uu = U(B1,G1,B2,G2,m,n,r1Start,r2Start,smax,Lu);
Umid = U(B1,G1,B2,G2,m,n,r1Start,r2Start,smax,Ld);
/* Search direction has U>0 for Lrt+dL */
Lrt = ( Uu < 0.0 ) ? (dL = Ld - Lu, Lu):
                     (dL = Lu - Ld, Ld);
k = 0;
while ( (fabs(dL) >= eps) && (Umid != 0.0) )
{
dL *= 0.5;
Lmid = Lrt+dL;
Umid = U(B1,G1,B2,G2,m,n,r1Start,r2Start,smax,Lmid);
if ( Umid <= 0.0 ) Lrt = Lmid;
k++;
if ( k > KMAX )
 { printf("\n!! > %i tries to bisect root",KMAX);
   Umid = 0.0; }
}
return  Lrt;
}
```

```
double LAS(int m, int n, double sign, double c)
/* Asymptotic series for eigenvalues */
{
double q,q2,q3,q4,q5,q6,fm,m2,m3,m4,m6,
 L0,L1,L2,L3,L4,L5,nu,cR,LLAS;
int nMm;

if ( sign > 0 )
{ /* prolate */
q = 2*(n-m)+1;
q2=q*q; q3=q2*q; q4=q3*q; q5=q4*q; q6=q5*q;
fm = m; m2 = fm*fm; m4 = m2*m2; m6 = m4*m2;

L0 = c*q+m2-(q2+5.0)/8.0;
L1 = -q*(q2+11.0-32.0*m2)/64;
L2 = -(5.0*(q4+26.0*q2+21.0) -
384.0*m2*(q2+1.0))/1024.0;
L3 = -q*((33.0*q4+1594.0*q2+5621.0)/128.0 -
 m2*(37.0*q2+167.0)+m4/8.0)/128.0;
L4 = -((63.0*q6+4940.0*q4+43327*q2+22470.0)/65536.0-
 m2*(115.0*q4+1310.0*q2+735.0)/512.0+3.0*m4*
 (q2+1.0)/8.0);
L5 =-q*((527.0*q6+61529.0*q4+1043961.0*
 q2+2241599.0)/1048576.0-
 m2*(5739.0*q4+127550.0*q2+298951.0)/32768.0+
 m4*(355.0*q2+1505.0)/512.0 - m6/16.0);
}
else
{
/* sign < 0; oblate */
nMm = n-m;
if ( nMm % 2 == 0 )
 { nu = nMm/2; q = n+1; }
else
 { nu = (nMm-1)/2; q = n; }
q2=q*q; q3=q2*q; q4=q3*q; q5=q4*q; q6=q5*q;
fm = m; m2 = fm*fm; m4 = m2*m2; m6 = m4*m2;

L0 = -c*c+2.0*c*(2.0*nu+fm+1.0)-2.0*nu*(nu+fm+1.0)-
 (fm+1.0);
L1 = -q*(q2+1.0 - m2)/8.0;
L2 = -(5.0*q4+10.0*q2+1.0-
 2.0*m2*(3.0*q2+1.0)+m4)/64.0;
L3 = -q*(33.0*q4+114.0*q2+37.0 -
 2.0*m2*(23.0*q2+25.0)+13.0*m4)/512.0;
L4 = -(63.0*q6+340.0*q4+239.0*q2+14.0 -
 10.0*m2*(10.0*q4+23.0*q2+3.0)+
 m4*(39.0*q2-18.0)-2.0*m6)/1024.0;
L5 = 0.0;
}

cR = 1.0/c;
LLAS = L0+(L1+(L2+(L3+(L4+L5*cR)*cR)*cR)*cR)*cR;
return  LLAS;
}
```

```
double LPS(int m, int n, double csq)
/* Power series for spherical wave function
   eigenvalues */
{
double
 tm,tn,tnM1,tnM3,tnM5,tnM7,tnP1,tnP3,tnP5,tnP7,tnP9,
 nMm,nMmM1,nMmM2,nMmM3,nMmP1,nMmP2,nMmP3,nMmP4,
 nPm,nPmM1,nPmM2,nPmM3,nPmP1,nPmP2,nPmP3,nPmP4,
 L0,L2,L4,L6,L8,A,B,C,D,LLPS;

L0 = n*(n+1);

tm = 2*m;  tn = 2*n;
tnM1 = tn-1.0; tnM3 = tn-3.0; tnM5 = tn-5.0;
tnM7 = tn-7.0;
tnP1 = tn+1.0; tnP3 = tn+3.0; tnP5 = tn+5.0;
tnP7 = tn+7.0; tnP9 = tn+9.0;
nMm = n - m;
nMmM1 = nMm-1.0; nMmM2 = nMm-2.0;
nMmM2 = nMm-2.0; nMmM3 = nMm-3.0;
nMmP1 = nMm+1.0; nMmP2 = nMm+2.0;
nMmP3 = nMm+3.0; nMmP4 = nMm+4.0;
nPm = n + m;
nPmM1 = nPm-1.0; nPmM2 = nPm-2.0;
nPmM2 = nPm-2.0; nPmM3 = nPm-3.0;
nPmP1 = nPm+1.0; nPmP2 = nPm+2.0;
nPmP3 = nPm+3.0; nPmP4 = nPm+4.0;

L2 = 0.5*(1.0 - ((tm-1.0)*(tm+1.0))/(tnM1*tnP3));

L4 = 0.5*(-nMmP1*nMmP2*nPmP1*nPmP2/
 (pow(tnP3,3.0)*tnP5) +
 nMmM1*nMm*nPmM1*nPm/(tnM3*pow(tnM1,3.0)))/tnP1;

L6 = (tm*tm-1.0)*(nMmP1*nMmP2*nPmP1*nPmP2/
 (pow(tnP3,4.0)*tnP5*tnP7) - nMmM1*nMm*nPmM1*nPm/
 (tnM5*tnM3*pow(tnM1,4.0)))/(tnM1*tnP1*tnP3);

A = (nMmM1*nMm*nPmM1*nPm/(pow(tnM5,2.0)*tnM3*
 pow(tnM1,5.0)) - nMmP1*nMmP2*nPmP1*nPmP2/
 (pow(tnP3,5.0)*tnP5*tnP7*tnP7))/
 (tnM1*tnM1*tnP1*tnP3*tnP3);

B = (nMmM3*nMmM2*nMmM1*nMm*nPmM3*nPmM2*nPmM1*nPm/
 (tnM7*tnM5*tnM5*pow(tnM3,3.0)*pow(tnM1,4.0)) -
 nMmP1*nMmP2*nMmP3*nMmP4*nPmP1*nPmP2*nPmP3*nPmP4/
 (pow(tnP3,4.0)*pow(tnP5,3.0)*tnP7*tnP7*tnP9))/tnP1;

C = (pow(nMmP1*nMmP2*nPmP1*nPmP2,2.0)/
 (pow(tnP3,7.0)*tnP5*tnP5) -
 pow(nMmM1*nMm*nPmM1*nPm,2.0)/
 (tnM3*tnM3*pow(tnM1,7.0)))/(tnP1*tnP1);

D = nMmM1*nMm*nMmP1*nMmP2*nPmM1*nPm*nPmP1*nPmP2/
 (tnM3*pow(tnM1*tnP3,4.0)*tnP1*tnP1*tnP5);

L8 = 2.0*pow(tm*tm-1.0,2.0)*A+B/16.0+C/8.0+D/2.0;
LLPS = L0+(L2+(L4+(L6+L8*csq)*csq)*csq)*csq;
return  LLPS;
}
```

```
double U(double B1[], double G1[], double B2[],
  double G2[],int m,int n,int r1Start,int r2Start,
  int smax, double LL)
{
/* Continued fraction with spherical wave function
   eigenvalue */
double  U1,U2;
int   sStop,s;

/* U1 series; finite */
if ( n == m+r1Start )  U1 = G1[r1Start] - LL;
else
  {
  U1 = B1[r1Start]/(G1[r1Start] - LL);
  sStop = (n-m-1)/2;
  s = r1Start+1;
  while ( s <=  sStop )
    { U1 = B1[s]/(G1[s] - LL - U1);   s++; }
  U1 = G1[s] - LL - U1;
  }

/* U2 series; infinite, but truncated
   after smax  terms  */
s = 1;
U2 = B2[1]/(G2[1] - LL);
while ( s <=  smax )
  { U2 = B2[s]/(G2[s] - LL - U2);   s++; }

return   U1 - U2;
}
```

Test Values

Check values of the spherical wave function eigenvalues, $\lambda_{mn}(c)$, can be generated by running the SWFeigenvalue cell in *Mathematica* notebook SWF, which is annotated in Section 20.13. Examples are given in Tables 13.1.1 (prolate) and 13.1.2 (oblate). These values can be used as benchmark values for the C functions and for optimizing the value of smax in a desired range of m, n, and c.

The values from our C and *Mathematica* programs agree with each other to 10 digits; they also agree with the values tabulated in Flammer [Fla57], Tables 10 (prolate) and 131 (oblate), although the latter tables give no more than 7 digits.

TABLE 13.1.1 Test values to 10 digits of the spheroidal wave function eigen-values for *prolate* coordinates with $c = 2$.

$m \setminus n$	0	1	2
0	1.127734065	4.287128544	8.225713001
1	0	2.734111026	7.653149562
2	0	0	6.542495274

TABLE 13.1.2 Test values to 10 digits of the spheroidal wave function eigenvalues for *oblate* coordinates with $c = 2$.

$m \setminus n$	0	1	2
0	-1.594493213	-0.5052439809	4.091509102
1	0	1.118553391	4.222747333
2	0	0	5.395010783

13.1.4 Auxiliary Functions for Eigenvalues

Keywords

auxiliary functions, spheroidal coordinates, spheroidal eigenvalues, spheroidal wave functions

C Function and *Mathematica* Notebook

SWFAngCoeff (C) in Section 13.2.1; function GTT in cell SWFAngCoefficient in *Mathematica* notebook SWF

Background

Computation of the spheroidal wave function coefficients (Sections 13.2, 13.3) is most accurate if an auxiliary function, t_{mn}, related to the eigenvalue λ_{mn} (Section 13.1.3) is used, as suggested by Little and Corbató [Stra56, page 57]. The reason for this is given in the following subsection. Also, by computing t_{mn} one may compare directly with the eigenvalue quantity $t = t_{mn}$ which is tabulated in Stratton et al [Stra56]. You may also find the *Mathematica* and C functions for t_{mn} to be useful for computing λ_{mn}.

Function Definition

The relation between the auxiliary function and the eigenvalue is

$$\lambda_{mn}(c) = n(n+1) + c^2 \left[\frac{2n(n+1) - 2m^2 - 1}{(2n-1)(2n+3)} + t_{mn}(c) \right] \tag{13.1.20}$$

The term that is added to t_{mn} is just l_2 in (13.1.19); thus, $t_{mn}(c)$ vanishes proportionally to c^2 as c tends to zero. A typical order of magnitude of $t_{mn}(c)$ is $-0.01c^2$, which we use as its starting estimate in the *Mathematica* function SWFevalTMN.

Numerical accuracy of the quantities $\gamma_r^m - \lambda_{mn}$ appearing in (13.1.15) and (13.1.16) is much improved by using the definitions (13.1.13) and (13.1.20) and making some algebraic simplifications to rewrite this quantity as

$$\gamma_{n-m+s}^m - \lambda_{mn} = s(2n+s+1) + $$
$$\frac{2s(2n+s+1)(4m^2-1)c^2}{(2n-1)(2n+3)(2n+2s-1)(2n+2s+3)} - c^2 t_{mn} \tag{13.1.21}$$

In the function U_1 given by (13.1.15), s ranges from zero to $-(n-m)$ or $-(n-m)+1$, depending on whether $n-m$ is even or odd. In U_2 given by (13.1.16), s ranges from 2 to the input parameter `smax`. Since the first few terms of U_1 usually dominate in determining the eigenvalue, it is advantageous to have the difference $\gamma_r^m - \lambda_{mn}$ small when s is small, as is satisfied by the s-dependent terms on the right side of (13.1.21), which vanish for $s = 0$. Note that in the above formulas the conversion from prolate to oblate cases is obtained just by a change of sign of c^2, as discussed below (13.1.12).

13.2 SPHEROIDAL ANGULAR FUNCTIONS

The spheroidal angular functions introduced in Section 13.1.2 are usually expanded in a basis of associated Legendre functions (Chapter 12). In Section 13.2.1 we give formulas and C functions for the expansion coefficients; Section 13.2.2 has results for spheroidal angular functions. Several analytical properties of the functions are given in [Abr64, Chapter 21] and [Erd53c, Sections 16.9–16.13].

13.2.1 Expansion Coefficients for Angular Functions

Keywords
expansion coefficients, spheroidal coordinates, spheroidal wave functions

C Function and *Mathematica* Notebook
SWFAngCoeff (C); cell SWFAngularCoefficient in *Mathematica* notebook SWF $\boxed{M}$

Function Definitions
Spheroidal angular functions are usually expanded into spherical Legendre functions, either of the first kind, $P_{m+r}^m(\eta)$ in Section 12.2.2, or of the second kind, $Q_{m+r}^m(\eta)$ in Section 12.2.3. For the spheroidal function of the first kind, which is regular at $\eta = \pm 1$, we follow Abramowitz and Stegun [Abr64, 21.7.1] and Flammer [Fla57, (3.1.4)] by writing

$$S_{mn}^{(1)}(c,\eta) = \sum_{r=0,1}^{\infty} d_r^{mn}(c) P_{m+r}^m(\eta) \tag{13.2.1}$$

in which the summation starts at $r = 0$ if $n - m$ is even, but it starts at $r = 1$ if $n - m$ is odd. In either case, r goes by steps of 2. As $c \to 0$, the spheroidal angular function must collapse to $P_{m+r}^m(\eta)$ with the same m and n values. The only nonzero angular coefficient is then $d_{n-m}^{mn}(c)$, corresponding to $r = n - m$, that is, to $s = 0$ in (13.1.21).

For the function of the second kind, which is irregular at $\eta = \pm 1$, we also follow [Abr64, 21.7.2] by writing

$$S_{mn}^{(2)}(c,\eta) = \sum_{r=0,1}^{\infty} d_r^{mn}(c) Q_{m+r}^m(\eta) \tag{13.2.2}$$

with the summation starting at $r = 0$ if $n - m$ is even but at $r = 1$ if $n - m$ is odd, and r goes by steps of 2. As $c \to 0$, this spheroidal angular function collapses to $Q_n^m(\eta)$, so that the only nonzero angular coefficient is $d_{n-m}^{mn}(c)$, that is, $r = n - m$ in (13.2.2) and $s = 0$ in

(13.1.21). Flammer [Fla57, Section 3.1] introduces but does not develop the spheroidal angular functions of the second kind.

The spheroidal angular coefficients $d_r^{mn}(c)$ can be computed numerically by using the recurrence relation between them, given in Abramowitz and Stegun [Abr64, 21.7.3]:

$$\alpha_r d_{r+2}^{mn} + (\beta_r - \lambda_{mn}) d_r^{mn} + \gamma_r d_{r-2}^{mn} = 0 \qquad (13.2.3)$$

All of these quantities depend on the parameter c; the terms α_r and γ_r are given by

$$\alpha_r = \frac{(2m+r+2)(2m+r+1)c^2}{(2m+2r+3)(2m+2r+5)} \qquad (13.2.4)$$

and

$$\gamma_r = \frac{r(r-1)c^2}{(2m+2r-3)(2m+2r-1)} \qquad (13.2.5)$$

For accurate numerical calculation of the d_r^{mn}, we write the term that involves λ_{mn} as

$$\beta_{n-m+s} - \lambda_{mn} = s(2n+s+1) + $$
$$c^2 \left[\frac{2s(2n+s+1)(4m^2-1)}{(2n-1)(2n+3)(2n+2s-1)(2n+2s+3)} - t_{mn} \right] \qquad (13.2.6)$$

Use of the recurrence relation (13.2.3) can procede in the direction of increasing r or of decreasing r. For modest values of $n-m$ the latter gives results that are more accurate; the details are discussed under Algorithm and Program.

Our normalization of the $d_r^{mn}(c)$ uses the Flammer scheme [Fla57; 21.7.15, and 21.7.16 in Abr64], for which the normalization is

$$\sum_{r=0}^{\infty} \frac{(-1)^{r/2}(r+2m)! \, d_r^{mn}}{2^r (r/2)!(r/2+m)!} = \frac{(-1)^{(n-m)/2}(n+m)!}{2^{n-m}[(n-m)/2]![(n+m)/2]!}$$
$$\sum_{r=1}^{\infty} \frac{(-1)^{(r-1)/2}(r+2m+1)! \, d_r^{mn}}{2^r [(r-1)/2]![(r+2m+1)/2]!} = \frac{(-1)^{(n-m-1)/2}(n+m+1)!}{2^{n-m}[(n-m-1)/2]![(n+m+1)/2]!} \qquad (13.2.7)$$

for $n-m$ even or odd, respectively. The spheroidal and spherical angular functions of the first kind then coincide at $\eta = 0$; $S_{mn}^{(1)}(c,0) = P_n^m(0)$.

If one uses alternatively to the Flammer scheme (13.2.7) the normalization in Stratton et al [Str56] that is given as [Abr64, 21.7.12], then the range of values of the coefficients becomes very large, especially for oblate coordinates. Coefficients in the Stratton scheme are tabulated extensively to 7-digit accuracy in [Stra56]. The values in the *Atlas* agree with these values to the accuracy quoted therein.

To remind readers of the *Atlas* of the progress in computing technology over 40 years, note that the roughly 75,000 numerical values in the tables of Stratton et al required two programmers "about six months of fairly intensive effort" and 10 hours of production time on the Whirlwind I computer at the Massachusetts Institute of Technology. The output was more than 5 km of paper tape, from which the tables were prepared on an electric typewriter. Such are the heroic deeds from which travelers' tales are spun!

Visualization

The expansion coefficients for the angular part of spheroidal wave functions, the $d_r^{mn}(c)$ in (13.2.1) and (13.2.2), depend upon the order n, the degree m, the parameter c in (13.1.5), and on whether one is using prolate or oblate spheroidal coordinates (Figure 13.1.1). We therefore display $d_r^{mn}(c)$ as surfaces made from plaquettes whose vertices are the coefficient values.

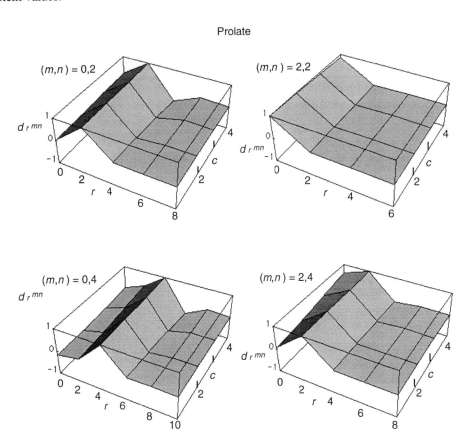

FIGURE 13.2.1 Coefficients (in the Flammer normalization) for expansion of spheroidal wave functions in a basis of Legendre functions according to (13.2.1) or (13.2.2). Shown for four pairs of m and n values as a function of parameter c that measures the prolateness having values 1 to 4. The coefficients peak at $r = n - m$, which is the unique value when $c = 0$. (Adapted from cell SWFAngCoefficient in *Mathematica* notebook SWF.)

As discussed below (13.2.1), the coefficients are expected to peak at $r = n - m$, at least for moderate c values; this is confirmed by the graphics in Figure 13.2.1.

For oblate coordinates (right side of Figure 13.1.1), the coefficients shown in Figure 13.2.2 behave quite similarly to those for prolate coordinates (Figure 13.2.1). For values of $-c^2$ much larger in magnitude than those shown here, however, the behavior of the $d_r^{mn}(c)$ is much more complicated, as can be explored by using the cell SWFcoefficient in *Mathematica* notebook SWF.

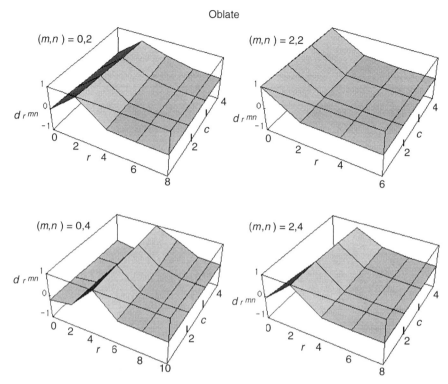

FIGURE 13.2.2 Coefficients (in the Flammer normalization) for expansion of spheroidal wave functions in a basis of Legendre functions according to (13.2.1) or (13.2.2). Shown for four pairs of m and n values as a function of the parameter c that measures the oblateness having values 1 to 4. The coefficients peak at $r = n - m$, which is the unique value when $c = 0$. (Adapted from cell SWFAngCoefficient in *Mathematica* notebook SWF.)

Algorithm and Program

Computing the expansion coefficients for spheroidal angular functions goes by two stages. First, the auxiliary functions for eigenvalues, $t_{mn}(c)$ in Section 13.1.4, are determined as the roots of function U in (13.1.17); the algorithm for this is the same as described for function SWFevalue in Section 13.1.3, except that the starting estimate for $t_{mn}(c)$ is much simpler than that for $\lambda_{mn}(c)$ and should usually be adequate. The coded function is SWF-evalTMN.

Second, the coefficients $d_r^{mn}(c)$ are determined by backward recurrence on r, starting with the coefficient for $r = m - n + 14$ being assigned the value unity. The range of r values from this value down to $r = 0$ or 1 typically gives 15 orders of magnitude range for the coefficients. Recurrence scheme (13.2.3) is used, expressed in terms of the functions aa for α_r in (13.2.4), bb for $\beta_r - \lambda_{mn}$ in (13.2.6), and cc for γ_r in (13.2.5). The C function SWFAngCoeff invokes SWFevalTMN. The driver program for the spheroidal functions given in Section 21.13 shows how the accuracy parameter for SWFAngCoeff, TMNeps, that determines the accuracy of the auxiliary eigenvalue function, is used. Normalization of the expansion coefficients is done by using the Flammer scheme [Fla57], given in (13.2.7). Factorials are computed in terms of logarithms of Γ functions, using the function LogGamma described in Section 6.1.1.

Here is the code for the primary functions SWFcoeff and SWFevalTMN.

```
void SWFAngCoeff(int m,int n,double sign,double c,
 int smax, double TMNeps, double dmn[], double *tmn)
/* Spherical angular function coefficients,
   using  tmn  of Little and Corbato in [Stra56]
   with backward iteration */


{
double SWFevalTMN(int m, int n, double csq,
                  int smax, double TMNeps),
 LogGamma( double eta), aa(int m,int r,double csq),
 bbtmn(int m,int n,int s,double csq,double tmn),
 cc(int m,int r,double csq);

double csq,tm,tmM1,fm,fnPm,fnMm,trSt,frSt,
 term,sum,fr,phase,Norm;
int nMm,rSt,rLarge,r;

csq = sign*c*c;
/* Modified eigenvalue */
*tmn = SWFevalTMN(m,n,csq,smax,TMNeps);

/* Make array of coefficients */
nMm = n - m;
rSt = ( nMm % 2 == 0 ) ?  0: 1;
rLarge = nMm + 14;
dmn[rLarge] = 1.0;

for ( r = rLarge; r >= rSt+2; r -= 2 )
  {
  if ( csq == 0.0 )
   { dmn[r] = ( r == nMm ) ? 1.0: 0.0; }
  else
    {
    dmn[r-2] = ( r == rLarge ) ?
                -bbtmn(m,n,r-nMm,csq,*tmn)*dmn[r]/
                 cc(m,r,csq):
                -(bbtmn(m,n,r-nMm,csq,*tmn)*dmn[r]+
                 aa(m,r,csq)*dmn[r+2])/cc(m,r,csq);

    }
  }
/* Normalization in  Flammer scheme; [Fla57] */
tm = m+m;
tmM1 = tm - 1.0;
fm = m;
fnPm = n + m;
fnMm = nMm;
trSt = 2*rSt;
frSt = rSt;

term = exp(LogGamma(tm+1.0+trSt) -
 LogGamma(fm+frSt+1.0))/pow(2.0,frSt);
sum = term*dmn[rSt];
```

```
for ( r = rSt+2; r <= rLarge; r += 2 )
  {
  fr = r;
  term *= -(fr+tmM1+frSt)/(fr - frSt);
  sum += term*dmn[r];
  }
phase = ( (nMm-rSt) % 4  == 0 ) ? 1.0 : -1.0;
Norm = phase*exp(LogGamma(fnPm+frSt+1.0) -
  LogGamma(0.5*(fnMm-frSt)+1.0) -
  LogGamma(0.5*(fnPm+frSt)+1.0))/(pow(2.0,fnMm)*sum);

for ( r = rSt; r <= rLarge; r += 2 )
  { dmn[r] *= Norm; }
return;
}


double aa(int m, int r, double csq)
/* Alpha term for SWF coefficient recurrence */
{
double tmPrP1,tmPtrP3;

tmPrP1 = 2*m+r+1;
tmPtrP3 = 2*(m+r)+3;
return
(tmPrP1+1.0)*tmPrP1*csq/(tmPtrP3*(tmPtrP3+2.0));
}


double bbtmn(int m, int n, int s, double csq,
             double tmn)
/* Beta term for SWF coefficient recurrence;
   modified for  tmn  auxiliary function  */
{
double fm,tn,ts,fs,tnPts;

fm = m;
tn = 2*n;
ts = 2*s;
fs = s;
tnPts = tn+ts;
return   fs*(tn+fs+1.0)*(1.0+csq*
  (2.0*(4.0*fm*fm-1.0)/((tn-1.0)*(tn+3.0)*
  (tnPts-1.0)*(tnPts+3.0)))) - csq*tmn;
}


double cc(int m, int r, double csq)
/* Gamma term for SWF coefficient recurrence */
{
double fr,tm,tmPtr;

fr = r;
tm = 2*m;
tmPtr = tm+2.0*fr;
return   fr*(fr-1.0)*csq/((tmPtr-3.0)*(tmPtr-1.0));
}
```

```
double SWFevalTMN(int m, int n, double csq,
                  int smax, double TMNeps)

/* Spheroidal wave function eigenvalue in tmn form
   for degree n, order m, argument csq;
   Retain  smax  terms in continued fraction and
   get modified eigenvalue to accuracy TMNeps */


{
double U(double B1[], double GTT1[], double B2[],
double GTT2[], int m,int n,int r1Start,int r2Start,
int smax, double cTMN),
c2D2,tcsq,c4,fm2M1,fr,tmPr,mPr,tmPtrM1,tn,tnM13,s,
tnPtsM1,tmn,cTMNest,cTMNu,cTMNd,Uu,Ud,cTstep,Umid,
cTMNrt,dcTMN,cTMNmid;

double B1[MAX],GTT1[MAX],B2[MAX],GTT2[MAX];
int nMm,r1Start,r2Start,r,t,k;

if ( csq == 0.0 )    return 0.0;

/* Make arrays of terms in continued fractions
   that are independent of eigenvalue */
c2D2 = 0.5*csq;
tcsq = 2.0*csq;
c4 = csq*csq;
nMm = n - m;
fm2M1 = 4.0*m*m - 1.0;


/* Elements for U1 function */
r1Start = ( nMm % 2 == 0 ) ?  0: 1;
r = r1Start;
fr = r;
tmPr = 2*m+r;
mPr = m+r;
tmPtrM1 = 2.0*mPr - 1.0;
tn = n+n;
tnM13 = (tn-1.0)*(tn+3.0);
s = fr - (double) nMm;

for ( t = r1Start; t <=  (nMm+1)/2; t++ )
  {
  B1[t] = (fr+2.0)*(fr+1.0)*(tmPr+2.0)*(tmPr+1.0)*c4/
    ((tmPtrM1+4.0)*(tmPtrM1+4.0)*(tmPtrM1+2.0)*
    (tmPtrM1+6.0));
  tnPtsM1 = tn+2.0*s-1.0;
  GTT1[t] = s*(tn+s+1.0)*(1.0+tcsq*fm2M1/
    (tnM13*tnPtsM1*(tnPtsM1+4.0)));
  fr += 2.0;
  tmPr += 2.0;
  tmPtrM1  += 4.0;
  mPr += 2.0;
  s += 2.0;
  }
```

```
/* Elements for U2 function;
   r starts at n-m+2smax for smax terms */
r2Start = nMm+2*smax;
fr = r2Start;
tmPr = 2*m+r2Start;
mPr = m+r2Start;
tmPtrM1 = 2*mPr - 1;
s = 2*smax;
for ( t = 1; t <=  smax; t++ )
 {
 B2[t] = fr*(fr-1.0)*tmPr*(tmPr-1.0)*c4/
  (tmPtrM1*tmPtrM1*(tmPtrM1-2.0)*(tmPtrM1+2.0));
 tnPtsM1 = tn+2.0*s-1.0;
 GTT2[t] = s*(tn+s+1.0)*(1.0+tcsq*fm2M1/
  (tnM13*tnPtsM1*(tnPtsM1+4.0)));
 fr -= 2.0;
 tmPr -= 2.0;
 tmPtrM1  -= 4.0;
 mPr -= 2.0;
 s -= 2.0;
 }
/* Estimate modified eigenvalue multiplied by c*c */
cTMNest =  -0.01*csq;

/* Bracket the modified eigenvalue */
cTMNu = cTMNest*1.1;
cTMNd = cTMNest*0.9;
Uu =
U(B1,GTT1,B2,GTT2,m,n,r1Start,r2Start,smax,cTMNu);
Ud =
U(B1,GTT1,B2,GTT2,m,n,r1Start,r2Start,smax,cTMNd);
k = 1;
while ( Uu*Ud > 0.0 )
 { cTstep = cTMNu - cTMNd;
   if ( fabs(Uu) < fabs(Ud) )
     { cTMNu += cTstep;
       Uu =
U(B1,GTT1,B2,GTT2,m,n,r1Start,r2Start,smax,cTMNu); }
   else
     { cTMNd -= cTstep;
       Ud =
U(B1,GTT1,B2,GTT2,m,n,r1Start,r2Start,smax,cTMNd); }
 if ( k > NTRY )
   { printf
   ("\n!! SWFevalTMN bracket > %i tries; 0 returned",
     NTRY);
     return 0.0; }
   k ++;
 }
/* Refine the root to fractional accuracy TMNeps by
   bisection method; start at cTMNu and cTMNd */
Uu =
 U(B1,GTT1,B2,GTT2,m,n,r1Start,r2Start,smax,cTMNu);
Umid =
 U(B1,GTT1,B2,GTT2,m,n,r1Start,r2Start,smax,cTMNd);
```

```
/* Search direction has U>0 for cTMNrt+dcTMN */
cTMNrt = ( Uu < 0.0 ) ? (dcTMN = cTMNd - cTMNu,
 cTMNu): (dcTMN = cTMNu - cTMNd, cTMNd);
k = 0;

while ( (fabs(dcTMN) >= TMNeps) && (Umid != 0.0) )
 {
 dcTMN *= 0.5;
 cTMNmid = cTMNrt+dcTMN;
 Umid =
 U(B1,GTT1,B2,GTT2,m,n,r1Start,r2Start,smax,cTMNmid);
 if ( Umid <= 0.0 ) cTMNrt = cTMNmid;
 k++;
 if ( k > KMAX )
  { printf("\n!! > %i tries to bisect root",KMAX);
    Umid = 0.0; }
 }
return  cTMNrt/csq;
}

double U(double B1[], double GTT1[], double B2[],
 double GTT2[],
 int m, int n, int r1Start, int r2Start, int smax,
double cTMN)
{
/* Continued fraction with modified spherical wave
   function eigenvalue TMN   times  csq */
double  U1,U2;
int  sStop,s;

/* U1 series; finite */
if ( n == m )    U1 = GTT1[0] - cTMN;
else
 {
 if ( n == m+1 )  U1 = GTT1[1] - cTMN;
 else
  {
  U1 = B1[r1Start]/(GTT1[r1Start] - cTMN);
  sStop = (n-m-1)/2;
  s = r1Start+1;
  while ( s <=  sStop )
   { U1 = B1[s]/(GTT1[s] - cTMN - U1);  s++; }
  U1 = GTT1[s] - cTMN - U1;
  }
 }

/* U2 series; infinite, but truncated
   after  smax  terms */
s = 1;
U2 = B2[1]/(GTT2[1] - cTMN);
while ( s <=  smax )
 { U2 = B2[s]/(GTT2[s] - cTMN - U2);  s++; }

return  U1 - U2;
}
```

Test Values

Test values of the auxiliary functions for eigenvalues, the $t_{mn}(c)$ defined in Section 13.1.4, can be generated by running the cell SWFAngCoefficient in *Mathematica* notebook SWF, which is annotated in Section 20.13. Samples are given in Tables 13.2.1 (prolate coordinates) and 13.2.2 (oblate coordinates) and can be used to benchmark the C functions.

The values from our C and *Mathematica* programs agree with each other to 10 digits; they also agree to 7 significant figures with the values tabulated in Stratton et al [Str56], wherein $t_{mn}(c)$ is denoted by t.

TABLE 13.2.1 Test values to 10 digits of the auxiliary functions for the spheroidal wave function eigenvalues for *prolate* coordinates with $c = 2$.

$m \backslash n$	0	1	2
0	–0.05139981712	–0.02821786401	0.03261872647
1	0	–0.0164722436	–0.01528403807
2	0	0	–0.00723332426

TABLE 13.2.2 Test values to 10 digits of the auxiliary functions for the spheroidal wave function eigenvalues for *oblate* coordinates with $c = 2$.

$m \backslash n$	0	1	2
0	0.06528996996	0.02631099522	–0.04668679935
1	0	0.02036165231	0.01574173809
2	0	0	0.00839016127

Check values of the expansion coefficients for angular functions, the $d_r^{mn}(c)$ in (13.2.1) and (13.2.2), can be obtained by running cell SWFAngCoefficient in *Mathematica* notebook SWF, annotated in Section 20.13. The examples given in Tables 13.2.3 (prolate coordinates) and 13.2.4 (oblate coordinates) can be used to benchmark the C functions.

TABLE 13.2.3 Test values to 10 digits of the spheroidal angular function expansion coefficients for *prolate* coordinates with $c = 2$.

$r / (m, n)$	$(0, 0)$	$(0, 1)$	$(2, 2)$
0	8.316189907, –01	0	9.626013828, –01
1	0	7.927602801, –01	0
2	–3.205879803, –01	0	–1.462189666, –02
3	0	–1.304916770,–01	0
4	2.106221614, –02	0	1.902542786, –04
5	0	5.979682966, –03	0

In Tables 13.2.3 and 13.2.4, powers of 10 are indicated by ",−01" for 10^{-1}, etc.

TABLE 13.2.4 Test values to 10 digits of the spheroidal angular function expansion coefficients for *oblate* coordinates with $c = 2$.

$r / (m, n)$	$(0, 0)$	$(0, 1)$	$(2, 2)$
0	1.302525566, 00	0	1.044919786, 00
1	0	1.278288017, 00	0
2	6.378139131, −01	0	1.841079559, −02
3	0	1.961926744, −01	0
4	4.475614052, −02	0	2.567034433, −04
5	0	8.750870761, −03	0

13.2.2 Spheroidal Angular Functions

Keywords

angular functions, spheroidal coordinates, spheroidal wave functions

C Functions and *Mathematica* Notebook

S1MN and S2MN (C); cell SWFAngWave in *Mathematica* notebook SWF

M

Function Definitions

The spheroidal angular functions are defined by (13.2.1) and (13.2.2) respectively for the functions that are regular at $\eta = \pm 1$, $S_{mn}^{(1)}(c, \eta)$, and those that irregular at $\eta = \pm 1$, $S_{mn}^{(2)}(c, \eta)$. These functions are expanded in series of regular Legendre functions, $P_{m+r}^{m}(\eta)$, or irregular Legendre functions, $Q_{m+r}^{m}(\eta)$, as described in Section 13.2.1. (Legendre functions are portrayed in Chapter 12.) The challenge in such an expansion is to compute the expansion coefficients, $d_r^{mn}(c)$, which are the same for both the functions of the first and second kind. These computations are described in Section 13.2.1.

Visualization

There are four arguments for each of the spheroidal angular functions of the first and second kind; m, n, c, and η. Also, we have to distinguish between prolate and oblate coordinates (Figure 13.1.1). To visualize $S_{mn}^{(1)}(c, \eta)$ and $S_{mn}^{(2)}(c, \eta)$ we choose $|c| = 2$ and superimpose three curves for each m and n; $c = 2$ (prolate case), $c = 0$ (spherical case). and $|c| = 2$ (oblate case). For $c = 0$ we have the spherical Legendre functions, $P_n^m(\eta)$ and $Q_n^m(\eta)$, which are also shown in Figures 12.2.2 and 12.2.5, respectively.

For the regular functions, $S_{mn}^{(1)}(c, \eta)$, we show η over the range $[0, 1]$, while for the irregular function, $S_{mn}^{(1)}$, we use the η range $[0, 0.8]$.

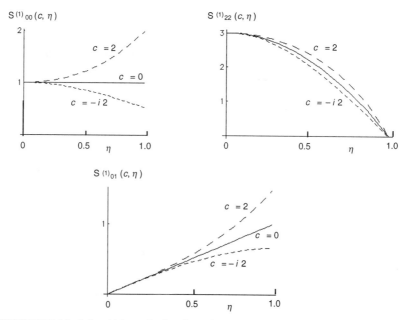

FIGURE 13.2.3 Spheroidal angular functions of the first kind (regular at $\eta = 1$) according to (13.2.1). Shown for three pairs of m and n values for the magnitude of parameter c equal to 2 in each graphic, for prolate (long dashes), spherical (solid curve), and oblate (short dashes). (Adapted from cell `SWFAngWave` in *Mathematica* notebook `SWF`.)

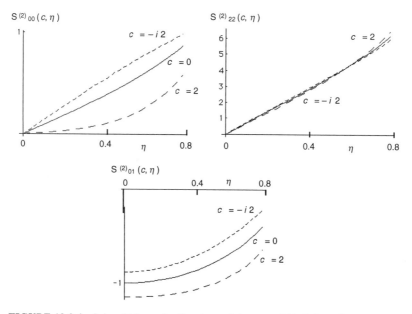

FIGURE 13.2.4 Spheroidal angular functions of the second kind (irregular at $\eta = \pm 1$) according to (13.2.2). Shown for three pairs of m and n values for the magnitude of parameter c equal to 2 in each graphic, for prolate (long dashes), spherical (solid curve), and oblate (short dashes). (Adapted from cell `SWFAngWave` in *Mathematica* notebook `SWF`.)

Notice that in both Figure 13.2.3 and Figure 13.2.4 the spherical angular function is approximately the average of the values for prolate and oblate coordinates, which indicates that the expansion coefficients $d_r^{mn}(c)$ are approximately even functions of c.

Algorithm and Program

The main task in computing the spheroidal angular functions $S_{mn}^{(1)}(c,\eta)$ and $S_{mn}^{(2)}(c,\eta)$ is to compute the expansion coefficients in (13.2.1) and (13.2.2), which is accomplished in Section 13.2.1. Assuming that the d_r^{mn} are available for a given choice of c for prolate or oblate coordinates, it is easy to sum the Legendre function series until the result is sufficiently accurate.

For simplicitly, we incorporate computing the coefficients and summing the series into the same driver program, SMN, which is annotated in Section 21.13. There we also show how the accuracy of the expansion is controlled by using the input parameter Seps.

The C code for the regular function, $S_{mn}^{(1)}(c,\eta)$, is in S1MN, as follows:

```
double S1MN(int m, int n, double eta,
            double dmn[], double Seps)
/* Spheroidal Angular Function of First Kind;
   degree n, order m, argument eta.
   Convergence to fractional acuracy  Seps  */

{
double PNM(int n, int m, double eta),sum,ratio,term;
int nMm,rStart,rLarge,r;

nMm = n - m;
rStart = ( nMm % 2 == 0 ) ?  0: 1;
rLarge = nMm +14;/* coefficient range about 10^15 */
sum = 0.0;
ratio = 10;
r = rStart;

while ( ratio > Seps )
  {
  if (  r > rLarge )
    { printf
      ("\n\n!!S1NM unconverged to %.6E in %i terms",
        Seps,rLarge);
     return  sum;
     }
  term = dmn[r]*PNM(m+r,m,eta);
  sum += term;
  ratio = fabs(term/sum);
  r += 2;
  }

return  sum;
}
```

The C function S2MN, which computes the irregular spheroidal angular function $S_{mn}^{(2)}(c,\eta)$, is very similar to that for the regular function, except for the irregular Legendre functions, $Q_{m+r}^m(\eta)$ in C function QNM, instead of the regular Legendre functions, $P_{m+r}^m(\eta)$ in C function PNM that appear in S1NM above. Here is the code for S2NM; note that it assumes that the coefficients dmn[r] have been precomputed; for how this is done, see Section 13.2.1 and the driver program in Section 21.13.

```
double S2MN(int m, int n, double eta,
            double dmn[], double Seps)
/* Spheroidal Angular Function of Second Kind;
   degree n, order m, argument eta.
   Convergence to fractional acuracy  Seps  */

{
double QNM(int n, int m, double eta),sum,ratio,term;
int nMm,rStart,rLarge,r;

nMm = n - m;
rStart = ( nMm % 2 == 0 ) ?  0: 1;
rLarge = nMm +14;/* coefficient range about 10^15 */
sum = 0.0;
ratio = 10;
r = rStart;

while ( ratio > Seps )
  {
  if (  r > rLarge )
   { printf
    ("\n\n!!S2NM unconverged to %.6E in %i terms",
       Seps,rLarge);
    return  sum;
   }
  term = dmn[r]*QNM(m+r,m,eta);
  sum += term;
  ratio = fabs(term/sum);
  r += 2;
  }

return  sum;
}
```

Test Values

Test values of the spheroidal angular functions— $S_{mn}^{(1)}(c,\eta)$ and $S_{mn}^{(2)}(c,\eta)$ in (13.2.1) and (13.2.2)—can be obtained by running cell SWFAngWave in *Mathematica* notebook SWF, annotated in Section 20.13. Examples are given in Tables 13.2.5 and 13.2.6—for $S_{mn}^{(1)}(c,\eta)$ in prolate and oblate coordinates, respectively—and in Tables 13.2.7 and 13.2.8 for $S_{mn}^{(2)}(c,\eta)$ in prolate and oblate coordinates. They can be used to benchmark the C functions S1MN and S2MN.

The regular spheroidal angular functions $S_{mn}^{(1)}(c,\eta)$ are tabulated to 4-digit accuracy by Flammer [Fla57] in Tables 60–69 for prolate coordinates and in Tables 153–165 for oblate coordinates. Abramowitz and Stegun [Abr64, Table 21.2] provide a 4-digit table for the regular function over very limited ranges of m, n, c, and η. To our knowledge, the irregular function, $S_{mn}^{(2)}(c,\eta)$, is not tabulated elsewhere. The function values computed by our C and *Mathematica* codes agree to at least 10 significant figures.

Here are the test values for $S_{mn}^{(1)}(c,\eta)$.

TABLE 13.2.5 Test values to 10 digits of the spheroidal angular functions of the first kind for *prolate* coordinates with $c = 2$.

η / (m, n)	$(0, 0)$	$(0, 1)$	$(2, 2)$
0.0	1	0	3
0.5	0.8654110232	0.4539778475	2.101460986
1.0	0.5315070076	0.6681191605	0

TABLE 13.2.6 Test values to 10 digits of the spheroidal angular functions of the first kind for *oblate* coordinates with $c = 2$.

η / (m, n)	$(0, 0)$	$(0, 1)$	$(2, 2)$
0.0	1	0	3
0.5	1.210281779	0.5541373857	2.425127624
1.0	1.986421857	1.483422681	0

Test values for $S_{mn}^{(2)}(c,\eta)$ (which is irregular at $\eta = \pm 1$) are as follows:

TABLE 13.2.7 Test values to 10 digits of the spheroidal angular functions of the second kind for *prolate* coordinates with $c = 2$.

η / (m, n)	$(0, 0)$	$(0, 1)$	$(2, 2)$
0.2	0.3035663912	−0.8285844306	1.671915984
0.5	0.7284492853	−0.5457394502	4.078013605
0.8	1.122281824	0.01259212405	7.615701178

TABLE 13.2.8 Test values to 10 digits of the spheroidal angular functions of the second kind for *oblate* coordinates with $c = 2$.

η / (m, n)	$(0, 0)$	$(0, 1)$	$(2, 2)$
0.2	0.03626069838	−1.128501496	1.516997335
0.5	0.2132070307	−0.9613846209	4.048404863
0.8	0.9584636797	−0.3230285639	8.528130842

13.3   SPHEROIDAL RADIAL FUNCTIONS

The spheroidal radial functions, $R_{mn}(c,\xi)$, that are introduced in Section 13.1.2 are usually expanded in a basis of spherical Bessel functions (Section 14.3.1). Such expansions turn out to be quite simple, since the major work of understanding and computing the expansion coefficients is done in Sections 13.1.3 and 13.1.4 (eigenvalues and their auxiliary functions) and in Section 13.2.1 (expansion coefficients for angular functions). The reason for this simplicity is that the radial expansion coefficients are proportional to the angular expansion coefficients.

For each choice of spherical-basis radial functions (Bessel, Neumann, or Hankel), there is a corresponding spheroidal radial function. For illustrative purposes and to allow ready comparison with tabulated functions in Flammer [Fla57] and in Stratton et al [Stra56], we present formulas, visualizations, and numerical values for the spheroidal radial function that is called $R_{mn}^{(1)}(c,\xi)$ in Flammer and in Abramowitz and Stegun [Abr64, 21.9.1], and is denoted by $je_{m\ell}(h,\xi)$ in Stratton et al, wherein $\ell = n$ and $h = c$ in our notation. When $c = 0$ $R_{mn}^{(1)}(c,\xi)$ collapses to $j_n(c\xi)$, the spherical Bessel function in Section 14.4.1.

Many analytical properties of radial functions are given in Section 11.3 of Morse and Feshbach [Mor53], in Erdélyi et al [Erd53c, Sections 16.9–16.13], in Meixner and Schäfke [Mei54], in Stratton et al [Stra56], in Flammer [Fla57, Chapter 4], and in Abramowitz and Stegun [Abr64, Section 21.9]. Section 13.3.1 has formulas, visualizations, and a C function for radial-function coefficients, while Section 13.3.2 describes $R_{mn}^{(1)}(c,\xi)$.

13.3.1 Expansion Coefficients for Radial Functions

Keywords
expansion coefficients, radial functions, spheroidal coordinates, spheroidal wave functions

C Function and *Mathematica* Notebook
SWFRadCoeff (C); cell SWFRadCoefficient in *Mathematica* notebook SWF

Function Definitions
The prototype spheroidal radial function is for the function that is regular at $\xi = \pm 1$ and is expanded in terms of spherical Bessel functions (Section 14.3.1), as follows:

$$R_{mn}^{(1)}(c,\xi) = \left(1 - 1/\xi^2\right)^{m/2} \sum_{r=0,1}^{\infty} a_r^{mn}(c) j_{m+r}(c\xi) \qquad (13.3.1)$$

in which the summation starts at $r = 0$ if $n - m$ is even, but it starts at $r = 1$ if $n - m$ is odd. In either case, r goes by steps of 2. The real coefficients $a_r^{mn}(c)$ are called the radial expansion coefficients. As $c \to 0$, the spheroidal radial function must collapse to $j_n(c\xi)$, so the only nonzero radial coefficient is then $a_{n-m}^{mn}(0)$, corresponding to $r = n - m$.

The spheroidal radial coefficients $a_r^{mn}(c)$ can be computed readily in terms of the angular coefficients d_r^{mn} in Section 13.2.1, as follows:

$$a_r^{mn} = \frac{(-1)^{(r-n+m)/2}(r+2m)! \, d_r^{mn}}{r! \, S_{mn}(c)} \qquad (13.3.2)$$

Here, the summation quantity, $S_{mn}(c)$, is given by

$$S_{mn}(c) = \sum_{r=0,1}^{\infty} \frac{(r+2m)! \, d_r^{mn}}{r!} \qquad (13.3.3)$$

with the sum starting at $r = 0$ if $n - m$ is even, but at $r = 1$ if $n - m$ is odd; either way, r goes by steps of 2. Thus, the exponents of -1 in (13.3.2) and in (13.3.4) following are integers. Although the normalization of the angular coefficients, d_r^{mn}, is different between the two primary sources, Stratton et al [Stra56] and Flammer [Fla57], the radial coefficients (a_r^{mn}) are the same, since the r-independent normalization difference between these two sources cancels when $S_{mn}(c)$ is applied. Radial coefficients satisfy the sum rule

$$\sum_{r=0,1}^{\infty} (-1)^{(r-n+m)/2} a_r^{mn}(c) = 1 \tag{13.3.4}$$

which could be used to check their accuracy and whether they have been computed to large enough r. In our *Mathematica* and C functions, however, we use (13.3.4) to force the normalization of the $a_r^{mn}(c)$. The asymptotic form of $R_{mn}^{(1)}(c,\xi)$ is

$$R_{mn}^{(1)}(c,\xi) \xrightarrow[c\xi \to \infty]{} \frac{\cos[c\xi - (n+1)\pi/2]}{c\xi} \tag{13.3.5}$$

Visualization

The radial expansion coefficients for spheroidal wave functions, the $a_r^{mn}(c)$ in (13.3.1), depend upon the order n, the degree m, the parameter c in (13.1.5), and on whether one is using prolate or oblate spheroidal coordinates (Figure 13.1.1). We display $a_r^{mn}(c)$ as surfaces made from plaquettes whose vertices are the values of the radial coefficients.

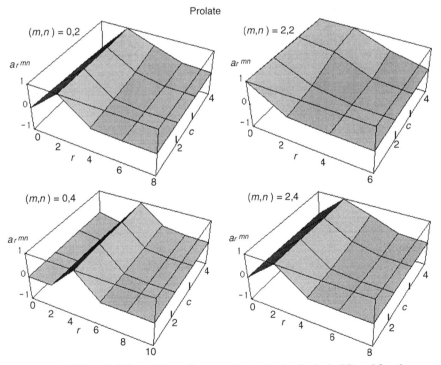

FIGURE 13.3.1 Radial coefficients for expansion in a basis of spherical Bessel functions according to (13.3.1), for four pairs of m and n values as a function of parameter c having values 1 to 4. The coefficients peak at $r = n - m$, which is the unique value when $c = 0$. (Adapted from cell `SWFRadCoefficient` in *Mathematica* notebook SWF.)

As discussed below (13.3.1), the radial coefficients are expected to peak at $r = n - m$, at least for moderate c values; this is confirmed by the graphics in Figure 13.3.1.

For oblate coordinates (right side of Figure 13.1.1), the radial coefficients shown in Figure 13.3.2 behave quite similarly to those for prolate coordinates (Figure 13.3.1). For values of $-c^2$ much larger in magnitude than those shown here, however, the behavior of the $a_r^{mn}(c)$ is much more complicated, as can be explored by using the cell SWFRad-Coefficient in *Mathematica* notebook SWF.

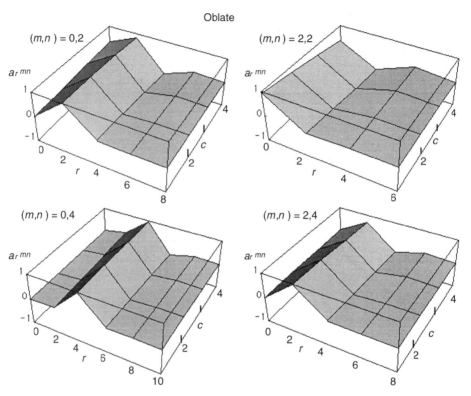

FIGURE 13.3.2   Radial coefficients for expansion in a basis of spherical Bessel functions according to (13.3.1), for four pairs of m and n values as a function of parameter c having values 1 to 4 for oblate coordinates. The coefficients peak at $r = n - m$, which is the unique value when $c = 0$. (Adapted from cell SWFRadCoefficient in *Mathematica* notebook SWF.)

Algorithm and Program

The expansion coefficients for spheroidal radial functions are computed in three stages. First, the auxiliary functions for eigenvalues, $t_{mn}(c)$ in Section 13.1.4, are determined as the roots of function U in (13.1.17), using the same algorithm as described for function SWFevalue in Section 13.1.3, except that the starting estimate for $t_{mn}(c)$ is much simpler than that for $\lambda_{mn}(c)$ and should usually be adequate. The coded function is SWFevalTMN.

Second, the angular coefficients $d_r^{mn}(c)$ are determined by backward recurrence on r, starting with $r = m - n + 14$ being assigned the value unity. The range of r values from this value down to $r = 0$ or 1 typically gives 15 orders of magnitude range for the coefficients.

The details are described in Section 13.2.1 under Algorithm and Program, while the resulting C function is `SWFAngCoeff`, which is listed in that subsection.

Third, the radial coefficients $a_r^{mn}(c)$ are computed in terms of the $d_r^{mn}(c)$ in a separate function, `SWFRadCoeff`, by using (13.3.2). Normalization of the $a_r^{mn}(c)$ is then done by using (13.3.3). Factorials are computed in terms of logarithms of Γ functions, using the function `LogGamma` described in Section 6.1.1.

Here is the code for the function `SWFRadCoeff`.

```
void SWFRadCoeff(int m, int n, double dmn[],
  double amn[])
/* Spherical radial function coefficients, amn[],
    in terms of angular coefficients  dmn[]
    previously computed by  SWFAngCoeff  */
{
double term,fr,frP1,frP1Ptm,sum,SNorm;
int nMm,rSt,rLarge,r;

/* Make array of unnormalized coefficients */
nMm = n - m;
rSt = ( nMm % 2 == 0 ) ?  0: 1;
rLarge = nMm + 14;
term = 1.0;
fr = rSt;
frP1 = fr+1.0;
frP1Ptm = frP1 + ( (double) (2*m) );

for ( r = rSt; r <= rLarge; r += 2 )
 {
 amn[r] = term*dmn[r];
 term *= - frP1Ptm*(frP1Ptm+1.0)/(frP1*(frP1+1.0));
 frP1 += 2.0;
 frP1Ptm += 2.0;
 }

/* Normalization */
term = 1.0;
sum = amn[rSt];

for ( r = rSt+2; r <= rLarge; r += 2 )
 {
 term = -term;
 sum += term*amn[r];
 }
SNorm = 1.0/sum;
for ( r = rSt; r <= rLarge; r += 2 )
  { amn[r] *= SNorm; }
return;
}
```

Test Values

Check values of the expansion coefficients for radial functions, the $a_r^{mn}(c)$ in (13.3.1) and (13.3.2), can be obtained by running cell `SWFRadCoefficient` in *Mathematica* notebook `SWF`, annotated in Section 20.13. The examples given in Tables 13.3.1 (for prolate

M

coordinates) and 13.3.2 (for oblate coordinates) can be used to benchmark the C function `SWFRadCoeff` given above; the driver program for this function is described in Section 21.13.

TABLE 13.3.1 **Test values to 10 digits of the spheroidal radial function expansion coefficients for *prolate* coordinates with $c = 2$.**

$r / (m, n)$	$(0, 0)$	$(0, 1)$	$(2, 2)$
0	1.564643512	0	1.272892899
1	0	1.186555224	0
2	0.6031679276	0	0.2900282832
3	0	0.1953119814	0
4	0.03962735361	0	0.01761075012
5	0	0.008950024666	0

TABLE 13.3.2 **Test values to 10 digits of the spheroidal radial function expansion coefficients for *oblate* coordinates with $c = 2$.**

$r / (m, n)$	$(0, 0)$	$(0, 1)$	$(2, 2)$
0	0.6557144754	0	0.7800400713
1	0	0.8617152976	0
2	−0.321086838	0	−0.2061568528
3	0	−0.1322567579	0
4	0.02253103507	0	0.01341416656
5	0	0.005899108107	0

13.3.2 Spheroidal Radial Functions

Keywords
radial functions, spheroidal coordinates, spheroidal wave functions

C Functions and *Mathematica* Notebook
R1MN (C) uses `SpherBess` (C); cell `SWFRadWave` in *Mathematica* notebook SWF

$\boxed{M}$

Function Definitions

The spheroidal radial function of the first kind that is regular at $\xi = 0$, $R_{mn}^{(1)}(c,\xi)$, is defined by (13.3.1). In this expansion it is understood that c and ξ are real, and that ξ is in the range for prolate coordinates, $\xi \geq 1$ according to (13.1.1). For oblate coordinates the transformation is $c \rightarrow -ic$ and $\xi \rightarrow i\xi$. Thus, one has the expansions

$$R_{mn}^{(1)}(c,\xi) = \left(1 - 1/\xi^2\right)^{m/2} \sum_{r=0,1}^{\infty} a_r^{mn}(c) j_{m+r}(c\xi) \qquad \text{prolate} \qquad (13.3.6)$$

$$R_{mn}^{(1)}(c,\xi) = \left(1 + 1/\xi^2\right)^{m/2} \sum_{r=0,1}^{\infty} a_r^{mn}(c) j_{m+r}(c\xi) \qquad \text{oblate} \qquad (13.3.7)$$

in which the radial expansion coefficients, $a_r^{mn}(c)$, depend upon whether the coordinates are prolate or oblate. This is shown in Figures 13.3.1 and 13.3.2, while numerical examples are given in Tables 13.3.1 and 13.3.2. How to compute the a_r^{mn} is described in Section 13.3.1.

Visualization

The spheroidal radial function of the first kind has four arguments—m, n, c, and ξ—and one has to distinguish between prolate and oblate coordinates (Figure 13.1.1). To visualize $R_{mn}^{(1)}(c,\xi)$ we choose $|c| = 2$ and show two curves for each m and n; $c = 2$ (prolate case) and $c = -i2$ (oblate case), as shown in Figure 13.3.3.

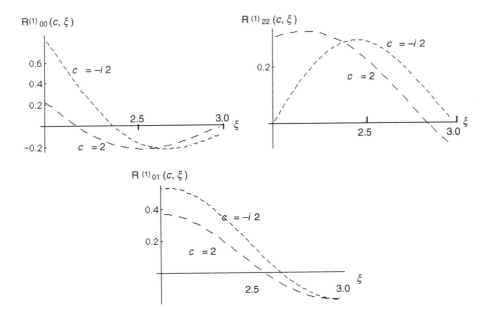

FIGURE 13.3.3 Spheroidal radial functions of the first kind (regular at $\xi = 1$) according to (13.3.6) or (13.3.7) for prolate and oblate coordinates, respectively. Shown for three pairs of m and n values for the magnitude of parameter c equal to 2 in each graphic, for prolate (long dashes) and oblate (short dashes). (Adapted from cell `SWFRadWave` in *Mathematica* notebook SWF.)

Algorithm and Program

The main task in computing the spheroidal radial functions $R_{mn}^{(1)}(c, \xi)$ is to compute the expansion coefficients a_r^{mn} in (13.3.1) and (13.3.2), which is described in Section 13.3.1. Assuming that the a_r^{mn} are available for a given choice of c for prolate or oblate coordinates, it is easy to sum the spherical Bessel function series until the result is sufficiently accurate.

For simplicitly, computing the coefficients and summing the series is incorporated into the same driver program, SMN, annotated in Section 21.13. There we also show how the accuracy of the expansion is controlled by using the input parameter Seps.

The C code for the function $R_{mn}^{(1)}(c, \xi)$ is in R1MN, as follows:

```
double R1MN(int m,int n,double sign,double c,
            double xi,
double amn[],double jn[],double Seps)
/* Spheroidal Radial Function of First Kind;
   degree n, order m, argument x = c*xi.
   Uses spherical Bessel function in table jn[].
   Convergence to fractional acuracy  Seps  */
{
double sum,ratio,term;
int nMm,rStart,rLarge,nMax,rMax,r;

nMm = n - m;
rStart = ( nMm % 2 == 0 ) ?  0: 1;
rLarge = nMm+14;/* coefficient range about 10^15 */
sum = 0.0;
ratio = 10;
nMax = NMAX - 8 - m;
rMax = ( rLarge < nMax ) ? rLarge: nMax;
r = rStart;
while ( ratio > Seps )
 {
 if (  r > rMax )
  { printf
     ("\n\n!!R1NM unconverged to %.6E in %i terms",
       Seps,rMax);
  return  sum;
  }
 term = amn[r]*jn[m+r];
 sum += term;
 ratio = fabs(term/sum);
 r += 2;
 }
if ( ( xi == 1.0 ) && ( m == 0 ) )
{ return  sum; }/* pow returns -INF for 0 to 0 ! */
return
   pow(1.0-sign/(xi*xi), 0.5*( (double) m) )*sum;
}
```

Regular spherical Bessel functions, $j_n(x)$, are needed in the expansion (13.3.1). Since the j_n are computed by backward recurrence on n, it is most efficient for the present use to make a table of the functions, as done in the following C function. Details of the algorithm are given in Section 14.3.1.

```
void SpherBess(int nMax, double x, double jn[])
/* Make a table of Spherical Bessel Functions
    for argument  x  from order  n = 0  to  nMax  */
{
double dblFctrl,tn,xs,jzero,jNorm;
int nLarge,n;

nLarge = nMax+6;
dblFctrl = 1.0; /* Make double factorial */
tn = 3.0;
for ( n = 1; n <= nLarge-1; n++ )
  {
  dblFctrl *= tn;
  tn += 2.0;
  }

/* Starting approximations */
jn[nLarge-1] = pow(x, (double) nLarge-1 )/dblFctrl;
jn[nLarge] = x*jn[nLarge-1]/tn;

/* Use backward iteration down to  n = 0    */
for ( n = nLarge-1; n >= 1; n-- )
  {
  tn -= 2.0;
  jn[n-1] = tn*jn[n]/x - jn[n+1];
  }

/* Normalize by comparison with  j0(x)   */
xs = x*x; /* Get  j0  accurate for small  x  */
jzero = ( fabs(x) > 0.001 ) ?  sin(x)/x:
        1.0-xs*(1.0-xs*(1.0-xs/42.0)/20.0)/6.0;
jNorm = jzero/jn[0];
for ( n = 0; n <= nMax; n++ )
  { jn[n] *= jNorm; }
return;
}
```

Test Values

Check values of the spheroidal radial functions $R_{mn}^{(1)}(c,\xi)$ in (13.3.6) and (13.3.7) can be obtained by running cell SWFRadWave in *Mathematica* notebook SWF, which is annotated in Section 20.13. Examples are given in Tables 13.3.3 and 13.3.4 on the next page for $R_{mn}^{(1)}(c,\xi)$ in prolate and oblate coordinates, respectively.

The regular spheroidal radial functions $R_{mn}^{(1)}(c,\xi)$ are tabulated over very limited ranges of m, n, c, and ξ to 4-digit accuracy by Flammer [Fla57] in Tables 71–85 for prolate coordinates and in Table 166 for oblate coordinates. These tables are reproduced in Abramowitz and Stegun [Abr64, Tables 21.3 and 21.4]. Function values computed by our C and *Mathematica* codes agree to at least 10 significant figures.

TABLE 13.3.3 Test values to 10 digits of the spheroidal radial functions of the first kind for *prolate* coordinates with c = 2.

ξ / (m, n)	(0, 0)	(0, 1)	(2, 2)
2.0	−0.1244173017	0.1830101679	0.2911576371
2.5	−0.2113445246	−0.06698264222	0.1903325072
3.0	−0.0874742011	−0.1709683875	0.009976946169

TABLE 13.3.4 Test values to 10 digits of the spheroidal radial functions of the first kind for *oblate* coordinates with c = 2.

ξ / (m, n)	(0, 0)	(0, 1)	(2, 2)
2.0	−0.2099701324	0.0700400961	0.2374981897
2.5	−0.1648342762	−0.1117075115	0.07793113436
3.0	−0.01417884923	−0.1617353351	−0.07603800096

REFERENCES ON SPHEROIDAL WAVE FUNCTIONS

[Abr64]   Abramowitz, M., and I. A. Stegun, *Handbook of Mathematical Functions*, Dover, New York, 1964.

[Erd53c]   Erdélyi, A., et al, *Higher Transcendental Functions*, Vol. 3, McGraw-Hill, New York, 1953; reprint edition, Krieger, Malabar, Florida, 1981.

[Fla57]   Flammer, C., *Spheroidal Wave Functions*, Stanford University Press, Stanford, 1957.

[Mei54]   Meixner, J., and F. W. Schäfke, *Mathieusche Funktionen und Sphäroidfunktionen*, Springer, Berlin, 1954.

[Mei80]   Meixner, J., F. W. Schäfke, and G. Wolf, *Mathieu Functions and Spheroidal Functions and Their Mathematical Foundations*, Springer, Berlin, 1980.

[Moo71]   Moon, P., and D. E. Spencer, *Field Theory Handbook*, Springer Verlag, Berlin, second edition, 1971.

[Mor53]   Morse, P. M., and H. Feshbach, *Methods of Theoretical Physics*, McGraw-Hill, New York, 1953.

[Str56]   Stratton, J. A., P. M. Morse, L. J. Chu, J. D. C. Little, and F. J. Corbató, *Spheroidal Wave Functions*, Technology Press of M.I.T. and Wiley, New York, 1956.

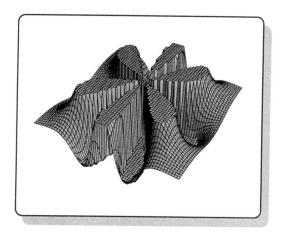

Chapter 14

BESSEL FUNCTIONS

Bessel functions are very widely used, especially in science and engineering. The chapter begins with an overview of these functions, then specializes in Sections 14.2 and 14.3 to various Bessel functions of integer and fractional order, respectively. As throughout the *Atlas*, functions are visualized by using *Mathematica* and C programs are constructed for them. Visualization is important because the irregular functions may be very complicated—as seen in the motif above, which shows $Y_6(z)$ in the complex plane.

The literature on Bessel functions is extensive and their analytical and computational properties are well understood. Watson's classic treatise on Bessel functions [Wat44] has a complete analytical treatment; we use his notation, which is that in Whittaker and Watson [Whi27] and in Chapters 9 and 10 of Abramowitz and Stegun [Abr64]. In the treatise on higher transcendental functions edited by Erdélyi et al [Erd53b], Chapter 7 has extensive lists of analytical properties. Formulas are also given in Sections 8.4 and 8.5 of Gradshteyn and Ryzhik [Gra94]. In Chapter 8 of the work by Jahnke and Emde [Jah45] there are formulas and an unusally large number of views of Bessel functions, clearly and painstakingly drawn by hand before the advent of quick and painless computer graphics.

The Bessel functions considered here are the following: the regular and irregular functions, $J_n(x)$ and $Y_n(x)$ respectively, that are of integer order n and real argument x (Sections 14.2.1 and 14.2.1), the modified regular and irregular functions, $I_n(x)$ and $K_n(x)$ (Section 14.2.3 and 14.2.4), and the Kelvin functions $ber_n(x)$, $bei_n(x)$, $ker_n(x)$, and $kei_n(x)$. Among the Bessel functions of fractional order, we cover the spherical Bessel functions $j_n(x)$ and $y_n(x)$ (Section 14.3.1), the modified spherical Bessel functions of the first and second kind $I_{n+1/2}(x)$ and $I_{-n-1/2}(x)$ (Section 14.3.2), and also the Airy functions $Ai(x)$ and $Bi(x)$ (Section 14.3.3).

14.1 OVERVIEW OF BESSEL FUNCTIONS

The mathematical properties and applications of Bessel functions of the varieties mentioned in the preceding paragraph are covered completely at both introductory and advanced levels in many text and reference books. What is generally missing in such books—except for the works by Jahnke and Emde [Jah45, Chapter 8] and by Spanier and Oldham [Spa87, Chapters 49–56]—is visualization of the functions that are to be computed numerically. This section therefore emphasizes graphics of the Bessel functions that are helpful for understanding their overall properties. In the sections that follow, the graphics are specific to the Bessel functions in the section.

Bessel's Differential Equation

Bessel functions are often defined in terms of regularized (standardized, normalized) solutions of the differential equation

$$z^2 w'' + z w' + (z^2 - v^2) w(z) = 0 \tag{14.1.1}$$

One basic solution is $J_v(z)$—which is regular near $z = 0$ for $v > 0$—as $J_v(z) \propto z^v$. The second basic solution is $Y_v(z)$—which is irrregular near $z = 0$ for $v > 0$—as $Y_v(z) \propto z^{-v}$. From these two basic solutions, rotations of z in the complex plane and linear combinations of the resulting functions produce a variety of Bessel functions.

When $v = n$, an integer, and z is a real variable, we have the Bessel functions described in this section; their names and our notation are summarized in Table 14.1.1. Note that notations vary widely between sources; a summary is given at the start of Chapter 9 in Abramowitz and Stegun [Abr64].

TABLE 14.1.1 **Basic Bessel functions of integer order.**

Function	Notation
Bessel function of the first kind, regular at $x = 0$.	$J_n(x)$
Bessel function of the second kind, irregular at $x = 0$); Weber's function.	$Y_n(x)$
Hyperbolic Bessel function, regular at $x = 0$.	$I_n(x) = e^{-n\pi i/2} J_n(ix)$
Hyperbolic Bessel function, irregular at $x = 0$; Bassett's function.	$K_n(x) = ie^{n\pi i/2}[J_n(ix) + iY_n(ix)]/2$
Kelvin functions, regular at $x = 0$.	$ber_n(x) = \mathrm{Re}\left[J_n(xe^{3\pi i/4})\right]$ $bei_n(x) = \mathrm{Im}\left[J_n(xe^{3\pi i/4})\right]$
Kelvin functions, irregular at $x = 0$.	$ker_n(x) = \mathrm{Re}\left[e^{-n\pi i/2} K_n(xe^{3\pi i/4})\right]$ $kei_n(x) = \mathrm{Im}\left[e^{-n\pi i/2} K_n(xe^{3\pi i/4})\right]$

Visualizing Bessel Functions

The figures that follow show many views of the Bessel functions $J_n(z)$ and $Y_n(z)$ from which the Bessel functions of integer order that are listed in Table 14.1.1 can be constructed. The 3D overviews show the behavior of $J_n(z)$ and $Y_n(z)$ as functions of two of the three parameters, n, x, y (where $z = x + iy$). Then transections of these surfaces (2D curves) are shown; either n or x is fixed ($y = 0$) while the other parameter varies. Views that are more detailed are shown for hyperbolic Bessel and Kelvin functions in the following sections.

$J_n(z)$ and $Y_n(z)$ in the Complex Plane

The Bessel function of the first kind, regular at $z = 0$, $J_n(z)$, is visualized as follows:

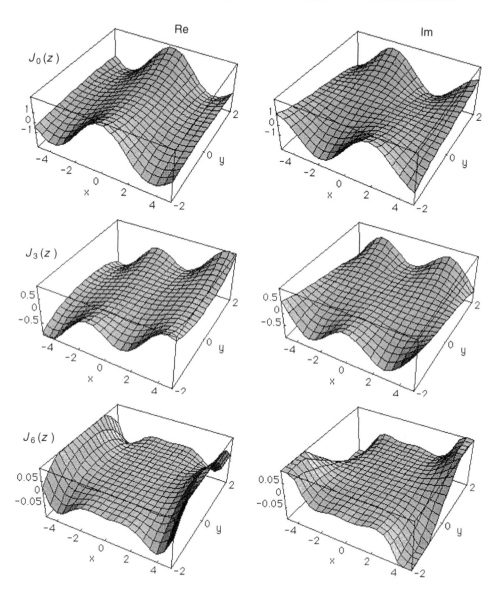

FIGURE 14.1.1 Regular Bessel functions in the complex plane for fixed values of $v = n = 0, 3, 6$; real parts on the left side and imaginary parts on the right side. Note that the vertical scales change very quickly with n; for $n = 6$ the extreme values of the Bessel function exceed 0.1 and have been truncated to this value. (Adapted from the output of cell BesselJY in *Mathematica* notebook BesselInteger.)

The Bessel function of the second kind, irregular at $z = 0$, $Y_n(z)$, can be visualized as follows:

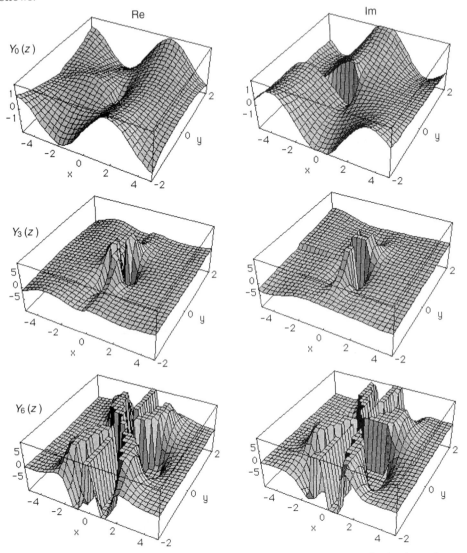

FIGURE 14.1.2 Irregular Bessel functions in the complex plane for fixed values of $v = n = 0, 3, 6$; real parts on the left side and imaginary parts on the right side. The range of values shown is -10 to 10; values larger in magnitude have been truncated. The bottom left graphic is shown in more detail as the motif at the start of this chapter. (Adapted from the output of cell `BesselJY` in the *Mathematica* notebook `BesselInteger`.)

Because of the singularity of type z^{-n} near $z = 0$ for $n \neq 0$, the behavior of $Y_n(z)$ is much more complicated than that of $J_n(z)$. In Figure 14.1.2 the top right graphic for $\mathrm{Im}[Y_0(z)]$ reminds us that there is also a branch cut along the negative real axis in order to make the functions holomorphic (regular). As n increases, the singularity structure becomes more complicated, as seen for $n = 6$ at the bottom of this figure. You can explore this aspect in detail by modifying and running the cell `BesselJY` in *Mathematica* notebook `BesselInteger`.

$J_n(x)$ and $Y_n(x)$ as Functions of x and n

If z is restricted to x, a real variable, then one can make 3D views of $J_n(x)$ and $Y_n(x)$ as x and n vary; such graphics are shown in Figures 14.1.3 and 14.1.4.

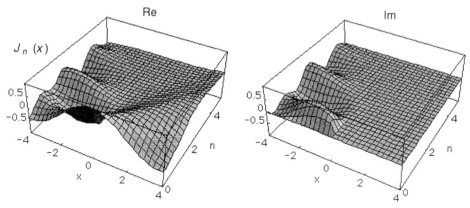

FIGURE 14.1.3 Regular Bessel functions in terms of of real variable x and order n; real parts on the left side, imaginary parts on the right side. The vertical scale ranges from -1 to 1. (Adapted from the output of cell `BesselJY` in the *Mathematica* notebook `Bessel-Integer`.)

Notice that when $x \geq 0$ the function $J_n(x)$ is purely real; the irregular function $Y_n(x)$ has the same property, as seen in Figure 14.1.4.

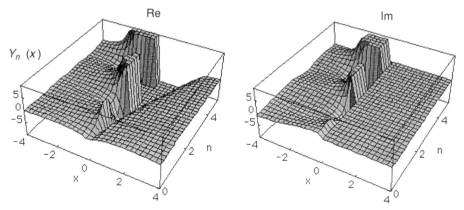

FIGURE 14.1.4 Irregular Bessel functions in terms of real variable x and order n; real parts on the left side, imaginary parts on the right side. The vertical scale ranges from -10 to 10. (Adapted from the output of cell `BesselJY` in *Mathematica* notebook `Bessel-Integer`.)

Transections of $J_n(x)$ and $Y_n(x)$

If a slice (transection) is cut parallel to a horizontal axis through one of the surfaces shown above, one gets the more usual (but less informative) curve of the Bessel function, allowing one to see how the function depends upon a single variable. Such curves are shown in Figures 14.1.5 and 14.1.6.

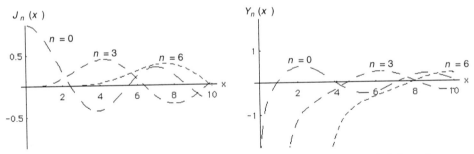

FIGURE 14.1.5 Bessel functions in terms of real variable x for order $n = 0$, 3,6; regular functions on the left side, irregular functions on the right side. (Adapted from the output of cell `BesselJY` in *Mathematica* notebook `BesselInteger`.)

Notice for $J_n(x)$ in the above figure that when x is fixed the function is generally increasing as n decreases; to compute numerical values of $J_n(x)$ accurately, this is therefore the appropriate direction for iterating on n. (See Section 3.6.) For $Y_n(x)$ the $x = 0$ singularity—for $n > 0$ of type x^{-n} and for $n = 0$ of type $\ln x$—persists out to larger x as n increases, so that the first zero of $Y_n(x)$ is near $x = n+1$.

If x is kept fixed and n is varied (slices into the page in Figures 14.1.3 and 14.1.4) then the curves in Figure 14.1.6 are obtained. The appropriate iteration direction for $J_n(x)$, discussed in the preceding paragraph, is also confirmed. Over the range of n and x shown below, $Y_n(x)$ should be computed by iterating towards larger n. Such details of numerical computation of Bessel functions are considered more fully in the following sections describing each function.

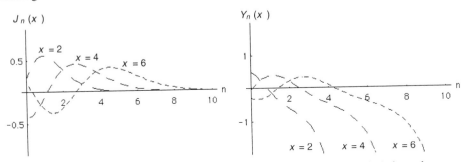

FIGURE 14.1.6 Bessel functions in terms of order n for real variable $x = 2$, 4, 6; regular functions on the left side, irregular functions on the right side. (Adapted from the output of cell `BesselJY` in *Mathematica* notebook `BesselInteger`.)

14.2   BESSEL FUNCTIONS OF INTEGER ORDER

Of the Bessel functions of integer order that we consider (Table 14.1.1), only the cylindrical functions $J_n(x)$ and $Y_n(x)$ have been displayed in much detail (Figures 14.1.5 and 14.1.6). In this section we therefore begin each subsection by visualizing the function, then we present the algorithm and C code for its efficient numerical computation.

14.2.1 Regular Cylindrical Bessel Function *J*

Keywords
cylinder functions, regular Bessel function

C Function and *Mathematica* Notebook
BessJN (C); cell BesselJY in notebook BesselInteger (*Mathematica*)   *M*

Background and Function Definition
The Bessel function $J_n(x)$ occurs most often in problems that use cylindrical 3D coordinates. Then the variable $x = kr$, in which k is a wavenumber and r is the radial coordinate. The regular function, $J_n(x)$, satisfies the differential equation

$$x^2 J_n'' + x J_n' + (x^2 - n^2) J_n(x) = 0 \tag{14.2.1}$$

A power series solution of this equation produces

$$J_n(x) = \left(\frac{x}{2}\right)^n \sum_{k=0}^{\infty} \frac{(-x^2/4)^k}{k!(n+k)!} \tag{14.2.2}$$

which we use in the Algorithm to generate an array of Bessel functions. An appropriate recurrence relation for accurate numerical computation of $J_n(x)$ is

$$J_{n-1}(x) = \frac{2n}{x} J_n(x) - J_{n+1}(x) \tag{14.2.3}$$

With the standardization indicated in (14.2.2), $J_n(x)$ satisfies the sum rule

$$J_0^2(x) + 2 \sum_{k=1}^{\infty} J_k^2(x) = 1 \tag{14.2.4}$$

which we use below to normalize our Bessel functions.

If $x \leq n_m^2/2$, where n_m is the largest n value of interest for given x, use of the above formulas generally gives accurate numerical values. If, however, $x > n_m^2/2$, another method is needed; we use Hankel's asymptotic expansions, derived in Section 7.2 of Watson's treatise [Wat44] and are cited as equations 9.2.5, 9.2.9, and 9.2.10 in Abramowitz and Stegun [Abr64], namely,

$$J_n(x) = \sqrt{\frac{2}{\pi x}} [P(n,x) \cos \chi - Q(n,x) \sin \chi]$$
$$\chi = x - (n + 1/2) \pi/2 \tag{14.2.5}$$

The auxiliary functions, P and Q, have the asymptotic expansions (see Section 3.4)

$$P(n,x) \sim 1 - \frac{(\mu-1)(\mu-9)}{2!(8x)^2} + \frac{(\mu-1)(\mu-9)(\mu-25)(\mu-49)}{4!(8x)^4} - \cdots \tag{14.2.6}$$

$$Q(n,x) \sim \frac{\mu-1}{8x} \left[1 - \frac{(\mu-9)(\mu-25)}{3!(8x)^2} + \frac{(\mu-9)(\mu-25)(\mu-49)}{5!(8x)^4} - \cdots \right. \tag{14.2.7}$$

in both of which $\mu = 4n^2$. The functions $P(n,x)$ and $Q(n,x)$ are shown in Figure 14.2.1 below for $n = 0$ and 1, which are used to start the upward iteration on n to compute $J_n(x)$.

Visualization

Views of Bessel functions of real argument, $J_n(x)$, are in Figures 14.1.3, 14.1.5 (left), and 14.1.6 (left). The latter two figures show that for modest values of x, $J_n(x)$ generally increases as n decreases; this is the clue for part of the accurate and efficient computational algorithm following.

The expression (14.2.5) requires asymptotic expansion formulas (14.2.6) and (14.2.7) for $P(n,x)$ and $Q(n,x)$. These functions are shown in Figure 14.2.1.

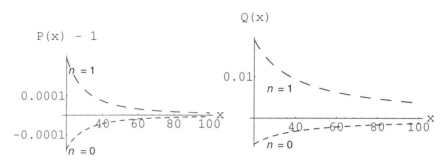

FIGURE 14.2.1 Asymptotic values of the auxiliary functions for computing Bessel functions, with $n = 0$ and 1. The left side shows $P-1$, and the right side shows Q. (Adapted from the output of cell `BesselPQ` in *Mathematica* notebook `BesselInteger`.)

Algorithm and Program

From the immediately preceding discussions one can devise a simple algorithm to compute an array of $J_n(x)$ with n varying and x fixed, which is often how the function values are needed, as in Fourier-Bessel expansions.

For a modest value of x, a sufficiently large n value, n_m, can be found such that the first term in the series (14.2.2) gives a fair approximation to both the magnitude and relative size of two adjacent Bessel functions:

$$J_{n_m-1}(x) := \left(\frac{x}{2}\right)^{\!\!\wedge} \Big/ (n_m - 1)! \qquad J_{n_m}(x) := x J_{n_m-1}(x)/(2n) \qquad (14.2.8)$$

in which the " := " stands for a value assignment. If one now iterates in the direction of decreasing n by using (14.2.3), then the relative error between any two adjacent computed values of $J_n(x)$ will be quickly damped out; typically, six iterations result in 10^{-10} relative accuracy of $J_n(x)$ values; these are stored in an array labelled by n. We terminate the iteration after $n = 1$. At this stage of the computation, the array elements are normalized by multiplying each by a common factor, `JNorm`, in function `BessJN` below, so that the sum rule (14.2.4) is satisfied through $k = n_m$. This produces an array holding $J_0(x)$, $J_1(x)$, ..., $J_{n_m}(x)$.

In function `BessJN` the value of `nLarge` = n_m+6 is input by the program user, as shown in the annotated C driver program in Section 21.14. We describe under Test Values how to optimize the choice of n_m to obtain a given accuracy in each $J_n(x)$.

For $x > n_m^2/2$ the above algorithm is not accurate for computing $J_n(x)$. We use instead the asymptotic expansions (14.2.6) and (14.2.7) for $P(n,x)$ and $Q(n,x)$ with $n = 0$ and 1 to start the iteration; this is tracked toward increasing n, which is now the stable direction.

The following are the C functions for computing the array of $J_n(x)$ values, `BessJN`, and $P(n,x)$ and $Q(n,x)$, functions `PP` and `QQ`, respectively.

```
/* maximum number of table entries
   allowing for 6 inaccurate starting values  */
#define NMAX 46

void BessJN(int nJMax, double x, double JN[])
/* Make a table of regular Bessel Functions
   from order  n = 0  to  nJMax; x is fixed  */
{
double PP(int n, double x),QQ(int n, double x),
 fnm2,fctrl,fn,xs,tn,tnDx,dtnDx,sum,JNorm,
 absx,chi0,chi1,Rtfact,sign;
const double PiD2 = 1.57079632679;
int nLarge,n;

if ( x == 0 )
 { /* treat this specially */
 JN[0] = 1.0;
 for ( n = 1; n <= nJMax; n++ )
  { JN[n] = 0.0; }
 return; }

nLarge = nJMax+6;
fnm2 = nJMax*nJMax;
absx = fabs(x);
if ( absx <=  0.5*fnm2 )
 {   /* Backward iteration method */
 fctrl = 1.0; /* Make factorial */
 fn = 2.0;
 for ( n = 2; n <= nLarge-1; n++ )
   { fctrl *= fn; fn += 1.0; }
/* Starting approximations */
 JN[nLarge-1] = pow(0.5*x, (double) nLarge-1)/fctrl;
 tn = 2.0*fn;
 JN[nLarge] = x*JN[nLarge-1]/tn;
 tnDx = tn/x;
 dtnDx = 2.0/x;
 /* Use backward iteration down to  n = 0   */
 for ( n = nLarge-1; n >= 1; n-- )
   {
   tnDx -= dtnDx;
   JN[n-1] = tnDx*JN[n] - JN[n+1];
   }
/* Normalize by comparing with sum rule  */
 sum = 0.0;
 for ( n = 1; n <= nJMax; n++ )
   { sum += JN[n]*JN[n]; }
 JNorm  = 1.0/sqrt(JN[0]*JN[0]+2.0*sum);
 for ( n = 0; n <= nJMax; n++ )
   { JN[n] *= JNorm; }
 return; }
else
 { /* forward iteration from P & Q series */
 /* Starting values */
 chi0 = absx - 0.5*PiD2;
 chi1 = chi0 - PiD2;
 Rtfact = 1.0/sqrt(PiD2*absx);
```

```
   JN[0] = Rtfact*
    (PP(0,absx)*cos(chi0) - QQ(0,absx)*sin(chi0));
   sign = ( x < 0.0 ) ? -1.0: 1.0;
   JN[1] = sign*Rtfact*
    (PP(1,absx)*cos(chi1) - QQ(1,absx)*sin(chi1));
  /* Use forward iteration up to  n = nJMax    */
   tnDx = 0.0;
   dtnDx = 2.0/x;
   for ( n = 1; n < nJMax; n++ )
    {
    tnDx += dtnDx;
    JN[n+1] = tnDx*JN[n] - JN[n-1];
    }
   return;
  } }

double PP(int n, double x)
/* Hankel's P series for asymptotic in  x > 0   */
/* Relative accuracy set to  eps = 10^(-12)   */
{
double mu,atxs,term,termold,fk,tk,sum,
 ratio,eps,termnew;
int k;

mu = 4*n*n;
atxs = 64.0*x*x;
k = 0;
term = 1.0;
termold = 1.0;
fk = -1.0;
tk = 0.0;
sum = 1.0;
ratio = 10.0;
eps = 1.0e-12;
while ( ratio > eps )
 {
 k++;
 fk += 4.0;
 tk += 2.0;
 term *=    -(mu-(fk-2.0)*(fk-2.0))*(mu-fk*fk)/
            (tk*(tk-1.0)*atxs);
 sum += term;
 ratio = fabs(term/sum);
 termnew = fabs(term);
 if ( termnew > termold )
  {
  printf
  ("\n\n!!In PP diverging after %i terms",k);
  ratio = 0.0;
  }
 else
  { termold = termnew; }
 }
return sum;
}
```

```
double QQ(int n, double x)
/* Hankel's Q series for asymptotic in  x > 0  */
/* Relative accuracy set to  eps = 10^(-12)  */
{
double mu,atxs,term,termold,fk,tk,sum,
 ratio,eps,termnew;
int k;

mu = 4*n*n;
atxs = 64.0*x*x;
k = 0;
term = 1.0;
termold = 1.0;
fk = 1.0;
tk = 1.0;
sum = 1.0;
ratio = 10.0;
eps = 1.0e-12;

while ( ratio > eps )
 {
 k++;
 fk += 4.0;
 tk += 2.0;
 term *=   -(mu-(fk-2.0)*(fk-2.0))*(mu-fk*fk)/
          (tk*(tk-1.0)*atxs);
 sum += term;
 ratio = fabs(term/sum);
 termnew = fabs(term);
 if ( termnew > termold )
  {
  printf
  ("\n\n!!In QQ diverging after %i terms",k);
  ratio = 0.0;
  }
 else
  { termold = termnew; }
 }
return (mu-1.0)*sum/(8.0*x);
}
```

Programming Notes

The parameter NMAX is used in the C driver program (Section 21.14) to assign the size of array JN that holds the values of $J_n(x)$. The driver program compares the maximum value of n requested, nJMax, with NMAX to ensure that enough storage is available. The rule relating nJMax, x, and the fractional accuracy, ε, is discussed under Test Values following, and sample values of these accuracy control parameters are given in Table 14.2.2.

Test Values

Check values of $J_n(x)$ can be generated by running the BesselZ cell in *Mathematica* notebook BesselInteger; examples are given in Table 14.2.1. Set nJMax = 20 for $x = 0$ to 8 to exercise the first method in the algorithm, and nJMax = 6 for $x = 20$ to exercise the second method.

M

TABLE 14.2.1 Test values to 10 digits of the regular Bessel functions.

$x \setminus n$	0	3	6
0	1.000000000	0	0
2.0	0.2238907790	0.1289432495	0.001202428972
4.0	−0.3971498099	0.4301714739	0.4301714739
6.0	0.1506452573	0.1147683848	0.2458368634
8.0	0.1716508071	−0.2911322071	0.3375759001
20.0	0.1670246643	−0.09890139456	−0.05508604956

To compute $J_n(x)$ efficiently and accurately it is worthwhile to optimize the parameter n_m, the maximum order used in the backward-iteration scheme (14.2.3). Note that elements of array JN are manipulated at least $3n_m$ times; one pass for iteration, one for the sum rule, and one for normalization. On the basis of many test runs, we find that six iterations are typically required to correct for inaccuracies in the ratio of the two starting values in (14.2.5); we therefore set nLarge = nJMax+6.

The other approximation made in our algorithm is that the sum (14.2.4) is carried out only to $k = n_m = $ nJMax. To estimate the fractional error in the normalization, ε, made by the finite summation limit, write (14.2.4) as

$$J_0^2(x) + 2 \sum_{k=1}^{n_m} J_k^2(x) = 1 - 2 \sum_{k=n_m+1}^{\infty} J_k^2(x)$$

$$\approx 1 - 2\left[(x/2)^{n_m+1}/(n_m+1)!\right]^2 \approx 1 - 2\varepsilon \tag{14.2.9}$$

Here, the first approximation is obtained if only the first remaining term is maintained and is approximated using (14.2.5). By approximating the factorial by the first term in its Stirling series, as in the first equation below (7.1.4), then simplifying by replacing $n_m + 1$ by n_m, one gets the condition

$$2n_m[\ln(x/2) - \ln(n_m) + 1] \approx \ln(2\pi\varepsilon) \tag{14.2.10}$$

Table 14.2.2 shows typical values of n_m predicted by (14.2.10); by running either our C program or the cell BesselZ in *Mathematica* notebook BesselInteger, you will find that these n_m values indeed give the required accuracy.

TABLE 14.2.2 Maximum order of the regular Bessel function, n_m, as a function of argument x and the fractional accuracy, ε. See also (14.2.10).

ε	x	n_m
10^{-6}	5	11
	10	19
	20	33
10^{-10}	5	14
	10	22
	20	36

14.2.2 Irregular Cylindrical Bessel Function Y

Keywords
cylinder functions, irregular Bessel function

C Function and *Mathematica* Notebook
BessYN (C); cell BesselYX in notebook BesselInteger (*Mathematica*)

M

Background and Function Definition
The Bessel function $Y_n(x)$ occurs most often in problems that use cylindrical 3D coordinates, with the variable $x = kr$, in which k is a wavenumber and r is the radial coordinate. The irregular function is most likely to arise in problems in which $r > 0$ only, such as application of an exterior boundary condition at $r = R$, a cylindrical bounding surface. Function $Y_n(x)$ satisfies the same equation as the regular solution $J_n(x)$, namely,

$$x^2 Y_n'' + x Y_n' + (x^2 - n^2) Y_n(x) = 0 \qquad (14.2.11)$$

but it is singular at $x = 0$. One expression for $Y_n(x)$—involving power series, logarithms, and regular Bessel functions—is [Abr64, 9.1.88]

$$Y_n(x) = -\frac{n!(x/2)^{-n}}{\pi} \sum_{k=0}^{n-1} \frac{(x/2)^k J_k(x)}{(n-k)k!} + \frac{2[\ln(x/2) - \psi(n+1)]J_n(x)}{\pi}$$
$$- \frac{2}{\pi} \sum_{k=1}^{\infty} \frac{(-1)^k (n+2k) J_{n+2k}(x)}{k(n+k)} \qquad (14.2.12)$$

where the ψ function is given in Section 6.2.1; this relation also holds if x is generalized to z. From (14.2.12) it is seen that for $n > 0$ a function that is nonsingular near $z = 0$ is

$$X_n(z) \equiv (z/2)^n \, Y_n(z) \qquad (14.2.13)$$

as shown in Figures 14.2.2 and 14.2.3. An appropriate recurrence relation for $Y_n(x)$ is

$$Y_{n+1}(x) = \frac{2n}{x} Y_n(x) - Y_{n-1}(x) \qquad (14.2.14)$$

applied with increasing n—the direction that is insensitive to subtractive cancellations. Values of $Y_0(x)$ and $Y_1(x)$ are required; we compute them using (14.2.12) with $n = 0$ and 1. The $J_n(x)$ up to high order are computed into array JN, as in Section 14.2.1.

If $x \le n_m^2/2$, where n_m is the largest n value of interest for given x, using the above formulas generally gives accurate numerical values. If, however, $x > n_m^2/2$, another method is needed; we use Hankel's asymptotic expansions derived in Section 7.2 of Watson's treatise [Wat44] and cited as equations 9.2.6, 9.2.9, and 9.2.10 in Abramowitz and Stegun [Abr64], namely

$$Y_n(x) = \sqrt{\frac{2}{\pi x}} [P(n,x)\sin\chi + Q(n,x)\cos\chi] \qquad \chi = x - (n+1/2)\pi/2 \quad (14.2.15)$$

The auxiliary functions, P and Q, have the asymptotic expansions given by (14.2.6) and (14.2.7); P and Q are depicted in Figure 14.2.1 for the starting n values of 0 and 1, over a relevant range of x, 20 to 100. Upward iteration on n using (14.2.14) is stable for large x.

Visualization
Views of $Y_n(z)$ and of $X_n(z)$ are shown in Figure 14.2.2; 3D views of $Y_n(z)$ are also shown in Figure 14.1.2 for $n = 0, 3, 6$ and for x over a slightly larger range.

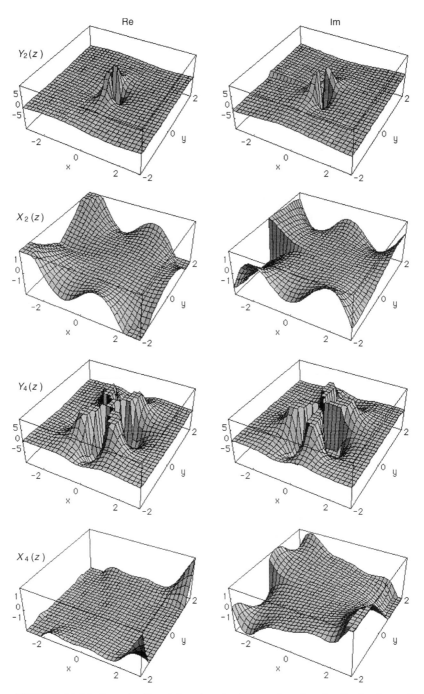

FIGURE 14.2.2 Irregular Bessel functions in the complex plane for fixed values of $v = n = 2$ and 4; real parts on the left and imaginary parts on the right. The range of Y_n values shown is -10 to 10, while that for X_n is -2 to 2; larger values have been truncated. (Adapted from the output of cell `BesselYX` in *Mathematica* notebook `Bessel-Integer`.)

Notice in Figure 14.2.2 that $X_n(z)$ has no singularity at $z = 0$ (or for $n > 0$), unlike $Y_n(z)$ whose singularity is of order n, in addition to the logarithmic singularity, as indicated in (14.2.12). This is also evident when $Y_n(x)$ and $X_n(x)$ are mapped as functions of x and n, as in Figure 14.2.3.

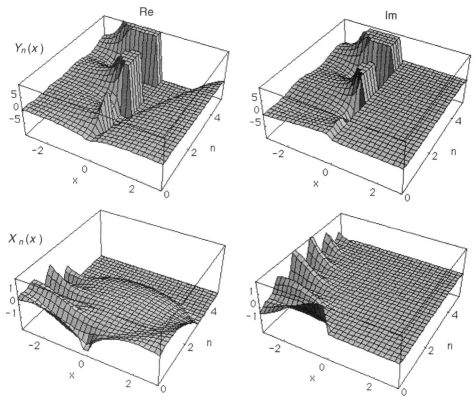

FIGURE 14.2.3 Irregular Bessel functions in terms of real variable x and order n; real parts on the left, imaginary parts on the right. The vertical scale ranges from -10 to 10 for Y_n and from -2 to 2 for X_n. (Adapted from the output of cell `BesselYX` in *Mathematica* notebook `BesselInteger`.)

From the above two figures, it is clear that if you need the irregular function near $x = 0$, it would be worthwhile to modify the starting formulas and iteration scheme in order to work with $X_n(x)$ rather than $Y_n(x)$. This territory we leave for users of the *Atlas* to explore on their own.

Algorithm and Program

In function `BessYN` the values of n_m and x are input by the program user, as shown in the annotated C driver program in Section 21.14. We describe in Section 14.2.1 under Test Values how to optimize the choice of n_m to obtain a given accuracy in each $J_n(x)$. Somewhat larger values of n_m should be used for computing the $Y_n(x)$, since J_k rather than J_k^L is required in (14.2.12).

The C function for computing the array YN of $Y_n(x)$ values is given on the following page.

```
void BessYN(int nYMax, double x, double YN[])
/* Make a table of irregular Bessel Functions
   from order  n = 0  to  nYMax; x>0 is fixed  */
{
double JN[NMAX], PP(int n, double x),
 QQ(int n, double x),
 fnm2,sum0,sum1,phase,fk,term0,logterm,
 chi0,chi1,Rtfact,tnDx,dtnDx;
const double TwoDPi = 0.636619772368,
 EulerG = 0.577215664902;
void BessJN(int nYMax, double x, double JN[]);
int k,n;

fnm2 = nYMax*nYMax;
if ( x <=  0.5*fnm2 )
 { /* YN[0] & YN[1] from table of regular
   Bessel functions Jn(x) order  n = 0 to nYMax */
 BessJN(nYMax,x,JN);
 sum0 = 0.0;
 sum1 = 0.0;
 phase = 1.0;
 for ( k = 1; k < nYMax/2; k++ )
  {
  phase = - phase;
  fk = k;
  term0 = phase/fk;
  sum0 += term0*JN[2*k];
  sum1 += term0*(0.5+fk)*JN[1+2*k]/(1.0+fk);
  }
 logterm = log(0.5*x)+EulerG;
 YN[0] = TwoDPi*(logterm*JN[0] - 2.0*sum0);
 YN[1] = TwoDPi*(-JN[0]/x+
         (logterm-1.0)*JN[1]-2.0*sum1);
 }
else
 { /* YN[0] & YN[1] from P & Q series */
 chi0 = x - 0.5/TwoDPi;
 chi1 = chi0 - 1.0/TwoDPi;
 Rtfact = sqrt(TwoDPi/x);
 YN[0] = Rtfact*
  (PP(0,x)*sin(chi0) + QQ(0,x)*cos(chi0));
 YN[1] = Rtfact*
  (PP(1,x)*sin(chi1) + QQ(1,x)*cos(chi1));
 }

/* Use forward iteration up to  n = nYMax  */
tnDx = 0.0;
dtnDx = 2.0/x;
for ( n = 1; n < nYMax; n++ )
 {
 tnDx += dtnDx;
 YN[n+1] = tnDx*YN[n] - YN[n-1];
 }
return;
}
```

Programming Notes

Function `BessYN` for the irregular Bessel function $Y_n(x)$ with $x > 0$ is slightly more complicated than `BessJN` for the regular Bessel function $J_n(x)$ in Section 14.2.1, since it usually needs this function for its computations. In addition, functions `PP` and `QQ` to compute the asymptotic expressions for P and Q in (14.2.15) are often needed. These three functions are given in the program listing for `BessJN` in Section 14.2.1; on the CD-ROM they are packaged with the C driver program for $Y_n(x)$, following our rule of "no assembly required" that is described in The Computer Interface in the introduction to the *Atlas*.

Parameter `NMAX` is used in the C driver program (Section 21.14) to assign the size of arrays `JN` and `YN` that hold the values of $J_n(x)$ and $Y_n(x)$. The driver program compares the maximum value of n requested, `nYMax`, with `NMAX` to ensure that enough storage is available. The rule relating `nYMax`, x, and the fractional accuracy of the $J_n(x)$ values, ε, is discussed under Test Values in Section 14.2.1; sample values of the accuracy control parameters are given in Table 14.2.2.

Test Values

Check values of $Y_n(x)$ can be generated by running the `BesselZ` cell in *Mathematica* notebook `BessInteger`. Examples are given in Table 14.2.3. Choose `nYMax = 30` for $x = 2$ to 10 to exercise the first method in the algorithm, and `nYMax = 6` for $x = 20$ to exercise the second method.

M

TABLE 14.2.3 Test values to 10 digits of the irregular Bessel functions.

$x \setminus n$	0	3	6
2.0	0.5103756726	−1.127783777	−46.91400242
4.0	−0.01694073933	−0.182022116	−1.500691784
6.0	−0.288194684	0.328248946	−0.4268258569
8.0	0.2235214894	0.0265421593	0.03755810693
10.0	0.05567116728	−0.2513626572	0.280352559
20.0	0.06264059681	0.1496732627	−0.174111621

14.2.3 Regular Hyperbolic Bessel Function *I*

Keywords

cylinder functions, hyperbolic Bessel function, modified Bessel function, regular Bessel function

C Function and *Mathematica* Notebook

M

`BessIN` (C); cell `BesselIK` in notebook `BesselInteger` (*Mathematica*)

Background and Function Definition

The Bessel function $I_n(x)$ occurs in problems using cylindrical 3D coordinates, with the variable $x = kr$, in which k is a wavenumber and r is the radial coordinate. The hyperbolic function that is regular at $x = 0$ often arises in problems in which there is a cylindrical bounding surface at some finite r. Analytical developments are discussed in Section 3.7 of Watson's treatise on Bessel functions [Wat44], as well as in Sections 7.2.2 and 7.2.5 of Erdélyi et al [Erd53b]. Numerous formulas are given in Section 9.6 of Abramowitz and Stegun [Abr64] and in Section 8.4 of Gradshteyn and Ryzhik [Gra94]. Some graphical properties are shown in Chapters 49 and 50 of Spanier and Oldham [Spa87].

The regular hyperbolic Bessel function $I_n(x)$ is sometimes called a Basset function or a Weber function, but the first nomenclature is also used for the irregular hyperbolic Bessel function $K_n(x)$ that is covered in Section 14.2.4.

Function $I_n(x)$ satisfies a modified Bessel's differential equation, namely,

$$x^2 I_n'' + x I_n' - (x^2 + n^2) I_n(x) = 0 \tag{14.2.16}$$

and it is regular at $x = 0$. Formally, one may write

$$I_n(z) = (-i)^n J_n(iz) \tag{14.2.17}$$

which disguises the fact that $I_n(x)$ is purely real when x is real. Indeed, if we display $I_n(z)$ in the complex z plane, as we do in the visualizations below, it is seen to be essentially the Bessel function $J_n(z)$. Along the real line, however, $I_n(x)$ is quite different from $J_n(x)$, so it is usual to discuss it separately. The relation (14.2.17) is useful to relate analytical properties of the two functions, as in the following equations.

A power series solution of (14.2.16) produces

$$I_n(x) = \left(\frac{x}{2}\right)^n \sum_{k=0}^{\infty} \frac{(x^2/4)^k}{k!(n+k)!} \tag{14.2.18}$$

which we use in the Algorithm to generate an array of regular hyperbolic Bessel functions. A recurrence relation for computing $I_n(x)$ accurately is

$$I_{n-1}(x) = \frac{2n}{x} I_n(x) + I_{n+1}(x) \tag{14.2.19}$$

With the standardization indicated in (14.2.18), $I_n(x)$ satisfies the sum rule

$$I_0(x) + 2 \sum_{k=1}^{\infty} I_k(x) = e^x \tag{14.2.20}$$

which we use below to normalize the Bessel functions. This series converges faster than the $I_n(x)$ analog of (14.2.4), which involves alternating signs for successive series terms.

If $x \leq n_m^2/2$, where n_m is the largest n value of interest for given x, the above formulas generally give accurate numerical values. If, however, $x > n_m^2/2$, then another method is needed; we use asymptotic expansions, which resemble (14.2.5)–(14.2.7) for $J_n(x)$ but are simpler. The asymptotic series for $I_n(x)$ is equation 9.7.1 in Abramowitz and Stegun [Abr64], namely,

$$I_n(x) = \frac{e^x P_I(n,x)}{\sqrt{2\pi x}} \qquad P_I(n,x) \sim 1 - \frac{(\mu-1)}{8x} + \frac{(\mu-1)(\mu-9)}{2!(8x)^2} - \cdots \tag{14.2.21}$$

with $\mu = 4n^2$.

Visualization

Many views of $I_n(z)$ and $I_n(x)$ are shown in Figures 14.2.4–14.2.6. Compare and contrast these with the views of $J_n(z)$ and $J_n(x)$ in Figures 14.1.1, 14.1.3, 14.1.5, and 14.1.6, where similar representations are made.

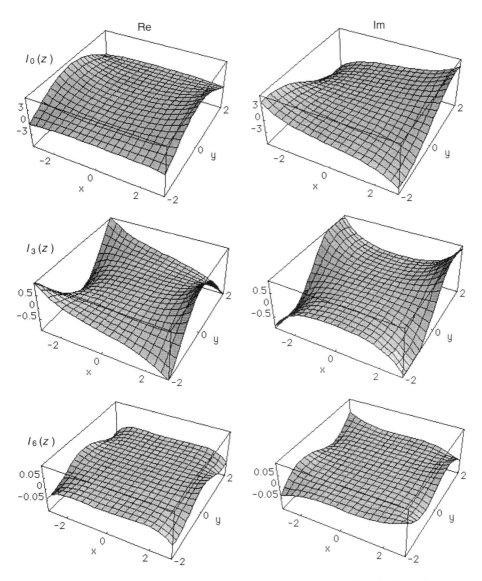

FIGURE 14.2.4 Regular hyperbolic Bessel functions in the complex plane for fixed values of $n = 0$, 3, and 6; real parts on the left and imaginary parts on the right. Note that the range of I_n values shown changes greatly as n changes. (Adapted from the output of cell BesselIK in *Mathematica* notebook BesselInteger.)

Notice in Figure 14.2.4 that the magnitude of $I_n(z)$ varies quite slowly as z varies over the x and y ranges shown; if the magnitude of x is increased much beyond 3, however, the lower-order functions begin to increase rapidly, as shown in more detail in Figure 14.2.6 below.

The behavior of $I_n(x)$ as x and n vary is shown in Figure 14.2.5. Note that although $I_n(x)$ is purely real when the order, n, is an integer, the surfaces in the figure range over noninteger orders.

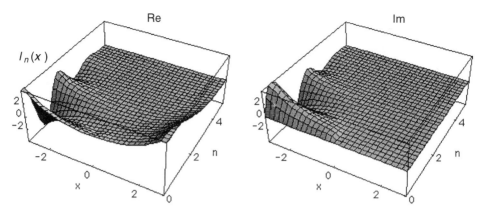

FIGURE 14.2.5 Regular hyperbolic Bessel functions in terms of real variable x and order n; real parts on the left, imaginary parts on the right. The vertical scale ranges from -4 to 4. (Adapted from the output of cell `BesselIK` in the *Mathematica* notebook `Bessel-Integer`.)

From the above two overview figures of the regular hyperbolic Bessel function, $I_n(x)$, we proceed to the two graphics in Figure 14.2.6, which help when devising the algorithm for our C function.

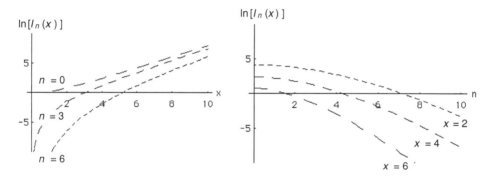

FIGURE 14.2.6 Logarithms of regular hyperbolic Bessel functions. Left side; I_n in terms of real variable x for order $n = 0, 3, 6$. Right side; I_n in terms of order n for $x = 2, 4, 6$. (Adapted from the output of cell `BesselIK` in the *Mathematica* notebook `Bessel-Integer`.)

Note on the left side of Figure 14.2.6, that for any positive x value $I_n(x)$ decreases as n increases; the direction of stable iteration is therefore toward decreasing n. Also in this graphic, note that $\ln[I_n(x)]$ eventually increases linearly with x, indicating an exponential increase of $I_n(x)$ with increase in x.

Algorithm and Program

The algorithm for function `BessIN` that computes $I_n(x)$ is similar to that for the regular cylinder Bessel function, $J_n(x)$, which is described in Section 14.2.1. In particular, `BessIN` computes an array of $I_n(x)$ values with n varying and x fixed. As justified immediately below Figure 14.2.6, the iteration is toward decreasing n; starting estimates for suitably large n are therefore required.

If $x \leq n_m^2$, where n_m is the largest n value of interest for given x, use of the leading term in (14.2.18) gives a good starting estimate of $I_n(x)$. The sum rule (14.2.20) is used to normalize the functions in this x range. Under Test Values we describe how to optimize the choice of n_m to obtain a given accuracy in each $I_n(x)$.

If $x > n_m^2$ then we use the asymptotic expansion (14.2.21), as coded in function PPI; in this function we set the fractional accuracy at eps $= 10^{-12}$. Downward iteration is used, but no normalization is needed. In function BessIN the values of n_m and x are input by the program user, as shown in the annotated C driver program in Section 21.14.

The C function for computing the array IN of $I_n(x)$ values is as follows:

```c
/* maximum number of table entries
    allowing for 6 inaccurate starting values   */
#define NMAX 46

void BessIN(int nIMax, double x, double IN[])
/* Make table of regular hyperbolic Bessel functions
    from order  n = 0  to  nIMax; x is fixed   */
{
double PPI(int n, double x),fnm2,absx,fctrl,fn,tn,
  sign,Rtfact,tnDx,dtnDx,sum,INorm;
const double TPi = 6.28318530718;
int nLarge,n;

if ( x == 0 )
 { /* treat this specially */
 IN[0] = 1.0;

 for ( n = 1; n <= nIMax; n++ )
  { IN[n] = 0.0; }

 return;
 }

nLarge = nIMax+6;
fnm2 = nIMax*nIMax;
absx = fabs(x);
if ( absx <=  fnm2 )
 {    /* Start with first term of power series */
 fctrl = 1.0; /* Make factorial */
 fn = 2.0;

 for ( n = 2; n <= nLarge-1; n++ )
  { fctrl *= fn; fn += 1.0; } /* then series term */
 IN[nLarge-1] = pow(0.5*x, (double) nLarge-1)/fctrl;
 tn = 2.0*fn;
 IN[nLarge] = x*IN[nLarge-1]/tn;

 /* Backward iteration down to  n = 0    */
 tnDx = ( (double) 2*nLarge )/x;
 dtnDx = 2.0/x;

 for ( n = nLarge-1; n >= 1; n-- )
  {
  tnDx -= dtnDx;
  IN[n-1] = tnDx*IN[n] + IN[n+1];
  }
```

```
    /* Normalize by linear sum rule  */
    sum = 0.0;

    for ( n = 1; n <= nLarge; n++ )
     { sum += IN[n]; }
    INorm  = exp(x)/(IN[0]+2.0*sum);

    for ( n = 0; n <= nIMax; n++ )
     { IN[n] *= INorm; }
    return;
    }


    /* PI series for larger  x  */
    Rtfact = exp(absx)/sqrt(TPi*absx);
    IN[nIMax-1] = Rtfact*PPI(nIMax-1,absx);
    IN[nIMax] = Rtfact*PPI(nIMax,absx);

    if ( x < 0 ) /* Check sign for  x < 0  */
     {
     if( (nIMax%2) > 0 )  IN[nIMax] = -IN[nIMax];
     else  IN[nIMax-1] = -IN[nIMax-1];
     }

    /* Use backward iteration down to  n = 0   */
   tnDx = ( (double) 2*nIMax )/x;
   dtnDx = 2.0/x;

   for ( n = nIMax-1; n >= 1; n-- )
    {
    tnDx -= dtnDx;
    IN[n-1] = tnDx*IN[n] + IN[n+1];
    }

   return;
   }


   double PPI(int n, double x)
   /* Series for  regular hyperbolic Bessel
      asymptotic in  x > 0 */
   /* Relative accuracy set to  eps = 10^(-12)   */
   {
   double mu,atex,term,termold,fk,tk,sum,
    ratio,eps,termnew;
   int k;

   mu = 4*n*n;
   atex = 8.0*x;
   k = 0;
   term = 1.0;
   termold = 1.0;
   fk = 0.0;
   tk = -1.0;
   sum = 1.0;
   ratio = 10.0;
   eps = 1.0e-12;
```

```
while ( ratio > eps )
 {
 k++;
 fk += 1.0;
 tk += 2.0;
 term *=   -(mu - tk*tk)/(fk*atex);
 sum += term;
 ratio = fabs(term/sum);
 termnew = fabs(term);

 if ( termnew > termold )
   {
   printf
   ("\n\n!!In PI diverging after %i terms",k);
   ratio = 0.0;
   }

 else
   { termold = termnew; }
 }

return sum;
 }
```

Programming Notes

Parameter NMAX is used in the C driver program (Section 21.14) to assign the size of array IN that holds the values of $I_n(x)$ with n varying from 0 to $n_m = $ nIMax. The driver program compares the maximum value of n requested, nIMax, with NMAX to ensure that enough storage is available. The rule relating nIMax, x, and the fractional accuracy of the $I_n(x)$ values, ε, is discussed under Test Values below; sample values of the accuracy control parameters are given in Table 14.2.5.

Test Values

Test values of $I_n(x)$ can be generated by running the BesselIK cell in *Mathematica* notebook BesselInteger; examples are given in Table 14.2.4. Choose nIMax = 20 for $x = 0$ to 8 to exercise the first method in the algorithm, and nIMax = 6 for $x = 20$ to exercise the second method.

$\boxed{M}$

TABLE 14.2.4 Test values to 10 digits of the regular hyperbolic Bessel functions.

$x \setminus n$	0	3	6
0	1.0	0	0
2.0	2.279585302	0.2127399592	0.001600173364
4.0	11.30192195	3.337275778	0.1544647999
6.0	67.23440698	30.1505403	3.355774847
8.0	427.5641157	236.0752210	43.61965919
20.0	4.355828256×10^7	3.459241634×10^7	1.742586421×10^7

To compute $I_n(x)$ efficiently and accurately it is worthwhile to optimize parameter n_m, the maximum order used in the backward-iteration scheme (14.2.19). For $x \leq n_m^2$, the elements of array IN are manipulated at least $3n_m$ times; once for iteration, once for the sum rule, and once for normalization. By making many test runs, we find that six iterations are typically required to correct for inaccuracies in the ratio of the two starting values; we therefore set nLarge = nIMax+6, where nIMax $= n_m$.

The other approximation made in our algorithm when $x \leq n_m^2$ is that the sum (14.2.20) is carried out only to $k = N =$ nLarge. To estimate the fractional error in the normalization, ε, made by the finite summation limit, write (14.2.20) as

$$I_0(x) + 2 \sum_{k=1}^{N} I_k(x) = e^x - 2 \sum_{k=N+1}^{\infty} I_k(x)$$
$$\approx e^x [1 - 2 e^{-x}(x/2)^{N+1}/(N+1)!] \approx e^x [1 - \varepsilon] \tag{14.2.22}$$

The first approximation is obtained if only the first remaining term is kept and if it is approximated using its leading term. The second approximation assumes that the error in each $I_n(x)$ comes only from the normalization error. If we replace the factorial by the first term in its Stirling series, as in the first equation below (7.1.4), then simplify by replacing $N+1$ by N, we get the condition

$$-x + N[\ln(x/2) - \ln N + 1] \approx \ln(\sqrt{\pi/2}\, \varepsilon) \tag{14.2.23}$$

Table 14.2.5 shows typical values of N predicted by (14.2.23); by running either our C program or the cell BesselIK in *Mathematica* notebook BesselInteger, you will find that these N values give the required accuracy.

TABLE 14.2.5 Maximum order of the regular hyperbolic Bessel function, N, as a function of argument x and the fractional accuracy, ε. See also (14.2.23).

ε	x	N
10^{-6}	5	13
	10	17
	20	20
10^{-10}	5	18
	10	24
	20	30
	30	33

14.2.4 Irregular Hyperbolic Bessel Function K

Keywords
hyperbolic functions, hyperbolic Bessel function, irregular Bessel function, modified Bessel function

C Function and *Mathematica* Notebook
BessKN (C); cell BesselIK in notebook BesselInteger (*Mathematica*)

Background and Function Definition

The Bessel function $K_n(x)$, sometimes called the Macdonald function, occurs most often in problems that use cylindrical 3D coordinates, with the variable $x = kr$, in which k is the wavenumber and r is the radial coordinate. The irregular function arises in problems in which $r > 0$ only, such as application of an exterior boundary condition for exponential decay beyond $r = R$, a cylindrical bounding surface. Analytical treatments are given in Section 3.7 of Watson's treatise on Bessel functions [Wat44] and in Sections 7.2.2 and 7.2.5 of Erdélyi et al [Erd53b]. Many formulas are given in Sections 9.6 and 9.7 of Abramowitz and Stegun [Abr64] and in Section 8.4 of Gradshteyn and Ryzhik [Gra94]. A few 2D graphs are shown in Chapter 51 of Spanier and Oldham [Spa87].

Function $K_n(x)$ satisfies the same equation as the regular solution $I_n(x)$, namely,

$$x^2 K_n'' + x K_n' - (x^2 + n^2) K_n(x) = 0 \tag{14.2.24}$$

but it is singular at $x = 0$. An expression for $K_n(x)$—involving power series, logarithms, and regular hyperbolic Bessel functions—is [Abr64, 9.6.53]

$$K_n(x) = -\frac{n! (x/2)^{-n}}{2} \sum_{k=0}^{n-1} \frac{(-x/2)^k I_k(x)}{(n-k)k!} +$$

$$(-1)^{n-1} [\ln(x/2) - \psi(n+1)] I_n(x) + (-1)^n \sum_{k=1}^{\infty} \frac{(n+2k) I_{n+2k}(x)}{k(n+k)} \tag{14.2.25}$$

where the ψ, is discussed in Section 6.2.1; this relation also holds if x is generalized to z.

An appropriate recurrence relation for $K_n(x)$—derived from [Abr64, 9.6.26]—is

$$K_{n+1}(x) = \frac{2n}{x} K_n(x) + K_{n-1}(x) \tag{14.2.26}$$

applied with increasing n—the direction insensitive to subtractive cancellation. Values of $K_0(x)$ and $K_1(x)$ are required; we compute them using (14.2.25) with $n = 0$ and 1. The $K_n(x)$ up to high order are then computed into array KN.

If $x \leq n_m^2$, where n_m is the largest n value of interest for given x, the above formulas give accurate numerical values. If, however, $x > n_m^2$, another method is needed; we use for $n = 0$ and 1 the asymptotic expansions in Abramowitz and Stegun [Abr64, 9.7.2], namely,

$$K_n(x) = \sqrt{\frac{\pi}{2x}} e^{-x} P_K(n,x) \qquad P_K(n,x) \sim 1 + \frac{(\mu - 1)}{8x} + \frac{(\mu - 1)(\mu - 9)}{2!(8x)^2} + \dots \tag{14.2.27}$$

Visualization

Several views of $K_n(z)$ and $K_n(x)$ are shown in the following graphics for small values of the parameters n and z or x and n. Figure 14.2.7 shows $K_n(z)$ in the complex plane for $n = 0$, 3, and 6, whereas Figure 14.2.8 is for $K_n(x)$ versus x and n, which is most relevant for devising our algorithm for the C function, BessKN.

Compare and contrast these visualizations of the irregular hyperbolic Bessel functions with the views of the irregular cylinder Bessel functions $Y_n(z)$ and $Y_n(x)$ in Figures 14.1.2, 14.1.4, 14.1.5, and 14.1.6, where similar representations are made. Note, however, that $K_n(z)$ is not related to $Y_n(z)$ by a simple rotation in the complex plane; the connection is given by equation 9.6.5 in Abramowitz and Stegun [Abr64].

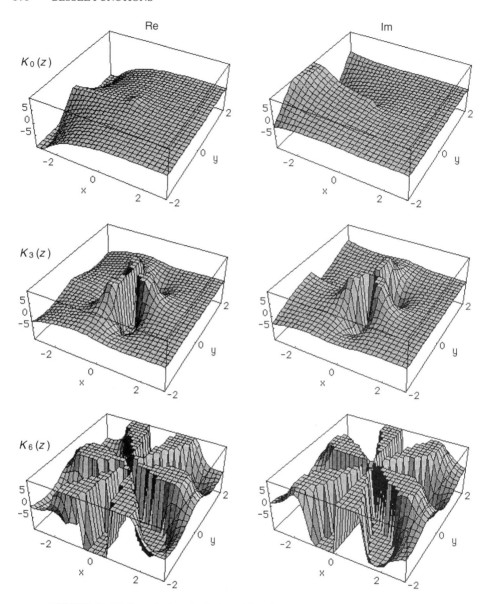

FIGURE 14.2.7 Irregular hyperbolic Bessel functions in the complex plane for fixed values of $n = 0$, 3, and 6; real parts on the left and imaginary parts on the right. The range of K_n values shown is -10 to 10; values larger in magnitude are truncated. (Adapted from the output of cell `BesselIK` in *Mathematica* notebook `BesselInteger`.)

Notice in Figure 14.2.7 that the behavior of $K_n(z)$ is dominated by the singularities at $z = 0$ for the x and y ranges shown; if the magnitude of x is increased much beyond 3 the lower-order functions begin to decrease rapidly, as shown in more detail in Figure 14.2.8 below. According to (14.2.27), the decay eventually becomes roughly exponential in x.

The behavior of $K_n(x)$ as x and n vary is shown in Figure 14.2.8. Although $K_n(x)$ is purely real when n is an integer, the surfaces in the figure range over noninteger orders.

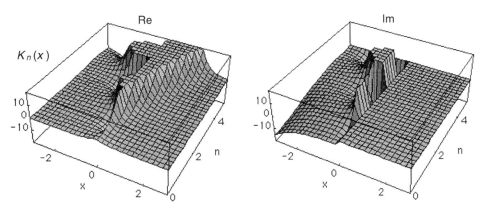

FIGURE 14.2.8 Irregular hyperbolic Bessel functions in terms of real variable x and order n; real parts on the left, imaginary parts on the right. The vertical scale ranges from -20 to 20. (Adapted from the output of cell `BesselIK` in *Mathematica* notebook `Bessel-Integer`.)

From the above two overview figures of the irregular hyperbolic Bessel function, $K_n(x)$, we proceed to the two graphics in Figure 14.2.9, which help us to devise the algorithm for our C function.

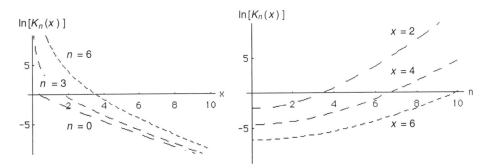

FIGURE 14.2.9 Logarithms of irregular hyperbolic Bessel functions. Left side; K_n in terms of real variable x for order $n = 0, 3, 6$. Right side; K_n in terms of order n for $x = 2, 4, 6$. (Adapted from the output of cell `BesselIK` in the *Mathematica* notebook `Bessel-Integer`.)

Note on the left side of Figure 14.2.9 that for any x value $K_n(x)$ increases as n increases; the direction of stable iteration is therefore towards increasing n. Also in this graphic, note that when n is given, $\ln[K_n(x)]$ eventually decreases nearly linearly with x, indicating an approximately exponential decrease of $K_n(x)$, consistent with (14.2.27).

Algorithm and Program

In function `BessKN` the values of n_m and x are input by the program user, as shown in the annotated C driver program in Section 21.14. We describe under Test Values how to optimize the choice of n_m to obtain a given accuracy in each $K_n(x)$.

The C function for computing the array KN of $K_n(x)$ values is given on the following page.

```
/* maximum number of table entries
   allowing for 6 inaccurate starting values  */
#define NMAX 47

void BessKN(int nKMax, double x, double KN[])
/* Table of irregular hyperbolic Bessel functions
   from order  n = 0  to  nKMax; x is fixed  */
{
double PPK(int n, double x),fnm2,IN[NMAX],
 sum0,sum1,fk,term,logterm,Rtfact,tnDx,dtnDx;
const double EulerG = 0.577215664902,
 TwoDPi = 0.636619772368;
void BessIN(int nKMax, double x, double IN[]);
int k,n;

fnm2 = nKMax*nKMax;
if ( x <=  fnm2 )
  {  /* KN[0] & KN[1] from table of regular
     hyperbolic Bessel functions In(x)
     order  n = 0 to nKMax    */
 BessIN(nKMax,x,IN);
 sum0 = 0.0;
 sum1 = 0.0;
 fk = 0.0;
 for ( k = 1; k < nKMax/2; k++ )
   {
   fk +=1.0;
   term = 1.0/fk;
   sum0 += term*IN[2*k];
   sum1 += term*(0.5+fk)*IN[1+2*k]/(1.0+fk);
   }
 logterm = log(0.5*x)+EulerG;
 KN[0] = - logterm*IN[0] + 2.0*sum0;
 KN[1] = IN[0]/x +(logterm-1.0)*IN[1]-2.0*sum1;
 }
else
  {   /* PK series for larger  x  */
 Rtfact = exp(-x)/sqrt(TwoDPi*x);
 KN[0] = Rtfact*PPK(0,x);
 KN[1] = Rtfact*PPK(1,x);
 }

 /* Forward iteration up to  n = nKMax   */
tnDx = 0.0;
dtnDx = 2.0/x;
for ( n = 1; n < nKMax; n++ )
 {
 tnDx += dtnDx;
 KN[n+1] = tnDx*KN[n] + KN[n-1];
 }
return;
}

double PPK(int n, double x)
/* Series for irregular hyperbolic Bessel
   asymptotic in  x > 0  */
/* Relative accuracy set to  eps = 10^(-12)  */
```

```
{
double mu,atex,term,termold,fk,tk,sum,
 ratio,eps,termnew;
int k;

mu = 4*n*n;
atex = 8.0*x;
k = 0;
term = 1.0;
termold = 1.0;
fk = 0.0;
tk = -1.0;
sum = 1.0;
ratio = 10.0;
eps = 1.0e-12;

while ( ratio > eps )
  {
  k++;
  fk += 1.0;
  tk += 2.0;
  term *=  (mu - tk*tk)/(fk*atex);
  sum += term;
  ratio = fabs(term/sum);
  termnew = fabs(term);
  if ( termnew > termold )
    {
    printf
    ("\n\n!!In PI diverging after %i terms",k);
    ratio = 0.0;
    }
  else
    { termold = termnew; }
  }

return sum;
}
```

Programming Notes

Our C function BessKN for the irregular hyperbolic Bessel function $K_n(x)$ with $x > 0$ is more complicated than BessIN for the regular function $I_n(x)$ in Section 14.2.3; it actually needs this function for its computation, using (14.2.25) with $n = 0$ and 1 to initiate the recurrence over n if $x \leq n_m^2$. In addition, the function PPK computes the asymptotic expressions for $P_K(x)$ in (14.2.27) if $x > n_m^2$ and is given in the program listing above. The function BessIN is given in Section 14.2.3. On the CD-ROM all the functions are packaged together, following our rule of "no assembly required" described in The Computer Interface in the Introduction to the *Atlas*.

Parameter NMAX is used in the C driver program (Section 21.14) to assign the size of arrays IN and KN that hold the values of $I_n(x)$ and $K_n(x)$. The driver program compares the maximum value of n requested, nKMax, with NMAX to ensure that enough storage is available. The rule relating nKMax, x, and the fractional accuracy of the $K_n(x)$ values, ε, is discussed under Test Values below; sample values of the accuracy control parameters are given in Table 14.2.7.

Test Values

Check values of $K_n(x)$ can be generated by running the BesselIK cell in *Mathematica* notebook BesselInteger. Examples are given in Table 14.2.6. Choose nKMax = 40 for $x = 2$ to 8 to exercise the first method in the algorithm, and nKMax = 6 for $x = 40$ to exercise the second method.

TABLE 14.2.6 Test values to 10 digits of the irregular hyperbolic Bessel functions.

$x \backslash n$	0	3	6
2.0	0.1138938727	0.6473853909	49.35116143
4.0	0.01115967609	0.02988492442	0.4480851840
6.0	0.001243994328	0.002471898096	0.01753673063
8.0	0.0001464707052	0.0002480257159	0.001145529809
40.0	$8.392861100 \times 10^{-19}$	$9.378903725 \times 10^{-19}$	$1.308050685 \times 10^{-18}$

To compute $K_n(x)$ efficiently and accurately it is worthwhile to optimize parameter n_m, the maximum order used in the forward-iteration scheme (14.2.26). In (14.2.25) for $x \leq n_m^2$, the elements of array IN are manipulated at least $3n_m$ times; once for iteration, once for the sum rule, and once for normalization. By making many test runs, we find that six iterations are typically required to correct for inaccuracies in the ratio of the two starting values; we therefore set nLarge = nIMax+6, where nIMax = n_m. Typical values of n_m are given in Table 14.2.5.

The other approximation made in our algorithm when $x \leq n_m^2$ is that the sum to infinity in (14.2.25) for the starting values $K_0(x)$ and $K_1(x)$ is actually carried out only to some $k = N$. To estimate the fractional error, ε, made by the finite summation limit, we use for $K_0(x)$ the first neglected term $2I_{N+2}(x)/(N/2+1)$. For this we use its leading term from (14.2.18) and we replace the factorial therein by the first term in its Stirling series, as in the first equation below (7.1.4). For purposes of estimating ε as a fractional error, we approximate $K_0(x)$ by the leading term of its asymptotic expansion (14.2.27), which is within 10% of its true value for $x \geq 1$. We then simplify by replacing $N+1$ by N, to obtain the condition

$$-x - \ln N + N[\ln(x/2) - \ln N + 1] \approx \ln(\pi \varepsilon / 4) \qquad (14.2.28)$$

Table 14.2.7 gives typical values of N predicted by (14.2.28); by running either our C program or cell BesselIK in *Mathematica* notebook BessInteger, you will find that these N values give the required accuracy for the given x values. Input parameter nKMax = n_m should be chosen at least as large as N.

The estimation formula (14.2.28) is not reliable for much larger x than shown in Table 14.2.7, since the approximations made in its derivation then break down. If you need $K_n(x)$ for large x, modify the function BessKN so that it is forced to use the asymptotic expansion (14.2.21); the fractional accuracy of this expansion is set to be 10^{-12}. Note, however, that it cannot be made of arbitrary accuracy for any given x, since it is an asymptotic expansion; see the discussion in Section 3.4.

TABLE 14.2.7 Maximum order of the irregular hyperbolic Bessel function, N, as a function of argument x and the fractional accuracy, ε. See also (14.2.8).

ε	x	N
10^{-6}	1	8
	5	17
	10	28
10^{-10}	1	11
	5	22
	10	33

14.3   KELVIN FUNCTIONS

The Kelvin functions occur frequently in electrical engineering when one uses cylindrical coordinates to investigate alternating-current problems, especially the skin effect, as discussed in detail in Chapters 8 and 9 of McLachlan's text [McL55] and in the text by Ramo et al [Ram84]. The functions were introduced in this context by Lord Kelvin (William Thomson) in 1889, so they are also called Thomson functions. The regular function arises in problems in which $r \geq 0$, while the irregular function occurs when $r > 0$ only, such as application of an exterior boundary condition at $r = R > 0$, a cylindrical bounding surface.

Analytical results on Kelvin functions are stated in Section 3.8 of Watson's treatise on Bessel functions [Wat44], in Section 7.2.3 of Erdélyi et al [Erd53b], and in Section 8.56 of Gradshteyn and Ryzhik [Gra94]. Graphs of the functions are given in Chapter 55 of Spanier and Oldham [Spa87]. Formulas relevant for numerical computing are in Sections 9.9 to 9.11 in Abramowitz and Stegun [Abr64].

The following two subsections consider the regular functions—ber and bei—first, then they develop the irregular functions—ker and kei.

14.3.1 Regular Kelvin Functions ber, bei

Keywords
Kelvin function, regular Bessel function, Thomson function

C Functions and *Mathematica* Notebook
BerBei (C); cell Kelvin in notebook BesselInteger (*Mathematica*)

$\boxed{M}$

Background and Function Definition
Regular Kelvin functions for integer order n and real argument x—$\mathrm{ber}_n x$ and $\mathrm{bei}_n x$—satisfy the Bessel-type equation

$$x^2 w_n'' + x w_n' - (ix^2 + n^2) w_n(x) = 0 \tag{14.3.1}$$

where the real functions $\mathrm{ber}_n x$ and $\mathrm{bei}_n x$ are related to $w(x)$ by

$$w(x) = \mathrm{ber}_n x + i\,\mathrm{bei}_n x = J_n(xe^{3\pi i/4}) \tag{14.3.2}$$

and $w(x)$ is regular at $x = 0$. When $n = 0$, it is common to write

$$\mathrm{ber}\, x = \mathrm{ber}_0 x \qquad\qquad \mathrm{bei}\, x = \mathrm{bei}_0 x \tag{14.3.3}$$

Power series for ber_n and bei_n are [Abr64, 9.9.9]

$$\mathrm{ber}_n x = \left(\frac{x}{2}\right)^n \sum_{k=0}^{\infty} \frac{\cos[(3n+2k)\pi/4](x/2)^{2k}}{k!\,(n+k)!}$$

$$\mathrm{bei}_n x = \left(\frac{x}{2}\right)^n \sum_{k=0}^{\infty} \frac{\sin[(3n+2k)\pi/4](x/2)^{2k}}{k!\,(n+k)!} \tag{14.3.4}$$

Appropriate recurrence relations for ber_n and bei_n—derived from [Abr64, 9.9.14]—are

$$\mathrm{ber}_{n-1} x = -\frac{n\sqrt{2}}{x}[\mathrm{ber}_n x - \mathrm{bei}_n x] - \mathrm{ber}_{n+1} x$$

$$\mathrm{bei}_{n-1} x = -\frac{n\sqrt{2}}{x}[\mathrm{bei}_n x + \mathrm{ber}_n x] - \mathrm{bei}_{n+1} x \tag{14.3.5}$$

applied with decreasing n—the direction insensitive to subtractive cancellation when x is small compared with n, as illustrated in Visualization.

A sum rule for the regular functions can be deduced from equations 9.1.46 and 9.9.1 in Abramowitz and Stegun [Abr64], namely,

$$\mathrm{ber}_0 x + 2\sum_{k=1}^{\infty} \mathrm{ber}_{2k} x + i\left[\mathrm{bei}_0 x + 2\sum_{k=1}^{\infty} \mathrm{bei}_{2k} x\right] = 1 \tag{14.3.6}$$

which is used in C function `BerBei` to normalize the Kelvin functions. If $x < 2\sqrt{2(n+2)}$ then the power series (14.3.4) converge quickly; for example, if $x < 4$ then $\mathrm{ber}\, x$ and $\mathrm{bei}\, x$ ($n = 0$) can readily be computed this way, as can ber_n and bei_n for larger n. If, however, convergence in (14.3.4) is slow then another method is needed.

Asymptotic expansions for ber_n and bei_n when $x > 0$ are obtained from [Abr64, 9.10.1, 9.10.2] as

$$\mathrm{ber}_n x = \frac{e^{x/\sqrt{2}}}{\sqrt{2\pi x}}[f_n(x)\cos\alpha_n + g_n(x)\sin\alpha_n] - \frac{\mathrm{kei}_n x}{\pi}$$

$$\mathrm{bei}_n x = \frac{e^{x/\sqrt{2}}}{\sqrt{2\pi x}}[f_n(x)\sin\alpha_n - g_n(x)\sin\alpha_n] + \frac{\mathrm{ker}_n x}{\pi} \tag{14.3.7}$$

In the asymptotic expansion formulas we include the irregular Kelvin functions ker_n and kei_n (Section 14.3.2). Although these terms are often ignored in discussions of the functions, they are required for accurate numerical values. For example, when $x = 15$ and $n = 0$ or 1, the irregular terms contribute about 10^{-9} relative accuracy to the regular functions; for $x = 20$ their contribution is about 10^{-12}. Asymptotic series for ker_n and kei_n are given in [Abr64, 9.10.3, 9.10.4] as

$$\mathrm{ker}_n x = \sqrt{\frac{\pi}{2x}}\, e^{-x/\sqrt{2}}\, [f_n(-x)\cos\beta_n - g_n(-x)\sin\beta_n]$$

$$\mathrm{kei}_n x = \sqrt{\frac{\pi}{2x}}\, e^{-x/\sqrt{2}}\, [-f_n(-x)\sin\beta_n - g_n(-x)\cos\beta_n]$$

(14.3.8)

In (14.3.7) and (14.3.8) the angles α_n and β_n are given by

$$\alpha_n = x/\sqrt{2} + (4n-1)\pi/8 \qquad \beta_n = \alpha_n + \pi/4$$

(14.3.9)

The occurrence of $e^{x/\sqrt{2}}$ in the asymptotic expansions suggests that such a scaling factor will enable ber_n and bei_n to be shown over a wide range of x values, as in the visualizations following.

The auxiliary functions f_n and g_n have the asymptotic expansions (Section 3.4) given by [Abr64, 9.10.6, 9.10.7] as

$$f_n(\pm x) \sim 1 + \sum_{k=1}^{\infty} \frac{(\mp 1)^k(\mu-1)(\mu-1)\ldots[(\mu-(2k-1)^2)]}{k!(8x)^k}\cos\left(\frac{k\pi}{4}\right)$$

$$g_n(\pm x) \sim \sum_{k=1}^{\infty} \frac{(\mp 1)^k(\mu-1)(\mu-1)\ldots[(\mu-(2k-1)^2)]}{k!(8x)^k}\sin\left(\frac{k\pi}{4}\right)$$

(14.3.10)

Visualization

We show several views of the regular functions ber_n and bei_n in the following graphics; first for small values of x, then for larger x values but with the functions modified to suppress the exponential growth with x. Figure 14.3.1 shows ber_n and bei_n versus x and n, which is relevant for devising our algorithm for the C function `BerBei`.

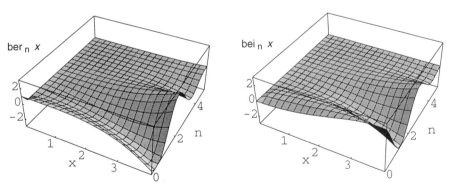

FIGURE 14.3.1 Regular Kelvin functions in the $x - n$ plane, with x ranging from 0 to 4 and n ranging from 0 to 5. The vertical scale ranges from –3 to 3. (Adapted from the output of cell `Kelvin` in *Mathematica* notebook `BesselInteger`.)

In order to illustrate the Kelvin functions for much larger x, we remove the exponential factor that appears in the large-x formula (14.3.7), thus obtaining the hills and dales seen in Figure 14.3.2.

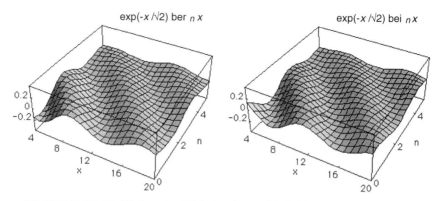

FIGURE 14.3.2 Modified regular Kelvin functions multiplied by an exponential damping factor. Shown as functions of x, which ranges from 4 to 20 for each value of n from 0 to 5. The vertical scale ranges from -0.4 to 0.4. (Adapted from the output of cell `Kelvin` in *Mathematica* notebook `BesselInteger`.)

The behavior of $\text{ber}_n x$ and $\text{bei}_n x$ as x varies with n fixed is shown in Figure 14.3.3 over the range of x for which the power-series expansion converges rapidly for any n.

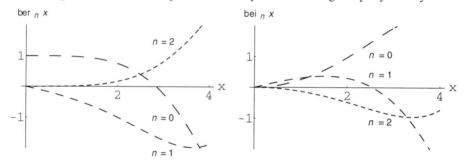

FIGURE 14.3.3 Regular Kelvin functions as functions of x, which ranges from 0 to 4 for each value of $n = 0, 1, 2$. The vertical scale ranges from -2 to 2. (Adapted from the output of cell `Kelvin` in *Mathematica* notebook `BesselInteger`.)

Figure 14.3.4 help us to devise the algorithm for our C function, since it shows that $\text{ber}_n x$ and $\text{bei}_n x$ generally increase as n decreases, which is thus the appropriate direction of iteration if relative errors from subtractive cancellation are to be minimized.

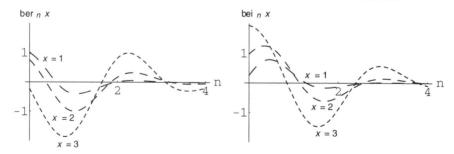

FIGURE 14.3.4 Regular Kelvin functions as functions of n, which ranges from 0 to 4 for each value of $x = 1, 2$, and 3. The vertical scale ranges from -2 to 2. (Adapted from the output of cell `Kelvin` in *Mathematica* notebook `BesselInteger`.)

The auxiliary functions $f_n(x)$ and $g_n(x)$ defined by (14.3.10) are required for computing the Kelvin functions for large x; they are shown for x in [20, 100] with $n = 0$ and $n = 1$ (the starting values for forward iteration) in Figure 14.3.5. Note that when x is large we have $f_1 - 1 \approx -3(f_0 - 1)$, $f_0 - 1 \approx g_0$, and $g_1 \approx -3g_0$.

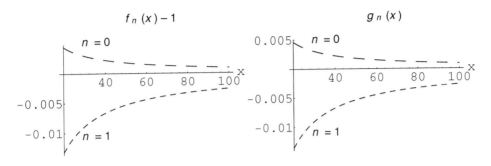

FIGURE 14.3.5 Auxiliary functions for asymptotic expansion of Kelvin functions. We show $n = 0$ and 1 with x ranging from 20 to 100. (Adapted from the output of cell `Kelvin` in *Mathematica* notebook `BesselInteger`.)

Algorithm and Program

The structure of function `BerBei` to compute into arrays `ber` and `bei` the regular Kelvin functions is very similar to the function `BessJN` for the regular Bessel functions in Section 14.2.1. The annotated driver program `BerBei.c` in Section 21.14 shows how to input the values of n_m (the maximum order of the Kelvin function) and x. We describe under Test Values how to optimize the choice of n_m to obtain a given accuracy in each $\text{ber}_n x$ and $\text{bei}_n x$ for n from zero up to n_m.

The algorithm uses one of two basic methods based upon the the formulas given above in Background and Function Definition. For $x \le n_m^2$, backward iteration—starting from the first term of the power series (14.3.4)—is used, following (14.3.5). The functions are normalized by using sum rule (14.3.6). Table 14.3.2 provides guidance for the appropriate choices of n_m for using the power-series method.

For a larger x value (that is, by choosing n_m small enough for practical use) the asymptotic expansions in (14.3.7)–(14.3.10) are used for $n = 0$ and 1; iteration then follows for increasing n up to n_m.

The C function for computing the arrays `ber` and `bei` of $\text{ber}_n x$ and $\text{bei}_n x$ values is as follows. The listing includes the functions `Fn` for $f_n(x)$ and `Gn` for $g_n(x)$.

```
/* maximum number of table entries
      allowing for 6 inaccurate starting values   */
#define NMAX 50

void BerBei(int nBMax, double x,
  double ber[], double bei[])
/* Make a table of regular Kelvin Functions
    ber n x  and  bei n x
    from order  n = 0  to  nBMax <= NMAX-1   */
```

```
{
double Fn(int n, double x),Gn(int n, double x),
 fnm2,absx,fctrl,fn,tn,tnDx,dtnDx,
 sumr,sumi,den,Normr,Normi,tempr,tempi,
 alpha,beta,bfact,kfact,bf0,bg0,bf1,bg1,
 kf0,kg0,kf1,kg1,kcos,ksin,ker0,kei0,ker1,kei1,
 bcos,bsin;
const double PiD2 = 1.570796326794897,
 PiD4 = 0.7853981633974483,
 PiD8 = 0.3926990816987241,
 /* root 2 to 16 digits for 10-digit accuracy */
 Rt2 = 1.414213562373095;
int nLarge,n;

if ( x == 0 )
 { /* treat this specially */
 ber[0] = 1.0;    bei[0] = 1.0;
 for ( n = 1; n <= nBMax; n++ )
  { ber[n] = 0.0;  bei[n] = 0.0; }
 return;
 }

nLarge = nBMax+6;
fnm2 = nBMax*nBMax;
absx = fabs(x);

if ( absx <= fnm2 )
 {   /* Backward iteration method */
 fctrl = 1.0; /* Make factorial */
 fn = 2.0;
 for ( n = 2; n <= nLarge-1; n++ )
  { fctrl *= fn; fn += 1.0; }

 /* Starting approximations */
 ber[nLarge-1] =
  0.5*pow(0.5*x,(double) nLarge-1)/fctrl;
 bei[nLarge-1] = ber[nLarge-1];
 tn = 2.0*fn;
 ber[nLarge] = x*ber[nLarge-1]/tn;
 bei[nLarge] = ber[nLarge];

 /* Use backward iteration down to  n = 0  */
 tnDx = - tn/(Rt2*x);
 dtnDx = - Rt2/x;
 for ( n = nLarge-1; n >= 1; n-- )
  {
  tnDx -= dtnDx;
  ber[n-1] = tnDx*(ber[n] - bei[n]) - ber[n+1];
  bei[n-1] = tnDx*(bei[n] + ber[n]) - bei[n+1];
  }

 /* Normalize by comparing with sum rule
    carried out to  nLarge = nBMax+6   */
 sumr = 0.0;  sumi = 0.0;
 for ( n = 2; n <= nBMax; n += 2 )
  {
  sumr += ber[n];   sumi += bei[n];
  }
```

```
  sumr = ber[0]+2*sumr;   sumi = bei[0]+2*sumi;
  den = sumr*sumr + sumi*sumi;
  Normr = sumr/den;   Normi = - sumi/den;
  for ( n = 0; n <= nBMax; n++ )
   {
   tempr = ber[n];  tempi = bei[n];
   ber[n] = Normr*tempr - Normi*tempi;
   bei[n] = Normr*tempi + Normi*tempr;
   }
  return;
  }
else
  {
  /* forward iteration from  f & g  series */
  /* Starting values */
  alpha = absx/Rt2 - PiD8;
  beta = alpha + PiD4;
  bfact = 0.5*exp(absx/Rt2)/sqrt(PiD2*absx);
  kfact = 0.5*exp(-absx/Rt2)/sqrt(PiD2*absx);
  /* Make and save auxiliary functions */
  bf0 = Fn(0,absx);      bg0 = Gn(0,absx);
  bf1 = Fn(1,absx);      bg1 = Gn(1,absx);
  kf0 = Fn(0,-absx);     kg0 = Gn(0,-absx);
  kf1 = Fn(1,-absx);     kg1 = Gn(1,-absx);
  /* Irregular Kelvin functions / Pi  */
  kcos = cos(beta);      ksin = sin(beta);
  ker0 =   kfact*(kf0*kcos - kg0*ksin);
  kei0 = - kfact*(kf0*ksin + kg0*kcos);
   /* For n=1 re-use n=0 cos & sin */
  ker1 = - kfact*(kf1*ksin + kg1*kcos);
  kei1 = - kfact*(kf1*kcos - kg1*ksin);
  /* Regular Kelvin functions to start */
  bcos = cos(alpha);     bsin = sin(alpha);
  ber[0] = bfact*(bf0*bcos + bg0*bsin) - kei0;
  bei[0] = bfact*(bf0*bsin - bg0*bcos) + ker0;
   /* For n=1 re-use n=0 cos & sin */
  ber[1] = bfact*(-bf1*bsin + bg1*bcos) - kei1;
  bei[1] = bfact*(bf1*bcos + bg1*bsin) + ker1;
  if ( x < 0 )
   { ber[1] = - ber[1];   bei[1] = - bei[1]; }

  /* Forward iteration up to  n = nBMax   */
  tnDx = 0.0;
  dtnDx = - Rt2/x;
  for ( n = 1; n < nBMax; n++ )
   {
   tnDx += dtnDx;
   ber[n+1] = tnDx*(ber[n] - bei[n]) - ber[n-1];
   bei[n+1] = tnDx*(bei[n] + ber[n]) - bei[n-1];
   }
  return;
  }
}
```

```c
double Fn(int n, double x)
/* Auxiliary  f  series for asymptotic in  x > 0
   of Kelvin functions  */
/* Relative accuracy set to  eps = 10^(-12)  */
{
double mu,phase,atex,term,termold,fk,tk,sum,
 ratio,eps,termnew;
const double PiD4 = 0.7853981633974483;
int k;

mu = 4*n*n;
if ( x > 0.0 )  { phase = -1.0; }
 else           { phase =  1.0; }

/* Make asymptotic series */
atex = 8.0*fabs(x);
k = 0;
term = 1.0;
termold = 1.0;
fk = 0.0;
tk = -1.0;
sum = 1.0;
ratio = 10.0;
eps = 1.0e-12;

while ( ratio > eps )
  {
  k++;
  fk += 1.0;
  tk += 2.0;
  term *=  phase*(mu-tk*tk)/(fk*atex);
  sum += term*cos(fk*PiD4);
  ratio = fabs(term/sum);
  termnew = fabs(term);
  if ( termnew > termold )
    {
    printf
    ("\n\n!!In Fn diverging after %i terms",k);
    ratio = 0.0;
    }
  else
    { termold = termnew; }
  }

return sum;
}


double Gn(int n, double x)
/* Auxiliary  g  series for asymptotic in  x > 0
   of Kelvin functions  */
/* Relative accuracy set to  eps = 10^(-12)  */
{
double mu,phase,atex,term,termold,fk,tk,sum,
 ratio,eps,termnew;
const double PiD4 = 0.7853981633974483;
int k;
```

```
mu = 4*n*n;
if ( x > 0.0 )  { phase = -1.0; }
  else          { phase =  1.0; }
/* Make asymptotic series */
atex = 8.0*fabs(x);
k = 0;
term = 1.0;
termold = 1.0;
fk = 0.0;
tk = -1.0;
sum = 0.0;
ratio = 10.0;
eps = 1.0e-12;

while ( ratio > eps )
 {
 k++;
 fk += 1.0;
 tk += 2.0;
 term *=  phase*(mu-tk*tk)/(fk*atex);
 sum += term*sin(fk*PiD4);
 ratio = fabs(term/sum);
 termnew = fabs(term);
 if ( termnew > termold )
   {
   printf
   ("\n\n!!In Gn diverging after %i terms",k);
   ratio = 0.0;
   }
 else
   { termold = termnew; }
 }
return sum;
}
```

Programming Notes

The functions $\text{ber}_n x$ and $\text{bei}_n x$ are the real and imaginary parts of a regular Bessel function of complex argument, as given in (14.3.2). We see this, for example, in the recurrence relation (14.3.5) and in the normalization rule (14.3.6). It would be easier to program function `BerBei` in complex arithmetic, but it is more efficient in execution to generate $\text{ber}_n x$ and $\text{bei}_n x$ in parallel, as we do.

In most of the C functions in the *Atlas*, numerical constants are given to 12 decimal digits, this usually being sufficient for 10-digit accuracy in the function values. In Ber-Bei, however, 16-digit accuracy of the $\sqrt{2}$ factor in (14.3.5) is required to assure 10 digits of accuracy in the Kelvin functions.

Auxiliary functions for the asymptotic series—f_n and g_n in (14.3.10), shown in Figure 14.3.5 for $n = 0$ and 1—are computed to 12-digit relative accuracy by Fn and Gn, respectively. If higher accuracy is needed, decrease parameter `eps` in these functions. As we do consistently for asymptotic series throughout the *Atlas* (see the discussion in Section 3.4), the convergence of the series is monitored and an error message is printed if the terms start to increase in magnitude before the requested tolerance is obtained. Functions

Fn and Gn are also used for computing the irregular Kelvin functions, $\ker_n x$ and $\kei_n x$ in Section 14.3.2.

Parameter NMAX is used in the C driver program (Section 21.14) to assign the size of the arrays ber and bei that hold the values of $\ber_n x$ and $\bei_n x$. The driver program compares the maximum value of n requested, nBMax, with NMAX to ensure that enough storage is available. The rule relating nBMax, x, and the fractional accuracy of the regular Kelvin function values computed by the backward-iteration method, ε, is discussed under Test Values below. Table 14.3.3 shows sample values of the accuracy control parameters.

Test Values

Test values of $\ber_n x$ and $\bei_n x$ can be generated by running the Kelvin cell in the *Mathematica* notebook BesselInteger. Examples are given in Table 14.3.1 for $\ber_n x$ and in Table 14.3.2 for $\bei_n x$. Choose nBMax = 30 for $x = 2$ to 8 to exercise the first method in the algorithm (backward iteration). Enter nBMax = 6 for $x = 40$ to exercise the second method (forward iteration starting with asymptotic series).

TABLE 14.3.1 Test values to 10 digits of the regular Kelvin function ber.

$x \setminus n$	0	3	6
2.0	0.7517341827	0.0856114485	$-1.979536185 \times 10^{-4}$
4.0	−2.563416557	−0.2826268017	−0.04892605288
6.0	−8.858315966	−6.430044105	−1.066577146
8.0	20.97395561	−15.42035815	−6.159793715
40.0	$-1.125966969 \times 10^{11}$	$3.359258003 \times 10^{10}$	$8.816320510 \times 10^{10}$

TABLE 14.3.2 Test values to 10 digits of the regular Kelvin function bei.

$x \setminus n$	0	3	6
2.0	0.9722916273	0.1442099416	$1.376499575 \times 10^{-3}$
4.0	2.292690323	1.437765103	0.07637784192
6.0	−7.334746541	1.901460523	0.3340501699
8.0	−35.01672516	−22.57504623	−4.426361358
40.0	$4.562813029 \times 10^{10}$	$1.070520684 \times 10^{11}$	$-5.444700992 \times 10^{9}$

To compute $\ber_n x$ and $\bei_n x$ efficiently and accurately, it is worthwhile to optimize parameter n_m, the maximum order used in the backward-iteration scheme (14.3.5). When $x \leq n_m^2$, the elements of arrays ber and bei are manipulated at least $3n_m$ times; once for iteration, once for the sum rule, and once for normalization. After many test runs we find that six iterations are typically required to correct for inaccuracies in the ratio of the two starting values, which are the first terms in (14.3.4); thus nLarge = nIMax+6, where nIMax = n_m. Table 14.3.3 gives typical values of n_m.

The other approximation made in our algorithm when $x \leq n_m^2$ is that the sum to infinity in (14.3.6) is actually carried out only to order $k = $ nLarge, which we abbreviate as N. To estimate the fractional error, ε, made by the finite summation limit, we use the first neglected term estimated to leading order from (14.3.4) and we replace the factorial therein by the first term in its Stirling series, as in the first equation below (7.1.4). We then simplify by replacing $N+1$ by N, to obtain the approximate condition

$$\ln(x^2/2) + N[\ln(x/2) - \ln N + 1] - 2\ln N \approx \ln(\sqrt{2\pi}\,\varepsilon) \qquad (14.3.11)$$

Table 14.3.3 gives typical values of N predicted by (14.3.11). By running either our C driver program, BerBei.c in Section 21.14 or the cell Kelvin in *Mathematica* notebook BesselInteger (Section 20.14), you will find that these N values give the required accuracy for the given x values.

Accuracy estimation formula (14.3.11) is not reliable for much larger x than shown in Table 14.3.3, since the approximations made in its derivation then break down. If you need $\text{ber}_n x$ and $\text{bei}_n x$ for large x, modify the function BerBei so that it is forced to use the asymptotic expansion (14.3.7); the fractional accuracy of this expansion is set to be 10^{-12}. Note, however, that it cannot be made of arbitrary accuracy for any given x, since it is an asymptotic expansion; see the discussion in Section 3.4.

TABLE 14.3.3 Maximum order of the regular Kelvin functions, N, as a function of argument x and the fractional accuracy, ε; see (14.3.11).

ε	x	N
10^{-6}	5	14
	10	22
	20	37
10^{-10}	5	19
	10	28
	20	43

14.3.2 Irregular Kelvin Functions ker, kei

Keywords
Kelvin function, irregular Bessel function, Thomson function

C Functions and *Mathematica* Notebook
KerKei (C); cell Kelvin in notebook BesselInteger (*Mathematica*)

Background and Function Definition
Irregular Kelvin functions for integer order n and real argument x, $\text{ker}_n x$ and $\text{kei}_n x$, satisfy a Bessel-type equation, as follows:

$$x^2 w_n'' + x K_n'' - (ix^2 + n^2) w_n(x) = 0 \qquad (14.3.12)$$

where the real functions $\text{ker}_n x$ and $\text{kei}_n x$ are related to $w(x)$ and the irregular hyperbolic Bessel function of complex argument, $K_n(z)$ in Section 14.2.4, by

$$w(x) = \text{ker}_n x + i\,\text{kei}_n x = e^{-n\pi i/2} K_n(xe^{\pi i/4}) \qquad (14.3.13)$$

and $w(x)$ is irregular at $x = 0$. When $n = 0$, it is usual to write

$$\ker x = \ker_0 x \qquad \kei x = \kei_0 x \qquad (14.3.14)$$

Because of relation (14.3.13), our numerical methods for the Kelvin functions are very similar to those for $K_n(x)$.

Series expansions for $\ker_n$ and $\kei_n$ for $x > 0$ are [Abr64, 9.9.11]

$$\ker_n x = \frac{1}{2}\left(\frac{x}{2}\right)^{-n} \sum_{k=0}^{n-1} \frac{(n-k-1)! \cos[(3n+2k)\pi/4](x/2)^{2k}}{k!}$$

$$+\frac{1}{2}\left(\frac{x}{2}\right)^{n} \sum_{k=0}^{\infty} \frac{[\psi(k+1)+\psi(n+k+1)]\cos[(3n+2k)\pi/4](x/2)^{2k}}{k!(n+k)!} \qquad (14.3.15)$$

$$-\ln(x/2)\ber_n x + \frac{\pi}{4}\bei_n x$$

$$\kei_n x = -\frac{1}{2}\left(\frac{x}{2}\right)^{-n} \sum_{k=0}^{n-1} \frac{(n-k-1)! \sin[(3n+2k)\pi/4](x/2)^{2k}}{k!}$$

$$+\frac{1}{2}\left(\frac{x}{2}\right)^{n} \sum_{k=0}^{\infty} \frac{[\psi(k+1)+\psi(n+k+1)]\sin[(3n+2k)\pi/4](x/2)^{2k}}{k!(n+k)!} \qquad (14.3.16)$$

$$-\ln(x/2)\bei_n x - \frac{\pi}{4}\ber_n x$$

where the ψ function is given in Section 6.2.1. For $n = 0$ these simplify to

$$\ker x = \ker_0 x = -\ln(x/2)\ber x + \frac{\pi}{4}\bei x + \sum_{k=0}^{\infty} \frac{(-1)^k \psi(2k+1)(x^2/4)^{2k}}{[(2k)!]^2}$$

$$\kei x = \kei_0 x = -\ln(x/2)\bei x - \frac{\pi}{4}\ber x + \sum_{k=0}^{\infty} \frac{(-1)^k \psi(2k+2)(x^2/4)^{2k+1}}{[(2k+1)!]^2} \qquad (14.3.17)$$

and for $n = 1$ we have

$$\ker_1 x = \frac{-1}{x\sqrt{2}} - \ln(x/2)\ber_1 x + \frac{\pi}{4}\bei_1 x$$

$$+\frac{x}{4}\sum_{k=0}^{\infty} \frac{\cos[(3+2k)\pi/4][\psi(k+1)+\psi(k+2)](x/2)^{2k}}{k!(1+k)!}$$

$$\kei_1 x = \frac{-1}{x\sqrt{2}} - \ln(x/2)\bei_1 x - \frac{\pi}{4}\ber_1 x \qquad (14.3.18)$$

$$+\frac{x}{4}\sum_{k=0}^{\infty} \frac{\sin[(3+2k)\pi/4][\psi(k+1)+\psi(k+2)](x/2)^{2k}}{k!(1+k)!}$$

These expressions are used to start the upward iteration on n when x is small enough.

Recurrence relations for $\ker_n$ and $\kei_n$, obtained from [Abr64, 9.9.14], are

$$\ker_{n+1} x = -\frac{n\sqrt{2}}{x}[\ker_n x - \kei_n x] - \ker_{n-1} x$$

$$\kei_{n+1} x = -\frac{n\sqrt{2}}{x}[\kei_n x + \ker_n x] - \kei_{n-1} x \qquad (14.3.19)$$

applied with increasing n, which is the direction insensitive to subtractive cancellation when x is small compared with n, as illustrated in Visualization. Since we need only $n = 0$ and $n = 1$ for starting values, the x values for rapid convergence of the power series in (14.3.17) and (14.3.18) to a given accuracy need be established only once; this is discussed below under Test Values.

Asymptotic series for $\ker_n$ and $\kei_n$ for $x > 0$ are given in [Abr64, 9.10.3, 9.10.4] as

$$\ker_n x = \sqrt{\frac{\pi}{2x}}\, e^{-x/\sqrt{2}}\, [f_n(-x)\cos\beta_n - g_n(-x)\sin\beta_n]$$

$$\kei_n x = \sqrt{\frac{\pi}{2x}}\, e^{-x/\sqrt{2}}\, [-f_n(-x)\sin\beta_n - g_n(-x)\sin\beta_n] \tag{14.3.20}$$

in which β_n is given by

$$\beta_n = x/\sqrt{2} + (4n+1)\pi/8 \tag{14.3.21}$$

The occurrence of $e^{-x/\sqrt{2}}$ in the asymptotic expansion suggests that removing such a scale factor will enable $\ker_n$ and $\kei_n$ to be shown over a wide range of x values, as in the following Figure 14.3.7. The auxiliary functions f_n and g_n have the asymptotic expansions (Section 3.4) given by [Abr64, 9.10.6, 9.10.7] as

$$f_n(\pm x) \sim 1 + \sum_{k=1}^{\infty} \frac{(\mp 1)^k (\mu - 1)(\mu - 1)\ldots[(\mu - (2k-1)^2)]}{k!(8x)^k} \cos\left(\frac{k\pi}{4}\right)$$

$$g_n(\pm x) \sim \sum_{k=1}^{\infty} \frac{(\mp 1)^k (\mu - 1)(\mu - 1)\ldots[(\mu - (2k-1)^2)]}{k!(8x)^k} \sin\left(\frac{k\pi}{4}\right) \tag{14.3.22}$$

For $n = 0$ and 1, convergence of these series to 10^{-10} relative accuracy is satisfactory for $x > 15$.

Visualization

In the following graphics we show the irregular functions, $\ker_n$ and $\kei_n$, from several viewpoints; first for small values of x, then for larger x values but with the functions modified to remove their exponential decay with x. Figure 14.3.6 shows $\ker_n$ and $\kei_n$ versus x and n, which helps when devising our algorithm for the C function `KerKei`.

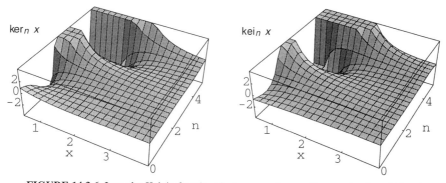

FIGURE 14.3.6 Irregular Kelvin functions in the $x - n$ plane, with x ranging from 0.5 to 4 and n ranging from 0 to 5. The vertical scale ranges from –4 to 4. Note the mesa-shaped parts of the surfaces; these are the truncated remnants of the singularities at $x = 0$. (Adapted from the output of cell `Kelvin` in *Mathematica* notebook `BesselInteger`.)

To illustrate the Kelvin functions for much larger x, we remove the exponential factor that appears in the large-x formula (14.3.22); this gives the billows seen in Figure 14.3.7.

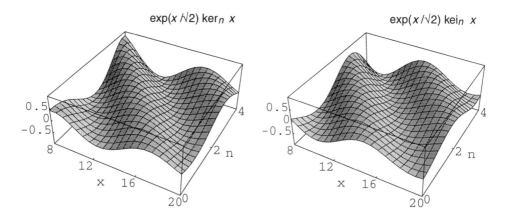

FIGURE 14.3.7 Modified irregular Kelvin functions multiplied by an exponential factor. Shown as functions of x, which ranges from 8 to 20 for each value of n from 0 to 4. The vertical scale ranges from –1 to 1. (Adapted from the output of cell `Kelvin` in *Mathematica* notebook `BesselInteger`.)

The behavior of $\ker_n$ and kei_n as x varies with n fixed is shown in Figure 14.3.8 over the range of x for which the power-series expansion converges rapidly for $n = 0$ and 1.

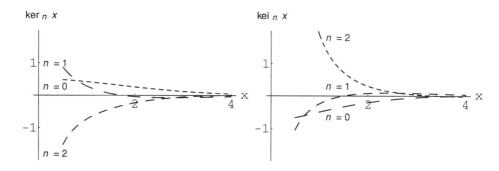

FIGURE 14.3.8 Irregular Kelvin functions versus x, which ranges from 0.5 to 4 for $n = 0$, 1, and 2. The vertical scale ranges from –2 to 2. (Adapted from the output of cell `Kelvin` in *Mathematica* notebook `BesselInteger`.)

Figure 14.3.9 on the next page justifies using the forward-iteration scheme (14.3.19) in the algorithm for our C function `KerKei`, since the figure shows that $\ker_n$ and kei_n generally increase as n increases, which is therefore the appropriate direction for iterating if relative errors from subtractive cancellation are to be minimized.

The auxiliary functions $f_n(x)$ and $g_n(x)$ defined in (14.3.22) are required for computing the Kelvin functions for large x; they are shown for $n = 0$ and $n = 1$ in Figure 14.3.5 for

x in [20, 100], as used for ber_n and bei_n (Section 14.3.1). We show in Figure 14.3.10 $f_n(-x)$ and $g_n(-x)$ over the same x range, as used for ker_n and kei_n.

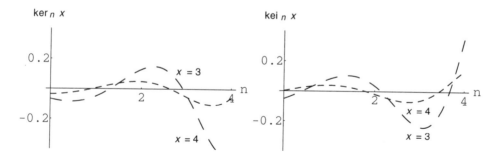

FIGURE 14.3.9 Irregular Kelvin functions versus n, which ranges from 0 to 4 for each value of $x = 3$ and 4. The vertical scale ranges from –0.4 to 0.4. (Adapted from the output of cell Kelvin in *Mathematica* notebook BesselInteger.)

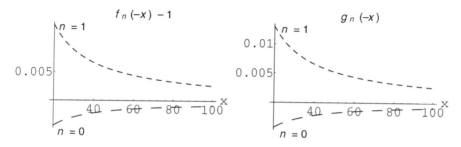

FIGURE 14.3.10 Auxiliary functions for asymptotic expansion of irregular Kelvin functions, as in (14.3.22) with negative arguments. We show $n = 0$ and 1 with x ranging from 20 to 100. (Adapted from the output of cell Kelvin in *Mathematica* notebook Bessel-Integer.)

Algorithm and Program

Function KerKei to compute into arrays ker and kei the irregular Kelvin functions is very similar in structure to the function BessKN for the irregular hyperbolic Bessel function $K_n(x)$ in Section 14.2.4, as you would surmise from the relation (14.3.13). Our annotated driver program KerKei.c in Section 21.14 shows how to input the values of n_m (the maximum order of the Kelvin function) and x. We describe under Test Values how to optimize the choice of n_m to obtain a given accuracy in each $\text{ker}_n x$ and $\text{kei}_n x$ for n from zero up to n_m.

The algorithm uses one of two basic methods based upon the formulas given above in Background and Function Definition. For $x \leq n_m^2$ we start with the series expansions (14.3.17) and (14.3.18) for $n = 0$ and 1. Table 14.3.2 provides guidance for the appropriate choices of n_m for using the power-series method. For larger x values (that is, by choosing n_m small enough for practical use) the asymptotic expansions in (14.3.20)–(14.3.22) are used for $n = 0$ and 1. In either case, iteration then follows for increasing n up to n_m.

The C function KerKei for computing the arrays ker and kei of $\text{ker}_n x$ and $\text{kei}_n x$ values is listed on the following pages; it uses functions BerBei, Fn, and Gn, which are described in Section 14.3.1.

```
/* maximum number of table entries
   allowing for 6 inaccurate starting values   */
#define NMAX 50

void KerKei(int nKMax, double x,
 double ker[], double kei[])
/* Make a table of irregular Kelvin Functions
   ker n x  and  kei n x
   from order  n = 0  to  nKMax <= NMAX-1  */

{
double Fn(int n, double x),Gn(int n, double x),
 ber[NMAX],bei[NMAX],psi[NMAX],fac[NMAX],fnm2,
 sumr,sumi,fk,xD2,xD22,xD24,lnx2,xmult,phase,
 angle,term,RxRt2,beta,kfact,kf0,kg0,kf1,kg1,
 kcos,ksin,tnDx,dtnDx;
const double PiD2 = 1.570796326794897,
 PiD4 = 0.7853981633974483,
 PiD8 = 0.3926990816987241,
 EulerG = 0.5772156649015329,
 /* root 2 to 16 digits for 10-digit accuracy */
 Rt2 = 1.414213562373095;
int k,n;
void BerBei(int nKMax, double x,
 double ber[], double bei[]);

fnm2 = nKMax*nKMax;

if ( x <=  fnm2 )
 { /* ker[0] & ker[1] from series and regular
      Kelvin functions  ber & bei
      order n = 0 & 1   */

 BerBei(nKMax,x,ber,bei);

 /* Tables of  psi  & factorial functions */
 psi[1] = - EulerG; /* is psi(1) */
 fac[0] = 1.0;
 fk = 1.0;
 for ( k = 1; k < nKMax; k++ )
  {
  psi[k+1] = psi[k] + 1.0/fk;
  fac[k] = fac[k-1]*fk;
  fk += 1.0;
  }

 xD2 = 0.5*x;
 xD22 = xD2*xD2;
 xD24 = xD22*xD22;
 lnx2 = log(xD2);

 /* ker & kei of order n = 0 */
 xmult = 1.0;
 sumr = 0.0;
 sumi = 0.0;
 phase = 1.0;
```

```
for ( k = 0; k < nKMax/2; k++ )
  { sumr += phase*psi[2*k+1]*xmult/
            (fac[2*k]*fac[2*k]);
    sumi += phase*psi[2*k+2]*xmult/
            (fac[2*k+1]*fac[2*k+1]);
    xmult *= xD24;
    phase = - phase; }
ker[0] = -lnx2*ber[0] + PiD4*bei[0]+sumr;
kei[0] = -lnx2*bei[0] - PiD4*ber[0]+xD22*sumi;

/* ker & kei of order n = 1 */
angle = 3.0*PiD4;
xmult = 1.0;
sumr = 0.0;
sumi = 0.0;
for ( k = 0; k < nKMax/2; k++ )
  { term = (psi[k+1]+psi[k+2])*xmult/
           (fac[k]*fac[k+1]);
    sumr += cos(angle)*term;
    sumi += sin(angle)*term;
    angle += PiD2;
    xmult *= xD22; }
RxRt2 = -1.0/(x*Rt2);
ker[1] = RxRt2 - lnx2*ber[1] +
         PiD4*bei[1] + 0.25*x*sumr;
kei[1] = RxRt2 - lnx2*bei[1] -
         PiD4*ber[1] + 0.25*x*sumi;

}

else
  {   /* f & g  series for larger  x  */
  kfact = sqrt(PiD2/x)*exp(-x/Rt2);
  beta = x/Rt2 + PiD8;
  kf0 = Fn(0,-x);   kg0 = Gn(0,-x);
  kf1 = Fn(1,-x);   kg1 = Gn(1,-x);
  kcos = cos(beta);   ksin = sin(beta);
  ker[0] =   kfact*(kf0*kcos - kg0*ksin);
  kei[0] = - kfact*(kf0*ksin + kg0*kcos);
   /* For n=1 re-use n=0 cos & sin */
  ker[1] = - kfact*(kf1*ksin + kg1*kcos);
  kei[1] = - kfact*(kf1*kcos - kg1*ksin);
  }

  /* Forward iteration up to  n = nKMax   */
  tnDx = 0.0;
  dtnDx = - Rt2/x;

  for ( n = 1; n < nKMax; n++ )
   {
   tnDx += dtnDx;
   ker[n+1] = tnDx*(ker[n] - kei[n]) - ker[n-1];
   kei[n+1] = tnDx*(kei[n] + ker[n]) - kei[n-1];
   }
return;
}
```

Programming Notes

Irregular Kelvin functions, $\ker_n x$ and $\ker_n x$, are the real and imaginary parts of an irregular Bessel function of complex argument, as in (14.3.13); the recurrence relation (14.3.19) shows this. It would be easier to program `KerKei` in complex arithmetic, but it is more efficient in execution to generate $\ker_n x$ and $\ker_n x$ in parallel, as we do.

In most C functions in the *Atlas*, numerical constants are given to 12 decimal digits, this usually being sufficient for 10-digit accuracy in function values. In `KerKei`, however, 16-digit accuracy of $\sqrt{2}$, as used in (14.3.19), is required to assure 10 digits of accuracy in the irregular Kelvin functions.

The auxiliary functions for the asymptotic series, f_n and g_n of negative argument $-x$ in (14.3.22), are shown in Figure 14.3.10 for $n = 0$ and 1. They are computed to 12-digit relative accuracy in functions `Fn` and `Gn`, respectively. If higher accuracy is needed, increase the parameter `eps` in these functions. As for asymptotic series throughout the *Atlas* (see the discussion in Section 3.4), the convergence of the series is monitored and an error message is printed if the terms start to increase in magnitude before the requested tolerance is obtained. Functions `Fn` and `Gn` are also used for computing the regular Kelvin functions, $\ker_n x$ and $\ker_n x$ in Section 14.3.1, where their programs are listed. If you need these functions at the same x value as the irregular functions, then include arrays `ber` and `bei` in the argument list of `KerKei`; the values of $\ker_n x$ and $\ker_n x$ will then be available to the function that calls `KerKei`.

Parameter `NMAX` is used in the C driver program (Section 21.14) to assign the size of the arrays `ker` and `kei` that hold the values of $\ker_n x$ and $\ker_n x$. The driver program compares the maximum value of n requested, `nKMax`, with `NMAX` to ensure that enough storage is available. The rule relating `nKMax`, x, and the fractional accuracy of the irregular Kelvin functions computed by the forward-iteration method, ε, is discussed under Test Values below. Table 14.3.6 gives sample values of the accuracy control parameters.

Test Values

Test values of $\ker_n x$ and $\ker_n x$ can be generated by running the `Kelvin` cell in the *Mathematica* notebook `BesselInteger`. Examples are given in Table 14.3.4 for $\ker_n x$ and in Table 14.3.5 for $\ker_n x$. Choose `nKMax = 40` for $x = 2$ to 8 to exercise the series method in the algorithm. Enter `nKMax = 6` for $x = 40$ to check the method for asymptotic series.

To compute $\ker_n x$ and $\ker_n x$ efficiently and accurately, we should optimize the choice of the largest n value for computing $\ker_n x$ and $\ker_n x$, as well as the upper limit, $k = N$, on the sums to infinity in (14.3.17) and (14.3.18) when computing the irregular functions for $n = 0$ and $n = 1$. Below Table 14.3.5 we discuss how to do this.

TABLE 14.3.4 Test values to 10 digits of the irregular Kelvin function ker.

$x \backslash n$	0	3	6
2.0	$-4.166451399 \times 10^{-2}$	$2.980215340 \times 10^{-1}$	$-1.183666000 \times 10^{1}$
4.0	$-3.617884790 \times 10^{-2}$	$-5.207111358 \times 10^{-2}$	$-6.147570225 \times 10^{-1}$
6.0	$-6.530375082 \times 10^{-4}$	$-1.144989901 \times 10^{-2}$	$-6.229629537 \times 10^{-2}$
8.0	$1.485834068 \times 10^{-3}$	$2.676688137 \times 10^{-4}$	$-3.151431990 \times 10^{-3}$
40.0	$-9.474811649 \times 10^{-14}$	$-5.130762219 \times 10^{-14}$	$1.070702912 \times 10^{-13}$

TABLE 14.3.5 Test values to 10 digits of the irregular Kelvin function kei.

$x \setminus n$	0	3	6
2.0	$-2.024000678 \times 10^{-1}$	$-8.868206985 \times 10^{-1}$	$-5.851979713 \times 10^{1}$
4.0	$2.198399295 \times 10^{-3}$	$-6.051816818 \times 10^{-2}$	$-6.218562685 \times 10^{-1}$
6.0	$7.216491544 \times 10^{-3}$	$4.511489668 \times 10^{-3}$	$5.322361059 \times 10^{-3}$
8.0	$3.695839557 \times 10^{-4}$	$2.263321265 \times 10^{-3}$	$7.027541105 \times 10^{-3}$
40.0	$4.011081399 \times 10^{-14}$	$-9.889096434 \times 10^{-14}$	$-9.250554124 \times 10^{-14}$

To optimize the efficiency of computing $\ker_n x$ and $\kei_n x$, first consider optimizing the choice of largest n value for $\ber_n x$ and $\bei_n x$, as in Table 14.2.7 and the discussion before it. Except when $\varepsilon = 10^{-6}$ and $x = 1$, the upper-limit values given there are smaller than those required for the k summation limit for $\ker_n x$ and $\kei_n x$ with $n = 0$ or 1, so accuracy is usually limited by the latter. For the summation limits, consider $\ker_0 x$ in (14.3.17) and make an analysis similar to that below Table 14.2.6 for $K_n(x)$. We thus obtain for $\ker_0 x$ the approximate condition for fractional accuracy ε, given by

$$2 N_0 \, [\, 2 \ln x - \ln(8 N_0) + 1] + \ln x / 2 + x / \sqrt{2} \approx \ln(\sqrt{2 \pi^3} \, \varepsilon) \qquad (14.3.23)$$

in which N_0 is the largest value of k needed. Noting that k in (14.3.17) corresponds to half that in (14.3.15) and (14.3.16), we take a typical N as twice N_0. Table 14.3.6 gives typical values of N predicted from (14.3.23). By running our C driver program `KerKei.c` in Section 21.14, you will find the values N_c, which are close to the N values predicted, but are pessimistic for $x > 3$.

TABLE 14.3.6 Maximum order of the irregular Kelvin functions, N, as a function of argument x and the fractional accuracy, ε; see (14.3.23).

ε	x	$N = 2N_0$	N_c
10^{-6}	1	6	8
	3	16	16
	5	30	22
	7	48	30
10^{-10}	1	10	12
	3	20	22
	5	36	28
	7	56	38

Accuracy formula (14.3.23) is not reliable for much larger x than in Table 14.3.6, since the approximations made in its derivation then break down. If you need $\ker_n x$ and $\kei_n x$ for large x, modify function `KerKei` so that it uses the asymptotic expansions in (14.3.20); the fractional accuracy of these expansions is set to be 10^{-12}. Note that it cannot be made of arbitrary accuracy for any given x, since it is an asymptotic expansion; see the discussion in Section 3.4.

14.4   BESSEL FUNCTIONS OF HALF-INTEGER ORDER

Section 14.1 of the *Atlas* provides a tour of Bessel functions, considering them mainly from graphical aspects. In particular, Figures 14.1.3, 14.1.4, and 14.1.6 show functions of order n, where n is treated as a continuous variable. This section develops visualizations and C functions for computing Bessel functions whose order is $v = n+1/2$, where n is an integer; these are spherical Bessel functions (Sections 14.4.1–14.4.4). The Airy functions, for which $v = 1/3$, are described in Section 14.5.

Both regular and irregular functions are considered for the spherical and modified spherical Bessel functions. Each subsection begins by giving relevant background and formulas for computing a function numerically, then we view the function from several perspectives using *Mathematica* graphics that you can also produce, before presenting the algorithm and the C function for its efficient numerical computation.

Analytical results for spherical Bessel functions are given in Section 3.4 of Watson's treatise [Wat44], in Chapter 8 of Jahnke and Emde [Jah45], in Section 7.2.6 of Erdélyi et al [Erd53b], and in Section 8.46 of Gradshteyn and Ryzhik [Gra94]. Formulas for numerical computing are in Sections 10.1–10.3 of Abramowitz and Stegun [Abr64].

14.4.1   Regular Spherical Bessel Function *j*

Keywords
regular spherical Bessel function

C Function and *Mathematica* Notebook

 SphereBessJ (C); cells `Bessel_j` and `SphereBesselPQ` in notebook `Bessel-Half` (*Mathematica*)

Background and Function Definition
The spherical Bessel function occurs most often in problems using spherical polar coordinates. In this context the variable $x = kr$, in which k is a wavenumber and r is the radial coordinate. The regular function, $j_n(x)$, satisfies the differential equation

$$x^2 j_n'' + 2x j_n' + [x^2 - n(n+1)] j_n(x) = 0 \qquad (14.4.1)$$

This can be solved in terms of the regular Bessel function of half-integer order as the standardized solution

$$j_n(x) = \sqrt{\frac{\pi}{2x}}\, J_{n+1/2}(x) \qquad (14.4.2)$$

A recurrence relation for $j_n(x)$ given by Abramowitz and Stegun [Abr64, 10.1.19] is

$$j_{n-1}(x) + j_{n+1}(x) = \frac{2n+1}{x}\, j_n(x) \qquad (14.4.3)$$

For analytical work, and for numerical applications when n is small, it is useful to have explicit formulas, which can be generated from (14.4.3) by starting with $j_0(x)$ and $j_1(x)$ to generate the expressions in Table 14.4.1. By running the cell `Bessel_j` in *Mathematica* notebook `BesselHalf` you can extend this table. Such explicit expressions are useful if you need numerical values for specific low-order functions. If they are coded efficiently they will usually produce accurate values faster than our general method described under Algorithm and Program.

Table 14.4.1 Algebraic expressions for regular spherical Bessel functions of order $n = 0$ to $n = 9$. The notation is $r = 1/x$, $s = \sin x$, and $c = \cos x$.

n	$j_n(x)$	n	$j_n(x)$
0	$rs = (1/x)\sin x$	5	$cr(-1 + 105r^2 - 945r^4) +$ $15r^2(1 - 28r^2 + 63r^4)s$
1	$-cr + r^2 s$	6	$21cr^2(-1 + 60r^2 - 495r^4) +$ $r(-1 + 210r^2 - 4725r^4 + 10395r^6)s$
2	$-3cr^2 + r(-1 + 3r^2)s$	7	$cr(1 - 378r^2 + 17325r^4 - 135135r^6) +$ $7r^2(-4 + 450r^2 - 8910r^4 + 19305r^6)s$
3	$cr(1 - 15r^2) +$ $3r^2(-2 + 5r^2)s$	8	$9cr^2(4 - 770r^2 + 30030r^4 - 225225r^6) +$ $r(1 - 630r^2 + 51975r^4 - 945945r^6 + 2027025r^8)s$
4	$5cr^2(2 - 21r^2) +$ $r(1 - 45r^2 + 105r^4)s$	9	$cr(-1 + 990r^2 - 135135r^4 + 4729725r^6 - 34459425r^8)$ $+45r^2(1 - 308r^2 + 21021r^4 - 360360r^6 + 765765r^8)s$

Note that the compact notation in Table 14.4.1—for example, $j_0(x) = rs$ rather than $j_0(x) = \sin x / x$—is appropriate for programming; one just makes the assignments $r = 1/x$, $s = \sin x$, and $c = \cos x$, then codes the algebraic statements in the table, which will be efficient of coding effort, memory, and execution time. However, beware of subtractive cancellations as x approaches zero; for small x, use the power series (14.4.4).

Equation (14.4.2) can also be solved by a power series solution, which produces— when normalized according to (14.4.2)—the expression [Abr64, 10.1.2]

$$j_n(x) = \frac{x^n}{(2n+1)!!}\left[1 - \frac{x^2/2}{1!(2n+3)} + \frac{(x^2/2)^2}{2!(2n+3)(2n+5)} - \cdots\right] \qquad (14.4.4)$$

which we use in the Algorithm for large n to initiate making an array of Bessel functions. With the standardization indicated in (14.4.2), $j_n(x)$ satisfies the sum rule [Abr64, 10.1.50]

$$\sum_{k=0}^{\infty} (2k+1)j_k^2(x) = 1 \qquad (14.4.5)$$

which may be used to check the spherical Bessel functions.

If $x^2 \le 2n_m + 4$, where n_m is the largest n value of interest for given x, use of (14.4.4) and backward iteration generally gives accurate numerical values. If $x^2 > 2n_m + 4$ another method is needed, however; we use equations 10.1.8 and 10.1.9 in Abramowitz and Stegun [Abr64], namely,

$$j_n(x) = \frac{1}{x}[p(n,x)\sin\chi_n + q(n,x)\cos\chi_n]$$

$$\chi_n = x - n\pi/2$$

(14.4.6)

The auxiliary functions, p and q, have the expansions

$$p(n,x) = \sum_{k=0}^{[n/2]} \frac{(-1)^k(n+2k)!}{(2k)!(n-2k)!(2x)^{2k}}$$

$$q(n,x) = \sum_{k=0}^{[(n-1)/2]} \frac{(-1)^k(n+2k+1)!}{(2k+1)!(n-2k-1)!(2x)^{2k+1}} \quad n>0 \qquad q(0,x)=0$$

(14.4.7)

in which $[i]$ denotes the integer part of i. The functions $p(n,x)$ and $q(n,x)$, which change quite slowly with x, are shown in Figure 14.4.3 below for $n = 0$ and 1, which are used to start upward iteration on n to compute $j_n(x)$ for higher n values.

Visualization

Views of regular spherical Bessel functions of real argument, $j_n(x)$, are seen in Figures 14.4.1 and 14.4.2. In the first figure note that $j_n(x)$ has an imaginary part if $x < 0$ and n is not an integer; we do not consider the latter possibility in our C function, so $j_n(x)$ is then purely real.

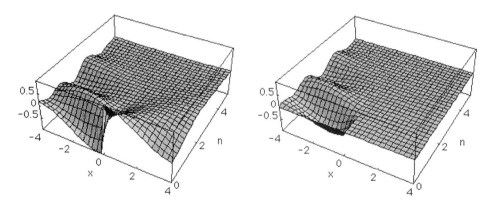

FIGURE 14.4.1 Regular spherical Bessel functions versus x and n. The left side shows the real part and the right side shows the imaginary part (Adapted from the output of cell `Bessel_j` in *Mathematica* notebook `BesselHalf`.)

Figure 14.4.2 shows transections of $j_n(x)$, as slices for constant n (left side) and as slices for constant x (right side). Note that for modest values of x, $j_n(x)$ generally increases as n decreases; this is the clue for part of the accurate and efficient computational algorithm following.

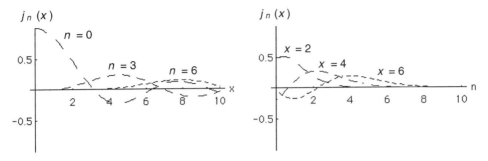

FIGURE 14.4.2 Regular spherical Bessel functions versus x with n fixed (right side) or versus n with x fixed (right side). (Adapted from the output of cell `Bessel_j` in *Mathematica* notebook `BesselHalf`.)

Expression (14.4.6) requires (14.4.7) for $p(n,x)$ and $q(n,x)$. These functions—which are polynomials in the variable $1/(2x)$—are shown in Figure 14.4.3.

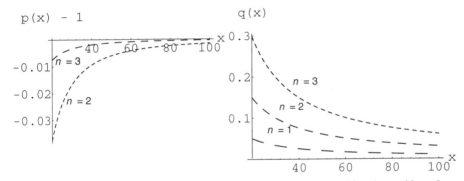

FIGURE 14.4.3 Auxiliary functions for computing spherical Bessel functions, with $n = 2$ and 3 for p, and $n = 1, 2, 3$ for q. (Adapted from the output of cell `BesselPQ` in *Mathematica* notebook `BesselHalf`.)

Convergence of $p(n,x)$ to unity and of $q(n,x)$ to zero is very slow if $n > 1$; agreement with the result $j_n(x) \xrightarrow[x \to \infty]{} \sin \chi_n / x$ to accuracy ε is attained if $x > n(n+1)/(2\varepsilon)$. Thus, for accuracy of 10^{-6} this simple formula holds for $n = 2$ only if $x > 3{,}000{,}000$.

Algorithm and Program

From the preceding discussion and visualizations we can devise a simple algorithm to compute an array of $j_n(x)$ with n varying and x fixed, which is often how the values are needed in Fourier-Bessel expansions. For a modest value of x, a sufficiently large n value, n_m, can be found such that the first two terms in series (14.4.4) give a fair approximation to the magnitude and relative size of two adjacent regular spherical Bessel functions:

$$j_{n_m-1}(x) := \frac{x^{n_m-1}}{(2n_m-1)!!}\left[1 - \frac{x^2}{2(2n_m+1)}\right]$$

$$j_{n_m}(x) := \frac{x^{n_m}}{(2n_m+1)!!}\left[1 - \frac{x^2}{2(2n_m+3)}\right]$$

(14.4.8)

in which the " := " stands for a value assignment. If we now iterate toward decreasing n by using (14.4.3), then the relative error between any two adjacent computed values of $j_n(x)$ is quickly damped out; typically, ten iterations result in 10^{-10} relative accuracy of values that are proportional to $j_n(x)$; these are stored in an array labelled by n. We terminate the iteration after $n = 1$. The array elements are then normalized by multiplying each by the common factor jNorm in function SphereBessJ below, so that agreement with $j_0(x)$ is obtained. This produces an array holding $j_0(x)$, $j_1(x)$, ..., $j_{n_m}(x)$.

In function SphereBessJ the value of nLarge $= n_m + 10$ is used after n_m is input by the program user, as shown in the annotated C driver program in Section 21.14. Under Test Values we show how to optimize n_m to obtain a given accuracy in each $j_n(x)$.

For $x^2 > 2n_m + 4$ the above algorithm is not accurate for computing $j_n(x)$. We use instead the expressions (14.4.6) and (14.4.7) with $n = 0$ and 1 to start the iteration; this is tracked toward increasing n, which is now the stable direction. Note that $p(n, x)$ and $q(n, x)$ are finite expansions; if you need the regular spherical Bessel function $j_n(x)$ for a single n value, it is thus efficient and accurate to use these expressions. For this reason, we program the general expressions in (14.4.7) rather than use the special values $p(0, x) = 1$, $p(1, x) = 1$, $q(0, x) = 0$, and $q(1, x) = 1/x$.

The following are the C functions for computing the array of $j_n(x)$ values, Sphere-BessJ, and the $p(n, x)$ and $q(n, x)$, functions pp and qq, respectively.

```
/* maximum number of table entries,
     allowing for 10 inaccurate starting values   */
#define NMAX 41

void SphereBessJ(int nJMax, double x, double jn[])
/* Make a table of regular spherical Bessel
     functions from order  n = 0  to  nJMax  */
{
double pp(int n, double x),qq(int n, double x);
double xs,dblFctrl,tn,tnDx,dtnDx,jzero,jNorm,
 chi0,chi1;
const double PiD2 = 1.57079632679;
int nLarge,n;

if ( x == 0.0 ) /* treat this specially */
  {
  jn[0] = 1.0;
  for ( n = 1; n <= nJMax; n++ )
    { jn[n] = 0.0; }
  return;
  }

xs = x*x;
if ( xs <= ( (double) (2*nJMax+4) )  )
  {
  nLarge = nJMax+10;
  dblFctrl = 1.0; /* Make double factorial */
  tn = 3.0;
  for ( n = 1; n <= nLarge-1; n++ )
    {
    dblFctrl *= tn;
    tn += 2.0;
    }
```

```
/* Starting approximations */
jn[nLarge-1] = pow(x, (double) (nLarge-1)
)/dblFctrl;
jn[nLarge] = x*jn[nLarge-1]*(1.0 -
0.5*xs/(tn+2.0))/tn;
jn[nLarge-1] *= (1.0 - 0.5*xs/tn);

/* Use backward iteration down to  n = 0   */
tnDx = ( (double) (2*nLarge+1) )/x;
dtnDx = 2.0/x;
for ( n = nLarge-1; n >= 1; n-- )
  {
  tnDx -= dtnDx;
  jn[n-1] = tnDx*jn[n] - jn[n+1];
  }

/* Normalize by comparison with  j0(x)  */
/*   Get  j0  accurate for small  x  */
jzero = ( fabs(x) > 0.001 ) ?  sin(x)/x:
        /* else series accurate to < 10^(-17) */
        1.0-xs*(1.0-xs*(1.0-xs/42.0)/20.0)/6.0;
jNorm = jzero/jn[0];
for ( n = 0; n <= nJMax; n++ )
  { jn[n] *= jNorm; }
return;
}

else

{  /* forward iteration from p & q series */

/* Starting values */
chi0 = x;
chi1 = x - PiD2;
jn[0] = (pp(0,x)*sin(chi0) + qq(0,x)*cos(chi0))/x;
jn[1] = (pp(1,x)*sin(chi1) + qq(1,x)*cos(chi1))/x;

/* Use forward iteration up to  n = nJMax    */
tnDx = 1.0/x;
dtnDx = 2.0/x;
for ( n = 1; n < nJMax; n++ )
  {
  tnDx += dtnDx;
  jn[n+1] = tnDx*jn[n] - jn[n-1];
  }
return;
}
}


double pp(int n, double x)
/* p series for spherical Bessel function  */
{
double fx2,nP2kM1,nMtkP2,tkM1,term,sum;
int kmax,k;

kmax = n/2;
if ( kmax == 0 )  { return  1.0; }
```

```
fx2 = 4.0*x*x;
nP2kM1 = n - 1;
nMtkP2 = n+2;
tkM1 = -1.0;
term = 1.0;
sum = 1.0;
for ( k = 1; k <= kmax; k++ )
  {
  nP2kM1 += 2.0;
  nMtkP2 -= 2.0;
  tkM1 += 2.0;
  term *= - nP2kM1*(nP2kM1+1.0)*nMtkP2*(nMtkP2-1.0)/
          (tkM1*(tkM1+1.0)*fx2);
  sum += term;
  }
return sum;
}


double qq(int n, double x)
/*  q series for spherical Bessel function  */
{
double fx2,nP2k,nMtk,tk,term,sum;
int kmax,k;

if ( n == 0 )   { return  0.0; }
if ( n == 1 )   { return  1.0/x; }
if ( n == 2 )   { return  3.0/x; }

kmax = (n-1)/2;
fx2 = 4.0*x*x;
nP2k = n;
nMtk = n;
tk = 0;
term = ( (double) (n*(n+1)) )/(2.0*x);
sum = term;
for ( k = 1; k <= kmax; k++ )
  {
  nP2k += 2.0;
  nMtk -= 2.0;
  tk += 2.0;
  term *= - nP2k*(nP2k+1.0)*nMtk*(nMtk+1.0)/
          (tk*(tk+1.0)*fx2);
  sum += term;
  }
return sum;
}
```

Programming Notes

The parameter NMAX is used in the C driver program (Section 21.14) to assign the size of array jn that holds the values of $j_n(x)$. The driver program compares the maximum value of n requested, nJMax, with NMAX to ensure that enough storage is available, after allowing for ten startup values in the backward recurrence. The rule relating nJMax and x is discussed under Test Values.

Test Values

Check values of $j_n(x)$ can be generated by running the `Bessel_j` cell in *Mathematica* notebook `BesselHalf`; examples are given in Table 14.4.2. Set `nJMax = 20` for $x = 0$ to 6 to exercise the first method in the algorithm, and `nJMax = 6` for $x = 20$ to exercise the second method.

TABLE 14.4.2 Test values to 10 digits of the regular spherical Bessel functions.

$x \setminus n$	0	3	6
0	1.000000000	0	0
2.0	0.4546487134	0.06072209766	0.0004140409734
4.0	−0.1892006238	0.2292438580	0.01746216868
6.0	−0.04656924970	0.1366851630	0.09379607104
20.0	0.04564726254	0.006030359081	−0.04129999977

To compute $j_n(x)$ efficiently and accurately one should optimize the parameter n_m, the maximum order used in the backward-iteration scheme. Noting that successive terms in the power series (14.4.4) are roughly in the ratio of powers of $x^2/[2(2n+4)]$ divided by factorials, suggests that $j_n(x)$ will change from predominantly a single power of x—the prefactor in (14.4.4)—to an oscillatory behavior when $x^2 > 2n+4$. We therefore use this as the break point between the backward iteration method ($x^2 \leq 2n_m+4$) and the forward iteration method starting with (14.4.6) for $n = 0$ and $n = 1$, used for $x^2 > 2n_m+4$. On the basis of many test runs, we find that in the first method ten iterations are typically required to correct for inaccuracies in the ratio of the two starting values in (14.4.8); we therefore set `nLarge = nJMax+10` when backward iteration is used. The choice of n_m is input by the function user as `nJMax`.

14.4.2 Irregular Spherical Bessel Function y

Keywords

irregular spherical Bessel function

C Function and *Mathematica* Notebook

`SphereBessY` (C); cells `Bessel_y`, `SphereBesselPQ` in notebook `BesselHalf` (*Mathematica*)

Background and Function Definition

The irregular spherical Bessel function, $y_n(x)$, occurs most often in problems involving spherical polar coordinates. In this context, $x = kr$, in which k is a wavenumber and r is the radial coordinate. The irregular function arises when solutions for $r > 0$ are required, as in a boundary condition on a sphere of radius R. Analytical results for $y_n(x)$ are given in Chapter 8 of Jahnke and Emde [Jah45], in Chapter 54 of Spanier and Oldham [Spa87], and in Section 8.46 of Gradshteyn and Ryzhik [Gra94]. Useful formulas for numerical computing are in Sections 10.1–10.3 of Abramowitz and Stegun [Abr64].

Function $y_n(x)$, satisfies the same differential equation as the regular solution, namely,

$$x^2 y_n'' + 2x y_n' + [x^2 - n(n+1)] y_n(x) = 0 \tag{14.4.9}$$

but the solution is singular at $x = 0$. This equation can be solved formally in terms of the irregular Bessel function of half-integer order (Section 14.2.2) as the standardized solution

$$y_n(x) = \sqrt{\frac{\pi}{2x}}\, Y_{n+1/2}(x) \tag{14.4.10}$$

A recurrence relation for $y_n(x)$ is given by Abramowitz and Stegun [Abr64, 10.1.19] as

$$y_{n-1}(x) + y_{n+1}(x) = \frac{2n+1}{x}\, y_n(x) \tag{14.4.11}$$

For analytical work, and for numerical applications when n is small, it is useful to have explicit formulas, which can be generated from (14.4.11) by starting with $y_0(x)$ and $y_1(x)$ to generate the expressions in Table 14.4.3, produced by running cell `Bessel_y` in *Mathematica* notebook `BesselHalf`. They can also be produced readily from the expressions in Table 14.4.1 for $j_n(x)$ by making the transpositions $s \to -c$ and $c \to s$.

Explicit expressions are useful if you need numerical values for specific low-order functions; if coded efficiently they often produce accurate values faster than our general methods described under Algorithm and Program.

TABLE 14.4.3 Algebraic expressions for irregular spherical Bessel functions of order $n = 0$ to $n = 9$. The notation is $r = 1/x$, $s = \sin x$, and $c = \cos x$.

n	$y_n(x)$	n	$y_n(x)$
0	$-cr = -\cos x / x$	5	$sr(-1 + 105r^2 - 945r^4) -$ $15r^2(1 - 28r^2 + 63r^4)c$
1	$-sr^2 - rc$	6	$21sr^2(-1 + 60r^2 - 495r^4) -$ $r(-1 + 210r^2 - 4725r^4 + 10395r^6)s$
2	$-3sr^2 - r(-1 + 3r^2)c$	7	$sr(1 - 378r^2 + 17325r^4 - 135135r^6) +$ $7r^2(-4 + 450r^2 - 8910r^4 + 19305r^6)c$
3	$sr(1 - 15r^2) -$ $3r^2(-2 + 5r^2)c$	8	$9sr^2(4 - 770r^2 + 30030r^4 - 225225r^6) -$ $r(1 - 630r^2 + 51975r^4 - 945945r^6 + 2027025r^8)c$
4	$5sr^2(2 - 21r^2) -$ $r(1 - 45r^2 + 105r^4)c$	9	$sr(-1 + 990r - 135135r^2 + 4729725r^4 - 34459425r^6)$ $-45r(1 - 308r^2 + 21021r^4 - 360360r^6 + 765765r^8)c$

Note that the compact notation in Table 14.4.3—for example, $y_0(x) = -cr$ rather than $y_0(x) = -\cos x / x$—is appropriate for programming; one just assigns $r = 1/x$, $s = \sin x$, and $c = \cos x$, then codes the algebraic statements in the table, which will be efficient of coding effort, memory, and execution time. However, beware of subtractive cancellations as x approaches zero; for small x, use the power series (14.4.12) below.

Equation (14.4.9) can be solved by a power-series solution, which produces—when normalized according to (14.4.10)—the expression [Abr64, 10.1.3]

$$y_n(x) = -\frac{(2n-1)!!}{x^n}\left[1 + \frac{x^2/2}{1!(2n-1)} + \frac{(x^2/2)^2}{2!(2n-1)(2n-3)} + \cdots\right] \quad (14.4.12)$$

which converges rapidly when $x^2 \le 2n$. Another formula that may be useful is equation 10.1.9 in Abramowitz and Stegun [Abr64], namely,

$$y_n(x) = \frac{(-1)^{n+1}}{x}[p(n,x)\cos\chi_n - q(n,x)\sin\chi_n] \quad (14.4.13)$$

$$\chi_n = x + n\pi/2$$

The auxiliary functions, p and q, are given in (14.4.7); they are also used for the regular spherical Bessel functions $j_n(x)$ and are depicted in Figure 14.4.3. Since both $p(n,x)$ and $q(n,x)$ are finite expansions, if you need the irregular spherical Bessel function for a single n value it is efficient and accurate to use these expressions. For this reason, functions pp and qq are given for the general expressions in (14.4.7) in the listing of function Sphere-BessJ in Section 14.4.1.

Visualization

Views of irregular spherical Bessel functions of real argument, $y_n(x)$, are shown in Figures 14.4.4 and 14.4.5. In the first figure note that $y_n(x)$ has an imaginary part if $x < 0$ and n is not an integer; we do not consider the latter possibility in our C function, so $y_n(x)$ is then purely real. The $(n+1)$th-order poles of the $y_n(x)$ as $x \to 0$ are quite evident.

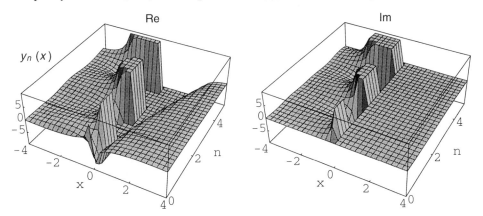

FIGURE 14.4.4 Irregular spherical Bessel functions versus x and n. The left side shows the real part and the right side shows the imaginary part Values whose absolute values are larger than 10 have been truncated. (Adapted from the output of cell Bessel_y in *Mathematica* notebook BesselHalf.)

Figure 14.4.5 shows slices of $y_n(x)$ for constant n (left side) and as slices for constant x (right side). For modest values of x, $y_n(x)$ generally increases as n increases; this suggests part of the accurate computational algorithm following.

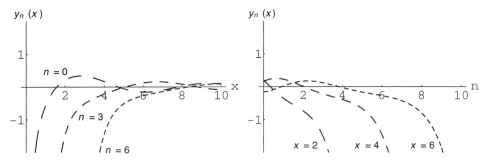

FIGURE 14.4.5 Irregular spherical Bessel functions versus x with n fixed (right side) or versus n with x fixed (right side). (Adapted from the output of cell Bessel_y in *Mathematica* notebook BesselHalf.)

Convergence of $p(n,x)$ to unity and of $q(n,x)$ to zero is very slow if $n > 1$; agreement with the asymptotic formula $y_n(x) \xrightarrow[x \to \infty]{} -\cos \chi_n / x$ to accuracy ε is attained only when $x > n(n+1)/(2\varepsilon)$. For accuracy of 10^{-6} this asymptotic formula holds for $n = 2$ only if $x > 3,000,000$!

Algorithm and Program

The algorithm to compute an array of irregular spherical Bessel functions— $y_n(x)$ with n varying and x fixed—is very simple. We start with

$$y_0(x) = -\cos x / x \qquad y_1(x) = [-\sin x + y_0(x)] / x \qquad (14.4.14)$$

then we iterate toward increasing n by using (14.4.11), to produce an array holding $y_0(x)$, $y_1(x)$, ..., $y_{n_m}(x)$. In the driver program for function SphereBessY the value n_m is input by the program user as nYMax, as shown in the annotated C driver program in Section 21.14. The C function, SphereBessY, computes the array of $y_n(x)$ values. In the program on the CD-ROM that comes with the *Atlas*, the function is given with its driver.

```
/* maximum number of table entries */
#define NMAX 41

void SphereBessY(int nYMax, double x, double yn[])
/* Make a table of irregular spherical Bessel
   functions from order  n = 0  to  nYMax  */
{
double tnDx,dtnDx;
int n;

yn[0] = - cos(x)/x;
yn[1] = ( - sin(x) + yn[0])/x;

/* Forward iteration up to  n = nYMax    */
tnDx = 1.0/x;
dtnDx = 2.0/x;
for ( n = 1; n < nYMax; n++ )
  {
  tnDx += dtnDx;
  yn[n+1] = tnDx*yn[n] - yn[n-1];
  }
return;
}
```

Programming Notes

The parameter NMAX is used in the C driver program (Section 21.14) to assign the size of array yn that holds the values of the $y_0(x)$. The driver program compares the maximum value of n requested, nYMax, with NMAX to ensure that enough storage is available.

Test Values

Test values of $y_0(x)$ can be generated by running the Bessel_y cell in *Mathematica* notebook BesselHalf; examples are given in Table 14.4.4. Set nYMax = 6 to obtain the irregular functions from $n = 0$ to $n = 6$.

TABLE 14.4.4 Test values to 10 digits of the irregular spherical Bessel functions.

$x \setminus n$	0	3	6
2.0	0.2080734183	−1.484366557	−97.79165769
4.0	0.1634109052	−0.2186419659	−1.430950935
6.0	−0.1600283811	0.1217499034	−0.2720957250
20.0	−0.02040410309	0.05001846343	−0.03059727053

14.4.3 Regular Modified Spherical Bessel Function *i*

Keywords

regular modified spherical Bessel function

C Function and *Mathematica* Notebook

SphereBessI (C); cells Bessel_i and BesselR in notebook BesselHalf (*Mathematica*)

Background and Function Definition

Modified spherical Bessel functions occur most often in problems using spherical 3D coordinates in which the variable $ix = ikr$, where k is the magnitude of a wavenumber and r is the radial coordinate. Relevant formulas are given in Section 10.2 of Abramowitz and Stegun [Abr64]. The regular function, which we call $i_n(x)$, satisfies the differential equation

$$x^2 i_n'' + 2x i_n' - [x^2 + n(n+1)]i_n(x) = 0 \qquad (14.4.15)$$

Formally, this can be solved in terms of either the regular hyperbolic Bessel function, I, of half-integer order or in terms of a regular spherical Bessel function of purely imaginary argument

$$i_n(x) \equiv \sqrt{\frac{\pi}{2x}}\, I_{n+1/2}(x) = (-i)^n j_n(ix) \qquad (14.4.16)$$

A recurrence relation for $i_n(x)$ is given by Abramowitz and Stegun [Abr64, 10.2.18] as

$$i_{n-1}(x) - i_{n+1}(x) = \frac{2n+1}{x}\, i_n(x) \qquad (14.4.17)$$

M

For analytical work, and for numerical applications when n is small, it is useful to have explicit formulas. These can be generated using (14.4.17) by starting with $i_0(x)$ and $i_1(x)$ to generate the expressions in Table 14.4.5. By running the cell SphereBessel_I in *Mathematica* notebook BesselHalf you can extend this table if necessary. Explicit expressions are useful if you need numerical values for a low-order function. If the function is coded efficiently it will usually produce accurate values faster than our general method described under Algorithm and Program.

TABLE 14.4.5 Algebraic expressions for regular modified spherical Bessel functions of order $n = 0$ to $n = 9$. The notation is $r = 1/x$, $s = \sinh x$, and $c = \cosh x$.

n	$i_n(x)$	n	$i_n(x)$
0	$rs = (1/x)\sinh x$	5	$cr(1 + 105r^2 + 945r^4) -$ $15r^2(1 + 28r^2 + 63r^4)s$
1	$cr - r^2 s$	6	$-21cr^2(1 + 60r^2 + 495r^4) +$ $r(1 + 210r^2 + 4725r^4 + 10395r^6)s$
2	$-3cr^2 + r(1 + 3r^2)s$	7	$cr(1 + 378r^2 + 17325r^4 + 135135r^6) -$ $7r^2(4 + 450r^2 + 8910r^4 + 19305r^6)s$
3	$cr(1 + 15r^2) -$ $3r^2(2 + 5r^2)s$	8	$-9cr^2(4 + 770r^2 + 30030r^4 + 225225r^6) +$ $r(1 + 630r^2 + 51975r^4 + 945945r^6 + 2027025r^8)s$
4	$-5cr^2(2 + 21r^2) +$ $r(1 + 45r^2 + 105r^4)s$	9	$cr(-1 + 990r + 135135r^2 + 4729725r^4 + 34459425r^6)$ $-45r(1 + 308r^2 + 21021r^4 + 360360r^6 + 765765r^8)s$

!

The compact notation in Table 14.4.5 is appropriate for programming; just make the assignments $r = 1/x$, $s = \sin x$, and $c = \cos x$, then code the algebraic statements in the table, which will be efficient of coding effort, memory, and execution time. Beware of subtractive cancellations as x approaches zero; for small x, use the power series (14.4.18).

Equation (14.4.15) can also be solved by a power-series solution, which produces— when normalized according to (14.4.16)—the expression [Abr64, 10.2.5]

$$i_n(x) = \frac{x^n}{(2n+1)!!}\left[1 + \frac{x^2/2}{1!(2n+3)} + \frac{(x^2/2)^2}{2!(2n+3)(2n+5)} + \ldots\right] \tag{14.4.18}$$

which we use in the Algorithm for large n to initiate making an array of modified Bessel functions. With the standardization in (14.4.16), $i_n(x)$ satisfies the sum rule [Abr64, 10.2.36]

$$\sum_{k=0}^{\infty}(2k+1)i_k(x) = 1 \tag{14.4.19}$$

which may be used to check the modified spherical Bessel functions.

If $x^2 \leq 2n_m + 4$, where n_m is the largest n value of interest for given x, use of (14.4.18) and backward iteration generally gives accurate numerical values. If $x^2 > 2n_m + 4$ another way is needed to compute $i_n(x)$; we use equations 10.2.9 and 10.2.11 given in Abramowitz and Stegun [Abr64]:

$$i_n(x) = \frac{1}{2x}[r(n,-x)\,e^x - (-1)^n\,r(n,x)e^{-x}] \tag{14.4.20}$$

The auxiliary function, r, has the expansion

$$r(n,x) = \sum_{k=0}^{n} \frac{(n+k)!}{k!(n-k)!(2x)^k} \tag{14.4.21}$$

and is shown in Figure 14.4.9 below for $n = 0$ and 1, which are used to start upward iteration on n to compute $i_n(x)$ for higher n values. If you need $i_n(x)$ for a single n value, then build a program using this formula, which is coded as function rr and listed in the following C program.

Visualization

Views of regular modified spherical Bessel functions of real argument, $i_n(x)$, are seen in Figures 14.4.6, 14.4.7, and 14.4.8. In the first figure note that $i_n(x)$ has an imaginary part if $x < 0$ and n is not an integer; we do not consider the latter possibility in our C function, so $i_n(x)$ is then purely real. The range of x displayed is unusually small because $i_n(x)$ grows nearly exponentially with x for large x; see Figure 14.4.8 below.

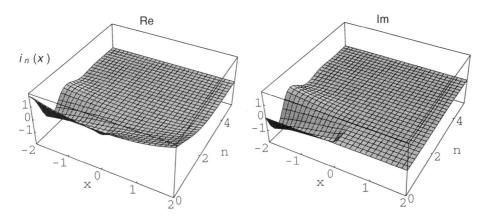

FIGURE 14.4.6 Regular modified spherical Bessel functions versus x and n. The left side shows the real part and the right side shows the imaginary part (Adapted from the output of cell `Bessel_i` in *Mathematica* notebook `BesselHalf`.)

Figures 14.4.7 and 14.4.8 show transections of $i_n(x)$, as slices for constant n or constant x. For modest values of x, $i_n(x)$ generally increases as n decreases; this suggests the accurate and efficient computational algorithm following.

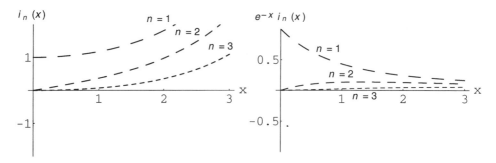

FIGURE 14.4.7 Regular modified spherical Bessel functions versus x with n fixed. Left side shows the function, and right side shows function with exponential of x removed. (Adapted from the output of cell `Bessel_i` in *Mathematica* notebook `BesselHalf`.)

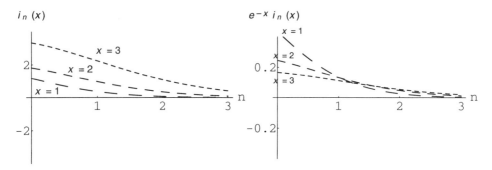

FIGURE 14.4.8 Regular modified spherical Bessel functions versus n with x fixed. Left side shows the function, and right side shows function with exponential of x removed. (Adapted from the output of cell `Bessel_i` in *Mathematica* notebook `BesselHalf`.)

The predominant behavior of $i_n(x)$ becomes exponential as x increases, as suggested by formula (14.4.20). This expression requires (14.4.21) for $r(n,x)$, which is a polynomial in variable $1/(2x)$; it is shown in Figure 14.4.9 for both $x > 0$ and $x < 0$, which are required in (14.4.20).

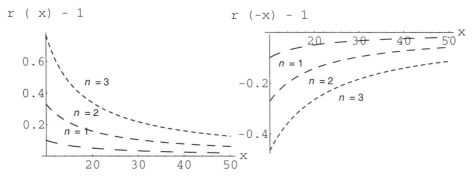

FIGURE 14.4.9 Auxiliary function for computing modified spherical Bessel functions, with $n = 1, 2, 3$. (Adapted from the output of cell `BesselR` in *Mathematica* notebook `BesselHalf`.)

Convergence of $r(n,x)$ to unity as x increases in magnitude is quite slow when $n > 0$; according to (14.4.21), $|r(n,x) - 1| \approx n(n+1)/(2|x|)$.

Algorithm and Program

From the preceding discussion and visualizations one can devise a simple algorithm to compute an array of $i_n(x)$ with n varying and x fixed. For a modest value of x, a sufficiently large n value, n_m, can be found such that the first two terms in series (14.4.18) give a fair approximation to the magnitude and relative size of two adjacent regular modified spherical Bessel functions:

$$i_{n_m-1}(x) := \frac{x^{n_m-1}}{(2n_m-1)!!}\left[1 + \frac{x^2}{2(2n_m+1)}\right]$$

$$i_{n_m}(x) := \frac{x^{n_m}}{(2n_m+1)!!}\left[1 + \frac{x^2}{2(2n_m+3)}\right]$$

(14.4.22)

in which ":=" stands for a value assignment. By iterating toward decreasing n by using (14.4.17), the relative error between any two adjacent computed values of $i_n(x)$ is quickly damped out; typically, six iterations result in 10^{-10} relative accuracy of values that are proportional to $i_n(x)$; these are stored in an array labelled by n. The iteration is terminated after $n = 1$. All array elements are then normalized by multiplying each by the factor iNorm in function SphereBessI below, so that agreement with $i_0(x)$ is obtained. This produces an array holding $i_0(x)$, $i_1(x)$, ..., $i_{n_m}(x)$. In SphereBessI the value of nLarge $= n_m+6$ is used after n_m is input by the program user, as shown in the annotated C driver program in Section 21.14.

For $x^2 > 2n_m + 4$ the above algorithm for $i_n(x)$ is not accurate. Instead, expressions (14.4.20) and (14.4.21) are used with $n = 0$ and 1 to start the iteration; this is tracked to increasing n, which is now the stable direction. Note that $r(n,x)$ is a finite expansion; if you need the regular modified spherical Bessel function $i_n(x)$ for a single n value, it is efficient and accurate to use these expressions. For this reason, we program the general expressions in (14.4.21) rather than use the special values $r(0,x) = 1$ and $r(1,x) = 1 + 1/x$.

The following are the C functions for computing the array of $i_n(x)$ values, SphereBessI, and the $r(n,x)$ function, called rr.

```
/* maximum number of table entries,
      allowing for 6 inaccurate starting values */
#define NMAX 37

void SphereBessI(int nIMax, double x, double in[])
/* Table of regular modified spherical Bessel
      functions from order   n = 0   to   nIMax   */


{
double rr(int n, double x);
double xs,dblFctrl,tn,tnDx,dtnDx,izero,iNorm,expX;
int nLarge,n;
```

```
if ( x == 0.0 ) /* treat this specially */
 {
 in[0] = 1.0;
 for ( n = 1; n <= nIMax; n++ )
  { in[n] = 0.0; }
 return;
 }

xs = x*x;
if ( xs <= ( (double) (2*nIMax+4) )  )
 {
 nLarge = nIMax+6;
 dblFctrl = 1.0; /* Make double factorial */
 tn = 3.0;

 for ( n = 1; n <= nLarge-1; n++ )
  {
  dblFctrl *= tn;
  tn += 2.0;
  }
 /* Starting approximations */
 in[nLarge-1] = pow(x, (double) (nLarge-1)
)/dblFctrl;
 in[nLarge] = x*in[nLarge-1]*(1.0 +
             0.5*xs/(tn+2.0))/tn;
 in[nLarge-1] *= (1.0 + 0.5*xs/tn);

 /* Use backward iteration down to  n = 0   */
 tnDx = ( (double) (2*nLarge+1) )/x;
 dtnDx = 2.0/x;

 for ( n = nLarge-1; n >= 1; n-- )
  {
  tnDx -= dtnDx;
  in[n-1] = tnDx*in[n] + in[n+1];
  }

 /* Normalize by comparison with  i0(x)   */
 /*   Get  i0  accurate for small  x   */
 izero = ( fabs(x) > 0.001 ) ?  sinh(x)/x:
          /* else series accurate to < 10^(-17) */
          1.0+xs*(1.0+xs*(1.0+xs/42.0)/20.0)/6.0;
 iNorm = izero/in[0];

 for ( n = 0; n <= nIMax; n++ )
  { in[n] *= iNorm; }
 return;
 }

else

 {  /* forward iteration from  r  series */

 /* Starting values */
 expX = exp(x);
 in[0] = 0.5*(rr(0,-x)*expX - rr(0,x)/expX)/x;
 in[1] = 0.5*(rr(1,-x)*expX + rr(1,x)/expX)/x;
```

```
    /* Use forward iteration up to   n = nIMax  */
    tnDx = - 1.0/x;
    dtnDx = 2.0/x;

    for ( n = 1; n < nIMax; n++ )
      {
      tnDx -= dtnDx;
      in[n+1] = tnDx*in[n] + in[n-1];
      }
    return;
    }
  }


double rr(int n, double x)
/*   r series for modified
      spherical Bessel function   */
  {
  double tx,nPk,nMkP1,fk,term,sum;
  int k;

  if ( n == 0 )  { return  1.0; }

  tx = 0.5/x;
  nPk = n;
  nMkP1 = n+1;
  fk = 0.0;
  term = 1.0;
  sum = 1.0;

  for ( k = 1; k <= n; k++ )
    {
    nPk += 1.0;
    nMkP1 -= 1.0;
    fk += 1.0;
    term *=  nPk*nMkP1*tx/fk;
    sum += term;
    }

  return  sum;
  }
```

Programming Notes

Parameter NMAX is used in the C driver program (Section 21.14) to assign the size of array in that holds the values of $i_n(x)$. The driver compares the maximum value of n requested, nIMax, with NMAX to ensure that enough storage is available, after allowing for six startup values in the backward recurrence.

Test Values

For most values of n and x, test values of $i_n(x)$ can be generated by running the Bessel_i cell in *Mathematica* notebook BesselHalf; examples are given in Table 14.4.6. Set nIMax = 20 for $x = 0$ to 6 to exercise the first method in the algorithm, and nIMax = 10 for $x = 10$ to exercise the second method.

TABLE 14.4.6 Test values to 10 digits of the regular modified spherical Bessel functions.

$x \setminus n$	0	3	6
1.0	1.175201194	$1.006509052 \times 10^{-2}$	$7.650333778 \times 10^{-6}$ *
2.0	1.813430204	$9.474252220 \times 10^{-2}$	$5.405952086 \times 10^{-4}$ *
3.0	3.339291642	0.4152875766	$7.245037363 \times 10^{-3}$
10.0	$1.101323287 \times 10^{3}$	$5.892079640 \times 10^{2}$	$1.309968827 \times 10^{2}$

In Table 14.4.6 the two values marked with $*$ disagree with those computed by *Mathematica*. The values computed by our C function `SphereBessI` are given here, since the values computed by this function agree those in Table 10.9 in Abramowitz and Stegun [Abr64] for $n = 9$ (8 digits) and for $n = 10$ (7 digits). The discrepancy with *Mathematica* values increases as x decreases; for $x = 1$ and $n = 10$ *Mathematica* calculates zero.

14.4.4 Irregular Modified Spherical Bessel Function k

Keywords
irregular modified spherical Bessel function

C Function and *Mathematica* Notebook
`SphereBessK` (C); cell `Bessel_k` in notebook `BesselHalf` (*Mathematica*)

Background and Function Definition
Irregular modified spherical Bessel functions, $k_n(x)$, occur in problems involving spherical polar coordinates, with the variable $ix = ikr$, in which k is the magnitude of a wavenumber and r is the radial coordinate. The modified functions occur when damped behavior with r is described, and the irregular function arises when solutions for $r > 0$ are required, as in a boundary condition on a sphere of radius R. Analytical properties of $k_n(x)$ are listed in Section 10.2 of Abramowitz and Stegun [Abr64], where it is referred to as the modified spherical Bessel function of the third kind.

Function $k_n(x)$ satisfies the same differential equation as the regular solution, namely

$$x^2 k_n'' + 2x k_n' + [x^2 - n(n+1)]k_n(x) = 0 \tag{14.4.23}$$

but the solution is singular at $x = 0$. This equation can be solved formally in terms of an irregular Bessel function of half-integer order (Section 14.2.4) as the standardized solution

$$k_n(x) = \sqrt{\frac{\pi}{2x}} \, K_{n+1/2}(x) \tag{14.4.24}$$

A recurrence relation for $k_n(x)$ can be inferred from Abramowitz and Stegun [Abr64, 10.2.18] as

$$k_{n-1}(x) - k_{n+1}(x) = -\frac{2n+1}{x} k_n(x) \tag{14.4.25}$$

For analytical work and for numerical applications when n is small, explicit formulas are useful. They can be generated by using (14.4.25), starting with $k_0(x)$ and $k_1(x)$, which are given by

$$k_0(x) = \frac{\pi}{2x} e^{-x} \qquad k_1(x) = \frac{\pi}{2x} e^{-x}[1 + 1/x] \qquad (14.4.26)$$

Expressions in Table 14.4.7 up to $k_9(x)$ can be produced by running cell `Bessel_k` in the *Mathematica* notebook `BesselHalf`. Explicit expressions are useful for numerical values of specific low-order functions; if coded efficiently they often produce accurate results faster than our general methods described under Algorithm and Program.

TABLE 14.4.7 Algebraic expressions for irregular modified spherical Bessel functions of order $n = 0$ to $n = 9$; the notation is $r = 1/x$ and $p = \pi e(-x)/(2x)$.

n	$k_n(x)$	n	$k_n(x)$
0	$p = \pi e^{-x}/(2x)$	5	$p(1 + 15r + 105r^2 + 420r^3 +$ $945r^4 + 945r^5)$
1	$p(1 + r)$	6	$p(1 + 21r + 210r^2 + 1260r^3 +$ $4725r^4 + 10395r^5 + 10395r^6)$
2	$p(1 + 3r + 3r^2)$	7	$p(1 + 28r + 378r^2 + 3150r^3 + 17325r^4 +$ $62370r^5 + 135135r^6 + 135135r^7)$
3	$p(1 + 6r + 15r^2 + 15r^3)$	8	$p(1 + 36r + 630r^2 + 6930r^3 + 51975r^4 + 270270r^5 +$ $945945r^6 + 2027025r^7 + 2027025r^8)$
4	$p(1 + 10r + 45r^2 +$ $105r^3 + 105r^4)$	9	$p(1 + 45r + 990r^2 + 13860r^3 + 135135r^4 + 945945r^5 +$ $4729725r^6 + 16216200r^7 + 34459425r^8 + 34459425r^9)$

The compact notation in Table 14.4.7 is appropriate for programming; just assign $r = 1/x$ and $p = \pi e^{-x}/(2x)$ then code the algebraic statements in the table, which will be efficient of coding effort, memory, and execution time.

Equation (14.4.23) can be solved by a power-series solution, which produces—when normalized according to (14.4.24)—the expression [Abr64, 10.2.15]

$$k_n(x) = -\frac{\pi e^{-x}}{2x} \sum_{k=0}^{n} \frac{(n+k)!}{k!(n-k)!(2x)^k} \qquad (14.4.27)$$

of which the formulas in Table 14.4.7 are examples.

Visualization

Views of irregular modified spherical Bessel functions of real argument, $k_n(x)$, are shown in Figures 14.4.10–14.4.12. In the first figure $k_n(x)$ has an imaginary part if $x < 0$ and n is not an integer; we do not consider the latter possibility in our C function, so $k_n(x)$ is then purely real. The $(n+1)$th-order poles of $k_n(x)$ as $x \to 0$ are quite evident.

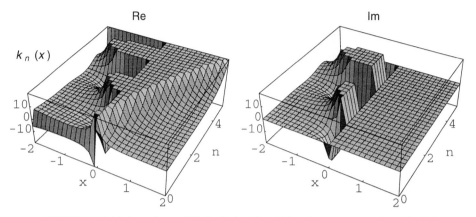

FIGURE 14.4.10 Irregular modified spherical Bessel functions versus x and n. The left side shows the real part and the right side shows the imaginary part. Values whose absolute values are larger than 20 have been truncated. (Adapted from the output of cell `Bessel_k` in *Mathematica* notebook `BesselHalf`.)

Figure 14.4.11 shows slices of $k_n(x)$ for constant n. The left side shows the rapid decline of the function; on the right side an x and an exponential factor to remove most of the x variation when n is small have been inserted. As n increases, $k_n(x)$ increases for any positive value of x, increasing n is therefore always the appropriate direction of iteration for numerical computation.

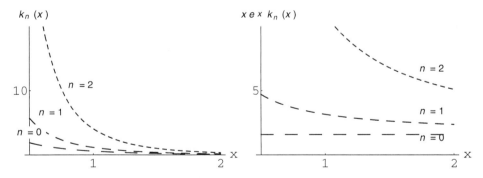

FIGURE 14.4.11 Irregular modified spherical Bessel functions versus x with n fixed. In the right-side view, factors of x and its exponential have been included to remove most of the x dependence. (Adapted from the output of cell `Bessel_k` in *Mathematica* notebook `BesselHalf`.)

In Figure 14.4.12 we show how $k_n(x)$ varies with n for constant x. As n increases, $k_n(x)$ increases for any positive value of x, confirming that increasing n is the appropriate direction of iteration.

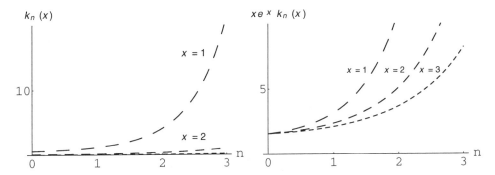

FIGURE 14.4.12 Irregular modified spherical Bessel functions versus n with x fixed. In the left-side view, $x = 1$ and 2. In the right-side view, factors of x and its exponential have been included to remove most of the x dependence, which is shown for $x = 1$, 2, and 3. (Adapted from the output of cell Bessel_k in *Mathematica* notebook BesselHalf.)

Algorithm and Program

The algorithm to compute an array of irregular modified spherical Bessel functions—$k_n(x)$ with n varying and x fixed—is very simple. We start with the values for $n = 0$ and 1, given in (14.4.26) then we iterate toward increasing n by using (14.4.25), to produce an array holding $k_0(x)$, $k_1(x)$, ..., $k_{n_m}(x)$.

In the driver program for function SphereBessK the value n_m is input by the user as nKMax, as shown in the annotated C driver program in Section 21.14. The C function computes an array of $k_n(x)$ values. In the program on the CD-ROM that comes with the *Atlas*, the function is given with its driver.

```c
/* maximum number of table entries */
#define NMAX 41

void SphereBessK(int nKMax, double x, double kn[])
/* Make a table of irregular spherical Bessel
   functions from order  n = 0  to  nKMax  */
{
double tnDx,dtnDx;
const double PiD2 = 1.57079632679;
int n;

kn[0] = PiD2*exp(-x)/x;
kn[1] = kn[0]*(1.0+1.0/x);
/* Forward iteration up to  n = nKMax   */
tnDx = 1.0/x;
dtnDx = 2.0/x;

for ( n = 1; n < nKMax; n++ )
  {
  tnDx += dtnDx;
  kn[n+1] = tnDx*kn[n] + kn[n-1];
  }
return;
}
```

Programming Notes

The parameter `NMAX` is used in the C driver program (Section 21.14) to assign the size of array `kn` that holds the values of the $k_n(x)$. The driver program compares the maximum value of n requested, `nKMax`, with `NMAX` to ensure that enough storage is available.

Test Values

Check values of $k_n(x)$ can be generated by running the `Bessel_k` cell in *Mathematica* notebook `BesselHalf`; examples are given in Table 14.4.4. Set `nKMax = 6` to obtain the irregular functions from $n = 0$ to $n = 6$.

TABLE 14.4.8 **Test values to 10 digits of the irregular modified spherical Bessel functions.**

$x \setminus n$	0	3	6
1.0	$5.778636749 \times 10^{-1}$	$2.138095597 \times 10^{1}$	$1.560636427 \times 10^{4}$
2.0	$1.062920829 \times 10^{-1}$	1.023061298	$1.067255553 \times 10^{2}$
3.0	$2.606844804 \times 10^{-2}$	$1.361352286 \times 10^{-1}$	5.040865452
10.0	$7.131404291 \times 10^{-6}$	$1.258692857 \times 10^{-5}$	$5.025390067 \times 10^{-5}$

14.5 AIRY FUNCTIONS

The Airy functions, Ai and Bi, occur widely in science, engineering, and mathematics, from the quantum mechanics of the linear potential, through the diffraction of waves, and the bending of plates under inclined loading, to asymptotic solutions of second-order differential equations. Since Airy functions are varieties of Bessel functions of order $v = 1/3$, they are covered in the following subsection. The derivatives of the functions with respect to their argument, Ai$'$ and Bi$'$, are also often used; these functions are described in Section 14.5.2.

Analytical results for Airy functions are given in Section 8.9 of Hochstadt [Hoc86], and in Chapter 56 of Spanier and Oldham [Spa87]. Formulas useful for numerical computing are given in Section 10.4 of Abramowitz and Stegun [Abr64].

14.5.1 Airy Functions Ai, Bi

Keywords
Airy functions

C Function and *Mathematica* Notebook
`AiryAiBi` (C); cells `Airy`, `Airy.f&g`, and `AiryAsymptote` in notebook `Bessel-Airy` (*Mathematica*)

Function Definitions
The Airy functions occur often as solutions of the differential equation,

$$w'' - zw(z) = 0 \tag{14.5.1}$$

of which Ai and Bi are two linearly-independent solutions. This equation can be solved in terms of the Bessel functions of order $v = 1/3$ as the standardized solutions

$$\text{Ai}(z) \equiv \frac{\sqrt{z}}{3}\left[I_{-1/3}(\zeta) - I_{1/3}(\zeta)\right] = \frac{\sqrt{z/3}}{\pi} K_{1/3}(\zeta)$$

$$\text{Bi}(z) \equiv \sqrt{z/3}\left[I_{-1/3}(\zeta) + I_{1/3}(\zeta)\right]$$

(14.5.2)

in terms of the variable

$$\zeta = 2z^{3/2}/3$$

(14.5.3)

One may also write

$$\text{Ai}(-z) = \frac{\sqrt{z}}{3}\left[J_{-1/3}(\zeta) + J_{1/3}(\zeta)\right]$$

$$\text{Bi}(-z) = \sqrt{z/3}\left[J_{-1/3}(\zeta) - J_{1/3}(\zeta)\right]$$

(14.5.4)

When z is real—as in our numerical applications below—the Bessel functions I (Section 14.2.3), K (Section 14.2.4), and J (Section 14.2.1) are real-valued functions of real arguments; the only novelty is that their orders are $v = \pm 1/3$ in the Airy functions.

For numerical work the above expressions are not practicable. We use instead—with $z = x$, a real variable—ascending series for x suitably small. The series are given in terms of auxiliary functions $f(x)$ and $g(x)$ by expressions 10.4.2 and 10.4.3 in Abramowitz and Stegun [Abr87], namely

$$\text{Ai}(x) = c_1 f(x) - c_2 g(x)$$

$$\text{Bi}(x) = \sqrt{3}\left[c_1 f(x) + c_2 g(x)\right]$$

(14.5.5)

Here the constants are given by

$$c_1 = 3^{-2/3} / \Gamma(2/3) = 0.35502\,80538\,87817$$

$$c_2 = 3^{-1/3} / \Gamma(1/3) = 0.25881\,94037\,92807$$

(14.5.6)

and the auxiliary functions are

$$f(x) = 1 + 1x^3/3! + 1.4x^6/6! + \ldots = \sum_{k=0}^{\infty} \frac{1.4.7.(3k-2)x^{3k}}{(3k)!}$$

$$g(x) = x\left[1 + 2x^3/4! + 2.5x^6/7! + \ldots\right] = x\sum_{k=0}^{\infty} \frac{2.5.8.(3k-1)x^{3k}}{(3k+1)!}$$

(14.5.7)

These functions are shown in Figure 14.5.4 below.

If the convergence of the series in (14.5.7) is too slow, one can use the asymptotic expansions. For $x \gg 0$ [Abr64, 10.4.59, 10.4.63] gives

$$\text{Ai}(x) \sim \frac{e^{-\zeta}}{2\sqrt{\pi}x^{1/2}} D(\zeta) \qquad D(\zeta) = \sum_{k=0}^{\infty} (-1)^k t_k / \zeta^k$$

$$\text{Bi}(x) \sim \frac{e^{\zeta}}{\sqrt{\pi}x^{1/2}} E(\zeta) \qquad E(\zeta) = \sum_{k=0}^{\infty} t_k / \zeta^k$$

(14.5.8)

showing that $Ai(x)$ eventually decays nearly exponentially and that $Bi(x)$ increases nearly exponentially, as shown in Figure 14.5.3 below. The t_k in this and following equations are given by

$$t_k = \frac{(2k+1)(2k+3)\ldots(6k-1)}{216^k \, k!} \qquad k > 0 \qquad t_0 = 1 \qquad (14.5.9)$$

For large negative arguments [Abr64, 10.4.60, 10.4.64] gives

$$Ai(-x) \sim \frac{1}{\sqrt{\pi x^{1/2}}} \left[\sin(\zeta + \pi/4)\,F(\zeta) - \cos(\zeta + \pi/4)\,G(\zeta) \right]$$

$$(14.5.10)$$

$$Bi(-x) \sim \frac{1}{\sqrt{\pi x^{1/2}}} \left[\cos(\zeta + \pi/4)\,F(\zeta) + \sin(\zeta + \pi/4)\,G(\zeta) \right]$$

which indicate oscillatory behavior for the Airy functions. The auxiliary functions, F and G, formally given by

$$F(\zeta) = \sum_{k=0}^{\infty} (-1)^k c_{2k} / \zeta^{2k} \qquad G(\zeta) = \sum_{k=0}^{\infty} (-1)^k c_{2k+1} / \zeta^{2k+1} \qquad (14.5.11)$$

are depicted in Figure 14.5.5 below. In these equations ζ is related to $z = x$ by (14.5.3).

Visualization

Perspective views of the Airy functions, Ai and Bi, are shown in Figures 14.5.1 and 14.5.2, respectively. Note that for $x < 0$ and $y \neq 0$ the functions grow rapidly with y, so that the behavior of the functions along the real line is not clearly visible.

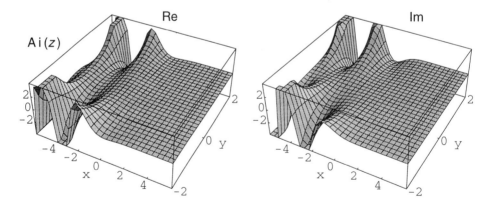

FIGURE 14.5.1 The Airy function, Ai, in the complex plane. The left side shows the real part and the right side shows the imaginary part for x in [-5, 5] and y in [–2, 2]. (Adapted from the output of cell `Airy` in *Mathematica* notebook `BesselAiry`.)

Figure 14.5.3 shows Ai and Bi along the real line, x —the argument for which we provide the C functions. The oscillations for $x < 0$ and the exponential-type behaviors for $x > 0$ are clear.

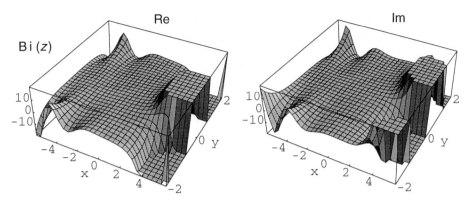

FIGURE 14.5.2 The Airy function, Bi, in the complex plane. The left side shows the real part and the right side shows the imaginary part for x in [-5, 5] and y in [-2, 2]. (Adapted from the output of cell `Airy` in *Mathematica* notebook `BesselAiry`.)

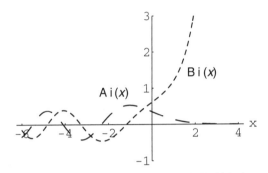

FIGURE 14.5.3 Airy functions along the real line in [-2, 2]; Ai is shown by long dashes and Bi is shown by short dashes. (Adapted from the output of cell `Airy` in *Mathematica* notebook `BesselAiry`.)

Finally, taking the breeze with the Airy family, one can view in Figures 14.5.4 and 14.5.5 the auxiliary functions for the power series and asymptotic expansions.

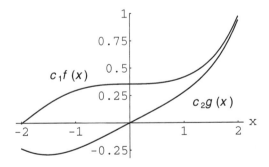

FIGURE 14.5.4 Auxiliary functions for computing Airy functions by power series expansions (14.5.7) in [-2, 2]. (Adapted from the output of cell `Airy` in *Mathematica* notebook `BesselAiry`.)

The *Mathematica* calculations for Figure 14.5.4 use expressions that are more compact than (14.5.7); following (5.17.3) in Lebedev's text on special functions [Leb65], we write

$$c_1 f(x) = \sum_{k=0}^{\infty} \frac{x^{3k}}{3^{2k+2/3} k! \Gamma(k+2/3)}$$

$$c_2 g(x) = \sum_{k=0}^{\infty} \frac{x^{3k+1}}{3^{2k+4/3} k! \Gamma(k+4/3)}$$

(14.5.12)

In cell `Airy.f&g` of the annotated *Mathematica* notebook `BesselAiry` (Section 20.14) we show how to use these expressions to generate these auxiliary functions. Note in Figure 14.5.4 that as x increases, the two functions approach each other; calculation of Ai(x) by the power series then becomes increasingly sensitive to subtractive cancellation when x exceeds 2.

When x is as small as 2, the auxiliary asymptotic functions F and G in (14.5.10) cannot reasonably be computed by using (14.5.11), either for visualizing them by *Mathematica* or for obtaining accurate numerical values by using a C function; only for very large x or ζ is the asymptotic expansion accurate enough. Instead, we use the built-in functions `AiryAi` and `AiryBi` in *Mathematica* to calculate accurate values of Ai and Bi, then from these determine F and G by inverting (14.5.10) to produce

$$F(\zeta) = \sqrt{\pi x^{1/2}} \left[\sin(\zeta + \pi/4) \mathrm{Ai}(-x) + \cos(\zeta + \pi/4) \mathrm{Bi}(-x) \right]$$

$$G(\zeta) = \sqrt{\pi x^{1/2}} \left[\sin(\zeta + \pi/4) \mathrm{Bi}(-x) - \cos(\zeta + \pi/4) \mathrm{Ai}(-x) \right]$$

(14.5.13)

For the C functions we find rational approximations to F and G thus obtained; we make use of functions in *Mathematica* package `NumericalMath`Approximations.`` The details of this procedure are given under Algorithm and Program. The auxiliary functions F and G in terms of variable ζ, (14.5.3), are shown in the following figure.

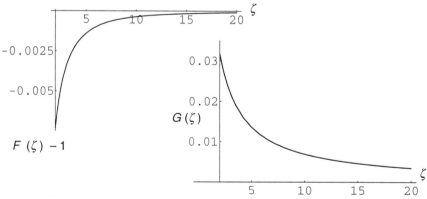

FIGURE 14.5.5 Auxiliary functions F and G for computing Airy functions by expansions (14.5.10). (Adapted from the output of cell `Airy.f&g` in *Mathematica* notebook `BesselAiry`.)

Algorithm and Program

From the preceding discussion and visualizations it is clear that computing the Airy functions Ai and Bi from the power series (14.5.7) when $|x| \le 2$ is straightforward, but for x

(or ζ) outside this range the algorithm is more challenging. We use the method summarized above Figure 14.5.5; that is, we fit Ai and Bi by rational approximations to D and E in (14.5.8) for $\zeta > 2$, and to F and G in (14.5.11) for $\zeta < -2$.

Each of the four rational approximations in ζ is of the form

$$R_{p/q}(\zeta) = \sum_{k=0}^{p} n_k \zeta^k \bigg/ \sum_{k=0}^{q} d_k \zeta^k \qquad (14.5.14)$$

in which the fitted coefficients n_k and d_k are determined by using the *Mathematica* function `RationalInterpolation`. The details of this can be seen in cell `AiryAsymptotes` annotated in Section 20.14. Table 14.5.1 summarizes the fitting parameters and the accuracy obtained when the fit is made over $|\zeta|$ in [2, 20], that is over $|x|$ in [2.080, 9.655].

TABLE 14.5.1 Fitting parameters for rational approximations to auxiliary functions to compute Airy functions of large argument.

Function, R	p	q	Maximum error, at ζ_m	ζ_m
D	5	6	-8.4×10^{-13}	3.6
E	9	10	-9.1×10^{-12}	5.2
F	5	6	-3.2×10^{-12}	3.1
G	5	6	-9.6×10^{-14}	4.4

The fitting coefficients are given in the C function `AiryAiBi` that follows. The fractional accuracy in the Airy functions described by the rational approximations for $\zeta > 2$ (the exponential region for Ai and Bi) is shown in Figure 14.5.6. Here $\varepsilon_{Ai}(\zeta)$ denotes the fractional error in Ai(x) and $\varepsilon_{Bi}(\zeta)$ denotes the fractional error in Bi(x); the fitting is done in ζ space, related to $z = x$ by (14.5.3). Note that the errors are much less than our accuracy goal of 10 digits for functions in the *Atlas*.

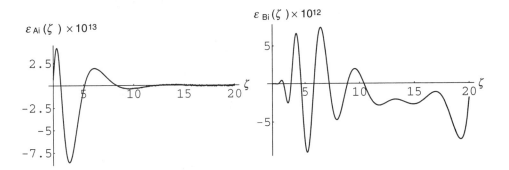

FIGURE 14.5.6 Fractional accuracy in computing Airy functions of positive argument by rational-fraction approximations for $\zeta < -2$. (Adapted from the output of cell `Airy-Asymptotes` in *Mathematica* notebook `BesselAiry`.)

Figure 14.5.7 shows the accuracy in the Airy functions of negative argument—the oscillatory region of $x < 0$ for Ai and Bi, as in (14.5.10)—described by the rational approximations for $\zeta > 2$. Fitting is done in ζ space, and the errors shown are actual errors. Note that they are two orders of magnitude less than our accuracy goal of 10 digits for functions in the *Atlas*.

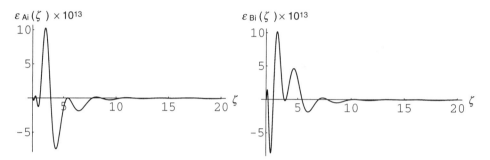

FIGURE 14.5.7 Accuracy in computing Airy functions of negative argument by rational-fraction approximations for $\zeta < -2$ (Adapted from the output of cell `AiryAsymptotes` in *Mathematica* notebook `BesselAiry`.)

The following is the C function `AiryAiBi` that computes Ai and Bi.

```
void AiryAiBi(double x, double eps,
  double *Ai, double *Bi)
/* Airy Functions  Ai  and  Bi  */
{
double x3,tk,fterm,fsum,fratio,gterm,gsum,gratio,
  zeta,nsum,dsum,DD,EE,expZeta,RSPiX,FF,GG,
  angle,Sin,Cos;
const double c1 = 0.355028053887817,
  c2 = 0.258819403792807, Rt3 = 1.73205080757,
  Pi = 3.14159265359, PiD4 = 0.785398163397,
  Dn[6] = {0.6223951529912393,18.406233696146327,
            90.62792720436782,136.22806191525729,
            72.26279462926934,11.691570161073225},
  Dd[7] = {1.0,22.093953355321673,98.14106331593061,
            140.86855055266716,73.07470903544198,
            11.691570165350335,4.249088927932299e-11},
  En[10] = {0.8256034744925347,2.883139497692721,
            2.1635619064914819,0.5401804200326157,
            0.6183717774780627,-0.03425732498078304,
            0.0493234227752643,-0.0026157454280489184,
            0.0008066890401987603,
            0.3380663415058702e-4},
  Ed[11] = {1.0,1.9602145386554288,2.4213535313953605,
            0.4230947450530819,0.6299743421502649,
            -0.0386826193746995,0.0495486788287488,
            -0.0026760752382712361,0.0008044364077155,
            0.3380498458461015e-4,
            1.2773999631503219e-11},
```

```
Fn[6] = {0.7087121120473341,7.760802902936213,
         10.923195558527917,6.956569411782221,
          1.9666881246587482,0.3104548161086819},
Fd[7] = {1.0,8.002143608605754,10.996184129739234,
         6.968099984002047,1.9666880434589649,
         0.3104548177549655,-1.4582338808688533e-11},
Gn[6] = {0.16704907184893067,0.3987010737972128,
         0.3027697203859155,0.10968156356297849,
         0.020300514040201847,0.0016257843492067754},
Gd[7] = {1.0,4.415035500383613,6.571238680327184,
         4.519929360200246,1.5922186312671058,
         0.2923274950430756,0.023411293712262138};
int k;

if ( fabs(x) <= 2.0 ) /* use power series */
 {
 x3 = x*x*x;
 tk = 0.0;
 fterm = 1.0;
 fsum = 1.0;
 fratio = 10.0;
 gterm = 1.0;
 gsum = 1.0;
 gratio = 10.0;

 while ( (fratio > eps) || (gratio > eps) )
  {
  tk += 3.0;
  fterm *= x3/(tk*(tk-1.0)); /* for  f  series */
  fsum += fterm;
  fratio = fabs(fterm/fsum);
  gterm *= x3/(tk*(tk+1.0)); /* for  g  series */
  gsum += gterm;
  gratio = fabs(gterm/gsum);
  }
 fsum *= c1;
 gsum *= c2*x;
 *Ai = fsum - gsum;
 *Bi = Rt3*(fsum + gsum);
 return; /* values for  |x| <= 2.5  */
 }

if ( x > 2.0 )
 { /* use rational approximations to D and E */
 zeta = 2.0*pow(x,1.5)/3.0;

 nsum = 0.0; /*  D approximation */
 dsum = Dd[6];

 for ( k = 5; k >= 0; k-- )
  {
  nsum = nsum*zeta+Dn[k];
  dsum = dsum*zeta+Dd[k];
  }
 DD = nsum/dsum;
```

```
      nsum = 0.0; /*   E approximation */
      dsum = Ed[10];
      for ( k = 9; k >= 0; k-- )
        {
        nsum = nsum*zeta+En[k];
        dsum = dsum*zeta+Ed[k];
        }
      EE = nsum/dsum;

      expZeta = exp(zeta);
      RSPiX = 1.0/sqrt(Pi*sqrt(x));
      *Ai = RSPiX/(2.0*expZeta)*DD;
      *Bi = RSPiX*expZeta*EE;
      return; /* values for  x > 2.0   */
      }
  /* x < -2.0;
       use rational approximations to F and G */
      zeta = 2.0*pow(-x,1.5)/3.0;

      nsum = 0.0; /*   F approximation */
      dsum = Fd[6];
      for ( k = 5; k >= 0; k-- )
        {
        nsum = nsum*zeta+Fn[k];
        dsum = dsum*zeta+Fd[k];
        }
      FF = nsum/dsum;

      nsum = 0.0; /*   G approximation */
      dsum = Gd[6];
      for ( k = 5; k >= 0; k-- )
        {
        nsum = nsum*zeta+Gn[k];
        dsum = dsum*zeta+Gd[k];
        }
      GG = nsum/dsum;

      RSPiX = 1.0/sqrt(Pi*sqrt(-x));
      angle = zeta+PiD4;
      Sin = sin(angle);
      Cos = cos(angle);
      *Ai = RSPiX*(Sin*FF - Cos*GG);
      *Bi = RSPiX*(Cos*FF + Sin*GG);
      return; /* values for  x < -2.0  */
      }
```

Programming Notes

Program `AiryAiBi` returns two values, `Ai` and `Bi`, through its argument list. The accuracy of these values for the x range $[-2, 2]$ in which the power series in (14.5.7) are used is controlled by the function user through input parameter `eps`, which gives the fractional accuracy required in `Ai` and `Bi`. The series is summed until both series meet this requirement; for `eps` $= 10^{-10}$ no more than ten terms are required.

For $|x| > 2$ the rational approximations described under Algorithm and Program are used. Although they require several high-accuracy coefficients to define them, the polynomials in (14.5.14) are evaluated efficiently by using the Horner algorithm. The approximations have hard-wired accuracy of about 10^{-11} or better, so they do not use the accuracy parameter eps; it must, however, be nonzero in the driver program shown in Section 21.14 in order that program execution continue.

Test Values
Check values of Ai and Bi can be generated by running the Airy cell in *Mathematica* notebook BesselAiry; examples are given in Table 14.5.2 for x values covering the three algorithms used.

TABLE 14.5.2 **Test values to 10 digits of the Airy functions Ai and Bi.**

x	-6	-4	-2
Ai	-0.3291451736	-0.07026553295	0.2274074282
Bi	-0.1466983767	0.3922347057	-0.412302588
x	2	4	6
Ai	0.03492413042	0.0009515638512	$9.94769436 \times 10^{-6}$
Bi	3.298095000	83.84707141	6536.446105

14.5.2 Derivatives of Airy Functions Ai′, Bi′

Keywords
Airy functions, derivatives of Airy functions

C Function and *Mathematica* Notebook
AiryPrimeAiBi (C); cells Airy, Airy.fp&gp, and AiryPrimeAsymptotes in the notebook BesselAiry (*Mathematica*)

Background and Function Definitions
Derivatives of Airy functions with respect to their arguments, Ai′ and Bi′, are often used in science and engineering, as summarized in the introduction to Section 14.5. Analytical results are given in Section 8.9 of Hochstadt [Hoc86], and in Chapter 56 of Spanier and Oldham [Spa87]. Formulas applicable to numerical computing are given in Section 10.4 of Abramowitz and Stegun [Abr64]. The Airy functions, Ai and Bi, are described in Section 14.5.1 of the *Atlas*.

Formally, Ai′ and Bi′ can be written in terms of Bessel functions of order $v = 1/3$, as in Abramowitz and Stegun [Abr64, 10.4.16, 10.4,17, 10.4.20, 10.4.21]

$$\text{Ai}'(z) = -\frac{z}{3}\left[I_{-2/3}(\zeta) - I_{2/3}(\zeta)\right] = \frac{z}{\pi\sqrt{3}} K_{2/3}(\zeta)$$

$$\text{Bi}'(z) = \frac{z}{\sqrt{3}}\left[I_{-2/3}(\zeta) + I_{2/3}(\zeta)\right]$$

$$(14.5.15)$$

in terms of the variable

$$\zeta = 2z^{3/2}/3 \qquad (14.5.16)$$

For numerical work with $z = x$, a real variable, we use methods similar to those for Ai and Bi. The power series expansions for small x are given in terms of the derivatives of the auxiliary functions $f(x)$ and $g(x)$ in (14.5.7) by

$$\mathrm{Ai}'(x) = c_1 f'(x) - c_2 g'(x)$$
$$\mathrm{Bi}'(x) = \sqrt{3}\left[c_1 f'(x) + c_2 g'(x)\right] \qquad (14.5.17)$$

with the constants given by

$$c_1 = 3^{-2/3}/\Gamma(2/3) = 0.35502\,80538\,87817$$
$$c_2 = 3^{-1/3}/\Gamma(1/3) = 0.25881\,94037\,92807 \qquad (14.5.18)$$

and the derivative functions are

$$f'(x) = \sum_{k=1}^{\infty} t_k \qquad t_1 = x^2/2 \qquad t_k/t_{k-1} = x^3/[(3k-1)(3k-3)]$$
$$g'(x) = \sum_{k=0}^{\infty} t_k \qquad t_0 = 1 \qquad t_k/t_{k-1} = x^3/[3k(3k-2)] \qquad (14.5.19)$$

These functions are shown in Figure 14.5.11 below.

If the convergence of these series is too slow, one could try using asymptotic expansions, but their convergence is poor, as described for Ai and Bi in Section 14.5.1. We therefore devise rational approximations for the derivative functions. For $x \gg 0$, following Abramowitz and Stegun [Abr64, 10.4.61, 10.4.66], one has

$$\mathrm{Ai}'(x) = -\frac{z^{1/4} e^{-\zeta}}{2\sqrt{\pi}} D_1(\zeta) \qquad \mathrm{Bi}'(x) = \frac{z^{1/4} e^{\zeta}}{\sqrt{\pi}} E_1(\zeta) \qquad (14.5.20)$$

For large negative arguments, following [Abr64, 10.4.62, 10.4.67], one writes

$$\mathrm{Ai}'(-x) = -\frac{x^{1/4}}{\sqrt{\pi}}\left[\cos(\zeta + \pi/4)F_1(\zeta) + \sin(\zeta + \pi/4)G_1(\zeta)\right]$$
$$\mathrm{Bi}'(-x) = \frac{x^{1/4}}{\sqrt{\pi}}\left[\sin(\zeta + \pi/4)F_1(\zeta) - \cos(\zeta + \pi/4)G_1(\zeta)\right] \qquad (14.5.21)$$

The auxiliary functions, D_1, E_1, F_1, and G_1, are to be obtained by fitting them as rational approximations similar to (14.5.14).

Visualization

Perspective views of the derivatives of the Airy functions, Ai' and Bi', are shown in Figures 14.5.8 and 14.5.9, respectively. Since when $x < 0$ and $y \neq 0$ the functions grow rapidly with y, the behavior of the these derivatives along the real line is not clear in these views.

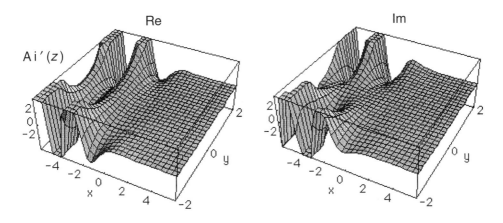

FIGURE 14.5.8 The derivative of the Airy function, Ai, in the complex plane. The left side shows the real part and the right side shows the imaginary part for x in [-5, 5] and y in [-2, 2]. (Adapted from the output of cell `Airy` in *Mathematica* notebook `BesselAiry`.)

In Figure 14.5.9 the approximately exponential increase of Bi′ along the positive real line, x —the argument for which we provide the C functions—makes it difficult to display the derivative beyond $x = 4$; we have truncated the derivative values at 20.

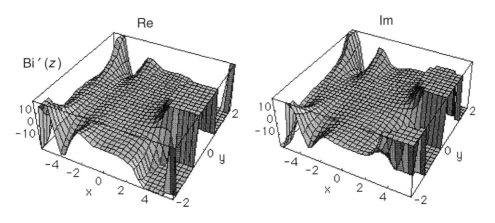

FIGURE 14.5.9 The derivative of the Airy function, Bi, in the complex plane. The left side shows the real part and the right side shows the imaginary part for x in [-5, 5] and y in [-2, 2]. Function values are truncated at 20. (Adapted from the output of cell `Airy` in *Mathematica* notebook `BesselAiry`.)

Figure 14.5.10 shows the derivatives of the Airy functions, Ai′ and Bi′, along the real line, x —the argument for which we provide the C functions. The oscillations for $x < 0$ and the exponential-type behaviors for $x > 0$ are clear.

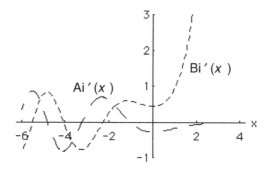

FIGURE 14.5.10 Derivatives of the Airy functions along the real line; Ai′ is shown by long dashes and Bi′ is shown by short dashes. (Adapted from the output of cell `Airy` in *Mathematica* notebook `BesselAiry`.)

Finally for the derivatives of the Airy functions, Figure 14.5.11 shows the auxiliary functions $f′$ and $g′$ for the power series in (14.5.19).

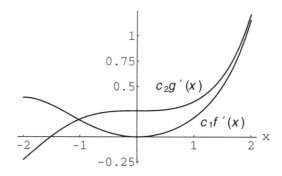

FIGURE 14.5.11 Auxiliary functions for computing derivatives of Airy functions by the power series in (14.5.19). (Adapted from the output of cell `Airy.fp&gp` in *Mathematica* notebook `BesselAiry`.)

The *Mathematica* calculations for Figure 14.5.11 use expressions that are easier to program than (14.5.19); differentiating in (14.5.12), produces

$$c_1 f'(x) = \sum_{k=1}^{\infty} \frac{x^{3k-1}}{3^{2k-1/3}(k-1)!\,\Gamma(k+2/3)}$$

$$c_2 g'(x) = \sum_{k=0}^{\infty} \frac{x^{3k}}{3^{2k+1/3}k!\,\Gamma(k+1/3)}$$

(14.5.22)

In cell `Airy.fp&gp` of the annotated *Mathematica* notebook `BesselAiry` (Section 20.14) we show how to use these expressions to generate the auxiliary derivative functions. Figure 14.5.11 shows that as x increases, the two functions approach each other. Calculation of Ai′(x) by the power series then becomes increasingly sensitive to subtractive cancellation when x exceeds 2; we then switch to another method.

When $|x|$ is as small as 2, the auxiliary asymptotic functions D and E in (14.5.8) or F and G in (14.5.11) cannot be differentiated with respect to x and used reliably in (14.5.21), either for visualizing the derivatives by *Mathematica* or for obtaining accurate numerical values by using a C function; only for very large x or ζ is the asymptotic expansion accurate enough. Rather, we use the built-in functions `AiryAiPrime` and `AiryBiPrime` in *Mathematica* to calculate accurate values of Ai$'$ and Bi$'$, then from these we determine D_1 and E_1 in (14.5.20). For F_1 and G_1 we invert (14.5.21) to produce

$$F_1(\zeta) = \sqrt{\pi / x^{1/2}}\left[\sin(\zeta + \pi/4)\text{Bi}'(-x) - \cos(\zeta + \pi/4)\text{Ai}'(-x)\right]$$

$$G_1(\zeta) = -\sqrt{\pi / x^{1/2}}\left[\sin(\zeta + \pi/4)\text{Ai}'(-x) + \cos(\zeta + \pi/4)\text{Bi}'(-x)\right]$$

(14.5.23)

For the C functions we find rational approximations to F_1 and G_1 by using functions in the *Mathematica* package `NumericalMath`Approximations`.` Details of this procedure are given under Algorithm and Program. The auxiliary functions F_1 and G_1 in terms of variable ζ, (14.5.16), are shown in the following figure. **M**

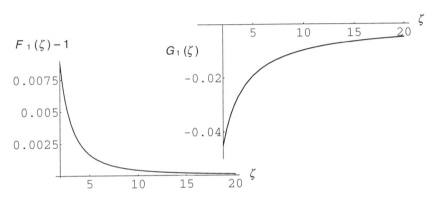

FIGURE 14.5.12 Auxiliary functions F_1 and G_1 for computing derivatives of Airy functions by expansions (14.5.21). (Adapted from the output of cell `Airy.fp&gp` in *Mathematica* notebook `BesselAiry`.)

Algorithm and Program

From the discussion and visualizations above, it is clear that computing the first derivatives of the Airy functions, Ai$'$ and Bi$'$, from the power series in (14.5.19) when $|x| \leq 2$ is straightforward, but for x (or ζ) outside this range the algorithm is more challenging. For $|x| > 2.5$ we use the method summarized above Figure 14.5.12; that is, we fit Ai$'$ and Bi$'$ by rational approximations to D_1 and E_1 in (14.5.20) for $\zeta > 2$, and to F_1 and G_1 in (14.5.21) for $\zeta < -2$. Each rational approximation in ζ is of the form

$$R_{p/q}^{(1)}(\zeta) = \sum_{k=0}^{p} n_k^{(1)} \zeta^k \Bigg/ \sum_{k=0}^{q} d_k^{(1)} \zeta^k$$

(14.5.24)

in which the coefficients $n_k^{(1)}$ and $d_k^{(1)}$ are determined by using the *Mathematica* function `RationalInterpolation`, as seen in the cell `AiryPrimeASymptotes` in Section 20.14. Table 14.5.3 gives the fitting parameters and the accuracy obtained when the fit is made over $|\zeta|$ in [2, 20], that is over $|x|$ in [2.080, 9.655].

TABLE 14.5.3 Fitting parameters for rational approximations to auxiliary functions to compute derivatives of Airy functions.

Function, R	p	q	Maximum error, at ζ_m	ζ_m
D_1	5	6	-2.0×10^{-13}	2.9
E_1	9	10	-2.0×10^{-11}	4.4
F_1	5	6	3.4×10^{-12}	3.1
G_1	5	6	-1.1×10^{-13}	2.5

The fitting coefficients are given in the following C function. The fractional accuracy in the Airy function derivatives described by the rational approximations for $\zeta > 2$ (the exponential region for Ai$'$ and Bi$'$) is shown in Figure 14.5.13. Here $\varepsilon_{\mathrm{Ai}'}(\zeta)$ denotes the fractional error in Ai$'(x)$ and $\varepsilon_{\mathrm{Bi}'}(\zeta)$ denotes the that in Bi$'(x)$; fitting is done in ζ space, related to $z = x$ by (14.5.16). Errors are much less than our accuracy goals of 10 digits.

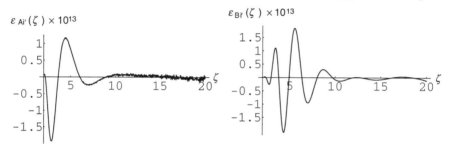

FIGURE 14.5.13 Fractional accuracy in computing derivatives of Airy functions of positive argument by rational-fraction approximations for $\zeta > 2$. (Adapted from the output of cell `AiryPrimeAsymptote` in *Mathematica* notebook `BesselAiry`.)

Figure 14.5.14 shows the fractional accuracy in the derivatives of Airy functions of negative argument—the oscillatory region of $x < 0$ for Ai$'$ and Bi$'$, as in (14.5.21)—described by the rational approximations for $\zeta > 2$. Fitting is done in ζ space, and the errors are actual differences; function minus fit. The errors are two orders of magnitude less than our accuracy goal of 10 digits for functions in the *Atlas*.

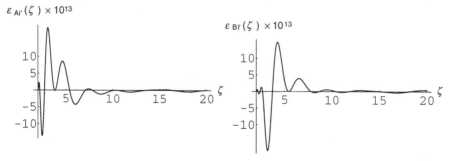

FIGURE 14.5.7 Accuracy in computing Airy functions of negative argument by rational-fraction approximations for $\zeta > 2$. (Adapted from the output of the cell `AiryPrime-Asymptote` in *Mathematica* notebook `BesselAiry`.)

The following is the C function for computing Ai′ and Bi′, `AiryPrimeAiBi`.

```
void AiryPrimeAiBi(double x, double eps,
  double *AiPrime, double *BiPrime)
/* Derivatives of Airy Functions, Ai' and  Bi' */
{
double x2,x3,tk,fterm,fsum,fratio,gterm,gsum,gratio,
  zeta,nsum,dsum,DD,EE,expZeta,RSPiX,FF,GG,angle,
  Sin,Cos;
const double c1 = 0.355028053887817,
  c2 = 0.258819403792807, Rt3 = 1.73205080757,
  Pi = 3.14159265359, PiD4 = 0.785398163397,
  D1n[6] = {1.765399083345135,32.51245961123978,
          128.96024013355196,171.29119587966386,
          83.83971685393736, 12.800523802235876},
  D1d[7] = {1.0,25.707051970697979,116.11422653072933,
          163.82285283357947,82.59522161455095,
          12.80052379932291,2.892647926744374e-11},
  E1n[10] = {1.3250584952011983,4.459554613245555,
            4.045748725309557,3.812506008078161,
            0.14004012797432785,0.6588213602856071,
            -0.07314276366636963,0.04138186753679998,
            -0.003535814704379303,
            0.0006209493121186009},
  E1d[11] = {1.0,7.066706001637072,3.644677059016622,
            4.039659022890137,0.17743464267023878,
            0.6560533854794217,-0.06941833554279864,
            0.04107726219940824,-0.003475610421847949,
            0.0006209519828011233,
            -1.9171296928438553e-11},
  F1n[6] = {1.4871127880501187,10.885581301222775,
          15.315193218337946,9.796686931111783,
          2.7857694521376417,0.4368670460997608},
  F1d[7] = {1.0,10.483175978046164,15.192999439883268,
          9.777511638153511,2.7857695725749192,
          0.4368670436222089,2.231025954257948e-11},
  G1n[6] = {-0.3913546968367662,-2.2240698457851491,
          -2.0683228910410607,-0.8705113304463556,
          -0.17738647071315156,
          -0.017711593671499819},
  G1d[7] = {1.0,11.414162467369463,26.583538550245219,
          22.071342536771844,9.033389065898527,
          1.8245467421976844,0.18217639035022146};
int k;

if ( fabs(x) <= 2.0 ) /* use power series */
  {
  x2 = x*x;
  x3 = x*x2;

  tk = 2.0;  /*  f1  series */
  fterm = 0.5*x2;
  fsum = fterm;
  fratio = 10.0;
```

```
    while ( fratio > eps )
      {
      tk += 3.0;
      fterm *= x3/(tk*(tk-2.0));
      fsum += fterm;
      fratio = fabs(fterm/fsum);
      }
    fsum *= c1;

    tk = 0.0;  /*  g1  series */
    gterm = 1.0;
    gsum = 1.0;
    gratio = 10.0;
    while ( gratio > eps )
      {
      tk += 3.0;
      gterm *= x3/(tk*(tk-2.0));
      gsum += gterm;
      gratio = fabs(gterm/gsum);
      }
    gsum *= c2;

    *AiPrime = fsum - gsum;
    *BiPrime = Rt3*(fsum + gsum);
    return; /* values for  |x| <= 2  */
    }

if ( x > 2.0 )
  { /* use rational approximations to D1 and E1 */
  zeta = 2.0*pow(x,1.5)/3.0;

  nsum = 0.0; /*  D1 approximation */
  dsum = D1d[6];
  for ( k = 5; k >= 0; k-- )
    {
    nsum = nsum*zeta+D1n[k];
    dsum = dsum*zeta+D1d[k];
    }
  DD = nsum/dsum;

  nsum = 0.0; /*  E1 approximation */
  dsum = E1d[10];
  for ( k = 9; k >= 0; k-- )
    {
    nsum = nsum*zeta+E1n[k];
    dsum = dsum*zeta+E1d[k];
    }
  EE = nsum/dsum;

  expZeta = exp(zeta);
  RSPiX = sqrt(sqrt(x)/Pi);
  *AiPrime = - RSPiX*DD/(2.0*expZeta);
  *BiPrime =   RSPiX*expZeta*EE;
  return; /* values for  x > 2  */
  }
```

```
/* x < -2;
    use rational approximations to F1 and G1 */
zeta = 2.0*pow(-x,1.5)/3.0;

nsum = 0.0; /*  F1 approximation */
dsum = F1d[6];

for ( k = 5; k >= 0; k-- )
  {
  nsum = nsum*zeta+F1n[k];
  dsum = dsum*zeta+F1d[k];
  }

FF = nsum/dsum;

nsum = 0.0; /*  G1 approximation */
dsum = G1d[6];

for ( k = 5; k >= 0; k-- )
  {
  nsum = nsum*zeta+G1n[k];
  dsum = dsum*zeta+G1d[k];
  }

GG = nsum/dsum;

RSPiX = sqrt(sqrt(-x)/Pi);
angle = zeta+PiD4;
Sin = sin(angle);
Cos = cos(angle);

*AiPrime = - RSPiX*(Cos*FF + Sin*GG);
*BiPrime =   RSPiX*(Sin*FF - Cos*GG);
return; /* values for  x < -2  */
}
```

Programming Notes

Program `AiryPrimeAiBi` returns two values—`AiPrime` and `BiPrime`—through its argument list. The accuracy of these values for the x range $[-2, 2]$ in which the power series in (14.5.22) are used is controlled by the function user through the input parameter `eps`, which gives the fractional accuracy required in Ai$'$ and Bi$'$. Each series is summed until both series meet this requirement; for `eps` $= 10^{-10}$ no more than ten terms are needed.

For $|x| > 2$ the rational approximations described under Algorithm and Program are used. Although they require several high-accuracy coefficients to define them, the polynomials in (14.5.24) are evaluated efficiently by using the Horner algorithm. The approximations have hard-wired accuracy of about 10^{-11} or better (Table 14.5.3), so they do not use the accuracy parameter `eps`; it must, however, be nonzero in the driver program shown in Section 21.14 in order that program execution continues.

Test Values

Check values of Ai$'$ and Bi$'$ can be generated by running the `Airy` cell in *Mathematica* notebook `BesselAiry`; examples are given in Table 14.5.4 for x values covering the three algorithms used.

M

TABLE 14.5.4 **Test values to 10 digits of derivatives of the Airy functions.**

x	-6	-4	-2
Ai$'$	0.3459354873	-0.7906285754	0.6182590207
Bi$'$	-0.8128987851	-0.1166705674	0.2787951669
x	2	4	6
Ai$'$	-0.05309038443	-0.001958640950	$-2.476520040 \times 10^{-5}$
Bi$'$	4.100682050	161.9266835	$1.572560262 \times 10^{4}$

REFERENCES ON BESSEL FUNCTIONS

[Abr64]   Abramowitz, M., and I. A. Stegun, *Handbook of Mathematical Functions,* Dover, New York, 1964.

[Erd53b]   Erdélyi, A., et al, *Higher Transcendental Functions*, Vol. 2, McGraw-Hill, New York, 1953; reprint edition, Krieger, Malabar, Florida, 1981.

[Gra94]   Gradshteyn, I. S., and I. M. Ryzhik, *Table of Integrals, Series, and Products*, Academic Press, Orlando, fifth edition, 1994.

[Hoc86]   Hochstadt, H., *The Functions of Mathematical Physics*, Dover, New York, 1986.

[Jah45]   Jahnke, E., and F. Emde, *Tables of Functions*, Dover, New York, fourth edition, 1945.

[Leb65]   Lebedev, N. N., *Special Functions and Their Applications*, Prentice-Hall, Englewood Cliffs, New Jersey, 1965.

[McL55]   McLachlan, N. W., *Bessel Functions for Engineers*, Clarendon Press, Oxford, second edition, 1955.

[Ram84]   Ramo, S., J. R. Whinnery, and T. van Duzer, *Fields and Waves in Communications Electronics*, Wiley, New York, second edition, 1984.

[Spa87]   Spanier, J., and K. B. Oldham, *An Atlas of Functions*, Hemisphere Publishing, Washington, D.C., 1987.

[Wat44]   Watson, G. N., *A Treatise on the Theory of Bessel Functions*, Cambridge University Press, New York, second edition, 1944.

[Whi27]   Whittaker, E. T., and G. N. Watson, *A Course of Modern Analysis*, Cambridge University Press, Cambridge, England, fourth edition, 1927.

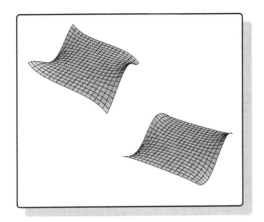

Chapter 15

STRUVE, ANGER, AND WEBER FUNCTIONS

The Struve, Anger, and Weber functions are three types of functions that are related to Bessel functions (Chapter 14) and have applications in science and engineering. The literature on the properties of these functions and on interrelations between them is quite sparse. Emphasizing mathematical analysis are the book by Watson [Wat44, Sections 10.1 and 10.4], and Sections 7.5.3 and 7.5.4 of the treatise on higher transcendental functions edited by Erdélyi et al [Erd53b], while there is a more practical emphasis in McLachlan's monograph on Bessel functions for engineers [McL55].

Formulas that are useful for numerical computations are given in Chapter 12 of the handbook of functions edited by Abramowitz and Stegun [Abr64], in Sections 8.9 and 8.10 of Jahnke and Emde [Jah45], and in Sections 6.81 and 8.58 of Gradshteyn and Ryzhik [Gra94]. A few graphs of the Struve functions are shown in Chapter 57 of Spanier and Oldham [Spa87].

15.1 STRUVE FUNCTIONS

The Struve functions are of two kinds, $\mathbf{H}_v(z)$, called the Struve function, and $\mathbf{L}_v(z)$, called the modified Struve function. Since $\mathbf{L}_v(z)$ is obtained from $\mathbf{H}_v(z)$ by a rotation in the z plane and a phase factor that depends on v, it is easiest to consider $\mathbf{H}_v(z)$ first. The use of bold-faced notation for the functions serves only to confuse scientists and engineers, it does not indicate a vector.

15.1.1 Struve Function H

Keywords
Struve functions

C Function and *Mathematica* Notebook

StruveH (C); cell StruveH in notebook StruveAngerWeber (*Mathematica*)

Background

The Struve functions occur in vibration problems involving disks, such as the fluid pressure on a vibrating surface—which is important when designing microphone diaphragms.

Function Definition

In the scientific context, the Struve function, $\mathbf{H}_v(z)$, is mostly readily defined in terms of solutions of the differential equation

$$z^2 w_v'' + z w_v' + [z^2 - v^2] w_v(z) = \frac{4(z/2)^{v+1}}{\sqrt{\pi}\, \Gamma(v+1/2)} \tag{15.1.1}$$

whose solution is a linear combination of solutions of the homogeneous equation, which are regular and irregular Bessel functions (Chapter 14), and a particular solution of the inhomogeneous equation, which is the Struve function. This has a power-series solution

$$\mathbf{H}_v(z) = (z/2)^{v+1} \sum_{k=0}^{\infty} \frac{(-1)^k (z/2)^{2k}}{\Gamma(k+3/2)\,\Gamma(k+v+3/2)} \tag{15.1.2}$$

and an integral representation [Abr64, 12.1.6]

$$\mathbf{H}_v(z) = \frac{2(z/2)^v}{\sqrt{\pi}\, \Gamma(v+1/2)} \int_0^1 (1-t^2)^{v-1/2} \sin(zt)\, dt \qquad \text{Re}\, v > -1/2 \tag{15.1.3}$$

The close relation of this Struve function to the regular Bessel function, $J_v(z)$, is evident from the integral representation for the latter [Abr64, 9.1.20]

$$J_v(z) = \frac{2(z/2)^v}{\sqrt{\pi}\, \Gamma(v+1/2)} \int_0^1 (1-t^2)^{v-1/2} \cos(zt)\, dt \qquad \text{Re}\, v > -1/2 \tag{15.1.4}$$

Recurrence relations for Struve functions are given by [Abr64, 12.1.9]

$$\mathbf{H}_{v-1}(z) + \mathbf{H}_{v+1}(z) = \frac{2v}{z} \mathbf{H}_v(z) + \frac{(z/2)^v}{\sqrt{\pi}\, \Gamma(v+3/2)} \tag{15.1.5}$$

and an asymptotic expansion when $|z|$ is large is [Abr64, 12.1.29]

$$\mathbf{H}_v(z) = Y_v(z) + \frac{1}{\pi} \sum_{k=0}^{m-1} \frac{\Gamma(k+1/2)}{\Gamma(v+1/2-k)(z/2)^{2k-v+1}} + R_m \tag{15.1.6}$$

in which $Y_v(z)$ is the irregular cylindrical Bessel function (Section 14.2.2). The remainder, R_m, is of order $|z|^{v-2m-1}$ and is of the same sign and smaller in magnitude than the first neglected term if v is real, if $z = x$ is positive, and if $m \geq v - 1/2$.

The power series, recurrence relations, and asymptotic expansion are used in the following algorithm for $v = n$, an integer, and $z = x$, a real variable.

Visualization

Since $\mathbf{H}_v(z)$ is a function of two variables, both of which may be complex, there are many possible viewpoints. However, since the Struve function resembles a cylindrical Bessel function in its definition—compare (15.1.3) and (15.1.4)—we make graphics that are similar to those in the overview of the Bessel functions (Section 14.1).

Figure 15.1.1 shows the surfaces of $\mathbf{H}_\nu(z)$ as a function of complex variable $z = x + iy$ with the order $\nu = n$, a nonnegative integer, being fixed.

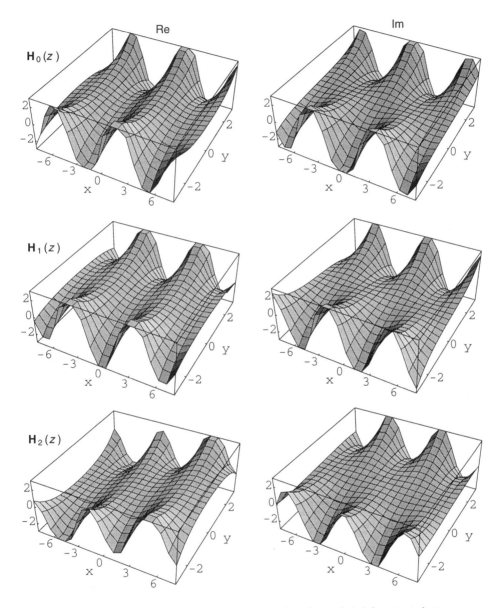

FIGURE 15.1.1 Struve functions in the complex plane for $n = 0, 1, 2$ from top to bottom; real parts left, imaginary parts right. (Adapted from cell `StruveH` in *Mathematica* notebook `StruveAngerWeber`.)

For insights that help when computing $\mathbf{H}_n(x)$, it is interesting to map the surfaces of $\mathbf{H}_n(x)$ in the x–n plane, as shown in Figure 15.1.2. For $x > 0$ the behavior of $\mathbf{H}_n(x)$ as n

increases depends upon the relative size of x and the maximum n of interest. The direction to iterate n in order to minimize subtractive cancellation must therefore be chosen carefully, as in our C function StruveH that follows.

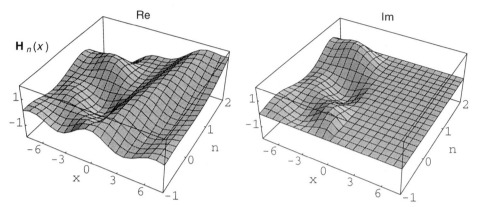

When transections of these surfaces are examined—either for fixed n and varying x (Figure 15.1.3 left) or for fixed x and varying n (Figure 15.1.3 right)—it is again seen that the appropriate direction of iteration for $x > 0$ depends upon n and x.

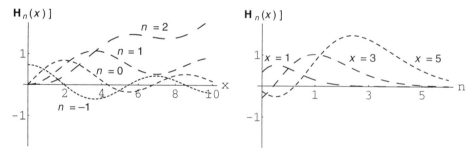

Algorithm, Program, and Programming Notes

Our algorithm to compute accurately and efficiently an array of Struve functions, $\mathbf{H}_n(x)$, from $n = 0$ up to $n = n_h$ over a broad range of $n_h \geq 0$ and $x \geq 0$ uses a variety of methods, summarized as follows.

For $x = 0$ all the elements of the array are set to zero. If $x < n_h/2$—where n_h is input in the driver program (Section 21.15) as variable nHMax—the function StruveH uses power series expansion (15.1.2) for $n = $ nLarge $ = $ nHMax+8, to compute $\mathbf{H}_n(x)$ and $\mathbf{H}_{n-1}(x)$; backward iteration by (15.1.5) then gives the functions down to $n = 0$. When $x \geq n_h/2$ there are two possible methods to compute $\mathbf{H}_0(x)$ and $\mathbf{H}_1(x)$ in preparation for forward iteration on n. The first method—for $x \leq 25$—uses the power series expansion, whereas the second method uses the asymptotic expansion (15.1.6), which also requires the irregular cylindrical Bessel functions $Y_0(x)$ and $Y_1(x)$, whose complicated method of computation is described in Section 14.2.2.

For $x \geq n_h/2$ the optimum switchover x value between the two methods depends upon the relative accuracy desired, which is the input parameter eps. The chosen value, $x = 25$, gives most efficient evaluation for eps $= 10^{-10}$ which is the accuracy goal for all functions in the *Atlas*. If higher accuracy is needed the switchover value must be increased to ensure convergence of the asymptotic series. For example, accuracy of better than 10^{-12} can be achieved if the switchover is postponed until $x = 30$, but more than 50 terms will be needed for the power series expansion for x up to this value.

In the following listing of C function StruveH, we show only functions not listed elsewhere in the *Atlas*. Function BessY01 is a simplified version of BessYN in Section 14.2.2, and the driver program for checking StruveH is in Section 21.15. The CD-ROM that comes with the *Atlas* has the complete program package, following our practice discussed under The C Functions: No Assembly Required, found in the Introduction to the *Atlas*.

```c
/* maximum number of table entries */
#define NMAX 39

void StruveH(int nHMax,double x,double eps,
 double HN[])
/* Make a table of StruveH functions
    from order  n = 0 to nHMax; x >= 0 is fixed */
{
double fnH,Mxs,sum0,term0,sum1,term1,ratio,
 MRxs,termold,tkM1,epsA,tkM3,termnew,YN[NMAX],
 tnDx,dtnDx,Tn,tkP1,tnP1;
double HHSeries(int n, double x, double eps);
const double TwoDPi = 0.636619772368;
void  BessY01(int nYMax, double x, double YN[]);
int nYMax,k,n,nLarge;

fnH = nHMax;
if ( x <   0.5*fnH )
  {
  if ( x == 0.0 )
    {
    for ( n = 0; n <= nHMax; n++ )
    { HN[n] = 0.0; }
    return;
    }
  /* 0<x<nHMax/2: power series & backward iteration */
  nLarge = nHMax+8;
  HN[nLarge]  = HHSeries(nLarge,x,eps);
  HN[nLarge-1] = HHSeries(nLarge-1,x,eps);
  /* Backward iteration from n=nHMax+8 down to n=0 */
  tkP1 = 1.0;/* Make extra term in recurrence */
  Tn = TwoDPi;
  for ( k = 1; k < nLarge; k++ )
    {
    tkP1 += 2.0;
    Tn *= x/tkP1;
    }
```

```
   tnDx = ( (double ) 2*nLarge )/x;
   dtnDx = 2.0/x;
   tnP1 = tkP1;
   for ( n = nLarge-1; n >=1; n-- )
    {
    tnDx -= dtnDx;
    HN[n-1] = tnDx*HN[n] - HN[n+1] + Tn;
    Tn *= tnP1/x;
    tnP1 -= 2.0;
    }
   return;
   }
  else  /*  x >= nHMax/2  */
   {
   if ( x <= 25.0 )
    { /* HN[0] & HN[1] from power series  */
    HN[0] = HHSeries(0,x,eps);
    HN[1] = HHSeries(1,x,eps);
    }
   else
    /* x>25, HN[0] & HN[1] by asymptotic expansion */
    {
    MRxs = -1.0/(x*x);
    k = 0;
    term0 = 1.0;  term1 = 1.0;
    termold = 1.0;
    tkM1 = -1.0;
    sum0 = 1.0;  sum1 = 1.0;
    ratio = 10.0;
    epsA = 1.0e-10;
   while ( ratio > epsA )
     {
     k++;
     tkM3 = tkM1;
     tkM1 += 2.0;
     term0 *=  tkM1*tkM1*MRxs;
     term1 *=  tkM1*tkM3*MRxs;
     sum0 += term0;
     sum1 += term1;  /* H0 converges slower */
     ratio = fabs(term0/sum0);
     termnew = fabs(term0);
     if( termnew > termold  )
     {
     printf
     ("\n\n!!In StruveH diverging after %i terms",k);
     ratio = 0.0;
     }
     else
       { termold = termnew; }
     }
    /* Irregular cylindrical Bessel functions,
       Y0 & Y1 */
    nYMax = 0;
    BessY01(nYMax,x,YN);
```

```
    HN[0] = YN[0] + TwoDPi*sum0/x;
    HN[1] = YN[1] + TwoDPi*sum1;
       }
    /* Use forward iteration up to  n = nHMax  */
    tnDx = 0.0;
    dtnDx = 2.0/x;
    Tn = TwoDPi;
    tnP1 = 1.0;
    for ( n = 1; n < nHMax; n++ )
      {
     tnDx += dtnDx;
     tnP1 += 2.0;
     Tn *= x/tnP1;
     HN[n+1] = tnDx*HN[n] - HN[n-1] + Tn;
      }
    return;
    }
}


double HHSeries(int n, double x, double eps)
/* Struve function of order n, argument x, to
    fractional accuracy  eps  by power series */
{
double Mxs,sum,term,ratio,tkP1,tkPnP1,hh;
const double TwoDPi = 0.636619772368;
int k;

Mxs = - x*x;
sum = 1.0;  term = 1.0;
ratio = 10.0;
tkP1 = 1.0;  tkPnP1 = 1+2*n;
while ( ratio > eps )
  {
  tkP1 += 2.0;  tkPnP1 += 2.0;
  term *= Mxs/(tkP1*tkPnP1);
  sum += term;
  ratio = fabs(term/sum);
  }
/* Make the prefactor */
hh = TwoDPi;
tkP1 = -1.0;
for ( k = 0; k <= n; k++ )
  {
  tkP1 += 2.0;
  hh *= x/tkP1;
  }
return   hh*sum;
}
```

Test Values

Check values of the Struve function $\mathbf{H}_n(x)$ can be generated by running the `StruveH` cell in *Mathematica* notebook `StruveAngerWeber`. In order to obtain values accurate to 10 decimal digits—the accuracy goal for all functions in the *Atlas*—the options that control

M

the accuracy of the *Mathematica* function NSum for infinite series must be adjusted carefully, as indicated in the annotated listing of the StruveH driver in Section 20.15.

Sample numerical values of $\mathbf{H}_n(x)$ that exercise all methods used in the C function StruveH are given in Table 15.1.1. Input nHMax = 6 in the driver program to exercise the three methods of computation, and use $\varepsilon = 10^{-11}$ to guarantee agreement in the 10th digit for all x.

TABLE 15.1.1 Test values to 10 digits of the Struve functions , H.

$x \setminus n$	0	3	6
1.0	0.5686566270	$5.842526535 \times 10^{-3}$	$4.607524757 \times 10^{-6}$
3.0	0.5743061488	0.3517111871	$8.446444418 \times 10^{-3}$
30.0	−0.09609842155	38.34349101	$1.507106020 \times 10^{3}$

15.1.2 Modified Struve Function L

Keywords
modified Struve functions, Struve functions

C Function and *Mathematica* Notebook
StruveL (C); cell StruveL in notebook StruveAngerWeber (*Mathematica*)

Background
The modified Struve functions, $\mathbf{L}_\nu(z)$, are related to the Struve functions, $\mathbf{H}_\nu(z)$ in Section 15.1.1 just as the hyperbolic Bessel functions $I_\nu(z)$ in Section 14.2.3 are related to the cylindrical Bessel functions $J_\nu(z)$ in Section 14.2.1.

Mathematical analysis of $\mathbf{L}_\nu(z)$ is mentioned in the treatise by Watson [Wat44, Section 10.4], and results are given in Section 7.5.4 of the work on higher transcendental functions edited by Erdélyi et al [Erd53b]. Practical formulas for numerical computations are in Section 12.2 of the handbook of functions edited by Abramowitz and Stegun [Abr64], while some graphs of $\mathbf{L}_\nu(z)$ are shown in Chapter 57 of Spanier and Oldham [Spa87].

Function Definition
The modified Struve function, $\mathbf{L}_\nu(z)$, is readily defined in terms of the Struve function $\mathbf{H}_\nu(z)$ by [Abr64, 12.2.1]

$$\mathbf{L}_\nu(z) \equiv -ie^{-i\nu\pi/2}\mathbf{H}_\nu(iz) \qquad (15.1.7)$$

the phase factors in which are chosen so that $\mathbf{L}_\nu(z)$ is real when z is real and positive. This is clear in the power-series expression obtained from power series (15.1.2) for $\mathbf{H}_\nu(z)$, thus

$$\mathbf{L}_\nu(z) = (z/2)^{\nu+1} \sum_{k=0}^{\infty} \frac{(z/2)^{2k}}{\Gamma(k+3/2)\,\Gamma(k+\nu+3/2)} \qquad (15.1.8)$$

Integral representations of $\mathbf{L}_\nu(z)$ are [Abr64, 12.2.3]

$$\mathbf{L}_\nu(x) = I_{-\nu}(x) - \frac{2(x/2)^\nu}{\sqrt{\pi}\,\Gamma(\nu+1/2)} \int_0^\infty (1+t^2)^{\nu-1/2}\sin(xt)dt \qquad \mathrm{Re}\,\nu < 1/2 \quad (15.1.9)$$

and also [Abr64, 12.2.2]

$$\mathbf{L}_v(z) = \frac{2(z/2)^v}{\sqrt{\pi}\,\Gamma(v+1/2)} \int_0^{\pi/2} \sin^{2v}\theta \sinh(z\cos\theta)d\theta \qquad \mathrm{Re}\,v > -1/2 \quad (15.1.10)$$

Recurrence relations for modified Struve functions are given by [Abr64, 12.2.4]

$$\mathbf{L}_{v-1}(z) - \mathbf{L}_{v+1}(z) = \frac{2v}{z}\mathbf{L}_v(z) + \frac{(z/2)^v}{\sqrt{\pi}\,\Gamma(v+3/2)} \qquad (15.1.11)$$

and an asymptotic expansion when $|z|$ is large is [Abr64, 12.2.6]

$$\mathbf{L}_v(z) \sim I_{-v}(z) + \frac{1}{\pi}\sum_{k=0}^{\infty}\frac{(-1)^{k+1}\,\Gamma(k+1/2)}{\Gamma(v+1/2-k)(z/2)^{2k-v+1}} \qquad |\arg z| < \pi/2 \quad (15.1.12)$$

in which $I_{-v}(z)$ is a regular hyperbolic Bessel function (Section 14.2.3). The power series, recurrence relations, and asymptotic expansion are used in the following algorithm for $v = n$, an integer, and $z = x \geq 0$, a nonnegative real variable.

Visualization

Function $\mathbf{L}_v(z)$ in the complex plane is, according to (15.1.7), just a rotation of $\mathbf{H}_v(z)$; however, we make new visualizations as in Figure 15.1.4. The function $\mathbf{L}_2(z)$ is shown as the motif at the start of this chapter.

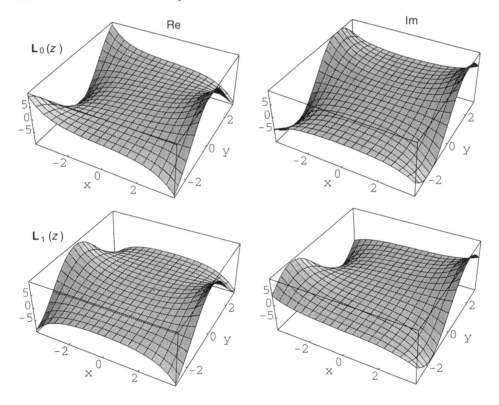

FIGURE 15.1.4 Modified Struve functions in the complex plane for $n = 0$ and 1; real parts left, imaginary parts right. (Adapted from cell `StruveL` in *Mathematica* notebook `StruveAngerWeber`.)

It is also interesting to map the surfaces of the modified Struve function, $\mathbf{L}_n(x)$ in the x–n plane, as shown in Figure 15.1.5. For $x > 0$ $\mathbf{L}_n(x)$ is a monotonically decreasing function of n. The direction to iterate n in order to minimize subtractive cancellation is therefore towards decreasing n, as in our C function `StruveL` that follows.

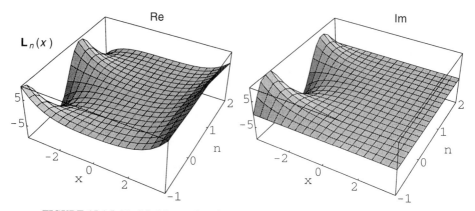

FIGURE 15.1.5 Modified Struve functions in the x–n plane; real part left, imaginary part right. Note that the imaginary part is zero for $x > 0$. (Adapted from cell `StruveL` in *Mathematica* notebook `StruveAngerWeber`.)

When transections of these surfaces are examined—either for fixed n and varying x (Figure 15.1.6 left) or for fixed x and varying n (Figure 15.1.6 right)—it is again seen that decreasing n is the appropriate direction of iteration for $x > 0$.

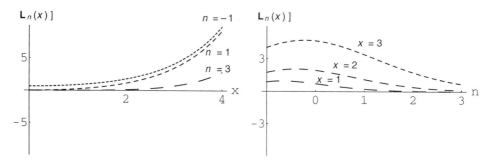

FIGURE 15.1.6 Modified Struve functions vs x for $n = -1, 0, 1, 2$ on left side and vs n for $x = 1, 3, 5$ on right side. (Adapted from cell `StruveL` in the *Mathematica* notebook `StruveAngerWeber`.)

Algorithm, Program, and Programming Notes

The algorithm to compute accurately and efficiently an array of modified Struve functions, $\mathbf{L}_n(x)$, from $n = 0$ to $n = n_h$ over a broad range of $n_h \geq 0$ and $x \geq 0$ is simpler than that for $\mathbf{H}_n(x)$. If $x > 0$ the algorithm uses either the power series (15.1.8) or the asymptotic expansion (15.1.12) to calculate $\mathbf{L}_{n_h}(x)$ and $\mathbf{L}_{n_h-1}(x)$. Backward iteration by (15.1.11) is then used to generate the rest of the array of $\mathbf{L}_n(x)$ values, down to $n = 0$. For $x = 0$ all the elements of the array are set to zero. The hyperbolic Bessel function $I_{-n}(x) = I_n(x)$, required for the asymptotic expansion (15.1.12) is computed as described for function `BessIN` in Section 14.2.3.

In the following listing of C function `StruveL`, we show only functions not listed elsewhere in the *Atlas*. Function `BessINLL` is a simplified version of `BessIN` in Section 14.2.3, and the driver program for checking `StruveL` is in Section 21.15; the CD-ROM that comes with the *Atlas* has the complete program package, following our rule of "no assembly required."

```c
/* maximum number of table entries */
#define NMAX 30

void StruveL(int nLMax,double x,double eps,double
LN[])
/* Make a table of modified Struve functions
   from order  n = 0 to nLMax; x >= 0 is fixed   */
{
double fnL,IN[NMAX],tnDx,dtnDx,Tn,tkP1,tnP1;
double LLSeries(int n, double x, double eps);
double LLAsymptote(int n, double x, double eps);
void BessINLL(int nIMax, double x, double IN[]);
const double TwoDPi = 0.636619772368;
int k,n;

if ( x == 0.0 )
  {
  for ( n = 0; n <= nLMax; n++ )
   { LN[n] = 0.0; }
  return;
  }
fnL = nLMax;
if ( ( x < 0.5*fnL ) || ( x <= 25.0 ) )
  {
  /*  0<x<nLMax/2: power series to start */
  LN[nLMax] = LLSeries(nLMax,x,eps);
  LN[nLMax-1] = LLSeries(nLMax-1,x,eps);
  }
else
  { /* x>25; use regular hyperbolic Bessel function
        In(x) plus asymptotic expansion */
  BessINLL(nLMax,x,IN);
  LN[nLMax] = IN[nLMax] + LLAsymptote(nLMax,x,eps);
  LN[nLMax-1] =
   IN[nLMax-1]+LLAsymptote(nLMax-1,x,eps);
  }

/* Backward iteration from n = nLMax down to n=0 */
tkP1 = 1.0;/* Make extra term in recurrence */
Tn = TwoDPi;
for ( k = 1; k < nLMax; k++ )
  {
  tkP1 += 2.0;
  Tn *= x/tkP1;
  }
tnDx = ( (double ) 2*nLMax )/x;
dtnDx = 2.0/x;
tnP1 = tkP1;
```

```
  for ( n = nLMax-1; n >=1; n-- )
    {
   tnDx -= dtnDx;
   LN[n-1] = tnDx*LN[n] + LN[n+1] + Tn;
   Tn *= tnP1/x;
   tnP1 -= 2.0;
    }
   return;
    }

double LLAsymptote(int n, double x, double eps)
/* Struve function of order n, argument x, to
  fractional accuracy eps by asymptotic expansion */
  {
double MRxs,term,termold,tkM1,tnMtkP1,sum,ratio,
  termnew,LL;
const double TwoDPi = 0.636619772368;
int k;

MRxs = - 1.0/(x*x);
k = 0;
term = 1.0;
termold = 1.0;
tkM1 = -1.0;
tnMtkP1 = 2*n+1;
sum = 1.0;
ratio = 10.0;
while ( ratio > eps )
  {
 k++;
 tkM1 += 2.0;
 tnMtkP1 -= 2.0;
 term *=  tkM1*tnMtkP1*MRxs;
 sum += term;
 ratio = fabs(term/sum);
 termnew = fabs(term);
 if( termnew > termold  )
   {
 printf
("\n\n!!In LLAsymptote diverging after %i terms",k);
  ratio = 0.0;
   }
 else
   { termold = termnew; }
  }
/* Prefactor */
LL = TwoDPi;
tkM1 = 1.0;
for ( k = 1; k < n; k++ )
  {
 tkM1 += 2.0;
 LL *= x/tkM1;
  }
return  - LL*sum;
  }
```

```
double LLSeries(int n, double x, double eps)
/* Modified Struve function of order n, argument x,
   to fractional accuracy   eps   by power series */
{
double xs,sum,term,ratio,tkP1,tkPnP1,LL;
const double TwoDPi = 0.636619772368;
int k;

xs = x*x;
sum = 1.0;   term = 1.0;
ratio = 10.0;
tkP1 = 1.0;   tkPnP1 = 1+2*n;
while ( ratio > eps )
  {
  tkP1 += 2.0;   tkPnP1 += 2.0;
  term *= xs/(tkP1*tkPnP1);
  sum += term;
  ratio = fabs(term/sum);
  }
/* Make the prefactor */
LL = TwoDPi;
tkP1 = -1.0;
for ( k = 0; k <= n; k++ )
  {
  tkP1 += 2.0;
  LL *= x/tkP1;
  }
return   LL*sum;
}
```

Test Values

Test values of the modified Struve function $\mathbf{L}_n(x)$ can be generated by running the cell StruveL in *Mathematica* notebook StruveAngerWeber. To obtain values accurate to 10 decimal digits—the accuracy goal for all functions in the *Atlas*—the options that control the accuracy of the *Mathematica* function NSum for infinite series must be adjusted carefully, as indicated in the annotated listing of the StruveL driver in Section 20.15.

M

Numerical values of $\mathbf{L}_n(x)$ that exercise both methods used in C function StruveL are given in Table 15.1.2; input nLMax = 6 in the driver program. If nLMax and x must both be large then normalization of $I_n(x)$ in the asymptotic expansion (15.1.12) becomes problematic, as discussed in relation to Table 14.2.5. Small coding changes in StruveL are necessary to include the case when both parameters are large.

TABLE 15.1.2 Test values to 10 digits of the modified Struve functions , L.

$x \setminus n$	0	3	6
1.0	0.7102431859	0.006291730750	$4.816920657 \times 10^{-6}$
3.0	4.648685732	0.6844313137	0.01259719752
50.0	$2.932553784 \times 10^{20}$	$2.677764139 \times 10^{20}$	$2.039389282 \times 10^{20}$

15.2 ANGER AND WEBER FUNCTIONS

15.2.1 Overview of Anger and Weber Functions

Anger functions (named after a 19th-century mathematician, not a state of mind), $\mathbf{J}_v(z)$, and Weber functions, $\mathbf{E}_v(z)$, are defined by integrals similar to those for Bessel functions, so their analytical properties are often considered together. The section begins with an overview of their common properties, then detailed properties are presented in Sections 15.2.2. and 15.2.3.

The bold-face notation for the two functions should not confuse physicists and electrical engineers into thinking that we are dealing with current density and electric field strength, respectively, even though the functions $\mathbf{J}_v(z)$ and $\mathbf{E}_v(z)$ date from the epoch of Maxwell. Also, don't confuse this Weber function with Weber's parabolic cylinder functions in Section 18.1.

Background

Mathematical analysis of Anger and Weber functions is considered in detail in Section 10.1 of Watson's monograph on Bessel functions [Wat44], many formulas are given in Section 7.5.3 of the compendium by Erdélyi et al [Erd53b], while Jahnke and Emde [Jah45] present in Section 8.9 a function $\Omega_p(z) \equiv -\mathbf{E}_p(z)$ as the Lommel-Weber function, show some curves of the function, and give some 4-figure numerical tables. Section 8.58 of Gradshteyn and Ryzhik [Gra94] has several formulas, as does Section 12.3 in Abramowitz and Stegun [Abr64]. The main results that we need are obtained from Watson.

Function Definitions

The Anger function is defined by

$$\mathbf{J}_v(z) \equiv \int_0^\pi \frac{\cos(v\theta - z\sin\theta)}{\pi} d\theta \tag{15.2.1}$$

When $v = n$, an integer, then $\mathbf{J}_n(z) = J_n(z)$, a regular Bessel function of integer order, as developed in Section 14.2.1. The Weber function is similarly defined by

$$\mathbf{E}_v(z) \equiv \int_0^\pi \frac{\sin(v\theta - z\sin\theta)}{\pi} d\theta \tag{15.2.2}$$

Generally, $\mathbf{E}_v(z)$ does not reduce to a Bessel function. The commonality of the Anger and Weber functions arises by combining the two definitions as

$$\mathbf{J}_v(z) + i\mathbf{E}_v(z) \equiv \int_0^\pi \frac{e^{i(v\theta - z\sin\theta)}}{\pi} d\theta \tag{15.2.3}$$

Thus, if both v and z are real then $\mathbf{J}$ and $\mathbf{E}$ are the real and imaginary parts, respectively, of the integral on the right—a result that helps understanding some the subsequent formulas.

The range of v and z over which Anger and Weber functions must be computed directly can be reduced by using the following symmetry properties that can be deduced by using trigonometric identities in the defining equations:

$$\mathbf{J}_v(-z) = \mathbf{J}_{-v}(z) \qquad \mathbf{E}_v(-z) = -\mathbf{E}_{-v}(z) \tag{15.2.4}$$

$$\mathbf{J}_{-v}(z) = \sin(v\pi)\mathbf{E}_v(z) + \cos(v\pi)\mathbf{J}_v(z)$$
$$\mathbf{E}_{-v}(z) = \cos(v\pi)\mathbf{E}_v(z) - \sin(v\pi)\mathbf{J}_v(z) \tag{15.2.5}$$

Therefore, only $v \geq 0$ needs to be considered. When $v = n \geq 0$ the results [Abr64, 12.3.2, 12.3.6] can be used, namely,

$$\mathbf{J}_n(z) = J_n(z)$$

$$\mathbf{E}_n(z) = \frac{1}{\pi} \sum_{k=0}^{[(n-1)/2]} \frac{\Gamma(k+1/2)(z/2)^{n-2k-1}}{\Gamma(n+1/2-k)} - \mathbf{H}_n(z) \tag{15.2.6}$$

In the following C functions, `AngerJ` and `WeberE`, the functions are computed only for noninteger positive orders, v. This avoids much complexity in the expansions that we use.

Expression of $\mathbf{J}_v(z)$ and $\mathbf{E}_v(z)$ in terms of power series is accomplished by using the following relations [Wat44, Section 10.1, (9) and (10)]

$$\mathbf{J}_v(z) = \frac{\sin(v\pi)}{\pi} s_{0,v}(z) - \frac{v \sin(v\pi)}{\pi} s_{-1,v}(z)$$

$$\mathbf{E}_v(z) = -\frac{1+\cos(v\pi)}{\pi} s_{0,v}(z) - \frac{v[1-\cos(v\pi)]}{\pi} s_{-1,v}(z) \tag{15.2.7}$$

in which $s_{\mu,v}(z)$ is a Lommel function—as described in [Wat44, Section 10.7]—defined by

$$s_{\mu,v}(z) \equiv \frac{z^{\mu+1}}{(\mu+1)^2 - v^2} \sum_{k=0}^{\infty} t_k \tag{15.2.8}$$

$$t_0 = 1 \qquad t_k/t_{k-1} = -z^2/[(\mu+2k+1)^2 - v^2]$$

This definition is valid unless $\mu \pm v$ is an odd integer, a situation that we avoid. When z is large in magnitude, asymptotic expansions for $\mathbf{J}_v(z)$ and $\mathbf{E}_v(z)$ can be developed. From Watson [Wat44, Section 10.14, (1) and (2)] we have that

$$\mathbf{J}_v(z) - J_v(z) \sim \frac{\sin(v\pi)}{\pi z} S_{0,v}(z) - \frac{v \sin(v\pi)}{\pi z^2} S_{-1,v}(z)$$

$$\mathbf{E}_v(z) + Y_v(z) \sim -\frac{1+\cos(v\pi)}{\pi z} S_{0,v}(z) - \frac{v[1-\cos(v\pi)]}{\pi z^2} S_{-1,v}(z) \tag{15.2.9}$$

where the second kind of Lommel function, $S_{\mu,v}(z)$, is defined as [Wat44, Section 10.71]

$$S_{\mu,v}(z) \equiv \sum_{k=0}^{\infty} u_k \tag{15.2.10}$$

$$u_0 = 1 \qquad u_k/u_{k-1} = -[(\mu-2k+1)^2 - v^2]/z^2$$

These formulas and use of the recurrence relations given below are sufficient for the algorithms described below.

This overview of the Anger and Weber functions establishes connections between them that are useful for numerical computations. In the following two sections each function is described in detail.

15.2.2 Anger Function J

Keywords
Anger functions, Bessel functions, Lommel functions

C Function and *Mathematica* Notebook
`AngerWeber` (C); cell `AngerJ` in notebook `StruveAngerWeber` (*Mathematica*)

M

Background and Function Definition

The analytical background to the Anger functions, $\mathbf{J}_\nu(z)$, is given in Section 15.2.1.

Visualization

Figure 15.2.1 shows the surfaces of $\mathbf{J}_\nu(z)$ as a function of complex variable $z = x + iy$ with the order ν being fixed.

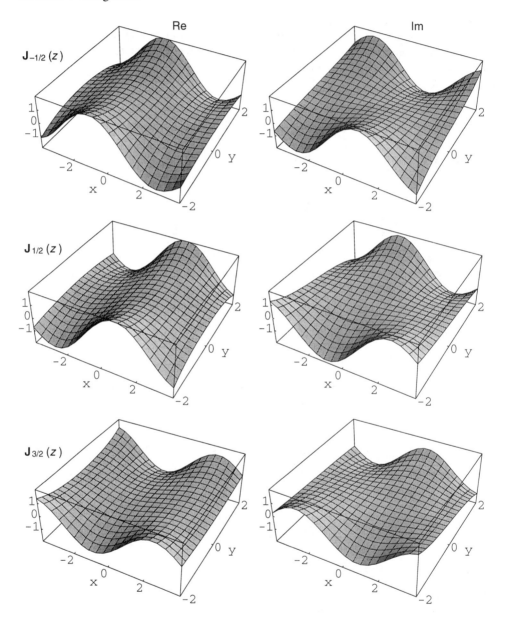

FIGURE 15.2.1 Anger functions in the complex plane for $\nu = -1/2$, $1/2$, $3/2$ from top to bottom; real parts left, imaginary parts right. (Adapted from cell `AngerJ` in *Mathematica* notebook `StruveAngerWeber`.)

It is also interesting to map the surface of $\mathbf{J}_\nu(x)$ in the x–ν plane, as shown in Figure 15.2.2. Also, the choice of direction to iterate n in order to minimize subtractive cancellation in our C function `AngerWeber` is helped by making transections of this surface, as seen on the right side of the figure.

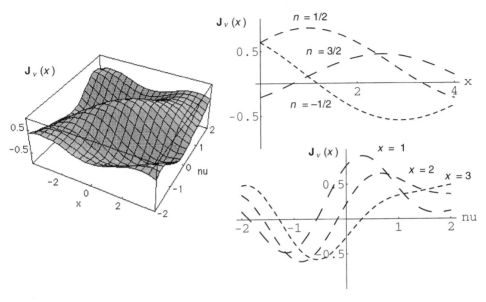

FIGURE 15.2.2 Anger functions in the x–n plane (left), vs x for $n = -1/2$, $1/2$, $3/2$ (right top) and vs n for $x = 1$, 2, 3 (right bottom).   (Adapted from cell `AngerJ` in *Mathematica* notebook `StruveAngerWeber`.)

Algorithm, Program, and Programming Notes

It might appear that brute-force numerical integration in the defining relation (15.2.1) for $\mathbf{J}_\nu(x)$ is appropriate for computing the Anger functions. The accuracy of the results obtained by this method is not easy to control, and the number of integration points needed for high accuracy may be large, particularly if ν or x is large, since the integrand then oscillates rapidly. For these reasons, we use numerical integration only for the *Mathematica* computations, which are relatively very slow and which need to be tweaked to obtain the accuracy required.

Our algorithm computes accurately and efficiently an array of Anger functions, $\mathbf{J}_\nu(x)$, beginning with $0 < \nu_m < 1$, called `nuMin` in C function `AngerWeber`. It first computes the starting values $\mathbf{J}_{\nu_m}(x)$ and $\mathbf{J}_{\nu_m+1}(x)$ by using the power series expressions in (15.2.7) if $x < 30$ and the asymptotic expansion (15.2.9) for larger x. Each function is obtained with fraction accuracy `eps`, which is an input parameter. In addition to the formulas in Section 15.2.1, the algorithm uses the recurrence relations for Anger functions given by [Wat44, Section 10.12]

$$\mathbf{J}_{\nu-1}(x) + \mathbf{J}_{\nu+1}(x) = \frac{2\nu}{x}\mathbf{J}_\nu(x) - \frac{2\sin(\nu\pi)}{\pi x} \tag{15.2.11}$$

Forward iteration is used to generate a total of `nAWMax` values, including the starting values. This direction of iteration is sensitive to subtractive cancellation when x is small; it is, however, straightforward to modify function `AngerWeber` to start at a different value of v, then to iterate downward if necessary. When $x = 0$, (15.2.11) cannot be used, but the integrals can be computed directly to give $\mathbf{J}_v(0) = \sin(v\pi)/(v\pi)$.

In the asymptotic region, $x \geq 30$, function `AngerWeber` returns $\mathbf{J}_v(x) - J_v(x)$, as in (15.2.9). If necessary, adapt the function `BessJN` in Section 14.2.1 to handle noninteger v values so that $J_v(x)$ can be added in.

The Lommel functions, $s_{\mu,v}(x)$ defined by (15.2.8) and $S_{\mu,v}(x)$ defined by (15.2.10), are computed by functions `LittleS` and `BigS`, respectively, which are included in the following listing. The switchover value of x at $x = 30$ between power series and asymptotic expansions is determined by our accuracy goal for all functions in the *Atlas*—a fractional accuracy of at least 10^{-10}. No more than 50 terms are then required in the power series, while only 15 terms in the asymptotic expansion are needed to achieve such convergence. If the accuracy you require exceeds this, the switchover must be moved to higher x. For less accuracy and faster computation, reduce the switchover value of x.

Function `AngerWeber` also computes the Weber functions, $\mathbf{E}_v(x)$, by using the same steps as for the Anger functions; more details are given in Section 15.2.3. An annotated listing of the C driver program for testing `AngerWeber` is in Section 20.15. Here is the C function for the Anger and Weber functions.

```c
/* maximum number of table entries   */
#define NMAX 10

void AngerWeber(int nAWMax, double nuMin, double x,
  double eps, double Anger[], double Weber[])
/* Makes a table of Anger and Weber Functions
    from noninteger order  nu > 0  for
    nAWMax < NMAX-1  values; fixed x >= 0.
    Fractional accuracy is    eps.
    If x >= 30 compute Anger-J, Weber+Y.  */
{
double nuPi,nu,nuMinP1,sZero,sMOne,PiX,CosNuPi,
  dtnuDx,tnuDx,tDPiX,
  LittleS(double mew, double nu, double x,
    double eps), BigS(double mew, double nu,
    double x, double eps);
const double Pi = 3.14159265359;
int n;

if ( x == 0 )
  { /* treat this specially */
  nuPi = nuMin*Pi;
  for ( n = 1; n <= nAWMax; n++ )
    {
    Anger[n] = sin(nuPi)/nuPi;
    Weber[n] = (1.0 - cos(nuPi))/nuPi;
    nuPi += Pi;
    }
  return;
  }

nuMinP1 = nuMin+1.0;
if ( x <   30.0 )
  {
```

```
    /* Starting values from power series;
       two lowest orders, nu  and  nu + 1  */
    sZero = LittleS( 0,nuMin,x,eps);
    sMOne = nuMin*LittleS(-1,nuMin,x,eps);
    Anger[1] = sin(nuMin*Pi)*(sZero - sMOne)/Pi;
    CosNuPi = cos(nuMin*Pi);
    Weber[1] = -((1.0+CosNuPi)*sZero+
     (1.0-CosNuPi)*sMOne)/Pi;
    sZero = LittleS( 0,nuMinP1,x,eps);
    sMOne = nuMinP1*LittleS(-1,nuMinP1,x,eps);
    Anger[2] = sin(nuMinP1*Pi)*(sZero - sMOne)/Pi;
    CosNuPi = cos(nuMinP1*Pi);
    Weber[2] = -((1.0+CosNuPi)*sZero+
     (1.0-CosNuPi)*sMOne)/Pi;
    }
 else
    {
    /* x > 30; start from asymptotic expansions;
       two lowest orders, nu  and  nu + 1  */
    sZero = BigS( 0,nuMin,x,eps);
    sMOne = BigS(-1,nuMin,x,eps);
    PiX = Pi*x;
    Anger[1] = sin(nuMin*Pi)*(sZero-nuMin*sMOne/x)/PiX;
    CosNuPi = cos(nuMin*Pi);
    Weber[1] = -((1.0+CosNuPi)*sZero+
     nuMin*(1.0-CosNuPi)*sMOne/x)/PiX;
    sZero = BigS( 0,nuMinP1,x,eps);
    sMOne = BigS(-1,nuMinP1,x,eps);
    Anger[2] = sin(nuMinP1*Pi)*(sZero-
     nuMinP1*sMOne/x)/PiX;
    CosNuPi = cos(nuMinP1*Pi);
    Weber[2] = -((1.0+CosNuPi)*sZero+
     nuMinP1*(1.0-CosNuPi)*sMOne/x)/PiX;
    }
    /* Use forward iteration for nAWMax-2 terms   */
    dtnuDx = 2.0/x;
    tnuDx = 2.0*nuMin/x;
    tDPiX = 2.0/(Pi*x);
    nuPi = nuMin*Pi;
    for ( n = 2; n < nAWMax; n++ )
       {
       tnuDx += dtnuDx;
       nuPi += Pi;
       Anger[n+1] = tnuDx*Anger[n]-Anger[n-1]-
        tDPiX*sin(nuPi);
       Weber[n+1] = tnuDx*Weber[n]-Weber[n-1]-
        tDPiX*(1.0-cos(nuPi));
       }
 }

double LittleS(double mew, double nu,
 double x, double eps)
/* Lommel function  s(mew,nu,x)  by power series
    to fractional accuracy  eps  */
```

```
{
double mewP1,nus,Mxs,mewPtkP1,term,sum,ratio;

mewP1 = mew+1.0;
nus = nu*nu;
Mxs = - x*x;
mewPtkP1 = mewP1;
term = 1.0;
sum = 1.0;
ratio = 10.0;
while ( ratio > eps )
  {
  mewPtkP1 += 2.0;
  term *=  Mxs/(mewPtkP1*mewPtkP1-nus);
  sum += term;
  ratio = fabs(term/sum);
  }
return  pow(x,mewP1)*sum/(mewP1*mewP1 - nus);
}
double BigS(double mew, double nu,
 double x, double eps)
/* Lommel function  S(mew,nu,x)  by asymptotic
 expansion to accuracy fractional accuracy eps */
{
double nus,MRxs,mewMtkP1,term,termold,
 sum,ratio,termnew;
int k;

nus = nu*nu;
MRxs = -1.0/(x*x);
mewMtkP1 = mew+1.0;
k = 0;
term = 1.0;
termold = 1.0;
sum = 1.0;
ratio = 10.0;
while ( ratio > eps )
  {
  k++;
  mewMtkP1 -= 2.0;
  term *=  MRxs*(mewMtkP1*mewMtkP1-nus);
  sum += term;
  ratio = fabs(term/sum);
  termnew = fabs(term);
  if ( termnew > termold )
    {
    printf
    ("\n\n!!In BigS diverging after %i terms",k);
    ratio = 0.0;
    }
  else
    { termold = termnew; }
  }
return  sum;
}
```

Test Values
Check values of the Anger function $\mathbf{J}_v(x)$ can be generated by running the AngerJ cell in *Mathematica* notebook StruveAngerWeber. In order to obtain values accurate to 10 decimal digits—the accuracy goal for all functions in the *Atlas*—the options controlling the accuracy of *Mathematica* function NIntegrate for numerical integration must be adjusted carefully, as indicated in the annotated listing of the driver for AngerJ in Section 20.15.

Sample numerical values of $\mathbf{J}_v(x)$ that exercise both methods used in the C function AngerWeber are given in Table 15.2.1. Input nAWMax = 3 in the driver program, and exercise both methods of computing initial values by using the indicated x values.

TABLE 15.2.1 Test values to 10 digits of the Anger functions, J. For $x = 30$ the regular cylindrical Bessel function has been subtracted, as indicated in (15.2.9).

$x \setminus v$	0.3	1.3	2.3
1.0	0.9047516990	0.2061459351	0.1462639471
3.0	0.006366655316	0.3816160596	0.4960460012
30.0	0.008489868474	−0.008219435284	0.007965654295

15.2.3 Weber Function E

Keywords
Bessel functions, Lommel functions, Weber functions

C Function and *Mathematica* Notebook
AngerWeber (C); cell WeberE in notebook StruveAngerWeber (*Mathematica*)

Background and Function Definition
Mathematical analysis of Weber functions is considered in detail in Section 10.1 of Watson's monograph on Bessel functions [Wat44], formulas are given in Section 7.5.3 of the compendium by Erdélyi et al [Erd53b], while Jahnke and Emde [Jah45] give in Section 8.9 a function $\Omega_p(z) \equiv -\mathbf{E}_p(z)$, the Lommel-Weber function, they show curves of the function and give 4-figure tables. Section 8.58 of Gradshteyn and Ryzhik [Gra94] has several formulas, as does Section 12.3 in Abramowitz and Stegun [Abr64]. The main results that we use are from Watson.

The Weber function, $\mathbf{E}_v(z)$, is defined by

$$\mathbf{E}_v(z) \equiv \int_0^\pi \frac{\sin(v\theta - z\sin\theta)}{\pi} d\theta \qquad (15.2.12)$$

Unlike the Anger function in Section 15.2.2, $\mathbf{E}_v(z)$ does not reduce to a Bessel function when v is an integer.

Do not confuse this Weber function with Weber's parabolic cylinder functions in Section 18.1.

Visualization

Figure 15.2.3 shows surfaces of $\mathbf{E}_\nu(z)$ as a function of the complex variable $z = x + iy$, the order, ν, being fixed.

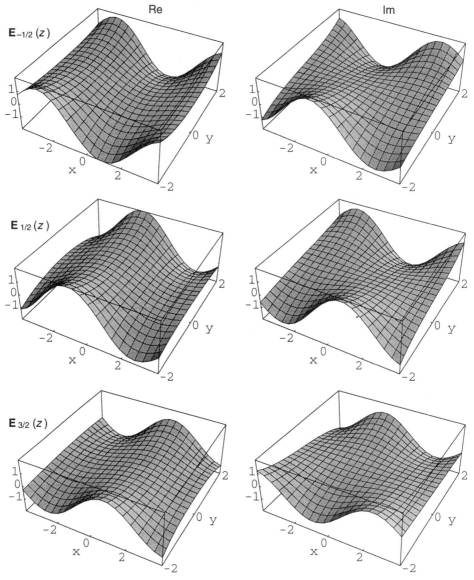

FIGURE 15.2.3 Weber functions in the complex plane for $\nu = -1/2$, $1/2$, $3/2$ from top to bottom; real parts left, imaginary parts right. (Adapted from cell `WeberE` in *Mathematica* notebook `StruveAngerWeber`.)

It is also interesting to map the surface of $\mathbf{E}_\nu(z)$ in the x–ν plane, as shown in Figure 15.2.4 on the following page. Also, the choice of direction to iterate n in order to minimize subtractive cancellation in C function `AngerWeber` is helped by making transections of this surface, as seen on the right side of the figure.

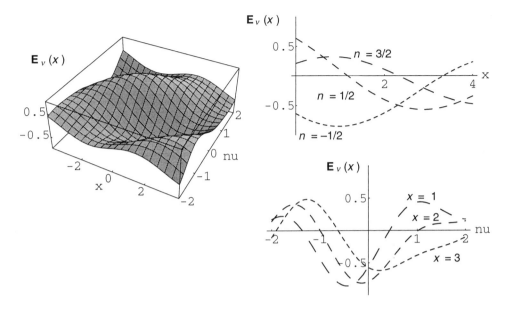

FIGURE 15.2.4 Weber functions in the x–n plane (left), vs x for $n = -1/2$, $1/2$, $3/2$ (right top) and vs n for $x = 1, 2, 3$ (right bottom).   (Adapted from cell `WeberE` in *Mathematica* notebook `StruveAngerWeber`.)

Algorithm, Program, and Programming Notes

As discussed for the Anger function in Section 15.2.2, it might appear that brute-force numerical integration in the relation (15.2.12) for $\mathbf{E}_\nu(x)$ is appropriate for computing the Weber functions. The accuracy of values obtained by this method is not easy to control, and the number of integration points needed for high accuracy may be large, particularly if ν or x is large, since the integrand then oscillates rapidly. We therefore use numerical integration only for the *Mathematica* computations, which are relatively very slow and need to be tweaked to obtain the accuracy required.

Our algorithm computes accurately and efficiently an array of Weber functions, $\mathbf{E}_\nu(x)$, beginning with $0 < \nu_m < 1$, called `numin` in C function `AngerWeber`. It first computes the starting values $\mathbf{E}_{\nu_m}(x)$ and $\mathbf{E}_{\nu_m+1}(x)$ by the power series expressions in (15.2.7) if $x < 30$ and the asymptotic expansion (15.2.9) for larger x. Each function is obtained with fraction accuracy `eps`, which is an input parameter. In addition to the formulas in Section 15.2.1, the algorithm uses the recurrence relations for Weber functions given by [Wat44, Section 10.12]

$$\mathbf{E}_{\nu-1}(x) + \mathbf{E}_{\nu+1}(x) = \frac{2\nu}{x} \mathbf{E}_\nu(x) - \frac{2[1 - \cos(\nu\pi)]}{\pi x} \tag{15.2.13}$$

Forward iteration is used to generate a total of `nAWMax` values, including the starting values. This direction of iteration is slightly sensitive to subtractive cancellation when x is small; it is, however, straightforward to modify function `AngerWeber` to start at a different value of ν, then to iterate downward if necessary. When $x = 0$, (15.2.13) cannot be used, but the integrals can be computed directly to give $\mathbf{E}_\nu(0) = [1 - \cos(\nu\pi)]/(\nu\pi)$.

In the asymptotic region, $x \geq 30$, function AngerWeber returns—as in (15.2.9)—the combination $\mathbf{E}_\nu(x) + Y_\nu(x)$, where $Y_\nu(x)$ is the irregular cylindrical Bessel function in Section 14.2.2. If necessary, adapt function BessYN in Section 14.2.2 to handle noninteger ν values so that $Y_\nu(x)$ can be subtracted out.

The Lommel functions, $s_{\mu,\nu}(x)$ in (15.2.8) and $S_{\mu,\nu}(x)$ in (15.2.10), are computed by functions LittleS and BigS, respectively, given in the program listing in Section 15.2.2. The switchover value of x at $x = 30$ between power series and asymptotic expansions is determined by our accuracy goal of at least 10^{-10} fractional accuracy for functions in the *Atlas*. No more than 50 terms are required in the power series, and only 15 terms in the asymptotic expansion are needed to achieve such convergence. If the accuracy you need is more than this, the switchover must be moved to higher x. For less accuracy and faster computation, reduce the switchover value.

Since the C function AngerWeber also computes the Anger functions, $\mathbf{J}_\nu(x)$, by using the same steps as for the Weber functions, the single program listed in Section 15.2.2 serves both functions. The C driver program for testing AngerWeber is in Section 20.15.

Test Values

Check values of the Weber function $\mathbf{E}_\nu(x)$ can be generated by running the WeberE cell in *Mathematica* notebook StruveAngerWeber. To obtain values accurate to 10 decimal digits—the accuracy goal for functions in the *Atlas*—the options controlling the accuracy of *Mathematica* function NIntegrate for numerical integration must be adjusted carefully, as indicated in the annotated listing of WeberE in Section 20.15.

Sample numerical values of $\mathbf{E}_\nu(x)$ that exercise both methods used in the C function AngerWeber are given in Table 15.2.2. Input nAWMax = 3 in the driver program, and exercise both methods of computing initial values by using the given x values.

TABLE 15.2.2 **Test values to 10 digits of the Weber functions, E. For $x = 30$ the irregular cylindrical Bessel function has been added, as indicated in (15.2.9).**

$x \backslash \nu$	0.3	1.3	2.3
1.0	−0.1631599870	0.3971390361	0.1849059951
3.0	−0.6151235107	−0.2970461334	0.02074503311
30.0	−0.01687360542	−0.005105249328	−0.01726269905

REFERENCES ON STRUVE, ANGER, AND WEBER FUNCTIONS

[Abr64]   Abramowitz, M., and I. A. Stegun, *Handbook of Mathematical Functions,* Dover, New York, 1964.

[Erd53b]   Erdélyi, A., et al, *Higher Transcendental Functions*, Vol. 2, McGraw-Hill, New York, 1953; reprint edition, Krieger, Malabar, Florida, 1981.

[Gra94]   Gradshteyn, I. S., and I. M. Ryzhik, *Table of Integrals, Series, and Products*, Academic Press, Orlando, fifth edition, 1994.

[Jah45]   Jahnke, E., and F. Emde, *Tables of Functions*, Dover, New York, fourth edition, 1945.

[McL55]   McLachlan, N. W., *Bessel Functions for Engineers*, Clarendon Press, Oxford, second edition, 1955.

[Spa87]   Spanier, J., and K. B. Oldham, *An Atlas of Functions*, Hemisphere Publishing, Washington, D.C., 1987.

[Wat44]   Watson, G. N., *A Treatise on the Theory of Bessel Functions*, Cambridge University Press, New York, second edition, 1944.

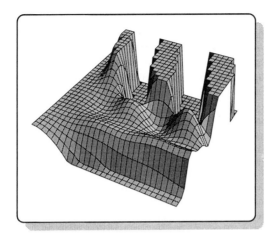

Chapter 16

HYPERGEOMETRIC FUNCTIONS
AND COULOMB WAVE FUNCTIONS

Hypergeometric functions of regular and confluent varieties are of great utility for analytical investigations, whereas Coulomb wave functions—involving special cases of confluent functions—are used to solve practical problems in the quantum mechanics of atoms and nuclei. This chapter shows how to visualize and to compute efficiently the hypergeometric functions $_2F_1$, $_1F_1$, and U, as well as the Coulomb wave functions—the regular function F_L and the irregular function G_L—plus their derivatives F_L' and G_L'.

Monographs by Gasper and Rahman [Gas90] and by Exton [Ext83] emphasize mathematical analysis of generalized hypergeometric functions, whereas physics applications are given in Seaborn's text [Sea91] and in Chapters 4 and 7 of Hochstadt's text on mathematical physics [Hoc86]. Many developments are summarized in Chapters 2, 4–6 of the treatise on higher transcendental functions edited by Erdélyi et al [Erd53a]. Practical formulas abound in Sections 9.1 and 9.2 of Gradshteyn and Ryzhik [Gra94], and in Chapters 13–15 of the handbook of functions edited by Abramowitz and Stegun [Abr64], which—as usual in the *Atlas*—is our primary reference. A few visualizations of confluent hypergeometric functions are given in Chapter 10 of Jahnke and Emde [Jah45], some confluent and Coulomb wave functions are shown in Abramowitz and Stegun, but there are no curves of hypergeometric functions in the atlas of Spanier and Oldham [Spa87].

16.1 HYPERGEOMETRIC FUNCTIONS

Keywords
Gauss hypergeometric functions, hypergeometric functions

C Function and *Mathematica* Notebook
Hyper2F1 (C); cell Hypergeometric2F1 in notebook Hyprgmtrc (*Mathematica*)   M

461

Background

Although hypergeometric functions of great generality have been defined and studied, and although they are sometimes useful to practitioners, attention in this section is restricted to the classical hypergeometric function $_2F_1(a,b,c,z)$, also denoted by $F(a,b;c;z)$. We use the former notation in order to conform with the notations for our C program and the *Mathematica* function.

The hypergeometric function is described extensively in Chapter 2 of Erdélyi et al [Erd53a] and formulas are summarized in Chapter 15 of Abramowitz and Stegun [Abr64]. Special cases of $_2F_1(a,b,c,z)$ include polynomials when either a or b is a negative integer, for example, the Chebyshev, Gegenbauer, Legendre, and Jacobi polynomials described in Sections 11.2.1, 11.3, 11.6, and 11.7, respectively, of the *Atlas*. It is generally not efficient, however, to compute these functions as special cases of $_2F_1$.

Legendre functions, of both the regular and irregular kinds as described in Chapter 12, are also related to hypergeometric functions. This viewpoint is emphasized in Chapter 3 of Erdélyi et al [Erd53a], where many formulas are given; some of these are listed in Section 15.4 of Abramowitz and Stegun [Abr64]. Again, for numerical work it is clearer and more efficient to use the specially-coded C functions rather than the related $_2F_1$ function.

Function Definition

The hypergeometric function can be defined as the solution of the differential equation

$$z(1-z)w'' + [c - (a+b+1)z]w' - abw(z) = 0 \tag{16.1.1}$$

that is regular at the origin and normalized (standardized) so that $w(0) = {}_2F_1(a,b,c,0) = 1$. The series solution can be expressed in terms of the rising-factorial functions (Pochhammer symbols, Section 7.1.2) of a, b, and c, as

$$_2F_1(a,b,c,z) = \sum_{n=0}^{\infty} \frac{(a)_n (b)_n z^n}{(c)_n n!} = 1 + \frac{abz}{c} + \frac{a(a+1)b(b+1)z^2}{c(c+1)\,2!} + \ldots \tag{16.1.2}$$

which converges for most values of the parameters a, b, c if $|z| < 1$. The series is not defined if $c = -m$, where m is a positive integer, unless either a or b is also a negative integer equal to $-n$, where n is a positive integer with $n < m$. Under these conditions, the series becomes a polynomial of order n.

Visualization

The hypergeometric function $_2F_1(a,b,c,z)$ could be visualized in a confusing variety of ways because it has three parameters—each of which can be complex—and an argument, z, which can also be complex. In most applications, however, each of a, b, and c, is a small integer or half integer, and $z = x$, a real variable. We therefore show mainly such examples, either as surfaces of the function (Figure 16.1.1) or as curves as functions of x with a, b, and c all fixed (Figures 16.1.2 and 16.1.3).

If you want to explore $_2F_1$ visually for other parameter and argument values, it is straightforward to modify the cell `Hypergeometric2F1` in *Mathematica* notebook `Hyprgmtrc`; its annotated listing is given in Section 20.16, and the cell is on the CD-ROM that comes with the *Atlas*.

Notice in each of Figures 16.1.1–16.1.3 that if a (or equivalently b) is small in magnitude then the hypergeometric function varies smoothly with x. This behavior changes rapidly as the magnitudes of a or b increase beyond unity or if c is close to a negative integer.

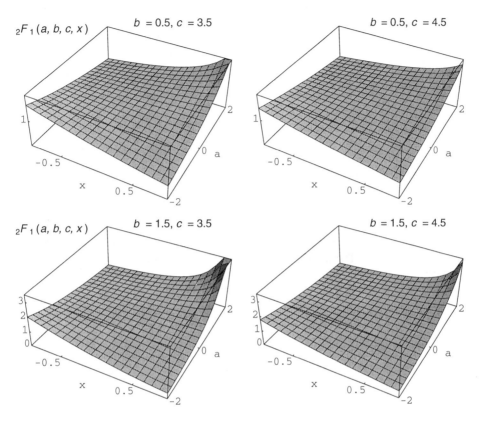

FIGURE 16.1.1 Hypergeometric functions as surfaces in x and a with b and c fixed. (Adapted from cell `Hypergeometric2F1` in *Mathematica* notebook `Hyprgmtrc`.)

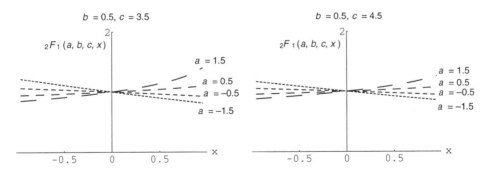

FIGURE 16.1.2 Curves of hypergeometric functions versus x, with $a, b,$ and c fixed. (Adapted from cell `Hypergeometric2F1` in *Mathematica* notebook `Hyprgmtrc`.)

More curves of the hypergeometric functions are shown on the following page.

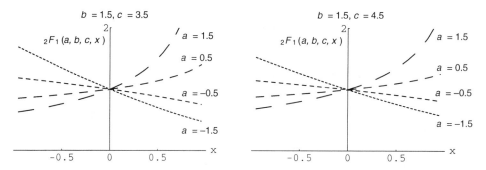

FIGURE 16.1.3 Curves of hypergeometric functions versus x, with a, b, and c fixed. (Adapted from cell `Hypergeometric2F1` in *Mathematica* notebook `Hyprgmtrc`.)

Algorithm, Program, and Programming Notes

For computing the hypergeometric function, $_2F_1(a,b,c,x)$, we use directly the defining power series (16.1.2). The driver program (Section 20.16) does not make tests to ensure that the series is convergent—users of the *Atlas* are assumed to have some travel experience in applied mathematics. The C function is simply as follows:

```
double Hyper2F1(double a, double b, double c,
          double x, double eps)
{
/* Hypergeometric function 2F1
    to fractional accuracy  eps  */
double  sum,ratio,aks,bks,kfact,cks,tk,xpow,
  eps10,term;

sum = 1.0;
ratio = 10.0;
aks = a;
bks = b;
kfact = 1.0;
cks = c;
tk = 1.0;
xpow = 1.0;
eps10 = 0.1*eps;
while ( ratio > eps10 )
  {
  tk *= aks*bks/(kfact*cks);
  xpow *= x;
  term = tk*xpow;
  sum += term;
  aks += 1.0;
  bks += 1.0;
  kfact += 1.0;
  cks += 1.0;
  ratio = fabs(term/sum);
  }
return  sum;
}
```

Test Values

Check values of $_2F_1(a,b,c,x)$ can be generated by running the `Hypergeometric2F1` cell in *Mathematica* notebook `Hyprgmtrc`; samples are given in Tables 16.1.1 and 16.1.2.

TABLE 16.1.1 **Exact test values of hypergeometric functions; $a = -2, b = -1$.**

$x \setminus c$	0.5	1.0	1.5
0.3	2.2	1.6	1.4
0.6	3.4	2.2	1.4
0.9	4.6	2.8	2.2

TABLE 16.1.2 **Test values to 10 digits of the hypergeometric functions with $a = 1, b = 1$.**

$x \setminus c$	0.5	1.0	1.5
0.3	1.970661834	1.428571429	1.264877612
0.6	5.213046033	2.500000000	1.808697355
0.9	47.47137317	10.00000000	4.163485908

16.2   CONFLUENT HYPERGEOMETRIC FUNCTIONS

Confluent hypergeometric functions can be defined in terms of solutions of

$$zw'' + (b-z)w' - aw(z) = 0 \qquad (16.2.1)$$

Two linearly-independent solutions are usually discussed, variously denoted by $_1F_1(a,b,z)$, $\Phi(a,b,z)$, or $M(a,b,z)$ for the first solution (regular near the origin), and by $U(a,b,z)$ or $\Psi(a,b,z)$ for the second solution (irregular near the origin). In Section 16.2.1 we explore the first solution and in Section 16.2.2 we develop the second solution.

16.2.1 Regular Function $_1F_1$

Keywords

confluent hypergeometric functions, hypergeometric functions, Kummer function

C Function and *Mathematica* Notebook

Hyper1F1 (C); cell `Confluent1F1` in the notebook `Hyprgmtrc` (*Mathematica*)

Background

The confluent hypergeometric function is described extensively in Chapter 16 of Whittaker and Watson [Whi27], in Chapter 6 of Erdélyi et al [Erd53a] and in Slater [Sla60], while formulas are summarized in Chapter 13 of Abramowitz and Stegun [Abr64]. Special cases of $_1F_1(a,b,z)$ that include many of the functions in the *Atlas* are listed in Section 13.6 of Abramowitz and Stegun. We use the notation $_1F_1(a,b,z)$ to conform with the notations for the *Mathematica* function and our C program.

Function Definition

The function $_1F_1(a,b,z)$ can be defined as the solution of (16.2.1) that is regular at the origin and is normalized (standardized) so that $w(0) = {}_1F_1(a,b,0) = 1$. The series solution can be expressed in terms of rising-factorial functions (Pochhammer symbols, Section 7.1.2) of a and b, as

$$_1F_1(a,b,z) = \sum_{n=0}^{\infty} \frac{(a)_n z^n}{(b)_n\, n!} = 1 + \frac{az}{b} + \frac{a(a+1)z^2}{b(b+1)\,2!} + \cdots \qquad (16.2.2)$$

which converges for most values of parameters a and b. The series is undefined if $b = -m$, where m is a positive integer, unless a is also a negative integer equal to $-n$, with n a positive integer with $n < m$. Under these conditions, the series is a polynomial of order n.

Visualization

The hypergeometric function $_1F_1(a,b,z)$ can be visualized in many ways, since a and b can be complex—as may be the argument z. In many applications, however, a and b are small integers or half integers, and $z = x$ or $z = 2ix$. We therefore show such examples, either as surfaces of the function (Figures 16.2.1 and 16.2.2) or as curves as functions of x with a and b fixed (Figure 16.2.3).

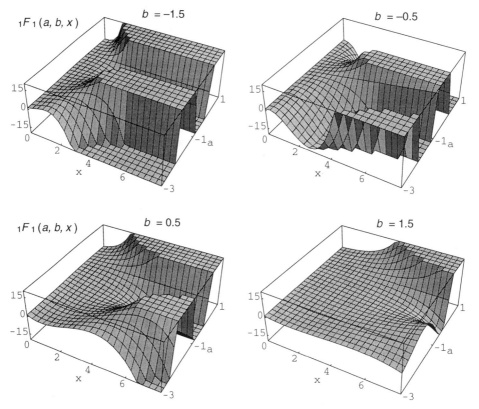

FIGURE 16.2.1 Confluent hypergeometric functions as surfaces in x in range [0, 8] and a in [−3, 1] with b fixed. (Adapted from cell `Confluent1F1` in *Mathematica* notebook `Hyprgmtrc`.)

In Figure 16.2.1 the functions are truncated when their magnitude exceeds 20. When the argument of $_1F_1(a,b,z)$ is purely imaginary, the functions become oscillatory when $a > 0$, as shown in Figure 16.2.2.

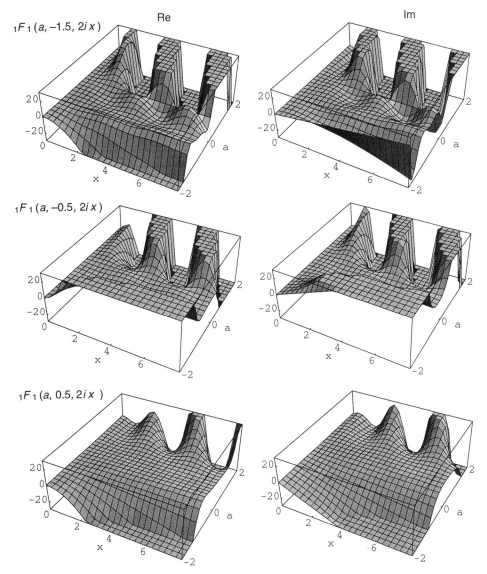

FIGURE 16.2.2 Confluent hypergeometric functions of purely imaginary argument as surfaces in x in range $[0, 8]$ and a in $[-2, 2]$ with b fixed. Real parts at left, imaginary parts at right. (Adapted from cell `Confluent1F1` in *Mathematica* notebook `Hyprgmtrc`.)

Curves of $_1F_1(a,b,x)$, with x varying while a and b are fixed, show how rapidly the function changes with a. Examples shown in Figure 16.2.3 can easily be increased by running the cell `Confluent1F1` in *Mathematica* notebook `Hyprgmtrc`.

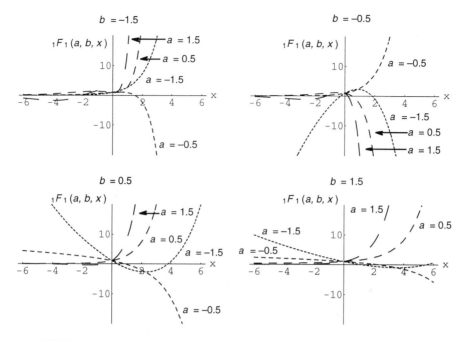

FIGURE 16.2.3 Curves of confluent hypergeometric functions versus x, with a and b fixed. Values are truncated at magnitude 20. (Adapted from cell `Confluent1F1` in the *Mathematica* notebook `Hyprgmtrc`.)

Algorithm and Program

For computing the confluent hypergeometric function, $_1F_1(a,b,x)$, we use the power series (16.2.2) for $|x| < 50$, which usually produces convergence to 10^{-10} relative accuracy in fewer than 100 terms. Whenever $b = a$, however, we use

$$_1F_1(a,a,z) = e^z \qquad (16.2.3)$$

For $|x| \geq 50$ we use asymptotic series derived from the integral representation given in Section 3.1 of Slater's monograph [Sla60] and cited as 13.2.1 in Abramowitz and Stegun [Abr64], namely,

$$\frac{\Gamma(b-a)\Gamma(a)}{\Gamma(b)} \, _1F_1(a,b,z) = \int_0^1 e^{zt} t^{a-1} (1-t)^{b-a-1} dt \qquad \mathrm{Re}\, b > \mathrm{Re}\, a > 0 \qquad (16.2.4)$$

From this one derives for $z = x$, a real variable, and for a and b real, the asymptotic expansion formulas [Abr64, 13.5.1]

$$_1F_1(a,b,x) \sim \left| \begin{array}{ll} \dfrac{\Gamma(b)e^x x^{a-b}}{\Gamma(a)} \displaystyle\sum_{k=0} \dfrac{(b-a)_k(1-a)_k}{k!\,x^k} & x \gg 0 \quad b > 0 \ a > 0 \\[3mm] \dfrac{\Gamma(b)(-x)^{-a}}{\Gamma(b-a)} \displaystyle\sum_{k=0} \dfrac{(a)_k(1+a-b)_k}{k!\,(-x)^k} & x \ll 0 \quad b > a \end{array} \right. \qquad (16.2.5)$$

in terms of rising factorials (Section 7.1.2). If the conditions on a and b are not satisfied, the following C function sets the asymptotic estimate of $_1F_1(a,b,x)$ to zero.

```
double Hyper1F1(double a, double b,
          double x, double eps)
/* Hypergeometric function 1F1
   to fractional accuracy  eps  */
{
double  sum,ratio,aks,bks,kfact,tk,xpow,
 eps10,term,fk,bMaPkM1,OneMaPkM1,termold,termnew,
 aPkM1,aMbPk,Mx,LogGamma(double x);
int k;
if ( a == b )  return  exp(x);
if ( b <= 0.0 )
  {
   printf
   ("\n\n!!In Hyper1F1  b<=0; 0 returned");
   return  0.0;
  }

eps10 = 0.1*eps;
if ( fabs(x) < 50.0 )
  {
  sum = 1.0;
  ratio = 10.0;
  aks = a;
  bks = b;
  kfact = 1.0;
  tk = 1.0;
  xpow = 1.0;
  while ( ratio > eps10 )
    {
    tk *= aks/(bks*kfact);
    xpow *= x;
    term = tk*xpow;
    sum += term;
    aks += 1.0;
    bks += 1.0;
    kfact += 1.0;
    ratio = fabs(term/sum);
    }
  return  sum;
  }

/* Use asymptotic series */
if ( x >= 50.0 )
  {
  if ( a <= 0.0 )
    {
    printf
    ("\n\n!!In Hyper1F1 asymptotic a<=0; 0 returned");
    return  0.0;
    }
  k = 0;
  fk = 1.0;
  bMaPkM1 = b - a;
  OneMaPkM1 = 1.0 - a;
```

```
 term = 1.0;
 termold = 100.0;
 sum = 1.0;
 ratio = 10.0;
 while ( ratio > eps10 )
  {
  k++;
  term *=  bMaPkM1*OneMaPkM1/(fk*x);
  sum += term;
  bMaPkM1 += 1.0;
  OneMaPkM1 += 1.0;
  fk += 1.0;
  ratio = fabs(term/sum);
  termnew = fabs(term);
  if ( termnew > termold )
    {
    printf
    ("\n\n!!In Hyper1F1 diverging after %i terms",k);
    ratio = 0.0; }
  else
    { termold = termnew; }
  }
 return  exp(LogGamma(b)+x-LogGamma(a))*
         pow(x,a-b)*sum;
 }
/*   x <= -50   asymptotic series    */
if ( b <= a )
 {
 printf
 ("\n\n!!In asymptotic Hyper1F1  b<=a; 0 returned");
  return  0.0; }
Mx = -x;
k = 0;
fk = 1.0;
aMbPk = a - b + 1.0;
aPkM1 = a;
term = 1.0;
termold = 100.0;
sum = 1.0;
ratio = 10.0;
while ( ratio > eps10 )
 {
 k++;
 term *= aPkM1*aMbPk/(fk*Mx);
 sum += term;
 aPkM1 += 1.0;
 aMbPk += 1.0;
 fk += 1.0;
 ratio = fabs(term/sum);
 termnew = fabs(term);
 if ( termnew > termold )
   {
   printf
   ("\n\n!!In Hyper1F1 diverging after %i terms",k);
   ratio = 0.0; }
```

```
    else
      { termold = termnew; }
    }
  return   exp(LogGamma(b)-LogGamma(b-a))*
           sum/pow(Mx,a);
  }
```

Programming Notes

The ratios of Γ functions that are needed in `Hyper1F1` from (16.2.5) are computed as the exponentials of differences of the logarithms of Γ functions, obtained from function `Log-Gamma` described in Section 6.1.1. The C function for the confluent hypergeometric function that is on the CD-ROM comes completely assembled with its driver program (Section 21.16) and with `LogGamma`.

Convergence of the power series (16.2.2) for ${}_1F_1(a,b,x)$ and of the asymptotic expansions (16.2.5) require a convergence parameter, input as the fractional accuracy `eps`. Thus, like all functions in the *Atlas*, `eps` $= 10^{-10}$ is a feasible accuracy for the confluent hypergeometric function.

Test Values

Check values of ${}_1F_1(a,b,x)$ can be generated by running the `Confluent1F1` cell in the *Mathematica* notebook `Hyprgmtrc`; samples are given in Table 16.2.1.

TABLE 16.2.1 Test values to 10 digits of confluent hypergeometric functions with $a = 2.0$.

$x \setminus b$	0.5	1.0	1.5
1.0	12.15039235	5.436563657	3.545117704
3.0	277.5103024	80.34214769	35.9550392
5.0	3823.379408	890.4789546	323.5090268
50.0	$3.346503339 \times 10^{24}$	$2.64419982 \times 10^{23}$	$3.281522692 \times 10^{22}$
100.0	$4.836032033 \times 10^{46}$	$2.714998313 \times 10^{45}$	$2.394193199 \times 10^{44}$

16.2.2 Irregular Function U

Keywords

confluent hypergeometric functions, hypergeometric functions, irregular confluent hypergeometric function, Kummer function, Whittaker function

C Function and *Mathematica* Notebook

`HyperU` (C); cell `ConfluentU` in the notebook `Hyprgmtrc` (*Mathematica*)

Background

The confluent hypergeometric function that may be irregular at the origin, $U(a,b,z)$ or $\Psi(a,b,z)$, is described extensively in Chapter 6 of Erdélyi et al [Erd53a] and formulas are summarized in Chapter 13 of Abramowitz and Stegun [Abr64]. Derivations of many of these formulas are given in Slater's monograph on the confluent hypergeometric function

[Sla60]. Special cases of $U(a,b,z)$ that include many of the functions in the *Atlas* are listed in Section 13.6 of Abramowitz and Stegun. Temme [Tem83] gives algorithms for computing $U(a,b,x)$, while Izarra et al [Iza95] use Padé approximants for the closely related Whittaker functions.

Function Definition

The function $U(a,b,z)$ can be defined as a solution of (16.2.1), but a definition that is more appropriate for our purpose is given by [Abr64, 13.2.5] as

$$\Gamma(a)\,U(a,b,z) = \int_0^{\widetilde{\infty}} e^{-zt} t^{a-1}(1+t)^{b-a-1}\,dt \qquad \mathrm{Re}\,z > 0 \quad \mathrm{Re}\,a > 0 \qquad (16.2.6)$$

which is similar to the integral definition of $_1F_1(a,b,z)$ in (16.2.4). It is convenient to re-write this as

$$\Gamma(a)\,U(a,b,z) = z^{-a}\int_0^{\widetilde{\infty}} e^{-w} w^{a-1}(1+w/z)^{b-a-1}\,dw \qquad \mathrm{Re}\,z > 0 \quad \mathrm{Re}\,a > 0 \qquad (16.2.7)$$

Visualization

As usual in an atlas, we show many maps of territories that may interest you, starting with the topography of the $U(a,b,x)$ function.

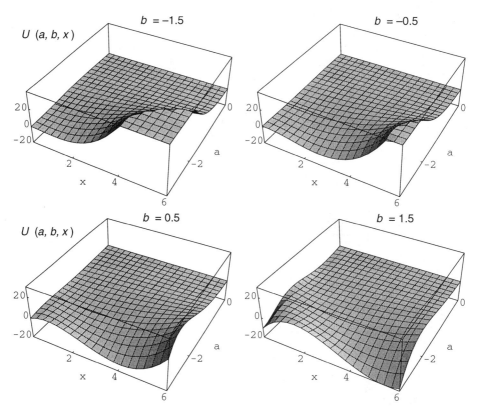

FIGURE 16.2.4 Confluent hypergeometric functions, U, as surfaces for x in $[0.1, 6]$ and a in $[-3, 0]$ with b fixed. Values exceeding 40 have been truncated. (Adapted from cell ConfluentU in *Mathematica* notebook Hyprgmtrc.)

The function $U(a,b,z)$ can be visualized in many ways, since a, b, and z can be complex. In typical applications, however, a and b are small integers or half integers, and $z = x$. We therefore show such examples, either as surfaces (Figures 16.2.4 and 16.2.5) or as curves versus x with a and b fixed (Figure 16.2.6).

The integral definition (16.2.7) shows that if $a < 0$ then $U(a,b,x)$ increases rapidly as x increases, and that the general behavior of the function is dominated by the x^{-a} outside the integral if $b - a$ is small. As $x \to 0$, on the other hand, $U(a,b,x)$ is divergent if $b \geq 1$, either logarithmically or as x^{1-b}; details are given in Section 6.8 of [Erd53a] and in Section 4.1.2 of Slater [Sla60]. The dominance of this x dependence for small x is also clear from (16.2.7).

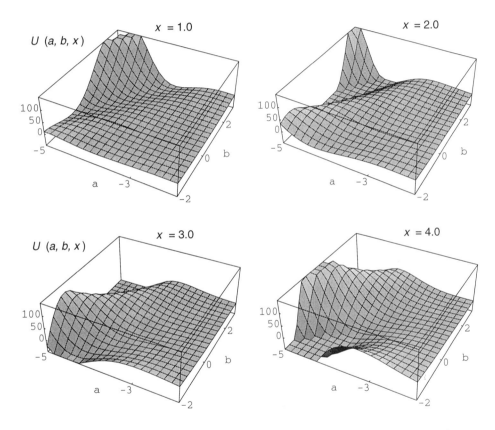

FIGURE 16.2.5 Confluent hypergeometric functions, U, as surfaces for a in $[-5, -2]$ and b in $[-2, 3]$ with x fixed. Values are truncated if less than -50 or greater than 150. (Adapted from cell `ConfluentU` in *Mathematica* notebook `Hyprgmtrc`.)

More detail of how $U(a,b,x)$ varies with x while a and b are fixed (transections of the surfaces in Figure 16.2.4), show how rapidly this confluent hypergeometric function changes with a. The examples shown in Figure 16.2.6 on the following page can easily be increased by running the cell `ConfluentU` in the *Mathematica* notebook `Hyprgmtrc`.

M

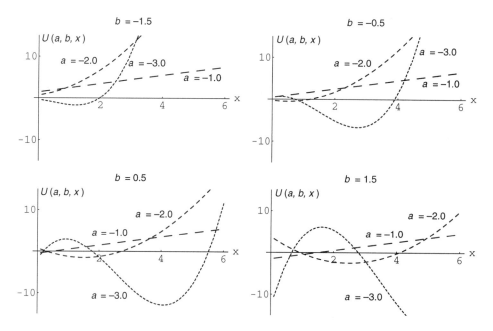

FIGURE 16.2.6 Curves of the confluent hypergeometric function, U, versus x, with a and b fixed. Values are truncated at magnitude 15. (Adapted from cell `ConfluentU` in the *Mathematica* notebook `Hyprgmtrc`.)

Algorithm and Program

For computing the confluent hypergeometric function $U(a,b,x)$ for $a > 0$, b real, and $x > 0$, we use the integral definition (16.2.7) for $x < 60$, with an asymptotic expansion for larger x values.

For the integration we use the nine-strip, equal-step, Newton-Cotes formula—equation 25.4.19 in Abramowitz and Stegun [Abr64]—namely,

$$\int_{x_0}^{x_9} f(x)\, dx \approx \frac{9h}{89600} [2857(f_0 + f_9) + 15741(f_1 + f_8) +$$

$$1080(f_2 + f_7) + 19344(f_3 + f_6) + 5778(f_4 + f_5)] \quad (16.2.8)$$

which has an error of order h^{11} multiplied by the 10th derivative of the integrand, where h is the step size. If the upper limit on the integral in (16.2.7) is set at $w = 80$ and we use $40 \times 27 = 1080$ integration points, then the integration can handle the range of x, a, and b values in Table 16.2.2 if we demand the usual accuracy goal for functions in the *Atlas* of 10^{-10}. To handle the programming conveniently, the C function `NineStripU` given below includes three applications of (16.2.8) for 28 points in each block of 27 strips.

For $|x| \geq 60$ we use an asymptotic expansion derived from the integral representation given in Section 4.1.2 of Slater's monograph [Sla60] and cited as 13.5.2 in Abramowitz and Stegun [Abr64], namely, in terms of rising factorials (Section 7.1.2)

$$U(a,b,x) \sim x^{-a} \sum_{k=0}^{\infty} \frac{(a)_k (1+a-b)_k}{k!(-x)^k} \qquad x \gg 0 \quad a > 0 \quad (16.2.9)$$

TABLE 16.2.2 Limits on a and b for given x to get 10-digit accuracy in the confluent hypergeometric function U.

x	a	b
0.01	10	20
0.1	20	30
1.0	25	30
10.0	25	35

Convergence of the expansion (16.2.9) is limited by the size of a and of $a - b$. For example, if we require 10-digit accuracy in $U(a,b,x)$ for $|x| \geq 60$ then $a = 8$ and $b = 5$ is an upper limit on these parameters. This topic is discussed further in the Programming Notes.

The C function for the confluent hypergeometric function $U(a,b,x)$ for $a > 0$, b real, and $x > 0$ is as follows:

```
double HyperU(double a, double b,
  double x, double eps)
/* Hypergeometric function U
    to fractional accuracy  eps   */
{
double  Dw,integral,wStart,
  sum,ratio,eps10,term,fk,termold,termnew,
  aPkM1,aMbPk,Mx,LogGamma(double x);
double NineStripU(double a, double b, double x,
  double wStart, double Dw);
const double PiD2 = 1.570796326794897;
int k;

if ( a <= 0.0 )
  {
  printf
  ("\n\n!!In HyperU  a<=0; 0 returned");
  return  0.0;
  }

if ( fabs(x) < 60.0 )
  { /* U  by integration */
  Dw = 2.0/27.0;
  integral = 0.0;
  wStart = 0.0;
  for ( k = 1; k <= 40; k++ )
    {
    integral += NineStripU(a,b,x,wStart,Dw);
    wStart += 2.0;
    }
  return   integral/exp(LogGamma(a)+a*log(x));
  }
```

```
/* Use asymptotic series for  x >= 60  */
eps10 = 0.1*eps;
Mx = -x;
k = 0;
fk = 1.0;
aMbPk = a - b + 1.0;
aPkM1 = a;
term = 1.0;
termold = 100.0;
sum = 1.0;
ratio = 10.0;
while ( ratio > eps10 )
  {
  k++;
  term *= aPkM1*aMbPk/(fk*Mx);
  sum += term;
  aPkM1 += 1.0;
  aMbPk += 1.0;
  fk += 1.0;
  ratio = fabs(term/sum);
  termnew = fabs(term);

  if ( termnew > termold )
    {
    printf
    ("\n\n!!In HyperU diverging after %i terms",k);
    ratio = 0.0;
    }
  else
    { termold = termnew; }
  }
return  pow(x,-a)*sum;
}


double NineStripU(double aa, double b, double x,
 double wStart, double Dw)
/* Nine-strip integral for hypergeometric U */
{
double aM1,bMaM1,w,f[28],integral;
double Power(double x, double p);
const double a[5] = {0.2869754464285714,
  1.581127232142857, 0.1084821428571429,
  1.943035714285714, 0.5803794642857144 };
int k;

aM1 = aa - 1.0;
bMaM1 = b - aa - 1.0;
w = wStart;

for ( k = 0; k <= 27; k++ )
  {
  f[k] = exp(-w)*Power(w,aM1)*Power(1.0+w/x,bMaM1);
  w += Dw;
  }
```

```
integral = Dw*(a[0]*(f[0]+f[27]+2.0*(f[9]+f[18]))+
    a[1]*(f[1]+f[8]+f[10]+f[17]+f[19]+f[26])+
    a[2]*(f[2]+f[7]+f[11]+f[16]+f[20]+f[25])+
    a[3]*(f[3]+f[6]+f[12]+f[15]+f[21]+f[24])+
    a[4]*(f[4]+f[5]+f[13]+f[14]+f[22]+f[23])));
return  integral;
}

double Power(double x, double p)
/* x to power p, including 0 to 0 = 1 */
{
if ( x == 0.0 && p == 0.0 )  return 1.0;
return  pow(x,p);
}
```

Programming Notes

Convergence of the asymptotic expansion (16.2.9) for $U(a,b,x)$ requires a convergence parameter, input as the fractional accuracy eps. Like all functions in the *Atlas*, a value of eps = 10^{-10} is feasible for this confluent hypergeometric function, provided that a and $a-b$ in (16.2.9) are not too large. For $x \geq 60$, 10-digit fractional accuracy is obtained with $0 < a \leq 8$ and $a-b \leq 3$. By increasing the lower bound on x for use of the asymptotic expansion or by reducing the accuracy demanded, convergence of the asymptotic expansion can be achieved for larger parameter values.

Three other functions are needed by HyperU. First is NineStripU, implementing integration formula (16.2.8) with f as the integrand in (16.2.7). Next is Power, which corrects the C library function pow for the error which sets $0^0 = 0$ rather than $0^0 = 1$; this result is needed in (16.2.7) at $w = 0$ if $a = 1$. Finally, the Γ function is used when HyperU is computed by (16.2.7); it is obtained from LogGamma described in Section 6.1.1. The C function for U on the CD-ROM accompanying the *Atlas* comes assembled with its driver program (Section 21.16), plus NineStripU, Power, and LogGamma.

Test Values

Some check values of $U(a,b,x)$ can be generated by running the ConfluentU cell in the *Mathematica* notebook Hyprgmtrc. Unfortunately, *Mathematica* sometimes returns zero for HypergeometricU for a and b about 4 whenever x exceeds about 48. Entries in Table 16.2.3 for such x values were checked against our C function, modified so that the same $U(a,b,x)$ is computed by integration and by asymptotic expansion.

TABLE 16.2.3 Test values to 10 digits of confluent hypergeometric functions with $a = 2.0$.

$x \backslash b$	3.0	4.0	5.0
0.1	100.0	2100.0	64100.0
5.0	0.04000000000	0.05600000000	0.08160000000
60.0	0.0002777777778	0.0002870370370	0.0002967592593

A useful result when testing the U function follows from (16.2.7), namely,

$$U(a,a+1,x) = x^{-a} \qquad x > 0 \qquad a \text{ positive integer} \qquad (16.2.10)$$

16.3   COULOMB WAVE FUNCTIONS

Coulomb wave functions are solutions of the Schrödinger equation in quantum mechanics for the motion of a particle with positive total energy E in the Coulomb potential of another particle when they have relative angular momentum $L = 0, 1$, etc. In dimensionless units

$$w'' + \left[1 - \frac{2\eta}{\rho} - \frac{L(L+1)}{\rho^2} \right] w(\rho) = 0 \qquad (16.3.1)$$

in which η is the Coulomb parameter ($\eta > 0$ for repulsion) and ρ is the particle separation, r, multiplied by the wavenumber, k. Note that $2\eta/\rho$ is the ratio of the Coulomb barrier at radius r to the total energy; thus, $2\eta/\rho > 1$ is the classically-forbidden region in which exponentially-damped behavior of the wave function is expected. Algorithms for negative total energy solutions in terms of Whittaker functions have been given by Izarra et al [Iza95].

Two linearly-independent solutions of (16.3.1) are defined; $F_L(\eta, \rho)$ which is regular near the origin $\rho = 0$, and $G_L(\eta, \rho)$ which is irregular near the origin. In Section 16.3.1 we explore the regular solution and its derivative with respect to ρ, denoted $F_L'(\eta, \rho)$. In Section 16.3.2 we describe the irregular solution and its derivative, $G_L'(\eta, \rho)$.

Our algorithm and the C function `CoulombFG` are designed to compute at the same time arrays of F_L, F_L', G_L, and G_L' for given η and ρ, with the array elements corresponding to $L = 0$, $L = 1$, etc, up to some maximum L value chosen by the program user. The methods used to compute all four functions are described under Algorithm and Program in Section 16.3.1.

16.3.1 Regular Functions and Derivatives F_L, F_L'

Keywords
Coulomb phase shifts, Coulomb wave functions, regular Coulomb wave functions

C Function and *Mathematica* Notebook
`CoulombFG` (C); cell `CoulombF` in the notebook `Hyprgmtrc`(*Mathematica*)

Background
The regular Coulomb wave function is described in a cursory manner in many nuclear and atomic physics texts. In general, computing accurate Coulomb wave functions efficiently is a challenging task, so many specialized methods have been developed. The methods we describe are suitable for nuclear collisions at energies not too far below the Coulomb barrier. For energies much below the barrier (typically in nuclear astrophysics applications) or for collisions between very heavy nuclei (when L is very large) special techniques are required. Some of the methods available are described by Thompson and Barnett [Tho86], Abad and Sesma [Aba92], and by Salvat et al [Sal95]. Our method is adapted from that described by Buck et al [Buc60]. Many formulas that may be useful when computing $F_L(\eta, \rho)$ and $F_L'(\eta, \rho)$ are summarized in Chapter 14 of Abramowitz and Stegun [Abr64].

Function Definition
The real function $F_L(\eta, \rho)$ is a solution of (16.3.1) that is regular at the origin $\rho = 0$. It can be defined in terms of the confluent hypergeometric function (Section 16.2.1) by equation 14.1.3 in Abramowitz and Stegun [Abr64], namely,

$$F_L(\eta, \rho) = C_L(\eta) \rho^{L+1} e^{-i\rho} \, _1F_1(L+1-i\eta, 2L+2, 2i\rho) \qquad (16.3.2)$$

in which, from [Abr64, 14.1.8, 14.1.10]

$$C_L(\eta) = \frac{\sqrt{L^2 + \eta^2}}{L(2L+1)} C_{L-1}(\eta) \qquad C_0(\eta) = \sqrt{\frac{2\pi\eta}{e^{2\pi\eta} - 1}} \qquad (16.3.3)$$

We use (16.3.2) for the *Mathematica* implementation of $F_L(\eta, \rho)$, since $_1F_1$ can be computed for complex arguments. This is not a practical method for efficient calculation of $F_L(\eta, \rho)$ by a C program.

The recurrence relation [Abr64, 14.2.3]

$$F_{L-1} = (2L+1)\left\{\left[\frac{\eta}{L(L+1)} + 1/\rho\right]F_L - \sqrt{\frac{\eta^2}{(L+1)^2} + 1}\, F_{L+1}\right\} \Big/ \sqrt{\frac{\eta^2}{L^2} + 1} \qquad (16.3.4)$$

allows a series of Coulomb functions to be computed by downward iteration once two adjacent values of L are available. The derivative relation that we use is [Abr64, 14.2.1]

$$F_L' = \sqrt{\eta^2 / L^2 + 1}\ F_{L-1} - [\eta / L + L/\rho]F_L \qquad (16.3.5)$$

We start the iteration in L for unnormalized F_L values at an L that is 20 beyond the largest L for which the Coulomb wave functions are required. Downward iteration is generally stable for F_L, thus unnormalized values of F_0 and F_0' are obtained. The Wronskian of the regular and irregular solution satisfies [Abr64, 14.2.4]

$$W_L \equiv F_L'G_L - F_LG_L' = 1 \qquad (16.3.6)$$

which we use to normalize F_0 and F_0' after accurate values of G_0 and G_0' are computed. All the other F_L and F_L' values are then normalized by the same value.

When (16.3.4) and (16.3.5) are used with their analogues for G_L and G_L' to generate the latter, the algebraic correctness of the Wronskian condition is assured; the numerical value of the Wronskian serves only to check the accuracy of the computer arithmetic. On the other hand, our *Mathematica* programs for the Coulomb wave functions—cells `CoulombF` and `CoulombG` in notebook `Hyprgmtrc` that is annotated in Section 20.16—use methods that are quite different from those for the C functions, therefore the Wronskian does serve as a useful check of the results obtained by *Mathematica*.

Accurate values of G_0 and G_0' are computed in a complicated way by using asymptotic expansions for $\rho \gg \eta$, as described in Section 16.3.2. Upward iteration on L for G_L and G_L' is generally stable, so the effort of computing G_0 and G_0' is worthwhile.

Visualization

The regular Coulomb wave function $F_L(\eta, \rho)$ can be visualized in many ways, depending upon the combinations of L, η, and ρ. The following figures show surfaces of $F_L(\eta, \rho)$ and $F_L'(\eta, \rho)$ in which two quantities vary and the other is fixed.

Figure 16.3.1 on the following page allows you to visualize how the regular Coulomb wave functions and their derivatives vary with distance (ρ) and with Coulomb field strength (η) for a fixed partial wave (L). Notice that for $L = 2$ the wave functions are small near $\rho = 0$, corresponding to a centripetal barrier. The classical turning point for a given partial wave L and Coulomb parameter η is the smallest ρ value, $\rho = \rho_L > 0$, for which $F_L''(\eta, \rho_L) = 0$. For $\rho < \rho_L$ the motion is classically forbidden and the quantum wave functions are exponentially damped. You can discern such behavior in the back-left corner of the surfaces for $F_0(\eta, \rho)$ and $F_2(\eta, \rho)$ in Figure 16.3.1.

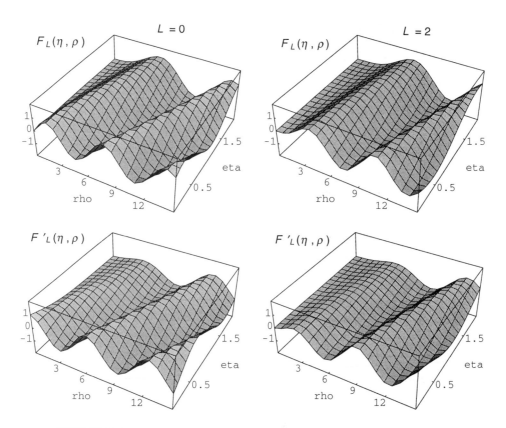

FIGURE 16.3.1 Regular Coulomb wave functions and their derivatives for ρ in $[0, 15]$ and η in $[0, 2]$ with L fixed. (Adapted from cell `CoulombF` in *Mathematica* notebook `Hyprgmtrc`.)

The classical turning point, $\rho < \rho_L$, for a repulsive potential ($\eta > 0$) is given by

$$\rho_L = \eta + \sqrt{\eta^2 + L(L+1)} \tag{16.3.7}$$

For example, $\rho_0 = 2\eta$. It can be seen in Figure 16.3.2 as the ρ value at which $F_L'(\eta,\rho)$ attains its first maximum.

If the regular Coulomb wave functions, $F_L(\eta,\rho)$, and their derivatives with respect to ρ, $F_L'(\eta,\rho)$, are shown as surfaces in the ρ–L plane with the Coulomb parameter, η, fixed, we can visualize how increasing the angular momentum (L) decreases the regular wave function at small radii (corresponding to small ρ) and how a larger Coulomb repulsion (increasing η) enhances this effect, as shown in Figure 16.3.2. Of course, you can also see what an attractive Coulomb potential does to these surfaces by running the cell `CoulombF` in *Mathematica* notebook `Hyprgmtrc` with a negative value of η.

Finally in the gallery of landscapes of $F_L(\eta,\rho)$ and $F_L'(\eta,\rho)$, one can portray η and L varying with ρ fixed. This corresponds, at a fixed energy, to varying the strength of the Coulomb interaction (η) and looking at different partial waves (L), but sampling the wave function at a fixed distance (ρ). Such surfaces are shown in Figure 16.3.3.

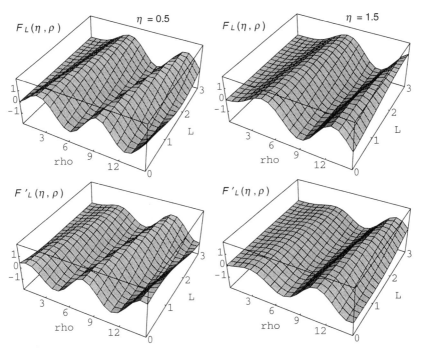

FIGURE 16.3.2 Regular Coulomb wave functions and derivatives for ρ in [0, 15] and L in [0, 3] with η fixed. (Cell `CoulombF` in *Mathematica* notebook `Hyprgmtrc`.)

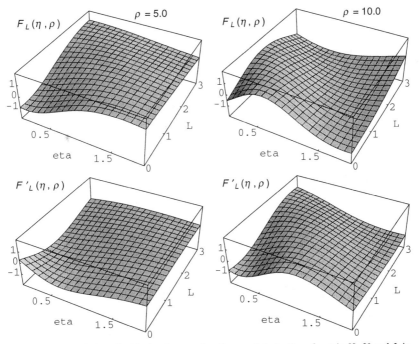

FIGURE 16.3.3 Regular Coulomb wave functions and derivatives for η in [0, 2] and L in [0, 3] with ρ fixed. (Cell `CoulombF` in *Mathematica* notebook `Hyprgmtrc`.)

Algorithm and Program

For computing the regular Coulomb wave function, $F_L(\eta, \rho)$, and its derivative with respect to ρ, $F'_L(\eta, \rho)$, we use the scheme outlined under Function Definition. The function CoulombFG also provides the absolute Coulomb phase shifts, $\sigma_L(\eta)$, given by [Abr64, 14.5.6]

$$\sigma_L(\eta) = \arg \Gamma(L + 1 + i\eta) \tag{16.3.8}$$

in terms of the Γ function (Section 6.1.1). The asymptotic expansion for large L is obtained as [Buc60, (21)]

$$\sigma_L(\eta) \sim (L + 1/2)\alpha + \eta(\ln \beta - 1) -$$

$$[\sin \alpha / 12 - \sin 3\alpha / (360\beta^2) + \sin 5\alpha / (1260\beta^4) -$$

$$\sin 7\alpha / (1680\beta^6) + \sin 9\alpha / (1188\beta^8) - ...]/\beta \tag{16.3.9}$$

$$\alpha = \arctan[\eta / (L+1)] \qquad \beta = \sqrt{\eta^2 + (L+1)^2}$$

then the recurrence relation [Buc60, (22)]

$$\sigma_{L-1}(\eta) = \sigma_L(\eta) - \arctan(\eta / L) \tag{16.3.10}$$

is applied until $L = 0$ is reached. The Coulomb phase shifts are stored in array elements Sigma[L].

Function CoulombFG uses function Gzero to compute $G_0(\eta, \rho)$ and $G'_0(\eta, \rho)$ for large ρ, then it uses Noumerov to integrate down to the ρ value of interest. The algorithms for these two functions are described in Section 16.3.2 but the C program is listed here.

```
#define MAX 51 /* for maximum  L value 30  */

void CoulombFG(int LMAX, double eta, double rho,
  double F[], double FP[], double G[], double GP[],
  double Sigma[])
{
/* Regular Coulomb Wave Functions F & G plus their
   derivatives F' & G' and the Coulomb phase shifts.
   Parameters L, eta, rho */
double  fL,fLP1,beta2,beta,alpha,norm,c3,c4;
void Gzero(double eta, double rho, double SigmaZ,
  double *Check, double *Gzero, double *GzeroP);
void Noumerov(double eta, double rhoN, double h,
  int Nrho, double gzero1, double gzero2,
  double *GzeroNew, double *GzeroH);
double SigmaZ,Check,h,rhoN,Gzero1,Gzero2,GzeroP1,
  GzeroP2,GzeroNew,GzeroH,h2D6,h3D24,TDrho,TDM1,
  A,B,Rrho,etaS;
int L,Nrho,n;

L = LMAX; /* Absolute Coulomb phase shifts */
fL = L;
fLP1 = fL+1.0;
alpha = atan(eta/fLP1);
beta2 = eta*eta+fLP1*fLP1;
beta = sqrt(beta2);
```

```
Sigma[L] = alpha*(fL+0.5)+eta*(log(beta) - 1.0) -
  (sin(alpha)/12.0 - (sin(3.0*alpha)/360.0 -
  (sin(5.0*alpha)/1260.0 - (sin(7.0*alpha)/1680.0 -
  sin(9.0*alpha)/(1188.0*beta2))/beta2)/beta2)/
  beta2)/beta;

for ( L = LMAX; L >= 1; L-- )
  { Sigma[L-1] = Sigma[L] - atan(eta/fL);
    fL -= 1.0; }
SigmaZ = Sigma[0];
/* Make  G[0] & G'[0]  */
Check = 1.0;
h = 0.01;
Nrho = -50;
rhoN = rho - 50.0*h;
while ( Check > 1.0e-5 )
  {
  Nrho += 50;
  rhoN += 50.0*h;
  /* Series for  G[0] & G'[0]  at rho = rhoN */
  Gzero(eta,rhoN,SigmaZ,&Check,&Gzero1,&GzeroP1);
  }

  /* then at one step further in */
Gzero(eta,rhoN-h,SigmaZ,&Check,&Gzero2,&GzeroP2);
if ( Nrho > 0 )
  {
  /* Iterate  g0  back from  rhoN  to  rho  */
  Noumerov(eta,rhoN,h,Nrho,Gzero1,Gzero2,
    &GzeroNew, &GzeroH);
  G[0] = GzeroNew;

  /* First derivative from last two  g0  values */
  h2D6 = h*h/6.0;
  h3D24 = 0.25*h*h2D6;
  TDrho = 2.0*eta/rho;
  TDM1 = TDrho - 1.0;
  A = 1.0+TDM1*h2D6 - 2.0*TDrho*h3D24/rho;
  B = 0.5*TDM1*h - TDrho*h2D6/rho +
    (TDM1*TDM1+2.0*TDrho/(rho*rho))*h3D24;
  GP[0] = ((GzeroH - GzeroNew)/h - GzeroNew*B)/A;
  }
else
  { G[0] = Gzero1;
    GP[0] = GzeroP1; }

/* Unnormalized F values */
F[LMAX+1] = 0.5e-16;
F[LMAX] = 1.0e-16;
Rrho = 1.0/rho;
etaS = eta*eta;
for ( L = LMAX; L >= 1; L-- )
  { fL = L;
    fLP1 = fL+1.0;
    F[L-1] = ((2.0*fL+1.0)*(eta/(fL*fLP1)+Rrho)*F[L] -
      sqrt(etaS/(fLP1*fLP1)+1.0)*F[L+1])/sqrt(etaS/
      (fL*fL)+1.0); }
```

```
FP[0] = (eta+Rrho)*F[0] - sqrt(etaS+1.0)*F[1];
/* Normalization */
norm = 1.0/(FP[0]*G[0]-F[0]*GP[0]);
FP[0] *= norm;

/* Normalize all the  F[L]  */
for ( L = 0; L<= LMAX; L++ )
 { F[L] *= norm; }

/* Compute  F'[L], G[L], G'[L]  by iteration */
for ( L = 1; L<= LMAX; L++ )
  {
  fL = L;
  c3 = eta/fL+fL*Rrho;
  c4 = sqrt(1.0+etaS/(fL*fL));
  FP[L] = c4*F[L-1] - c3*F[L];
  G[L] = (c3*G[L-1] - GP[L-1])/c4;
  GP[L] = c4*G[L-1] - c3*G[L];
  }
}

void Gzero(double eta, double rho, double SigmaZ,
 double *Check, double *Gzero, double *GzeroP)
{
/*  G0  and  G'0  from asymptotic estimates */
double  Trho,Teta,etaS,nP1,etaDrho,TnP1,s,t,S,T,
 sSum,tSum,SSum,TSum,TnP1Trho,TnP1eta,Rrho,stemp,
 ttemp,Stemp,Ttemp,fn,etaSMnTnP1,theta,Cos,Sin;
int n;

Trho = 2.0*rho; /* Series for  G[0] & G'[0]  */
Rrho = 1.0/rho;
Teta = 2.0*eta;
etaS = eta*eta;
etaDrho = eta*Rrho;
nP1 = 1.0;
TnP1 = 1.0;
s = eta/Trho;
t = etaS/Trho;
S = t*((eta-1.0)*Rrho - 1.0);
T = etaDrho*(-etaDrho+0.5);
sSum = 1.0 + s;
tSum = t;
SSum = S;
TSum = 1.0 - etaDrho + T;
TnP1Trho = Trho;
TnP1eta = eta;
for ( n = 1; n <= 15; n++ )
{
stemp = s;
ttemp = t;
Stemp = S;
Ttemp = T;
TnP1Trho += Trho;
TnP1eta += Teta;
fn = n;
```

```
etaSMnTnP1 = etaS - fn*(fn+1.0);
s = (TnP1eta*stemp-etaSMnTnP1*ttemp)/TnP1Trho;
t = (TnP1eta*ttemp+etaSMnTnP1*stemp)/TnP1Trho;
S = (TnP1eta*Stemp-etaSMnTnP1*Ttemp)/TnP1Trho-s*Rrho;
T = (TnP1eta*Ttemp+etaSMnTnP1*Stemp)/TnP1Trho-t*Rrho;
sSum += s;
tSum += t;
SSum += S;
TSum += T;
}

theta = -eta*log(Trho)+rho+SigmaZ;
Cos = cos(theta);
Sin = sin(theta);

*Check = fabs(1.0 - (sSum*TSum - SSum*tSum));
*Gzero = sSum*Cos - tSum*Sin;
*GzeroP = SSum*Cos - TSum*Sin;
}


void Noumerov(double eta, double rhoN, double h,
 int Nrho, double gzero1, double gzero2,
 double *GzeroNew, double *GzeroH)
{
/* Noumerov algorithm for Coulomb equation reduces G0
   from value at  rho+Nrho*h  to value at  rho  */
double Teta,hS,y1,y2,Y,YM,dYk,Yk,rho,Fh,Fh12,y;
int k;

Teta = 2.0*eta;
hS = h*h;

/* Starting values from asymptotic series */
y1 = gzero1;
y2 = gzero2;

/* Noumerov algorithm; iteration down in  rho  */
/* Starting values */
Y = (1.0 - hS*(Teta/rhoN - 1.0)/12.0)*y1;
YM = (1.0 - hS*(Teta/(rhoN-h) - 1.0)/12.0)*y2;
dYk = YM - Y;
Yk = YM;
rho =  rhoN - h;

for ( k = Nrho-2; k >= 0; k--)
  {
  Fh = hS*(Teta/rho - 1.0);
  Fh12 = Fh/12.0;
  dYk += Fh*(1.0+Fh12)*Yk; /* decrement in  Y  */
  Yk += dYk; /* new  Y  value */
  y = Yk/(1.0 - Fh12); /* next  y  value */
  rho = rho - h;
  if ( k == 1 )  { *GzeroH = y; }
  }

*GzeroNew = y;
}
```

Test Values

Check values of the regular Coulomb wave functions and their derivatives can be generated by running the `CoulombF` cell in *Mathematica* notebook `Hyprgmtrc`. The method used there is formula (16.3.2) for $F_L(\eta, \rho)$ and (16.3.5) for its derivatives with respect to ρ, $F_L'(\eta, \rho)$. Entries in Tables 16.3.1–16.3.6 were checked against our C function `Coulomb-FG`, run by using the driver program that is annotated in Section 21.16. Note that the accuracies of the values of the regular functions in this section and the irregular functions in Section 16.3.2 become intertwined because of our use of the Wronskian relation (16.3.6) for $L = 0$ to normalize the regular functions and their derivatives.

We depart from the standard accuracy goal for functions in the *Atlas*—a fractional accuracy of 10^{-10}—because such accuracy is difficult to obtain over a wide range of L, η, and ρ. Instead, the accuracy controls in `CoulombFG` have been adjusted to produce values of $F_L(\eta, \rho)$, $F_L'(\eta, \rho)$, $G_L(\eta, \rho)$, and $G_L'(\eta, \rho)$ that have at least 7 significant digits of accuracy, and usually 8 digits, unless $\eta \geq 1.25\rho$, corresponding to a total energy of less than 40% of the Coulomb barrier. Consult references [Tho86], [Aba92], and [Sal95] for other methods that extend the range for accurate calculations. Note also that, as a matter of practical quantum mechanics, an accuracy of 10^{-7} in a wave function indicates about the same fractional accuracy in the energy. However, relativistic kinematic effects are of this order for a proton with 0.1 keV energy or for an electron with 0.1 eV energy, which are unlikely to be appropriate energies at which to calculate Coulomb wave functions.

TABLE 16.3.1 Test values to 7 digits of the regular Coulomb wave functions for $\eta = 0$.

$L \setminus \rho$	2.0	10.0	18.0
0	0.9092974	-0.5440211	-0.7509872
4	0.02815879	-1.055893	-0.2924804
8	1.336641×10^{-5}	1.255780	0.9864487

TABLE 16.3.2 Test values to 7 digits of the regular Coulomb wave functions for $\eta = 2.5$.

$L \setminus \rho$	2.0	10.0	18.0
0	0.05755952	-0.3035138	-0.3650427
4	7.614019×10^{-4}	1.320764	1.009386
8	3.182492×10^{-7}	0.4380604	-1.065645

TABLE 16.3.3 Test values to 7 digits of the regular Coulomb wave functions for $\eta = 5.0$.

$L \setminus \rho$	2.0	10.0	18.0
0	0.0002862203	0.9179449	-1.222578
4	6.664777×10^{-6}	0.4168626	-0.5433925
8	3.818830×10^{-9}	0.04485260	1.404230

As a special check, note that when $\eta = 0$ one has $F_L(0, \rho) = \rho \, j_L(\rho)$ in terms of the spherical Bessel functions (Section 14.4.1).

TABLE 16.3.4 Test values to 7 digits of the derivatives of the regular Coulomb wave functions for $\eta = 0$.

$L \setminus \rho$	2.0	10.0	18.0
0	-0.4161468	-0.8390715	0.6603167
4	0.06512662	0.02739869	0.9431397
8	5.872748×10^{-5}	0.1292381	0.3450782

TABLE 16.3.5 Test values to 7 digits of the derivatives of the regular Coulomb wave functions for $\eta = 2.5$.

$L \setminus \rho$	2.0	10.0	18.0
0	0.08280378	-0.8085763	-0.8663576
4	0.002131978	-0.1267255	0.3647206
8	1.486002×10^{-6}	0.2711457	-0.3502354

TABLE 16.3.6 Test values to 7 digits of the derivatives of the regular Coulomb wave functions for $\eta = 5.0$.

$L \setminus \rho$	2.0	10.0	18.0
0	6.200837×10^{-4}	0.3310321	0.01832682
4	2.145410×10^{-5}	0.2383407	-0.7003106
8	1.882787×10^{-8}	0.04160134	-0.2038695

16.3.2 Irregular Functions and Derivatives G_L, G'_L

Keywords

Coulomb phase shifts, Coulomb wave functions, irregular Coulomb wave functions

C Function and *Mathematica* Notebook

CoulombFG (C); cell CoulombG in the notebook Hyprgmtrc (*Mathematica*) $\boxed{M}$

Background

Coulomb wave functions are described in many nuclear and atomic physics texts. To compute accurate Coulomb wave functions efficiently is challenging, and many specialized methods have been developed. The methods we describe for the regular functions in Section 16.3.1 and for the irregular functions in this section are suitable for collisions at energies not too far below the Coulomb barrier. For energies much below the barrier or for collisions between very heavy nuclei special techniques are required. Methods available are summarized by Thompson and Barnett [Tho86], Abad and Sesma [Aba92], and by Salvat et al [Sal95]. Our method—described in detail in Section 16.3.1—is adapted from that given by Buck et al [Buc60]. Formulas that may be used when computing $G_L(\eta, \rho)$ and $G'_L(\eta, \rho)$ are summarized in Chapter 14 of Abramowitz and Stegun [Abr64].

Function Definition

The function $G_L(\eta,\rho)$ is a solution of (16.3.1) that is irregular at the origin $\rho = 0$. For present purposes, it can be defined in terms of the integral given in Abramowitz and Stegun [Abr64, 14.3.3], written as

$$G_L(\eta,\rho) = \frac{e^{-\pi\eta}\rho^{L+1}}{(2L+1)!\,C_L(\eta)}$$

$$\left[\int_0^\infty \left\{-\left[\frac{2}{e^t + e^{-t}}\right]^{2(L+1)} \sin(\rho\tanh t - 2\eta t) + (1+t^2)^L e^{-\rho t + 2\eta\arctan t}\right\}dt\right] \qquad (16.3.11)$$

in which, from [Abr64, 14.1.8, 14.1.10]

$$C_L(\eta) = \frac{\sqrt{L^2 + \eta^2}}{L(2L+1)} C_{L-1}(\eta) \qquad C_0(\eta) = \sqrt{\frac{2\pi\eta}{e^{2\pi\eta} - 1}} \qquad (16.3.12)$$

The L-dependent term in (16.3.11) simplifies the term containing $1 - \tanh^2 t$ in [Abr64, 14.3.3] so that subtractive cancellation is minimized as t increases. Our definition of $G_L(\eta,\rho)$ is quite different from the corresponding one for $F_L(\eta,\rho)$ given as (16.3.2), even though [Abr64, 14.3.3] gives a matching integral definition for the regular function. We use (16.3.11) and (16.3.12) for the *Mathematica* implementation of $G_L(\eta,\rho)$. This is not a practical method for efficient calculation by a C program.

The recurrence relation [Abr64, 14.2.2] used for $G_L(\eta,\rho)$ is

$$G_{L+1} = \left[(\eta/L + L/\rho)G_{L-1} - G'_{L-1}\right]/\sqrt{\eta^2/L^2 + 1} \qquad (16.3.13)$$

which allows a series of irregular Coulomb functions to be computed by upward iteration once two adjacent values of L are available. The derivative relation is [Abr64, 14.2.1]

$$G'_L = \left[\sqrt{\eta^2/L^2 + 1}\, G_{L-1} - (\eta/L + L/\rho)G_L\right] \qquad (16.3.14)$$

As discussed below (16.3.6), the Wronskian of the regular and irregular solution cannot be used with our C function to check the accuracy of function values. It can be used with our *Mathematica* programs in cells `CoulombF` and `CoulombG` of notebook `Hyprgmtrc` to check the results obtained; typically, the Wronskian is unity to within the order of the computer accuracy.

Accurate values of G_0 and G'_0 are computed by asymptotic expansions given by Fröberg and by Buck et al [Frö55, Buc60] for $\rho \gg \eta$, as follows:

$$G_0(\eta,\rho) = s\cos\theta - t\sin\theta \qquad G'_0(\eta,\rho) = S\cos\theta - T\sin\theta \qquad (16.3.15)$$

in which, in terms of the $L = 0$ Coulomb phase shift σ_0 whose computation is described below (16.3.8), we have

$$\theta = -\eta\ln 2\rho + \rho + \sigma_0 \qquad (16.3.16)$$

and the asymptotic expansions for the auxiliary function s, t, S, and T are given by

$$s \sim \sum_{n=0}^N s_n \qquad t \sim \sum_{n=0}^N t_n \qquad S \sim \sum_{n=0}^N S_n \qquad T \sim \sum_{n=0}^N T_n \qquad (16.3.17)$$

The terms in this expansion are computed by iteration as

$$s_{n+1} = p_n s_n - q_n t_n \qquad\qquad t_{n+1} = p_n t_n + q_n s_n$$

$$S_{n+1} = p_n S_n - q_n T_n - s_{n+1}/\rho \qquad T_{n+1} = p_n T_n + q_n S_n - t_{n+1}/\rho$$

$$p_n = (2n+1)\eta/[2\rho(n+1)] \qquad q_n = [\eta^2 - n(n+1)]/[2\rho(n+1)]$$

$$s_0 = 1 \qquad s_1 = \frac{\eta}{2\rho} \qquad S_0 = 0 \qquad S_1 = \frac{\eta^2}{2\rho}\left(\frac{\eta-1}{\rho}-1\right) \qquad (16.3.18)$$

$$t_0 = 0 \qquad t_1 = \frac{\eta^2}{2\rho} \qquad T_0 = 1 - \frac{\eta}{\rho} \qquad T_1 = \frac{\eta}{\rho}\left(-\frac{\eta}{\rho}+1/2\right)$$

As usual in programming, these formulas are used in the opposite order to their specification. In (16.3.17) an upper limit to the sums of $N = 15$ is sufficiently accurate. Since these are asymptotic expansions (Section 3.4), they need to be controlled. We use the conditions on the auxiliary functions that

$$sT - St = 1 \qquad (16.3.19)$$

In our C function Gzero we require that (16.3.15) be computed at $\rho = \rho_N$ which is large enough that condition (16.3.19) is satisfied to 10^{-5}. This ρ value is often much larger than the ρ of interest, especially if the latter is small or if η is large. Therefore $G_0(\eta, \rho_N)$ usually has to be extrapolated back to $G_0(\eta, \rho)$.

We relate $G_0(\eta, \rho)$ to $G_0(\eta, \rho_N)$ by solving numerically the Coulomb equation (16.3.1) with $L = 0$, starting with $G_0(\eta, \rho_N)$ and $G_0(\eta, \rho_N - h)$, where h is a suitable step size. We use the Noumerov algorithm for linear second-order differential equations that have no first-derivative terms; this efficient algorithm—whose error per step is of order $h^6/240$—is derived in Section 8.3 of Thompson's text on computing [Tho92] and its implementation as a C function is given in Section 8.5 therein. For Coulomb wave functions and derivatives with at least 10^{-7} accuracy we require $h = 0.01$.

If no extrapolation in ρ is required then we have $G_0(\eta, \rho)$ and $G_0'(\eta, \rho)$ from (16.3.15). Otherwise, given $G_0(\eta, \rho)$ and $G_0(\eta, \rho + h)$ from the Noumerov algorithm extrapolation, we can use a Taylor expansion of $G_0(\eta, \rho + h)$ in terms of $G_0(\eta, \rho)$ and its derivatives. Each derivative can be expressed in terms of $G_0(\eta, \rho)$ and $G_0'(\eta, \rho)$ by using the defining equation (16.3.1) with $L = 0$. By solving for $G_0'(\eta, \rho)$ we obtain

$$G_0'(\eta, \rho) \approx \left\{[G_0(\eta, \rho + h) - G_0(\eta, \rho)]/h - G_0(\eta, \rho)B\right\}/A$$

$$A = 1 + (2\eta/\rho - 1)h^2/6 - 4\eta h^3/(24\rho^2) \qquad (16.3.20)$$

$$B = (2\eta/\rho - 1)h/2 - 2\eta h^2/(6\rho^2) + [(2\eta/\rho - 1)^2 + 4\eta/\rho^3]h^3/24$$

with an error of order $h^4/120$, which is about 10^{-10} for our choice of h. To check the algorithm and coding of this formula in function Noumerov that is listed at the end of Section 16.3.1, the cautious traveler using the *Atlas* will try $\eta = 0$, when $G_0(\eta, \rho) = \cos\rho$; indeed $G_0'(0, \rho) \approx -\sin\rho$ according to (16.3.20) and the error is as indicated.

Given $G_0(\eta, \rho)$ and $G_0'(\eta, \rho)$, the normalization of the regular functions is established through the Wronskian (16.3.6), plus upward iteration for the $G_L(\eta, \rho)$ and $G_L'(\eta, \rho)$ using (16.3.13) and (16.3.14) can begin. Note that the accuracy of $F_L(\eta, \rho)$ and $F_L'(\eta, \rho)$ depends upon the accuracy of the irregular functions because of the Wronskian connection.

Visualization

The irregular Coulomb function $G_L(\eta, \rho)$ and its derivative $G_L'(\eta, \rho)$ can be visualized in many ways. We show surfaces in which two quantities vary while the other is fixed.

Figure 16.3.4 shows how the irregular Coulomb wave functions and their derivatives vary with distance (ρ) and with Coulomb field strength (η) for a fixed partial wave (L). Notice that the singularities at $\rho = 0$ have effects even at $\rho = 3$ where the displays start, especially as η and L increase.

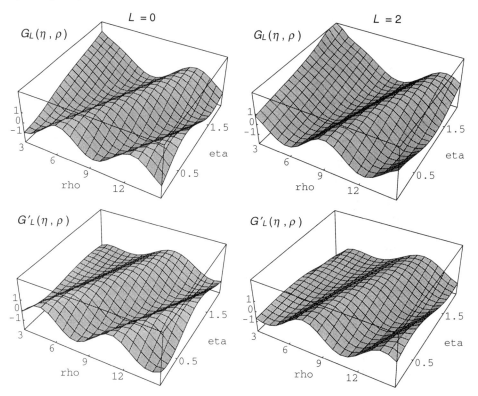

FIGURE 16.3.4 Irregular Coulomb wave functions and their derivatives for ρ in [3, 15] and η in [0, 2] with L fixed. (Adapted from cell `CoulombG` in the *Mathematica* notebook `Hyprgmtrc`.)

When the irregular Coulomb wave functions, $G_L(\eta,\rho)$, and their derivatives with respect to ρ, $G_L'(\eta,\rho)$, are displayed as surfaces in the ρ–L plane with the Coulomb parameter, η, fixed, we can visualize how increasing the angular momentum (L) increases the irregular wave function at small radii (corresponding to small ρ) and how a larger Coulomb repulsion (increasing η) enhances this effect, as shown in Figure 16.3.5. You can also see what an attractive Coulomb potential does to these surfaces by running the cell `CoulombG` in *Mathematica* notebook `Hyprgmtrc` with a negative value of η.

In the vistas of $G_L(\eta,\rho)$ and $G_L'(\eta,\rho)$, one can also view them with η and L varying and ρ fixed. This corresponds, at a fixed energy, to varying the strength of the Coulomb interaction (η) and considering different partial waves (L), but sampling the wave function at a fixed distance (ρ). Such surfaces are shown in Figure 16.3.6. Note that the formula used in our *Mathematica* cell, corresponding to the integral definition (16.3.11) can be used for Figures 16.3.5 and 16.3.6, in which L is a continuous variable, but the method used for our C implementation is not suitable because it requires integer L values.

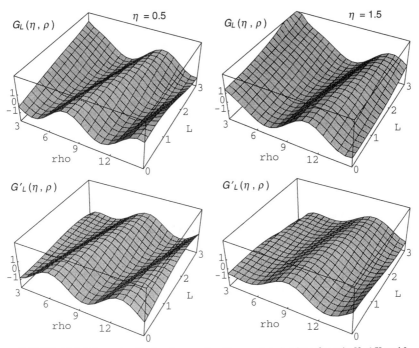

FIGURE 16.3.5 Irregular Coulomb wave functions and derivatives for ρ in [3, 15] and L in [0, 3] with η fixed. (Cell CoulombG in *Mathematica* notebook Hyprgmtrc.)

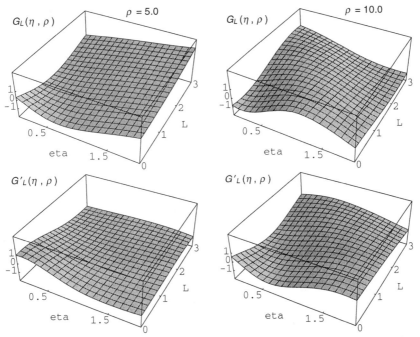

FIGURE 16.3.6 Irregular Coulomb wave functions and derivatives for η in [0, 2] and L in [0, 3] with ρ fixed. (Cell CoulombG in *Mathematica* notebook Hyprgmtrc.)

Algorithm and Program

The regular and irregular Coulomb functions, $F_L(\eta,\rho)$ and $G_L(\eta,\rho)$, and their derivatives with respect to ρ, $F'_L(\eta,\rho)$ and $G'_L(\eta,\rho)$, are computed together in our algorithm. The formulas used are in the Function Definition parts of Section 16.3.1 and of this section, where the details of computing $G_0(\eta,\rho)$ and $G'_0(\eta,\rho)$ are given. The complete C function is listed at the end of Algorithm and Program in Section 16.3.1.

The C driver program for CoulombFG is listed and annotated in Section 21.16. The complete program—driver, function, and the functions Gzero and Noumerov—is given as a single unit on the CD-ROM that comes with the *Atlas*.

Test Values

Check values of the regular Coulomb wave functions and their derivatives can be generated by running the CoulombG cell in *Mathematica* notebook Hyprgmtrc. The method used is formula (16.3.11) for $G_L(\eta,\rho)$ and (16.3.14) for its derivative with respect to ρ. Entries in Tables 16.3.7–16.3.12 were checked against our C function CoulombFG, run using the driver program in Section 21.16.

As justified under Test Values in Section 16.3.1, for the Coulomb wave functions we depart from the usual accuracy goal for functions in the *Atlas* (fractional accuracy 10^{-10}) because such accuracy is difficult to obtain over a wide range of L, η, and ρ. Instead, the accuracy controls have been adjusted to produce $F_L(\eta,\rho)$, $F'_L(\eta,\rho)$, $G_L(\eta,\rho)$, and $G'_L(\eta,\rho)$ with at least 7 significant digits of accuracy and usually 8 digits, unless $\eta \geq 1.25\rho$, corresponding to a total energy of less than 40% of the Coulomb barrier.

TABLE 16.3.7 Test values to 7 digits of the irregular Coulomb wave functions for $\eta = 0$.

$L \backslash \rho$	2.0	10.0	18.0
0	-0.4161468	-0.8390715	0.6603167
4	8.922583	0.01659930	0.9729608
8	9.060232×10^3	0.4111733	0.3995828

TABLE 16.3.8 Test values to 7 digits of the irregular Coulomb wave functions for $\eta = 2.5$.

$L \backslash \rho$	2.0	10.0	18.0
0	7.131768	-1.145628	-1.021203
4	256.6578	-0.08818117	0.4585935
8	3.552596×10^5	2.360854	-0.5219067

TABLE 16.3.9 Test values to 7 digits of the irregular Coulomb wave functions for $\eta = 5.0$.

$L \backslash \rho$	2.0	10.0	18.0
0	8.696901×10^2	1.608525	-0.003247673
4	2.492323×10^4	2.667598	-1.144669
8	2.787400×10^7	13.30351	-0.2715383

As a check, note that when $\eta = 0$ one has $G_L(0,\rho) = -\rho\, y_L(\rho)$ in terms of the irregular spherical Bessel functions (Section 14.4.2).

TABLE 16.3.10 Test values to 7 digits of the derivatives of the irregular Coulomb wave functions for $\eta = 0$.

$L \setminus \rho$	2.0	10.0	18.0
0	-0.9092974	0.5440211	0.7509872
4	-14.87643	0.9466351	0.2815986
8	-3.500682×10^4	-0.7540019	-0.8739559

TABLE 16.3.11 Test values to 7 digits of the derivatives of the irregular Coulomb wave functions for $\eta = 2.5$.

$L \setminus \rho$	2.0	10.0	18.0
0	-7.113726	0.2427318	0.3157803
4	-5.947073×10^6	-0.7486767	-0.8249982
8	-1.483377×10^6	-0.8214957	0.7668686

TABLE 16.3.12 Test values to 7 digits of the derivatives of the irregular Coulomb wave functions for $\eta = 5.0$.

$L \setminus \rho$	2.0	10.0	18.0
0	-1.609667×10^3	-0.5093189	0.8179925
4	-6.981396×10^4	-0.8736758	0.3650695
8	-1.244339×10^8	-9.956081	-0.6727113

REFERENCES ON HYPERGEOMETRIC FUNCTIONS AND COULOMB WAVE FUNCTIONS

[Aba92]   Abad, J., and J. Sesma, Computer Physics Communications, **71**, 110 (1992).

[Abr64]   Abramowitz, M., and I. A. Stegun, *Handbook of Mathematical Functions,* Dover, New York, 1964.

[Buc60]   Buck, B., R. N. Maddison, and P. E. Hodgson, Philosophical Magazine, **5**, 1181 (1960).

[Erd53a] Erdélyi, A., et al, *Higher Transcendental Functions*, Vol. 1, McGraw-Hill, New York, 1953; reprint edition, Krieger, Malabar, Florida, 1981.

[Ext83]   Exton, H., *q-Hypergeometric Functions and Applications*, Ellis Horwood, Chichester, 1983.

[Frö55]   Fröberg, C.-E., Reviews of Modern Physics, **27**, 399 (1955).

[Gas90]   Gasper, G., and M. Rahman, *Basic Hypergeometric Series*, Cambridge University Press, Cambridge, England, 1990.

[Gra94]   Gradshteyn, I. S., and I. M. Ryzhik, *Table of Integrals, Series, and Products*, Academic Press, fifth edition, 1994.

[Hoc86]   Hochstadt, H., *The Functions of Mathematical Physics*, Dover, New York, 1986.

[Iza95]   de Izarra, C., O. Vallee, J. Picart, and N. Tran Minh, Computers in Physics, **9**, 318 (1995).

[Jah45]   Jahnke, E., and F. Emde, *Tables of Functions*, Dover, New York, 1945.

[Sal95]   Salvat, F., J. M. Fernández-Varea, and W. Williamson, Jr., Computer Physics Communications, **90**, 151 (1995).

[Sea91]   Seaborn, J. B., *Hypergeometric Functions and Their Applications*, Springer-Verlag, New York, 1991.

[Sla60]   Slater, L. J., *Confluent Hypergeometric Functions*, Cambridge University Press, Cambridge, England, 1960.

[Spa87]   Spanier, J., and K. B. Oldham, *An Atlas of Functions,* Hemisphere Publishing, Washington, D.C., 1987.

[Tem83]   Temme, N. M., Numerische Mathematik, **41**, 63 (1983).

[Tho86]   Thompson, I. J., and A. R. Barnett, Journal of Computational Physics, **64**, 490 (1986).

[Tho92]   Thompson, W. J., *Computing for Scientists and Engineers*, Wiley, New York, 1992.

[Whi27]   Whittaker, E. T., and G. N. Watson, *A Course of Modern Analysis*, Cambridge University Press, Cambridge, England, fourth edition, 1927.

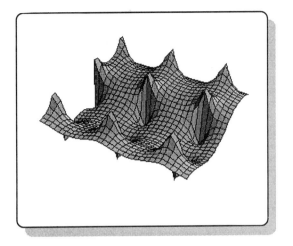

Chapter 17

ELLIPTIC INTEGRALS AND
ELLIPTIC FUNCTIONS

Elliptic integrals and functions are used in mathematics, statistics, science, and engineering—in problems as diverse as lengths of arcs, symmetric random walks on lattices, oscillations of pendulums, and trajectories of charged particles in nonuniform magnetic fields. This chapter begins with an overview of the functions, then the elliptic integrals and the elliptic functions are presented, beginning—as throughout the *Atlas*—with formulas and figures, then developing C programs for accurate and efficient computation. Users should beware of getting lost in the badlands of elliptic functions, such as the Jacobi cn function shown in the motif above.

17.1   OVERVIEW OF ELLIPTIC INTEGRALS AND ELLIPTIC FUNCTIONS

Elliptic integrals and functions were first encountered more than three centuries ago, so there is an extensive literature on them. Analytical properties are emphasized in works by Whittaker and Watson [Whi27], Cayley [Cay56], Neville [Nev44], and Greenhill [Gre59], in Chapter 13 of the treatise edited by Erdélyi et al [Erd53b], and in Chapter 5 of the text on complex analysis by González [Gon92]. Practical emphasis is given in Chapter 5 of the text on special functions by Carlson [Car77] and in the article by Sidhu [Sid95]. Extensive formulas are in the monograph by Byrd and Friedman [Byr71], in Chapters 5 and 6 of Jahnke and Emde [Jah45], Chapters 16 and 17 of the handbook of functions of Abramowitz and Stegun [Abr64], as well as in Section 8.1 of Gradshteyn and Ryzhik [Gra94].

   Analytical properties of elliptic integrals and functions are very well developed, so there is a wide variety of notations used for them. For computational purposes, however, simpler notations are usually the best. Here we summarize the notations used in the *Atlas*, whereas detailed views of the functions and programs for computing them are in the following sections. Table 17.1.1 gives the notation for elliptic integrals, while that for the elliptic functions is in Table 17.1.2.

TABLE 17.1.1 **Notation for elliptic integrals and complete elliptic integrals.**

Name	Elliptic integral	Complete elliptic integral
First kind	$F(\phi \mid m) \equiv \int_0^\phi \dfrac{1}{\sqrt{1 - m \sin^2 \theta}} \, d\theta$	$K(m) = F(\pi/2 \mid m)$
Second kind	$E(\phi \mid m) \equiv \int_0^\phi \sqrt{1 - m \sin^2 \theta} \, d\theta$	$E(m) = E(\pi/2 \mid m)$
Third kind	$\Pi(n; \phi \mid m) \equiv \int_0^\phi \dfrac{1}{(1 - n \sin^2 \theta)\sqrt{1 - m \sin^2 \theta}} \, d\theta$	$\Pi(n \mid m) = \Pi(n; \pi/2 \mid m)$

M

This notation is consistent with that in Chapter 17 of Abramowitz and Stegun [Abr64] and is the notation used in *Mathematica*. The following notation for Jacobi elliptic functions is also consistent with [Abr64] and with *Mathematica*. Let

$$u = F(\phi \mid m) = \int_0^\phi \frac{1}{\sqrt{1 - m \sin^2 \theta}} \, d\theta \qquad (17.1.1)$$

then ϕ is the *amplitude* of the elliptic integral of the first kind, so we write

$$\phi = \mathrm{am}(u \mid m) \qquad (17.1.2)$$

in which the $\mid m$ part is often omitted. The Jacobi elliptic functions are trigonometric functions of ϕ. The basic functions are summarized as follows.

TABLE 17.2.2 **Notation for basic Jacobi elliptic functions.**

Name	Jacobi elliptic function	$m = 0$	$m = 1$
sn	$\mathrm{sn}(u \mid m) = \sin \phi$	$\sin u$	$\tanh u$
cn	$\mathrm{cn}(u \mid m) = \cos \phi$	$\cos u$	$\mathrm{sech}\, u$
dn	$\mathrm{dn}(u \mid m) = \sqrt{1 - m \sin^2 \phi} \equiv \Delta\phi$	1	$\mathrm{sech}\, u$

From these basic functions Jacobi functions analogous to trigonometric functions can be generated, as described in detail in Section 120 of Byrd and Friedman [Byr71].

17.2   ELLIPTIC INTEGRALS

We discuss the standard forms of the elliptic integrals, as given in Table 17.1.1. Although some of the visualizations are given for an extensive range of the parameters m and n, our C functions are designed for m and n both less than unity.

17.2.1 Elliptic Integrals of the First Kind

Keywords
complete elliptic integrals, elliptic functions, elliptic integrals, incomplete elliptic integrals

C Function and *Mathematica* Notebook

EllptcFK (C); cell EllipticF in notebook Elliptics (*Mathematica*)

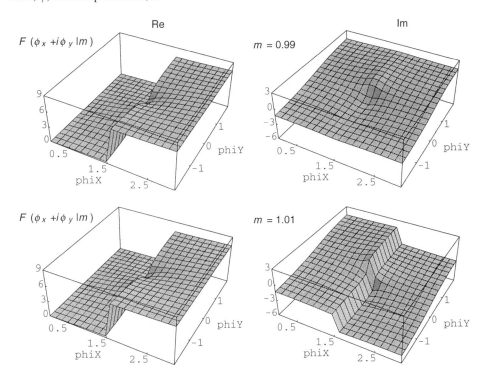 (partial — logo M top right)

Background

Background to the elliptic integrals is given in Section 17.1.

Function Definition

The incomplete elliptic integral of the first kind, $F(\phi \,|\, m)$, is defined by

$$F(\phi \,|\, m) \equiv \int_0^{\phi} \frac{1}{\sqrt{1 - m \sin^2 \theta}}\, d\theta \qquad (17.2.1)$$

in which m is called the parameter. Alternative expressions are given in Section 110 of Byrd and Friedman [Byr71] and in Section 17.2 of Abramowitz and Stegun [Abr64]. The complete elliptic integral of the first kind is $K(m) = F(\pi/2 \,|\, m)$. Note that, in general, $F(\phi \,|\, m)$ is real only if ϕ is real and if m is real and less than unity.

Visualization

Here are several views of the elliptic intgrals of the first kind—as functions of the argument, ϕ, and the parameter, m.

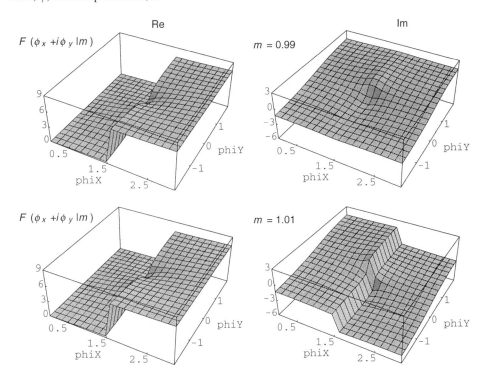

FIGURE 17.2.1 Elliptic integrals of the first kind with complex argument $\phi = \phi_x + i\,\phi_y$ and parameter m; left side shows real part and right side shows imaginary part. Top surfaces for $m = 0.99$ and bottom surfaces for $m = 1.01$; notice the rapid change in the imaginary parts. (Adapted from cell EllipticF in *Mathematica* notebook Elliptics.)

Note in Figure 17.2.1 that there are branch cuts in $F(\phi \mid m)$ because of the branch cuts of the square root in the integrand in (17.2.1).

For the ϕ and m ranges that produce real elliptic integrals of the first kind, it is enlightening to envision $F(\phi \mid m)$ either as a surface depending on ϕ and on $m < 1$ or as curves of functions of ϕ with m fixed or vice versa. Such surfaces and curves are shown in Figure 17.2.2.

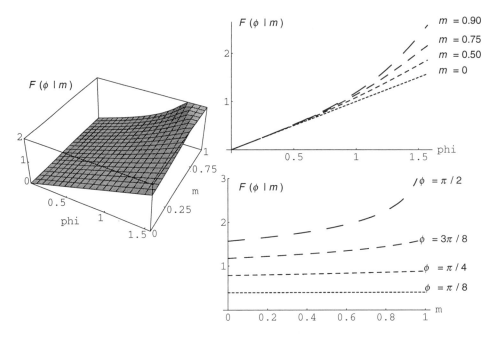

FIGURE 17.2.2 Elliptic integrals of the first kind; left side as a surface generated by parameters ϕ and m; right side top curves for ϕ varying and m fixed; right side botom curves for m varying and ϕ fixed. The curve for $\phi = \pi/2$ shows the complete elliptic integral of the first kind. (Adapted from cell `EllipticF` in *Mathematica* notebook `El-liptics`.)

In Figure 17.2.2 the integral has been truncated when its value exceeds 2, which occurs for ϕ near $\pi/2$ and m approaching unity, where the integral diverges. For example, if $\phi = \pi/2$ then $F(\phi \mid m) > 2$ for $m > 0.8$, while for $m = 1$ $F(\phi \mid m) > 2$ for $\phi > 1.3$.

The right top curve in Figure 17.2.2 shows that $F(\phi \mid m)$ becomes a more linear function of ϕ as m approaches zero. A transformation of $F(\phi \mid m)$ that steadily reduces m will therefore result in the computation of m being elementary—just ϕ in the limit $m = 0$. This is the basis of the descending Landen transform described in the following.

Algorithm and Program

Our algorithm for computing $F(\phi \mid m)$ uses the transform discovered by Landen in 1775 [Lan75] to reduce the parameter m so that the remaining integral is more nearly that of a circular function than that of an elliptic function. Because the transformation is not clearly explained in most previous works, we describe it in some detail from a geometric viewpoint.

The Landen Transform. The descending Landen transform that we use is derived clearly by Sidhu [Sid95] using the geometry of ellipses. Sidhu shows that one can write

$$F(\phi \mid m) = F(a_0, b_0, \phi_0) = \int_0^{\phi_0} \frac{1}{\sqrt{a_0^2 \cos^2 \theta_0 + b_0^2 \sin^2 \theta_0}} d\theta_0 \qquad (17.2.2)$$

with $a_0 = 1$, $b_0 = \sqrt{1-m}$, and $\phi_0 = \phi$. The transformation property is

$$F(a_n, b_n, \phi_n) = F(a_{n+1}, b_{n+1}, \phi_{n+1})/2 \qquad (17.2.3)$$

in which

$$a_{n+1} = (a_n + b_n)/2 \qquad b_{n+1} = \sqrt{a_n b_n} \qquad n \geq 0 \qquad (17.2.4)$$

and the angle ϕ_n increases monotonically according to

$$\phi_{n+1} = 2\phi_n - \delta_n \qquad \tan \delta_n = \frac{(a_n - b_n)\tan\phi_n}{a_n + b_n \tan^2 \phi_n} \qquad n \geq 0 \qquad (17.2.5)$$

which is numerically stable as n increases. In (17.2.4), each successive a_n value is the arithmetic mean of the previous a_n and b_n values, while each b_n value is the geometric mean of the preceding values, so that the algorithm is often called the arithmetic-geometric mean (AGM) algorithm.

Figure 17.2.3 shows geometrically the successive stages in the Landen transform for parameter $m = 0.9975$. If we draw ellipses with semi-major and semi-minor axes a_n and b_n, respectively, then their shapes for $n = 0$, 1, and 2 are as shown at left; alternatively, as shown at right, we can rescale the ellipses at each stage. The denominator in (17.2.2) is then the length P_1, which is integrated over the angle θ from 0 to ϕ_1.

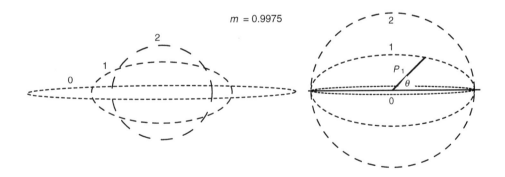

FIGURE 17.2.3 Successive stages in the Landen transform for $m = 0.9975$, thus $a_0 = 1$, $b_0 = 0.05$; each stage reduces the ellipticity greatly. On the left side the actual ellipses are drawn to scale, while on the right side the scale is changed at each stage so that the major axis is of constant length.

The arithmetic-geometric mean converges very rapidly, so that the ellipse approaches a circle and the integral becomes elementary. For example, choosing even the extreme case $m = 0.9975$ and writing $m_n = 1 - (b_n / a_n)^2$, we have the sequence given in Table 17.2.1.

Table 17.2.1 Convergence of the arithmetic-geometric mean (AGM) in the Landen transform of the elliptic integral of the first kind.

n	a_n	b_n	m_n
0	1.0000	0.0500	0.9975
1	0.5250	0.2236	0.8186
2	0.3743	0.3426	0.1622
3	0.3585	0.3581	0.0022

Our algorithm for computing the incomplete elliptic integral of the first kind, $F(\phi \mid m)$, therefore consists of making the transformations (17.2.3)–(17.2.5) N times, resulting in

$$F(\phi \mid m) \approx F(a_N, b_N, \phi_N) = \frac{\phi_N}{2^N a_N} \qquad (17.2.6)$$

For the complete integral of the first kind, $K(m)$, we have $\phi_N = 2^{N-1}\pi$, so that

$$K(m) \approx \frac{\pi}{2 a_N} \qquad (17.2.7)$$

The choice of N determines the accuracy and efficiency of the result.

Convergence of the Landen Transform. After N stages of the descending transform the integral that remains to be evaluated is

$$F_N = \int_0^{\phi_N} \frac{1}{\sqrt{1 - m_N \sin^2 \theta}}\, d\theta \qquad (17.2.8)$$

Since m_N is presumed to be small, the integrand can be expanded as a Taylor series in m_N then the integral can be estimated from its first two terms. The result is

$$F_N \approx \phi_N \{1 + m_N[1 - \sin(2\phi_N)/(2\phi_N)]/4\} \qquad (17.2.9)$$

The term proportional to m_N is therefore an estimate of the fractional error in the integral incurred by stopping after N stages. Since this term is less than $m_N / 2$, if we require a fractional accuracy of ε then we can terminate at an N such that $m_N < 2\varepsilon$. Note that

$$m_N = (a_N - b_N)(a_N + b_N)/a_N^2 \approx 2(a_N - b_N)/a_N \qquad (17.2.10)$$

Therefore do not attempt to set ε to less than the computational accuracy of your computer! For 10-digit accuracy—thus $\varepsilon = 10^{-10}$, which is the design goal for functions in the *Atlas*— $N \leq 6$ even for $m = 0.9999$, so the convergence of the Landen transform is very rapid even for high accuracy.

In the following C function, `EllptcFK`, both the incomplete elliptic integral of the first kind—$F(\phi \mid m)$ in (17.2.1)—and the complete integral of the first kind—$K(m)$ in Table 17.2.1—are computed. The first is returned as the value of function `EllptcFK`, while the second is returned as the argument Km. This is illustrated by the driver program in Section 21.17.

```
double EllptcFK(double phi, double m,
 double eps, double *Km)
/* Elliptic integrals of first kind by AGM method;
   F(phi|m)   and  K(m)   */

{
double  a,b,psi,c,epsD2,tanpsi,d,temp;
double const PiD2 = 1.57079632679;
int n;

a = 1.0;
b = sqrt(1.0 - m);
psi = phi;
c = 1.0;
temp = 1.0;
epsD2 = 0.5*eps;
n = 0;

while ( fabs(c) > epsD2*temp )
  {
  tanpsi = tan(psi);
  d = atan((a-b)*tanpsi/(a+b*tanpsi*tanpsi));
  psi = 2.0*psi-d;
  temp = a;
  a = 0.5*(a+b);
  c = 0.5*(temp-b);
  b = sqrt(temp*b);
  n++;
  }

*Km =   PiD2/a;

return psi/(pow(2.0, (double) n )*a);
}
```

Test Values

Check values of $F(\phi|m)$ and $K(m)$ can be generated by running the `EllipticF` cell in *Mathematica* notebook `Elliptics`. The values for $K(m)$ are just those for $F(\phi|m)$ with $\phi = \pi/2$. Examples are given in Table 17.2.2.

$\boxed{M}$

TABLE 17.2.2 **Test values of the elliptic integrals of the first kind.**

$\phi \backslash m$	0.25	0.50	0.75
0.5	0.5050887276	0.5104671356	0.5161698213
1.0	1.037356120	1.083216773	1.142468797
1.5	1.604024521	1.754036917	2.015275104
$\pi/2$	1.685750355	1.854074677	2.156515647
2.0	2.176587705	2.444382636	2.952569674

17.2.2 Elliptic Integrals of the Second Kind

Keywords

complete elliptic integrals, elliptic functions, elliptic integrals, incomplete elliptic integrals

C Function and *Mathematica* Notebook

$\boxed{M}$ EllptcEE (C); cell EllipticE in notebook Elliptics (*Mathematica*)

Background

Background to the elliptic integrals is given in Section 17.1.

Function Definition

The incomplete elliptic integral of the second kind, $E(\phi \mid m)$, is defined by

$$E(\phi \mid m) \equiv \int_0^{\phi} \sqrt{1 - m \sin^2 \theta} \, d\theta \qquad (17.2.11)$$

in which m is the parameter. Alternative expressions are given in Section 110 of Byrd and Friedman [Byr71] and in Sections 17.3–17.4 of Abramowitz and Stegun [Abr64]. The complete elliptic integral of the second kind is $E(m) = E(\pi/2 \mid m)$. In general, $E(\phi \mid m)$ is real only if ϕ is real and if m is real and not more than unity.

Visualization

Look at several views of the elliptic integrals of the second kind—as functions of argument, ϕ, and parameter, m.

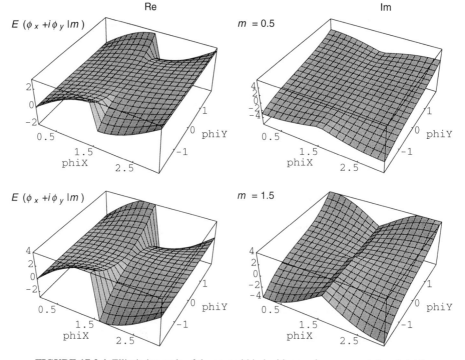

FIGURE 17.2.4 Elliptic integrals of the second kind with complex argument $\phi = \phi_x + i\,\phi_y$ and parameter m; left side shows real part and right side shows imaginary part. Top surfaces for $m = 0.5$ and bottom surfaces for $m = 1.5$. (Adapted from cell EllipticE in *Mathematica* notebook Elliptics.)

In Figure 17.2.4 there are branch cuts in $E(\phi \,|\, m)$, since there are branch cuts of the square root in the integrand in (17.2.11).

For the ϕ and m ranges that produce real elliptic integrals of the second kind, it is interesting to envision $E(\phi \,|\, m)$ either as a surface depending on ϕ and on $m \le 1$ or as curves of functions of ϕ with m fixed or vice versa. Such surfaces and curves are shown in Figure 17.2.5.

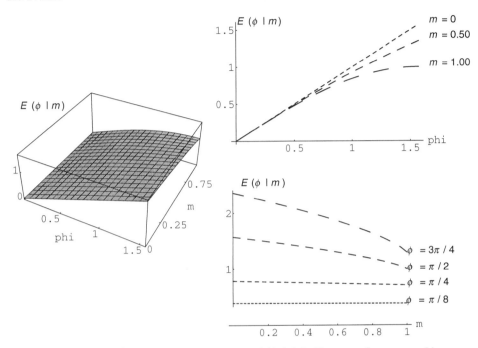

FIGURE 17.2.5 Elliptic integrals of the second kind; left side as a surface generated by parameters ϕ and m; right side top curves for ϕ varying and m fixed; right side bottom curves for m varying and ϕ fixed. The curve for $\phi = \pi/2$ shows the complete elliptic integral of the second kind. (Adapted from cell `EllipticE` in *Mathematica* notebook `El-liptics`. The truncated vertical axis is an artifact of *Mathematica*.)

The right top curve in Figure 17.2.5 shows that $E(\phi \,|\, m)$ becomes a more linear function of ϕ as m approaches zero. A transformation of $E(\phi \,|\, m)$ that steadily reduces m will therefore result in the computation of m being elementary—just ϕ in the limit $m = 0$. This is the basis of the descending Landen transform described in the following.

Algorithm and Program
Our algorithm for computing $E(\phi \,|\, m)$ uses Landen's transform [Lan75] to reduce the parameter m so that the remaining integral is more nearly that of a circular function than that of an elliptic function. The transformation is described from a geometric viewpoint in Section 17.2.1 for the elliptic integral of the first kind, following Sidhu's analysis [Sid95]. Here we give only the changes needed for the integral of the second kind.

The Landen Transform. The descending Landen transform for the elliptic integral of the second kind is shown by Sidhu to be expressible as

$$E(\phi \mid m) = E(a_0, b_0, \phi_0) = \int_0^{\phi_0} \sqrt{a_0^2 \cos^2 \theta_0 + b_0^2 \sin^2 \theta_0} \, d\theta_0 \qquad (17.2.12)$$

with $a_0 = 1$, $b_0 = \sqrt{1-m}$, and $\phi_0 = \phi$. The transformed integral can be written as

$$E(\phi \mid m) = \left[1 - m/2 - \sum_{k=1}^{N} 2^{k-1} c_k^2 \right] F(\phi \mid m) + \sum_{k=1}^{N} c_k \sin \phi_k \qquad (17.2.13)$$

in which, just as for the integral of the first kind in Section 17.2.1, we have

$$a_{n+1} = (a_n + b_n)/2 \qquad b_{n+1} = \sqrt{a_n b_n} \qquad n \geq 0 \qquad (17.2.14)$$

and additionally

$$c_{n+1} = (a_n - b_n)/2 \qquad n \geq 0 \qquad (17.2.15)$$

The angle ϕ_n increases monotonically according to

$$\phi_{n+1} = 2\phi_n - \delta_n \qquad \tan \delta_n = \frac{(a_n - b_n) \tan \phi_n}{a_n + b_n \tan^2 \phi_n} \qquad n \geq 0 \qquad (17.2.16)$$

which is numerically stable as n increases. Since in (17.2.14) each successive a_n value is the arithmetic mean of the previous a_n and b_n values, while each b_n value is the geometric mean of the preceding values, the algorithm is often called the arithmetic-geometric mean (AGM) algorithm. Figure 17.2.3 in Section 17.2.1 shows successive stages in the Landen transform. The arithmetic-geometric mean converges rapidly even for m near unity, as shown in Table 17.2.1.

Convergence of the Landen Transform. After N stages of the descending transform the integral remaining to be evaluated is

$$E_N = \int_0^{\phi_N} \sqrt{1 - m_N \sin^2 \theta} \, d\theta \qquad (17.2.17)$$

Since m_N is assumed to be small, the integrand can be expanded as a Taylor series in m_N and the integral estimated from its first two terms. The result is

$$E_N \approx \phi_N \{ 1 - m_N [1 - \sin(2\phi_N)/(2\phi_N)]/4 \} \qquad (17.2.18)$$

The term proportional to m_N is then an estimate of the fractional error incurred by stopping after N stages. Since this term is less in magnitude than $m_N/2$, if we require fractional accuracy of ε then we can terminate at an N such that $m_N < 2\varepsilon$. Noting that

$$m_N = (a_N - b_N)(a_N + b_N)/a_N^2 \approx 2(a_N - b_N)/a_N \qquad (17.2.19)$$

the termination condition for fractional accuracy ε can be estimated as

$$c_N \leq \varepsilon a_N/2 \qquad (17.2.20)$$

Therefore do not attempt to set ε to less than the computational accuracy of your computer! For 10-digit accuracy—thus $\varepsilon = 10^{-10}$, which is the design goal for functions in the *Atlas*— $N \leq 6$ even for $m = 0.9999$, so the convergence of the Landen transform is very rapid even for high accuracy.

In the following C function, `EllptcEE`, both the incomplete elliptic integral of the second kind—$E(\phi \,|\, m)$ in (17.2.11)—and the complete integral of the second kind—$E(m)$ in Table 17.2.1—are computed. The first is returned as the value of function `EllptcEE`, and the second is returned as the argument `Em`. This is illustrated by the driver program in Section 21.17. If the angle ϕ exceeds $\pi/2$, just compute the complete integral $E(m)$ then add in this into the incomplete integral for the part that is less than $\pi/2$.

```c
double EllptcEE(double phi, double m,
  double eps, double *Em)
/* Elliptic integrals of second kind by AGM method;
    E(phi|m)   for phi <= Pi/2  and  E(m)  */
{
double  a,b,psi,c,eps2,temp,sum1,sum2,pow2,tanpsi,d;
double const PiD2 = 1.57079632679;
int n;

if ( m == 1.0 )
 {
 *Em = 1.0;
 return  sin(phi);
 }

a = 1.0;
b = sqrt(1.0 - m);
psi = phi;
c = 1.0;
n = 0;
eps2 = 0.5*eps;
temp = 1.0;
sum1 = 1.0 - 0.5*m;
sum2 = 0.0;
pow2 = 1.0;

while ( fabs(c) > eps2*temp )
 {
 tanpsi = tan(psi);
 d = atan((a-b)*tanpsi/(a+b*tanpsi*tanpsi));
 psi = 2.0*psi-d;
 c = 0.5*(a-b);
 temp = a;
 a = 0.5*(a+b);
 b = sqrt(temp*b);
 sum1 -= pow2*c*c;
 sum2 += c*sin(psi);
 pow2 *= 2.0;
 n++;
 }

*Em = sum1*PiD2/a;

return  sum1*psi/(pow(2.0, (double) n )*a)+sum2;
}
```

Programming Notes

If you make frequent use of both elliptic integrals, $F(\phi\,|\,m)$ and $E(\phi\,|\,m)$, it would be efficient to merge the two functions `EllptcFK` and `EllptcEE`, since they have many statements in common.

Test Values

Check values of $E(\phi\,|\,m)$ and $E(m)$ can be generated by running the `EllipticE` cell in *Mathematica* notebook `Elliptics`. The values for $E(m)$ are just those for $E(\phi\,|\,m)$ with $\phi = \pi/2$. Examples are given in Table 17.2.3.

TABLE 17.2.3   **Test values of the elliptic integrals of the second kind.**

$\phi \setminus m$	0.25	0.50	0.75
0.5	0.4950017030	0.4899109598	0.4847213374
1.0	0.9648764543	0.9273298836	0.8866251235
1.5	1.406133741	1.300541573	1.175569442
$\pi/2$	1.467462209	1.350643881	1.211056028

17.2.3 Jacobi Zeta Function

Keywords

elliptic functions, elliptic integrals, Jacobi zeta function

C Function and *Mathematica* Notebook

`JacobiZeta` (C); cell `JacobiZeta` in notebook `Elliptics` (*Mathematica*)

Background

The Jacobi zeta function, $Z(\phi\,|\,m)$, bridges the elliptic integrals and elliptic functions. It is described in Sections 140–144 of Byrd and Friedman [Byr71] and in a cursory way in Section 17.4 of Abramowitz and Stegun [Abr64]. The zeta function can also be used to compute the elliptic integral of the third kind.

Do not confuse Jacobi's zeta function, $Z(\phi\,|\,m)$, with Riemann's zeta function, ζ, described in Section 8.3.

Function Definition

The Jacobi zeta function can be defined in terms of elliptic integrals of the first and second kinds (Sections 17.2.1 and 17.2.2, respectively) by

$$Z(\phi\,|\,m) \equiv E(\phi\,|\,m) - E(m)F(\phi\,|\,m)/K(m) \qquad (17.2.21)$$

Alternatively, in terms of integrals of Jacobi elliptic functions sn (Section 17.3.1),

$$Z(\phi \mid m) = \frac{m \sin\phi \cos\phi \sqrt{1 - m\sin^2\phi}}{K(m)} \int_0^{K(m)} \frac{\operatorname{sn}^2 u}{1 - m\sin^2\phi \operatorname{sn}^2 u}\, du \qquad (17.2.22)$$

in which m is the parameter. The complete integral, $Z(\pi/2 \mid m)$, is not interesting, since the zeta function has the special values

$$Z(\pi/2 \mid m) = 0 \qquad Z(0 \mid m) = 0 \qquad (17.2.23)$$

Other special values are listed in Section 141 of [Byr71].

Visualization

Here are several views of the Jacobi zeta function—as functions of the argument, ϕ, and the parameter, m.

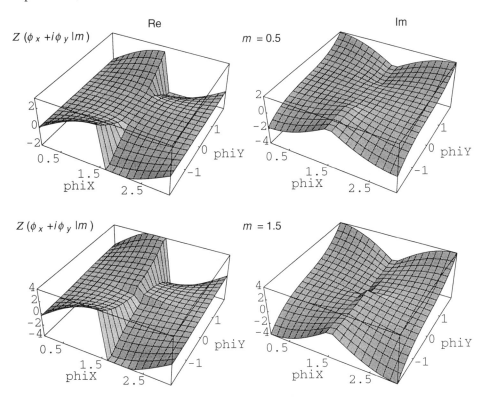

FIGURE 17.2.6 Jacobi zeta function with complex argument $\phi = \phi_x + i\,\phi_y$ and parameter m; left side shows real part and right side shows imaginary part. (Adapted from cell JacobiZeta in *Mathematica* notebook Elliptics.)

Figure 17.2.6 has branch cuts in $Z(\phi \mid m)$, since there are branch cuts in the square roots appearing in the integrands of $E(\phi \mid m)$ and $F(\phi \mid m)$.

For ϕ and m ranges that produce Jacobi zeta functions, it is interesting to view $Z(\phi \mid m)$ either as a surface depending on ϕ and on $m \le 1$ or as curves of functions of ϕ with m fixed or vice versa. Such surfaces and curves are shown in Figure 17.2.7.

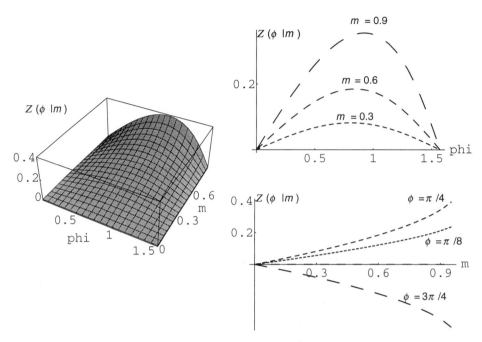

FIGURE 17.2.7 Jacobi zeta function with real argument ϕ and parameter $m < 1$; left side shows the surface of the function, right side top shows curves of the function with ϕ varying and m fixed, while right side bottom shows curves with m varying and ϕ fixed. Note that the function vanishes at $\phi = \pi/2$. (Adapted from cell `JacobiZeta` in *Mathematica* notebook `Elliptics`.)

Algorithm and Program

Our algorithm for computing $Z(\phi \,|\, m)$ uses a descending Landen transformation similar to those used for $F(\phi \,|\, m)$ and $E(\phi \,|\, m)$. Such transformations are described from a geometric viewpoint in Section 17.2.1 for the elliptic integral of the first kind. For $Z(\phi \,|\, m)$ the appropriate arithmetic-geometric-mean (AGM) scale is obtained by combining the algorithms in Sections 17.2.1 and 17.2.2 for $F(\phi \,|\, m)$ and $E(\phi \,|\, m)$, respectively. Start with

$$a_0 = 1 \qquad b_0 = \sqrt{1-m} \qquad c_0 = 1 \qquad (17.2.24)$$

Then compute iteratively

$$a_{n+1} = (a_n + b_n)/2 \qquad b_{n+1} = \sqrt{a_n b_n} \qquad c_{n+1} = (a_n - b_n)/2 \qquad n \geq 0 \quad (17.2.25)$$

and the angle ϕ_n increases monotonically according to

$$\phi_{n+1} = 2\phi_n - \delta_n \qquad \tan\delta_n = \frac{(a_n - b_n)\tan\phi_n}{a_n + b_n \tan^2\phi_n} \qquad n \geq 0 \qquad (17.2.26)$$

which is numerically stable as n increases. Terminate after N steps when c_N is negligible for the accuracy required of $Z(\phi \,|\, m)$. For 10-digit accuracy—the design goal for functions in the *Atlas*—$N \leq 6$ even for $m = 0.9999$, so convergence of this Landen transform is very rapid even for high accuracy. We then have

$$Z(\phi \,|\, m) = \sum_{k=1}^{N} c_k \sin\phi_k \qquad (17.2.27)$$

Use of the following C function, `JacobiZeta`, is illustrated by the driver program in Section 21.17. If the angle ϕ exceeds $\pi/2$, just compute the part that corresponds to the excess beyond an integer multiple of $\pi/2$.

```
double JacobiZeta(double phi, double m, double eps)
/* Jacobi Zeta function by combining AGM methods
    for elliptic integrals of first & second kinds */
{
double  a,b,psi,c,temp,sum,epsD2,tanpsi,d;
double const PiD2 = 1.57079632679;
int n;

a = 1.0;
b = sqrt(1.0 - m);
psi = phi;
c = 1.0;
temp = 1.0;
sum = 0.0;
epsD2 = 0.5*eps;
n = 0;

while ( fabs(c) > epsD2*temp )
  {
  tanpsi = tan(psi);
  d = atan((a-b)*tanpsi/(a+b*tanpsi*tanpsi));
  psi = 2.0*psi-d;
  temp = a;
  a = 0.5*(a+b);
  c = 0.5*(temp-b);
  b = sqrt(temp*b);
  sum += c*sin(psi);
  n++;
  }

return  sum;
}
```

Test Values

Check values of $Z(\phi \,|\, m)$ can be generated by running the `JacobiZeta` cell in *Mathematica* notebook `Elliptics`. Examples are given in Table 17.2.4.

M

TABLE 17.2.4 **Test values to 10 digits of the Jacobi zeta function.**

$\phi \setminus m$	0.25	0.50	0.75
0.5	0.05531701426	0.1180492858	0.1948506963
1.0	0.06184778257	0.1382353968	0.2450375121
1.5	0.009814670909	0.02277252871	0.04383127658

17.2.4 Heuman Lambda Function

Keywords
elliptic functions, elliptic integrals, Heuman lambda function

C Function and *Mathematica* Notebook
HeumanLambda (C); cell HeumanLambda in notebook Elliptics(*Mathematica*)

Background
The Heuman lambda function, $\Lambda_0(\phi \mid m)$, helps to bridge the elliptic integrals and elliptic functions. It is described in Sections 150–154 of Byrd and Friedman [Byr71] and it makes a cameo appearance in Section 17.4 of Abramowitz and Stegun [Abr64]. The lambda function—first tabulated by Heuman [Heu41]—can be used to compute the elliptic integral of the third kind.

Function Definition
The Heuman lambda function can be defined in terms of elliptic integrals of the first and second kinds (Sections 17.2.1 and 17.2.2, respectively) by

$$\Lambda_0(\phi \mid m) \equiv \frac{2}{\pi}\left\{[E(m)-K(m)]F(\phi \mid 1-m)+K(m)E(\phi \mid 1-m)\right\} \qquad (17.2.28)$$

Alternatively, in terms of integrals of Jacobi elliptic functions sn (Section 17.3.1),

$$\Lambda_0(\phi \mid m) = \frac{-2\sin\phi\sqrt{1+m\tan^2\phi}}{\pi}\int_0^{K(m)}\frac{1-m\sin^2 u}{1+m\tan^2\phi\,\mathrm{sn}^2 u}\,du \qquad (17.2.29)$$

in which m is the parameter. The complete integral, $\Lambda_0(\pi/2 \mid m)$, is not particularly interesting, since the lambda function has the special values

$$\Lambda_0(\phi \mid 0) = \sin\phi \qquad \Lambda_0(\phi \mid 1) = 2\phi/\pi$$
$$\Lambda_0(0 \mid m) = 0 \qquad \Lambda_0(\pi/2 \mid m) = 1 \qquad (17.2.30)$$

Other special values are listed in Section 151 of [Byr71].

Visualization
We show in Figures 17.2.8 and 17.2.9 several views of $\Lambda_0(\phi \mid m)$ as functions of the argument, ϕ, and the parameter, m. You can explore the lambda function in more detail by modifying then running the *Mathematica* cell HeumanLambda that is on the CD-ROM and whose annotated listing is in Section 20.17. Figure 17.2.8 has branch cuts in $\Lambda_0(\phi \mid m)$, since there are branch cuts in the square roots appearing in the integrands of $E(\phi \mid m)$ and $F(\phi \mid m)$.

For ϕ and m ranges that produce real $\Lambda_0(\phi \mid m)$, it is interesting to view $\Lambda_0(\phi \mid m)$ either as a surface depending on ϕ and on $0 \leq m \leq 1$ or as curves of functions of ϕ with m fixed or vice versa. Such surfaces and curves are shown in Figure 17.2.9.

Algorithm and Program
Our algorithm for computing the Heuman lambda function adapts the descending Landen transformations used for the elliptic integrals of the first and second kind that appear in the defining (17.2.28) and which are described geometrically in Section 17.2.1 for the elliptic integral of the first kind.

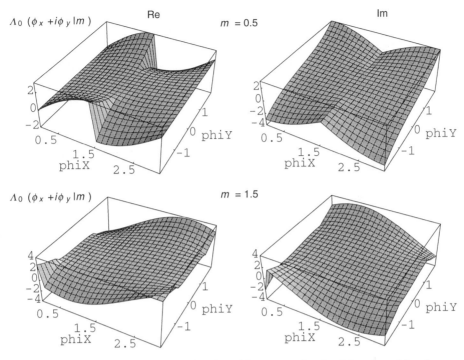

FIGURE 17.2.8 Heuman lambda function with argument $\phi = \phi_x + i\phi_y$ and $m = 0.5$ and 1.5; left shows real part and right shows imaginary part. (Adapted from cell `Heuman-Lambda` in notebook `Elliptics`.)

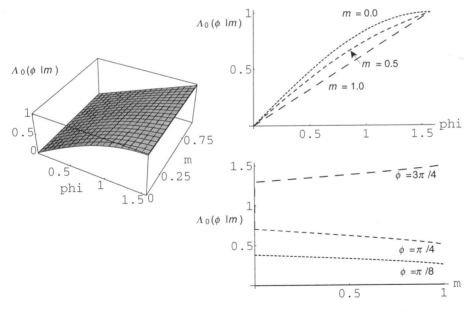

FIGURE 17.2.9 Heuman lambda function with argument ϕ and parameter m; left is the surface; right top shows curves with ϕ varying and m fixed, while right bottom shows m varying and ϕ fixed. (Adapted from cell `HeumanLambda` in notebook `Elliptics`.)

The driver program in Section 21.17 shows how to use the following C function. Input parameter `eps` controls the fractional accuracy of the AGM algorithm for the elliptic integrals in (17.2.28), as discussed in Sections 17.2.1 and 17.2.2 in relation to convergence of the Landen transform. Set `eps` $= 10^{-10}$ for 10-digit accuracy, as in the test values below.

```c
double HeumanLambda(double phi, double m, double eps)
/* Heuman Lambda function by combining AGM methods
   for elliptic integrals of first & second kinds */
{
double  a,b,psi,c,temp,sum1,sum2,pow2,epsD2,tanpsi,d,
 Fphimm,Ephimm,sum;
double const PiD2 = 1.57079632679;
int n;

if ( m == 0.0 )   return  sin(phi);
if ( m == 1.0 )   return  phi/PiD2;
if ( phi == 0.0 ) return  0.0;

/* AGM for E(phi|1-m) - F(phi|1-m)   */
a = 1.0;
b = sqrt(m);
psi = phi;
c = 1.0;
temp = 1.0;
sum1 = 0.5*(1.0+m);
sum2 = 0.0;
pow2 = 1.0;
epsD2 = 0.5*eps;
n = 0;
while ( fabs(c) > epsD2*temp )
  {
  tanpsi = tan(psi);
  d = atan((a - b)*tanpsi/(a+b*tanpsi*tanpsi));
  psi = 2.0*psi - d;
  temp = a;
  a = 0.5*(a+b);
  c = 0.5*(temp - b);
  b = sqrt(temp*b);
  sum1 -= pow2*c*c;
  sum2 += c*sin(psi);
  pow2 *= 2.0;
  n++;
  }
Fphimm = psi/(pow(2.0, (double) n )*a);
Ephimm = sum1*Fphimm+sum2;

/* AGM for K(m)  and  E(m)  */
a = 1.0;
b = sqrt(1.0 - m);
psi = phi;
c = 1.0;
temp = 1.0;
sum = -0.5*m;
pow2 = 1.0;
n = 0;
```

```
while ( fabs(c) > epsD2*temp )
  {
  tanpsi = tan(psi);
  d = atan((a - b)*tanpsi/(a+b*tanpsi*tanpsi));
  psi = 2.0*psi - d;
  temp = a;
  a = 0.5*(a+b);
  c = 0.5*(temp - b);
  b = sqrt(temp*b);
  sum -= pow2*c*c;
  pow2 *= 2.0;
  n++;
  }
return (sum*Fphimm+Ephimm)/a;
}
```

Test Values
Test values of the Heuman lambda function, $\Lambda_0(\phi \mid m)$, can be generated by running the
HeumanLambda cell in *Mathematica* notebook Elliptics. Examples are given in
Table 17.2.5.

TABLE 17.2.5 **Test values to 10 digits of the Human lambda function.**

$\phi \setminus m$	0.25	0.50	0.75
0.5	0.4484638788	0.4146601421	0.3755661454
1.0	0.7927451830	0.7474006351	0.7002771969
1.5	0.9815441455	0.9729237139	0.9649939810

17.2.5 Elliptic Integrals of the Third Kind

Keywords
complete elliptic integrals, elliptic functions, elliptic integrals, incomplete elliptic integrals

C Function and *Mathematica* Notebook
EllptcPi (C); cell EllipticPi in notebook Elliptics (*Mathematica*)

Background
Background to the elliptic integrals is given in Section 17.1.

Function Definition
The incomplete elliptic integral of the third kind with parameters n and m, $\Pi(n;\phi \mid m)$, is
defined by

$$\Pi(n;\phi \mid m) \equiv \int_0^{\phi} \frac{1}{(1 - n\sin^2\theta)\sqrt{1 - m\sin^2\theta}} \, d\theta \qquad (17.2.31)$$

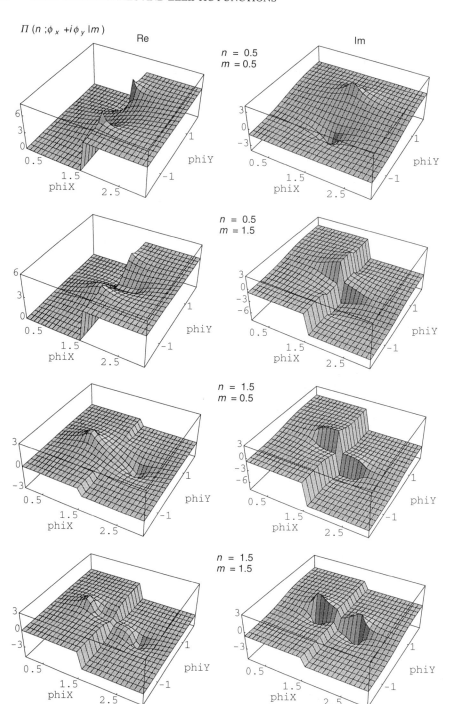

FIGURE 17.2.10 Elliptic integrals of the third kind with complex argument $\phi = \phi_x + i\,\phi_y$ and parameters n and m; right side shows real part and left side shows imaginary part. (Adapted from cell `EllipticPi` in *Mathematica* notebook `Elliptics`.)

Note that n sometimes appears with the opposite sign in the defining (17.2.31). Other expressions are given in Section 110 of Byrd and Friedman [Byr71] and in Section 17.7 of Abramowitz and Stegun [Abr64]. The complete elliptic integral of the third kind is just $\Pi(n\,|\,m) = \Pi(n;\pi/2\,|\,m)$. In general, $\Pi(n;\phi\,|\,m)$ is real only if ϕ is real and if m and n are real and less than unity.

Visualization

Scan the panorama of the elliptic integrals of the third kind—as functions of the complex argument, ϕ, and the parameters, n and m—in Figure 17.2.10, which shows branch cuts in $\Pi(n;\phi\,|\,m)$, since there are branch cuts of the square root in the integrand in (17.2.31).

For the ϕ, n, and m ranges that produce real elliptic integrals of the third kind, it is also interesting to envision $\Pi(n;\phi\,|\,m)$ as a surface depending on ϕ and on $m < 1$ with $n < 1$ fixed, as shown in Figure 17.2.11.

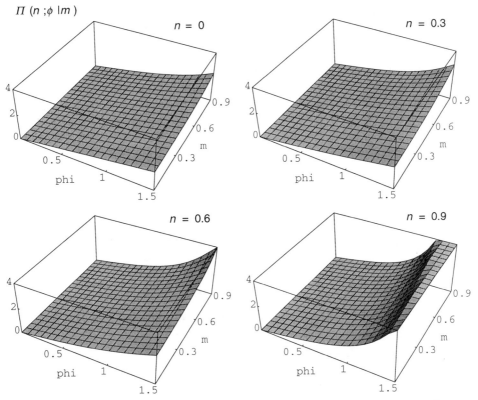

FIGURE 17.2.11 Elliptic integrals of the third kind as surfaces generated by parameters ϕ and m, with n fixed at 0, 0.3, 0.6, or 0.9. For $n = 0.9$, function values exceeding 4 have been truncated. (Adapted from cell `EllipticPi` in *Mathematica* notebook `Elliptics`.)

Algorithm and Program

The algorithms for computing $\Pi(n;\phi\,|\,m)$ are very complicated if one attempts to use the formulas listed in Section 17.7 of Abramowitz and Stegun [Abr64], and there are four cases that depend upon n and m. Travelers reading the *Atlas* should beware that the smooth terrain of $\Pi(n;\phi\,|\,m)$ seen in Figure 17.2.11 hides quicksands of computation if these formulas are used.

It is practical, accurate, and efficient to compute $\Pi(n;\phi\,|\,m)$ by direct numerical integration in (17.2.31) for $n < 1$ and $m < 1$. A robust integration method is needed to handle with sufficient accuracy the development of singularities (Figure 17.2.11 bottom right) as n or m approaches unity. The complete integral— $\Pi(n\,|\,m)$ in Table 17.2.1—can be computed just by setting $\phi = \pi/2$.

For integration we use the nine-strip, equal-step, Newton-Cotes formula given as 25.4.19 in Abramowitz and Stegun [Abr64], namely

$$\int_{x_0}^{x_9} f(x)\,dx \approx \frac{9h}{89600}[2857(f_0 + f_9) + 15741(f_1 + f_8) +$$
$$1080(f_2 + f_7) + 19344(f_3 + f_6) + 5778(f_4 + f_5)] \tag{17.2.32}$$

which has an error of order h^{11} multiplied by the 10th derivative of the integrand, where h is the step size. Until n or m approaches unity and ϕ is near $\pi/2$, this derivative is relatively small, but it grows quickly if these conditions are not met. If one demands the usual accuracy goal for functions in the *Atlas*, 10^{-10}, then for $\phi = \pi/2$ one needs $h \leq 0.01$ in the nine-strip formula for $n \leq 0.94$ and $m \leq 0.94$.

To handle the programming conveniently, the C function `NineStrip` given below includes three applications of (17.2.32) for a total of 28 points in a block of 27 strips. Function `EllptcPi` uses 6 such blocks for a total of $27 \times 6 = 162$ points. If the maximum n and m that you use are smaller than 0.94, or if less accuracy is satisfactory, you can make the program more efficient by reducing the number of blocks and correspondingly increasing the step size.

```c
double EllptcPi(double n, double phi, double m)
/* Elliptic integral of third kind by integration */

{
double Dtheta,integral1,integral2,
  integral3,integral4,integral5,integral6;
double NineStrip(double n, double m,
  double thetaStart, double Dtheta);
int k;

Dtheta = phi/162.0;

integral1 = NineStrip(n,m,0.0,Dtheta);
integral2 = NineStrip(n,m,phi/6.0,Dtheta);
integral3 = NineStrip(n,m,phi/3.0,Dtheta);
integral4 = NineStrip(n,m,phi/2.0,Dtheta);
integral5 = NineStrip(n,m,4.0*phi/6.0,Dtheta);
integral6 = NineStrip(n,m,5.0*phi/6.0,Dtheta);

return  integral1+integral2+integral3+
        integral4+integral5+integral6;

}
```

```
double NineStrip(double n, double m,
  double thetaStart, double Dtheta)
/* Nine-strip integral for Elliptic Pi */
{
double theta,s,s2,f[28],integral;
const double a[5] = {0.2869754464285714,
    1.581127232142857, 0.1084821428571429,
    1.943035714285714, 0.5803794642857144 };
int k;

theta = thetaStart;

for ( k = 0; k <= 27; k++ )
  {
  s = sin(theta);   s2 = s*s;
  f[k] = 1.0/((1.0 - n*s2)*sqrt(1.0 - m*s2));
  theta += Dtheta;
  }

integral =
Dtheta*(a[0]*(f[0]+f[27]+2.0*(f[9]+f[18]))+
        a[1]*(f[1]+f[8]+f[10]+f[17]+f[19]+f[26])+
        a[2]*(f[2]+f[7]+f[11]+f[16]+f[20]+f[25])+
        a[3]*(f[3]+f[6]+f[12]+f[15]+f[21]+f[24])+
        a[4]*(f[4]+f[5]+f[13]+f[14]+f[22]+f[23]));

return  integral;
}
```

Test Values

Check values of $\Pi(n;\phi\,|\,m)$ can be generated by running the `EllipticPi` cell in *Mathematica* notebook `Elliptics`. Examples are given in Table 17.2.6.

M

TABLE 17.2.6 Test values to 10 digits of elliptic integrals of the third kind.

$\phi \setminus (n, m)$	(0.25, 0.30)	(0.25, 0.90)	(0.75, 0.90)	(0.90, 0.90)
0.5	0.5166356028	0.5307696141	0.5554795280	0.5637222218
1.0	1.128965797	1.292346614	1.632335009	1.807930721
1.5	1.879018105	2.799552239	5.342645197	8.857772174

17.3 JACOBI ELLIPTIC FUNCTIONS AND THETA FUNCTIONS

In this section we develop visualizations and C functions for the Jacobi elliptic functions and theta functions. Standard forms of Jacobi elliptic functions are given in Table 17.1.1 in the overview of the elliptic integrals and functions.

17.3.1 Jacobi Elliptic Functions

Keywords

elliptic functions, Jacobi elliptic functions

C Function and *Mathematica* Notebook

JacobiSnCnDn (C); cells JacobiSN, JacobiCN, and JacobiDN in notebook Elliptics (*Mathematica*)

Function Definitions and Background

Suppose that we have an elliptic integral of the first kind (Section 17.2.1)

$$u = F(\phi \,|\, m) = \int_0^\phi \frac{1}{\sqrt{1 - m\sin^2\theta}} \, d\theta \tag{17.3.1}$$

then ϕ is called the *amplitude* of the elliptic integral, so one writes

$$\phi = \mathrm{am}(u \,|\, m) \tag{17.3.2}$$

The basic Jacobi elliptic functions are trigonometric functions of ϕ, summarized as

$$\mathrm{sn}(u \,|\, m) = \sin\phi \qquad \mathrm{cn}(u \,|\, m) = \cos\phi \tag{17.3.3}$$

The first function is often called the *sine amplitude*, while the second function is the *cosine amplitude*. They have the special values

$$\mathrm{sn}(u \,|\, 0) = \sin u \qquad \mathrm{sn}(u \,|\, 1) = \tanh u$$
$$\mathrm{cn}(u \,|\, 0) = \cos u \qquad \mathrm{cn}(u \,|\, 1) = \mathrm{sech}\, u \tag{17.3.4}$$

It is also usual to define the function dn by

$$\mathrm{dn}(u \,|\, m) = \sqrt{1 - m\sin^2\phi} \equiv \Delta\phi$$
$$\mathrm{dn}(u \,|\, 0) = 1 \qquad \mathrm{dn}(u \,|\, 1) = \mathrm{sech}\, u \tag{17.3.5}$$

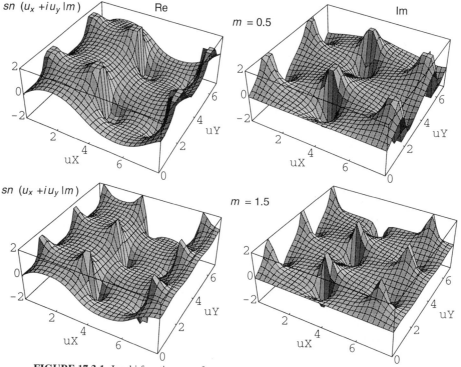

FIGURE 17.3.1 Jacobi functions, sn, for argument $u_x + i\,u_y$ with $m = 0.5$ (top) and $m = 1.5$ (bottom). (Adapted from cell JacobiSN in *Mathematica* notebook Elliptics.)

From the basic functions (17.3.3) and (17.3.5) many Jacobi functions analogous to the trigonometric functions can be generated, as described in Chapter 6 of Jahnke and Emde [Jah45], in great detail in Sections 13.9—13.23 of Erdélyi et al [Erd53b], in the monograph on applications by Greenhill [Gre59], and in Sections 120–129 of Byrd and Friedman [Byr71]. Summaries of properties are given in Chapter 16 of Abramowitz and Stegun [Abr64], whose notation we use, and in Section 8.1 of Gradshteyn and Ryzhik [Gra94]. Some graphics are shown in Chapter 63 of Spanier and Oldham [Spa87].

Visualization

In the adjacent figures we show views of the basic Jacobi elliptic functions, sn, cn and dn, as functions of the complex argument, $u = u_x + iu_y$, with the parameter, m, fixed (Figures 17.3.1–17.3.3). Although in the complex plane cn and dn are not essentially different from sn, it is useful to be able to compare transections of each surface with the curves in the figures that follow.

All three Jacobi functions are doubly periodic and they have doubly-periodic behavior in the complex u plane. Such behavior is illustrated and explained clearly in Sections 6.1–6.3 of Jahnke and Emde [Jah45].

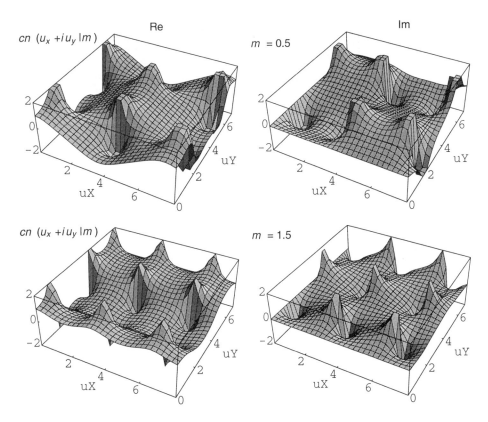

FIGURE 17.3.2 Jacobi functions, cn, as surfaces generated for complex argument $u_x + i\,u_y$ for parameter $m = 0.5$ (top) and $m = 1.5$ (bottom). (Adapted from cell `JacobiCN` in *Mathematica* notebook `Elliptics`.)

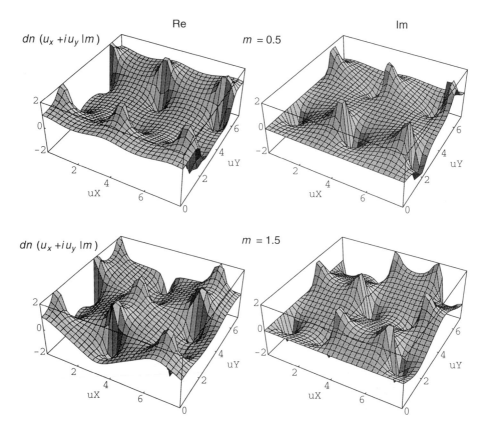

FIGURE 17.3.3 Jacobi functions, dn, as surfaces for argument $u_x + i\,u_y$ for $m = 0.5$ (top) and $m = 1.5$ (bottom). (Cell `JacobiDN` in *Mathematica* notebook `Elliptics`.)

The following three figures show the Jacobi elliptic functions—sn, cn, and dn—as surfaces for real argument u and real parameter m. The front edges of these surfaces, which have $m = 0$, are then the sine and cosine functions for sn and cn, respectively, and the constant value of unity for dn.

By making a transection of each surface, one can hold m constant and vary u (top right curves) or u can be constant and m can vary (bottom right curves). For each Jacobi function, we show three transection curves for each choice of constant value; thus we cover the range of interesting parameter values for the surfaces in Figures 17.3.3–17.3.6.

The cells in *Mathematica* notebook `Elliptics` that are used for the visualizations of the Jacobi elliptic functions are very easy to understand and to adapt to display sn, cn, or dn in regions of parameter space that are of interest to you. The ready-to-run notebook is on the CD-RO; the annotated listings of the cells in Section 20.17 have suggestions for how to explore further.

The convergence of the Jacobi functions to trigonometric functions as m tends to zero, suggests the possibility of expanding them in terms of sines and cosines. Such expansions are given, for example, in Section 16.13 of Abramowitz and Stegun [Abr64]. Similarly, as m approaches unity, expansions in terms of hyperbolic functions can be made, as given in Section 127.02 of Byrd and Friedman [Byr71].

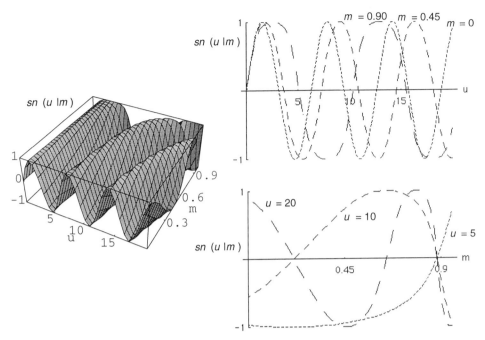

FIGURE 17.3.4 Jacobi functions, sn. Left: Surface for arguments u from 0 to 20 and m from 0 to 0.95. Right top: Curves for u from 0 to 20 and $m = 0$ (sine), 0.45, and 0.90. Right bottom: Curves for m from 0 to 0.95 and $u = 5$, 10, and 20. (Adapted from cell `JacobiSN` in *Mathematica* notebook `Elliptics`.)

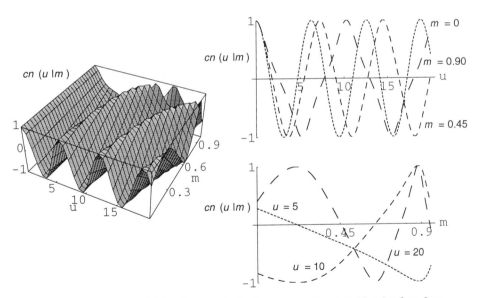

FIGURE 17.3.5 Jacobi functions, cn. Left: Surface for u from 0 to 20 and m from 0 to 0.95. Right top: Curves for u from 0 to 20 and $m = 0$ (sine), 0.45, and 0.90. Right bottom: Curves for m from 0 to 0.95 and $u = 5$, 10, and 20. (Adapted from cell `JacobiCN` in *Mathematica* notebook `Elliptics`.)

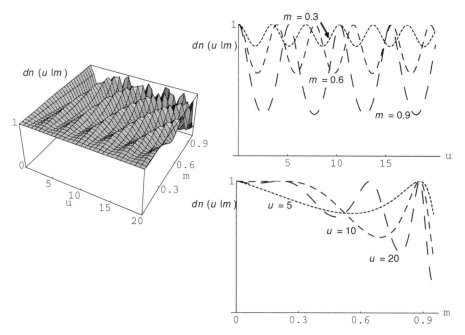

FIGURE 17.3.6 Jacobi functions, dn. Left: Surface generated for u from 0 to 20 and m from 0 to 0.95. Right top: Curves for u from 0 to 20 and $m = 0$ (sine function), 0.45, and 0.90. Right bottom: Curves for m from 0 to 0.95 and $u = 5$, 10, and 20. (Adapted from cell `JacobiDN` in *Mathematica* notebook `Elliptics`.)

Algorithm and Program

For computing the Jacobi elliptic functions we use the descending Landen transforms—also called the arithmetic-geometric mean (AGM)—as given in Section 16.4 of Abramowitz and Stegun [Abr64]. (Landen transforms are explained in Section 17.2.1 of the *Atlas* under Algorithm and Program.) The transform procedes as follows. Set

$$a_0 = 1 \qquad b_0 = \sqrt{1-m} \qquad c_0 = \sqrt{m} \qquad (17.3.6)$$

Then form in succession the arithmetic and geometric means of the preceding values, as

$$a_{n+1} = (a_n + b_n)/2 \qquad b_{n+1} = \sqrt{a_n b_n} \qquad n \geq 0 \qquad (17.3.7)$$

and additionally

$$c_{n+1} = (a_n - b_n)/2 \qquad n \geq 0 \qquad (17.3.8)$$

Continue this until $c_n < \varepsilon$, where ε is some user-specified accuracy control that is not less than the accuracy of the computer's arithmetic—else the subtraction to form c_n is not meaningful. Figure 17.2.3 in Section 17.2.1 shows successive stages in the Landen transform; the AGM converges rapidly even for m near unity, as shown in Table 17.2.1.

Now compute the angle

$$\phi_n = 2^n a_n u \qquad (17.3.9)$$

then compute by downward iteration the decreasing angles

$$\phi_{n-1} = [\phi_n + \arcsin(c_n \sin \phi_n / a_n)]/2 \qquad n \geq 1 \qquad (17.3.10)$$

The three primary Jacobi elliptic functions are then given by

$$\text{sn}(u \mid m) = \sin \phi_0 \qquad \text{cn}(u \mid m) = \cos \phi_0$$
$$\text{dn}(u \mid m) = \cos \phi_0 / \cos(\phi_1 - \phi_0) \qquad (17.3.11)$$

The algorithm for our C function `JacobiSnCnDn` is just a straightforward coding of (17.3.6)—(17.3.11). A call to this void function (the analogue of a Fortran subroutine) returns three values, sn, cn, and dn in the argument list of the function. The annotated listing of the C driver program in Section 21.17 shows how `JacobiSnCnDn` can be used.

```c
#define MAX 10   /* Size of array */

void JacobiSnCnDn(double u, double m,
 double eps, double *sn, double *cn, double *dn)
/* Jacobi elliptic functions by AGM method;
    sn, cn, and  dn  */
{
double a[MAX],b[MAX],c[MAX],phi[MAX];
int n;

if ( m == 0.0 )
 { *sn = sin(u); *cn = cos(u); *dn = 1.0;
 return; }

if ( m == 1.0 )
 { *sn = tanh(u); *cn = 1.0/cosh(u); *dn = *cn;
 return; }

a[0] = 1.0;
b[0] = sqrt(1.0 - m);
c[0] = sqrt(m);

n = 0;
while ( fabs(c[n]) > eps )
 {
 a[n+1] = 0.5*(a[n]+b[n]);
 b[n+1] = sqrt(a[n]*b[n]);
 c[n+1] = 0.5*(a[n]-b[n]);

 if ( n >= MAX-2 )
   {
   printf
   ("\n\n!!In JacobiSnCnDn  n=%i  too large",n);
   printf
   ("\nIncrease MAX=%i or decrease  eps",MAX);
   c[MAX-1] = 0.0;
   }

 n++;
 }
```

```
n--;
phi[n] = pow(2.0, (double) n )*a[n]*u;

while ( n > 0 )
  {
  phi[n-1] =
    0.5*(phi[n]+asin(c[n]*sin(phi[n])/a[n]));
  n--;
  }

*sn = sin(phi[0]);
*cn = cos(phi[0]);
*dn = *cn/cos(phi[1]-phi[0]);
return;
}
```

Programming Notes

In the above coding for the Jacobi elliptic functions the coefficients a_n, b_n, and c_n are stored in arrays of size MAX, which we have set to 10, thus allowing a maximum n value of 9. In practice, however, n does not exceed 6, even for $\varepsilon = 10^{-12}$, which is two orders of magnitude more accurate than the design goal for functions in the *Atlas*. If you require accuracy greater than this—and if the machine precision is sufficient—just increase MAX.

Test Values

Test values of sn, cn, and dn can be generated by running the JacobiSN, JacobiCN, and JacobiDn cells, respectively, in *Mathematica* notebook Elliptics. Examples are given in Tables 17.3.1–17.3.3.

TABLE 17.3.1 Test values to 10 digits of the Jacobi elliptic function sn($u \mid m$).

$u \setminus m$	0.3	0.6	0.9
1.0	0.8187707145	0.7949388393	0.7700857249
3.0	0.4114226606	0.7432676861	0.9906305999
5.0	−0.9929697113	−0.8398103136	0.1549888206

TABLE 17.3.2 Test values to 10 digits of the Jacobi elliptic function cn($u \mid m$).

$u \setminus m$	0.3	0.6	0.9
1.0	0.5741206467	0.6066895761	0.6379404175
3.0	−0.9114446743	−0.6689941306	−0.136568717
5.0	−0.1183687139	−0.5428799472	−0.9879162239

TABLE 17.3.3 Test values to 10 digits of the Jacobi elliptic function dn($u \mid m$).

$u \setminus m$	0.3	0.6	0.9
1.0	0.8938033090	0.7879361300	0.6828405221
3.0	0.9742789222	0.8176379933	0.3417395397
5.0	0.8391682464	0.7594940304	0.9891312446

17.3.2 Theta Functions

Keywords
elliptic functions, Jacobi theta functions, theta functions

C Function and *Mathematica* Notebook
JacobiTheta (C); cell Theta in notebook Elliptics (*Mathematica*)

M

Background
Jacobi theta functions—usually just called theta functions—are simply-periodic functions that can be represented by rapidly convergent power series. In the context of elliptic integrals and functions, they provide a direct way to compute these functions. Murty et al [Mur93] give a survey of theta functions and their applications. Clear presentations are in Chapter 21 of Whittaker and Watson [Whi27], as well as in Sections 13.19 and 13.20 of Erdélyi et al [Erd53b]. Many formulas are presented in Sections 1050–1053 of Byrd and Friedman [Byr71] in Chapter 4 of Jahnke and Emde [Jah45], and in Sections 8.18–8.19 of Gradshteyn and Ryzhik [Gra94].

Formulas relevant for computing the theta functions are in Sections 16.27–16.34 of Abramowitz and Stegun [Abr64], whose notation we use. Beware that the notation for theta functions is quite variable.

Theta functions satisfy diffusion-type equations, so they may be useful in heat-flow problems. In the context of elliptic functions, the argument q in the following is the nome function, and therefore has a range of (0, 1).

Function Definitions and Visualizations
The definition of the four theta functions in terms of rapidly-convergent Fourier series is quite simple. We therefore define each function and immediately show several views of it.

Theta Function of the First Kind. This function is defined by

$$\theta_1(u,q) \equiv 2q^{1/4} \sum_{n=0}^{\infty} (-1)^n q^{n(n+1)} \sin[(2n+1)u] \qquad (17.3.12)$$

This function is convergent for $|q| < 1$ for any choice of u. It is periodic, with period 2π and half period π..

Note that when $u = \pi/2$ the n-dependent phase is removed by the sine value, so the theta function of the first kind attains its maximum value. In the neighborhood of $u = \pi/2$ the function therefore grows very rapidly as $q \to 1$. In the following 3D visualization we truncate the display at $\theta_1(u,q) = 4$ and do not allow q to exceed 0.99.

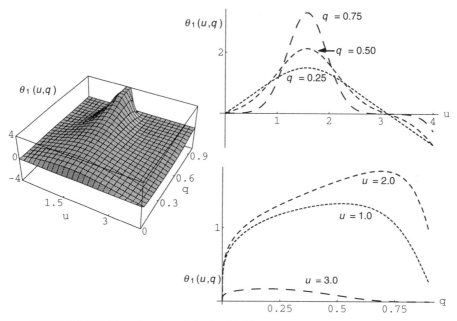

FIGURE 17.3.7 Theta functions of the first kind for arguments u and q (the nome). Left: Surface for u in $[0, 4]$ and q in $[0, 0.99]$, truncated for values > 4. Right top: u varying and q fixed. Right bottom: q varying in $[0, 0.9]$ and u fixed. (Cell Theta in *Mathematica* notebook Elliptics.)

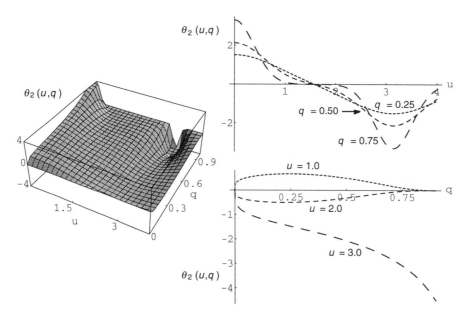

FIGURE 17.3.8 Theta functions of the second kind for arguments u and q (the nome). Left: Surface for u in $[0, 4]$ and q in $[0, 0.99]$, truncated for absolute values > 4. Right top: u varying and q fixed. Right bottom: q varying in $[0, 0.9]$ and u fixed. (Cell Theta in the *Mathematica* notebook Elliptics.)

Theta Function of the Second Kind. The Jacobi theta function of the second kind—shown in Figure 17.3.8—is defined by

$$\theta_2(u,q) \equiv 2q^{1/4} \sum_{n=0}^{\infty} q^{n(n+1)} \cos[(2n+1)u] \qquad (17.3.13)$$

This function is convergent for $|q| < 1$ for all u; it has period 2π and half period π.. It has the symmetry $\theta_2(u+\pi,q) = -\theta_2(u,q)$, as seen in the figure. In the neighborhood of $u = 0$ or $u = \pi$ the function grows very rapidly as $q \to 1$. In Figure 17.3.8 we truncate the display at $\theta_2(u,q) = 4$ and we stop q at 0.99; even so, convergence to accuracy $\varepsilon = 10^{-10}$ is not reached until $n \approx 50$.

Theta Function of the Third Kind. The Jacobi theta function of the third kind is defined by a series that is convergent for $|q| < 1$ for all u, namely,

$$\theta_3(u,q) \equiv 1 + 2\sum_{n=1}^{\infty} q^{n^2} \cos(2nu) \qquad (17.3.14)$$

The theta function of the third kind has period 2π and half period π.. It has the symmetry $\theta_3(u+\pi,q) = \theta_3(u,q)$, as seen in Figure 17.3.9. Near $u = 0$ or $u = \pi$ the function grows very rapidly as $q \to 1$., therefore in the figure we truncate the display at $\theta_3(u,q) = 4$ and we stop at $q = 0.99$; convergence to accuracy $\varepsilon = 10^{-10}$ requires $n \approx 50$.

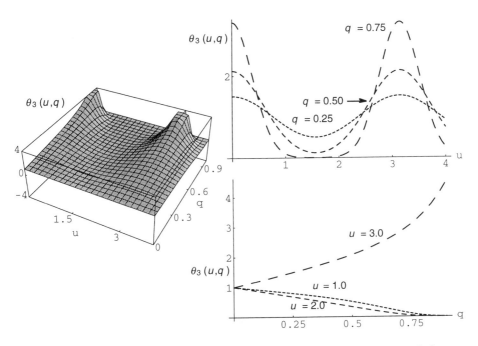

FIGURE 17.3.9 Theta functions of the third kind for arguments u and q (the nome). Left: Surface for u in $[0, 4]$ and q in $[0, 0.99]$, truncated for values exceeding 4. Right: Top shows curves with u varying and q fixed, while bottom shows curves with q varying in $[0, 0.9]$ with u fixed. (Adapted from cell Theta in *Mathematica* notebook Elliptics.)

Theta Function of the Fourth Kind. The Jacobi theta function of the fourth kind is defined by a series that is convergent for $|q| < 1$ for all u. It is given by

$$\theta_4(u,q) \equiv 1 + 2\sum_{n=1}^{\infty} (-1)^n q^{n^2} \cos(2nu) \qquad (17.3.15)$$

The function of the fourth kind has period 2π and half period π., and the symmetry $\theta_4(u+\pi,q) = \theta_4(u,q)$, as seen in Figure 17.3.10. Near $u = \pi/2$ the function grows very rapidly as $q \to 1$, therefore we truncate the display at $\theta_4(u,q) = 4$ and we use $q \le 0.99$; convergence to accuracy $\varepsilon = 10^{-10}$ still requires $n \approx 50$ in this case.

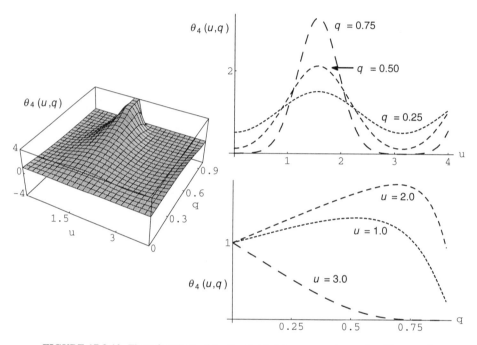

FIGURE 17.3.10 Theta functions of the fourth kind for arguments u and q (the nome). Left: Surface for u in [0, 4] and q in [0, 0.99], truncated for values exceeding 4. Right: Top shows curves with u varying and q fixed, while bottom shows curves with q varying in [0, 0.9] with u fixed. (Adapted from cell `Theta` in *Mathematica* notebook `Elliptics`.)

⚠ Note that the fourth Jacobi theta function, θ_4, is often denoted by θ_0, by θ, or by ϑ.

Algorithms and Program

A single C function, `JacobiTheta`, is devised to compute accurately and efficiently the four Jacobi theta functions. The program is a direct implementation of the defining equations (17.3.12)–(17.3.15), with u and q as input arguments, and an accuracy control parameter, `eps`, to estimate the absolute accuracy of the function that converges slowest.

If you need only one or two of the theta functions and many of these have to be computed accurately, it is straightforward to remove the statements that compute the partial sums for the unneeded functions.

```
void JacobiTheta(double u, double q, double eps,
 double *theta1, double *theta2,
 double *theta3, double *theta4)
/* Jacobi theta functions by series */
{
double qS,rn,sn,tn,phase,TU,TnU,TnP1U,
 sum1,sum2,sum3,sum4,small,eps10,term,factor;
int n;

qS = q*q;
rn = q;
sn = q;
tn = qS;
phase = -1.0;
TU = 2.0*u;
TnU = TU;
TnP1U = TnU+u;
sum1 = sin(u);
sum2 = cos(u);
sum3 = 0.0;
sum4 = 0.0;
small = 1.0;
eps10 = 0.1*eps;
n = 1;
while ( small > eps10 )
  {
  sum1 +=  phase*tn*sin(TnP1U);
  sum2 +=  tn*cos(TnP1U);
  term = sn*cos(TnU);
  sum3 += term;
  sum4 +=  phase*term;
  rn *= qS;
  sn *= rn;
  tn *= rn*q;
  phase = -phase;
  TnP1U += TU;
  TnU += TU;
  small = sn;
  n++;
  }
factor = 2.0*pow(q,0.25);

*theta1 = factor*sum1;
*theta2 = factor*sum2;
*theta3 = 1.0+2.0*sum3;
*theta4 = 1.0+2.0*sum4;

return;
}
```

Programming Notes

The convergence of the series for the four theta functions is tested only for what is likely to be the slowest-converging series, namely θ_4, given by (17.3.15). For this series, the C function tests only the q^{n^2} term, ignoring the cosine term, which may vanish. The oscillating sign in this series for θ_4 is accounted for by demanding that the series continue until the q-dependent term is a factor of 10 smaller than the input eps.

Note that we test for absolute accuracy, rather than the fractional accuracy that is usual in most of the *Atlas*. There is generally no way to specify fractional accuracy, since some of the theta functions have zeros whenever u is a multiple of $\pi/2$. Convergence of the series changes rapidly with q. For example, if $\varepsilon = 10^{-10}$, when $q = 0.5, 0.9, 0.95$, and 0.99, the number of terms required is 6, 15, 22, and 50, respectively.

Test Values

Check values of θ_1, θ_2, θ_3, and θ_4 can be generated by running the Theta cell in *Mathematica* notebook Elliptics. Examples are given in Tables 17.3.4–17.3.7.

TABLE 17.3.4 Test values to 10 digits of the Jacobi theta function of the first kind.

$u \setminus q$	0.25	0.50	0.75
1.0	1.177215174	1.330378498	1.064797946
2.0	1.310449911	1.632025903	1.741887940
3.0	0.1633718098	0.08080641825	0.002602152561

TABLE 17.3.5 Test values to 10 digits of the Jacobi theta function of the second kind.

$u \setminus q$	0.25	0.50	0.75
1.0	0.6766970505	0.5001981385	0.1022076506
2.0	−0.5039423273	−0.3181628217	−0.03561855843
3.0	−1.480856455	−2.068233820	−3.082142269

TABLE 17.3.6 Test values to 10 digits of the Jacobi theta function of the third kind.

$u \setminus q$	0.25	0.50	0.75
1.0	0.7868273164	0.5058938857	0.1022084382
2.0	0.6720479082	0.3314359783	0.03562460647
3.0	1.486682790	2.068244348	3.082142269

TABLE 17.3.7 Test values to 10 digits of the Jacobi theta function of the fourth kind.

$u \setminus q$	0.25	0.50	0.75
1.0	1.202959502	1.330686328	1.064797947
2.0	1.325678653	1.632130562	1.741887940
3.0	0.5265024281	0.1427450318	0.002849602040

17.3.3 Logarithmic Derivatives of Theta Functions

Keywords
elliptic functions, logarithmic derivatives of theta functions, theta functions

C Function and *Mathematica* Notebook $\boxed{M}$
LogDerTheta (C); cell LogDerTheta in notebook Elliptics (*Mathematica*)

Background
Logarithmic derivatives of the Jacobi theta functions (Section 17.3.2) are useful for computing elliptic integrals and elliptic functions (Sections 17.1, 17.2). The derivatives can be represented by power series obtained from product expansions for theta functions. Expressions for the derivatives are in Section 16.29 of Abramowitz and Stegun [Abr64].

Developments of the above ideas are given in Neville's monograph [Nev44] and in Sections 13.19 and 13.20 of Erdélyi et al [Erd53b]. For consistency with Section 17.3.2, we use the notation in Chapter 16 of Abramowitz and Stegun. Beware that the notation for theta functions is quite variable. In the following, we consider the logarithmic derivatives of all four theta functions and we give a single C function that computes them, just as for the theta functions in Section 17.3.2.

Function Definitions and Visualizations
Definition of the logarithmic derivatives of theta functions in terms of Fourier series is simple. In the following we define each of the four derivative functions then show views of it as surfaces and curves.

Logarithmic Derivative of the Theta Function of the First Kind. This is given by

$$\theta_1'(u,q)/\theta_1(u,q) = \cot u + 4\sum_{n=1}^{\infty} \frac{q^{2n}\sin(2nu)}{1-q^{2n}} \qquad (17.3.16)$$

Near $u=\pi$ the function grows very rapidly, especially as $q\to1$. In the 3D visualization (Figure 17.3.11 on the following page) we display the logarithmic derivative in the q range $[0.1, 0.9]$ only. In the curves of q varying and u fixed we also use q in $[0.1, 0.9]$, since the logarithmic derivative becomes very large as $u\to\pi$.

Logarithmic Derivative of the Theta Function of the Second Kind. This derivative—shown in Figure 17.3.12 following—is given by

$$\theta_2'(u,q)/\theta_2(u,q) = -\tan u + 4\sum_{n=1}^{\infty} \frac{(-1)^{n-1}q^{2n}\sin(2nu)}{1-q^{2n}} \qquad (17.3.17)$$

Near $|u|=\pi/2$ the derivative grows very rapidly, so we show the logarithmic derivative in the q range $[0.1, 0.9]$ only. In curves with q varying and u fixed we show q in $[0.1, 0.9]$ only, since the functions become very large as $u\to\pi/2$.

Logarithmic Derivative of the Theta Function of the Third Kind. This derivative (Figure 17.3.13 following) has a series expansion convergent for all u if $|q|<1$, namely,

$$\theta_3'(u,q)/\theta_3(u,q) = 4\sum_{n=1}^{\infty} \frac{(-1)^{n-1}q^{n}\sin(2nu)}{1-q^{2n}} \qquad (17.3.18)$$

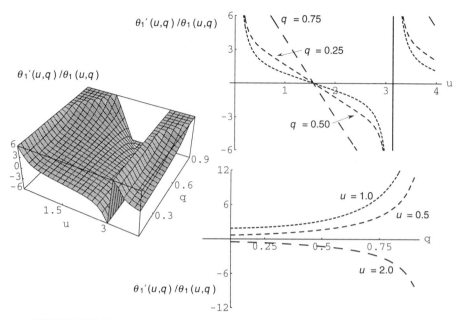

FIGURE 17.3.11 Logarithmic derivatives of theta functions of the first kind for arguments u and q. Left: Surface for u in $[0.1, 4]$ and q in $[0.1, 0.9]$. Right top: u varying and q fixed. Right bottom: q varying in $[0.1, 0.9]$ and u fixed. (Cell `LogDerTheta` in *Mathematica* notebook `Elliptics`.)

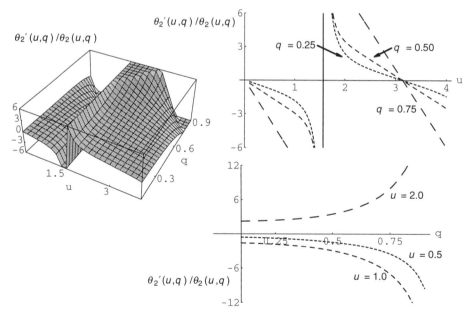

FIGURE 17.3.12 Logarithmic derivatives of theta functions of the second kind for arguments u and q. Left: Surface for u in $[0.1, 4]$ and q in $[0.1, 0.9]$. Right top: u varying and q fixed. Right bottom: q varying in $[0.1, 0.9]$ and u fixed. (Cell `LogDerTheta` in *Mathematica* notebook `Elliptics`.)

For the logarithmic derivative of the theta function of the third kind the derivative is very large for some values of u and q, so we display the logarithmic derivative in the range $[-6, 6]$ or $[-12, 12]$. In the curve with q varying and u fixed (Figure 17.3.13 right bottom) we show q in $[0.1, 0.9]$ only, since the functions become very large as $q \to 1$.

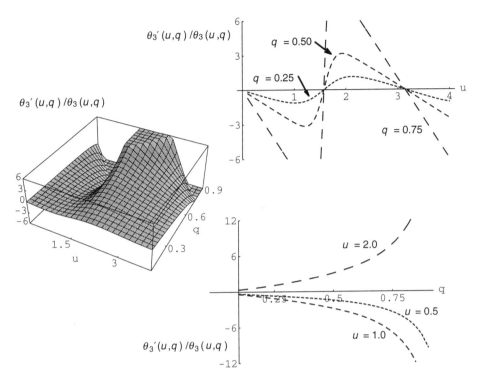

FIGURE 17.3.13 Logarithmic derivatives of theta functions of the third kind for arguments u and q. Left: Surface for u in $[0.1, 4]$ and q in $[0.1, 0.9]$. Right top: u varying and q fixed. Right bottom: q varying in $[0.1, 0.9]$ and u fixed. (Cell `LogDerTheta` in *Mathematica* notebook `Elliptics`.)

Logarithmic Derivative of the Theta Function of the Fourth Kind. This logarithmic derivative is given by a series that converges for all u if $|q| < 1$. Following 16.29.4 in Abramowitz and Stegun [Abr64], it is given by

$$\theta_4'(u,q)/\theta_4(u,q) = 4\sum_{n=1}^{\infty} \frac{q^n \sin(2nu)}{1-q^{2n}} \tag{17.3.19}$$

Note that the fourth Jacobi theta function, θ_4, is often denoted by θ_0, by θ, or by ϑ. The logarithmic derivatives are shown by the surface and curves in Figure 17.3.14 on the following page. Since the magnitude of the function is very large for some values of u and q, we display it in the range $[-6, 6]$ or $[-12, 12]$. In the curve with q varying and u fixed (Figure 17.3.14 right bottom) we show q in $[0.1, 0.9]$ only, since the functions become very large as $q \to 1$.

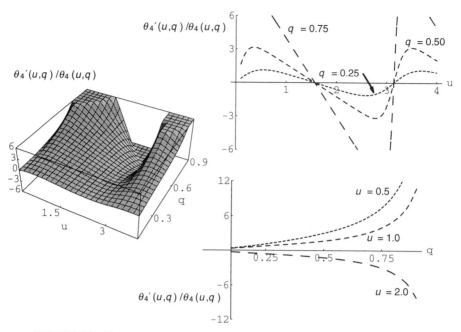

FIGURE 17.3.14 Logarithmic derivatives of theta functions of the fourth kind for arguments u and q. Left: Surface for u in $[0.1, 4]$ and q in $[0.1, 0.9]$. Right top: u varying and q fixed. Right bottom: q varying in $[0.1, 0.9]$ and u fixed. (Cell LogDerTheta in *Mathematica* notebook Elliptics.)

Algorithms and Program

A single C function, LogDerTheta, is devised to compute accurately and efficiently the logarithmic derivatives of the four Jacobi theta functions in Section 17.3.2. The program is a direct implementation of the defining equations (17.3.16)–(17.3.19), with u and q as input arguments. Also input is an accuracy control parameter, eps, to estimate the absolute accuracy of the function that converges slowest.

If you need only one or two of the theta function logarithmic derivatives and many of these have to be computed efficiently, it is straightforward to remove the statements that compute the partial sums for the unneeded functions.

```
void LogDerTheta(double u, double q, double eps,
  double *LDtheta1, double *LDtheta2,
  double *LDtheta3, double *LDtheta4)
/*  Log derivatives of Jacobi theta functions
    by series */
{
double qS,sn,tn,phase,TU,TnU,sum1,sum2,sum3,sum4,
  small,eps10,sinTerm,sumTerm;
int n;

qS = q*q;
sn = q;
tn = qS;
phase = -1.0;
```

```
TU = 2.0*u;
TnU = TU;
sum1 = 0.0;
sum2 = 0.0;
sum3 = 0.0;
sum4 = 0.0;
small = 1.0;
eps10 = 0.1*eps;
n = 1;
while ( small > eps10 )
  {
  sinTerm = sin(TnU)/(1.0-tn);
  sumTerm = tn*sinTerm;
  sum1 +=  sumTerm;
  sum2 +=  phase*sumTerm;
  sumTerm = sn*sinTerm;
  sum3 += phase*sumTerm;
  sum4 +=  sumTerm;
  sn *= q;
  tn *= qS;
  phase = -phase;
  TnU += TU;
  small = sn;
  n++;
  }
*LDtheta1 =   1.0/tan(u)+4.0*sum1;
*LDtheta2 =  -tan(u)+4.0*sum2;
*LDtheta3 =   4.0*sum3;
*LDtheta4 =   4.0*sum4;

return;
}
```

Programming Notes

Convergence of the series for the four logarithmic derivatives of the theta functions is tested only for the slowest-converging series, namely the series for the derivative of θ_3, which is given by (17.3.18). For this series, the C function tests only the q^n term, ignoring the sine term (which may vanish) and the q dependence in the denominator. The oscillating sign in this series is accounted for by demanding that the series continue until the q-dependent term is a factor of 10 smaller than the input eps.

We test for absolute accuracy, rather than the fractional accuracy usual in most of the *Atlas*. There is generally no way to specify fractional accuracy because the logarithmic-derivative functions have zeros and poles, as seen in Figures 17.3.11–17.3.14. Convergence of the series changes rapidly with q. For example, if $\varepsilon = 10^{-10}$, as q increases from 0.5 to 0.9 the number of terms required increases from 33 to 218. If modest accuracy of $\varepsilon = 10^{-4}$ suffices, the number of terms required is just 13 at $q = 0.5$ and 87 at $q = 0.9$.

Test Values

Check values for the logarithmic derivatives of θ_1, θ_2, θ_3, and θ_4 can be generated by running the LogDerTheta cell in *Mathematica* notebook Elliptics. Examples are given in Tables 17.3.8–17.3.11.

TABLE 17.3.8 Test values to 10 digits of logarithmic derivatives of the Jacobi theta function of the first kind.

$u \setminus q$	0.25	0.50	0.75
1.0	0.8724857825	1.648018715	3.968243992
2.0	−0.6444906343	−1.238710433	−2.983875009
3.0	−7.098973039	−7.597891856	−10.97445661

TABLE 17.3.9 Test values to 10 digits of logarithmic derivatives of the Jacobi theta function of the second kind.

$u \setminus q$	0.25	0.50	0.75
1.0	−1.811423198	−2.936999984	−6.952203151
2.0	2.402876058	3.483021691	7.938342250
3.0	0.2093223502	0.4085698189	0.9843689764

TABLE 17.3.10 Test values to 10 digits of logarithmic derivatives of the Jacobi theta function of the third kind.

$u \setminus q$	0.25	0.50	0.75
1.0	−1.125576786	−2.834361244	−6.952034837
2.0	1.080145633	3.112430779	7.934634004
3.0	0.1992474687	0.4085302579	0.9843689764

TABLE 17.3.11 Test values to 10 digits of logarithmic derivatives of the Jacobi theta function of the fourth kind.

$u \setminus q$	0.25	0.50	0.75
1.0	0.7755329328	1.645921647	3.968243978
2.0	−0.5942197984	−1.238129629	−2.983875008
3.0	−0.4989188171	−2.157170006	−8.987707858

REFERENCES ON ELLIPTIC INTEGRALS AND ELLIPTIC FUNCTIONS

[Abr64]   Abramowitz, M., and I. A. Stegun, *Handbook of Mathematical Functions,* Dover, New York, 1964.

[Byr71]   Byrd, P. F., and M. D. Friedman, *Handbook of Elliptic Integrals for Engineers and Scientists*, Springer-Verlag, New York, second edition, 1971.

[Car77]   Carlson, B. C., *Special Functions of Applied Mathematics*, Academic Press, New York, 1977.

[Cay56]   Cayley, A., *An Elementary Treatise on Elliptic Functions*, Dover, New York, 1956.

[Erd53b]   Erdélyi, A., et al, *Higher Transcendental Functions*, Vol. 2, McGraw-Hill, New York, 1953; reprint edition, Krieger, Malabar, Florida, 1981.

[Gon92]   González, M. O., *Complex Analysis*, Marcel Dekker, New York, 1992.

[Gra94]   Gradshteyn, I. S., and I. M. Ryzhik, *Table of Integrals, Series, and Products*, Academic Press, fifth edition, 1994.

[Gre59]   Greenhill, A. G., *The Applications of Elliptic Functions*, Dover, New York, 1959.

[Heu41]   Heuman, C., Journal of Mathematics and Physics, **20**, 127 (1941).

[Jah45]   Jahnke, E., and F. Emde, *Tables of Functions*, Dover, New York, 1945.

[Lan75]   Landen, J., Philosophical Transactions of the Royal Society, **65**, 283 (1775).

[Mur93]   Murty, M. R. (editor), *Theta Functions from the Classical to the Modern*, American Mathematical Society, Providence, Rhode Island, 1993.

[Nev44]   Neville, E. H., *Jacobian Elliptic Functions*, Clarendon Press, Oxford, 1944.

[Sid95]   Sidhu, S., Computers in Physics, **9**, 268 (1995).

[Spa87]   Spanier, J., and K. B. Oldham, *An Atlas of Functions,* Hemisphere Publishing, Washington, D.C., 1987.

[Whi27]   Whittaker, E. T., and G. N. Watson, *A Course of Modern Analysis*, Cambridge University Press, Cambridge, England, fourth edition, 1927.

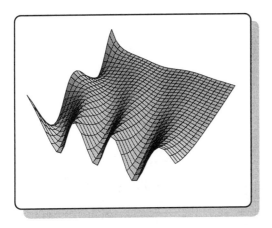

Chapter 18

PARABOLIC CYLINDER FUNCTIONS

The Weber parabolic cylinder functions are solutions of the wave equation in parabolic cylinder coordinates. Tabulations of analytical and numerical results for parabolic cylinder functions are considered in Chapters 19 of our guidebook, Abramowitz and Stegun [Abr64]. This chapter emphasizes visualizing the functions by using *Mathematica*, then using the insight obtained to devise accurate and efficient algorithms for computing the functions by C programs. Section 18.1 reviews the parabolic cylinder coordinates, then Section 18.2 describes the functions U and V.

18.1 PARABOLIC CYLINDER COORDINATES

Weber's parabolic cylinder functions—not to be confused with Weber functions in Section 15.2—occur in many contexts. For example, in quantum mechanics, statistical mechanics, and mathematical statistics. We discuss the coordinate system and solution of the Laplace equation in this system.

When solving differential equations in which parabolic cylinders provide surfaces with particularly simple properties, such as the vanishing of the electrostatic potential on the surfaces when solving the Laplace equation, it is practical to use parabolic cylinder coordinates. These coordinates, (α, β, z), are related to Cartesian coordinates by

$$x = c(\alpha^2 - \beta^2)/2 \qquad y = c\alpha\beta \qquad z = z \qquad (18.1.1)$$

in which $c > 0$ is a scale factor.

This triply-orthogonal system of surfaces is composed of parabolic cylinders with focii at the origin and of planes $z =$ constant, as shown in Figure 18.1.1 on the following page. You can explore these coordinates from several viewpoints by running cell Parabolic-Coords in *Mathematica* notebook ParCyl, which is annotated in Section 20.18.

Separation of the Laplacian in this coordinate system is discussed in the monograph by Moon and Spencer on field theory for engineers [Moo61], in their field-theory handbook

539

[Moo71], and in Lebedev's text on special functions [Leb65]. They show that the solution of the Laplace equation reduces to solving the differential equation

$$w'' - (z^2/4 + a)w(z) = 0 \qquad (18.1.2)$$

in which variable z is proportional to α or to β and a is related to the separation constant (eigenvalue) of the Laplace equation. The solutions of this equation are obtained as parabolic cylinder functions, to which we direct our attention in Section 18.2.

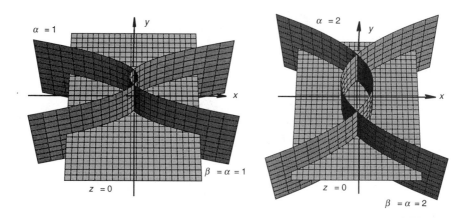

FIGURE 18.1.1 Parabolic cylinder coordinates, showing surfaces of constant α and β (both parabolic cylinders), and of constant z (planes perpendicular to the z axis, which is out of the plane of the paper). Left view $\alpha = \beta = 1$, right view $\alpha = \beta = 2$. (Adapted from cell `ParabolicCoords` in *Mathematica* notebook `ParCyl`.)

18.2   PARABOLIC CYLINDER FUNCTIONS

Two types of parabolic cylinder functions, $U(a, x)$ and $V(a, x)$, are considered. The latter can be obtained as linear combinations of the former, but it is most convenient to consider it separately—as in Section 18.2.2.

18.2.1 Parabolic Cylinder Functions U

Keywords
Hermite function, parabolic cylinder function, Weber function, Whittaker function

C Function and *Mathematica* Notebook
`ParCylU` (C); cell `ParabolicCylinderU` in notebook `ParCyl` (*Mathematica*)

Background
The parabolic cylinder function $U(a, x)$, also denoted by $D_{-a-1/2}(x)$ in Whittaker's notation [Whi27, Section 16.5], is the solution of (18.1.2) for $z = x$ that has asymptotic behavior for $x \gg |a|$

$$D(a, x) \sim e^{-x^2/4} / x^{a+1/2} \qquad (18.2.1)$$

The analysis background to $U(a,x)$ is given in Chapter 8 of Erdélyi et al [Erd53b]. Lebedev [Leb65, Section 10.2] denotes it as the Hermite function, since if $a = -n - 1/2$ with n a nonnegative integer then $U(a,x)$ is related simply to the Hermite polynomial $H_n(x/\sqrt{2})$ described in Section 11.4 of the *Atlas*. Many formulas are given in Section 7.7 of Gradshteyn and Ryzhik [Gra94] and in Chapter 19 of Abramowitz and Stegun [Abr64]. Some curves of $U(a,x)$ are shown in Chapter 46 of Spanier and Oldham [Spa87].

Background to numerical computation of U is given in Miller's introduction to the tables in [NPL55]. In spite of the title of this work, the function tabulated is $W(a,x)$, related to the solutions of (18.1.2) when $z = ix$, a purely imaginary argument. Other methods for $U(a,x)$ than those that we use are described in the works of Latham and Redding [Lat74, Red76], Schulten et al [Sch81], and Taubmann [Tau92].

Function Definition

Two definitions are used when computing the parabolic cylinder function $U(a,x)$ for moderate values of a and x, namely, definition in terms of confluent hypergeometric functions $_1F_1$ (Section 16.2.1) and in terms of an integral. The first definition can be deduced from equations 19.3.1–19.3.4 in [Abr64], or (through several changes of notation) from equation (4) in Section 8.2 of [Erd53b]. It results in

$$U(a,x) = \frac{\sqrt{\pi}\, e^{-x^2/4}}{2^{(a+1/2)/2}}$$

$$\times \left[\frac{{}_1F_1(a/2 + 1/4, 1/2, x^2/2)}{\Gamma(3/4 + a/2)} - \frac{\sqrt{2}\, x\, {}_1F_1(a/2 + 3/4, 3/2, x^2/2)}{\Gamma(1/4 + a/2)} \right] \tag{18.2.2}$$

This result is practical when a is in the range $[-5, 10]$ and x is in the range $[-10, 2]$; for larger x there is significant subtractive cancellation between the two terms in (18.2.2).

As a and x increase, it becomes more accurate—but less efficient—to use an integral definition of $U(a,x)$. We use

$$U(a,x) = \frac{e^{-x^2/4} \int_0^\infty I(x,w)\,dw}{\Gamma(1/2 + a)\, x^{a+1/2}} \qquad I(x,w) = e^{-w - w^2/(2x^2)} w^{a-1/2} \tag{18.2.3}$$

which is valid for $a > -1/2$ and $x > 0$. This expression is derived from [Erd53b, 8.3 (3)] by using a change of variable from v to a and from t to $w = xt$. As written, the asymptotic behavior of $U(a,x)$ in (18.2.1) is clear. The integrand $I(x,w)$ is investigated under Visualization, while methods for evaluating the integral are described under Algorithm and Program.

Astute readers have noticed that the definitions (18.2.2) and (18.2.3) must fail when a satisfies $a = -n - 1/2$ with n a nonnegative integer, since the Γ functions (Section 6.1.1) then diverge. This special case is handled by [Abr64, 19.13.1]

$$U(-n - 1/2, x) = e^{-x^2/4} H_n(x/\sqrt{2})/2^{n/2} \qquad n = 0, 1, 2, \dots \tag{18.2.4}$$

in terms of Hermite polynomials (Section 11.4].

Visualization

As functions of two variables, a and x, the parabolic cylinder functions $U(a,x)$ are ideal for visualizing in surface plots. The large range of variation of the function and its oscillatory behavior in some regions make it difficult, however, to display the details of $U(a,x)$ if the range of a or x is large.

Figure 18.2.2 shows the function from several viewpoints, as surfaces then as transections through these surfaces with either a fixed and x varying (top right) or with x fixed and a varying (bottom right). If you want to see more detail in these views, just modify the plotting parameters in the cell `ParabolicCylinderU` that is in *Mathematica* notebook `ParCyl`; this notebook is listed and annotated in Section 20.18, as well as being available off the CD-ROM that comes with the *Atlas*.

M

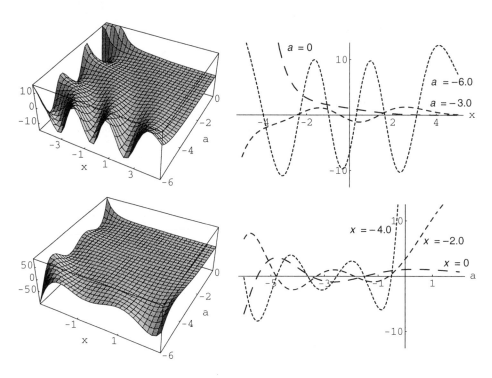

FIGURE 18.2.2 Parabolic cylinder functions, U. Top left; surface of the function for x in $[-5, 5]$ and a in $[-6, 0]$. Bottom left; U function with exponential factor in (18.2.2) or (18.2.3) removed. Top right; curves of U for x varying in $[-5, 5]$ and a fixed. Bottom right; curves of U for a varying in $[-6, 2]$ and x fixed. (Adapted from cell `Parabolic-CylinderU` in *Mathematica* notebook `ParCyl`.)

The integrand $I(x,w)$ in (18.2.3) is also interesting to visualize, along with the corresponding integrand that is applicable for $a < 1/2$, namely,

$$U(a,x) = e^{x^2/4} \, | \, x \, |^{a-1/2} \int_0^\infty J(x,w)dw$$

$$J(x,w) = e^{-w^2/(2x^2)} w^{-(a+1/2)} \cos[\text{sgn}(x)w + (a+1/2)\pi/2]$$

(18.2.5)

which is valid for $a > -1/2$ and $x > 0$. This expression derives from [Erd53b, 8.3 (4)] by making a change of variable from v to a and from t to $w = xt$. Figure 18.2.3 shows $J(x,w)$ (top row) and $I(x,w)$ (bottom row).

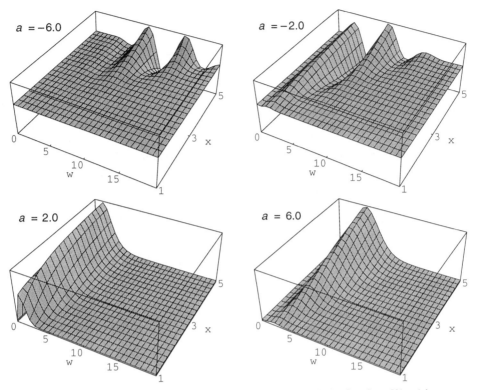

FIGURE 18.2.3 Integrands for computing the parabolic cylinder functions $U(a, x)$ by direct numerical integration. The variable w in (18.2.3) or (18.2.5) ranges from 0 to 20, and the argument x ranges from 1 to 5. (Adapted from cell `ParabolicCylinderU` in *Mathematica* notebook `ParCyl`.)

Note in Figure 18.2.3 that for $a < 0$ (top row) the integrands become highly oscillatory as x increases for given w. From this observation we infer that accurate and efficient numerical integration of (18.2.5) to at least 8 digits is likely to be very difficult; in the *Mathematica* program the heavy-duty function `NIntegrate` is invoked, and even it sometimes fails to achieve the desired accuracy. For $a > 0$ (bottom row) the integrand has its maximum at $w = w_m$, given by

$$w_m = x \left[\sqrt{x^2 + 4a - 2} - x \right] / 2 \qquad (18.2.6)$$

which is nearly $a - 1/2$ if $x \gg |a|$. When a is small, as seen in the bottom left graphic in Figure 18.2.3, the numerical integration therefore has to be handled carefully; although the integrand in (18.2.5) is analytically integrable if $a > -1/2$, the integrand is singular at $w = 0$ unless $a > 1/2$. How this is handled is discussed in the following.

Algorithm and Program
The algorithm for the parabolic cylinder function $U(a, x)$ and the corresponding C function `ParCylU` are designed to give numerical accuracy of at least 8 decimal digits and usually 10 digits. To do this, we use expression (18.2.2) for x in the range $[-10, 3]$ with a in the range $[-5$ to $10]$. If $x > 3$ and $a > 1$, we use the integral expression (18.2.3).

The integration method used is the nine-strip, equal-step, Newton-Cotes formula—25.4.19 in Abramowitz and Stegun [Abr64]—namely,

$$\int_{x_0}^{x_9} f(x)\,dx \approx \frac{9h}{89600}[2857(f_0 + f_9) + 15741(f_1 + f_8) +$$
$$1080(f_2 + f_7) + 19344(f_3 + f_6) + 5778(f_4 + f_5)] \tag{18.2.7}$$

which has an error of order h^{11} multiplied by the 10th derivative of the integrand, where h is the step size. If the upper limit on the integral in (18.2.3) is set to be at least $w = 25$ and we use $90 \times 27 = 2430$ integration points, then the integration can handle the range of x and a just discussed. For convenient programming, the C function `Uax9Strip` given below includes three applications of (18.2.7) for 28 points in each block of 27 strips.

Although we use the integration only for $a > 1/2$, the rapid variation of the integrand in (18.2.3) near $w = 0$ requires that the integral be evaluated analytically in some ε neighborhood of the origin; we use $\varepsilon = 0.05$ and make a Maclaurin expansion of the exponential factor in (18.2.3) through terms in w^3. This results in

$$\int_0^\varepsilon I(w)dw \approx \varepsilon^{a+1/2}\left[\frac{1}{a+1/2} - \frac{\varepsilon}{a+3/2} + \frac{(1-1/x^2)\varepsilon^2/2}{a+5/2} + \frac{(1/3-1/x^2)\varepsilon^3/2}{a+7/2}\right] \tag{18.2.8}$$

to accuracy of order 10^{-10} or better.

For negative a, one has to check for the special case covered by (18.2.4); we deem that this condition is met if a is within 10^{-6} of a negative half-integer value. If so, the function `Hermite`, listed in Section 11.4, is invoked.

The confluent hypergeometric function is computed by using the function `Hyper1F1` in Section 16.2.1, but is modified to use Γ functions rather than their logarithms, which is not permissible when the argument is negative. Further, to allow in some way for effects from subtractive cancellation in (18.2.2), an input accuracy parameter `eps` produces an accuracy parameter of 10^{-3} times this for use in `Hyper1F1`. We do not list the modified functions `Hyper1F1GG` or `Gamma`, since the changes are minor. Following our philosophy in the *Atlas* of "no assembly required" (see The Computer Interface in the Introduction), the CD-ROM version of the function embedded in its driver (Section 21.18) includes all these necessary functions.

```
double ParCylU(double a, double x, double eps)
/* Parabolic Cylinder function U(a,x).
   Fractional accuracy goal is  eps  for series. */
{
double aPh,aD2,Dw,integral,epsI,wStart,xS,xSD2,xSD4,
  wMax,DwStart,Uax,Ma,rem,epsK;
double Gamma(double x),Hyper1F1GG(double a,double b,
  double x, double eps),Uax9Strip(double aa,double x,
  double wStart, double Dw),Hermite(int n, double x);
const double RRt2 = 0.707106781187,
  RtPi = 1.77245385091;
int kMax,k,flMa;

aPh = a + 0.5;
aD2 = 0.5*a;
xS = x*x;
xSD2 = 0.5*xS;
xSD4 = 0.5*xSD2;
```

```
if ( ( x > 3.0 ) && ( a > 1.0 ) )
 {  /* Use 9-strip integrals */
 wMax = 0.5*x*(sqrt(x*x+4.0*a-2.0) - x);
 wMax = ( wMax > 2.5 ) ? 8.0*wMax : 25.0;
 kMax = 90;   /* is number of blocks */
 DwStart = wMax/( (double) kMax );
 Dw = DwStart/27.0;  /* is step size in a block */
  /* Analytical integral near origin */
 epsI = 0.05;
 integral = pow(epsI,aPh)*(1.0/aPh-epsI/(aPh+1.0)+
  0.5*epsI*epsI*((1.0-1.0/xS)/(aPh+2.0)+
  epsI*(1.0/xS-1.0/3.0)/(aPh+3.0)));
 wStart =  epsI;
 for ( k = 1; k <= kMax; k++ )
  { integral += Uax9Strip(a,x,wStart,Dw);
    wStart +=  DwStart; }
 Uax = integral/(exp(xSD4+aPh*log(x))*Gamma(aPh));
 return  Uax;
 }
if ( a < 0.0 )
 {    /* Hermite polynomials */
 Ma = -a;
 flMa = floor(Ma);
 rem = Ma - (double) flMa ;
 if ( fabs(rem - 0.5) < 1.0e-6 )
   /* Hermite polynomial */
  { return  Hermite(flMa,RRt2*x)/
    (pow(2.0,0.5*((double)flMa))*exp(xSD4)); }
 }
/* Hypergeometric series */
epsK = 0.001*eps;
Uax = RtPi*exp(-xSD4)*
(pow(2.0,-0.25- aD2)*
 Hyper1F1GG(aD2+0.25,0.5,xSD2,epsK)/
 Gamma(aD2+0.75) -  x*pow(2.0,0.25 - aD2)*
 Hyper1F1GG(aD2+0.75,1.5,xSD2,epsK)/
 Gamma(aD2+0.25));
return  Uax;
}

double Uax9Strip(double aa, double x,
 double wStart, double Dw)
/* Nine-strip integral for parabolic cylinder U
   with  aa > 0  */
{
double aMh,R2xS,w,f[28],integral;
double Power(double x, double p);
const double a[5] = {0.2869754464285714,
   1.581127232142857, 0.1084821428571429,
   1.943035714285714, 0.5803794642857144 };
int k;

aMh = aa - 0.5;
R2xS = 1.0/(2.0*x*x);
w = wStart;
```

```
for ( k = 0; k <= 27; k++ )
  {
  f[k] = exp(-w*(1.0+w*R2xS))*Power(w,aMh);
  w += Dw;
  }
integral = Dw*(a[0]*(f[0]+f[27]+2.0*(f[9]+f[18]))+
   a[1]*(f[1]+f[8]+f[10]+f[17]+f[19]+f[26])+
   a[2]*(f[2]+f[7]+f[11]+f[16]+f[20]+f[25])+
   a[3]*(f[3]+f[6]+f[12]+f[15]+f[21]+f[24])+
   a[4]*(f[4]+f[5]+f[13]+f[14]+f[22]+f[23])));
return  integral;
}
```

Test Values

Check values for the parabolic cylinder function $U(a,x)$ can be generated by running the ParabolicCylinderU cell in *Mathematica* notebook ParCyl. Examples are given in Table 18.2.1. Outside the ranges of a in [–5, 10] and x in [–10, 5] the accuracy of both the C and *Mathematica* functions will decrease.

TABLE 18.2.1 Test values to 10 digits of the parabolic cylinder function U.

$a \backslash x$	−5.0	0.0	5.0
−5.0	−9.615606270	3.052183664	1.879976816
0.0	3.330632780×10^2	1.216280214	$8.513565221 \times 10^{-4}$ *
5.0	4.599828923×10^4	0.1033543675	$1.552271295 \times 10^{-7}$
10.0	1.770340625×10^4	$5.911895930 \times 10^{-4}$	$1.549656896 \times 10^{-11}$

* The accuracy of the C function is 9 digits.

18.2.2 Parabolic Cylinder Functions V

Keywords
Hermite function, parabolic cylinder function, Weber function, Whittaker function

C Function and *Mathematica* Notebook
ParCylV (C); cell ParabolicCylinderV in notebook ParCyl (*Mathematica*)

Background
The second parabolic cylinder function, $V(a,x)$, is the solution of (18.1.2) for $z = x$ that has asymptotic behavior for $x \gg |a|$

$$V(a,x) \sim \sqrt{2/\pi}\, e^{x^2/4} x^{a-1/2} \qquad (18.2.9)$$

   Since V diverges for large x, it is generally of less use when solving scientific problems, in which x is usually proportional to distance. It may, however, be used to construct solutions of the Laplace equation in parabolic-cylinder coordinates (Section 18.2.1) when the region includes the origin. Formulas for $V(a,x)$ are given in Chapter 19 of Abramowitz and Stegun [Abr64]. Background to numerical computation of V is given in Miller's introduction to the tables in [NPL55], where the function tabulated is $W(a,x)$, related to the solutions of (18.2.2) when $z = ix$. Other methods for $V(a,x)$ are described in the articles by Redding and Latham [Red76], Schulten et al [Sch81], and Weniger [Wen96].

Function Definition

We have from Section 18.2.1 considerable understanding of the other parabolic cylinder function, $U(a,x)$, so it is convenient to relate $V(a,x)$ to it. Their relation is given by [Abr64, 19.3.8] as

$$V(a,x) = \Gamma(a+1/2)\big[\sin(\pi a)U(a,x)+U(a,-x)\big]/\pi \qquad (18.2.10)$$

This can be used directly, unless $a = -n-1/2$ with n a nonnegative integer. In this case, by manipulating formulas in Sections 19.2 and 19.3 of [Abr64] one obtains

$$V(-n-1/2,x) = \frac{2^{n/2}e^{-x^2/4}}{\sqrt{\pi}\,\Gamma(1+n)}\left|\begin{matrix}(-1)^{n/2}\sqrt{2}\,\Gamma[1+n/2]\,_1F_1(-n/2+1/2,3/2,x^2/2)\\(-1)^{(n+1)/2}\Gamma[(1+n)/2]\,_1F_1(-n/2,1/2,x^2/2)\end{matrix}\right. \qquad (18.2.11)$$

Visualization

The $V(a,x)$ parabolic cylinder functions are ideal for visualizing in surface plots as functions of the two variables, a and x. The function has a modest range of variation for the a and x values of interest for our C function `ParCylV`.

Figure 18.2.4 shows the function from several viewpoints, as surfaces and as transections with either a fixed and x varying or with x fixed and a varying. To see more detail, modify the plotting parameters in cell `ParabolicCylinderV` in *Mathematica* notebook `ParCyl`, which is listed in Section 20.18, and is available off the CD-ROM.

M

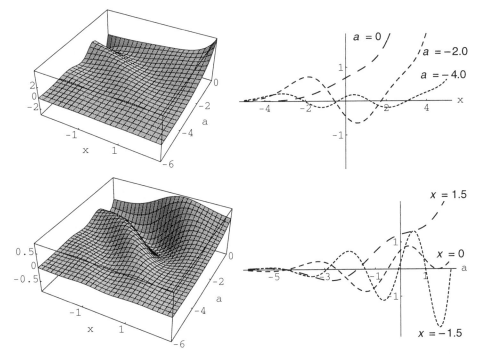

FIGURE 18.2.4 Parabolic cylinder functions, V. Top left: Surface of the function for x in $[-3, 3]$ and a in $[-6, 0]$. Bottom left; V function with the exponential factor removed. Top right: Curves of V for x varying in $[-5, 5]$ and a fixed. Bottom right: Curves of V for a varying in $[-6, 2]$ and x fixed. (Adapted from cell `ParabolicCylinderV` in *Mathematica* notebook `ParCyl`.)

The behavior of the second parabolic cylinder function, $V(a,x)$, for negative x when a is nearly an integer is important for obtaining accurate numerical values from (18.2.10). For such arguments, $U(a,x)$ is much larger in magnitude than $U(a,-x)$, especially when $a > 0$ (Table 18.2.1). The factor $\sin(\pi a)$ in (18.2.10) must therefore be calculated very accurately. To visualize this, the top of Figure 18.2.5 shows a general view of $V(a,x)$ for such a and x values; the function zeros are nearly independent of x. A closer look (lower panel) reveals, however, that the vanishing of $\sin(\pi a)$ for a an integer and the fact that $U(a,x) \gg U(a,-x)$ produces the closeness of the zeros.

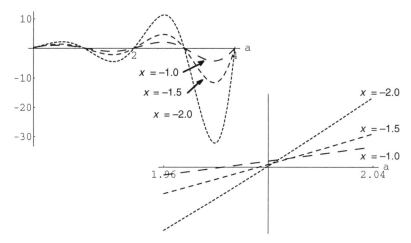

FIGURE 18.2.5 Variation of the parabolic cylinder functions V near integer values of a for $x < 0$. Left: Curves of V vs a for fixed negative x. Right: Detail of V near $a = 2$, where the numerical value of V is sensitive to the accuracy of the factor $\sin(\pi a)$ in (18.2.10). (Adapted from cell `ParabolicCylinderV` in *Mathematica* notebook `ParCyl`.)

In our C function `ParCylV` the value of π is given to 16 digits in order to ensure that the sine function is computed accurately for a near an integer.

Algorithm and Program
The algorithm for $V(a,x)$ is very simple compared with that for $U(a,x)$ in Section 18.2.1, which function it uses in (18.2.10). For negative a, the special case $a = -n - 1/2$ in (18.2.11) is checked; this condition is considered as met if a is within 10^{-6} of a negative half-integer, n. Function `Hypergeometric1F1`—used for $U(a,x)$—is then invoked.

```
double ParCylV(double a, double x, double eps)
/* Parabolic Cylinder function V(a,x).
   Fractional accuracy goal is  eps  for series
   calculation of  U(a,x)  */
{
double Hyper1F1GG(double a, double b, double x,
 double eps),ParCylU(double a,double x,double eps),
 Gamma(double x);
double Ma,rem,fn,xSD2,xSD4,factor,epsK,sign,Vax;
const double Pi=3.141592653589793 /* 16 digits */,
 RtPi = 1.77245385091, Rt2 = 1.41421356237;
int n;
```

```
if ( a < 0.0 )
{/* Check if a is within 10(-6) of half integer */
 Ma = -a;
 n = floor(Ma);
 rem = Ma - (double) n ;
 if ( fabs(rem - 0.5) < 1.0e-6 )
   {
    fn = n; /* integer part */
    xSD2 = 0.5*x*x;
    xSD4 = 0.5*xSD2;
    factor = pow(2.0,0.5*fn)/
      (RtPi*exp(xSD4)*Gamma(1.0+fn));
    epsK = 0.001*eps;
    if ( 2*(n/2) == n )
      { /*   n  even */
      sign = ( 4*(n/4) == n ) ? 1.0: -1.0;
      return  factor*Rt2*sign*Gamma(1.0+0.5*fn)*x*
              Hyper1F1GG(0.5*(1.0-fn),1.5,xSD2,epsK);
      }
         /*   n  odd */
      sign = ( 4*((n+1)/4) == (n+1) ) ? 1.0: -1.0;
      return  factor*sign*Gamma(0.5*(1.0+fn))*
              Hyper1F1GG(-0.5*fn,0.5,xSD2,epsK);
   }
 }
/* Use relation to parabolic cylinder function
   U(a,x) */
Vax = Gamma(a+0.5)*
  (sin(Pi*a)*ParCylU(a,x,eps)+ParCylU(a,-x,eps))/Pi;
return  Vax;
}
```

Test Values

Test values of the second parabolic cylinder function, $V(a,x)$, can be generated by running `ParabolicCylinderV` cell in *Mathematica* notebook `ParCyl`. Examples are given in Table 18.2.2. Outside the ranges of a in [–5, 10] and x in [–10, 5] the accuracy of both the C and *Mathematica* functions decreases.

M

TABLE 18.2.2 Test values to 10 digits of the second parabolic cylinder function V.

$a \backslash x$	−5.0	0.0	5.0
−5.0	−0.03591664210	−0.05831145754	0.1837045468
−2.4	−4.945524636	−0.08263818370	5.207088341
0.1	67.91908959	0.7673145573	2.197897891×10^2
9.5	-2.524979215×10^9	0.0	2.524979215×10^9

REFERENCES ON PARABOLIC CYLINDER FUNCTIONS

[Abr64]   Abramowitz, M., and I. A. Stegun, *Handbook of Mathematical Functions,* Dover, New York, 1964.

[Erd53b]   Erdélyi, A., et al, *Higher Transcendental Functions,* Vol. 2, McGraw-Hill, New York, 1953; reprint edition, Krieger, Malabar, Florida, 1981.

[Gra94]   Gradshteyn, I. S., and I. M. Ryzhik, *Table of Integrals, Series, and Products,* Academic Press, fifth edition, 1994.

[Jah45]   Jahnke, E., and F. Emde, *Tables of Functions,* Dover, New York, fourth edition, 1945.

[Lat74]   Latham, W. P., and R. W. Redding, Journal of Computational Physics, **16**, 66 (1974).

[Leb65]   Lebedev, N. N., *Special Functions and Their Applications,* Prentice-Hall, Englewood Cliffs, New Jersey, 1965.

[Moo61]   Moon, P., and D. E. Spencer, *Field Theory for Engineers,* Van Nostrand, Princeton, New Jersey, 1961.

[Moo71]   Moon, P., and D. E. Spencer, *Field Theory Handbook,* Springer-Verlag, Berlin, second edition, 1971.

[NPL55]   National Physical Laboratory, *Tables of Weber Parabolic Cylinder Functions,* Her Majesty's Stationery Office, London, 1955.

[Red76]   Redding, R. W., and W. P. Latham, Journal of Computational Physics, **20**, 256 (1976).

[Sch81]   Schulten, Z., R. G. Gordon, and D. G. M. Anderson, Journal of Computational Physics, **42,** 213 (1981).

[Spa87]   Spanier, J., and K. B. Oldham, *An Atlas of Functions,* Hemisphere Publishing, Washington, D.C., 1987.

[Tau92]   Taubmann, G., Computer Physics Communications, **69**, 415 (1992).

[Wen96]   Weniger, E. J., Computers in Physics, **10**, 496 (1996).

[Whi27]   Whittaker, E. T., and G. N. Watson, *A Course of Modern Analysis,* Cambridge University Press, Cambridge, England, fourth edition, 1927.

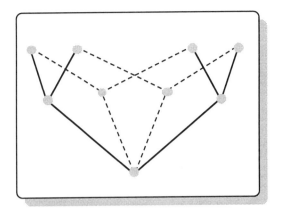

Chapter 19

MISCELLANEOUS FUNCTIONS FOR SCIENCE AND ENGINEERING

This chapter provides miscellaneous functions that are used often in science and engineering but that are not trivial to compute. The functions in Sections 19.1 through 19.5 are also described and tabulated in Chapter 27 of Abramowitz and Stegun [Abr64]. The Voigt function (Section 19.6) and the coupling coefficients (Section 19.7)—the 3–j, 6–j, and 9–j coefficients—are not given in [Abr64], except for a brief algebraic and numerical table of Clebsch-Gordan coefficients, which are essentially 3–j coefficients.

Mathematica notebooks for visualizing the functions and for quick calculations are provided as usual in the *Atlas*, except for the coupling coefficients (Section 19.7). For these we indicate *Mathematica* built-in functions.

19.1   DEBYE FUNCTIONS

Keywords
Debye functions

C Function and *Mathematica* Notebook
Debye (C); cell Debye in notebook Miscellaneous (*Mathematica*)

M

Background
Debye functions occur in many applications. They originated with Debye's theory of the specific heats of materials. Their properties and values are tabulated in Section 27.1 of Abramowitz and Stegun [Abr64].

Function Definition
The Debye functions, which we denote by $D_n(x)$ with $n \geq 1$ and $x > 0$, are defined as in Section 27.1 of [Abr64] by

$$D_n(x) = \int_0^x \frac{t^n}{e^t - 1} dt \tag{19.1.1}$$

For $x \to 0$ we obtain, by expanding the exponential into powers of t then integrating by parts, the limit

$$D_n(x) \xrightarrow[x \to 0]{} \frac{x^n}{n} \left[1 - \frac{x}{2(n+1)} \right] \tag{19.1.2}$$

The modified Debye function $D_n'(x)$, given by

$$D_n'(x) \equiv \frac{n}{x^n} D_n(x) \tag{19.1.3}$$

is therefore of order unity for $x \ll n$, which provides a useful way of visualizing and tabulating the functions. As $x \to \infty$ the Debye function is related to the Riemann zeta function, ζ (Section 8.3), by

$$D_n(x) \xrightarrow[x \to \infty]{} n! \, \zeta(n+1) \tag{19.1.4}$$

This can be used as another test limit for $D_n(x)$, as in Table 19.1.1 below.

The Debye function is monotonically increasing with x, as is clear from the positive integrand in (19.1.1). The function is a continuous function of n for $n \geq 1$.

Visualization

The relief map of $D_n(x)$ shown in Figure 19.1.1 is very simple because of the monotonicity in x and the continuity in n. The modified function $D_n'(x)$ in (19.13) varies less rapidly for $x \ll n$ is also shown in the figure.

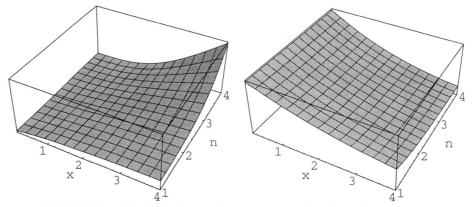

FIGURE 19.1.1 Left: Debye function (19.1.1) vs x and n for x in $(0, 4]$ and n in $[1, 4]$. Right: Modified Debye function (19.1.3) vs x and n for x in $(0, 4]$ and n in $[1, 4]$. (Cell Debye in *Mathematica* notebook Miscellaneous.)

Algorithm and Program

A simple, accurate, and efficient algorithm for $D_n(x)$ is to use the integral definition (19.1.1) directly. The integration is performed by using Simpson's formula, coded in the function SimpInt. The function Power is needed to compute integer powers correctly, since the C library function pow does not compute 0^0 as 1.

```
#define MAX 31

double Debye(int n, double x, int Nint)
/* Debye function of order n, argument x,
   using Nint Simpson integration points */
{
double h,t,Power(double, int),Intgrnd[MAX],
  SimpInt(double Intgrnd[], int Nint, double h);
int N2,tj;

N2 = Nint/2;
Nint = 2*N2; /* forces to even number */
h = x/Nint; /* is integration step size */

/* Make Debye function integrand */
Intgrnd[0] = Power(0,n-1);
t = h;
for ( tj = 1; tj <= Nint; tj++ )
  {
  Intgrnd[tj] = Power(t,n)/(exp(t)-1);
  t += h;
  }
/* then integrate by Simpson formula */
return   SimpInt(Intgrnd, Nint, h);
}

double Power(double x, int i)
/* x  to integer power i */
{
double floati;
/* always return 1.0 for i=0 */

if ( i == 0 )  return 1.0;

floati = i;  return pow(x,floati);
}

double SimpInt(double Intgrnd[], int Nint,
               double h)
/* Simpson integrate Nint (even) points
   of integrand in array Intgrnd */
{
double sum;
int N2,j;

N2 = Nint/2;
sum = (Intgrnd[0]+Intgrnd[Nint])/2
      +2*Intgrnd[Nint-1];
for ( j = 1; j < N2; j++ )
  {
  sum += Intgrnd[2*j]+2*Intgrnd[2*j-1];
  }
return  2*h*sum/3;
}
```

Programming Notes

The input to function Debye consists of n, x, and the number of integration points, Nint. If this is not input as even (which is necessary for Simpson's formula), then it is truncated to the nearest smaller integer. The parameter MAX gives the number of array elements for storing the integrand of the Debye function. The values MAX = 31 and Nint = 30 give at least 6 decimal digit accuracy for $n \leq 4$ and $x \leq 10$. The stepsize h is computed from x and Nint. At the first integration point, t = h, the integrand is computed as t^{n-1} in order to avoid the near singularity of the defining integrand in (19.1.1).

In order to obtain the asymptotic values in Table 19.1.1 by using function Debye, it is necessary to choose x about 40 and to define MAX (the array size) to be at least 501.

Test Values

Table 19.1.1 gives test values of the modified Debye functions (19.1.3) to 10 decimal digits, for x in [0, 10] and the value of the Debye functions (19.1.1) as $x \to \infty$.

TABLE 19.1.1 Values to 10 digits of the modified Debye function defined by (19.1.3) for x in [0, 10], and the asymptotic value of the Debye function, (19.1.4), to 7 digits.

$x \setminus n$	1	2	3	4
1.0	0.7765048841	0.7078774760	0.6744155631	0.6548740689
2.0	0.6064474096	0.4930823941	0.4411284736	0.4118927367
3.0	0.4801019696	0.3426138495	0.2835798281	0.2518786364
4.0	0.3878980838	0.2405536250	0.1817369138	0.1518546126
$D_n(\infty)$	1.644934	2.404113	6.493939	24.886266

19.2   SIEVERT INTEGRAL

Keywords

attenuation integral, Sievert integral

C Function and *Mathematica* Notebook

Sievert (C); cell Sievert in notebook Miscellaneous (*Mathematica*)

Background

The Sievert integral, $S(\theta, x)$, occurs in problems involving exponential attenuation averaged over angles, as in the attenuation of radiation and signals through uniform media. The integral gives the total amount of radiation reaching depth x for angles of emission between zero and θ. It is presented and tabulated in Section 27.4 of Abramowitz and Stegun [Abr64]. In such uses, θ is the maximum angle at which the radiation impinges on the attenuator and x is the dimensionless attenuation depth in units of an attenuation length. The situation is sketched in Figure 19.2.1.

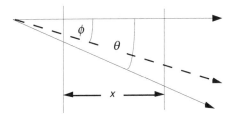

FIGURE 19.2.1 A typical problem in which the Sievert integral appears. Radiation from a point in the upper left is attenuated as it passes through a uniform medium of thickness x, in units of an attenuation length.

Function Definition

The Sievert integral, $S(\theta,x)$, is defined, as in [Abr64, Section 27.4], by

$$S(\theta,x) \equiv \int_0^\theta e^{-x\sec\phi}\, d\phi \qquad 0 \le \theta < \pi/2 \qquad x \ge 0 \qquad (19.2.1)$$

It is also convenient to define a modified Sievert integral by

$$S'(\theta,x) \equiv \int_0^\theta e^{-x\sec\phi}\, d\phi \Big/ \theta = S(\theta,x)/\theta \qquad \theta > 0 \qquad (19.2.2)$$

which gives the average value of the exponential attenuation factor over the range 0 to θ. In particular

$$S'(\theta,0) = 1 \qquad (19.2.3)$$

which serves as a check on the Sievert integral and helps to visualize it.

Visualization

The function $S(\theta,x)$ is monotonic in both θ and x, as shown in Figure 19.2.2. The modified integral, $S'(\theta,x)$, is more uniform as a function of θ, as also seen in the figure.

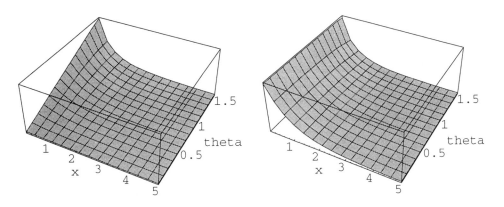

FIGURE 19.2.2 Left: Sievert integral (19.2.1) as a function of angle θ and attenuation depth x. Right: Modified Sievert integral (19.2.2). (Cell `Sievert` in *Mathematica* notebook `Miscellaneous`.)

Algorithm and Program

A simple, accurate, and efficient algorithm for the Sievert integral $S(\theta, x)$ is to use the integral definition (19.2.1) directly. The integration is performed by using Simpson's formula, coded in the function SimpInt.

```c
#define MAX 31

double Sievert(double theta, double x, int Nint)
/* Sievert integral of arguments theta & x,
   using Nint Simpson integration points */
{
double h,phi,Intgrnd[MAX],
  SimpInt(double Intgrnd[], int Nint, double h);
int N2,jphi;

N2 = Nint/2;
Nint = 2*N2; /* forces to even number */
h = theta/Nint; /* integration step size */

/* Make Sievert function integrand */
Intgrnd[0] = exp(-x);
phi = h;

for ( jphi = 1; jphi <= Nint; jphi++ )
  {
  Intgrnd[jphi] = exp(-x/cos(phi));
  phi += h;
  }
/* then integrate by Simpson formula */
return   SimpInt(Intgrnd, Nint, h);
}


double SimpInt(double Intgrnd[], int Nint,
          double h)
/* Simpson integrate Nint (even) points
   of integrand in array Intgrnd */
{
double sum;
int N2,j;

N2 = Nint/2;
sum = (Intgrnd[0]+Intgrnd[Nint])/2
 +2*Intgrnd[Nint-1];
for ( j = 1; j < N2; j++ )
  {
  sum += Intgrnd[2*j]+2*Intgrnd[2*j-1];
  }
return 2*h*sum/3;
}
```

Programming Notes

The input to function `Sievert` consists of `theta` (for angle variable θ in radian), `x` (in units of attenuation lengths), and the number of integration points, `Nint`. If `Nint` is not input as even (which is necessary for Simpson's formula), then it is truncated to the nearest smaller integer.

The parameter `MAX` gives the number of array elements for storing the integrand of the Sievert integral. The values `MAX` = 31 and `Nint` = 30 give at least 6 decimal digit accuracy for $\theta < \pi/2$ and $x \leq 5$. The stepsize `h` is computed from `theta` and `Nint`.

Test Values

Check values of the Sievert integral $S(\theta, x)$ are given in Table 19.2.1. Note that, with θ in radian, $S(\theta, 0) = \theta$. The test-table angles are thus 20°, 40°, and 60°.

$\boxed{M}$

TABLE 19.2.1 Values to 10 digits of the Sievert integral (19.2.1) for x in [0, 4] and $\theta < \pi/2$.

$x \backslash \theta$	$\pi/9 = 0.3490658504$	$2\pi/9 = 0.6981317008$	$\pi/3 = 1.047197551$
0	0.3490658504	0.6981317008	1.047197551
2.0	0.0453349018	0.0796444183	0.0953424480
4.0	0.0058961436	0.0093297977	0.0101275408

19.3   ABRAMOWITZ FUNCTION

Keywords

Abramowitz function

C Function and *Mathematica* Notebook

`Abrmwtz` (C); cell `Abramowitz` in notebook `Miscellaneous` (*Mathematica*)

$\boxed{M}$

Background

The Abramowitz function occurs in several problems arising in physics and engineering, such as atomic physics, radio-frequency spectra, and thermal-neutron absorption.

Function Definition

The function $f_m(x)$ that we call the Abramowitz function, since it was investigated extensively by Abramowitz [Abr53], is defined in Section 27.5 of Abramowitz and Stegun [Abr64] by

$$f_m(x) \equiv \int_0^\infty t^m e^{-t^2 - x/t} \, dt \qquad m \geq 0 \qquad x \geq 0 \qquad (19.3.1)$$

It satisfies the differential relation [Abr64, 27.5.2]

$$\frac{df_m}{dx} = -f_{m-1} \qquad m > 0 \tag{19.3.2}$$

and the recurrence relation

$$2f_m(x) = -(m-1)f_{m-2}(x) + x f_{m-3}(x) \qquad m > 2 \tag{19.3.3}$$

Visualization

The Abramowitz function, $f_m(x)$, is a smooth function of its variables m (usually taken to be a nonnegative integer) and x (nonnegative), as shown in Figure 19.3.1.

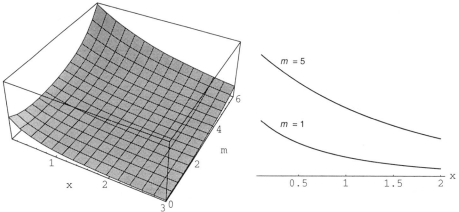

FIGURE 19.3.1 Abramowitz functions as defined by (19.3.1). Left: Surface for m in [0, 6] and x in [0, 3]. Right: Curves for $m = 1$ and $m = 5$. (Adapted from cell Abramowitz in *Mathematica* notebook Miscellaneous.)

Algorithm and Program

It would be tempting to evaluate the Abramowitz function directly from (19.3.1) as a numerical integral. The integrand is, however, quite ill-behaved when m is small, as shown in Figures 19.3.2 and 19.3.3.

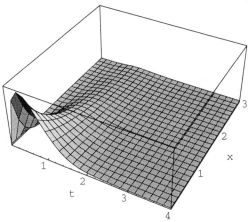

FIGURE 19.3.2 Integrand of the Abramowitz function (19.3.1) for $m = 0$ as a function of parameter x. (Cell Abramowitz in *Mathematica* notebook Miscellaneous.)

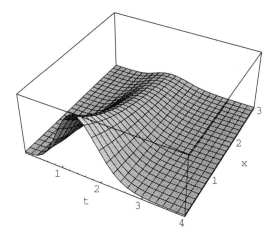

FIGURE 19.3.3 Integrand of the Abramowitz function (19.3.1) for $m = 4$ as a function of parameter x. (Cell Abramowitz in *Mathematica* notebook Miscellaneous.)

An alternative way of evaluating $f_m(x)$ is required. Series expansions for $m = 0, 1,$ and 2 are rapidly convergent. Iteration formulas can then be used if $m > 2$. The power series for $f_1(x)$ is given by [Abr64, 27.5.4], and this can be differentiated and (19.3.2) used with $m = 1$ to obtain the series for $f_0(x)$. Similarly, the power series for $f_2(x)$ can be obtained by using (19.3.2) with $m = 3$, in which $f_3(x)$ is obtained from (19.3.3) with $m = 3$. Resulting from this tedious algebra are the series expansions

$$f_0(x) = -\frac{1}{2} \sum_{k=1}^{\infty} \left[a_k + k(a_k \ln x + b_k) \right] x^{k-1} \tag{19.3.4}$$

$$f_1(x) = \frac{1}{2} \sum_{k=0}^{\infty} (a_k \ln x + b_k) x^k \tag{19.3.5}$$

and

$$f_2(x) = \frac{1}{4} \sum_{k=1}^{\infty} \left[(2k-2)a_k + k(k-2)(a_k \ln x + b_k) \right] x^{k-1} \tag{19.3.6}$$

The expansion coefficients are given by

$$a_0 = a_1 = 0 \qquad a_2 = -1$$
$$b_0 = 1 \qquad b_1 = -\sqrt{\pi} \qquad b_2 = 3(1-\gamma)/2$$
$$a_k = \frac{-2a_{k-2}}{k(k-1)(k-2)} \qquad k \geq 3 \tag{19.3.7}$$
$$b_k = \frac{-2b_{k-2} - (3k^2 - 6k + 2)a_k}{k(k-1)(k-2)} \qquad k \geq 3$$

in which γ is Euler's constant.

For $m \geq 3$ the recurrence relation (19.3.3) is used to generate f_m from preceding values by using the above expansions for f_0, f_1, and f_2.

```c
#define KMAX 21
#define MMAX 7

double Abrmwtz(int m, double x,
          double eps, int first)
/* Abramowitz function of arguments m & x,
   to fractional accuracy  eps  */
{
/* Euler's constant */
const double Gamma = 0.5772156649,
 RtPi = 1.77245385091;
static double a[KMAX],b[KMAX];
double f[MMAX],xk,term,next0,
 next1,next2,rat0,rat1,rat2,bigrat;
int k,denk,kkeep,n;

/* If this is the first call, make tables
   of the a and b coefficients */
if ( first == 1 )
  {
  a[0] = 0;     a[1] = 0;    a[2] = -1;
  b[0] = 1;    b[1] = -sqrt(4*atan(1.0));
  b[2] = 3*(1-Gamma)/2;
  for ( k = 3; k <= KMAX-1; k++ )
    {
    denk = k*(k-1)*(k-2);
    a[k] = -2*a[k-2]/denk;
    b[k] = (-2*b[k-2]-(k*(3*k-6)+2)*a[k])/denk;
    } /* end  k  loop for coefficients */
  }/* end first call */
if ( x==0 )
    {
    if ( m==0 )  return  RtPi/2;
    if ( m==1 )  return  0.5;
    if ( m==2 )  return  RtPi/4;
    f[0] = RtPi/2; f[1] = 0.5; f[2] = RtPi/4;
    /* Recurrence for  m >= 3  */
    for ( n = 3; n <= m; n++ )
      { f[n] = ((n-1)*f[n-2]+x*f[n-3])/2; }
    return f[m];
    }
/* Series expansions for m = 0, 1, & 2 */
f[0] = 0;     f[1] = b[0];    f[2] = 0;
xk = 1;
for ( k = 1; k <= KMAX-1; k++ )
  {
  term = a[k]*log(x)+b[k];
  /* Add in next terms in series */
  next0 = (a[k]+k*term)*xk;
  f[0] += next0;
  next1 = term*xk*x;
  f[1] += next1;
  next2 = (2*(k-1)*a[k]+k*(k-2)*term)*xk;
  f[2] += next2;
  xk *= x; /* updates x to k-th power */
```

```
/* Test convergences */
kkeep = k;
rat0 = fabs(next0/f[0]);
rat1 = fabs(next1/f[1]);
rat2 = fabs(next2/f[2]);
bigrat = ( rat0 > rat1 ) ? rat0 : rat1;
bigrat = (bigrat > rat2 ) ? bigrat : rat2;

if ( bigrat < eps )   k = KMAX;
} /* end  k  loop for m = 0, 1, & 2 */

if ( kkeep == KMAX-1 )
  {
  printf
  ("\n!!Abrmwtz not converged to eps=%6.4le by",
    eps);
  printf("KMAX=%i terms.\nIncrease
    KMAX\n",KMAX);
  }

f[0] = -f[0]/2;
f[1] = f[1]/2;
f[2] = f[2]/4;
if ( m == 0 )     return f[0];
if ( m == 1 )     return f[1];
if ( m == 2 )     return f[2];

/* Recurrence for  m >= 3   */
for ( n = 3; n <= m; n++ )
  { f[n] = ((n-1)*f[n-2]+x*f[n-3])/2; }

return f[m];
}
```

Programming Notes

The formulas in (19.3.7) for the series expansion coefficients are fairly complicated, therefore the coefficients are expensive to compute. For increased efficiency, all the coefficients that are likely to be needed for a given x are stored in arrays a[] and b[] if the parameter first, which is passed to Abrmwtz, is set to unity. Thereafter, if there are enough array elements precomputed and if x is not changed, first can be set to be other than unity, so that the coefficients (which are kept in static arrays) can just be reused.

Convergence of the series for $m = 0$, 1, and 2 is controlled by input parameter eps, which measures the largest relative error in the first neglected term compared with the partial series for $m = 0$, 1, and 2. Typically, fewer than a half dozen terms are required for 6-decimal accuracy. If not enough array elements are available for a given eps, the series is truncated with an error message. Recompile the function after increasing the array-size parameter KMAX.

For $m > 2$ the accuracy of the iterated values of f_m is not checked. The value of f_m is stored in array element f[m]. The program calling Abrmwtz should check that the array size parameter MMAX is larger than the maximum m value for which the function will be used. Otherwise, the function should be recompiled with larger MMAX.

Test Values

From the series (19.3.4) – (19.3.6) the explicit numerical approximations given as 27.5.5 – 27.5.7 in [Abr64] can be obtained readily. In particular, making use of the recurrence relation (19.3.3), we obtain that at $x = 0$

$$f_m(0) = \left| \begin{array}{ll} \dfrac{(m-1)!!}{2^{(m+1)/2}} & m \quad \text{odd} \\[3mm] \dfrac{(m-1)!!\sqrt{\pi}}{2^{m/2+1}} & m \quad \text{even}, \ m \geq 2 \end{array} \right. \tag{19.3.8}$$

and that $f_0(0) = \sqrt{\pi}/2$. These expressions can be used to generate the $x = 0$ values in Table 19.3.1.

TABLE 19.3.1 Values of the Abramowitz function to 10 digits for x in [0, 1.5] and m in [0, 3].

$x \setminus m$	0	1	2	3
0	0.8862269255	0.5000000000	0.4431134627	0.5000000000
0.5	0.2987173528	0.2531761739	0.2653820441	0.3278555121
1.0	0.1500459645	0.1465633814	0.1684873484	0.2215863636
1.5	0.0839095672	0.0900263074	0.1107033003	0.1529584827

19.4   SPENCE INTEGRAL

Keywords

Spence integral, dilogarithm, polylogarithm

C Function and *Mathematica* Notebook

Spence (C); cell Spence in notebook Miscellaneous (*Mathematica*)

Background

The Spence integral or dilogarithm, is one of the polylogarithms (Section 8.5), and is related to the Debye function with $n = 1$ (Section 19.1) by a simple transformation of variables [Abr64, 27.7.7].

Function Definition

We define Spence's integral function, $S(x)$, following the discussion in Section 27.7 of Abramowitz and Stegun [Abr64]. Namely,

$$S(x) \equiv -\int_1^x \frac{\ln t}{t-1}\, dt = -\int_0^{x-1} \frac{\ln(1+u)}{u}\, du \qquad x \geq 0 \tag{19.4.1}$$

Spence's integral is related to the dilogarithm Li_2 (Section 8.5) by

$$S(x) = \text{Li}_2(1-x) \tag{19.4.2}$$

Visualization

Spence's integral, being a function of the single nonnegative real variable x, is easily visualized, as shown in Figure 19.4.1.

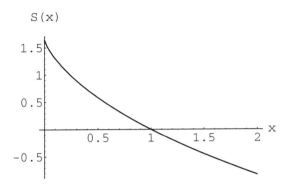

FIGURE 19.4.1 The Spence integral as a function of variable x in $[0, 2]$. (Cell `Spence` in *Mathematica* notebook `Miscellaneous`.)

Algorithm and Program

For most values of x, Spence's integral is readily calculated by using the second form of the integral in (19.4.1). The region around $x = 0$ is troublesome, however, while that near $x = 1$ (at which point the integral is zero) requires some care if relative accuracy is to be maintained.

As $x \to 0$ the integration of (19.4.1) becomes difficult because of divergence of the logarithm. Using the identity (27.7.3) in [Abr64] and integrating the remaining integral term by term produces

$$S(x) = \frac{\pi^2}{6} - \ln x \ln(1-x) - \sum_{k=0}^{\infty} x^{k+1}/(k+1)^2 \tag{19.4.3}$$

to within 10-digit relative error for $0.001 \le x \le 0.025$ if $k \le 5$ is used. For $x = 0$ we have that

$$S(0) = \frac{\pi^2}{6} \approx 1.644934067 \tag{19.4.4}$$

As $x \to 1$ the integral tends to zero, so relative accuracy is difficult to achieve by direct integration. Use the second form of the integral in (19.4.1), expansion of the logarithm term by term, integration, then substitution of the endpoint values, obtains

$$S(x) = \sum_{k=0}^{\infty} (1-x)^{k+1}/(k+1)^2 \tag{19.4.5}$$

This gives 10-digit relative accuracy for $S(x)$ if $k \le 5$ is used for x within 0.025 of unity.

If x is beyond the two danger zones just discussed and if $x \le 2$, then integration using the second form in (19.4.1) and Simpson's formula gives at least 10-digit accuracy if 140 integration points are used. For larger values of x or higher accuracy, the number of integration points needs to be increased. Function `Spence` is listed on the following pages.

```
#define MAX 141

double Spence (double x, int Nint)
/* Spence integral  by series & integration  */
{
double sum,term,fkP1,OneMx,h,u,Intgrnd[MAX],
 SimpInt(double Intgrnd[],int Nint,double h);
const double PiSD6 = 1.64493406685;
int k,N2,ju;

if ( fabs(x) < 0.025 )
  { /* For 10^-10 accuracy if x within 0.025
       of 0 start with Pi^2/6 for x = 0 */
  sum = 0.0;
  term = 1.0;
  for ( k = 0; k <= 5; k++ )
    {
    fkP1 = k+1;
    term *= x;
    sum += term/(fkP1*fkP1);
    }
  return  PiSD6 - log(x)*log(1-x) - sum;
  }

OneMx = 1.0 - x;
if ( fabs(OneMx) < 0.025 )
  { /* Direct integration of power series
       for 10^-10 relative accuracy if
       x within 0.025 of 1 */
  sum = 0.0;
  term = 1.0;
  for ( k = 0; k <= 5; k++ )
    {
    fkP1 = k+1;
    term *= OneMx;
    sum += term/(fkP1*fkP1);
    }
  return  sum;
  }

/* Full integration of  u  form */
N2 = Nint/2;
Nint = 2*N2; /* forces to even number */
h = -OneMx/Nint; /* is integration step size */

/* Make integrand of Spence's integral */
/* First, at u  origin */
Intgrnd[0] = 1;
u = h;
for ( ju = 1; ju <= Nint; ju++ )
  {
  Intgrnd[ju] = log(1+u)/u;
  u += h;
  }
/* then integrate by Simpson formula */
return   -SimpInt(Intgrnd, Nint, h);
}
```

```
double SimpInt(double Intgrnd[], int Nint,
          double h)
/* Simpson integrate Nint (even) points
   of integrand in array Intgrnd */
{
double sum;
int N2,j;
N2 = Nint/2;
sum =
(Intgrnd[0]+Intgrnd[Nint])/2+2*Intgrnd[Nint-1];
for ( j = 1; j < N2; j++ )
  {
  sum += Intgrnd[2*j]+2*Intgrnd[2*j-1];
  }
return  2.0*h*sum/3.0;
}
```

Programming Notes

It should be noted that there are three regions and different methods by which $S(x)$ is computed in `Spence`, as detailed in the preceding section . If fractional accuracy of better than 10 digits is required, formulas (19.4.3) and (19.4.5) should be used with more terms. For the direct integration, a larger number of integration points should be allowed for by recompiling `Spence` after increasing the array size parameter `MAX` beyond 141. This value of `MAX` gives 10-digit accuracy for x up to 2.

Test Values

Check values of $S(x)$ are given in Table 19.4.1 for the special regions near $x = 0$ and $x = 1$, as well as for a value obtained by integration—that at $x = 2$.

TABLE 19.4.1 **Values to 10 digits of Spence's integral for x in [0.001, 2].**

x	0	0.001	0.99
$S(x)$	1.644934067	1.637022605	0.01002511174
x	1.00	1.01	2.00
$S(x)$	0	−0.009975110490	−0.8224670334

19.5   CLAUSEN INTEGRAL

Keywords

Clausen integral, series

C Function and *Mathematica* Notebook

Clausen (C); cell Clausen in notebook Miscellaneous(*Mathematica*)

Background

Clausen's integral occurs most often in the context of summable series involving trigonometric functions. For example, we have [Abr64, 27.8.6] that

$$\sum_{k=1}^{\infty} \frac{\cos kt}{k} = -\ln\left(2\sin\frac{t}{2}\right) \qquad 0 < t < 2\pi \qquad (19.5.1)$$

from which it follows that

$$C(\theta) \equiv -\int_0^{\theta} \ln\left(2\sin\frac{t}{2}\right)dt = \sum_{k=1}^{\infty} \frac{\sin k\theta}{k^2} \qquad 0 < \theta < \pi \qquad (19.5.2)$$

in which $C(\theta)$ is called Clausen's integral. The understanding is that this series can not be summed by usual methods, indeed that evaluating the integral is a means of summing it. There are several related series, however, such as in [Abr64, 27.8.6] and in Jolley [Jol61], which have simple closed expressions.

Function Definition
Clausen's integral is defined by the integral expression in (19.5.2).

Visualization
Because the integrand in (19.5.2) contains an integrable singularity as $t \to 0$, the integral is not trivial to evaluate numerically. The behavior of the integrand of $C(\theta)$, including its sign, is shown in Figure 19.5.1.

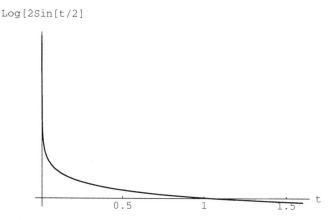

FIGURE 19.5.1 Integrand of the Clausen integral (19.5.1) as a function of variable t. (Cell `Clausen` in *Mathematica* notebook `Miscellaneous`.)

Note in Figure 19.5.1 that the integrand is positive until $t = \pi/3$. Therefore $C(\theta)$ is monotonically increasing until $\theta = \pi/3$, then it is monotonically decreasing until $\theta = \pi$, where it is zero. The resulting behavior of Clausen's integral is shown in Figure 19.5.2.

Algorithm and Program
Direct numerical evaluation of the integral expression for $C(\theta)$ in (19.5.1) is made difficult by the singularilty near $t = 0$, as shown in Figure 19.5.1. The trigonometric series in

(19.5.1) is not very practical either; its convergence is only of order $1/k^2$ because the numerator of each term in the series oscillates between ± 1. For example, if accuracy of order 6 decimal digits were required then at least 1000 terms would be needed.

C(theta)

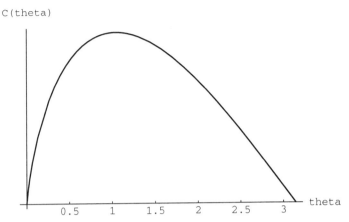

theta

0.5   1   1.5   2   2.5   3

FIGURE 19.5.2 The Clausen integral (19.5.1) as a function of variable θ. (Cell Clausen in *Mathematica* notebook Miscellaneous.)

Development of a practical algorithm for $C(\theta)$ involves expanding the integrand in (19.5.1) in two Maclaurin series, integrating term by term, then rearranging the series into a power series in θ. The resulting expression can be written, following [Abr64, 27.8.2],

$$C(\theta) = -\theta \ln \theta + \theta + \sum_{k=1}^{\infty} D_k \theta^{2k+1} \qquad 0 < \theta \le \pi/2 \qquad (19.5.3)$$

in which the coefficients are given by

$$D_k = \frac{(-1)^{k-1} F_k}{2k(2k+1)} \qquad D_1 = 1/72 \qquad (19.5.4)$$

Here the F_k are related to the Bernoulli numbers (Section 7.5) by

$$F_k = \frac{B_{2k}}{(2k)!} \qquad (19.5.5)$$

Unlike the Bernoulli numbers, B_k, the F_k decrease very rapidly with k. For example, $F_{10} \approx 2 \times 10^{-7}$ They can be generated by the following formula, derived from (7.5.4), as

$$F_k = \frac{k-1/2}{(2k+1)!} - \sum_{s=1}^{k-1} \frac{F_s}{(2k+1-2s)!} \qquad k > 1 \qquad (19.5.6)$$

$$F_0 = 1 \qquad F_1 = 1/12$$

in which the factorials of odd integers starting at $k = 3$ are generated by the recurrence relation

$$(2k+1)! = (2k+1)2k(2k-1)! \qquad (19.5.7)$$

The factorials can therefore be labelled by array index k, rather than by $2k+1$. Because generating the D_k coefficients is the major cost of using the function Clausen, these coefficients should be computed the first time that the function is called, then they can be reused subsequently.

For $\theta > \pi/2$ an expression similar to (19.5.3) can be obtained, as in [Abr64, 27.8.3], namely,

$$C(\theta) = (\pi - \theta)\ln 2 - \sum_{k=1}^{\infty} E_k (\pi - \theta)^{2k+1} \qquad \theta > \pi/2 \qquad (19.5.8)$$

in which the coefficients are given by

$$E_k = \left(2^{2k} - 1\right) D_k \qquad E_1 = 1/24 \qquad (19.5.9)$$

These coefficients are to be computed and stored the first time that Clausen is called.

Function Clausen for Clausen's integral, $C(\theta)$, is now expressed as follows:

```
#define KMAX 21

double Clausen
  (double theta, double eps, int first)
/* Clausen integral of argument  theta,
   by Bernoulli series to accuracy  eps */
{
/* pi/2, pi, & natural log of 2 */
const double pid2 = 1.5707963267949,
  pi = 3.1415926535898, ln2 = 0.69314718055995;

double tk,phase,pk,f[KMAX],F[KMAX],
thetaSq,thetaPow,sum,term,t,tSq,tPow;
static double D[KMAX],E[KMAX];
int k,s;

/* At first call make tables of coefficients */
if ( first == 1 )
  {
  /* Factorials of odd integers, starting at 3.
    Labelling is by ordinal, starting at 1 */
  f[1] = 6.0;
  tk = 4.0;

  for ( k = 2; k <= KMAX; k++ )
    {
    f[k] = (tk+1.0)*tk*f[k-1];
    tk += 2.0;
    }

  /* Modified Bernoulli numbers */
  F[0] = 1.0;    F[1] = 1.0/12.0;
  D[1] = 1.0/72.0;    E[1] = 1.0/24.0;
  tk = 4.0;    phase = -1.0;    pk = 16.0;
```

```
for ( k = 2; k <= KMAX; k++ )
  {
 ·F[k] = (k-0.5)/f[k];

  for ( s = 1; s < k; s++ )
  { F[k] -= F[s]/f[k-s]; }

 /* Coefficients for C(theta) */
 D[k] = phase*F[k]/(tk*(tk+1.0));
 E[k] = (pk-1.0)*D[k];
 tk += 2.0;
 phase =  - phase;
 pk *= 4.0;
 }/* end  k  loop */

 } /* end  first == 1  loop */

if ( theta == 0 )    return 0;


if ( theta <= pid2 )
 {
 thetaSq = theta*theta;
 thetaPow = theta;
 sum = theta*(1.0-log(theta));

 for ( k = 1; k <= KMAX-1; k++ )
   {
   thetaPow *= thetaSq;
   term = D[k]*thetaPow;
   sum += term;
   if ( fabs(term/sum) < eps ) break;
   }

 /* Convergence check */
 if ( k == KMAX )
 printf
 ("\n!!Clausen unconverged: %i terms\n",KMAX);

 return  sum;
 } /* end theta <= pi/2 loop */


if ( theta > pid2 )
  {
 t = pi - theta;
 tSq = t*t;
 tPow = t;
 sum = t*ln2;

 for ( k = 1; k <=KMAX-1; k++ )
   {
   tPow *= tSq;
   term = E[k]*tPow;
   sum -= term;
   if ( fabs(term/sum) < eps ) break;
   }
```

```
/* Convergence check */
if ( k == KMAX )
printf
("\n!!Clausen unconverged: %i terms\n",KMAX);
return sum;
} /* end theta > pi/2 loop */
}
```

Programming Notes

Function `Clausen` for $C(\theta)$ follows closely the algorithm for the power series expansions (19.5.3) and (19.5.9). If the argument `first` is set to 1, then the expansion coefficients D_k and E_k are computed up through $k = $ KMAX–1 and are stored in static arrays. If `Clausen` is then called with `first` not equal to unity, the coefficients are used directly.

The accuracy of Clausen's integral is controlled by input argument `eps`, which is used to check the series convergence by testing the ratio of the last term included to the series partial sum. Typically, 10-digit fractional accuracy is obtained with fewer than 10 terms in the series expansion (19.5.3) or (19.5.9). If convergence is not obtained within KMAX–1 terms, then a warning message is printed by `Clausen`. The function should be recompiled using a larger value of KMAX.

Test Values

Check values of Clausen's integral, $C(\theta)$, are given in Table 19.5.1. Note also that $C(0) = 0$.

TABLE 19.5.1 **Values to 10 digits of Clausen's integral (19.5.1) for θ in $[0, \pi)$.**

θ	$\pi/3$	$2\pi/3$	π
$C(\theta)$	1.014941606	0.6766277376	0

19.6   VOIGT (PLASMA DISPERSION) FUNCTION

Keywords

Cauchy distribution, convolutions, Gauss distribution, Hjerting function, Lorentz distribution, normal distribution, plasma dispersion function, Voigt function

C Function and *Mathematica* Notebook

`Voigt` (C); cell `Voigt` in notebook `Miscellaneous` (*Mathematica*)

Background

The Voigt function, V, is the convolution of a Gauss distribution (Section 9.3.1), G, and a Lorentz (Cauchy) function, L. For example, in nuclear fission reactors, to obtain reaction rates for thermal neutrons the distribution of neutron velocities (G) is folded into Breit-Wigner functions (L) describing neutron resonance reactions with reactor materials. In astronomy and optical spectroscopy, Gaussians describe Doppler shifts from random thermal velocities of atoms, and Lorentzians describe atomic transition rates for absorption or emission of light. Optical spectra then contain effects of Doppler broadening expressed through V. The Voigt function is essentially the plasma dispersion function. A survey of methods for computing V is given in an article by Thompson [Tho93].

Function Definition

Consider convolution of Gaussian (normal) and a Lorentzian (Cauchy) distributions as a function of parameter v:

$$I(v) = \int_{-\infty}^{\infty} G(v')L(v - v')dv' \tag{19.6.1}$$

where G is the normalized Gaussian centered on $v = 0$

$$G(v) = \frac{1}{\sqrt{\pi} \, \Delta v'_G} \exp\left[-v^2 / \left(\Delta v'_G\right)^2\right] \tag{19.6.2}$$

in terms of its width parameter

$$\Delta v'_G = \Delta v_{GF} / \sqrt{4\ln 2} = \sqrt{2}/\sigma \tag{19.6.3}$$

where Δv_{GF} is the Full Width at Half Maximum (FWHM) and σ is the standard deviation. The normalized Lorentzian in (19.6.1), centered on v_0, is given by (Section 9.4.2)

$$L(v) = \frac{\Gamma}{2\pi} \frac{1}{(v - v_0)^2 + \Gamma^2/4} \tag{19.6.4}$$

in terms of its FWHM value Γ.

By appropriate changes of variables, (19.6.1) can be written as

$$I(v) = \sqrt{\frac{4\ln 2}{\pi}} \frac{1}{\Delta v_{GF}} V(u, a) \tag{19.6.5}$$

in which V is the Voigt function, defined by

$$V(u, a) \equiv \frac{1}{\pi} \int_{-\infty}^{\infty} \frac{\exp\left[-(ay - u)^2\right]}{y^2 + 1} dy \tag{19.6.6}$$

The dimensionless parameters a and u are given in terms of the Gaussian and Lorentzian width parameters by the positive width-ratio parameter

$$a = \sqrt{\ln 2} \, \Gamma / \Delta v_{GF} = 0.8326 \, \Gamma / \Delta v_{GF} \geq 0 \tag{19.6.7}$$

and the parameter measuring the displacement from the center of the convolution

$$u = 2\sqrt{\ln 2} \, (v - v_0) / \Delta v_{GF} = 1.6651(v - v_0) / \Delta v_{GF} \tag{19.6.8}$$

Sometimes a is denoted by y, u is called v or x, and $V(u, a) = H(a, u)$. As $a \to 0$ the integral in (19.6.6) tends to the normalization over a Lorentzian, with a Gaussian of variable u as a factor. If $a \to \infty$ then V tends to a Lorentzian.

The Voigt function can be manipulated into an expression in terms of the complementary error function of complex argument, erfc(z), as

$$V(u,a) = \mathrm{Re}\Big[\exp\!\big(z^2\big)\mathrm{erfc}(z)\Big] \qquad z = a + iu \qquad (19.6.9)$$

Here are several graphics to help visualizing the Voigt function. In Figure 19.6.1 the surface has as its front face the Gaussian and as its left face ($u = 0$) a Lorentzian.

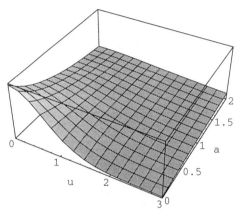

FIGURE 19.6.1 Surface of the Voigt function (19.6.6) as a function of u and a. (Cell Voigt in *Mathematica* notebook Miscellaneous.)

Figure 19.6.2, shows the evolution of the Voigt function from a Gaussian to a Lorentzian as the parameter a in (19.6.7) increases.

FIGURE 19.6.2 Left: Voigt profile function for five values of the ratio of the FWHM of the Lorentzian to that of the Gaussian by equal steps. Rear shows V for $a = 0$ (a Gaussian), while front shows $a = 1.6651$, nearly a Lorentzian. Right: Comparison of a Gaussian (solid curve) and a Lorentzian (dashed curve) having the same Full Width at Half Maximum and peak value. (Cell Voigt in the *Mathematica* notebook Miscellaneous.)

Algorithm and Program

Most algorithms for the Voigt function, V, begin with either the integral representation, (19.6.6), or the error-function relation (19.6.9) to derive analytical approximations for V. Many algorithms are discussed in the review article by Thompson [Tho93]. The algorithm used here is the expansion of V as a power series in a, written as

$$V(u,a) = \sum_{n=0}^{\infty} a^n H_n(u) \qquad (19.6.10)$$

in terms of functions H_n, sometimes called Hjerting functions. As shown in detail in Section 10.3 of Thompson's computing text [Tho92], the H_n can be obtained analytically from

$$H_{n+2}(u) = -\frac{1}{(n+1)(n+2)}\frac{d^2 H_n(u)}{du^2} \tag{19.6.11}$$

with starting values

$$H_0(u) = \exp(-u^2) \qquad H_1(u) = \frac{2}{\sqrt{\pi}}[2uF(u)-1] \tag{19.6.12}$$

in terms of Dawson's integral, $F(u)$, described in Section 10.3. For example;

$$H_2(u) = (1-2u^2)\exp(-u^2)$$

$$H_3(u) = \frac{4}{3\sqrt{\pi}}\left[(3-2u^2)uF(u)+u^2-1\right]$$

$$H_4(u) = \frac{1}{6}(3-12u^2+4u^4)\exp(-u^2) \tag{19.6.13}$$

$$H_5(u) = \frac{2}{15\sqrt{\pi}}\left[(15-20u^2+4u^4)uF(u)-2u^4+9u^2-4\right]$$

$$H_6(u) = \frac{1}{90}(15-90u^2+60u^4-8u^6)\exp(-u^2)$$

The $H_n(u)$ for higher $n > 6$ can be produced by successive differentiation, for example by using *Mathematica*. This technique is used for the Hjerting functions through $n = 10$ in our C function `Voigt`.

Convergence of the series (19.6.10) is very rapid, especially for $a < 0.5$ and when u increases beyond unity, that is, in the wings of the line profile. Note that V is an even function of u, being the convolution of two functions that are even functions of their arguments. This result is useful for programming and debugging. The exponentially damped behavior of F and of the H_n as u increases is shown in Figure 19.6.3.

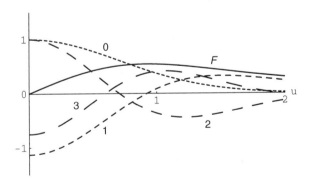

FIGURE 19.6.3 Hjerting functions, H_n, in (19.6.10) for $n = 0, 1, 2, 3$, and the Dawson integral F in (19.6.12) and (19.6.13), each as a function of u. (Adapted from cell `Voigt` in *Mathematica* notebook `Miscellaneous`.)

The program for the Voigt function is a straightforward implementation of the formulas in (19.6.10) – (19.6.13). Function `FDawson` (Section 10.3) is provided with the machine-readable program.

```
double Voigt(double u, double a)
/* Voigt function by series expansion thru n = 10 */
{
double eps,asq,even,FD,odd;
double FDawson(double u,double eps),H0(double u),
  H2(double u),H4(double u),H6(double u),
  H8(double u),H10(double u),
  H1(double RtPi, double FD, double u),
  H3(double RtPi, double FD, double u),
  H5(double RtPi, double FD, double u),
  H7(double RtPi, double FD, double u),
  H9(double RtPi, double FD, double u);
const double RtPi = 1.77245385091;

eps = 1.0e-10; /* accuracy of Dawson integral */
asq = a*a;
/* Even terms in series */
even = H0(u)+asq*(H2(u)+asq*(H4(u)+
  asq*(H6(u)+asq*(H8(u)+asq*H10(u)))));
FD = FDawson(u,eps);
/* Odd terms in series have Dawson integral, FD */
odd = a*(H1(RtPi,FD,u)+asq*(H3(RtPi,FD,u)+
  asq*(H5(RtPi,FD,u)+asq*(H7(RtPi,FD,u)+
  asq*H9(RtPi,FD,u)))));

return even+odd;
}

double H0(double u)
/* Zeroth-order H function */
{ return  exp(-u*u); }

double H1(double RtPi, double FD, double u)
/* First-order H function */
{ return  2.0*(2.0*u*FD-1.0)/RtPi; }

double H2(double u)
/* Second-order H function */
{ return  (1.0-2.0*u*u)*exp(-u*u); }

double H3(double RtPi, double FD, double u)
/* Third-order H function */
{return 4.0*((3.0-2.0*u*u)*u*FD+u*u-1.0)/(3.0*RtPi);}

double H4(double u)
/* Fourth-order H function */
{ double usq;
  usq = u*u;
  return  (3.0-(12.0-4.0*usq)*usq)*exp(-usq)/6.0; }

double H5(double RtPi, double FD, double u)
/* Fifth-order H function */
{ double usq;
  usq = u*u;  return
          2.0*((15.0-usq*(20.0-4.0*usq))*u*FD-
          usq*(2.0*usq-9.0)-4.0)/(15.0*RtPi);
}
```

```
double H6(double u)
/* Sixth-order H function */
{
double usq;
usq = u*u;
return
  (15.0-usq*(90.0-usq*(60.0-8.0*usq)))*exp(-u*u)/90.0;
}

double H7(double RtPi, double FD, double u)
/* Seventh-order H function */
{
double usq;
usq = u*u;
return 2.0*((105.0-usq*(210.0-usq*(84.0-
8.0*usq)))*u*FD-
  (24.0-usq*(87.0-usq*(40.0-4.0*usq))))/(315.0*RtPi);
}

double H8(double u)
/* Eighth-order H function */
{
double usq;
usq = u*u;
return  (105.0-usq*(840.0-usq*(840.0-
  usq*(224.0-16.0*usq))))*exp(-u*u)/2520.0;
}

double H9(double RtPi, double FD, double u)
/* Ninth-order H function */
{
double usq;
usq = u*u;
return
  ((945.0-usq*(2520.0-usq*(1512.0-288.0*usq)))*u*FD-
(192.0-usq*(975.0-usq*(690.0-usq*(140.0-8.0*usq)))))/
(5670.0*RtPi);
}

double H10(double u)
/* Tenth-order H function */
{
double usq;
usq = u*u;
return  (945.0-usq*(9450.0-usq*(12600.0-
  usq*(5040.0-usq*(720.0-32.0*usq)))))*
  exp(-u*u)/113400.0;
}
```

Programming Notes

When computing the series expansion (19.6.10) for the Voigt function, $V(u,a)$, in function Voigt, each value that is obtained by calling FDawson is used several times, so we improve efficiency by invoking the function F only once, then saving its value. The programming for odd n accommodates this sharing of F.

Test Values

Test values of $V(u,a)$—to the number of digits and over a range of values of a for which the series expansion (19.6.10) through $n = 10$ is adequate—are given in Table 19.6.1.

TABLE 19.6.1 Values of the Voigt function (19.6.6).

$a \setminus u$	0	0.50	1.00
0	1.000000000	0.7788007831	0.3678794412
0.1	0.896456980	0.7175877422	0.3731701483
0.2	0.809019520	0.663222625	0.373152914

19.7   ANGULAR MOMENTUM COUPLING COEFFICIENTS

The 3–j, 6–j, and 9–j coefficients from angular momentum theory are described in, for example, Thompson's text on rotational symmetries [Tho94]. As coded here, these functions are easy to understand and are moderately efficient. Algorithms and programs that may provide more efficient execution are described in [Tho94] and by Fack et al [Fac95]. First, common features of the functions are described, then Sections 19.7.1–19.7.3 describe the functions individually. The three functions are prepared uniformly, with all angular momentum numbers entered as decimals. Thus, $a = 3/2$ is entered as 1.5 and $m_b = -2$ is entered as –2., where the decimal point may be optional in some computers.

Because the coupling coefficients have either six or nine arguments, it is not practical to provide visualizations of them. We therefore do not provide a *Mathematica* notebook. For restricted values of the arguments, pictorial representations are feasible. Examples are given in [Tho94] for 3-j coefficients (Section 7.3.2) and for 6-j coefficients (Section 9.3.2). This reference has *Mathematica* notebooks for understanding these coupling coefficients.

Before a C function is executed, its arguments should be checked within the program calling it, in order to see whether the function is automatically zero because of violation of selection rules on magnetic substates m ($|m| \le j$) or between the total angular momenta j values (triangle selection rules). No such testing is done within a function itself. Because such checking is complicated, the driver program illustrating this checking for each angular momentum function is included here, rather than in Section 21.19.

We use the following conventions for coding the functions. A variable name beginning with "t" indicates *twice* an angular momentum, such as `tj` for 2j or `te` for 2e. After this has been done, there is a one-to-one correspondence between program variables and formula variables, with the exception that m' is replaced by mp.

Multi-Use Functions

Six multi-use functions are used by more than one angular momentum function; `Fctrl`, `Max`, `Min`, `Power`, `TriangleBroken`, and `Twice`. Following our philosophy of no assembly required (see The Computer Interface in the introduction to the *Atlas*), these functions are included with the functions on the CD-ROM when needed. The functions are as follows:

`Fctrl`. This computes the factorial function as a floating-point variable for non-negative integer arguments. There is no overflow check for this function.

```
double Fctrl(int N)
/* Factorial function;
   assumes N >= 0 */
{
double product;
int i;

product = 1;
if ( N <= 1 )  return product;
else
  {
  for ( i = 2; i <= N; i++ )
    { product = product*i; }
  return product;
  }
}
```

Max. This function returns the biggest of the N elements that are stored in array. The [0] element of the array is not used.

```
int Max(int array[], int N)
/* Maximum of first N elements
   of array[1] … [N] */
{
int big,i;

big = array[1];
for ( i = 2; i <= N; i++ )
  {
  if ( array[i] > big ) big = array[i];
  }
return big;
}
```

Min. Function Min returns the smallest of the N elements that are stored in array. The [0] element of the array is not used.

```
int Min(int array[], int N)
/* Minimum of first N elements
   of array [1]...[N] */
{
int small,i;

small = array[1];
for ( i = 2; i <= N; i++ )
  {
  if ( array[i] < small ) small = array[i];
  }
return small;
}
```

Power. The Power function returns x^i, in which i is an integer. The C library function $pow(x, y)$ assumes that x and y are double variables. Also, if $x = 0$ and $y \leq 0$, the function returns negative infinity. However, since $x^0 \equiv 1$, Power traps for $i = 0$, in which case it returns 1. Hence, the function:

```
double Power(double x, double i)
/* x to integer power */
{

/* Always return 1.0 for i=0 */
if ( i == 0 )  return 1.0;
 return pow(x, (double) i);
}
```

TriangleBroken. This function tests the triangle condition, using *twice* each of the three angular momenta. Therefore, the sum of the three function arguments must be even. A TRUE (=1) is returned if either this condition or the triangle rule is broken, else a FALSE (=0) is returned. The calling functions then test for a 1 or a 0 being returned.

```
int TriangleBroken(int tj1, int tj2, int tj3)
/* tests the triangle condition */
{
int sum;

sum = tj1+tj2+tj3;
if ( sum != 2*(sum/2) ||
    tj2 < abs(tj1-tj3) || tj2 > tj1+tj3 )
  { printf
  ("\n!! (%i,%i,%i) breaks triangle rule:",
   tj1,tj2,tj3);
  return TRUE; }
else return FALSE;
}
```

Twice. Function Twice converts a decimal half integer, such as 1.5 or −2.0, to an integer that is twice this. The possibility that the decimal representation is imprecise is allowed for by adding the roundoff term 0.01.

```
int Twice(double x)
/* Converts to integer(2*x)  */
{
int int2x;

if ( x < 0 ) int2x = -2*(fabs(x)+0.01);
        else int2x = 2*(x+0.01);
return int2x;
}
```

19.7.1 3–*j* Coefficients

Keywords

3–*j* coefficients, angular momentum, Clebsch-Gordan coefficients, Wigner coefficients

C and *Mathematica* Functions

_3j (C); ThreeJSymbol (*Mathematica*)

Background

Wigner's 3-*j* coefficient is used to combine three angular momenta, a, b, c, with the sum of the magnetic substate values, α, β, γ, being zero. Applications of 3-*j* coefficients are discussed in Chapters 7, 8, and 9 of Thompson's text on rotational symmetries [Tho94].

Function Definition

We use a standard formula for the 3-*j* coefficient, as derived in Section 7.3.1 of [Tho94];

$$\begin{pmatrix} a & b & c \\ \alpha & \beta & \gamma \end{pmatrix} = \delta_{\alpha+\beta+\gamma,0}(-1)^{a-b-\gamma}$$

$$\times \sqrt{\frac{(c+a-b)!(c-a+b)!(a+b-c)!(c-\gamma)!(c+\gamma)!}{(a+b+c+1)!(a-\alpha)!(a+\alpha)!(b-\beta)!(b+\beta)!}} \qquad (19.7.1)$$

$$\times \sum_{k} \frac{(-1)^{k+b+\beta}(b+c+\alpha-k)!(a-\alpha+k)!}{k!(c-a+b-k)!(c-\gamma-k)!(k+a-b+\gamma)!}$$

in which the sum is over all values of k such that the arguments of the factorials in the denominator are nonnegative.

Visualization

The simplest way to envision the 3-*j* coefficient (19.7.1) is in terms of a triangle graph:

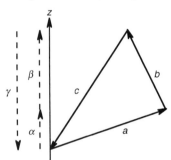

FIGURE 19.7.1 Triangle relations for the 3-*j* coefficient (19.7.1).

Algorithm and Program

The algorithm used for the 3-*j* coefficient is a direct coding of formula (19.7.1), except that in the driver program below the formula is not computed if the sum of the *m* values is nonzero or if the magnitude of any *m* value exceeds its matching *j* value.

```
#include <stdio.h>
#include <math.h>

#define TRUE  1
#define FALSE 0

main() /* 3-j coupling coefficients */
/* Complete program uses _3j,EvenOrOdd given here,
     plus Fctrl,Max,Min,Power,TriangleBroken,Twice */
```

```
{
double a,b,c,ma,mb,mc,Three_j;
int ta,tb,tc,tma,tmb,tmc;
double _3j(int ta,int tb,int tc,int tma,int tmb,int
  tmc);
int EvenOrOdd(int tj,int tm);

printf("Angular Momentum\n");
printf("3-j coupling coefficients\n");
ta = 0;
while ( ta >= 0 )
  {
  printf("\n\nInput  a  as 0.5, etc (<0 to end): ");
  scanf("%lf",&a);
  if ( a < 0 )
    { printf("\nEnd 3-j coupling coefficients");
    exit(0); }
  printf("\nInput b, c, ma, mb, mc: ");
  printf("\n(Use only spaces to separate values) ");
  scanf("%lf%lf%lf%lf%lf",&b,&c,&ma,&mb,&mc);
  /* Convert to integers as twice input values */
  ta = Twice(a);tb = Twice(b); tc = Twice(c);
  tma = Twice(ma); tmb = Twice(mb);  tmc = Twice(mc);
  /* Testing for zero coefficient */
  if ( tma+tmb+tmc != 0 )
    {printf("\n!!ma+mb+mc not zero; try again"); }
  else
    {
    if ( TriangleBroken(ta,tb,tc) != 0 )
      printf(" try again");
    else
      {
      if ( abs(tma) > ta ||
          abs(tmb) > tb ||
          abs(tmc) > tc )
      printf("\n!!An m value is too big; try again");
        else
        {
        if ( EvenOrOdd(ta,tma) ||
            EvenOrOdd(tb,tmb) ||
            EvenOrOdd(tc,tmc) > 0 )
        printf
          ("\n!!An m doesn't match a j; try again");
        else
          { Three_j = _3j(ta,tb,tc,tma,tmb,tmc);
          printf("\n 3-j coefficient = %.9E",Three_j);
          } } } } } }

double _3j(int ta, int tb, int tc,
          int tma, int tmb, int tmc)
/* 3-j coupling coefficient */
{
double n1,n2,n3,d1,d2,d3,norm,sum,phase,term;
int minmax[3],tkmin,tkmax,tk;
```

```
double Fctrl(int N),Power(double x, int i);
int Max(int minmax[],int N),Min(int minmax[],int N);

/* Normalization factorials */
n1 = Fctrl((tc+ta-tb)/2)*Fctrl((tc-ta+tb)/2);
n2 = Fctrl((ta+tb-tc)/2);
n3 = Fctrl((tc-tmc)/2)*Fctrl((tc+tmc)/2);
d1 = Fctrl((ta+tb+tc)/2+1);
d2 = Fctrl((ta-tma)/2)*Fctrl((ta+tma)/2);
d3 = Fctrl((tb-tmb)/2)*Fctrl((tb+tmb)/2);
norm = Power(-1.0,(ta-tb-tmc)/2)*
  sqrt(n1*n2*n3/(d1*d2*d3));

minmax[1] = 0; minmax[2] = tb-ta-tmc;
tkmin = Max(minmax,2);
minmax[1] = tc-ta+tb;  minmax[2] = tc-tmc;
tkmax = Min(minmax,2);
sum = 0;
phase = Power(-1.0,(tkmin+tb+tmb)/2);

for ( tk = tkmin; tk <= tkmax; tk = tk+2 )
  {
  n1 = Fctrl((tb+tc+tma-tk)/2);
  n2 = Fctrl((ta-tma+tk)/2);
  d1 = Fctrl(tk/2)*Fctrl((tc-ta+tb-tk)/2);
  d2 = Fctrl((tc-tmc-tk)/2);
  d3 = Fctrl((tk+ta-tb+tmc)/2);
  term = phase*n1*n2/(d1*d2*d3);
  phase = -phase;
  sum = sum+norm*term;
  }
return sum;
}

int EvenOrOdd(int tj,int tm)
/* tests for tj+tm even */
{
int sum;

sum = tj+tm;
if ( sum != 2*(sum/2) ) return 1;
else return 0;
}
```

Programming Notes

If many 3-j coefficients are to be calculated, the efficiency of _3j can be improved by storing a table of factorials up to the largest argument likely to be used, then using table lookup whenever a factorial is needed. If this straightforward reprogramming is done, the driver program should have a check that the factorial array will not be overflowed.

Test Values

Spot checks for a range of angular momenta, but with the smallest value ≤ 1, can be made by comparison with numerical results from the algebraic expressions in Table 19.7.1 on the following page. A testing method that is more complete is to check the orthogonality conditions for 3-j coefficients. Such tests are also useful for checking the numerical accuracy of the coefficients.

Computer-generated numerical tables of 3–j coefficients as rational fractions were made by Rotenberg et al [Rot59]. Within the limitations of the largest angular momentum equal to 8, these are the most useful numerical tables for testing 3–j coefficients.

TABLE 19.7.1 3-j coefficients with smallest angular momentum ≤ 1. Related coefficients can be obtained by permutation of columns and sign changes of lower row elements.

$$\begin{pmatrix} 0 & a & a \\ 0 & \alpha & -\alpha \end{pmatrix} = (-1)^{a-\alpha} \frac{1}{\sqrt{1+2a}}$$

$$\begin{pmatrix} 1/2 & a & a+1/2 \\ 1/2 & \alpha & -1/2-\alpha \end{pmatrix} = (-1)^{1-a+\alpha} \sqrt{\frac{1+a+\alpha}{2(1+a)(1+2a)}}$$

$$\begin{pmatrix} 1 & a & a \\ 1 & \alpha & -1-\alpha \end{pmatrix} = \sqrt{\frac{(a-\alpha)(1+a+\alpha)}{2a(1+a)(1+2a)}}$$

$$\begin{pmatrix} 1 & a & a \\ 0 & \alpha & -\alpha \end{pmatrix} = (-1)^{a-\alpha} \frac{\alpha}{\sqrt{a(1+a)(1+2a)}}$$

$$\begin{pmatrix} 1 & a & a+1 \\ 1 & \alpha & -\alpha-1 \end{pmatrix} = (-1)^{a-\alpha} \sqrt{\frac{(a+\alpha+1)(a+\alpha+2)}{(2a+1)(2a+2)(2a+3)}}$$

$$\begin{pmatrix} 1 & a & a+1 \\ 0 & \alpha & -\alpha \end{pmatrix} = (-1)^{1-a+\alpha} \sqrt{\frac{(1+a-\alpha)(1+a+\alpha)}{(1+a)(1+2a)(3+2a)}}$$

19.7.2   6–j Coefficients

Keywords
6–j coefficients, angular momentum, Racah coefficients, Wigner coefficients

C and *Mathematica* Functions
M _6j (C); SixJSymbol (*Mathematica*)

Background
Wigner's 6-j coefficient is used when combining six total angular momenta, which we label as $a, b, c, d, e,$ and f. See, for example, Section 9.1 of Thompson's text on angular momentum [Tho94].

Function Definition and Visualization
A 6-j coefficient can be expressed as a sum over products of four 3-j coefficients [Tho94, (9.16)], which is desirable in formal manipulations but is inefficient for numerical work. Racah [Rac42] reduced the equivalent relation between Clebsch-Gordan and Racah coefficients to a single algebraic sum. This is written in terms of the Δ function defined by

$$\Delta(abc) \equiv \sqrt{\frac{(a+b-c)!(a+c-b)!(b+c-a)!}{(a+b+c+1)!}} \qquad (19.7.2)$$

In this and the following expression it is assumed that the six arguments satisfy the triangle relations shown in Figure 19.7.2, and these conditions should be checked before function _6j is called, as illustrated in the driver program under Algorithm and Program.

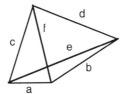

FIGURE 19.7.2 Triangle relations among the arguments of the 6-*j* coefficient (19.7.3).

The 6-*j* coefficient is expressed as

$$\begin{Bmatrix} a & b & e \\ d & c & f \end{Bmatrix} = (-1)^{a+b+c+d} \Delta(abe)\Delta(acf)\Delta(bdf)\Delta(cde)$$
$$\times \sum_{k}(-1)^k (a+b+c+d+1-k)!$$
$$\times \left[k!(e+f-a-d+k)!(e+f-b-c+k)! \right]^{-1}$$
$$\times \left[(a+b-e-k)!(c+d-e-k)!(a+c-f-k)!(b+d-f-k)! \right]^{-1}$$

(19.7.3)

The range of summation over k is chosen so that no factorial argument in the sum is negative. If the initial phase factor is omitted, one has the Racah coefficient $W(abcd;ef)$.

Algorithm and Program

The algorithm used for the 6-*j* coefficient is a direct coding of (19.7.3), except that in the driver program the formula is not computed either if the sum of the *m* values is nonzero or if the magnitude of any *m* value exceeds its matching *j* value.

```
#include <stdio.h>
#include <math.h>

#define TRUE  1
#define FALSE 0

main() /* 6-j recoupling coefficients */
/* Complete program uses _6j,Delt given here
   plus Fctrl,Max,Min,Power,TriangleBroken,Twice  */
{
double a,b,c,d,e,f,Six_j;
double _6j(int ta,int tb,int tc,int td,int te,int tf);
int ta,tb,tc,td,te,tf;

printf("Angular Momentum\n");
printf("6-j recoupling coefficients\n");
ta = 0;
while ( ta >= 0 )
```

```
 {
 printf("\n\nInput a  as 0.5, etc (<0 to end): ");
 scanf("%lf",&a);ta = Twice(a);
 if ( ta < 0 )
  {
  printf("\nEnd 6-j recoupling coefficients");
  exit(0);
  }
 printf("\nInput b, c, d, e, f: ");
 printf("\n(Use only spaces to separate values) ");
 scanf("%lf%lf%lf%lf%lf",&b,&c,&d,&e,&f);
 /* Convert to integers as twice input values */
 tb = Twice(b);tc = Twice(c);
 td = Twice(d);te = Twice(e); tf = Twice(f);

 /* Testing for zero coefficient */
 if (  TriangleBroken(ta,tb,te) != 0 ||
     TriangleBroken(ta,tc,tf) != 0 ||
     TriangleBroken(tb,td,tf) != 0 ||
     TriangleBroken(tc,td,te) != 0 )
  printf(" try again");
 else
  {
  Six_j = _6j(ta,tb,tc,td,te,tf);
  printf("\n 6-j coefficient = %.9E",Six_j);
  } } }

double _6j(int ta, int tb, int tc,
          int td, int te, int tf)
/* 6-j recoupling coefficient */
{
double norm,sum,phase,n1,d1,d2,d3,d4,d5,d6,term;
int minray[4],maxray[6],tkmin,tkmax,tk;
double Delt(int ta,int tb,int tc),Fctrl(int N),
 Power(double x,int i);
int Max(int minmax[],int N),Min(int minmax[],int N);

/* Normalization factorials */
norm =  sqrt(Delt(ta,tb,te)*Delt(ta,tc,tf)*
        Delt(tb,td,tf)*Delt(tc,td,te));
/* Minimum summation index */
minray[1] = 0;                minray[2] = ta+td-te-tf;
minray[3] = tb+tc-te-tf; tkmin = Max(minray,3);
/* Maximum summation index */
maxray[1] = ta+tb+tc+td+2; maxray[2] = ta+tb-te;
maxray[3] = tc+td-te;         maxray[4] = ta+tc-tf;
maxray[5] = tb+td-tf;         tkmax = Min(maxray,5);
sum = 0;                     /* Phase of 6-j */
phase = Power(-1.0,(ta+tb+tc+td+tkmin)/2);
for ( tk = tkmin; tk <= tkmax; tk = tk+2 )
 {
 n1 = Fctrl((ta+tb+tc+td-tk)/2+1);
 d1 = Fctrl(tk/2)*Fctrl((te+tf-ta-td+tk)/2);
 d2 = Fctrl((te+tf-tb-tc+tk)/2);
 d3 = Fctrl((ta+tb-te-tk)/2);
 d4 = Fctrl((tc+td-te-tk)/2);
```

```
    d5 = Fctrl((ta+tc-tf-tk)/2);
    d6 = Fctrl((tb+td-tf-tk)/2);
    term = phase*n1/(d1*d2*d3*d4*d5*d6);
    phase = -phase;
    sum = sum+norm*term;
    }
 return sum;
 }

 double Delt(int ta, int tb, int tc)
   /* Square of delta factor */
 { return  Fctrl((ta+tb-tc)/2)*
     Fctrl((ta+tc-tb)/2)*Fctrl((tb+tc-ta)/2)/
     Fctrl((ta+tb+tc)/2+1); }
```

Programming Notes

When many 6-*j* coefficients are to be calculated, the efficiency of the function _6j can be improved by storing a table of factorials up to the largest argument that is likely to be used, then using table lookup whenever a factorial is needed. If such reprogramming is done, the driver program should have a check that the factorial array will not overflow.

Test Values

Tabulations of Racah or 6-*j* coefficients, both algebraic—as in Tables 19.7.2 and 19.7.3—and numerical, can be used to provide test values. You can also use the *Mathematica* function SixJSymbol. Beware that this function does not check whether the four triangle conditions in Figure 19.7.2 and (19.7.3) are satisfied. Extensive tables of algebraic expressions for Clebsch-Gordan and Racah coefficients are available in Biedenharn and Louck [Bie81] and in Varshalovich et al [Var88]. Such tables were developed by hand calculation and the formulas were typeset by hand. Both procedures can have errors in the complicated expressions involved.

Extensive computer-generated numerical tables of 6-*j* coefficients as exact rational fractions were prepared by Rotenberg et al [Rot59]. Within the limitations of the tabulation (largest angular momentum equal to 8), these are the most useful numerical tables for testing. A more complete testing method is to check orthogonality conditions . Such tests are also useful for checking numerical accuracy.

TABLE 19.7.2 The 6-*j* coefficients with the smallest angular momentum *a* = 0 or 1/2. Related coefficients are obtained by permuting columns and by pairwise row and column interchanges. Variables in the coefficients must be chosen to satisfy the triangle rules and must be non-negative.

$$\begin{Bmatrix} 0 & b & b \\ d & c & c \end{Bmatrix} = \frac{(-1)^{b+c+d}}{\sqrt{(1+2b)(1+2c)}}$$

$$\begin{Bmatrix} 1/2 & b & b+1/2 \\ d & c & c+1/2 \end{Bmatrix} = (-1)^{1/2+b+c+d}\sqrt{\frac{(1/2+b-c+d)(1/2-b+c+d)}{(1+2b)(2+2b)(1+2c)(2+2c)}}$$

$$\begin{Bmatrix} 1/2 & b & b+1/2 \\ d & c & c-1/2 \end{Bmatrix} = (-1)^{1/2+b+c+d}\sqrt{\frac{(1/2+b+c-d)(3/2+b+c+d)}{(1+2b)(2+2b)2c(1+2c)}}$$

TABLE 19.7.3 The 6-j coefficients with the smallest angular momentum $a = 1$. Related coefficients are obtained by permuting columns and by pairwise row and column interchanges. Variables in the coefficients must satisfy the triangle rules and must be non-negative.

$$\begin{Bmatrix} 1 & b & b+1 \\ d & c & c+1 \end{Bmatrix} = (-1)^{1+b+c+d} \sqrt{\frac{(b-c+d)(1+b-c+d)(-b+c+d)(1-b+c+d)}{(1+2b)(2+2b)(3+2b)(1+2c)(2+2c)(3+2c)}}$$

$$\begin{Bmatrix} 1 & b & b+1 \\ d & c & c \end{Bmatrix} = (-1)^{1+b+c+d} \sqrt{\frac{(1+b+c-d)(1+b-c+d)(-b+c+d)(2+b+c+d)}{(1+2b)(2+2b)(3+2b)c(1+2c)(2+2c)}}$$

$$\begin{Bmatrix} 1 & b & b+1 \\ d & c & c-1 \end{Bmatrix} = (-1)^{b+c+d} \sqrt{\frac{(b+c-d)(1+b-c+d)(-b+c+d)(1+b+c+d)}{2b(1+2b)(2+2b)\,c(-1+2c)(1+2c)}}$$

$$\begin{Bmatrix} 1 & b & b \\ d & c & c \end{Bmatrix} = (-1)^{1+b+c+d} \frac{b(1+b)+c(1+c)-d(1+d)}{\sqrt{b(1+2b)(2+2b)c(1+2c)(2+2c)}}$$

19.7.3   9-j Coefficients

Keywords

9-j coefficients, angular momentum, Wigner coefficients

C Function

_9j

Background

The Wigner 9-j coefficients occur in angular momentum when four angular momenta are combined in different orders, for example, in transformations between L–S and j–j coupling in quantum mechanics. See, for example, the treatment in Section 9.5 of Thompson's angular momentum text [Tho94].

Function Definition

An expansion in terms of 6-j coefficients is used for the 9-j coefficients. The formula is not computed if any of the six independent triangle conditions is not satisfied. The function _6j is obtained from Section 19.7.2.

$$\begin{Bmatrix} a & b & c \\ d & e & f \\ g & h & i \end{Bmatrix} = \sum_k (-1)^{2k}(2k+1) \begin{Bmatrix} a & i & k \\ h & d & g \end{Bmatrix} \begin{Bmatrix} b & f & k \\ d & h & e \end{Bmatrix} \begin{Bmatrix} a & i & k \\ f & b & c \end{Bmatrix} \qquad (19.7.4)$$

Visualization

    The tree representation describing the transformation between the L–S and j–j coupling schemes is shown in Figure 19.7.3.

Algorithm and Program

Formula (19.7.4) is used directly, with the range of summation over k being determined within _9j from the triangle selection rules on the 6-j coefficients appearing in the sum.

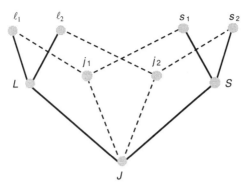

FIGURE 19.7.3 Tree visualization of transformation between L-S and j-j coupling schemes, in which solid lines represent one coupling scheme and dashed lines represent the other scheme.

```
#include <stdio.h>
#include <math.h>

#define TRUE  1
#define FALSE 0

main() /* 9-j recoupling coefficients */
/* Complete program uses _9j & Delt given here,
plus Fctrl,Max,Min,Power,TriangleBroken,Twice */
{
double Nine_j;
int ta,tb,tc,td,te,tf,tg,th,ti;
double a,b,c,d,e,f,g,h,i,
_6j(int ta,int tb,int tc,int td,int te,int tf),
_9j(int ta,int tb,int tc,
    int td,int te,int tf,int tg,int th,int ti);

printf("Angular Momentum\n");
printf("9-j recoupling coefficients\n");
ta = 0;
while ( ta >= 0 )
  {
  printf("\n\nInput a (<0 to end): ");
  scanf("%lf",&a);ta = Twice(a);
  if ( ta < 0 )
    {
    printf("\nEnd 9-j recoupling coefficients");
    exit(0);
    }
  printf("\nInput b, c: ");
  printf("\n(Use only spaces to separate values) ");
  scanf("%lf%lf",&b,&c);
  printf("\nInput d, e, f: ");
  scanf("%lf%lf%lf",&d,&e,&f);
  printf("\nInput g, h, i: ");
  scanf("%lf%lf%lf",&g,&h,&i);
  tb = Twice(b);tc = Twice(c);
  td = Twice(d);te = Twice(e); tf = Twice(f);
  tg = Twice(g);th = Twice(h); ti = Twice(i);
```

```
/* Testing for zero coefficient */
if (  TriangleBroken(ta,tb,tc) != 0 ||
    TriangleBroken(td,te,tf) != 0 ||
    TriangleBroken(tg,th,ti) != 0 ||
    TriangleBroken(ta,td,tg) != 0 ||
    TriangleBroken(tb,te,th) != 0 ||
    TriangleBroken(tc,tf,ti) != 0  )
  printf(" try again");

else
  {
  Nine_j = _9j(ta,tb,tc, td,te,tf, tg,th,ti);
  printf("\n 9-j coefficient = %.9E",Nine_j);
  } } }

double _9j(int ta, int tb, int tc,
 int td, int te, int tf, int tg, int th, int ti)
/* 9-j recoupling coefficient */
{
double sum,term,_6j(),Power(double x,int i);
double term1,term2,term3;
int minmax[4],tkmin,tkmax,tk;

/* Minimum summation index */
minmax[1] = abs(ta-ti); minmax[2] = abs(th-td);
minmax[3] = abs(tb-tf); tkmin = Max(minmax,3);

/* Maximum summation index */
minmax[1] = ta+ti;  minmax[2] = th+td;
minmax[3] = tb+tf;  tkmax = Min(minmax,3);

sum = 0;
for ( tk = tkmin; tk <= tkmax; tk = tk+2 )
  {
  term = _6j(ta,ti,td, th,tk,tg)*
      _6j(tb,tf,th, td,tk,te)*
      _6j(ta,ti,tb, tf,tk,tc);
  sum = sum+(tk+1)*term;
  }
return  Power(-1.0,tkmin)*sum;
}
```

Programming Notes

If many 9-*j* coefficients are to be computed, the efficiency of the function _9j can be improved by storing—for computation of the 6–*j* coefficients needed—a table of factorials up to the largest argument that is likely to be used, then using table lookup for the factorials. The driver program should then include a check that the factorial array will not overflow.

A variety of algorithms for 9-*j* coefficients—both for serial and for parallel processing—is discussed in Section 9.5.1 of [Tho94].

Test Values

Spot checks can be made by comparing the results from _9j with those in the tabulation by Matsunobu and Takebe [Mat55]. For example,

$$\begin{Bmatrix} 2. & 1. & 2. \\ 1.5 & 1.5 & 1. \\ 0.5 & 0.5 & 1. \end{Bmatrix} = -\frac{1}{2.5\sqrt{6}} \approx -0.04082482905$$

$$\begin{Bmatrix} 4. & 2. & 5. \\ 3.5 & 1.5 & 4. \\ 0.5 & 0.5 & 1. \end{Bmatrix} = \frac{\sqrt{7}}{4.3.5\sqrt{3}} \approx 0.02545875386$$

A testing method that is more complete is to check sum rules for 9–j coefficients, which is also useful for checking the numerical accuracy of the coefficients. For example, the orthogonality sum rule is [Tho94, (9.65)]

$$\sum_{cf} (2c+1)(2f+1) \begin{Bmatrix} a & b & c \\ d & e & f \\ g & h & i \end{Bmatrix} \begin{Bmatrix} a & b & c \\ d & e & f \\ j & k & i \end{Bmatrix} = \frac{\delta_{gj}\,\delta_{hk}}{(2g+1)(2h+1)} \qquad (19.7.5)$$

Because of the high symmetry of the 9–j coefficients, any one argument can be moved to a special position—such as the top-left corner—which will at most introduce a phase change. For $a = 0$, using expansion (19.7.4) and simplifying gives

$$\begin{Bmatrix} 0 & b & c \\ d & e & f \\ g & h & i \end{Bmatrix} = \frac{(-1)^{b+d+f+h}}{\sqrt{(2b+1)(2d+1)}} \begin{Bmatrix} d & e & f \\ b & i & h \end{Bmatrix} \delta_{bc}\,\delta_{dg} \qquad (19.7.6)$$

In terms of a 6–j coefficient (Section 19.7.2), a particular case of this coefficient is for $b = c = 1/2$;

$$\begin{Bmatrix} 0 & 1/2 & 1/2 \\ d & e & f \\ g & h & i \end{Bmatrix} = \frac{(-1)^{1/2+d+f+h}}{\sqrt{2(2d+1)}} \begin{Bmatrix} 1/2 & i & f \\ d & e & h \end{Bmatrix} \delta_{dg} \qquad (19.7.7)$$

Another such formula that is useful for testing is

$$\begin{Bmatrix} 1 & b & b \\ d & e & f \\ d & h & i \end{Bmatrix} = \frac{[e(e+1)+i(i+1)-f(f+1)-h(h+1)]}{\sqrt{4b(b+1)(2b+1)d(d+1)(2d+1)}}$$
$$\times (-1)^{b+d+f+h} \begin{Bmatrix} d & e & f \\ b & i & h \end{Bmatrix} \qquad (19.7.8)$$

REFERENCES ON MISCELLANEOUS FUNCTIONS
FOR SCIENCE AND ENGINEERING

[Abr53]   Abramowitz, M., Journal of Mathematical Physics, **32**, 188 (1953).

[Abr64]   Abramowitz, M., and I. A. Stegun, *Handbook of Mathematical Functions,* Dover, New York, 1964.

[Bie81]   Biedenharn, L. C., and J. D. Louck, *Angular Momentum in Quantum Physics,* Addison-Wesley, Reading, Massachusetts, 1981.

[Fac95]   Fack, V., S. N. Pitre, and J. van der Jeugt, Computer Physics Communications, **86**, 105 (1995).

[Jol61]   Jolley, L. B. W., *Summation of Series*, Dover, New York, 1961.

[Mat55]   Matsunobu, H., and H. Takebe, Progress of Theoretical Physics, **14**, 589 (1955).

[Rac42]   Racah, G., Physical Review, **62**, 186 (1942); ibid, **62**, 438 (1942).

[Rot59]   Rotenberg, M., R. Bivins, N. Metropolis, and J. K. Wooten, Jr., *The 3–j and 6–j Symbols*, Technology Press, M.I.T., Cambridge, Massachusetts, 1959.

[Tho92]   Thompson, W. J., *Computing for Scientists and Engineers*, Wiley, New York, 1992.

[Tho93]   Thompson, W. J., Computers in Physics, **7**, 627 (1993).

[Tho94]   Thompson, W. J., *Angular Momentum*, Wiley, New York, 1994.

[Var88]   Varshalovich, D. A., A. N. Moskalev, and V. K. Khersonskii, *Quantum Theory of Angular Momentum*, World Scientific, Singapore, 1988.

PART II
THE COMPUTER INTERFACE

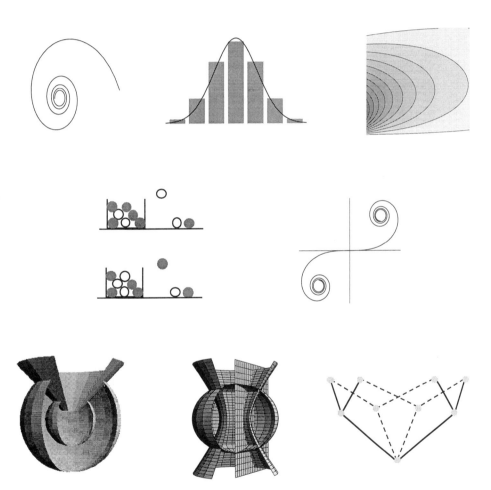

PART II

THE COMPUTER INTERFACE

Chapter 20

THE *MATHEMATICA* NOTEBOOKS

This chapter contains all the notebooks referred to in the text and graphics of the *Atlas*. The notebooks are annotated extensively so that you can match what is being performed within a *Mathematica* cell to the material in Part I of the *Atlas*. In addition, for each notebook cell there is a description suggesting how adventuresome travelers can use or modify each cell to explore further multifaceted properties of the special functions—analytical, graphical, and numerical.

20.1   INTRODUCTION TO THE NOTEBOOKS

The notebooks are designed to be used interactively, as an essential aid for you to understand the properties of functions. Thereby you can travel in the realms of applied mathematics, rather than making the journey vicariously by merely looking at static figures in the text.

Note also that the version of the notebooks on the CD-ROM accompanying the *Atlas* contains cells that have been executed, providing sample output against which you should compare the output from executing that cell on your computer system. After making such tests, you should then be well prepared to adapt the *Mathematica* programs to your particular needs.

Graphical output for three-dimensional objects produced by *Mathematica* is illuminated by red, green, and blue lights, and the polygonal surfaces on the objects reflect light diffusely. The default lighting and reflection directives are used. By displaying the graphic on a color monitor and printing it in color, you will enhance your understanding of the function properties. The cover of the *Atlas* shows a colored graphic—the gamma function in the complex plane.

Many of the notebooks have cells, or parts of cells, that are not used directly in Part I. For example, there are program segments used to generate algebraic expressions for functions in the text. These are identified and annotated in each notebook.

20.2   EXPLORING WITH THE NOTEBOOK CELLS

Because the *Atlas* is designed to be used interactively, it is important that you go exploring on your own. For every *Mathematica* cell in a notebook, there are suggestions in the following sections on how to adapt the programs to see the functions from different viewpoints, or to compute different numerical and analytical properties.

For example, the Γ function is depicted in the complex plane in the adjacent motif. Its graphic object is computed by running cell GammaFunction in notebook GmBt. In Section 6.1 there are seven different views of the gamma function—slices of surfaces, contour plots, and curves—and these may have stirred your interest to explore. Therefore in Section 20.6, just before the annotated cell, there are suggestions for how to modify the program to see other aspects of the function.

20.3   THE ANNOTATED NOTEBOOKS

Each cell in a *Mathematica* notebook in the *Atlas* is listed and annotated, as shown in the sample at right for the Γ function in Section 6.1; the matching notebook is in Section 20.6, since section numbers in this chapter match chapter numbers in Part I.

The main function of the annotations is to key the *Mathematica* code to equations, figures, and tables in the text. Explanations of what the code is doing are given if appropriate.

Typically, definitions are distinguished from objects, and from surfaces, curves, or tables. Usually, all the definitions in a given cell appear before they are used. The order of definitions is graphics of surfaces and curves, followed by numerics, just as in Part I.

```
(* GammaFunction cell *)

(* Gamma Function in 3D *)
GammaPlot[ReOrIm] :=
  Plot3d{ReOrIm[Gamma[x + I * y]],
  {x,-4,4},{y,0.0001,1}}


RealPlot =
  Show[GammaPlot[Re]];
ImaginaryPlot =
  Show[GammaPlot[Im]];

(* Gamma Function contours *)
GammaAbsContour = ContourPlot[
  Log[Abs[Gamma[x + I * y]]],
  {x,-4,4},{y,-2,2}];
```

Define real or imaginary part of surface of Γ; start above real axis.

Show real surface, then imaginary surface, Figure 6.1.1.

Contour plot of log of absolute value of Γ, as in Figure 6.1.2.

20.4 ELEMENTARY TRANSCENDENTAL FUNCTIONS

Notebook and Cells
Notebook `ElemTF`; cells `Exp&Log`,
`Circular`, `InverseCircular`,
`Hyperbolic`, `InverseHyperbolic`

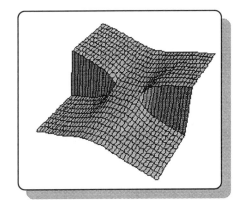

Overview
The *Mathematica* notebook for the elementary transcendental functions includes
exponential and logarithm functions (Section 4.1), the circular functions (Section
4.2), and hyperbolic functions in Section 4.3. It makes 3D views of surfaces of
functions of the *complex* variable
$z = x + iy$, which is feasible because these
functions have no other arguments. We also generate transections of these surfaces—
usually along the $x = 0$ or $y = 0$ planes—to provide the usual curves of the functions.
These curves are made for the direct functions—such as cosine, sine, and tangent—as well
as for their reciprocals—such as secant, cosecant, and cotangent.

Surfaces and curves of the inverse functions are also generated by the `ElemTF` notebook. For example, we visualize the surface of the arctanh of a complex variable as well
as the hyperbolic tangent, tanh. Such visualizations are especially useful for showing
branch-cut discontinuities of the logarithmic, inverse circular, and hyperbolic functions, for
example, the logarithm discontinuity shown in the motif immediately below.

This notebook also serves as an introduction to the *Mathematica* notebooks for subsequent chapters, since (in principle, even if not in fact) you should already understand the
analytical properties of the elementary transcendental functions. In particular, the branch-
cut discontinuities for the inverse functions are not trivial to understand and are often quite
tricky to account for in C functions.

The `Exp&Log` cell [4.1]
Executing this cell produces Figures 4.1.1
and 4.1.2 for the exponential and logarithm (principal value) of $z = x + iy$ with x
in $[-4, 4]$ and y in $[-1, 1]$.

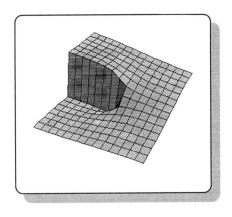

Exploring with `Exp&Log`
The surface of the exponential in
Figure 4.1.1 is quite flat because of the
small range of x and y that we used.
Extend both the x and y ranges in the surface plots to reveal the oscillations that
occur as the y range is increased. On the
contrary, by shrinking the x and y ranges
around the origin of the complex plane,
you can investigate the singularity structure of the logarithm in more detail.

```
(* Exp&Log cell *)

(* Exponentials *)
(* Exponentials of Complex Variables in 3D *)
ExpPlot[ReOrIm_] := Plot3D[ReOrIm[Exp[x+I*y]],
 {x,-1,1},{y,-1,1},
 PlotRange->All,PlotPoints->{20,20},
 Ticks->{Automatic,Automatic,None},
 AxesLabel->{"x"," y",""}]
ExpPlot[Re]; ExpPlot[Im];
```

Define function for surface of exponential, real or imaginary part.

Show real and imaginary exponential surfaces in Figure 4.1.1.

```
(* Exponentials in 2D  *)
(* For  y = 0  *)
Exp2DPlotXX[ReOrIm_,y_] := Plot[ReOrIm[Exp[x+I*y]],
 {x,-1,1},PlotPoints->40,
 Ticks->{{-1,0,1},None},AxesLabel->{"x",""}]
Exp2DPlotXX[Re,0]; Exp2DPlotXX[Im,0];
```

Define exponential curve for *y* fixed.

Show exponential curve for *y* = 0 in Figure 4.1.1.

```
(* For  x = 0  *)
Exp2DPlotYY[ReOrIm_,x_] := Plot[ReOrIm[Exp[x+I*y]],
 {y,-1,1},PlotPoints->40,
 Ticks->{{-1,0,1},None},AxesLabel->{"y",""}]
Exp2DPlotYY[Re,0]; Exp2DPlotYY[Im,0];
```

Define exponential curve for *x* fixed.

Show exponential curve for *x* = 0 in Figure 4.1.1.

```
(* Exponentials of Complex Numbers: Numerical *)
ExpTable = N[Table[
 {x,y,N[Re[Exp[x+I*y]]]+I*N[Im[Exp[x+I*y]]]},
 {x,-1,1,2},{y,-1,1,2}],10]
```

Numerical complex exponentials to 10 digits, Table 4.1.1.

```
(* Logarithms  *)
(* Logarithms of Complex Variables in 3D *)
LogPlot[ReOrIm_] := Plot3D[ReOrIm[Log[x+I*y]],
 {x,-1,1},{y,-1,1},
 PlotRange->All,PlotPoints->{20,20},
 Ticks->{Automatic,Automatic,None},
 AxesLabel->{"x"," y",""}]
LogPlot[Re]; LogPlot[Im];
```

Define function for surface of logarithm, real or imaginary part.

Show real and imaginary logarithm surfaces in Figure 4.1.2.

```
(* Logarithms in 2D  *)
(* For  y = 0  *)
Log2DPlotXX[ReOrIm_,y_] := Plot[ReOrIm[Log[x+I*y]],
 {x,-1,1},PlotPoints->40,
 Ticks->{{-1,0,1},None},AxesLabel->{"x",""}]
Log2DPlotXX[Re,0]; Log2DPlotXX[Im,0];
```

Define logarithm curve for *y* fixed.

Show logarithm curve for *y* = 0 in Figure 4.1.2.

```
(* For  x = 0  *)
Log2DPlotYY[ReOrIm_,x_] := Plot[ReOrIm[Log[x+I*y]],
 {y,-1,1},PlotPoints->40,
 Ticks->{{-1,0,1},None},AxesLabel->{"y",""}]
Log2DPlotYY[Re,0]; Log2DPlotYY[Im,0];
```

Define logarithm curve for *x* fixed.

Show logarithm curve for *x*= 0 in Figure 4.1.2.

```
(* Principal Value Logarithms: Numerical *)
LogTable = N[Table[
   {x,y,N[Re[Log[x+I*y]]]+I*N[Im[Log[x+I*y]]]},
   {x,-1,1,2},{y,-1,1,2}],10]
```

Numerical complex logarithms to 10 digits, Table 4.1.2.

The `Circular` cell [4.2.1]

Executing this cell produces Figure 4.2.1 for surfaces of the circular functions cosine, sine, and tangent for complex variable z, as well as curves of these functions and their reciprocals as functions of real variable x, as shown in Figure 4.2.2.

Exploring with `Circular`

The surface of the real parts of the cosine and sine surfaces in Figure 4.2.1 do not vary much over the small range of y that we used. Extend both the x and y ranges in the surface plots to reveal the oscillations that occur as the y range is

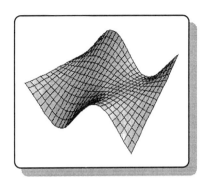

increased. By contrast, if you shrink the x and y ranges around the origin of the complex plane, you can investigate the singularity structure of the tangent in detail.

```
(* Circular cell *)

(* Circular functions: cosine, sine, tangent  *)

Off[Plot::plnr,General::spell1,General::stop];
(* suppresses irrelevant error messages *)

(* Circular Functions of Complex Variables in 3D *)
SurfPlot[CST_,ReOrIm_] := Plot3D[
  ReOrIm[CST[x+I*y]],{x,-4,4},{y,-1,1},
  PlotRange->All,PlotPoints->{40,20},
  Ticks->{Automatic,Automatic,None},
  AxesLabel->{"x","  y",""}]

SurfPlot[Cos,Re];   SurfPlot[Cos,Im];
SurfPlot[Sin,Re];   SurfPlot[Sin,Im];
SurfPlot[Tan,Re];   SurfPlot[Tan,Im];

(* Circular Functions in 2D, x-axis views  *)
CSTPlot[CSTorRec_,AutoDash_] := Plot[CSTorRec[x],
   {x,-4,4},PlotPoints->100,AxesLabel->{"x",""},
   Ticks->{{-4,-3,-2,-1,0,1,2,3,4},None},
   PlotRange->{-2,2},PlotStyle->AutoDash]
```

Define function for surfaces of circular functions in the complex plane.

Plor real and imaginary parts of cosine, sine, tangent, as in Figure 4.2.1.

Define curves of circular functions.

```
(* Circular cell continued  *)


Auto := Automatic
Dash := {Dashing[{0.02,0.02}]}
CSTPlot[Cos,Auto];   CSTPlot[Sec,Dash];
CSTPlot[Sin,Auto];   CSTPlot[Csc,Dash];
CSTPlot[Tan,Auto];   CSTPlot[Cot,Dash];


(* Cosine, Sine, & Tangent functions: Numerical *)
CosTable = N[Table[
  {x,y,N[Re[Cos[x+I*y]]]+I*N[Im[Cos[x+I*y]]]},
  {x,-1,1,2},{y,-1,1,2}],10]
SinTable = N[Table[
  {x,y,N[Re[Sin[x+I*y]]]+I*N[Im[Sin[x+I*y]]]},
  {x,-1,1,2},{y,-1,1,2}],10]
TanTable = N[Table[
  {x,y,N[Re[Tan[x+I*y]]]+I*N[Im[Tan[x+I*y]]]},
  {x,-1,1,2},{y,-1,1,2}],10]
```

Two abbreviations.

Plot curves of cosine, sine, tangent (solid) and reciprocals (dashed), as in Figure 4.2.2.

Numerical values to 10 digits of cosine, sine, and tangent for complex arguments, Table 4.2.1.

The `InverseCircular` cell [4.2.2]

Execute this cell to produce Figure 4.2.3 for surfaces of the principal values of the arccosine, arcsine, and arctangent as functions of a complex variable. Also, curves of these functions and their reciprocals for real arguments are produced, as in Figure 4.2.4.

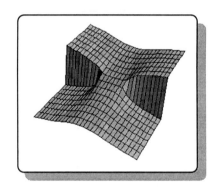

Exploring with `InverseCircular`

By extending the y range in the surface plots you will discover the larger range of variation that occurs as the y range is increased. On the other hand, by focusing on x and y ranges near the discontinuities (Table 4.2.2), you can investigate the singularity structure of the inverse circular functions in more detail.

```
(* InverseCircular cell *)

(* Inverse Circular Functions:
   arccos, arcsin, arctangent *)


Off[Plot::plnr,General::spell1,General::stop];
(* suppresses irrelevant error messages *)
```

Turn off nuisance messages.

```
(* Inverse Circular Functions
   of Complex Variables in 3D *)
InvSurfPlot[CSTinv_,ReOrIm_] := Plot3D[
 ReOrIm[CSTinv[x+I*y]],{x,-4,4},{y,-1,1},
 PlotRange->All,PlotPoints->{40,20},
 Ticks->{Automatic,Automatic,None},
 AxesLabel->{"x"," y",""}]

(* Inverse Circular Functions in 3D *)
InvSurfPlot[ArcCos,Re]; InvSurfPlot[ArcCos,Im];
InvSurfPlot[ArcSin,Re]; InvSurfPlot[ArcSin,Im];
InvSurfPlot[ArcTan,Re]; InvSurfPlot[ArcTan,Im];

(* Inverse Circular Functions in 2D, x-axis views *)
InvCSTPlot[CSTinv_,AutoDash_,Range_] :=
 Plot[CSTinv[x],{x,-Range,Range},
 PlotPoints->100,AxesLabel->{"x",""},
 Ticks->{{-4,-3,-2,-1,0,1,2,3,4},None},
 PlotRange->All,PlotStyle->AutoDash]

Auto := Automatic
Dash := {Dashing[{0.02,0.02}]}
InvCSTPlot[ArcCos,Auto,1]; InvCSTPlot[ArcSec,Dash,4];
InvCSTPlot[ArcSin,Auto,1]; InvCSTPlot[ArcCsc,Dash,4];

InvCSTPlot[ArcTan,Auto,4]; InvCSTPlot[ArcCot,Dash,4];

(* Inverse Circular Functions: Numerical *)
ArcCosTable = N[Table[
 {x,y,N[Re[ArcCos[x+I*y]]]+I*N[Im[ArcCos[x+I*y]]]},
 {x,-4,4,8},{y,-1,1,2}],10]
ArcSinTable = N[Table[
 {x,y,N[Re[ArcSin[x+I*y]]]+I*N[Im[ArcSin[x+I*y]]]},
 {x,-4,4,8},{y,-1,1,2}],10]
ArcTanTable = N[Table[
 {x,y,N[Re[ArcTan[x+I*y]]]+I*N[Im[ArcTan[x+I*y]]]},
 {x,-4,4,8},{y,-1,1,2}],10]
```

Define function for surface of inverse circular functions, real or imaginary part.

Show surfaces of inverse cosine, sine, and tangent, Figure 4.2.3.

Define curves of inverse circular functions versus x.

Abbreviate.

Show curves in Figure 4.2.4, with reciprocal functions dashed.

Numerical values to 10 digits of inverse cosin, sine, and tangent, as in Table 4.2.3.

The Hyperbolic cell [4.3.1]

Execute this cell to produce Figure 4.3.1 (surfaces) and Figure 4.3.2 (curves) for hyperbolic functions of $z = x + iy$. For the surfaces x is in $[-4,4]$ and y is in $[-1,1]$. For the curves the function values must be in $[-2,2]$. Numerical values (Table 4.3.1) are also generated.

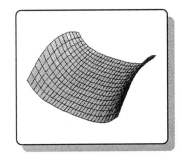

Exploring with Hyperbolic

The surfaces of the imaginary parts of the hyperbolic cosine and sine in Figure 4.3.1 are quite flat because of the small range of y used. Extend the y range in these plots to reveal oscillations that occur as the y range is increased. To investigate the hyperbolic tangent in detail, shrink the x range around the origin then vary y.

```
(* Hyperbolic cell *)


(* Hyperbolic functions: cosh, sinh, tanh  *)


Off[Plot::plnr,General::spell1,General::stop];
(* suppresses irrelevant error messages *)


(* Hyperbolic Functions of Complex Variables in 3D *)
HyperPlot[CSTh_,ReOrIm_] := Plot3D[
 ReOrIm[CSTh[x+I*y]],{x,-4,4},{y,-1,1},
 PlotRange->All,PlotPoints->{40,20},
 Ticks->{Automatic,Automatic,None},
 AxesLabel->{"x"," y",""}]


HyperPlot[Cosh,Re];  HyperPlot[Cosh,Im];
HyperPlot[Sinh,Re];  HyperPlot[Sinh,Im];
HyperPlot[Tanh,Re];  HyperPlot[Tanh,Im];


(* Hyperbolic Functions in 2D, x-axis views *)
HypCSTPlot[CSThorRec_,AutoDash_] :=
 Plot[CSThorRec[x],
 {x,-2,2},PlotPoints->100,AxesLabel->{"x",""},
 Ticks->{{-2,-1,0,1,2},None},
 PlotRange->{-2,2},PlotStyle->AutoDash]


Auto := Automatic
Dash := {Dashing[{0.02,0.02}]}


HypCSTPlot[Cosh,Auto];  HypCSTPlot[Sech,Dash];
HypCSTPlot[Sinh,Auto];  HypCSTPlot[Csch,Dash];
HypCSTPlot[Tanh,Auto];  HypCSTPlot[Coth,Dash];


(* Cosh, Sinh, & Tanh functions: Numerical *)


CoshTable = N[Table[
 {x,y,N[Re[Cosh[x+I*y]]]+I*N[Im[Cosh[x+I*y]]]},
 {x,-1,1,2},{y,-1,1,2}],10]


SinhTable = N[Table[
 {x,y,N[Re[Sinh[x+I*y]]]+I*N[Im[Sinh[x+I*y]]]},
 {x,-1,1,2},{y,-1,1,2}],10]


TanhTable = N[Table[
 {x,y,N[Re[Tanh[x+I*y]]]+I*N[Im[Tanh[x+I*y]]]},
 {x,-1,1,2},{y,-1,1,2}],10]
```

Define function for surface of hyperbolic function, real or imaginary part.

Show cosh, sinh, tanh, as in Figure 4.3.1.

Define curves of hyperbolic functions, aolid or dashed.

Abbreviations.

Show cosh, sinh, tanh (solid) and sech, csch, coth, as in Figure 4.3.2.

Numerical values to 10 digits of cosh, sinh, tanh for complex arguments (x, y), as in Table 4.3.1.

The `InverseHyperbolic` cell [4.3.2]
Execute this cell to produce the graphics for
Figures 4.3.3 and 4.3.4 for the principal value
of the inverse hyperbolic functions, arccosh,
arcsinh, and arctanh for $z = x + iy$ with x in
$[-4, 4]$ and y in $[-1, 1]$. Numerical values for
Table 4.3.3 will also be produced.

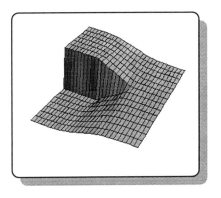

Exploring with `InverseHyperbolic`
The most interesting features of the surfaces
of the inverse hyperbolic functions in Figure 4.3.3 are the branch-cut discontinuities,
which are specified in Table 4.3.2. Explore
the neighborhoods of these discontinuities in
detail by zooming in on the x and y values near the branch cuts. You can check out this
behavior for the C functions `CoshZ`, `SinhZ`, and `TanhZ`.

```
(* InverseHyperbolic cell *)


(* Inverse Hyperbolic Functions:
   arccosh, arcsinh, arctanh *)

Off[Plot::plnr,General::spell1,General::stop];
(* suppresses irrelevant error messages *)


(* Hyperbolic Functions of Complex Variables in 3D *)
InvHypPlot[CSThinv_,ReOrIm_] := Plot3D[
  ReOrIm[CSThinv[x+I*y]],{x,-4,4},{y,-1,1},
  PlotRange->All,PlotPoints->{40,20},
  Ticks->{Automatic,Automatic,None},
  AxesLabel->{"x"," y",""}]


(* Inverse Hyperbolic Functions in 3D *)
InvHypPlot[ArcCosh,Re];   InvHypPlot[ArcCosh,Im];
InvHypPlot[ArcSinh,Re];   InvHypPlot[ArcSinh,Im];
InvHypPlot[ArcTanh,Re];   InvHypPlot[ArcTanh,Im];


(* Inverse Hyperbolic Functions in 2D:
   x-axis views *)
InvHypCSTPlot[CSThinv_,AutoDash_,Range_] :=
  Plot[CSThinv[x],{x,-Range,Range},
  PlotPoints->100,AxesLabel->{"x",""},
  Ticks->{{-4,-3,-2,-1,0,1,2,3,4},None},
  PlotRange->{-2,2},PlotStyle->AutoDash]

Auto := Automatic
Dash := {Dashing[{0.02,0.02}]}
```

Turn off error
messages.

Define function for
surface of invese
hyperbolic function,
real or imaginary.

Show surfaces of
inverse cosh, sinh,
tanh, as in Figure
4.3.3.

Define curves of
inverse functions of
real variable x.

Abbreviations.

```
(* InverseHyperbolic cell continued *)


InvHypCSTPlot[ArcCosh,Auto,4];
InvHypCSTPlot[ArcSech,Dash,4];


InvHypCSTPlot[ArcSinh,Auto,4];
InvHypCSTPlot[ArcCsch,Dash,4];


InvHypCSTPlot[ArcTanh,Auto,4];
InvHypCSTPlot[ArcCoth,Dash,4];


(* Inverse Hyperbolic Functions: Numerical *)
ArcCoshTable = N[Table[
  {x,y,N[Re[ArcCosh[x+I*y]]]+I*N[Im[ArcCosh[x+I*y]]]},
  {x,-4,4,8},{y,-1,1,2}],10]

ArcSinhTable = N[Table[
  {x,y,N[Re[ArcSinh[x+I*y]]]+I*N[Im[ArcSinh[x+I*y]]]},
  {x,-4,4,8},{y,-1,1,2}],10]

ArcTanhTable = N[Table[
  {x,y,N[Re[ArcTanh[x+I*y]]]+I*N[Im[ArcTanh[x+I*y]]]},
  {x,-4,4,8},{y,-1,1,2}],10]
```

Show curves of inverse functions, solid for inverse cosh, sinh, tanh, dashed for inversesech, csch, and coth.

Numerical values to 10 digits of inverse cosh, sinh, tanh, for complex arguments (x, y), Table 4.3.3.

20.5   EXPONENTIAL INTEGRALS AND RELATED FUNCTIONS

Notebook and Cells
Notebook `ExpInt`; cells `ExpIntFirst`, `ExpIntSecond`, `LogIntegral`, `Cos&-SinIntegral`

Overview
The *Mathematica* notebook for the exponential integrals of first and second kinds (Section 5.1) and related functions—the logarithmic integrals (Section 5.1.3) and the cosine and sine integrals (Section 5.2)—is simple to use. It provides 3D and 2D views of the functions, the parametric form of the Sici spiral (Section 5.2) shown at

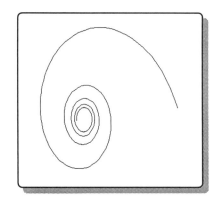

right, as well as numerical values of the integrals, to be used with the C functions to test program correctness and accuracy.

The basic definitions of all the functions in this chapter are in terms of integrals, and some of these integrals have singular regions (Figures 5.1.4 and 5.1.7). Alternative expressions in terms of series expansions are therefore used in the C functions. Cell `ExpIntSecond` in notebook `Expint` enables these series to be explored through visualizations, as in Figure 5.1.6.

The `ExpIntFirst` cell [5.1.1]

Executing this cell produces Figures 5.1.1–5.1.3
for values of the parameter n in the defining integral (5.1.1) between $n = 1$ and $n = 4$. Numerical tables are generated for $n = 1$ and $n = 3$.

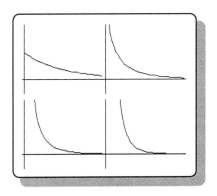

Exploring with `ExpIntFirst`

Adventuresome users will push on to larger values of n and x. Since the integrands become
singular as x approaches zero (especially for
larger n), a combination of small x and large n
challenges the robustness of the *Mathematica*
functions.

It is also interesting to plot surfaces of the modified exponential integral of the first
kind, $E'_n(z)$ defined analogously to (5.1.3). This needs only a simple modification of the
function `ExpIntFirstPlot` defined at the top of the cell.

```
(* ExpIntFirst cell *)


(* Exponential Integrals of First Kind in 3D *)
ExpIntFirstPlot[n_,ReOrIm_] := Plot3D[
  ReOrIm[ExpIntegralE[n,x+I*y]],
  {x,0.001,2},{y,0,0.5},
  PlotRange->All,PlotPoints->{30,10},
  Ticks->{Automatic,Automatic,None},
  AxesLabel->{"x"," y",""},
  DisplayFunction->Identity]


ExpIntFirstComplexPlot[n_] :=
  Show[GraphicsArray[{ExpIntFirstPlot[n,Re],
  ExpIntFirstPlot[n,Im]}],
  DisplayFunction->$DisplayFunction]


ExpIntFirstComplexPlot[1];

ExpIntFirstComplexPlot[4];


(* Modified Exponential Integrals
   of First Kind in 2D   *)
Cnvrt[n_] := ToString[n]

ExpIntFirstPlot[n_] := Plot[
  ExpIntegralE[n,x]/x^(n-1),{x,0.002,2},
  PlotRange->{-0.5,2.0},AxesLabel->
  {"x","E"<>Cnvrt[n]<>"(x)/x^"<>Cnvrt[n-1]<>""},
  DisplayFunction->Identity];
```

Define function for surface
of exponential integral of
first kind, order n. Start at
$x = 0.001$, away from singularity of integrand.

Show real and imaginary
surfaces side by side.

Surfaces for $n = 1$.

Surfaces for $n = 4$.

Brevity is the soul of wit.

Define function for curves of
modified surface integrals
(5.1.3), with axis labels.
Start at $x = 0.002$, away from
singularity of integrand.

```
(* ExpIntFirst cell continued *)

Show[GraphicsArray[
  {{ExpIntFirstPlot[0],ExpIntFirstPlot[1]},
   {ExpIntFirstPlot[2],ExpIntFirstPlot[3]}},
  GraphicsSpacing->{0,0.1}],
  DisplayFunction->$DisplayFunction];

(* Exponential Integrals
   of First Kind: Numerical *)
ExpIntFirstNum[n_] := N[Table[
  {x,ExpIntegralE[n,x]},{x,0.5,2,0.5}],10]
n = 1
ExpIntFirstNum[n]
n = 3
ExpIntFirstNum[n]
```

Curves are stacked in pairs for $n = 0$ and 1, then for $n = 2$ and 3, Figure 5.1.3.

Numerical values to 10 digits in Table 5.1.1 for $n = 1$ and $n = 3$.

The ExpIntSecond cell [5.1.2]

By executing this cell you will produce the graphics in Figure 5.1.6 for terms of the power series (5.1.10) and asymptotic series (5.1.11). Numerical values in Table 5.1.2 are also generated.

Exploring with ExpIntSecond

To explore further with ExpIntSecond, derive an improved empirical relation between the convergence parameter eps and the switchover x value for use in the C function ExpIntSecond, as discussed below (5.1.11).

The root of $Ei(x)$ that occurs near $x = 0.3725$ (Figure 5.1.5) can be refined, either by manual cut-and-try search or by using the *Mathematica* function FindRoot with the search confined to a small range about 0.3725.

```
(* ExpIntSecond cell *)

(* Exponential Integral of Second Kind *)

(* Integrand Contours in the Complex Plane *)
ContourPlot[Abs[Exp[t+I*u]/(t+I*u)],{t,-2,2},
  {u,-1,1},
  PlotPoints->60,Contours->7,ContourShading->False,
  ContourSmoothing->Automatic,AspectRatio->Automatic];

(* Exponential Integral of Second Kind: 2D Plot *)
ExpIntSecondPlot = Plot[
ExpIntegralEi[x],{x,0.001,4},
  PlotPoints->50,
  AxesLabel->{"x","Ei(x)"}];
```

Contours of integrand in Figure 5.1.4.

Curve in Figure 5.1.5 for function defined by (5.1.9).

```
(* Exponential Integral of Second Kind: Series *)
(*  Terms in Power Series *)
EiPowerTerms[x_] := Table[
  {m,Log[10,x^m/(m*(m!))]}],{m,1,50}]


EiPowPic[x_] := ListPlot[EiPowerTerms[x],
  AxesOrigin->{0,0},PlotJoined->True,
  PlotStyle->{AbsoluteThickness[0.1*x]},
  DisplayFunction->Identity]


Show[EiPowPic[10],EiPowPic[15],
  DisplayFunction->$DisplayFunction];


(*  Terms in Asymptotic Series *)
EiAsymTerms[x_] := Table[
  {m,Log[10,m!/(x^m)]}],{m,0,25}]


EiAsymPic[x_] := ListPlot[EiAsymTerms[x],
  AxesOrigin->{0,0},PlotJoined->True,
  PlotStyle->{AbsoluteThickness[0.1*x]},
  DisplayFunction->Identity]


Show[EiAsymPic[15],EiAsymPic[20],
  DisplayFunction->$DisplayFunction];


(* Exponential Integrals of Second Kind: Numerical *)
ExpIntSecondSmall = N[Table[
  {x,N[ExpIntegralEi[x]]},{x,0.1,0.5,0.1}],10]
ExpIntSecondBig = N[Table[
  {x,N[ExpIntegralEi[x]]},{x,1,5,1}],10]
```

Terms in the power series (5.1.10).

Power-series curve for given x as a plot in m.

Show curves on left side of Figure 5.1.6 for $x = 10$ and 15.

Terms in the asymptotic series (5.1.11).

Asymptotic-series curve for given x as a plot in m.

Show curves on right side of Figure 5.1.6 for $x = 15$ and 20.

Numerical values to 10 digits in tables for small then large x values.

The `LogIntegral` cell [5.1.3]

Executing this cell produces Figure 5.1.8 and the numerical values in Table 5.1.3. Note that we regard the logarithmic integral as the exponential integral of the second kind (Section 5.1.2) with transformed variable.

Exploring with `LogIntegral`

Explorations discussed for cell `ExpIntSecond` are also appropriate for cell `LogIntegral`, since it is related by the transformation (5.1.13).

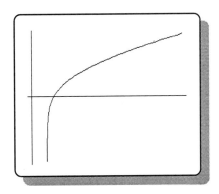

The annotated listing of cell `LogIntegral` is given on the following page.

```
(* LogIntegral cell *)

(* Logarithmic Integral *)

LogIntPlot = Plot[
  LogIntegral[x],{x,1.001,10},
  PlotPoints->50,
  AxesLabel->{" x","li(x)"}];

(* Logarithmic Integrals: Numerical *)
LogIntTable = N[Table[
  {x,N[LogIntegral[x]]},{x,2,10,2}],10]
```

Curve in Figure 5.1.5 for the function defined by (5.1.12).

Numerical values to 10 digits, as in Table 5.1.3.

The Cos&SinIntegral cell [5.2]
Execute this cell to produce Figures 5.2.1–5.2.3, the Sici spiral Figure 5.2.4, and numerical values of cosine and sine integrals (Tables 5.2.1, 5.2.2).

Exploring with Cos&SinIntegral
Optimization of the rational-fraction approximations, (5.2.9) and (5.2.10), is worth exploring if you make extensive use of C functions CosIntegral or SinIntegral.

It is also interesting to investigate the shape of the Sici spiral as a function of the range of the parameter *t* in (5.2.4).

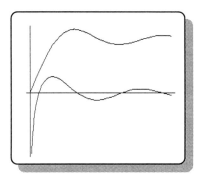

```
(* Cos&SinIntegral cell *)

(* Cosine & Sine Integrals *)
CosPlot[ReOrIm_] := Plot3D[
  ReOrIm[CosIntegral[x+I*y]],
  {x,0.1,10},{y,0,0.5},PlotRange->All,
  PlotPoints->{20,10},
  Ticks->{Automatic,Automatic,None},
  AxesLabel->{"x"," y",""}]

CosPlot[Re];
CosPlot[Im];

SinPlot[ReOrIm_] := Plot3D[
  ReOrIm[SinIntegral[x+I*y]],
  {x,0,10},{y,0,0.5},PlotRange->All,
  PlotPoints->{20,10},
  Ticks->{Automatic,Automatic,None},
  AxesLabel->{"x"," y",""}];
```

Define real and imaginary parts of surfaces for cosine integral. Start at *x* = 0.1 to avoid singularity.

Surfaces in Figure 5.2.1 for function (5.2.1).

Define real and imaginary parts of surfaces for sine integral.

```
SinPlot[Re];
SinPlot[Im];
(* Integrals in 2D; fronts of the 3D plots *)
CosIntPlot = Plot[CosIntegral[x],{x,0.1,10},
 DisplayFunction->Identity];
SinIntPlot = Plot[SinIntegral[x],{x,0,10},
 DisplayFunction->Identity];
CosSinPlot = Show[CosIntPlot,SinIntPlot,
 AxesLabel->{"x",""},
 DisplayFunction->$DisplayFunction];


(* Cos & Sin Integrals: Numerical *)
 CosNum = Table[
  {x,N[CosIntegral[x],10]},{x,0.5,2.0,0.5}]
 CosNum =Table[
  {x,N[CosIntegral[x],10]},{x,5,20,5}]
 SinNum = Table[
  {x,N[SinIntegral[x],10]},{x,0.5,2.0,0.5}]
 SinNum = Table[
  {x,N[SinIntegral[x],10]},{x,5,20,5}]
(*  Sici Spiral *)
Sici =
ParametricPlot[{CosIntegral[t],SinIntegral[t]},
 {t,2,30},PlotPoints->100,
 AspectRatio->1,Ticks->None];
```

Surfaces in Figure 5.2.2 for function (5.2.2).

Lower curve in Figure 5.2.3.
Upper curve in Figure 5.2.3.
Show curves together.

Numerical values to 10 digits of cosine and sine integrals, Tables 5.2.1 and 5.2.2.

Sici spiral with *t* in range 2 to 30, Figure 5.2.4.

20.6   GAMMA AND BETA FUNCTIONS

Notebook and Cells

Notebook GmBt; cells GammaFunction, BetaFunction, PsiFunction, PolygammaFunction, IncompleteGammaFunction, IncompleteBetaFunction

Overview

This *Mathematica* notebook for the gamma and beta function (Section 6.1), the psi (digamma) and polygamma functions (Section 6.2), and the incomplete gamma and beta functions (Section 6.3), will reward users of the *Atlas* who like to explore mountains, such as the rocky landscape of the Γ function in the complex plane,

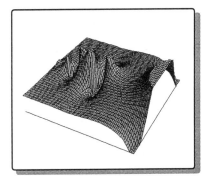

whose real part is shown in the insert. Notebook GmBt provides 3D and 2D views of the functions, as well as numerical values of the integrals that may be used with the C functions to test program correctness.

Given the appropriate choice of algorithms for real variables in the above functions, the C functions are robust and accurate. It is therefore not appropriate to show the substructure of the various algorithms used (power series, asymptotic series, and continued fractions. If you wish to go beyond the parameter ranges that we indicate for these functions, it would be worthwhile for you to modify some of the following cells to investigate the algorithms in more detail before modifying the corresponding C functions.

The `GammaFunction` cell [6.1.1]

By executing this cell you will get Figures 6.1.1–6.1.3 for the Γ function, as surfaces and contours in the complex plane, or as curves along the *x* and *y* axes. Numerical values (Table 6.1.1) for testing the Γ function for real arguments are also generated.

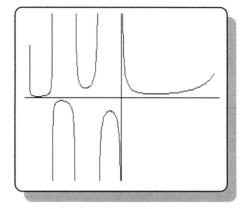

Exploring with `GammaFunction`

When drawing Figures 6.1.1 and 6.1.3, we chose slices of the Γ function along the axes. By modifying the function `GammaPlot` below, you can readily look at slices along other directions in the complex plane. As an extension of the contour plot shown in Figure 6.1.2, try plotting the phase of Γ(*z*) over the same region of the complex plane.

```
(* GammaFunction cell *)


(* Gamma Function in 3D *)
GammaPlot[ReOrIm_] := Plot3D[ReOrIm[Gamma[x+I*y]],
  {x,-4,4},{y,0.0001,1},PlotPoints->{80,20},
  Ticks->{Automatic,None,None},
  AxesEdge->{{-1,-1},{1,-1},{-1,-1}},
  PlotRange->{-6,6},ViewPoint->{0.0,-3.0,0.5},
  DisplayFunction->Identity]


RealPlot = Show[GammaPlot[Re],
  DisplayFunction->$DisplayFunction];
ImaginaryPlot = Show[GammaPlot[Im],
  DisplayFunction->$DisplayFunction];


(* Gamma Function contour plot; log scale *)
GammaAbsContour =
ContourPlot[Log[Abs[Gamma[x+I*y]]],
  {x,-4,4},{y,-2,2},PlotRange->{0,3},
  PlotPoints->{120,60},AspectRatio->Automatic,
  Contours->15,ContourShading->False,
  ContourSmoothing->Automatic];
```

Define function for the real or imaginary part of the surface of the Γ function, starting just above the real axis.

Show real surface, then show imaginary surface, as in Figure 6.1.1.

Contour plot of log of absolute value of the Γ function in the complex plane, Figure 6.1.2.

```
(* Gamma Function in 2D *)
(* Real-axis plot; poles at integers < 0  *)
GammaXXPlotRe = Plot[Re[Gamma[x]],
  {x,-4,4},PlotPoints->80,Ticks->{Automatic,None},
  AxesLabel->{"x",""}];
```

Curves of the Γ function along the *x* axis, Figure 6.1.3.

```
(* Real-axis plot; reciprocal of gamma *)
RecGammaXXPlotRe = Plot[1/Re[Gamma[x]],
  {x,-4,4},PlotPoints->80,
  Ticks->{Automatic,None},
  AxesLabel->{"x",""},
  PlotStyle->Dashing[{0.02,0.02}]];
```

Curves of 1/Γ along the *x* axis, Figure 6.1.3.

```
GammaYYPlotRe = Plot[Re[Gamma[I*y]],
  {y,-2,2},PlotPoints->60,Ticks->Automatic,
  AxesLabel->{"y",""}];
```

Curves of the ReΓ function along the *y* axis, Figure 6.1.3.

```
GammaYYPlotIm = Plot[Im[Gamma[I*y]],
  {y,-2,2},PlotPoints->60,Ticks->Automatic,
  AxesLabel->{"y",""}];
```

Curves of the ImΓ function along the *y* axis, Figure 6.1.3.

```
(* Gamma Function: Numerical *)
  GammaTable1 = Table[{x,N[Gamma[x],10]},
    {x,0.5,2,0.5}]
  GammaTable2 = Table[{x,N[Gamma[x],10]},
    {x,5,20,5}]
```

Numerical values to 10 digits of Γ along *x* axis, Table 6.1.1.

The BetaFunction cell [6.1.2]

By executing this cell you will produce Figures 6.1.4 and 6.1.5 for the *B* function, as surfaces and contours in the complex plane, or as curves along the *w* axis of $B(x,w)$ in Figure 6.1.4. Numerical values for Table 6.1.2 are also produced.

Exploring with BetaFunction

When drawing Figures 6.1.4 and 6.1.5, we chose the arguments of the *B* function as real. With some trepidation, you might try visualizing the beta function throughout the complex plane, in terms of the complex

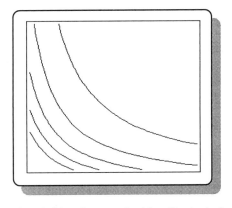

arguments *z* and *w*. Since you now have four real variables, how to do this effectively is not so clear. As an extension of the contour plot shown in Figure 6.1.5, draw the phase of *B* over the same region of the complex plane.

The annotated listing of the BetaFunction cell is given on the following page.

```
(* BetaFunction cell *)

(* Beta Function with real arguments in 3D *)
Beta3DPlot = Plot3D[Beta[x,w],
  {x,1,4},{w,1,4},PlotPoints->{20,20},
  PlotRange->{0,1},
  Ticks->{Automatic,Automatic,None},
  AxesEdge->{{-1,-1},{1,-1},{-1,-1}},
  AxesLabel->{"x","w",""}];

(* Beta Function contour plot *)
BetaContour = ContourPlot[Beta[x,w],
  {x,1,4},{w,1,4},PlotRange->Automatic,
  PlotPoints->{20,20},AspectRatio->1,
  Contours->5,ContourShading->False,
  ContourSmoothing->Automatic];

(* Beta Function with real arguments in 2D *)
Beta2D[x_] := Plot[Beta[x,w],
  {w,1,4},PlotPoints->20,PlotRange->{0,1},
  Ticks->{Automatic,Automatic},
  AxesLabel->{"w","B(x,w)"},
  DisplayFunction->Identity]

Beta2D13 = Show[Beta2D[1],Beta2D[3],
  DisplayFunction->$DisplayFunction];

(* Beta Function: Numerical *)
Beta10[x_,w_] := N[Beta[x,w],10]
Beta10[0.5,2.0]
Beta10[2.0,5.0]
Beta10[2.5,7.5]
Beta10[3.0,11.0]
```

Define graphics object for the *B* function surface with real arguments *x* and *w*, Figure 6.1.4 left.

Contours of the *B* function with real arguments *x* and *w*, Figure 6.1.5.

Define *B* function curve for fixed argument *x*, Figure 6.1.4 right.

Show curves for *x* = 1 and 3.

Numerical values to 10 digits of test values in Table 6.1.2.

The `PsiFunction` cell [6.2.1]

Executing this cell will produce Figures 6.2.1–6.2.3 for the psi (digamma) function, as surfaces and contours in the complex plane, or as curves along the *x* axis. Numerical values of $\psi(x)$ are produced for Table 6.2.3. Since ψ is the simplest of the derivatives of the Γ function (Table 6.2.1), if you understand where you are going with it then you will be well on your way to understanding the polygamma functions described in Section 6.2.2.

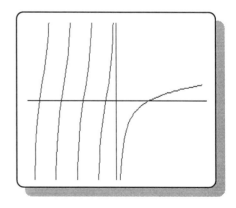

Exploring with `PsiFunction`

Go beyond the contour plot of the logarithm of the absolute value of $\psi(z)$ shown in Figure 6.2.2 to draw the phase of $\psi(z)$ over the same region of the complex plane.

```
(* PsiFunction cell *)

(* Psi Function in 3D; Real *)

RePsiPlot = Plot3D[Re[PolyGamma[x+I*y]],
  {x,-4,4},{y,0.0001,1},PlotPoints->{80,20},
  PlotRange->{-6,6},BoxRatios->{1,1,0.4},
  Ticks->{Automatic,None,None},
  AxesEdge->{{-1,-1},{1,-1},{-1,-1}},
  ViewPoint->{0.0,-3.0,0.5}];
```

Define graphics object for Reψ function surface just above the x axis, top of Figure 6.2.1.

```
(* Psi Function in 3D; Im is >= 0 *)

ImPsiPlot = Plot3D[Im[PolyGamma[x+I*y]],
  {x,-4,4},{y,0.0001,1},PlotPoints->{80,20},
  PlotRange->{0,6},BoxRatios->{1,1,0.2},
  Ticks->{Automatic,None,None},
  AxesEdge->{{-1,-1},{1,-1},{-1,-1}},
  ViewPoint->{0.0,-3.0,0.5}];
```

Define graphics object for Imψ function surface just above the x axis, bottom of Figure 6.2.1.

```
(* Psi Function contour plot; log scale *)
PsiAbsContour = ContourPlot[
  Log[Abs[PolyGamma[x+I*y]]],
  {x,-4,4},{y,-2,2},PlotRange->{0,3},
  PlotPoints->{120,60},AspectRatio->Automatic,
  Contours->15,ContourShading->False,
  ContourSmoothing->Automatic];
```

Contour plot for log of magnitude of ψ in complex plane, Figure 6.2.2.

```
(* Psi Function in 2D *)

(* Real-axis plot; poles at integers < 0  *)
PsiXXPlotRe = Plot[Re[PolyGamma[x]],
  {x,-4,4},PlotRange->{-6,6},
  PlotPoints->80,Ticks->{Automatic,None},
  AxesLabel->{"x",""}];
```

Curves of $\psi(x)$ along the real axis, including values near the poles, Figure 6.2.3 top.

```
(* Real-axis plot; reciprocal of psi *)

RecPsiXXPlotRe = Plot[1/Re[PolyGamma[x]],
  {x,-4,4},PlotRange->{-6,6},
  PlotPoints->80,Ticks->{Automatic,None},
  AxesLabel->{"x",""},
  PlotStyle->Dashing[{0.02,0.02}]];
```

Curves of $1/\psi(x)$ along the real axis, Figure 6.2.3 bottom.

```
(* Psi Function: Numerical *)
PsiCheck[x_] := N[PolyGamma[x],10]
PsiCheck[0.5]
PsiCheck[1.5]
PsiCheck[2.0]
PsiCheck[3.0]
PsiCheck[15.0]
```

Test values to 10 digits of $\psi(x)$, Table 6.2.3.

The `PolygammaFunction` cell [6.2.2]

By executing this cell, Figure 6.2.4 for polygamma functions of order $n = 1$ to 3 will be
produced, as surfaces for real variable x in
$\psi^{(n)}(x)$, or as curves along the x axis for
$n = 1$ and $n = 3$. Numerical values of $\psi^{(n)}$
are produced for part of Table 6.2.5.

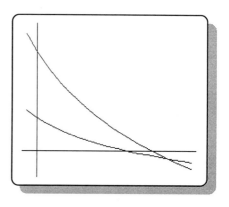

The series (6.2.10) was found to be faster
for computing $\psi^{(n)}(x)$ than the builtin *Mathematica* function `PolyGamma`. Rather
than the awkward sign reversal with n that
occurs because of definition (6.2.9), we dis-
card the sign. Thus, for z real, the function
`AbsPolyGam` at the beginning of the cell is
positive.

Algebraic expressions for $\psi^{(n)}(z)$, starting with (6.2.3) for $\psi(z)$, can be developed by
successive symbolic differentiation, as in function `PolyGam[n,z]` in the cell below.
Note that `r[k]` stands for $\varsigma(2k+1)-1$ in terms of the Riemann zeta function (Section 8.1).
The numerical value of this object decreases rapidly with k, thereby assuring rapid numer-
ical convergence of the series (6.2.3) if $|z|<1$.

Exploring with `PolygammaFunction`

Our C functions are provided only for the trigamma ($n = 1$) and tetragamma ($n = 2$) functions, (6.2.18) and (6.2.20), respectively. For higher n values, `PolyGam[n,z]` may be
converted to C code by applying *Mathematica* function `CForm`. If care is taken to code
around subtractive cancellations in the terms before the series—as discussed for using
(6.2.14)—then a rapidly convergent series may be obtained. For much larger n values, the
simple series (6.2.10) becomes practical for numerical work. For example, when $n = 5$ and
$z = 1$, part-per-million accuracy is obtained after $k = 10$. Try it and see.

```
(* PolygammaFunction cell *)


(* Polygamma: Speedup by using series *)
AbsPolyGam[n_,x_] := n!*NSum[
   (x+k)^(-(n+1)), {k,0,Infinity},
   AccuracyGoal->3, PrecisionGoal->3,
   WorkingPrecision->6]


(* Real-axis plot of absolute values
   truncated at 5  *)
PolyGamPlot = Plot3D[AbsPolyGam[n,x],
   {x,0.5,2}, {n,1,3}, PlotRange->{0,5},
   PlotPoints->{50,20},
   Ticks->{Automatic,Automatic,None},
   AxesLabel->{"x","n",""}];
```

Define polygamma function by
(6.2.10), ignoring phase.

Make graphics object for surface of polygamma function,
Figure 6.2.4 left.

```
(* Polygamma: Log display in 2D *)
PolyGam2DPlot[n_] := Plot[Log[AbsPolyGam[n,x]],
   {x,0.5,2},PlotRange->{-1,5},
  PlotPoints->20,
  Ticks->{Automatic,Automatic}]


PolyGam2DPlot[1]; (* Trigamma *)
PolyGam2DPlot[3]; (* Pentagamma *)


(* Polygamma Functions: Numerical *)
PG[n_,x_] := N[PolyGamma[n,x],10]
PolyGTable[n_] := {n,PG[n,0.5],PG[n,1.5],
 PG[n,2.0],PG[n,15.0]}


PolyGTable[1]   (* Trigamma *)
PolyGTable[2]   (* Tetragamma *)


(* Polygamma Functions by Derivatives *)
(* Define phi(z) = psi(1+z):   *)
phi[z_] := 1/(2*z) - (Pi/2)/Tan[Pi*z] -
   1/(1 - z^2) + 1 - Gamma -
   Sum[r[k]*z^(2*k),{k,1,Infinity}]


(* Its n-th derivative gets
    polygamma  n  at  1+z    *)
PolyGam[n_,z_] := D[phi[z],{z,n}]


PolyGam[2,z]
PolyGam[3,z]
```

Define curves for log of magnitude of polygamma on real axis.

Curves for $n = 1$ and for $n = 3$, Figure 6.2.4 right.

Define table of numerical values for given n.

Numerical values to 10 digits for $n = 1$ and for $n = 2$, Table 6.2.5.

Series for psi (digamma) function, (6.2.3).

Successive differentiation produces polygamma functions by using (6.2.9).

The `IncompleteGammaFunction` cell [6.3.1]

Execute this cell to produce Figures 6.3.1 and 6.3.2 for the incomplete gamma functions, $\gamma(a,x)$, then Figures 6.3.3 and 6.3.4 for regularized incomplete gamma functions, $P(a,x)$. Variable a corresponds to z in the full gamma function of Section 6.1.1 and x is the upper limit of the integral, becoming infinite to obtain the full gamma function. We restrict the visualizations and programs to real values of a because of the complexity of displays for $\gamma(a,x)$ when a, and thereby γ, is complex-valued.

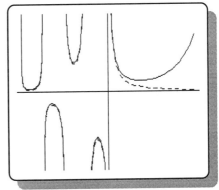

Many numerical values of $\gamma(a,x)$ and of $P(a,x)$ are produced. Some of the latter are in Table 6.3.1. Since we have chosen the a values therein as positive integers, the conversion to $\gamma(a,x)$ requires only multiplication by the factorial $(a-1)!$.

Exploring with `IncompleteGammaFunction`

Adventuresome computers of mathematical functions will explore $\gamma(a,x)$ and $P(a,x)$ for complex-valued a, as done in Section 6.1.1 for $\Gamma(z)$. When doing this, it is probably best to keep the upper limit of the integral (x) fixed, in order to avoid confusing visualizations.

```
(* IncompleteGammaFunction cell *)

(* Incomplete gamma function in 3D plots *)
IncGamma[a_,x_] := Gamma[a,0,x]
IncGamPlot = Plot3D[IncGamma[a,x],
  {a,-4,4},{x,0,10},PlotRange->{-6,6},
  PlotPoints->{80,40},
  Ticks->{Automatic,Automatic,None},
  AxesLabel->{"a","x",""}];

(* Incomplete gamma in 2D plots *)
(* As function of  a  for  x=1,5  fixed  *)
IncGam2DPlotX1 = Plot[IncGamma[a,1],
  {a,-4,4},PlotRange->{-6,6},
  PlotPoints->40,Ticks->{Automatic,None},
  AxesLabel->{"a",""},
  PlotStyle->Dashing[{0.02,0.02}],
  DisplayFunction->Identity];
IncGam2DPlotX5 = Plot[IncGamma[a,5],
  {a,-4,4},PlotRange->{-6,6},PlotPoints->40,
  Ticks->{Automatic,None},AxesLabel->{"a",""},
  DisplayFunction->Identity];
Show[IncGam2DPlotX1,IncGam2DPlotX5,
  DisplayFunction->$DisplayFunction];

(* As function of  x  for  a=1,4  fixed  *)
IncGam2DPlotAA1 = Plot[IncGamma[1,x],
  {x,0,10},PlotRange->{-6,6},
  PlotPoints->40,Ticks->{Automatic,None},
  PlotStyle->Dashing[{0.02,0.02}],
  AxesLabel->{"x",""},
  DisplayFunction->Identity];
IncGam2DPlotAA4 = Plot[IncGamma[4,x],
  {x,0,10},PlotRange->{-6,6},
  PlotPoints->40,Ticks->{Automatic,None},
  AxesLabel->{"x",""},
  DisplayFunction->Identity];
Show[IncGam2DPlotAA1,IncGam2DPlotAA4,
  DisplayFunction->$DisplayFunction];

(* Incomplete gamma function: Numerical *)
IncGamTable[x_]:=Table[{a,N[IncGamma[a,x],10]},
  {a,1,5,1}]
IncGamTable[1]
IncGamTable[5]
IncGamTable[10]
```

Incomplete gamma function

Incomplete gamma function surface as a function of real a and of x, Figure 6.3.1.

Curves for a varying and upper limit $x = 1$ in Figure 6.3.2 left.

Curves for a varying and upper limit $x = 5$ in Figure 6.3.2 left.

Show above curves superimposed.

Curves for x varying and $a = 1$ in Figure 6.3.2 right.

Curves for x varying and $a = 4$ in Figure 6.3.2 right.

Show above curves superimposed.

Numerical values to 10 digits of incomplete function defined as tables for x fixed. Make tables for $x = 1, 5, 10$.

```
(* Regularized Incomplete Gamma Function *)
RegIncGamma[a_,x_] := 1 - GammaRegularized[a,x]

RegIncGamPlot = Plot3D[RegIncGamma[a,x],
  {a,-4,4},{x,0.1,10},
  PlotPoints->{60,30},PlotRange->{0,2},
  Ticks->{Automatic,Automatic,None},
  AxesLabel->{"a","x",""}];


(* Regularized incomplete gamma in 2D plots *)
(* As function of  a  for  x=1,5  fixed  *)

RegIncGam2DPlotX1 = Plot[RegIncGamma[a,1],
  {a,-4,4},AxesLabel->{"a",""},
  PlotRange->{0,1.5},
  PlotPoints->40,Ticks->{Automatic,Automatic},
  PlotStyle->Dashing[{0.02,0.02}],
  DisplayFunction->Identity];

RegIncGam2DPlotX5 = Plot[RegIncGamma[a,5],
  {a,-4,4},AxesLabel->{"a",""},
  PlotRange->{0,1.5},
  PlotPoints->40,Ticks->{Automatic,Automatic},
  DisplayFunction->Identity];
Show[RegIncGam2DPlotX1,RegIncGam2DPlotX5,
  DisplayFunction->$DisplayFunction];

(* As function of  x  for  a=1,4  fixed  *)
RegIncGam2DPlotAA1 = Plot[RegIncGamma[1,x],
  {x,0,10},AxesLabel->{"x",""},
  PlotRange->{0,1.5},
  PlotPoints->40,Ticks->{Automatic,Automatic},
  PlotStyle->Dashing[{0.02,0.02}],
  DisplayFunction->Identity];

RegIncGam2DPlotAA4 = Plot[RegIncGamma[4,x],
  {x,0,10},AxesLabel->{"x",""},
  PlotRange->{0,1.5},
  PlotPoints->40,Ticks->{Automatic,Automatic},
  DisplayFunction->Identity];

Show[RegIncGam2DPlotAA1,RegIncGam2DPlotAA4,
  DisplayFunction->$DisplayFunction];


(* Regularized incomplete gamma function:
   Numerical *)
RegIncGamTable[x_] := Table[
                {a,x,N[RegIncGamma[a,x],10]},
                {a,1,5,1}]
RegIncGamTable[1]
RegIncGamTable[5]
RegIncGamTable[10]
```

Regularized incomplete gamma function

Regularized incomplete gamma function surface as a function of real a and of x, Figure 6.3.3.

Curves for a varying and upper limit $x = 1$ in Figure 6.3.4 left.

Curves for a varying and upper limit $x = 5$ in Figure 6.3.4 left.

Show above curves superimposed.

Curves for x varying and $a = 1$ in Figure 6.3.4 right.

Curves for x varying and $a = 4$ in Figure 6.3.4 right.

Show above curves superimposed.

Numerical values to 10 digits of regularized incomplete function defined as tables for x fixed. Make tables for $x = 1, 5, 10$.

The `IncompleteBetaFunction` cell [6.3.2]
Execute this cell to produce visualizations of the incomplete beta functions $B_x(a,b)$, as well as Figures 6.3.5 and 6.3.6 for the regularized incomplete functions, $I_x(a,b)$. Variable x is the upper limit of the integral, with $x = 1$ producing unity for the regularized function. Visualizations and programs are for real values of a and b, since the displays are confusing for complex a and b.

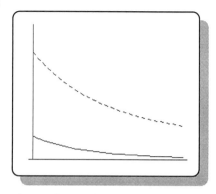

Many numerical values of $B_x(a,b)$ and of $I_x(a,b)$ are produced. Some of the latter are in Table 6.3.2. Because the a and b values therein are positive integers, conversion to $B_x(a,b)$ requires only multiplication by the factor $(a+b-1)!/[(a-1)!(b-1)!]$.

Exploring with `IncompleteBetaFunction`
Try computing $B_x(a,b)$ and $I_x(a,b)$ for complex-valued a and b. It is probably best to keep the upper limit of the integral (x) fixed, in order to avoid very confusing pictures.

```
(* IncompleteBetaFunction cell *)

(* Incomplete beta function in 3D plots *)
IncBetaPlot[x_] := Plot3D[Beta[x,a,b],
  {a,1,4},{b,1,4},PlotRange->{0,1},
  PlotPoints->{20,20},
  Ticks->{Automatic,Automatic,None},
  AxesLabel->{"a","b","B("<>ToString[x]<>",a,b)"}]

(* Incomplete beta in 2D plots *)
(* As function of  b  for  a & x  fixed  *)
IncBeta2D[a_,x_,DashIt_] := Plot[
  Beta[x,a,b],{b,1,4},PlotRange->{0,1},
  PlotPoints->30,Ticks->{Automatic,Automatic},
  AxesLabel->{"b","B("<>ToString[x]<>",a,b)"},
  PlotStyle->Dashing[{0.02,0.02*DashIt}],
  DisplayFunction->Identity]

(*  x = 0.5  *)
IncBetaPlot[0.5];
Beta1P5 = IncBeta2D[1,0.5,1]; (* dashed *)
Beta3P5 = IncBeta2D[3,0.5,0]; (* solid *)
Show[Beta1P5,Beta3P5,
  DisplayFunction->$DisplayFunction];

(*  x = 0.8  *)
IncBetaPlot[0.8];
Beta1P8 = IncBeta2D[1,0.8,1]; (* dashed *)
Beta3P8 = IncBeta2D[3,0.8,0]; (* solid *)
Show[Beta1P8,Beta3P8,
  DisplayFunction->$DisplayFunction];
```

Incomplete beta function

Define incomplete beta function surface as function of x.

Define incomplete function curves for a and upper limit x varying.

Surface for a varying and $x = 0.5$, Figure 6.3.5.
Curve for $a = 1$, $x = 0.5$, then for $a = 3$, $x = 0.5$.

Show curves together.

Surface for a varying and $x = 0.8$, Figure 6.3.5.
Curve for $a = 1$, $x = 0.8$, then for $a = 3$, $x = 0.8$.

Show curves together.

```
(* Incomplete beta function: Numerical *)
IncBetaTable[x_,a_] :=
 Table[{x,a,b,N[Beta[x,a,b],10]},{b,1,5,1}]


IncBetaTable[0.5,1]
IncBetaTable[0.5,3]
IncBetaTable[0.8,1]
IncBetaTable[0.8,3]
```

Define numerical table to
10 digits with x and a
varying.

$x = 0.5$, $a = 1$, then
$x = 0.5$, $a = 3$.
$x = 0.8$, $a = 1$, then
$x = 0.8$, $a = 3$.

```
(* Regularized Incomplete Beta Function *)
(* Regularized incomplete beta in 3D plots *)
RegIncBetaPlot[x_] :=
Plot3D[BetaRegularized[x,a,b],
 {a,1,4},{b,1,4},PlotRange->{0,1},
 PlotPoints->{20,20},
 Ticks->{Automatic,Automatic,None},
 AxesLabel->{"a","b","I("<>ToString[x]<>",a,b)"}]
```

**Regularized incomplete
beta function**

Define regularized incom-
plete beta function surface
as function of x.

```
(* Regularized incomplete beta in 2D plots *)
(* As function of  b  for  a & x  fixed  *)
RegIncBeta2D[a_,x_,DashIt_] := Plot[
 BetaRegularized[x,a,b],{b,1,4},PlotRange->{0,1},
 PlotPoints->30,Ticks->{Automatic,Automatic},
 AxesLabel->{"b","I("<>ToString[x]<>",a,b)"},
 PlotStyle->Dashing[{0.02,0.02*DashIt}],
 DisplayFunction->Identity]
```

Define regularized incom-
plete function curves for a
and upper limit x varying.

```
(*  x = 0.5  *)
RegIncBetaPlot[0.5];
RegBeta1P5 = RegIncBeta2D[1,0.5,1];(* dashed *)
RegBeta3P5 = RegIncBeta2D[3,0.5,0];(* solid *)
Show[RegBeta1P5,RegBeta3P5,
 DisplayFunction->$DisplayFunction];
```

Surface for a and b vary-
ing, $x = 0.5$, Figure 6.3.6.
Curve for $a = 1$, $x = 0.5$,
then for $a = 3$, $x = 0.5$.
Show curves together,
Figure 6.3.6 left.

```
(*  x = 0.8  *)
RegIncBetaPlot[0.8];
RegBeta1P8 = RegIncBeta2D[1,0.8,1];(* dashed *)
RegBeta3P8 = RegIncBeta2D[3,0.8,0];(* solid *)
Show[RegBeta1P8,RegBeta3P8,
 DisplayFunction->$DisplayFunction];
```

Surface for a and b vary-
ing, $x = 0.8$, Figure 6.3.6.
Curve for $a = 1$, $x = 0.8$,
then for $a = 3$, $x = 0.8$.
Show curves together,
Figure 6.3.6 right.

```
(* Regularized incomplete beta: Numerical *)
RegIncBetaTable[x_,a_] := Table[
 {x,a,b,N[BetaRegularized[x,a,b],10]},{b,1,5,1}]


RegIncBetaTable[0.5,1]
RegIncBetaTable[0.5,3]
RegIncBetaTable[0.8,1]
RegIncBetaTable[0.8,3]
```

Define numerical table to
10 digits with x and a
varying.
$x = 0.5$, $a = 1$, then
$x = 0.5$, $a = 3$.
$x = 0.8$, $a = 1$, then
$x = 0.8$, $a = 3$.

20.7   COMBINATORIAL FUNCTIONS

Notebook and Cells
Notebook `Combinatorial`; cells `Factorial`, `RisingFactorial`, `Binomial`, `Multinomial`, `StirlingNumbers`, `Fibonacci`, `Lucas`

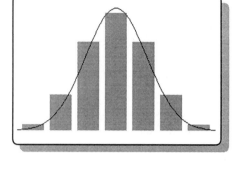

Overview
This *Mathematica* notebook for the combinatorial functions includes factorials and rising factorials (Section 7.1), binomial and multinomial coefficients (Section 7.2), Stirling numbers (Section 7.3), as well as the Fibonacci and Lucas polynomials (Section 7.4).

Calculations in this notebook produce discrete-valued functions; thus, most of the visualizations are 2- and 3-D bar charts (histograms). Notebook `Combinatorial` also produces numerical values that can be used with the C functions to test program correctness.

The `Factorial` cell [7.1.1]
By executing this cell you will get most of Figures 7.1.1 and 7.1.2 for the factorial function, $n!$, except that the labelling is omitted. Tables of approximations to Stirling's asymptotic expansion (7.1.4) are also produced. Numerical values (Table 7.1.1) for testing the factorial function are generated.

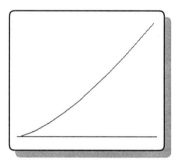

Exploring with `Factorial`
If you wish to use the Stirling asymptotic expansion (7.1.4) for large n, explore how the number of terms used should be chosen for a given accuracy. Note that—since this is an asymptotic expansion that eventually diverges (see Section 3.4)—the accuracy obtainable by using it is not arbitrarily great.

```
(* Factorial cell *)

(* Growth of factorials *)
LogFctr = N[Log[10,E]];
nmax = 10;
Log10Fctrl = Table[{n,N[Log[10,Factorial[n]]]},
               {n,1,nmax}];
StatMechStirl = Table[{n,LogFctr*N[n*Log[n/E]
  +0.5*Log[2*N[Pi]]]},{n,1,nmax}];
FctrlPlot = ListPlot[Log10Fctrl,PlotJoined->True];
StirlPlot = ListPlot[StatMechStirl];
```

Tables of $\log_{10} n!$ up to `nmax`, for the statistical-mechanics approximation, then graphic objects of these.

```
StirlForm =
  Normal[Series[LogGamma[n],{n,Infinity,8}]]
StirlThru7 = Table[{n-1,N[LogFctr*StirlForm]},
                {n,2,nmax+1}];
StirlThru7Plot = ListPlot[StirlThru7];

Show[FctrlPlot,StirlPlot,StirlThru7Plot];


(* Word-size limits for factorials *)
WordSize =
Table[{n,N[Log[2,Factorial[n]]]},{n,0,35}];
SizePlot = ListPlot[WordSize,Frame->True,
               GridLines->Automatic];


(* Factorials: Numerical *)
Fctrl[n_] := Factorial[n]
Table[{Fctrl[0],Fctrl[1],Fctrl[10],Fctrl[20],
  Fctrl[30]}]
```

First 7 terms of the Stirling series (7.1.4). Numerical table of the values, then graphics object for this.

Graphics for Figure 7.1.1.

Tables of $\log_2 n!$ up to 35, for word size at which $n!$ overflows. Curve for Figure 7.1.2.

Test values for $n!$ in Table 7.1.1.

By executing this cell you will get Figure 7.1.3 for the rising factorial function, $(n)_m$, except that its labelling is omitted.

Exploring with RisingFactorial
Follow up on the cautionary remark made under Algorithm, Program, and Programming Notes in Section 7.1.2 by exploring the ranges of n and m that produce overflow of computer words of various sizes for exact, integer values of the rising factorial $(n)_m$.

If approximate values of $(n)_m$ are suitable for your applications, develop and test the algorithms for $\ln(n)_m$ in *Mathematica*, then code them in C or another procedural language for greater efficiency.

```
(* RisingFactorial cell *)

<<Graphics`Graphics3D`

(* Rising Factorials: 3-D Bar Charts *)
RsngFctrl[n_,m_] :=
    Factorial[n+m-1]/Factorial[n-1]

RFTable[n_] := Table[RsngFctrl[n,m],{m,0,3}]
RFTable2 = RFTable[2];
RFTable4 = RFTable[4];
RFTable6 = RFTable[6];
RFTable8 = RFTable[8];
```

Need this package for BarChart3D

Define rising factorial in terms of factorials, (7.1.7).

Make tables for $n = 2,4,6,8$.

```
(* RisingFactorial cell continued *)

BarChart3D[{RFTable8,RFTable6,RFTable4,RFTable2},
  XSpacing->0.5,Boxed->False, Axes->False,
  SolidBarEdgeStyle->GrayLevel[0.5],
  ViewPoint->{2.94,-1.66,0.245}];
```

Stack bar charts for Figure 7.1.3.

The `Binomial` cell [7.2.1]

Executing this cell produces most of Figure 7.2.1 for the binomial coefficients, except that labelling is omitted. Comparing the Gaussian (normal) distribution (7.2.4) with the binomial distribution for $n = 6$ produces the graphic motif at the start of Chapter 7.

Exploring with `Binomial`

Run the barts of the cell that make the comparison of the binomial-coefficient distribution with the Gauss distribution for values of n that are both smaller and larger than our example, $n = 6$.

Calculate and plot the percentage residuals between the two functions against n. For what value of n does the Gaussian approximation become accurate enough for practical purposes in your mathematical calculations?

Follow up the remark made under Programming Notes in Section 7.2.1 by mapping out the ranges of n and m that produce overflow of computer words of various sizes for exact, integer values of the binomial coefficient. If approximate logarithms of binomial coefficients are suitable for your applications, develop and test algorithms for $\ln(n)_m$ in *Mathematica*, such as the Stirling series (7.1.4), code them in C or another procedural language for greater efficiency, then use (7.2.1) for the binomial coefficients in logarithmic form.

```
(* Binomial cell *)

<<Graphics`Graphics`
(* Binomials: Comparison with Gaussian *)
n = 6;
BinTable = Table[Binomial[n,m]/2^n,{m,0,n}];
BinChart = BarChart[BinTable,Axes->False,
  BarStyle->{GrayLevel[0.6]},BarEdges->False,
  DisplayFunction->Identity];


(* Gauss distribution *)
MidValue = 2/(Sqrt[2*N[Pi]*n]);
Gauss = Plot[MidValue*Exp[-(x-1-n/2)^2/Sqrt[n]],
  {x,0,n+2},Axes->False,DisplayFunction->Identity];


Show[BinChart,Gauss,
  DisplayFunction->$DisplayFunction];
```

Needs this package for `BarChart`.

Table of binomials, then in a bar chart.

Gaussian for comparison, (7.2.4).

Compare binomial and Gaussian.

```
<<Graphics`Graphics3D`

BinTable2 = Table[Binomial[2,m],{m,0,6}];
BinTable4 = Table[Binomial[4,m],{m,0,6}];
BinTable6 = Table[Binomial[6,m],{m,0,6}];


BarChart3D[{BinTable6,BinTable4,BinTable2},
  XSpacing->0.5,Boxed->False, Axes->False,
  SolidBarEdgeStyle->GrayLevel[0.5],
  ViewPoint->{2.94,-1.66,0.245}];
```

Needs this package for BarChart3D.

Tables of binomials for $n = 2,4,6$.

Stack bar charts for Figure 7.2.1.

The Multinomial cell [7.2.2]

Executing this cell produces most of Figure 7.2.2 for the trinomial coefficients with $n \le 6$, except that axes labelling is omitted. Numerical tables of trinomial coefficients are produced. Formula (7.2.6) is used with $m = 3$.

Exploring with Multinomial

Generalize the program in this cell so that it can produce bar charts and tables for any multinomial, rather than only for $m = 3$ used in the present cell. This is not a simple programming task in *Mathematica* and the resulting program is not likely to be clear.

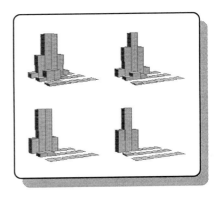

Under the C Programming Notes in Section 7.2.2 we mention the possibility of overflows, in spite of our strategy (7.2.8) that is implemented in the C function Multinomial. Map out the ranges of the m parameters for a given choice of n that produce overflow of computer words of various sizes for exact, integer values of the multinomial coefficient. Develop and test in *Mathematica* an algorithm for the logarithm of the multinomial, code this in C or another procedural language for greater efficiency, then use the tabular output part of cell Multinomial (or its generalization to m other than 3) to obtain test values.

```
(* Multinomial cell *)

<<Graphics`Graphics3D`

(* Multinomials: 3-D Bar Charts *)

Fctrl[n_] := Factorial[n]

Trinml[n_,n1_,n2_] :=
  If[ n-n1-n2 < 0, 0,
  Fctrl[n]/(Fctrl[n1]*Fctrl[n2]*Fctrl[n-n1-n2])]
TrnTable[n_,n2_] := Table[
  Trinml[n,n1,n2],{n1,0,6}]
BarTable[n2_] :=
  {TrnTable[6,n2],TrnTable[4,n2],TrnTable[2,n2]}
```

Needs this package for BarChart3D.

Abbreviation.

Define trinomial, $m = 3$ in (7.2.6).

Define table; n_1 varying.
Define table for bar chart with $n = 2, 4, 6$, and n_2 varying.

```
(* Multinomial cell continued *)

TrnChart3D[n2_] := BarChart3D[BarTable[n2],
  PlotRange->All,XSpacing->0.5,Boxed->False,
  Axes->False,SolidBarEdgeStyle->GrayLevel[0.5],
  ViewPoint->{2.94,-1.66,0.245},
  DisplayFunction->Identity]


For[ n2 = 1, n2 <= 6, n2++, TrnChart3D[n2] ];
Show[GraphicsArray[
  {{TrnChart3D[1],TrnChart3D[2],TrnChart3D[3]},
   {TrnChart3D[4],TrnChart3D[5],TrnChart3D[6]}}],
   DisplayFunction->$DisplayFunction];


(* Multinomials: Numerical *)
TrnTable4[n_,n2_] := Table[
  Trinml[n,n1,n2],{n1,0,4}]


BarTable4[n2_] :=
  {TrnTable4[1,n2],TrnTable4[2,n2],
   TrnTable4[3,n2],TrnTable4[4,n2]}


BarTable4[1]
BarTable4[2]
BarTable4[3]
BarTable4[4]
```

Define 3D bar chart for n_2 varying.

Stacked arrays of bar charts for n_2 from 1 to 6, Figure 7.2.2, for trinomials.

Define table of trinomial coefficients with n and n_2 varying, then combine the tables for the allowed n values.

Tables of binomials for n_2 from 1 to 4, Table 7.2.2.

The `StirlingNumbers` cell [7.3]

By executing this cell you can produce the 3D bar charts Figures 7.3.1 and 7.3.2 for the Stirling numbers of the first and second kind, respectively, but labeling of axes is omitted. We consider $n = 2, 4, 6$, with $m = 0 \ldots 6$. For Stirling numbers of the first kind a phase $(-1)^{n-m}$ is omitted in order that the values be nonnegative. All Stirling numbers of the second kind are nonnegative.

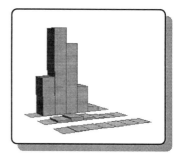

Exploring with `StirlingNumbers`

As written, the `StirlingNumbers` cell assumes that $n > 6$ will not be used, since the tables are generated with $m \leq 6$. Generalize the program so that it can produce 3D bar charts and tables for any range of the Stirling numbers.

For large n values and small word sizes for integer arithmetic, as in a C or Fortran program, the usual overflow problems of combinatorial functions will arise for the Stirling numbers. These are not easily avoided by working with logarithms of the numbers. Use the generalized cell in *Mathematica* with arbitrary-precision arithmetic to map out the ranges of n and m that produce overflow of computer words of sizes that you need.

```
(* StirlingNumbers cell *)

<<Graphics`Graphics3D`

(* Stirling Numbers: Numerical & 3-D Bar Charts *)

(* Stirling Numbers of First Kind: Magnitude *)
StrlTab[1,n_] := Table[(-1)^(n-m)*StirlingS1[n,m],
                  {m,0,6}]

(* Stirling Numbers of First Kind: Numerical *)
StrlTab[1,2]
StrlTab[1,4]
StrlTab[1,6]

(* Bar Chart *)
s = 1;
StirlPlot1 = BarChart3D[
  {StrlTab[s,6],StrlTab[s,4],StrlTab[s,2]},
  XSpacing->0.5,Boxed->False,Axes->False,
  SolidBarEdgeStyle->GrayLevel[0.5],
  ViewPoint->{2.94,-1.66,0.245}];

(* Stirling Numbers of Second Kind: Numerical *)
StrlTab[2,n_] := Table[StirlingS2[n,m],{m,0,6}]

StrlTab[2,2]
StrlTab[2,4]
StrlTab[2,6]

(* Bar Chart *)
s = 2;
StirlPlot2 = BarChart3D[
  {StrlTab[s,6],StrlTab[s,4],StrlTab[s,2]},
  XSpacing->0.5,Boxed->False,Axes->False,
  SolidBarEdgeStyle->GrayLevel[0.5],
  ViewPoint->{2.94,-1.66,0.245}];
```

Needs this package for BarChart3D.

Define table of magnitude of Stirling numbers of first kind and order n, m from 0 to 6.

Output tables for $n = 2, 4, 6$.

For Stirling numbers of first kind, stack bar charts for $n = 2, 4, 6$.

Define table of Stirling numbers of second kind and order n, m from 0 to 6.

Output tables for $n = 2, 4, 6$.

For Stirling numbers of second kind, stack bar charts for $n = 2, 4, 6$.

The Fibonacci cell [7.4.1]
By executing this cell you can produce Figure 7.4.1 for the Fibonacci polynomials and Fibonacci numbers, except that most labeling of axes is omitted. Test values of the Fibonacci polynomials in Table 7.4.1 are also generated.

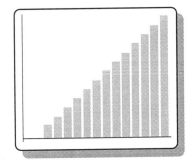

Exploring with Fibonacci
If the order of the Fibonacci polynomial, n, is large and the word size for integer arithmetic is small, as often in a C or Fortran program, the common over-flow problems of combinatorial functions will arise for the Fibonacci numbers, F_n, which are just the polynomials evaluated at $x = 1$. Use the Fibonacci cell in *Mathematica*—

whose integer arithmetic is exact—to map out the range of n that will produce overflow of computer words of sizes relevant for your needs.

   If approximate Fibonacci numbers are adequate for your needs, then you can adapt the golden-ratio formula (7.4.4) to work with logarithms of the numbers and to use floating-point approximations to F_n.

```
(* Fibonacci cell *)


<<Graphics`Graphics`
<<Graphics`Graphics3D`


(* Fibonacci Numbers *)
(* Definition by recursion *)
FibNum[n_] := FibNum[n-1]+FibNum[n-2]
FibNum[1] = 1;   FibNum[2] = 1;


FibNumbers = Table[{n,N[FibNum[n]]},{n,1,10}]


(* Fibonacci Numbers: Log-base-10 Plot *)
nmax = 15;
LogFibNum =
Table[Log[10,N[FibNum[n]]],{n,1,nmax}];
FibNumChart = BarChart[LogFibNum,
 BarStyle->{GrayLevel[0.8]},BarEdges->False,
 AxesLabel->{"n","log10[F(n)]"}];


(* Fibonacci Polynomials *)
(* Definition by recursion *)
FibPoly[n_,x_] := x*FibPoly[n-1,x]+FibPoly[n-2,x]
FibPoly[1,x_] = 1;   FibPoly[2,x_] = x;


(* Fibonacci Polynomials: Algebraic *)
nmax = 6
Clear[x];
FibPolyAlg =
Table[Expand[FibPoly[n,x]],{n,1,nmax}]


(* Fibonacci Polynomials: Numerical *)
 FibTable[x_,nmax_] := Table[
  N[FibPoly[n,x],10],{n,1,nmax}]

nmax = 10;
FibTable000 = FibTable[0.00,nmax]
FibTable025 = FibTable[0.25,nmax]
FibTable050 = FibTable[0.50,nmax]
FibTable075 = FibTable[0.75,nmax]
FibTable100 = FibTable[1.00,nmax]
```

Needs these two packages.

Define Fibonacci numbers by recursion.

Table of Fibonacci numbers, n from 1 to 10.

Logs of Fibonacci numbers, n from 1 to nmax.

Bar chart, Figure 7.4.1 lower left.

Define Fibonacci polynomials by recursion.

Table of algebraic Fibonacci polynomials through $n = 6$, (7.4.3).

Define table of numerical Fibonacci polynomials to 10 digits through $n = $ max with x fixed.

Tables for $x = 0$ (0.25) 1 for nmax = 10. Used in Table 7.4.1.

```
(*  Fibonacci Polynomials: 3D Bar Charts *)
BarChart3D[{FibTable100,FibTable075,FibTable050,
  FibTable025,FibTable000},
  XSpacing->0.5,Boxed->False, Axes->False,
  SolidBarEdgeStyle->GrayLevel[0.5],
  ViewPoint->{2.94,-1.66,0.245}];
```

Bar charts of above
Fibonacci polynomials,
Figure 7.4.1 right.

```
(*  Fibonacci Polynomials: 2D Curves *)
FibPoly2D[n_] := Plot[FibPoly[n,x],{x,0,1},
  AxesLabel->{" x","Fn(x)"},PlotRange->All,
  DisplayFunction->Identity];
```

Define curve of polynom-
ial with x varying and n
fixed.

```
Show[FibPoly2D[2],FibPoly2D[4],FibPoly2D[6],
  DisplayFunction->$DisplayFunction];
```

Show polynomials
together for $n = 2, 4, 6$.

The Lucas cell [7.4..2]

If you execute this cell you will produce the three
graphics in Figure 7.4.2 for the Lucas polynomials
and Lucas numbers, except that most labelling of
axes is omitted. Test values of Lucas polynomials
that are in Table 7.4.1 are also generated.

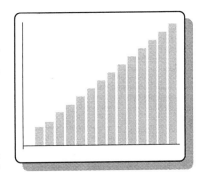

Exploring with Lucas

Like the Fibonacci polynomials in the Fibonacci
cell immediately above, if n is large and the word
size for integer arithmetic is small, as often in a C or
Fortran program, the overflow problems of combina-
torial functions will arise for the Lucas numbers, L_n, which are just the polynomials evaluated at $x = 1$. By using the Lucas cell in *Mathematica*—whose integer arithmetic is
exact—you can map out the range of n that will produce overflow of computer words of
sizes relevant for your needs.

```
(* Lucas cell *)

<<Graphics`Graphics3D`
<<Graphics`Graphics`

(* Lucas Polynomials *)

(* Definition by recursion *)
LucasPoly[n_,x_] := x*LucasPoly[n-1,x]+
                    LucasPoly[n-2,x]
LucasPoly[0,x_] = 2;
LucasPoly[1,x_] = x;
```

Needs these two packages.

Define Lucas polynomials
by recursion.

```
(* Lucas cell continued *)


(* Lucas Polynomials: Algebraic *)
nmax = 6

Clear[x];


LucasPolyAlg =
Table[Expand[LucasPoly[n,x]],{n,0,nmax}]


(* Lucas Numbers *)
LucasNumbers = Table[{n,N[LucasPoly[n,1]]},
                    {n,0,10}]


(* Lucas Numbers: Log-base-10 Plot *)
nmax = 15;

LogLucasNum =
Table[Log[10,N[LucasPoly[n,1]]],{n,1,nmax}];
LucasNumChart = BarChart[LogLucasNum,
  BarStyle->{GrayLevel[0.8]},BarEdges->False,
  AxesLabel->{"n","log10[L(n)]"}];


(* Lucas Numbers: Table *)
LucasTable[x_,nmax_] := Table[
  N[LucasPoly[n,x],10],{n,0,nmax}]


(* Lucas Polynomials: 3D Bar Charts *)
nmax = 10;
LucasTable000 = LucasTable[0.00,nmax]
LucasTable025 = LucasTable[0.25,nmax]
LucasTable050 = LucasTable[0.50,nmax]
LucasTable075 = LucasTable[0.75,nmax]
LucasTable100 = LucasTable[1.00,nmax]


BarChart3D[{LucasTable100,LucasTable075,
  LucasTable050,LucasTable025,LucasTable000},
  XSpacing->0.5,Boxed->False, Axes->False,
  SolidBarEdgeStyle->GrayLevel[0.5],
  ViewPoint->{2.94,-1.66,0.245}];


(* Lucas Polynomials: 2D Curves *)
LucasPoly2D[n_] := Plot[LucasPoly[n,x],{x,0,1},
  AxesLabel->{" x","Ln(x)"},PlotRange->{0,17},
  DisplayFunction->Identity];


Show[LucasPoly2D[2],LucasPoly2D[4],LucasPoly2D[6],
  DisplayFunction->$DisplayFunction];
```

Table of algebraic Lucas polynomials through $n = 6$, (7.4.6).

Table of Lucas numbers, n from 0 to 10.

Logs of Lucas numbers, n from 1 to nmax.

Bar chart, Figure 7.4.2 lower left.

Define table of numerical Lucas polynomials to 10 digits, $n = 0$ to max with x fixed.

Tables for $x = 0$ (0.25) 1 for nmax = 10. Used for Table 7.4.2.

Bar charts of above Lucas polynomials, Figure 7.4.2 right.

Define curve of polynomial with x varying and n fixed.

Show polynomials together for $n = 2, 4, 6$.

20.8   NUMBER THEORY FUNCTIONS

Notebook and Cells
Notebook `NumberTheory`; cells `Bernoulli`, `Euler`, `RiemannZeta`, `SumRecPower`, `PolyLogarithm`

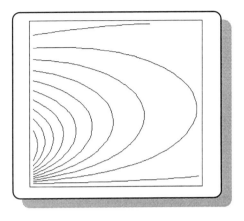

Overview
This *Mathematica* notebook for number theory functions includes Bernoulli numbers and polynomials (Section 8.1), Euler numbers and polynomials (Section 8.2), the Riemann zeta function (Section 8.3), other sums of reciprocal powers (Section 8.4), and the polylogarithms (Section 8.5).

Calculations in this notebook produce both discrete-valued functions and functions of complex variables, so that the visualizations include bar charts (histograms), 2D curves, and 3D surfaces. Notebook `NumberTheory` also produces numerical values that can be used with the C functions to test program correctness.

The `Bernoulli` cell [8.1]
By executing this cell you produce the three graphics in Figures 8.1.1 and 8.1.2 for the Bernoulli numbers and Bernoulli polynomials, except that labelling of axes and curves is omitted. Test values of Bernoulli numbers and polynomials in Tables 8.1.1 and 8.1.2 are also generated.

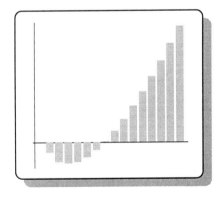

Exploring with `Bernoulli`
Like most combinatorial functions, overflow problems can arise if n is large and if the nonzero Bernoulli numbers, B_n (the polynomials evaluated at $x = 1$), are expressed exactly by rational fractions and the word size for integer arithmetic is small. This may occur in C or Fortran programs if the formulas (8.1.4) are coded in integer arithmetic. (Our C function `BernoulliNum` in Section 8.1.1 is coded in floating-point approximation.)

The rapid growth of B_n for even n with $n > 6$ is shown in Figure 8.1.2. If you need to use rational-fraction Bernoulli numbers, map out the range of n that will produce overflow of computer words of sizes relevant for your needs by extending the object `TabNmbr` in `Bernoulli` cell to a larger maximum n value. Note, as derived below (7.5.5), that $(n+1)!\,B_n$ is an integer. It is straightforward to modify the C function `BernoulliNum` to use integer arithmetic, provided that integer overflow can be accommodated.

```
(* Bernoulli cell *)

<<Graphics`Graphics`

(* Bernoulli Numbers *)
TabNmbr = Table[BernoulliB[n],{n,1,14}]
TabDecimal = N[TabNmbr,10]

(* Bernoulli Numbers: Bar chart *)
BrnllNmbrPlot = BarChart[TabNmbr,
  BarStyle->{GrayLevel[0.8]},BarEdges->False];

(* Bernoulli Numbers: Log scale bar chart *)
LogBrnll = BarChart[
  Table[Log[10,Abs[N[BernoulliB[n]]]],
  {n,0,30,2}],Ticks->{None,Automatic},
  AspectRatio->Automatic,PlotRange->All,
  BarStyle->{GrayLevel[0.8]},BarEdges->False];

(* Bernoulli Polynomials: Algebraic *)
Clear[x];
Table[BernoulliB[n,x],{n,0,5}]

(* Bernoulli Polynomials: Curves *)
BrnllPlot[n_] := Plot[BernoulliB[n,x],{x,0,1},
  DisplayFunction->Identity];

BrnllPolys = Show[
  BrnllPlot[2],BrnllPlot[3],BrnllPlot[4],BrnllPlot[5],
  DisplayFunction->$DisplayFunction];

(* Bernoulli Polynomials: Numerical *)
BrnllTable[x_] := N[Table[BernoulliB[n,x],{n,2,5}],10]
BrnllTable[0]
BrnllTable[0.5]
BrnllTable[1]
```

Table of Bernoulli numbers (10 digits) from 1 to 14, in fraction and decimal forms.

Bar chart of above numbers, Figure 8.1.1 top.

Bar chart of logs of Bernoulli numbers, Figure 8.1.2.

Table of algebraic Bernoulli polynomials through $n = 5$, as in (8.1.8).

Define curve of polynomial with x varying, n fixed. Show polynomials together for $n = 2$ to 5, Figure 8.1.1 bottom.

Numerical values of polynomials to 10 digits, Table 8.1.2.

The `Euler` cell [8.2]

Executing this cell produces the three graphics in Figures 8.2.1 and 8.2.2 for the Euler numbers and Euler polynomials, except that labelling of the axes and curves is omitted. Test values of Euler numbers and polynomials in Table 8.2.1 and 8.2.2 are also generated.

Exploring with `Euler`

As with most combinatorial functions, overflow problems can arise if n is large and if the nonzero Euler numbers, E_n—the polynomials evaluated according to (8.2.7)—are expressed exactly as integers and the word size for integer

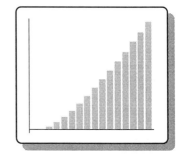

arithmetic is small. This may occur in C or Fortran programs if the formula is coded in integer arithmetic, as in our C function `EulerNum` in Section 8.2.1.

The very rapid growth of E_n for even n with $n > 6$ is shown in Figure 8.2.2. If you need Euler numbers for fairly large n, first map out the range of n that will produce overflow of computer words of sizes relevant for your needs by extending object `TabNmElr` in `Euler` cell to a larger maximum n value. Note that E_n is an integer. It is straightforward to modify the C function `EulerNum` in Section 8.2.1 to use floating-point arithmetic, thereby postponing the overflow, but at the expense of exactness.

```
(* Euler cell *)


<<Graphics`Graphics`

(* Euler Numbers: Numerical *)
TabNmElr1 = Table[EulerE[n],{n,1,4}]
TabNmElr2 = Table[EulerE[n],{n,14,16,2}]


(* Euler numbers scaled by 1/4^n *)
TabDnP2 = N[Table[EulerE[n]/(4^n),{n,1,12}],10]


(* Euler Numbers: Graphical *)
EulerNmbrPlot = BarChart[TabDnP2,
  BarStyle->{GrayLevel[0.8]},BarEdges->False];


LogEuler = BarChart[
  Table[Log[10,N[(-1)^(n/2)*EulerE[n]]],
  {n,0,30,2}],Ticks->{None,Automatic},
  BarStyle->{GrayLevel[0.8]},BarEdges->False];


(* Euler Polynomials: Algebraic *)
Table[EulerE[n,x],{n,0,5}]



(* Euler Polynomials: Graphical *)
ElrPlot[n_] := Plot[EulerE[n,x],{x,0,1},
  DisplayFunction->Identity]


ElrPolys = Show[ElrPlot[2],ElrPlot[3],
  ElrPlot[4],ElrPlot[5],
  DisplayFunction->$DisplayFunction];


(* Euler Polynomials: Numerical *)
EulerTable[x_] := N[Table[EulerE[n,x],
  {n,2,5}],10]
EulerTable[0]
EulerTable[0.5]
EulerTable[1]
```

Table of Euler numbers from 1 to 12, in fraction and decimal forms.

Bar chart of modified Euler numbers, Figure 8.2.1 top.

Bar chart of logs of Euler numbers, Figure 8.2.2.

Table of algebraic Euler polynomials through $n = 5$, as in (8.2.10).

Define curve of polynomial with x varying and n fixed. Show polynomials together for $n = 2$–5, Figure 8.2.1 bottom.

Numerical values of Euler polynomials to 10 digits, Table 8.2.2.

The `RiemannZeta` cell [8.3]

Execute this cell to get surfaces of the real and imaginary parts of the zeta function in the complex plane (Figure 8.3.1), as well as magnitude and phase surfaces, in addition to the magnitude of the function along the critical line (Figure 8.3.2). Test values of the function along the real line (Table 8.3.1) are also made.

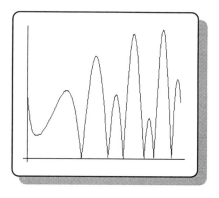

Exploring with `RiemannZeta`

An unproven conjecture in number theory is Riemann's hypothesis (1859) that all nonreal zeros of the zeta function lie on the line $\text{Re}\,z = 1/2$. Use the root-finding capabilities of *Mathematica* to check this out for a limited region of the z plane. Extensive calculations to test the hypothesis are described in Chapter 3 of Varga's book, reference [Var90] in Chapter 8.

```
(* RiemannZeta cell *)

(* Riemann zeta function in the complex plane *)

(* Riemann zeta; 3D views *)
Riemann3D[ReIm_] := Plot3D[ReIm[Zeta[x+I*y]],
  {x,1.5,4},{y,0,4},PlotRange->All,
  PlotPoints->20,AxesLabel->{"x","y",""},
  Ticks->{{2,3,4},Automatic,None}]
  Riemann3D[Re];   Riemann3D[Im];

(* Riemann zeta: Absolute value in 3D *)
Riemann3DAbs = Plot3D[Abs[Zeta[x+I*y]],
  {x,1.5,4},{y,0,4},PlotRange->All,
  PlotPoints->20,AxesLabel->{"x","y",""},
  Ticks->{{2,3,4},Automatic,None}];

(* Riemann zeta: Phase contour plot *)
Riemann3DPhase = ContourPlot[
  ArcTan[Im[Zeta[x+I*y]],Re[Zeta[x+I*y]]],
  {x,1.5,4},{y,0,4},PlotRange->All,
  PlotPoints->30,FrameTicks->{{2,3,4},Automatic},
  ContourShading->False,ContourSmoothing->6];

(* Riemann zeta on the critical line; 2D view *)
RZ2D = Plot[Abs[Zeta[1/2+I*y]],{y,0,40}];

(* Riemann zeta minus unity: Numerical *)
(* Scaled zeta calculation accuracy *)
ZetaM1[s_] := N[N[Zeta[s],12+0.3*s]-1,10]
ZetaM1Small = Table[{s,ZetaM1[s]},{s,2.5,4,0.5}]
ZetaM1Large = Table[{s,ZetaM1[s]},{s,10,40,10}]
```

Define surface of zeta function of complex variable. Show surfaces of real and imaginary parts, Figure 8.3.1.

Surface of absolute value of function, Figure 8.3.2 left top.

Contour plot of zeta function, Figure 8.3.2 left bottom.

Curve of Riemann zeta on the critical line, Figure 8.3.2 right.

Zeta function minus unity to 10 digits, Table 8.3.1, for small and large real arguments.

The `SumRecPower` cell [8.4]

Execute this cell to produce the four bar charts in Figure 8.4.1 for the logarithms of the magnitudes of the sums of reciprocal powers minus unity. Test values of the sums of reciprocal powers other than the zeta function are also generated for Table 8.4.2.

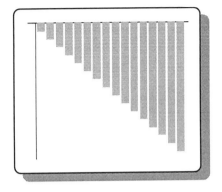

Exploring with `SumRecPower`

Check out the statement made in the discussion of Test Values in Section 8.4 about the number of digits accuracy that is required in the *Mathematica* calculations to ensure 10-digit relative accuracy in the sums minus unity. Explain this result analytically for the case that *s* is large.

```
(* SumRecPower cell *)

(* Sums of Reciprocal Powers *)

<<Graphics`Graphics`

Off[General::spell1]; Unprotect[Beta];
```
Disable spelling and conflicting `Beta` use.

```
(* Functions of reciprocal powers *)
Eta[n_] := (1-2^(1-n))*Zeta[n]
Lambda[n_] := (1-2^(-n))*Zeta[n]
Beta[n_] := N[NSum[(-1)^k/(2*k+1)^n,{k,0,Infinity},
 AccuracyGoal->30],30]
```
Define functions by (8.4.4) and (8.4.3).

```
(* Sums of Reciprocal Powers: Numerical *)
RecTable = Table[
 {n,N[Zeta[n],11],Eta[n],Lambda[n],N[Beta[n],11]},
 {n,2,10}]
```
Tables for Figure 8.4.1.

```
(* Sums of Reciprocal Powers: Graphical *)
RecPowMOne[PowSum_] := BarChart[
 Table[Log[10,Abs[N[PowSum[n]-1]]],
 {n,2,32,2}],Ticks->{None,Automatic},
 BarStyle->{GrayLevel[0.8]},BarEdges->False,
 AxesLabel->
 {"n","log[ |"<>ToString[PowSum]<>"(n)-1|]"},
 PlotRange->{-16,0}]
```
Define bar charts for each power sum.

```
ZetaChart = RecPowMOne[Zeta];
EtaChart = RecPowMOne[Eta];
LambdaChart = RecPowMOne[Lambda];
BetaChart = RecPowMOne[Beta];
```
Mark charts as in Figure 8.4.1.

```
(* Sums of Reciprocal Powers - 1 : Numerical *)
(* Scaled calculation accuracy *)
EtaM1[s_] := N[N[Eta[s],12+0.3*s]-1,10]
EtaM1Small = Table[{s,EtaM1[s]},{s,2.5,3.5,0.5}]
EtaM1Large = Table[{s,EtaM1[s]},{s,10,20,5}]

LambdaM1[s_] := N[N[Lambda[s],12+0.3*s]-1,10]
LambdaM1Small = Table[{s,LambdaM1[s]},{s,2.5,3.5,0.5}]
LambdaM1Large = Table[{s,LambdaM1[s]},{s,10,20,5}]

BetaM1[s_] := N[N[Beta[s],12+0.3*s]-1,10]
BetaM1Small = Table[{s,BetaM1[s]},{s,2.5,3.5,0.5}]
BetaM1Large = Table[{s,BetaM1[s]},{s,10,20,5}]

Protect[Beta];
```

Tables with umber of significant digits depending upon *s*, for each sum of reciprocal powers.

Restore `Beta`.

The `PolyLogarithm` cell [8.5]

By executing this cell you will obtain for the polylogarithms the four surfaces in Figure 8.5.1, as well as the surface and curves in Figure 8.5.2. Test values of polylogarithms, as in Table 8.5.2, are also generated.

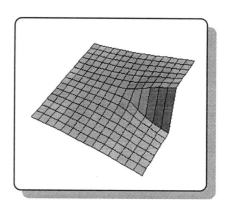

Exploring with `PolyLogarithm`

The dependence on order n of the polylogarithm, $Li_n(x)$, can be mapped out by making a 2D plot similar to that on the right side of Figure 8.5.2, but with n varying and x fixed. By doing this, you can see more clearly the n dependence of the surface of $Li_n(x)$ that is on the left side of figure 8.5.2.

```
(* PolyLogarithm cell *)


(* PolyLogarithm for fixed n vs z: 3D views *)
ComplexPolyLog[ReIm_,n_] := Plot3D[
 ReIm[PolyLog[n,x+I*y]],
 {x,-4,4},{y,-4,4},PlotRange->All,
 PlotPoints->20,AxesLabel->{"x","y",""},
 Ticks->{{-4,-3,-2,-1,0,1,2,3,4},
 {-4,-3,-2,-1,0,1,2,3,4},None}]

ComplexPolyLog[Re,2]; ComplexPolyLog[Im,2];
ComplexPolyLog[Re,4]; ComplexPolyLog[Im,4];
```

Define surface of real or imaginary part of polylogarithm of order n.

Show surfaces for $n = 2$ and 4, Figure 8.5.1.

```
(* PolyLogarithm vs x & n: 3D views *)
PolyLog3D = Plot3D[PolyLog[n,x],
  {x,-0.95,0.95},{n,2,6},PlotRange->All,
  PlotPoints->15,AxesLabel->{"x","n",""},
  Ticks->{{-0.5,0,0.5},Automatic,None}];
```

Surface for real argument x and order n, Figure 8.5.2.

```
(* PolyLogarithm for fixed n vs x: 2D views *)
PolyPlot[n_] := Plot[PolyLog[n,x],
  {x,-0.95,0.95},PlotRange->{-1,1},
  PlotPoints->20,AxesLabel->{"x",""},
  Ticks->{{-0.5,0,0.5},{-1,-0.5,0,0.5,1}},
  DisplayFunction->Identity]
```

Define curves for fixed order n and varying argument x.

```
Poly1 = PolyPlot[1]; Poly5 = PolyPlot[5];
Show[Poly1,Poly5,
  DisplayFunction->$DisplayFunction];
```

Make curves for $n = 1$ and 5, then show together in Figure 8.5.2.

```
(* PolyLogarithm: Numerical *)
PolyLogTable = Table[{n,x,N[PolyLog[n,x],10]},
  {x,-1,1,0.5},{n,1,6,1}]
```

Numerical values to 10 digits, as in Table 8.5.2.

20.9   PROBABILITY DISTRIBUTIONS

Notebooks

Notebooks DiscretePDF, NormalPDF, Other-ContinuousPDF
(Cells are listed at the start of the description of each notebook.)

Overview

There are three *Mathematica* notebooks for the probability distributions in Chapter 9, corresponding to discrete distributions (Section 9.2), normal distributions (Section 9.3), and other probability distributions (Section 9.4).

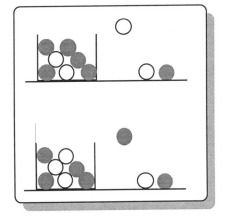

All functions in this chapter are straightforward to compute, both as C functions (Chapter 9 and Section 20.9) and as *Mathematica* functions. Their visualization is also usually simple, although graphics of these distributions are seldom provided in statistics books. Since the cumulative distribution is monotonic with respect to its random variable, the density distributions (or discrete values) are usually more interesting to visualize. For example, the graphical motifs at the beginning of the cell descriptions all show densities or probabilities.

Notebook `DiscretePDF` [9.2]

Cells `BinomialPDF`, `NegativeBinom-ialPDF`, `GeometricPDF`, `Hypergeo-metricPDF`, `LogSeriesPDF`, `Poisson-PDF`

The following cell at the head of notebook `DiscretePDF` must be executed first in order to load two packages and to define functions `DPDFunc` and `DPDFChart2`.

```
(* Discrete Probability Distribution Functions *)

(* You MUST execute this cell first *)

<<Statistics`DiscreteDistributions`
<<Graphics`Graphics3D`

(*  Discrete PDFs; names having 2 arguments *)
DPDFunc[NN_,p_,DPDFname_] := Table[
  PDF[DPDFname[NN,p],k],{k,0,NN}]

DPDTable[NN_,p_,DPDFname_] :=
  N[Table[{NN,p,DPDFunc[NN,p,DPDFname]}],10]

DPDFChart2[NN_,DPDFname_] := BarChart3D[
  {DPDFunc[NN,0.1,DPDFname],
    DPDFunc[NN,0.3,DPDFname],
    DPDFunc[NN,0.5,DPDFname]},
  Frame->False,Ticks->None,
  PlotLabel->"N = "<>ToString[NN]<>" ",
  AxesEdge->{Automatic,{1,-1},Automatic},
  AxesLabel->{"p","k"," "},
  DisplayFunction->Identity]
```

Load statistics and graphics packages.

Abbreviate functions for probability densities.

Define numerical tables to 10 digits.

Define 3D bar charts for $p = 0.1, 0.3, 0.5$.

The `BinomialPDF` cell [9.2.1]

If you execute this cell you will produce for the binomial distribution Figure 9.2.1 with values of the parameter N in the definition (9.2.1) of $N = 4$ and $N = 10$.

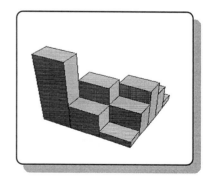

Exploring with `BinomialPDF`

Many approximations have been derived for the binomial distribution, in order to simplify calculations that use it. Such approximations are discussed in the references cited in Section 9.2.1. Investigate the approximations by programming them into cell `BinomialPDF` and overlaying their graphics with the objects from the binomial distribution bar charts.

```
(* BinomialPDF cell *)

(* Binomial Distribution: 3D views *)
Show[GraphicsArray[
  {DPDFChart2[4,BinomialDistribution],
   DPDFChart2[10,BinomialDistribution]},
   GraphicsSpacing->0.3],
   DisplayFunction->$DisplayFunction];

(* Bimomial Distribution: Numerical *)
NN = 4;
DPDTable[NN,0.1,BinomialDistribution]
DPDTable[NN,0.3,BinomialDistribution]
```

Show functions in 3D bar charts for $N = 4$ and 10, Figure 9.2.1.

Numerical binomial distribution values, Table 9.2.1.

The `NegativeBinomialPDF` cell [9.2.2]
Execute this cell to produce Figure 9.2.2 for values of the negative-binomial (Pascal) distribution with parameter N in the definition (9.2.1) having $N = 1$, 4, and 10. Check values for Table 9.2.2 are also generated.

Exploring with `NegativeBinomialPDF`
The negative binomial distribution has three parameters—N, p, and k. Therefore, there is a wide range of parameter space to explore. For example, various approximations to the distribution that are discussed in the references in Section 9.2.2 can be checked for reliability. Numerical test values for the C function `NegBinomDist` can also be obtained.

```
(* NegativeBinomialPDF cell *)

(* Negative Binomial Distribution: 3D views *)
Show[GraphicsArray[
  {DPDFChart2[1,NegativeBinomialDistribution],
   DPDFChart2[4,NegativeBinomialDistribution],
   DPDFChart2[10,NegativeBinomialDistribution]},
   GraphicsSpacing->0.2],
   DisplayFunction->$DisplayFunction];
```

Show functions in 3D bar charts for $N = 1$, 4, and 10, Figure 9.2.2.

```
(* NegativeBinomialPDF cell continued *)

(* Negative Binomial Distribution: Numerical *)
p = 0.3
N[Table[PDF[NegativeBinomialDistribution[1,p],k],
  {k,0,3}],10]
N[Table[PDF[NegativeBinomialDistribution[2,p],k],
  {k,0,3}],10]
N[Table[PDF[NegativeBinomialDistribution[3,p],k],
  {k,0,3}],10]
```

$p = 0.3$ in Table 9.2.2 for check values of negative binomial distribution.

The GeometricPDF cell [9.2.3]

By executing this cell you will produce the 3D bar chart and the two curves in Figure 9.2.3. Test values for Table 9.2.3 are generated and convergence of the sum rule (9.2.6) is investigated by summing the series up to $k = 15$.

Exploring with GeometricPDF

The geometric distribution has two parameters that can be varied, p and k. Investigate how closely the exponential distribution in Section 9.4.2 is approached as p approaches unity. Check convergence of sum rule (9.2.6) by using the Apply functions, as in the following cell.

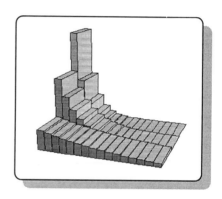

```
(* GeometricPDF cell *)

DPDGeometric[p_] := Table[
  PDF[GeometricDistribution[p],k],{k,0,15}]
```

Simplify geometric density distribution function.

```
(* Geometric Distribution; 3D views *)
BarChart3D[{DPDGeometric[0.5],DPDGeometric[0.3],
  DPDGeometric[0.1]},Frame->False,
  Ticks->{None,Automatic,None},
  PlotLabel->"Geometric distribution",
  AxesEdge->{{1,-1},{1,-1},Automatic},
  AxesLabel->{"p","k"," "},
  ViewPoint->{3,1,1}];
```

3D bar charts for $p = 0.1, 0.3, 0.5$, Figure 9.2.3 center.

```
(* Geometric Distribution; 2D views *)
GeoPlot[k_] := Plot[PDF[GeometricDistribution[p],k],
  {p,0,1},PlotRange->All,Ticks->{Automatic,None},
  AxesLabel->{"p",""}]

GeoPlot[3]; GeoPlot[12];
```

Define curves as a function of k for p varying, then show curves for $k = 3$ and 12, Figure 9.2.3 left and right.

```
(* Geometric Distribution PDF; Numerical *)
p = 0.1;
GDlist1 = N[Table[PDF[GeometricDistribution[p],k],
   {k,0,15}],10]
N[Apply[Plus,GDlist1],10] (* to check the sum rule *)
p = 0.3;
GDlist2 = N[Table[PDF[GeometricDistribution[p],k],
   {k,0,15}],10]
N[Apply[Plus,GDlist2],10]
p = 0.5;
GDlist3 = N[Table[PDF[GeometricDistribution[p],k],
   {k,0,15}],10]
N[Apply[Plus,GDlist3],10]
```

Entries in Table 9.2.3 (and more) for check values of geometric distribution. Sum the table entries by using `Apply[Plus, …]` to see convergence of sum (9.2.6) up to $k = 15$ for $p = 0.1$, 0.3, 0.5.

The `HypergeometricPDF` cell [9.2.4]
Executing this cell produces the 3D bar charts of the hypergeometric distribution in Figure 9.2.4. Test values for Table 9.2.4 are generated and convergence of the sum rule (9.2.9) is investigated by summing the series up to $k = 4$.

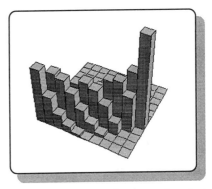

Exploring with `HypergeometricPDF`
The hypergeometric distribution has the three parameters, N, N_1, n and k, that can be varied. In our visualizations (Figure 9.2.4) we have kept the total number of objects, N, fixed at 10. How do the distributions vary if you change only N? To find out, all you need to do is to change the 10 in the `Show[]` below, and to change `NN` at the start of the numerical section of the cell.

```
(* HypergeometricPDF cell *)

(* NN=total; N1=number white; n=number of trials *)

DPDHyper[NN_,N1_] := Table[
  PDF[HypergeometricDistribution
    [n,N1,NN],k],{n,1,NN},{k,0,NN}]
```

Define table of hypergeometric distribution function.

```
(* Hypergeometric Distribution; 3D views *)
DPDFChartHyper[NN_,N1_] := BarChart3D[
  DPDHyper[NN,N1],Frame->False,Ticks->None,
  PlotLabel->"N1 = "<>ToString[N1]<>" ",
  AxesEdge->{Automatic,{1,-1},Automatic},
  AxesLabel->{"n","k"," "},ViewPoint->{3,-1,1},
  DisplayFunction->Identity]
```

Define 3D bar charts for N and N_1 fixed.

```
(* HypergeometricPDF cell continued *)


Show[GraphicsArray[
  {DPDFChartHyper[10,4],DPDFChartHyper[10,8]},
    GraphicsSpacing->0.3],
    DisplayFunction->$DisplayFunction];


(* Hypergeometric Distribution; Numerical *)
HyperTable[n_,N1_,NN_] := N[Table[
  PDF[HypergeometricDistribution[n,N1,NN],k],
  {k,0,4}],10]


NN = 10;  N1 = 4;
HGDlist1 = HyperTable[1,N1,NN]
Apply[Plus,HGDlist1] (* checks sum rule *)
HGDlist2 = HyperTable[2,N1,NN]
Apply[Plus,HGDlist2]
HGDlist3 = HyperTable[3,N1,NN]
Apply[Plus,HGDlist3]
HGDlist4 = HyperTable[4,N1,NN]
Apply[Plus,HGDlist4]
```

Show bar charts for
$N = 10$, $N_1 = 4$ and
8, as in Figure 9.2.4.

Numerical entries in
Table 9.2.4 to 10
digits for check
values of hypergeo-
metric distribution.
Sum table entries us-
ing `Apply[Plus]`
to see convergence of
sum (9.2.9) up to
$k = 4$.

The `LogSeriesPDF` cell [9.2.5]
Executing this cell produces the 3D bar charts
of the logarithmic series distribution in the
center of Figure 9.2.5, as well as the curves of
the distribution as functions of θ for fixed k
that are shown at the left and right sides of the
figure.

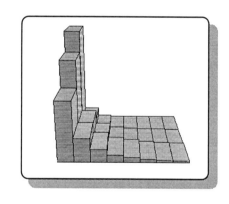

Exploring with `LogSeriesPDF`
The logarithmic series distribution has just two
parameters, θ and k, which can be varied. Try
some more values of k and see how well the approximations described in the references given
in Section 9.2.5 are satisfied. Add a section to the cell to check the sum rule (9.2.11),
similarly to that at the end of `HyperGeometricPDF`.

```
(* LogSeriesPDF cell *)


(* Logarithmic Series Distribution *)

DPDLogSeries[theta_,kmax_] := N[Table[
  PDF[LogSeriesDistribution[theta],k],{k,1,kmax}],10]
```

Define table of log
series distribution
function to 10
digits.

```
(* Logarithmic Series: 3D views *)
BarChart3D[{DPDLogSeries[0.2,8],DPDLogSeries[0.5,8],
  DPDLogSeries[0.8,8]},Frame->False,
  Ticks->{None,Automatic,None},
  PlotLabel->"Logarithmic series distribution",
  AxesEdge->{{1,-1},{1,-1},Automatic},
  AxesLabel->{" theta"," k",""},ViewPoint->{3,0,1}];
```

Make 3D bar charts for $\theta = 0.2$, 0.5, 0.8, Figure 9.2.5 center.

```
(* Logarithmic Series Distribution; 2D views *)
LogSeriesPlot[k_] :=
Plot[PDF[LogSeriesDistribution[theta],k],
  {theta,0.01,0.99},PlotRange->All,Ticks-
>{Automatic,None},
  AxesLabel->{"theta",""}]
LogSeriesPlot[2]; LogSeriesPlot[6];
```

Define curves for log series as function of θ for k fixed. Show curves for $k = 2$ and 6, Figure 9.2.5 right.

```
(*  Logarithmic Series Distribution: Numerical *)
DPDLogSeries[0.2,3]
DPDLogSeries[0.5,3]
DPDLogSeries[0.8,3]
```

Numerical values to 10 digits, Table 9.2.5.

The `PoissonPDF` cell [9.2.6]

By executing this cell you will produce the 3D bar charts of the Poisson distribution at the left of Figure 9.2.6, as well as the curves of the distribution as functions of m for fixed k, as shown at the right side of the figure. Numerical values in Table 9.2.6 are also produced.

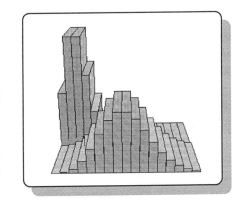

Exploring with `PoissonPDF`

As m increases, the Poisson distribution tends to a Gauss (normal) distribution whose mean value and variance are both equal to m. Check this out graphically and numerically by comparing the Poisson distributions generated by this cell with Gaussians (Section 9.3.1) having the same variance. Do this as a function of m, and decide an m value appropriate for your applications at which the Poisson distribution could be replaced by the normal distribution. Use the `Apply[Plus, ...]` function in *Mathematica* to sum the numerical values of the distribution, in order to check the sum rule (9.2.13).

```
(* PoissonPDF cell *)

(*  Poisson Distribution *)
DPDPoisson[m_] := Table[
  PDF[PoissonDistribution[m],k],{k,0,30}]
```

Define table of Poisson series distribution for fixed mean value m.

```
(* PoissonPDF cell continued *)


(* Poisson Distribution: 3D views *)
BarChart3D[{DPDPoisson[2],DPDPoisson[10],
  DPDPoisson[18]},Frame->False,
  Ticks->{None,Automatic,None},
  PlotLabel->"Poisson distribution",
  AxesEdge->{{1,-1},{1,-1},Automatic},
  AxesLabel->{"  m","  k"," "},ViewPoint->{3,0,1}];


(* Poisson Distribution; 2D views *)
PoissonPlot[k_] := Plot[PDF[PoissonDistribution[m],k],
  {m,1,20},PlotRange->All,Ticks->{Automatic,None},
  Axes->{True,False},AxesLabel->{"m ",""}]
PoissonPlot[6]; PoissonPlot[14];


(* Poisson Distribution: Numerical *)
m = 2;
Table[N[PDF[PoissonDistribution[m],k],10],{k,0,20,10}]
m = 10;
Table[N[PDF[PoissonDistribution[m],k],10],{k,0,20,10}]
```

Make 3D bar chart
for *m* = 2,10, 18,
as in Figure 9.2.6
left.

Define curves for
Poisson distribu-
tion as function of
k for *m* varying.
Show curves for
k = 6, 14, Figure
9.2.6 right.

Numerical values
to 10 digits as in
Table 9.2.6.

Notebook NormalPDF [9.3]

Cells NormalPDF, BivariatePDF,
ChiSquarePDF, F-Distribution-
PDF, StudentT-PDF, Lognormal-
PDF

Properties of distributions that are related to the normal distribution and are discussed in Section 9.3 are summarized in
Table 9.3.1.   Note also Figure 9.3.1,
showing interrelations between various
distributions and the normal (Gauss) distribution.

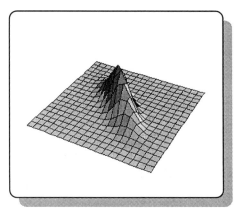

   The following cell at the head of the
notebook NormalPDF must be executed first in order to load the package for normal
distributions.

```
(* You MUST execute this cell first *)


<<Statistics`NormalDistribution`
```

Load statistics package for normal
distributions.

The `NormalPDF` cell [9.3.1]

If you execute this cell you will produce for the normal distribution the surface and curves in Figure 9.3.2, as well as the numerical check values in Table 9.3.2.

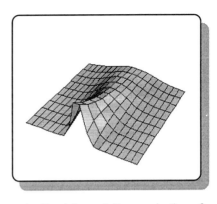

Exploring with `NormalPDF`

The normal distribution is often expressed as the standardized form (9.3.3), having $\sigma = 1$. This usage presupposes, however, that the parameters m and σ have been estimated well enough that the origin and scale can be removed as parameters. To explore how uncertainties in the parameters affect construction of a standardized form, follow up in the references cited in Section 9.3.1 the techniques for estimating m and σ. Use typical uncertainties for your field of applications, then run cell `NormalPDF` to produce the corresponding graphics and numerical values.

```
(* NormalPDF cell *)


(* Normal Distribution *)
NormPDF[sigma_,x_] :=
PDF[NormalDistribution[0,sigma],x]


(* Normal Distribution: 3D views *)
NPDFChart = Plot3D[NormPDF[sigma,x],{x,-10,10},
  {sigma,1,4},PlotRange->All,PlotPoints->{40,20},
  Ticks->{Automatic,Automatic,None},
  PlotLabel->"Gauss (normal) distribution",
  AxesEdge->{Automatic,{1,-1},Automatic},
  AxesLabel->{"x","sigma"," "}];


(* Normal Distribution: 2D views *)
NormPlot[x_] := Plot[NormPDF[sigma,x],
  {sigma,0.5,2},PlotRange->All,
  Ticks->{{0.5,1.0,1.5,2.0},None},
  Axes->{True,False},AxesOrigin->{0.6,0.0},
  AxesLabel->{"sigma",""},DisplayFunction->Identity]


Show[NormPlot[0],NormPlot[1],NormPlot[2],
  DisplayFunction->$DisplayFunction];


(* Normal Distribution: Numerical *)
sigma = 1
N[Table[NormPDF[sigma,x],{x,0,10,2}],10]
```

Show functions as a surface for x in -10 to 10 and σ in 1 to 4, Figure 9.3.2.

Define curves for σ varying and x fixed.

Show curves for $x = 0$, 1, 2.

Numerical values of the normal distribution in Table 9.3.2.

The `BivariatePDF` cell [9.3.2]
Execute this cell to produce graphics for understanding the bivariate normal distributions (9.3.5) and (9.3.7). The surfaces in Figure 9.3.3 are displayed, as are the surfaces in Figure 9.3.4 for rotated bivariate distributions.

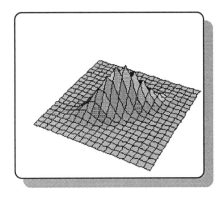

To emphasize that rotation produces ellipses whose major axes coincide with the x- and y-variable axes, contour plots of the rotated bivariate distributions are made for Figure 9.3.5. Cell `BivariatePDF` also makes the numerical values of B_ρ, for Table 9.3.3.

Exploring with `BivariatePDF`
Explore how bivariate normal probability density surfaces—as in Figure 9.3.3—change as the correlation coefficient between the x and y random variables changes. For example, a correlation coefficient that produces a visible change in the surface contours from circles ($\rho = 0$) to ellipses ($\rho \neq 0$) will often be of statistical significance.

Modify cell `BivariatePDF` so that graphics for ρ values of interest in your applications are produced. What is a significant non-zero value of ρ? Similarly, investigate how a change in ρ (such as a change equal to its estimated standard deviation) changes the bivariate distribution in relevant regions of x and y.

```
(* BivariatePDF cell *)

(* Bivariate Normal Distribution
   in terms of correlation coefficient rho *)

del[rho_] := 1-rho^2

BnPDF[x_,y_,rho_] :=
 Exp[-(x^2-2*rho*x*y+y^2)/(2*del[rho])]/
 (2*N[Pi]*Sqrt[del[rho]])

(* Bivariate Normal Distribution: 3D views *)
BnPDFChart[rho_] := Plot3D[BnPDF[x,y,rho],{x,-4,4},
 {y,-4,4},PlotRange->All,PlotPoints->{30,30},
 Ticks->{Automatic,Automatic,None},
 AxesEdge->{Automatic,{1,-1},Automatic},
 PlotLabel->"rho = "<>ToString[rho]<>" ",
 AxesLabel->{"x","y"," "}]

BnPDFChart[-0.8]; BnPDFChart[0]; BnPDFChart[0.8];

(* Rotated Bivariate Normal Distribution *)
del[rho_] := 1-rho^2
BPrimePDF[xp_,yp_,rho_] :=
 Exp[-(xp^2/(1-rho)+yp^2/(1+rho))]/
    (N[Pi]*Sqrt[1-rho^2])
```

Define bivariate pdf, (9.3.5).

Define surface of bivariate density for x and y in –4 to 4, as a function of ρ.

Show surfaces for $\rho = -0.8, 0, 0.8$, Figure 9.3.3.

Define rotated bivariate pdf, (9.3.7).

```
(* Rotated Bivariate Normal Distribution: 3D views *)
BPrimeChart[rho_] := Plot3D[BPrimePDF[xp,yp,rho],
  {xp,-4,4},{yp,-4,4},PlotRange->All,
  PlotPoints->{30,30},Ticks->None,
  AxesEdge->{Automatic,{1,-1},Automatic},
  PlotLabel->"rho = "<>ToString[rho]<>" ",
  AxesLabel->{"x'","y'"," "},
DisplayFunction->Identity]
```

Define surface of rotated bivariate density for x and y in –4 to 4, as function of ρ.

```
Show[GraphicsArray[{BPrimeChart[-0.8],
  BPrimeChart[0.8]},GraphicsSpacing->0.4,
  PlotLabel->"Rotated bivariate distributions"],
  DisplayFunction->$DisplayFunction];
```

Show surfaces for $\rho = -0.8$ and 0.8, Figure 9.3.4.

```
(* Rotated Bivariate Normal Distribution:
   Contour plots *)
BPcontour[r_] := ContourPlot[
  Log[Evaluate[BPrimePDF[xp,yp,r]]],
  {xp,-6,6},{yp,-6,6},PlotRange->All,
  PlotPoints->{50,50},Ticks->None,
  AxesLabel->{"x'","y'"},DisplayFunction->Identity]
```

Define contours of rotated bivariate density for x and y in –6 to 6, as a function of ρ.

```
Show[GraphicsArray[
  {BPcontour[-0.8],BPcontour[0.8]},
  GraphicsSpacing->0.4,
  PlotLabel->
  "Contours of logs of rotated bivariate PDFs",
  AxesLabel->{"x'","y'"}],
  DisplayFunction->$DisplayFunction];
```

Show contours for $\rho = -0.8$ and 0.8, Figure 9.3.5.

```
(* Bivariate Normal Distribution: Numerical *)
r = 0.8
N[Table[BnPDF[x,y,r],{x,-2,2,1},{y,-2,2,1}],10]
```

Numerical values of bivariate normal distribution (9.3.5).

The `ChiSquarePDF` cell [9.3.3]

By executing this cell you will produce several interesting graphics that help you understand the chi-square distributions defined by (9.3.8) and (9.3.9). The surfaces and curves for Figure 9.3.6 and 9.3.7 are displayed, and numerical values of the density and cumulative distributions (Tables 9.3.4 and 9.3.5) are produced.

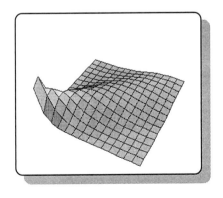

Exploring with `ChiSquarePDF`

The chi-square distribution is widely used for estimating confidence intervals in statistics based upon the normal distribution (Section 9.3.1). It is therefore interesting to explore the density and cumulative distributions in more detail. For example, how much confidence should you place in the rule of thumb used by many physicists and other practitioners of the "exact" sciences that when $\chi^2/v = 1$ one has obtained a good fit of a model to data? This can be investigated easily by modifying object `CumDFChart` in the `ChiSquarePDF` cell.

```
(* ChiSquarePDF cell *)

(*  Chi-Square Probability Density *)
CsPDF[chisq_,nu_] :=
PDF[ChiSquareDistribution[nu],chisq]

(* Chi-Square Probability Densities: 3D views *)
CsPDFChart =
Plot3D[CsPDF[chisq,nu],{chisq,1,20},{nu,1,20},
  PlotRange->All,PlotPoints->{20,20},
  Ticks->{Automatic,Automatic,None},
  PlotLabel->"Chi-square probability density",
  AxesEdge->{Automatic,{1,-1},Automatic},
  AxesLabel->{"chisq","nu"," "}];

(*  Chi-Square Probability Densities: 2D views *)
ChiSPlot[nu_] := Plot[CsPDF[chisq,nu],
  {chisq,0.5,14},PlotRange->All,
  Ticks->{{2,4,6,8,10,12,14},None},
  Axes->{True,False},AxesOrigin->{0.0,0.0},
  AxesLabel->{"chisq",""},DisplayFunction->Identity]

Show[ChiSPlot[5],ChiSPlot[10],ChiSPlot[15],
  DisplayFunction->$DisplayFunction];

(* Chi-Square Probability Densities: Numerical *)
nu = 1
N[Table[CsPDF[chisq,nu],{chisq,2,10,2}],10]
nu = 2
N[Table[CsPDF[chisq,nu],{chisq,2,10,2}],10]
nu = 5
N[Table[CsPDF[chisq,nu],{chisq,2,10,2}],10]
nu = 10
N[Table[CsPDF[chisq,nu],{chisq,2,10,2}],10]

(*  Chi-Square Cumulative Distribution *)
CsCumDF[chisq_,nu_] :=
  CDF[ChiSquareDistribution[nu],chisq]

(*  Chi-Square Cumulative Distribution: 3D views *)
CsCumDFChart = Plot3D[
  CsCumDF[chisq,nu],{chisq,1,20},{nu,1,20},
  PlotRange->All,PlotPoints->{20,20},
  Ticks->{Automatic,Automatic,None},
  PlotLabel->"(chisq,nu)",
  AxesEdge->{Automatic,{1,-1},Automatic},
  AxesLabel->{"chisq"," nu"," "}];
```

Define chi-square density, (9.3.8).

Make surface of chi-square density for χ^2 and v in 1 to 20, Figure 9.3.6 left.

Show curves for $v = 5$, 10, and 15, Figure 9.3.6 right.

Numerical values of the χ^2 density, as in Table 9.3.4.

Define chi-square cumulative, (9.3.9).

Make surface of chi-square cumulative for χ^2 and v in 1 to 20, Figure 9.3.7 left.

```
(* Chi-Square Cumulative Distribution: chisq/nu *)
CumDFChart = Plot3D[
  CsCumDF[chisqNu*nu,nu],{chisqNu,1,5},{nu,1,4},
  PlotRange->All,PlotPoints->{20,20},
  Ticks->{Automatic,Automatic,None},
  PlotLabel->"(chisq/nu,nu)",
  AxesEdge->{Automatic,{1,-1},Automatic},
  AxesLabel->{"chisq/nu"," nu"," "}];
```

Surface of chi-square cumulative in terms of χ^2/ν and ν, Figure 9.3.7 right.

```
(* Chi-Square Cumulative Distribution: Numerical *)
nu = 1
N[Table[CsCumDF[chisq,nu],{chisq,2,10,2}],10]
nu = 2
N[Table[CsCumDF[chisq,nu],{chisq,2,10,2}],10]
nu = 5
N[Table[CsCumDF[chisq,nu],{chisq,2,10,2}],10]
```

Numerical values to 10 digits of χ^2 cumulative, as in Table 9.3.5.

The `F-DistributionPDF` cell [9.3.4]

Execute this cell to produce several interesting graphics that help you understand the F-ratio distributions defined by (9.3.14) and (9.3.15, (9.3.16). The surfaces and curves for Figures 9.3.8 and 9.3.9 are displayed, and numerical values of the density and cumulative distributions (Tables 9.3.6 and 9.3.7) are produced.

Exploring with `F-DistributionPDF`

The F-ratio distribution (also known as the variance-ratio distribution) is widely used when estimating the probability that two

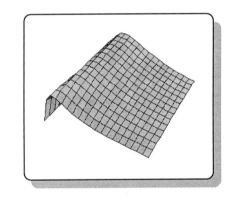

samples based upon the normal distribution (Section 9.3.1) have the same variance. It is interesting to explore the F-ratio density and cumulative distributions in more detail.

We show in Figures 9.3.8 and 9.3.9 curves of the density and cumulative density for $\nu_1 = \nu_2 = \nu$. Investigate what happens as the number of degrees of freedom is moved away from equality. This can be done easily by modifying the object `FratPlot` in the `F-DistributionPDF` cell so that it has the two degrees as freedom as arguments. To visualize the changes when the two degrees of freedom move away from equality, it might be better to plot surfaces (`Plot3D` function) rather than curves (`Plot` function).

```
(* F-DistributionPDF cell *)


(* F Distribution Probability Density *)
FratPDF[f_,nu1_,nu2_] :=
  PDF[FRatioDistribution[nu1,nu2],f]
```

Define F-ratio density, (9.3.14).

```
(* F-DistributionPDF cell continued *)

(*  F Distribution Probability Density: 3D views *)
FratPDFChart[nu1_] := Plot3D[FratPDF[f,nu1,nu2],
  {f,0,2},{nu2,1,6},
  PlotRange->All,PlotPoints->{20,20},
  Ticks->{Automatic,Automatic,None},
  PlotLabel->"F density: nu1 = "<>ToString[nu1]" ",
  AxesEdge->{Automatic,{1,-1},Automatic},
  AxesLabel->{"f","nu2"," "}];


FratPDFChart[6];


(* F Probability Densities: 2D views, nu1=nu2=nu *)
FratPlot[nu_] := Plot[FratPDF[f,nu,nu],
  {f,0.05,2},PlotRange->All,
  Ticks->{{0,0.5,1.0,1.5,2.0},None},
  Axes->{True,False},AxesOrigin->{0.1,0.0},
  AxesLabel->{"f",""},DisplayFunction->Identity]


Show[FratPlot[2],FratPlot[6],FratPlot[10],
  DisplayFunction->$DisplayFunction];

(* F Probability Density: Numerical *)
nu1 = 5
nu2 = 3
Table[N[FratPDF[f,nu1,nu2],10],{f,0,2,0.5}]
nu1 = 10
nu2 = 3
Table[N[FratPDF[f,nu1,nu2],10],{f,0,2,0.5}]


(*  F Cumulative Distribution in Q Form  *)
FratioCDF[f_,nu1_,nu2_] := 1 - CDF[
  FRatioDistribution[nu1,nu2],f]


(* F Cumulative Distribution in Q Form: 3D views *)
FratCDFGraf[nu1_] := Plot3D[FratioCDF[f,nu1,nu2],
  {f,0,2},{nu2,1,6},
  PlotRange->All,PlotPoints->{20,20},
  Ticks->{Automatic,Automatic,None},PlotLabel->
  "F cumulative (Q form): nu1 = "<>ToString[nu1]" ",
  AxesEdge->{Automatic,{1,-1},Automatic},
  AxesLabel->{"f","nu2"," "}];


FratCDFGraf[6];

(* F Cumulative Distribution: 2D views, nu1=nu2=nu *)
FratCDFPlot[nu_] := Plot[FratioCDF[f,nu,nu],
  {f,0.05,2},PlotRange->All,
  Ticks->{{0,0.5,1.0,1.5,2.0},None},
  Axes->{True,False},AxesOrigin->{0.1,0.0},
  AxesLabel->{"f",""},DisplayFunction->Identity]
```

Define surface of *F*-ratio density as function of ν_1.

Density surface for $\nu_1 = 6$, Figure 9.3.8 left.

Define curves for equal ν values.

Show curves for $\nu = 2, 6, 10$, Figure 9.3.8 right.

Numerical values of density to 10 digits, as in Table 9.3.6.

Define *F*-ratio cumulative function in *Q* form, (9.3.15).

Define *F*-ratio cumulative surface in *Q* form with ν fixed.

Surface of *F*-ratio cumulative, $\nu = 6$, Figure 9.3.8 left.

Define cumulative curves for fixed ν.

```
Show[FratCDFPlot[2],FratCDFPlot[6],FratCDFPlot[10],
   DisplayFunction->$DisplayFunction];
```

Show cumulative
curves for $v = 2$,
6, 10; Figure 9.3.9
right.

```
(* F Cumulative Distribution in Q Form: Numerical *)
nu1 = 5
nu2 = 3
Table[N[FratioCDF[f,nu1,nu2],10],{f,0,2,0.5}]
nu1 = 10
nu2 = 3
Table[N[FratioCDF[f,nu1,nu2],10],{f,0,2,0.5}]
```

Numerical values
to 10 digits of *F*-
ratio cumulative,
as in Table 9.3.7.

The `StudentT-PDF` cell [9.3.5]

By executing this cell you will produce several interesting graphics that will help you understand the Student-*t* distributions defined by (9.3.17) and (9.3.18). Surfaces and curves for Figures 9.3.10–9.3.12 are displayed when the `StudentT-PDF` cell is run. Both density distributions and cumulative distributions are visualized. Numerical values of the density and cumulative distributions in Tables 9.3.8 and 9.3.9 are also produced.

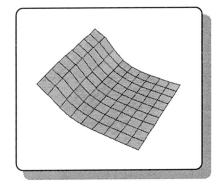

Comparison of the Student-*t* distribution density with the standardized Gauss (normal) distribution density (9.3.3) is prepared by the function `GandT[]`, which is used to produce Figure 9.3.11 for $v = 2$ and $v = 6$.

Exploring with `StudentT-PDF`

The Student-*t* distribution is often used when estimating the probability that two samples based upon the normal distribution (Section 9.3.1) have the same mean. It is therefore interesting to explore its density and cumulative distributions in more detail.

Explore how quickly the Student-*t* distribution density approaches the normal distribution density as the number of degrees of freedom, *v*, increases. This can be done easily by modifying the function `GandT[]` in cell `StudentT-PDF`. To visualize the approach to the Gaussian, it might be useful to plot the surface of the fractional difference between the two densities as a function of *t* and *v* by using the `Plot3D` function. A similar comparison can be made for the cumulative distributions by comparing the *A* distribution—related to the incomplete beta function by (9.3.19)—to the error function (Section 10.1).

```
(* StudentT-PDF cell *)

(* Student's t  & Gauss Probability Densities *)
StudentTPDF[t_,nu_] :=
PDF[StudentTDistribution[nu],t]
```

Define Student-*t*
density, (9.3.17).

```
(* StudentT-PDF cell continued *)

StudT[nu_] := Plot[StudentTPDF[t,nu],{t,-5,5},
  PlotRange->All,PlotPoints->20,
  Ticks->{Automatic,None},AxesLabel->{"t"," "},
  DisplayFunction->Identity]
```

Define curve of *t* density as function of v.

```
Gauss = Plot[N[Exp[-t^2/2]/Sqrt[2*Pi]],{t,-5,5},
  PlotRange->All,PlotPoints->20,
  PlotStyle->{Dashing[{0.02,0.02}]},
  Ticks->{Automatic,None},AxesLabel->{"t"," "},
  DisplayFunction->Identity];
```

Curve of standardized Gaussian, Figure 9.3.11 dashed.

```
GandT[nu_] := Show[Gauss,StudT[nu],
  PlotLabel->"nu = "<>ToString[nu]<>"",
  DisplayFunction->Identity]

GT2 = GandT[2];    GT6 = GandT[6];

Show[GraphicsArray[{GT2,GT6},GraphicsSpacing->0],
  DisplayFunction->$DisplayFunction];
```

Define curves for comparing Gaussian and Student-*t* at same v value.

Show comparison at $v = 2$ and 6 in Figure 9.3.11.

```
(* Student's t Probability Density in A Form:
     3D views *)
StudTPDFGraf = Plot3D[StudentTPDF[t,nu],
  {t,0,5},{nu,1,5},
  PlotRange->All,PlotPoints->{20,20},
  Ticks->{Automatic,Automatic,None},
  PlotLabel->"Student's t: Density (A form)",
  AxesEdge->{Automatic,{1,-1},Automatic},
  AxesLabel->{"t","nu"," "}];
```

Surface of Student-*t* density, Figure 9.3.10 left.

```
(* Student t Probability Densities: 2D views *)
StudPlot[nu_] := Plot[StudentTPDF[t,nu],
  {t,0,5},PlotRange->All,
  Ticks->{{0,1,2,3,4,5},None},
  Axes->{True,False},AxesOrigin->{0.1,0.0},
  AxesLabel->{"t",""},DisplayFunction->Identity]
```

Define curve of density as function of v.

```
Show[StudPlot[2],StudPlot[10],
  DisplayFunction->$DisplayFunction];
```

Show curves for $v = 2$ and 10, Figure 9.3.10 right.

```
(* t Probability Density: Numerical *)
nu = 1
Table[N[StudentTPDF[t,nu],10],{t,0,6,2}]
nu = 2
Table[N[StudentTPDF[t,nu],10],{t,0,6,2}]
nu = 5
Table[N[StudentTPDF[t,nu],10],{t,0,6,2}]
```

Numerical values to 10 digits of Student-*t* density, Table 9.3.8.

```
(* Student's t Cumulative Distribution:
   A form, 3D views *)
StudsTCDF[t_,nu_] :=
     2*CDF[StudentTDistribution[nu],t]-1


StudentCDFGraf = Plot3D[StudsTCDF[t,nu],
   {t,0,5},{nu,1,5},
   PlotRange->All,PlotPoints->{20,20},
   Ticks->{Automatic,Automatic,None},
   PlotLabel->"Student's t: Cumulative (A form)",
   AxesEdge->{Automatic,{1,-1},Automatic},
   AxesLabel->{"t","nu"," "}];


(*  Student t Cumulative Distribution: 2D views *)
StudACDF[nu_] := Plot[StudsTCDF[t,nu],
   {t,0,5},PlotRange->All,
   Ticks->{{0,1,2,3,4,5},None},
   Axes->{True,False},AxesOrigin->{0.1,0.0},
   AxesLabel->{"t",""},DisplayFunction->Identity]


Show[StudACDF[2],StudACDF[10],
   DisplayFunction->$DisplayFunction];


(* t Cumulative Distribution: A form, numerical *)
nu = 1
Table[N[StudsTCDF[t,nu],10],{t,0,6,2}]
nu = 2
Table[N[StudsTCDF[t,nu],10],{t,0,6,2}]
nu = 5
Table[N[StudsTCDF[t,nu],10],{t,0,6,2}]
```

Define Student-*t* cumulative, (9.3.18).

Surface of Student-*t* cumulative, Figure 9.3.12 left.

Define cumulative curve as function of *v*.

Show curves for $v = 2$ and 10, Figure 9.3.12 right.

Numerical values of Student-*t* cumulative, Table 9.3.9.

The `LognormalPDF` cell [9.3.6]

Executing this cell produces graphics that help you understand the lognormal distributions defined by (9.3.21) and (9.3.23). When `LognormalPDF` cell is run, surfaces for Figures 9.3.13 (densities) and 9.3.14 (cumulative distributions) are displayed. Numerical values of the density and cumulative distributions in Tables 9.3.10 and 9.3.11 are also produced.

Exploring with `LognormalPDF`

Since the lognormal distribution is a mapping of the variable of the normal (Gauss) distribution in Section 9.3.1 according to the transformation rule (9.3.20), it might seem not worth much further investigation. It is, however, interesting to explore its density and cumulative distributions in more detail, since the mapping is nonlinear and therefore distorts the distributions away from the symmetric shapes that they have for the normal distribution.

One possibility is to take slices through the surfaces in Figures 9.3.13 and 9.3.14. That is, apply the graphics function Plot rather than Plot3D as we do. Thereby you will produce curves of the lognormal distributions as functions of x, σ, or ξ.

```
(* LognormalPDF cell *)

(* Lognormal Distribution *)
LogNormalPDF[zeta_,sigma_,x_] :=
  PDF[NormalDistribution[zeta,sigma],Log[x]]/x

LogNPDF[zeta_,sigma_,x_] :=
   LogNormalPDF[zeta,sigma,x]

(* Lognormal Density Distribution: 3D views *)
  LogNPDFChart[zeta_] := Plot3D[
  LogNPDF[zeta,sigma,x],{x,0.1,3},{sigma,0.1,0.5},
  PlotRange->All,PlotPoints->{40,20},
  Ticks->{Automatic,Automatic,None},
  PlotLabel->"zeta = "<>ToString[zeta]<>"",
  AxesEdge->{Automatic,{1,-1},Automatic},
  AxesLabel->{"x","   sigma"," "}]

LogNPDFChart[-0.7];    LogNPDFChart[0.9];

(* Lognormal Density Distribution: Numerical *)
zeta = 0.5
sigma = 0.3
N[Table[LogNPDF[zeta,sigma,x],{x,1,2,0.5}],10]

(* Lognormal Cumulative Distribution *)
LogNormCDF[zeta_,sigma_,x_] :=
  Erfc[(zeta-Log[x])/(Sqrt[2]*sigma)]/2;

(* Lognormal Cumulative Distribution: 3D views *)
LogNCDFGraf[zeta_] := Plot3D[
  LogNormCDF[zeta,sigma,x],
   {x,0.1,3},{sigma,0.1,0.5},
  PlotRange->All,PlotPoints->{40,20},
  Ticks->{Automatic,Automatic,None},
  PlotLabel->"zeta = "<>ToString[zeta]<>"",
  AxesEdge->{Automatic,{1,-1},Automatic},
  AxesLabel->{"x","   sigma"," "}]

LogNCDFGraf[-0.7];  LogNCDFGraf[0.9];

(* Lognormal Cumulative Distribution: Numerical *)
zeta  = 0.5
sigma = 0.3
N[Table[LogNormCDF[zeta,sigma,x],{x,1,2,0.5}],10]
```

Define lognormal density, (9.3.21), with a choice of function names.

Define surfaces of density as function of ξ.

Show surfaces for ξ = –0.7 and 0.9, Figure 9.3.13.

Numerical lognormal densities as in Table 9.3.10.

Define lognormal cumulative, (9.3.23).

Define surfaces of cumulative distribution as function of ξ.

Show surfaces for ξ = –0.7 and 0.9, Figure 9.3.14.

Numerical lognormal cumulative distributions as in Table 9.3.11.

Notebook ocPDF [9.4]
Cells CauchyPDF, Exponential-
PDF, ParetoPDF, WeibullPDF,
LogisticPDF, LaplacePDF, Kol-
mogorovSmirnovPDF, BetaPDF

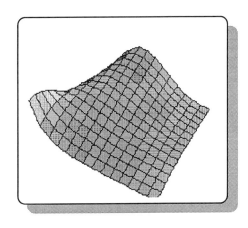

Continuous distributions not directly related to the normal distribution (Section
9.3) are discussed in Section 9.4 and
their properties are summarized in
Table 9.4.1. This *Mathematica* notebook
includes eight univariate continuous
probability distributions. (The notebook
for normal distributions is NPDF, given
immediately above, and the notebook for
discrete distributions is DPDF described at the start of Section 20.9.)

The following cell at the head of notebook ocPDF must be executed first in order to
load the package for continuous distributions.

```
(* You MUST execute this cell first *)

<<Statistics`ContinuousDistributions`
```
Load statistics package for
continuous distributions.

The CauchyPDF cell [9.4.1]
Execute this cell to produce for the Cauchy
(Lorentz) distribution the surface and curves
in Figures 9.4.2 and 9.4.3. Numerical values
of the density and cumulative distributions
are also produced.

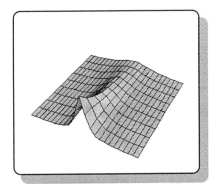

Exploring with CauchyPDF
The Cauchy distribution resembles the Gauss
(normal) distribution, as discussed in more
detail in Section 19.6 where the Voigt (plasma dispersion) function—which is the convolution of a Cauchy with a Gauss distribu-
tion—is described completely. It is therefore of interest to compare the Cauchy and Gauss
distributions. To do this, either run the appropriate parts of cell Voigt in notebook
Miscellaneous or adapt the CauchyPDF cell to include a Gauss distribution (see also
Section 9.3.1). Make graphical comparisons of the two distributions.

```
(* CauchyPDF cell *)

(* Cauchy Distribution Density *)
CauchyPDF[lambda_,x_] :=
PDF[CauchyDistribution[0,lambda],x]
```

Define Cauchy density distribution with $m = 0$, (9.4.1).

```
(* Cauchy Distribution Density: 3D views *)
CauchyChart = Plot3D[
  CauchyPDF[lambda,x],{x,-10,10},{lambda,1,4},
  PlotRange->All,PlotPoints->{40,20},
  Ticks->{Automatic,Automatic,None},
  PlotLabel->"Cauchy distribution: density",
  AxesEdge->{Automatic,{1,-1},Automatic},
  AxesLabel->{"x","lambda"," "}];
```

Make surface of Cauchy density distribution, Figure 9.4.2 left.

```
(* Cauchy Distribution Density: 2D views *)
CauchyPlot[x_] := Plot[CauchyPDF[lambda,x],
  {lambda,0.5,3},PlotRange->All,
  Ticks->{{0.5,1.0,1.5,2.0,2.5,3.0},None},
  Axes->{True,False},AxesOrigin->{0.6,0.0},
  AxesLabel->{"lambda",""},DisplayFunction->Identity]
```

Define curve of Cauchy density for x fixed and λ varying.

```
Show[CauchyPlot[0],CauchyPlot[0.5],CauchyPlot[1],
  DisplayFunction->$DisplayFunction];
```

Show curves of density for $x = 0$, 0.5, 1, Figure 9.4.2 right.

```
(* Cauchy Distribution Density; Numerical values *)
lambda = 1
N[Table[CauchyPDF[lambda,x],{x,0,10,2}],10]
```

Cauchy density numerical values to 10 digits, Table 9.4.2.

```
(*  Cauchy Distribution Cumulative *)
CauchyCum[lambda_,x_] :=
  CDF[CauchyDistribution[0,lambda],x]
```

Define Cauchy cumulative distribution.

```
(*  Cauchy Distribution Cumulative: 3D views *)
CauchyCDFGraf = Plot3D[
  CauchyCum[lambda,x],{x,-10,10},{lambda,1,4},
  PlotRange->All,PlotPoints->{40,20},
  Ticks->{Automatic,Automatic,None},
  PlotLabel->"Cauchy distribution: cumulative",
  AxesEdge->{Automatic,{1,-1},Automatic},
  AxesLabel->{"x","lambda"," "}];
```

Make surface of Cauchy cumulative distribution, Figure 9.4.3 left.

```
(*  Cauchy Distribution Cumulative: 2D views *)
CauchyCDFPlot[x_] := Plot[CauchyCum[lambda,x],
  {lambda,0.5,3},PlotRange->All,
  Ticks->{{0.5,1.0,1.5,2.0,2.5,3.0},None},
  Axes->{True,False},AxesOrigin->{0.6,0.0},
  AxesLabel->{"lambda",""},DisplayFunction->Identity]
```

Define curve of Cauchy cumulative for x fixed and λ varying.

```
Show[CauchyCDFPlot[0],CauchyCDFPlot[0.5],
  CauchyCDFPlot[1],
  DisplayFunction->$DisplayFunction];
```

Show curves of cumulative for $x = 0$, 0.5, 1, Figure 9.4.3 right.

```
(* Cauchy Distribution Cumulative; Numerical *)
lambda = 1
N[Table[CauchyCum[lambda,x],{x,0,10,2}],10]
```

Cauchy cumulative numerical values to 10 digits, Table 9.4.2.

The ExponentialPDF cell [9.4.2]

By executing cell ExponentialPDF you will produce the surfaces and curves in Figures 9.4.4 and 9.4.5 that provide visualizations of the exponential density and cumulative distributions.

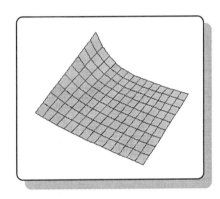

Exploring with ExponentialPDF

The exponential density distribution (9.4.4) is often expressed in the standardized form (9.4.5), which has $\lambda = 1$. This assumes, however, that λ has been estimated well enough that the scale can be removed as a parameter. To explore how uncertainties in λ affect construction of a standardized form, follow up in the references cited in Section 9.4.2 the techniques for estimating λ. Use typical uncertainties for your field of applications, then run cell ExponentialPDF to produce the corresponding graphics and numerical values.

```
(* ExponentialPDF cell *)


(* Exponential Distribution Density *)
ExpPDF[lambda_,x_] :=
  PDF[ExponentialDistribution[lambda],x]
```

Define exponential density distribution, (9.4.4).

```
(* Exponential Distribution Density: 3D views *)
ExpChart = Plot3D[
  ExpPDF[lambda,x],{x,0,2},{lambda,1,3},
  PlotRange->All,PlotPoints->{40,20},
  Ticks->{Automatic,Automatic,None},
  PlotLabel->"Exponential distribution: density",
  AxesEdge->{Automatic,{1,-1},Automatic},
  AxesLabel->{"x","lambda"," "}];
```

Surface of exponential density, Figure 9.4.4 left.

```
(* ExponentialPDF cell continued *)

(* Exponential Distribution Density: 2D views *)
ExpPlot[x_] := Plot[ExpPDF[lambda,x],
  {lambda,0.5,3},PlotRange->All,
  Ticks->{{0.5,1.0,1.5,2.0,2.5,3.0},None},
  Axes->{True,False},AxesOrigin->{0.6,0.0},
  AxesLabel->{"lambda",""},DisplayFunction->Identity]

Show[ExpPlot[0],ExpPlot[0.5],ExpPlot[1],
  DisplayFunction->$DisplayFunction];


(* Exponential Distribution Density; Numerical *)
lambda = 1
N[Table[ExpPDF[lambda,x],{x,0,4,1}],10]


(* Exponential Distribution Cumulative *)
ExpCum[lambda_,x_] :=
  CDF[ExponentialDistribution[lambda],x]


(* Exponential Distribution Cumulative: 3D views *)
ExpGraf = Plot3D[
  ExpCum[lambda,x],{x,0,2},{lambda,1,3},
  PlotRange->All,PlotPoints->{40,20},
  Ticks->{Automatic,Automatic,None},
  PlotLabel->"Exponential distribution: cumulative",
  AxesEdge->{Automatic,{1,-1},Automatic},
  AxesLabel->{"x","lambda"," "}];


(* Exponential Distribution Cumulative: 2D views *)
ExpCDFPlot[x_] := Plot[ExpCum[lambda,x],
  {lambda,0.5,3},PlotRange->All,
  Ticks->{{0.5,1.0,1.5,2.0,2.5,3.0},None},
  Axes->{True,False},AxesOrigin->{0.6,0.0},
  AxesLabel->{"lambda",""},DisplayFunction->Identity]

Show[ExpCDFPlot[0.5],ExpCDFPlot[1.0],ExpCDFPlot[1.5],
  DisplayFunction->$DisplayFunction];


(* Exponential Distribution Cumulative; Numerical *)
lambda = 1
N[Table[ExpCum[lambda,x],{x,0,4,1}],10]
```

Define curve of exponential density for x fixed.

Show curves for $x = 0, 0.5, 1$, Figure 9.4.4 right.

Numerical values to 10 digits of exponential density, Table 9.4.3.

Define exponential cumulative distribution, (9.4.6).

Surface of exponential cumulative, Figure 9.4.5 left.

Define curve of exponential cumulative for x fixed.

Show curves for $x = 0.5, 1.0, 1.5$, Figure 9.4.5 right.

Numerical values of exponential cumulative, Table 9.4.3.

The ParetoPDF cell [9.4.3]

If you execute this cell the computer will produce for the Pareto distribution the surface and curves in Figures 9.4.6 and 9.4.7. This will certainly make you wiser, but will probably not make you wealthier.

Exploring with `ParetoPDF`

Both the Pareto density and cumulative distributions are discontinuous, according to (9.4.8) and (9.4.9). This can be confusing, so it is worthwhile to explore the surfaces and curves of the two distributions.

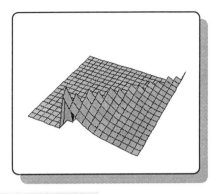

For example, look at the surfaces of the distributions from a different viewpoint to see the the sudden rise of the density and cumulative functions.

```
(* ParetoPDF cell *)


(* Pareto Distribution Density *)
ParetoPDF[a_,k_,x_] := 0 /; x < k
ParetoPDF[a_,k_,x_] := a*k^a/x^(a+1) /; x >= k


(* Pareto Distribution Density: 3D views *)
ParChart[a_] := Plot3D[
  ParetoPDF[a,k,x],{x,0,1.5},{k,0.5,1.5},
  PlotRange->All,PlotPoints->{40,30},
  Ticks->{Automatic,Automatic,None},
  PlotLabel->"a = "<>ToString[a]<>"",
  AxesEdge->{Automatic,{1,-1},Automatic},
  AxesLabel->{"x","k"," "}]

ParChart[1.5];


(* Pareto Distribution Density: 2D views *)
ParPlot[a_,x_] := Plot[ParetoPDF[a,k,x],
  {k,0.5,1.5},PlotRange->All,
  Ticks->{{0.5,1.0,1.5},None},
  Axes->{True,False},AxesOrigin->{0.6,0.0},
  AxesLabel->{"k",""},DisplayFunction->Identity]

Show[ParPlot[1.0,1.0],ParPlot[1.0,1.5],
  ParPlot[1.0,2.0],
  DisplayFunction->$DisplayFunction];
Show[ParPlot[2.0,1.0],ParPlot[2.0,1.5],
  ParPlot[2.0,2.0],
  DisplayFunction->$DisplayFunction];


(* Pareto Distribution Density; Numerical *)
a = 1.5
k = 1
N[Table[ParetoPDF[a,k,x],{x,1.0,2.5,0.5}],10]
```

Define Pareto density in two regions of *x*.

Define surface of Pareto density distribution as a function of parameter *a*.

Surface for *a* = 1.5, Figure 9.4.6 left.

Define curve of Pareto density over *k* for *a* and *x* fixed.

Show density curves for *a* = 1 then for *a* = 2, Figure 9.4.6 right.

Pareto distribution numerical densities to 10 digits for Table 9.4.4.

```
(* ParetoPDF cell continued *)

(* Pareto Distribution Cumulative *)
ParCDF[a_,k_,x_] := 0 /; x < k
ParCDF[a_,k_,x_] := 1 - (k/x)^a /; x >= k


(* Pareto Distribution Cumulative: 3D views *)
ParGraf[a_] := Plot3D[
  ParCDF[a,k,x],{x,0,1.5},{k,0.5,1.5},
  PlotRange->All,PlotPoints->{40,30},
  Ticks->{Automatic,Automatic,None},
  PlotLabel->"a = "<>ToString[a]<>"",
  AxesEdge->{Automatic,{1,-1},Automatic},
  AxesLabel->{"x","k"," "}]

ParGraf[1.5];


(* Pareto Distribution Cumulative: 2D views *)
ParCDFPlot[a_,x_] := Plot[ParCDF[a,k,x],
  {k,0.5,1.5},PlotRange->All,
  Ticks->{{0.5,1.0,1.5},None},
  Axes->{True,False},AxesOrigin->{0.6,0.0},
  AxesLabel->{"k",""},DisplayFunction->Identity]

Show[ParCDFPlot[1.0,1.0],ParCDFPlot[1.0,1.5],
  ParCDFPlot[1.0,2.0],
  DisplayFunction->$DisplayFunction];
Show[ParCDFPlot[2.0,1.0],ParCDFPlot[2.0,1.5],
  ParCDFPlot[2.0,2.0],
  DisplayFunction->$DisplayFunction];


(* Pareto Distribution Cumulative; Numerical *)
a = 1.5
k = 1
N[Table[ParCDF[a,k,x],{x,1.0,2.5,0.5}],10]
```

Define Pareto cumulative in two regions of x.

Define surface of Pareto cumulative distribution as a function of a.

Surface for $a = 1.5$, Figure 9.4.7 left.

Define curve of Pareto cumulative over k for a and x fixed.

Show cumulative curves for $a = 1$ then for $a = 2$, Figure 9.4.7 right.

Pareto distribution numerical cumulative distributions to 10 digits, Table 9.4.4.

The WeibullPDF cell [9.4.4]

Execute this cell to produce for the Weibull distribution defined by the density and cumulative distributions (9.4.12) and (9.4.13), respectively, the surfaces and curves in Figures 9.4.8 and 9.4.9.

Exploring with WeibullPDF

The Weibull density distribution is either expressed in the general form (9.4.11) or in the centralized and standardized form (9.4.12), which has $\alpha = 1$ and $\beta = 0$. This presupposes that both α and β have been estimated well

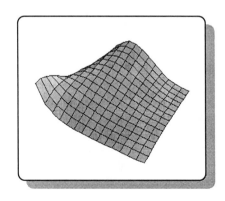

enough that the scale and origin can be removed as parameters, even if the third parameter, c, is not well characterized.

To explore how uncertainties in α and β affect construction of a standardized form, follow up in the references cited in Section 9.4.4 the techniques for estimating α and β Use typical uncertainties for your field of applications, then run cell `WeibullPDF` to produce the corresponding graphics and numerical values. Do this for a range of c values that are appropriate to your investigation.

```
(* WeibullPDF cell *)


(* Weibull Standardized Distribution: Density *)
WbllPDF[c_,x_] := PDF[WeibullDistribution[c,1],x]
```
Define standardized Weibull density (9.4.12).

```
(* Weibull Standardized Distribution Density:
   3D views *)
WbllChart = Plot3D[
  WbllPDF[c,x],{x,10^(-1),2},{c,0.5,2.5},
  PlotRange->All,PlotPoints->{30,30},
  Ticks->{Automatic,Automatic,None},
  PlotLabel->"Weibull distribution: density",
  AxesEdge->{Automatic,{1,-1},Automatic},
  AxesLabel->{"x"," c"," "}];
```
Surface of standardized Weibull density, Figure 9.4.8 left.

```
(* Weibull Distribution Density: 2D views *)
WbllPlot[x_] := Plot[WbllPDF[c,x],
  {c,0.5,2.5},PlotRange->All,
  Ticks->{{0.5,1.0,1.5,2.0,2.5},None},
  Axes->{True,False},AxesOrigin->{0.6,0.0},
  AxesLabel->{"c",""},DisplayFunction->Identity]

Show[WbllPlot[0.5],WbllPlot[1.5],
  DisplayFunction->$DisplayFunction];
```
Define curve of standardized Weibull density with c varying and x fixed. Show density curves for $x = 0.5, 1.5$, Figure 9.4.8 right.

```
(* Weibull Distribution Density; Numerical *)
c = 1
N[Table[WbllPDF[c,x],{x,0.5,2,0.5}],10]
c = 2
N[Table[WbllPDF[c,x],{x,0.5,2,0.5}],10]
```
Numerical tables of standardized Weibull density to 10 digits, Table 9.4.5.

```
(* Weibull Standardized Distribution: Cumulative *)
WbllCumDF[c_,x_] := CDF[WeibullDistribution[c,1],x]
```
Define standardized Weibull cumulative distribution (9.4.12).

```
(* Weibull Distribution Cumulative: 3D views *)
WbllGraf = Plot3D[
  WbllCumDF[c,x],{x,10^(-1),2},{c,0.5,2.5},
  PlotRange->All,PlotPoints->{30,30},
  Ticks->{Automatic,Automatic,None},
  PlotLabel->"Weibull distribution: cumulative",
  AxesEdge->{Automatic,{1,-1},Automatic},
  AxesLabel->{"x"," c"," "}];
```
Surface of standardized Weibull cumulative distribution, Figure 9.4.9 left.

```
(* WeibullPDF cell continued *)

(* Weibull Distribution Cumulative: 2D views *)
WbllCDFPlot[x_] := Plot[WbllCumDF[c,x],
  {c,0.5,2.5},PlotRange->All,
  Ticks->{{0.5,1.0,1.5,2.0,2.5},None},
  Axes->{True,False},AxesOrigin->{0.6,0.0},
  AxesLabel->{"c",""},DisplayFunction->Identity]

Show[WbllCDFPlot[0.5],WbllCDFPlot[1.5],
  DisplayFunction->$DisplayFunction];

(* Weibull Distribution Cumulative; Numerical *)
c = 1
N[Table[WbllCumDF[c,x],{x,0.5,2,0.5}],10]
c = 2
N[Table[WbllCumDF[c,x],{x,0.5,2,0.5}],10]
```

Define curve of standardized Weibull cumulative distribution with c varying x fixed. Show cumulative distribution curves for $x = 0.5$, 1.5, Figure 9.4.9 right.

Numerical tables of standardized Weibull cumulative distribution, Table 9.4.6.

The LogisticPDF cell [9.4.5]

By executing this cell you will produce the surface and curves shown in Figures 9.4.10 and 9.4.11 for the standardized logistic density and cumulative distributions, (9.4.16) and (9.4.15), respectively.

Exploring with LogisticPDF

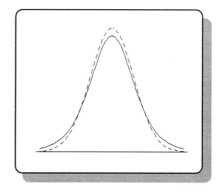

The logistic distribution is often substituted for the normal (Gauss) distribution presented in Section 9.3.1 or for the Cauchy (Lorentz) distribution in Section 9.4.1. We compare in Figure 9.4.10, and in the graphic at right, centralized logistic (solid curve) and normal distributions (dashed curve) having the same full width at half maximum (FWHM). Here are suggestions for exploring the similarities and differences between the logistic and normal probability distributions.

First, do the necessary mathematical analysis to find the relations between the parameters of these two centralized distributions if they have the same variance. The tables of moments—Tables 9.3.1 and 9.4.1—are useful for this. Then modify cell LogisticPDF so that graphical and numerical comparisons are made. To explore further, especially if the wings of the distributions are of consequence, repeat the analysis and calculations for the two distributions under the condition that their fourth moments (or the coefficients γ_2) are the same. Note that one of the distributions has to be not standardized for such conditions to be satisfied.

```
(* LogisticPDF cell *)

(* Logistic Standardized Distribution: Density *)
LgstcPDF[x_] := PDF[LogisticDistribution[0,1],x]
```

Define standardized logistic density, (9.4.16).

```
(* Logistic Density: 2D view *)
LgstcChart = Plot[LgstcPDF[x],{x,-5,5},
  PlotRange->All,PlotPoints->50,
  Ticks->{Automatic,None},
  PlotLabel->"Logistic distribution: density",
  AxesLabel->{"x"," "}];

(* Logistic & Gauss Probability Densities *)
Logistic = Plot[LgstcPDF[x],{x,-5,5},
  PlotRange->All,PlotPoints->50,
  Ticks->{Automatic,None},AxesLabel->{"x"," "},
  DisplayFunction->Identity];
Gauss = Plot[0.266470*Exp[-0.223072*x^2],{x,-5,5},
  PlotRange->All,PlotPoints->50,
  PlotStyle->{Dashing[{0.02,0.02}]},
  Ticks->{Automatic,None},AxesLabel->{"x"," "},
  DisplayFunction->Identity];
GandL = Show[Gauss,Logistic,
  DisplayFunction->$DisplayFunction];


(* Logistic Distribution Density; Numerical *)
N[Table[LgstcPDF[x],{x,-2,2,2}],10]


(* Logistic Standardized Distribution: Cumulative *)
LgstcCumDF[x_] := CDF[LogisticDistribution[0,1],x]


(* Logistic Distribution Cumulative; 2D view *)
LgstcGraf = Plot[LgstcCumDF[x],{x,-5,5},
  PlotRange->All,PlotPoints->50,
  Ticks->{Automatic,None},
  PlotLabel->"Logistic distribution: cumulative",
  AxesLabel->{"x"," "}];

(*  Logistic Distribution Cumulative; Numerical *)
LgstcCumDF[x_] := CDF[LogisticDistribution[0,1],x]
N[Table[LgstcCumDF[x],{x,-2,2,2}],10]
```

Curve of logistic density, Figure 9.4.10 left.

Compare logistic and Gaussian densities having same FWHM, Figure 9.4.10 right.

Numerical standardized logistic density in Table 9.4.7.

Define standardized logistic cumulative, (9.4.15).

Curve of logistic cumulative, Figure 9.4.11.

Numerical standardized logistic cumulative in Table 9.4.7.

The `LaplacePDF` cell [9.4.6]

Execute this cell to produce for the Laplace (double-sided exponential) distribution the surface and curves displayed in Figures 9.4.12 and 9.4.13.

Exploring with `LaplacePDF`

We give the Laplace distribution in the standardized form (9.4.18). This presupposes that any scaling parameter for the variable x has been estimated well enough that the scale parameter can be removed. To explore how uncertainties in the parameters affect construction of a standardized

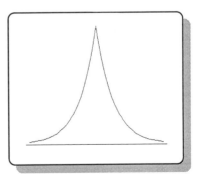

form, follow up in the references cited in Section 9.4.6 the techniques for estimating the scale parameter. Use typical uncertainties for your field of applications, then run cell `LaplacePDF` to produce the corresponding graphics and numerical values.

```
(* LaplacePDF cell *)


(* Laplace Distribution Density *)
LaplacePDF[x_] := PDF[LaplaceDistribution[0,1],x]


(* Laplace Distribution Density: 2D view *)
LaplaceChart = Plot[ LaplacePDF[x],{x,-4,4},
  PlotRange->All,PlotPoints->40,
  Ticks->{Automatic,None},
  PlotLabel->"Laplace distribution: density",
  AxesLabel->{"x"," "}];


(* Laplace Distribution Density; Numerical *)
N[Table[LaplacePDF[x],{x,-2,2,2}],10]


(* Laplace Distribution Cumulative*)
LaplaceCumDF[x_] := CDF[LaplaceDistribution[0,1],x]


(* Laplace Distribution Cumulative: 2D view *)
LaplaceGraf = Plot[ LaplaceCumDF[x],{x,-4,4},
  PlotRange->All,PlotPoints->40,
  Ticks->{Automatic,None},
  PlotLabel->"Laplace distribution: cumulative",
  AxesLabel->{"x"," "}];


(* Laplace Distribution Cumulative; Numerical *)
N[Table[LaplaceCumDF[x],{x,-2,2,2}],10]
```

Define centralized and standardized Laplace distribution density, (9.4.18).

Curve of Laplace density, Figure 9.4.12.

Numerical densities to 10 digits, Table 9.4.8.

Define Laplace cumulative distribution, (9.4.19).

Curve of Laplace cumulative, Figure 9.4.13.

Numerical cumulative for Table 9.4.8.

The `KolmogorovSmirnovPDF` cell [9.4.7] Execute this cell to produce for the distribution of Kolmogorov and Smirnov the curves of density and cumulative distributions in Figure 9.4.15.

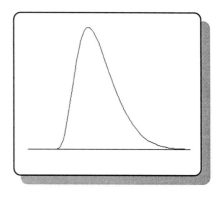

Exploring with `KolmogorovSmirnovPDF` Investigate the rate of convergence of the density and cumulative distributions, (9.4.21) and (9.4.22), by using the cell `KolmogorovSmirnovPDF`. To do this, change the upper limit on j in functions `KSPDF` and `KSCumDF` from 20 to a variable that you control. Since convergence of both the density and cumulative distribution sums depends on the product jx rather than on only j, it is appropriate for the upper limit in `NSum[]` to depend on x.

Another way to explore the convergence of the two series is to use the C functions `KSDnsty` and `KSQCum` (Section 9.4.7) and to output the values of `fj` for a given accuracy criterion `eps` that is input to the functions.

```
(* KolmogorovSmirnovPDF cell *)


(* Kolmogorov-Smirnov PDF: Density *)
KSPDF[x_] := 8*x*NSum[
    Re[(-1)^(j-1)*j^2*Exp[-2*j^2*x^2]],{j,1,20}];


(* Kolmogorov-Smirnov Density: 2D view *)
KSChart = Plot[Evaluate[KSPDF[x]],{x,0.1,2},
    Ticks->{Automatic,None},
    PlotLabel->"Kolmogorov-Smirnov: density"];


(* Kolmogorov-Smirnov Density: Numerical *)
N[Table[KSPDF[x],{x,0.5,2.0,0.5}],10]


(* Kolmogorov-Smirnov PDF: 1 - Cumulative *)
KSCumDF[x_] := 2*NSum[
    Re[(-1)^(j-1)*Exp[-2*j^2*x^2]],{j,1,20}];


(* Kolmogorov-Smirnov: 1 - Cumulative: 2D view *)
KSGraf = Plot[Evaluate[KSCumDF[x]],{x,0.1,2},
    Ticks->{Automatic,None},
    PlotLabel->"Kolmogorov-Smirnov: cumulative"];


(* Kolmogorov-Smirnov Cumulative: Numerical *)
N[Table[KSCumDF[x],{x,0.5,2.0,0.5}],10]
```

Define K-S density distribution by (9.4.21).

Curve of K-S density, Figure 9.4.15 left.

Numerical values of K-S density in Table 9.4.9.

Define K-S Q-cumulative distribution by (9.4..22).

Curve of K-S Q-cumulative distribution, Figure 9.4.15 right.

Numerical K-S Q-cumulative distribution in Table 9.4.9.

The `BetaPDF` cell [9.4.8]

Execute this cell to produce for the beta probability distribution the surfaces in Figures 9.4.15 and 9.4.16. Similar graphics are produced for the regularized incomplete beta functions (Section 6.3.2) and the F-distribution functions (Section 9.3.4).

Exploring with `BetaPDF`

The beta distribution depends upon two parameters, p and q, so it can be visualized from many viewpoints. In `BetaPDF` we chose p as the variable that is held fixed, while surfaces as functions of x and q are drawn.

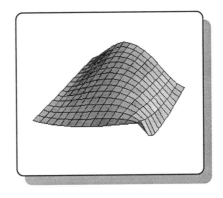

Explore how to envision the symmetry relations $B_{qp}(1-x) = B_{pq}(x)$ for the density distribution and $C_{qp}(1-x) = 1 - C_{pq}(x)$ for the cumulative distribution, which follow from the symmetries of the regularized incomplete beta functions in Section 6.3.2.

```
(* BetaPDF cell *)

(* Beta Distribution Probability Density *)
BetaPDF[p_,q_,x_] :=
  PDF[BetaDistribution[p,q],x]
```
Define beta distribution density, (9.4.24).

```
(* Beta Distribution Density: 3D views *)
BetaChart[p_] := Plot3D[BetaPDF[p,q,x],
  {x,0,0.99},{q,1.01,3},
  PlotRange->All,PlotPoints->{20,20},
  Ticks->{Automatic,Automatic,None},
  PlotLabel->"Beta density: p = "<>ToString[p]" ",
  AxesEdge->{Automatic,{1,-1},Automatic},
  AxesLabel->{"x"," q"," "}]
```
Define density surface as a function of p.

```
BetaChart[1.5];  BetaChart[2.5];
```
Show surfaces for $p = 1.5, 2.5$, Figure 9.4.15.

```
(* Beta Probability Density: Numerical *)
q = 2
p = 1.5
N[Table[N[BetaPDF[p,q,x]],{x,0.25,0.75,0.25}],10]
q = 2
p = 2.5
N[Table[N[BetaPDF[p,q,x]],{x,0.25,0.75,0.25}],10]
```
Beta density numerical to 10 digits, Table 9.4.10.

```
(* Beta Distribution Cumulative *)
BetaCumDF[p_,q_,x_] :=
  CDF[BetaDistribution[p,q],x]
```
Define beta distribution cumulative, (9.4.25).

```
(* Beta Distribution Cumulative: 3D views *)
BetaCumGraf[p_] := Plot3D[BetaCumDF[p,q,x],
  {x,0,0.99},{q,1.01,3},
  PlotRange->All,PlotPoints->{20,20},
  Ticks->{Automatic,Automatic,None},
  PlotLabel->"Beta cumulative: p = "<>ToString[p]" ",
  AxesEdge->{Automatic,{1,-1},Automatic},
  AxesLabel->{"x"," q"," "}]
```
Define cumulative distribution surface as a function of p.

```
BetaCumGraf[1.5];  BetaCumGraf[2.5];
```
Show surfaces for $p = 1.5, 2.5$, Figure 9.4.16.

```
(* Beta Distribution Cumulative: Numerical *)
q = 2
p = 1.5
N[Table[N[BetaCumDF[p,q,x]],{x,0.25,0.75,0.25}],10]
q = 2
p = 2.5
N[Table[N[BetaCumDF[p,q,x]],{x,0.25,0.75,0.25}],10]
```
Beta cumulative numerical to 10 digits, Table 9.4.11.

20.10 ERROR FUNCTION, FRESNEL AND DAWSON INTEGRALS

Notebook and Cells
Notebook `ErFD`; cells `Erf`, `Fresnel`,
`FresnelApprox`, `Dawson`

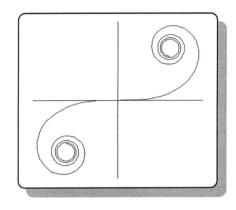

Overview
This *Mathematica* notebook—including
the error function (Section 10.1), Fresnel
cosine and sine integrals (Section 10.2),
and Dawson integral (Section 10.3)—provides relatively simple visualizations of
these functions. Since the functions are
univariate (or at most bivariate for the error function) with no parameters, they can
usually be represented in 2D curves, as we
show in the adjacent motif for the Cornu spiral and the motifs for each of the three cells
that follow.

The `Erf` cell [10.1]
Execute this cell to produce for the error function the surfaces in Figures 10.1.1, the curve in
Figure 10.1.2, and test values in Table 10.1.1.
Note that the curve on the front surface of Figure 10.1.1 (real part of the error function) gives
Figure 10.1.2, which is the cumulative distribution function of the normal (Gauss) distribution in Section 9.3.1.

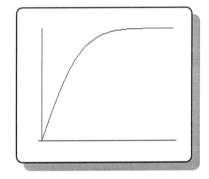

Exploring with `Erf`
Use the high accuracy of arithmetic in *Mathematica* to check the accuracy of the rational-
fraction approximations given in equations 7.1.26–7.1.28 of Abramowitz and Steguns's
handbook. Develop a rational-fraction expansion to about 10-digit accuracy by using Padé
approximants as implemented in the *Mathematica* package `Calculus`Pade``. This provides an efficient alternative to computing the error function as an incomplete gamma
function, as we do in Section 10.1.

```
(* Erf cell *)

(* Error Function of Complex Argument *)

ErfRealPlot = Plot3D[Re[Erf[x+ I*y]],
    {x,0.001,3},{y,0,1},PlotRange->All,PlotPoints->30,
    Ticks->{Automatic,Automatic,None},
    AxesLabel->{"x"," y",""}];
```

Surface of real part of
error function of a
complex variable,
Figure 10.1.1 left.

```
(* Erf cell continued *)

ErfImagPlot = Plot3D[Im[Erf[x+ I*y]],
  {x,0.001,3},{y,0,1},PlotRange->All,PlotPoints->30,
  Ticks->{Automatic,Automatic,None},
  AxesEdge->{Automatic,{1,-1},None},
  AxesLabel->{"x"," y",""}];


(* Error Function in 2D; front of 3D Real plot *)
ErfPlot = Plot[Erf[x],{x,0,3},PlotRange->{0,1},
  AxesOrigin->{0,0},AxesLabel->{"x","erf(x)"}];


(* Error Function: Numerical *)
ErfNum = N[Table[{x,Erf[x]},{x,0.5,1.5,0.5}],10]
```

Surface of imaginary part of error function of a complex variable, Figure 10.1.1 right.

Curve of error function, Figure 10.1.2.

Numerical values to 10 digits, Table 10.1.1.

The `Fresnel` cell [10.2]

By executing this cell you will produce surfaces of the Fresnel integrals in the complex plane (Figure 10.2.1), curves of the conventional Fresnel integrals (Figure 10.2.2), and of the Cornu spiral (Figure 10.2.3). Numerical values of the Fresnel integrals (Tables 10.2.1 and 10.2.2)will also be obtained.

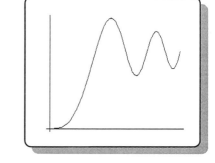

Exploring with `Fresnel`

Fresnel is not known for his geographical explorations, but rather for his discoveries in optics. You can discover similar visual delights by using the `Fresnel` cell. For example, how do the inner parts of the Cornu spiral depend upon the range of the parameter *t* in (10.2.6)?

For numerical applications, check the accuracy that is claimed for our C functions in Section 10.2 by generating the table objects `FresnelCosNum` and `FresnelInNum` over appropriate ranges of x and with controlled accuracy in the arithmetic done by *Mathematica*. If you need more accuracy than our rational-fraction approximations developed in `FresnelApprox` cell provide (better than 10^{-12}), follow up the references given in Section 10.2 and check out their algorithms by using the numerical parts of this cell.

```
(* Fresnel cell *)

(* Error Function in Form for Fresnel Integrals *)
FresnelErf[z_] := (1+I)*Erf[Sqrt[N[Pi]]*(1-I)*z/2]/2


FresnelRealPlot = Plot3D[Re[FresnelErf[x+ I*y]],
  {x,0.001,3},{y,0,1},PlotRange->All,PlotPoints->30,
  Ticks->{Automatic,Automatic,None},
  AxesLabel->{"x"," y",""}];
```

Define Fresnel integral by (10.2.5).

Surface of real part of Fresnel integral, Figure 10.2.1 left.

```
FresnelImagPlot = Plot3D[Im[FresnelErf[x+ I*y]],
  {x,0.001,3},{y,0,1},PlotRange->All,PlotPoints->30,
  Ticks->{Automatic,Automatic,None},
  AxesEdge->{Automatic,Automatic,None},
  AxesLabel->{"x"," y",""}];
```

Surface of imaginary part of Fresnel integral, Figure 10.2.1 right.

```
(* Fresnel Integrals in 2D; fronts of 3D plots *)
FresnelCos[x_] := Re[FresnelErf[x]]
FresnelSin[x_] := Im[FresnelErf[x]]
CosIntPlot = Plot[FresnelCos[x],{x,0,3},
  AxesLabel->{"x","C(x)"}];
SinIntPlot = Plot[FresnelSin[x],{x,0,3},
  AxesLabel->{"x","S(x)"}];
```

Curves of Fresnel cosine and sine integrals, Figure 10.2.2.

```
(* Fresnel Integrals: Numerical *)
FresnelCosNum = N[Table[
  {x,FresnelCos[x]},{x,0.5,1.5,0.5}],10]
FresnelSinNum = N[Table[
  {x,FresnelSin[x]},{x,0.5,1.5,0.5}],10]
```

Numerical integrals to 10 digits, Tables 10.2.1 and 10.2.2.

```
(*  Cornu Spiral *)
Cornu = ParametricPlot[{FresnelCos[t],FresnelSin[t]},
  {t,-4,4},PlotPoints->40,AspectRatio->1,Ticks->None];
```

Cornu spiral, Figure 10.2.3.

The FresnelApprox cell [10.2]
Execute this cell to investigate the auxiliary functions *f* and *g* for the Fresnel integrals (Figure 10.2.4), such as *f* shown at right. Rational-fraction approximations to these functions can also be obtained.

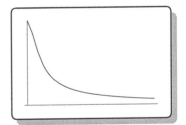

Exploring with FresnelApprox
Use the FresnelApprox cell to explore how the accuracy of the Fresnel integrals depends on the order of the polynomials used for the rational-fraction expressions. The *Atlas* uses 10th-order numerator polynomials and 11th-order denominator polynomials to give *f* and *g* to fractional accuracy of at least 15 digits. If you need less accuracy than this, run FresnelApprox with polynomials of lower order to find the polynomial coefficients. After inserting these in C function FresnelCS, you will have a function that executes more efficiently but with decreased accuracy.

```
(* FresnelApprox cell *)

 (* Approximations for Fresnel integral
    auxiliary functions f and g in x range 1 to 6 *)

<<NumericalMath`Approximations`;

(* Error Function in Form for Fresnel Integrals *)
FErf[z_] := (1+I)*Erf[Sqrt[N[Pi]]*(1-I)*z/2]/2
FCos[x_] := Re[FErf[x]]
FSin[x_] := Im[FErf[x]]
```

Load package for rational interpolation.
Define Fresnel cosine and sine integrals.

```
(* Auxiliary Functions: Definitions *)
FF[x_] :=
(1/2-FSin[x])*Cos[Pi*x*x/2]-(1/2-FCos[x])*Sin[Pi*x*x/2]
GG[x_] :=
(1/2-FCos[x])*Cos[Pi*x*x/2]+(1/2-FSin[x])*Sin[Pi*x*x/2]

(* Auxiliary Functions: 2D views *)
FFPlot = Plot[FF[x],{x,0,10},PlotRange->{0,0.5}];
GGPlot = Plot[GG[x],{x,0,10},PlotRange->{0,0.5}];

(* Rational approximation of  f  in 0.5 to 6 *)
FFPadeBias = RationalInterpolation[
FF[x],{x,10,11},{x,0.5,6},Bias->-0.5]


(* Check near  x = 1 *)
FFTableBias1 = Table[
  {x,N[FFPadeBias - FF[x],12]},{x,0.5,1.5,0.1}]
(* Check near  x = 5 *)
FFTableBias5 = Table[
  {x,N[FFPadeBias - FF[x],12]},{x,4,6,0.2}]


(* Rational approximation of  g  in 0.5 to 6 *)
GGPadeBias = RationalInterpolation[
  GG[x],{x,10,11},{x,0.5,6},Bias->-0.5]


(* Check near  x = 1 *)
GGTableBias1 = Table[
  {x,N[GGPadeBias - GG[x],12]},{x,0.5,1.5,0.1}]
  (* Check near  x = 5 *)
GGTableBias5 = Table[
  {x,N[GGPadeBias - GG[x],12]},{x,4,6,0.2}]
```

Define auxiliary functions f and g by (10.2.9) and (10.2.10).

Curves of f and g, Figure (10.2.4).

Approximate f by fraction with polynomials of order 10 and 11.

Numerical values of differences at $x = 1$ and $x = 5$.

Approximate g by fraction with poly nomials of order 10 and 11.

Numerical values of differences at $x = 1$ and $x = 5$.

The Dawson cell [10.3]

Execute this cell to produce for the Dawson integral (10.3.1) the curves in Figures 10.3.1 and 10.3.2, as well as the numerical values in Table 10.3.1.

Exploring with Dawson

To optimize our C function FDawson (Section 10.3) find an appropriate x value at which to switch from expansion (10.3.5) to the asymptotic expansion (10.3.4) depending upon the accuracy parameter eps. To do this, modify the parts of the Dawson cell that generate terms in the asymptotic series and add coding for (10.3.4) so that the accuracy of both expansions can be mapped out as functions of x and eps.

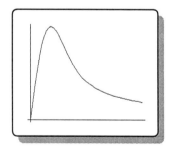

```
(* Dawson cell *)

(* Dawson Integral as a complex Error Function *)
Dawson[x_] := I*Sqrt[N[Pi]]*Erf[-I*x]/(2*Exp[x^2])
```

Define Dawson integral by (10.3.1).

```
(* Dawson cell continued *)

DawsonPlot = Plot[Dawson[x],{x,0,5},
  AxesLabel->{"x",""},
  DisplayFunction->Identity];
```

Curve of Dawson integral, solid curve in Figure 10.3.1.

```
(* First two terms of asymptotic expansion *)
DawsonAsymptote = Plot[(0.5/x)*(1+0.5/x^2),{x,1.5,5},
  PlotStyle->{Dashing[{0.02,0.02}]},
  DisplayFunction->Identity];
```

Curve of two terms of asymptotic expansion (10.3.5), dashed segment in Figure 10.3.1.

```
DawsonBoth = Show[DawsonPlot,DawsonAsymptote,
  DisplayFunction->$DisplayFunction];
```

Show both curves.

```
(* Dawson Integral: Numerical *)
DawsonNum = Table[{x,N[Dawson[x],10]},{x,2,8,2}]
```

Numerical values to 10 digits, Table 10.3.1.

```
(* Dawson Integral: Terms in the Asymptotic Series *)
DawAsymTerms[x_] := Table[
  {j,Log[10,(2*j-1)!!/((2*x^2)^j)]},{j,0,20}]
```

Define table of terms in asymptotic series as function of x, then define plot for each x value.

```
DawAsymPic[x_] := ListPlot[DawAsymTerms[x],
  AxesOrigin->{0,0}, PlotJoined->True,
  PlotStyle->{AbsoluteThickness[0.4*x]},
  DisplayFunction->Identity]
```

```
Show[DawAsymPic[2],DawAsymPic[4],DawAsymPic[6],
  AxesLabel->{"j",""},
  DisplayFunction->$DisplayFunction];
```

Show curves together, Figure 10.3.2.

20.11 ORTHOGONAL POLYNOMIALS

Notebooks and Cells
Notebooks OrthoPIter, OrthoP; cells ChebyT, ChebyU, Gegenbauer, Hermite, Laguerre, Legendre, Jacobi

Overview
There are two *Mathematica* notebooks for orthogonal polynomials. The first notebook, OrthoPIter, generates iteratively (Section 3.2) formulas for the seven orthogonal polynomials in Chapter 11; Chebyshev polynomials of the first kind (Section 11.2.1) and of the second kind (Section 11.2.2), the Gegenbauer (ultra-

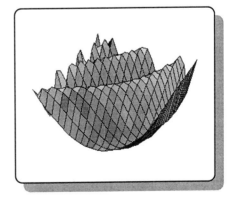

spherical) polynomials (Section 11.3), Hermite polynomials (Section 11.4), generalized Laguerre polynomials (Section 11.5), Legendre polynomials (Section 11.6), and Jacobi polynomials (Section 11.5). Table 11.1.3 summarizes the recurrence relations used in this notebook.

The second *Mathematica* notebook is for visualizations and numerical values of the orthogonal polynomials. Being polynomials, the functions in this chapter are all straightforward to compute numerically, both as C functions (Chapter 11) and as *Mathematica* functions. Their visualization is interesting because they must oscillate in sign with their argument in order to be orthogonal, as seen in the adjacent motif of the surface of the Chebyshev polynomial of the first kind, $T_n(x)$, as a function of x and n. Since there is at least one parameter—the order of the polynomial, n —the visualizations are usually best shown as surfaces. For the more general polynomials—Gegenbauer, Laguerre, and Jacobi—there are additional parameters, so that ingenuity is needed to display them.

Notebook `OrthoPIter`

By executing this notebook you will obtain algebraic expressions for the orthogonal polynomials of lowest order, n, that are given explicitly in Chapter 11. You can extend the expressions to arbitrary high n by invoking `TableOP[name_,n]` with the n of your choice.

```
(* Orthogonal Polynomials by Iteration *)

(* Iteration to get  OrthoP  up to order maxn_
   as a function of x, given the starting
   array elements OrthoP[0] and OrthoP[1] *)

OrthoIter[maxn_] :=
 Do[
  (
  OrthoP[i] = Simplify[((b2[i]+b3[i]*x)*OrthoP[i-1]-
        b4[i]*OrthoP[i-2])/b1[i]]), {i,2,maxn,1}]

(* Table of polynomial formulas *)
TableOP[name_,maxn_] :=
  (
  OrthoIter[maxn];
  TableForm[Table[{n,Collect[OrthoP[n],x]},{n,0,maxn}],
  TableHeadings->{None,{"n",""<>ToString[name]<>"(x)"}}]
  )


(* Chebyshev polynomials of first kind, T *)
(* Starting elements *)
OrthoP[0] = 1;  OrthoP[1] = x;
(* Rules for the b_n *)
b1[n_] := 1; b2[n_] := 0; b3[n_] := 2; b4[n_] := 1;
(* Chebyshev polynomials of first kind up to n = 7 *)
TableOP[Tn,7]

(* Chebyshev polynomials of second kind, U *)
(* Starting elements *)
OrthoP[0] = 1;  OrthoP[1] = 2*x;
(* Rules for the b_n *)
b1[n_] := 1; b2[n_] := 0; b3[n_] := 2; b4[n_] := 1;
(* Chebyshev polynomials of second kind up to n = 7 *)
TableOP[Un,7]
```

Define general iteration function as function of maximum *n*. Iteration relation (11.1.3) for *i*th element.

Arrange polynomials into a labelled table.

See Table 11.1.3.

Table 11.2.1.

See Table 11.1.3.

Table 11.2.3.

```
(* Orthogonal Polynomials by Iteration continued *)


(* Gegenbauer (ultraspherical) polynomials, C(a)  *)
(* Starting elements *)
Clear[a]; OrthoP[0] = 1; OrthoP[1] = 2*a*x;
(* Rules for the b_n *)
b1[n_] := n;  b2[n_] := 0;
b3[n_] := 2*(n+a-1); b4[n_] := n+2*a-2;
(* Gegenbauer polynomials up to n = 3 *)
TableOP[Cna,3]


(* Hermite polynomials, H  *)
(* Starting elements *)
OrthoP[0] = 1; OrthoP[1] = 2*x;
(* Rules for the b_n *)
b1[n_] := 1; b2[n_] := 0;
b3[n_] := 2; b4[n_] := 2*n-2;
(* Hermite polynomials up to n = 9 *)
TableOP[Hn,9]


(* Laguerre polynomials, L(a)  *)
(* Starting elements *)
Clear[a]; OrthoP[0] = 1; OrthoP[1] = 1+a-x;
(* Rules for the b_n *)
b1[n_] := n; b2[n_] := 2*n+a-1;
b3[n_] := -1; b4[n_] := n+a-1;
(* Laguerre polynomials up to n = 3 *)
TableOP[Lna,3]


(* Legendre polynomials, P  *)
(* Starting elements *)
OrthoP[0] = 1; OrthoP[1] = x;
(* Rules for the b_n *)
b1[n_] := n; b2[n_] := 0;
b3[n_] := 2*n-1; b4[n_] := n-1;
(* Legendre polynomials up to n = 7 *)
TableOP[Pn,7]


(* Jacobi polynomials, P(a,b)  *)
(* Starting elements *)
Clear[a,b]; OrthoP[0] = 1;
OrthoP[1] = (a-b+(2+a+b)*x)/2;
(* Rules for the b_n *)
b1[n_] := 2*n*(n+a+b)*(2*n+a+b-2);
b2[n_] := (2*n+a+b-1)*(a^2-b^2);
b3[n_] := (2*n+a+b-2)*(2*n+a+b-1)*(2*n+a+b);
b4[n_] := 2*(n+a-1)*(n+b-1)*(2*n+a+b);
(* Jacobi polynomials up to n = 2 *)
TableOP[Pnab,2]
```

See
Table 11.1.3.

Table 11.3.1.

See
Table 11.1.3.

Table 11.4.1.

See
Table 11.1.3.

Table 11.5.1.

See
Table 11.1.3.

Table 11.6.1.

See
Table 11.1.3.

Table 11.7.1.

Notebook OrthoP

This notebook contains the cells for visualizing and obtaining numerical values of the orthogonal polynomials in Chapter 11.

The following cell at the head of notebook OrthoP must be executed first in order to define a function that builds numerical tables of the polynomials with argument x and a single parameter—the polynomial order n. This is used for both Chebyshev polynomials (cells ChebyT and ChebyU), for the Hermite polynomials (cell Hermite), and for the Legendre polynomials (cell Legendre).

```
(* You MUST execute this cell first *)

(* For polynomials depending on just n and x *)
NumTableOP[poly_,name_,maxn_,xmin_,xmax_,xstep_] :=
   N[TableForm[Table[{n,poly[n,x]},
   {x,xmin,xmax,xstep},{n,0,maxn}],
   TableHeadings->
     {None,{"n",""<>ToString[name]<>"(x)"}}],10]
```

Define numerical table to 10 digits of polynomials depend–ing on just the order n and the argument x.

The ChebyT cell [11.2.1]

Execute this cell to produce (Figure 11.2.1) the surface of the Chebyshev polynomial of the first kind and the curves as a function of x for n fixed. Also, a table of numerical values to check the C function is produced for Table 11.2.2.

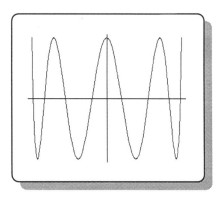

Exploring with ChebyT

How does the Chebyshev polynomial of the first kind behave if x is fixed and n is varied, giving a transect of the surface in Figure 11.2.1? To investigate this, make a function like ChebyFirst2D[] in the ChebyT cell, but let its variable be x and the variable in Plot be n.

```
(* ChebyT cell *)
(* Chebyshev Polynomials of First Kind, Tn *)

ChebyFirstChart = Plot3D[
   ChebyshevT[n,x],{x,-1,1},{n,2,8},
   PlotRange->All,PlotPoints->{30,30},
   Ticks->{Automatic,Automatic,None},
   PlotLabel->"Chebyshev Tn(x)",
   AxesEdge->{Automatic,{1,-1},Automatic},
   AxesLabel->{"x"," n"," "}];
```

Surface of Chebyshev polynomial, Figure 11.2.1 center.

```
(* ChebyT cell continued *)

ChebyFirstChart = Plot3D[
  ChebyshevT[n,x],{x,-1,1},{n,2,8},
  PlotRange->All,PlotPoints->{30,30},
  Ticks->{Automatic,Automatic,None},
  PlotLabel->"Chebyshev Tn(x)",
  AxesEdge->{Automatic,{1,-1},Automatic},
  AxesLabel->{"x"," n"," "}];


(*  Chebyshev Tn: 2D plot *)
ChebyFirst2D[n_] := Plot[
  ChebyshevT[n,x],{x,-1,1},
  PlotRange->All,PlotPoints->40,Ticks->None];
ChebyFirst2D[3];
ChebyFirst2D[8];


(* Chebyshev Tn: Numerical *)
NumTableOP[ChebyshevT,Tn,4,-1,1,0.5]
```

Surface of Chebyshev polynomial, Figure 11.2.1 center.

Define curve of polynomial for fixed order *n*.
Curves for *n* = 3, 8, Figure 11.2.1 left and right.

Numerical values; Table 11.2.2.

The ChebyU cell [11.2.2]

By executing this cell you will produce (Figure 11.2.2) the surface of the Chebyshev polynomial of the second kind and curves as a function of *x* for *n* fixed. A table of numerical values that can be used to check the C function is produced for Table 11.2.4.

Exploring with ChebyU

Investigate how the Chebyshev polynomial of the second kind behaves if *x* is fixed and *n* is varied, giving a transect of the surface in Figure 11.2.2. To do this, make a function like ChebySecond2D[] in the ChebyU cell, but let its variable be x and the variable in Plot be n.

```
(* ChebyU cell *)
(* Chebyshev Polynomials of Second Kind, Un *)

ChebySecondChart = Plot3D[
  ChebyshevU[n,x],{x,-0.9,0.9},{n,2,8},
  PlotRange->All,PlotPoints->{30,30},
  Ticks->{Automatic,Automatic,None},
  PlotLabel->"Chebyshev Un(x)",
  AxesEdge->{Automatic,{1,-1},Automatic},
  AxesLabel->{"x"," n"," "}];
```

Surface of Chebyshev polynomial, Figure 11.2.2 center.

```
ChebySecondChart = Plot3D[
  ChebyshevU[n,x],{x,-0.9,0.9},{n,2,8},
  PlotRange->All,PlotPoints->{30,30},
  Ticks->{Automatic,Automatic,None},
  PlotLabel->"Chebyshev Un(x)",
  AxesEdge->{Automatic,{1,-1},Automatic},
  AxesLabel->{"x"," n"," "}];

(* Chebyshev Un: 2D plot *)
ChebySecond2D[n_] := Plot[
  ChebyshevU[n,x],{x,-0.8,0.8},
  PlotRange->All,PlotPoints->40,Ticks->None];
ChebySecond2D[3];
ChebySecond2D[8];

(* Chebyshev Un: Numerical *)
NumTableOP[ChebyshevU,Un,4,-1,1,0.5]
```

Surface of Chebyshev polynomial, Figure 11.2.2 center.

Define curve of polynomial for fixed order *n*.
Curves for *n* = 3, 8, Figure 11.2.2 left and right.

Numerical values; Table 11.2.4.

The `Gegenbauer` cell [11.3]

Execute this cell to generate surfaces of the Gegenbauer polynomials divided by α^2 as a function of x and α with the order, n, fixed, as in Figure 11.3.1. Numerical values, some of which are in Table 11.3.2, are also generated.

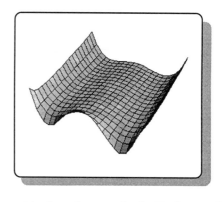

Exploring with `Gegenbauer`

Explore the relations to other special functions, such as those in (11.3.5) for Legendre polynomials (Section 11.6), Chebyshev polynomials of the second kind (Section 11.2.2), and associated Legendre functions (Section 12.1). Relations to other functions are discussed in the references in the Background for Gegenbauer polynomials.

```
(* Gegenbauer cell *)
(* Gegenbauer Polynomials, C(a)n *)

GegenbauerChart[n_] := Plot3D[
  GegenbauerC[n,a,x]/(a^2),{x,-1.0,1.0},{a,0.1,2},
  PlotRange->All,PlotPoints->{30,30},
  Ticks->{Automatic,Automatic,None},
  PlotLabel->"n = "<>ToString[n]<>"",
  AxesEdge->{Automatic,{1,-1},Automatic},
  AxesLabel->{"x"," a"," "}];

GegenbauerChart[2];    GegenbauerChart[3];
GegenbauerChart[4];    GegenbauerChart[5];
```

Define surface of Gegenbauer polynomial for order *n* fixed.

Show surfaces for *n* = 2, 3, 4, 5, Figure 11.3.1.

```
(* Gegenbauer cell continued *)

(* Gegenbauer Polynomials: Numerical *)
NumTableGegen[a_,maxn_] :=
  N[TableForm[Table[{n,GegenbauerC[n,a,x]},
  {x,-1,1,0.5},{n,0,maxn}],
  TableHeadings->
  {None,{"n","C("<>ToString[a]<>")n(x)"}}],10]


NumTableGegen[0.5,5]   (*  = Pn(x)   *)
NumTableGegen[1.5,5]
```

Define numerical table to 10 digits of Gegenbauer polynomials for α and maximum n fixed.

Show tables for $\alpha = 1/2$, 3/2, and n up to 5, Table 11.3.2.

The `Hermite` cell [11.4]

By executing this cell you will see the surface of the modified Hermite polynomial as in the adjacent motif, as well as curves for n fixed and tables that can be used to check the C function `Hermite` in Section 11.4.

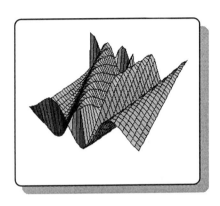

Exploring with `Hermite`

Search out Hermite's cell in the wilderness of special functions. It is illuminating to modify cell `Hermite` to make transects of the rugged landscape seen in the surface plot of the modified polynomials, (11.4.1), slicing in the n direction in addition to the x direction shown in Figure 11.4.1 and generated by running the cell.

```
(* Hermite cell *)
(* Hermite Polynomials, Hn *)

(* Modified Hermite Polynomials, H'n *)
HermiteHPrime[n_,x_] :=
  HermiteH[n,x]/(Sqrt[Exp[x*x]*2^n*n!])


HermiteChart = Plot3D[
  HermiteHPrime[n,x],{x,-3,3},{n,2,6},
  PlotRange->All,PlotPoints->{40,40},
  Ticks->{Automatic,Automatic,None},
  PlotLabel->"Hermite  H'n(x)",
  AxesEdge->{Automatic,{1,-1},Automatic},
  AxesLabel->{"x"," n"," "}];


(*  Hermite H'n: 2D plot *)
HermitePrime2D[n_] := Plot[
  HermiteHPrime[n,x],{x,-3,3},
  PlotRange->All,PlotPoints->40,Ticks->None];
HermitePrime2D[3];     HermitePrime2D[6];
```

Define modified Hermite polynomials, (11.4.1).

Surface of modified Hermite polynomials as a function of x and n, Figure 11.4.1 center.

Define curve of modified Hermite polynomial for n fixed.
Curves for $n = 3$ and 6, Figure 11.4.1 left and right

```
(* Hermite Polynomials: Numerical *)
NumTableOP[HermiteH,Hn,4,0.5,2,0.5]
```

Numerical Hermite polynomials, Table 11.4.2.

The Laguerre cell [11.5]

Execute this cell to obtain the (generalized) Laguerre polynomials as surfaces (Figure 11.5.1) and numerical values (as in Tables 11.5.2 and 11.5.3).

Exploring with Laguerre

Check the numerical accuracy of the C function Laguerre for high n (which is generated by upward recurrence in the order of the polynomial) by running the cell with corresponding n and x values for its numerical output.

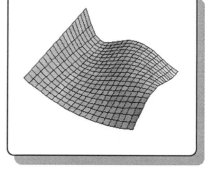

If your interests are statistical, follow up the references in the Background to devise a program that produces probability densities for pairs of correlated random variables.

```
(* Laguerre cell *)
(* Laguerre Polynomials, L(a)n *)

LaguerreGraph[a_] := Plot3D[
  LaguerreL[n,a,x],{x,0,5},{n,2,6},
  PlotRange->All,PlotPoints->{30,30},
  Ticks->{Automatic,Automatic,None},
  PlotLabel->"a = "<>ToString[a]<>"",
  AxesEdge->{Automatic,{1,-1},Automatic},
  AxesLabel->{"x"," n"," "}];
```

Define surface of generalized Laguerre polynomial for α fixed.

```
LaguerreGraph[0];
LaguerreGraph[1];
```

Show surfaces for $\alpha = 0$ and 1, Figure 11.5.1.

```
(* Laguerre Polynomials: Numerical *)
NumTableLaguerre[a_,maxn_] :=
  N[TableForm[Table[{n,LaguerreL[n,a,x]},
  {x,0,2,1},{n,1,maxn}],
  TableHeadings->
  {None,{"n","L("<>ToString[a]<>")n(x)"}}],10]
```

Define numerical tables to 10 digits of generalized polynomials for α and maximum n fixed.

```
NumTableLaguerre[0,4]
NumTableLaguerre[1.5,4]
```

Make tables; $\alpha = 0$, $n \leq 4$, and $\alpha = 1.5$, $n \leq 4$, Tables 11.5.2 and 11.5.3.

The `Legendre` cell [11.6]

Execute this cell to obtain the surface for the Legendre polynomials (center in Figure 11.6.1), the curves with n fixed and x varying (sides of Figure 11.6.1), and numerical values of the function (Table 11.6.2).

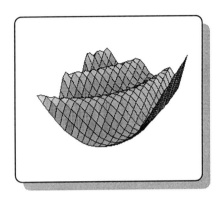

Exploring with `Legendre`

What does a slice through the Legendre polynomial surface look like if x is fixed and n varies? To answer this, make a function like `Legendre2D[]` in cell `Legendre` but let it be a function of x with the variable in `Plot` running over n. Such a view is useful for understanding partial-wave expansions of the Helmholtz wave equation.

```
(* Legendre cell *)
(* Legendre Polynomials, Pn *)

LegendreChart = Plot3D[
  LegendreP[n,x],{x,-1,1},{n,2,8},
  PlotRange->All,PlotPoints->{30,30},
  Ticks->{Automatic,Automatic,None},
  PlotLabel->"Legendre  Pn(x)",
  AxesEdge->{Automatic,{1,-1},Automatic},
  AxesLabel->{"x"," n"," "}];

(* Legendre Pn: 2D plot *)
Legendre2D[n_] := Plot[LegendreP[n,x],{x,-1,1},
  PlotRange->All,PlotPoints->40,Ticks->None];
Legendre2D[3];    Legendre2D[8];

(* Legendre Polynomials: Numerical *)
NumTableOP[LegendreP,Pn,3,-1,1,0.5]
```

Surface of Legendre polynomial as function of x and n, Figure 11.6.1 center.

Define curve of Legendre polynomial for fixed n. Show curves for $n = 3$ and 8, Figure 11.6.1 left and right.

Numerical values to 10 digits in Table 11.6.2.

The `Jacobi` cell [11.7]

By executing this cell you will be able to visualize the Jacobi polynomial in all its generality, as shown in Figure 11.7.2. Tables of numerical values will also be generated, as seen in Tables 11.7.2 and 11.7.3.

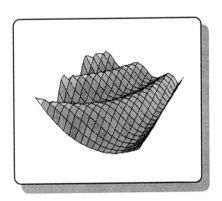

Exploring with `Jacobi`

Since the Jacobi polynomial is the most general of the classical orthogonal polynomials (Figures 11.7.1 and 11.7.2), it is interesting to investigate further its relation to the other orthogonal polynomials. For example, is it efficient for algebraic or numerical work to program only the Jacobi polynomial then to derive the Gegenbauer, Chebyshev, and Legendre polynomials as special cases of it, as Figure 11.7.1 indicates?

```
(* Jacobi cell *)
(* Jacobi Polynomials, P(a,b)n *)

JacobiChart[a_,b_] := Plot3D[
  JacobiP[n,a,b,x],{x,-1,1},{n,2,8},
  PlotRange->All,PlotPoints->{20,20},Ticks->None];

JacobiChart[-1/2,-1/2];
JacobiChart[-1/2,0];
JacobiChart[-1/2,1/2];

JacobiChart[0,-1/2];
JacobiChart[0,0];
JacobiChart[0,1/2];

JacobiChart[1/2,-1/2];
JacobiChart[1/2,0];
JacobiChart[1/2,1/2];

(* Jacobi Polynomials: Numerical *)
NumTableJacobi[a_,b_,maxn_] :=
  N[TableForm[Table[{n,JacobiP[n,a,b,x]},
  {x,0,1,0.5},{n,1,maxn}],TableHeadings->
  {None,{"n",
  "C("<>ToString[a]<>","<>ToString[b]<>")n(x)"}}],10]

NumTableJacobi[-0.5,-0.5,4]
NumTableJacobi[0.5,0,4]
```

Define surface of Jacobi polynomial with *a* and *b* fixed.

Show three surfaces for $\alpha = -1/2$, Figure 11.7.2 top row.

Show three surfaces for $\alpha = 0$, Figure 11.7.2 middle row.

Show three surfaces for $\alpha = 1/2$, Figure 11.7.2 bottom row.

Define numerical table to 10 digits of Jacobi polynomials for *a*, *b*, and maximum *n* fixed.

Numerical values, Tables 11.7.2 and 11.7.3.

20.12 LEGENDRE FUNCTIONS

Notebooks and Cells

Notebook LegendreCoords; with cells SphereCoords, TorusCoords, Cone-Coords. Notebook Legendre; cells LegendreP, LegendreQ, ToroidP, ToroidQ, Conical.

Overview

There are two *Mathematica* notebooks for the Legendre functions. The first notebook, LegendreCoords, generates graphics for two coordinate systems—spherical polar and toroidal—and for the part of spherical coordinates appropriate for conical functions (Section 12.4), which we show in the adjacent motif.

The second *Mathematica* notebook is for Legendre functions, excluding the Legendre polynomials, which are in Section 11.6 with other orthogonal polynomials. This notebook invites you to visualize and to compute both analytical and numerical values of Legendre functions—spherical, toroidal, and conical. Since there is at least one parameter—the order of the polynomial, n —in addition to the function argument, most of the visualizations are best shown as surfaces.

For the spherical functions (Section 12.2) and the toroidal functions (Section 12.3) we explore both regular functions and irregular functions. One complication is that the functions on the cut in the complex plane from $x = -1$ to $x = 1$ (the conventional Legendre functions, here denoted by Roman P for the regular function and Q for the irregular function) differ in definition from those in the rest of the complex plane (Italic P and Q), a distinction that is clarified in discussing (12.1.3) and (12.1.4).

Notebook `LegendreCoords` [12.2]

Use this notebook to obtain surfaces that clarify the coordinates used for the three classes of Legendre functions we consider; spherical (conventional, with integer degree n and order m), toroidal (half-integer degree and integer order), and conical (degree of -1/2 plus imaginary, zeroth order).

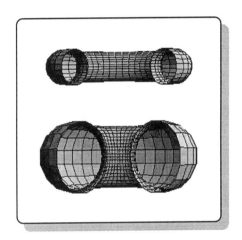

Of the three cells in this notebook, `TorusCoords` is probably the least familiar. Check how spherical polar coordinates are generated. Understand toroidal coordinates better by varying the parameter η, as discussed in Section 12.3.1 in relation to Figure 12.3.2, part of which is shown in the adjacent motif, for a narrow ring ($\eta = 2$, top) and for a wider ring ($\eta = 1$, bottom).

The `SphereCoords` cell [12.2.1]

By executing this cell you get a 3D view of the surfaces of constant r (spheres), constant θ (cones) and constant ϕ (half planes though the z axis). A sphere and a cone are shown in the adjacent motif, but the half plane in Figure 12.3.1 has been omitted.

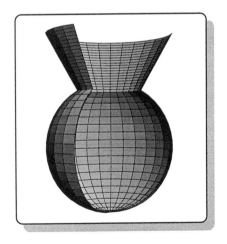

Exploring with `SphereCoords`

Execute this cell to develop your insight into the computation of surfaces in spherical polar coordinates. This will be useful for understanding the toroidal coordinates in the next cell, which are more complicated.

```
(* SphereCoords cell *)

SetOptions[ParametricPlot3D,
  PlotPoints->{20,20},Boxed->False,Axes->False,
  ViewPoint->{0,-3,0},DisplayFunction->Identity];


(* Spherical Coordinates; (r,theta,phi) *)
xs[r_,theta_,phi_] := r*Sin[theta]*Cos[phi]
ys[r_,theta_,phi_] := r*Sin[theta]*Sin[phi]
zs[r_,theta_,phi_] := r*Cos[theta]

SphereSurf[r_] := ParametricPlot3D[
  {xs[r,theta,phi],ys[r,theta,phi],zs[r,theta,phi]},
  {theta,Pi/6,Pi},{phi,0,4*Pi/3}]

ConeSurf[theta_] := ParametricPlot3D[
  {xs[r,theta,phi],ys[r,theta,phi],zs[r,theta,phi]},
  {r,1,2},{phi,0,4*Pi/3}]

PlaneSurf[phi_] := ParametricPlot3D[
  {r*Cos[phi],r*Sin[phi],z},{r,0,2.5},{z,-1,1.8}]

SphericalCoords = Show[
  SphereSurf[1],ConeSurf[Pi/6],PlaneSurf[Pi/3],
  DisplayFunction->$DisplayFunction];
```

By setting the options, we avoid repetition below.

Define Cartesian coordinates in terms of polar coordinates.

Define sphere surface as function of *r*.

Define cone surface as function of θ

Define half-plane surface as function of ϕ.

Show the three surfaces together.

The TorusCoords cell [12.3.1]

Execute this cell for toroidal coordinates to get a 3D view of the surfaces of constant η (toroids, or rings), constant θ (spheres) and constant ϕ (half planes through the *z* axis), as shown in the adjacent motif.

Exploring with TorusCoords

Develop your insight into computing surfaces in toroidal coordinates by running this cell with various values of the parameters η, θ, and ϕ. Note that the fourth parameter, *c* in (12.3.1) provides only an overall scale factor for the coordinates, analogous to the *r* in spherical polar coordinates. For purposes of visualization, it is therefore not interesting to vary *c*.

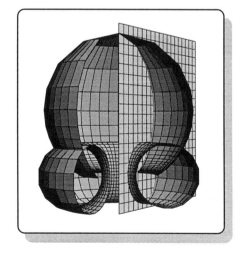

By varying η, you can see the various toroidal shapes, as in the motif for this notebook and in Figure 12.3.2. Formulas for the radii of the cross-section circles and for the circle in the $x - y$ plane that is traced out by the torus center, are given above that figure.

```
(* TorusCoords cell *)

(* Toroidal Coordinates; (eta,theta,phi) *)

SetOptions[ParametricPlot3D,
  PlotPoints->{30,30},Boxed->False,Axes->False,
  ViewPoint->{0,-3,0},DisplayFunction->Identity];

xt[eta_,theta_,phi_] :=
  c*Sinh[eta]*Cos[phi]/(Cosh[eta]-Cos[theta])
yt[eta_,theta_,phi_] :=
  c*Sinh[eta]*Sin[phi]/(Cosh[eta]-Cos[theta])
zt[eta_,theta_,phi_] :=
  c*Sin[theta]/(Cosh[eta]-Cos[theta])

c = 2;
TorusSurf[eta_] := ParametricPlot3D[
{xt[eta,theta,phi],yt[eta,theta,phi],zt[eta,theta,phi]},
{theta,-Pi,Pi},{phi,0,4*Pi/3}]

TorusSphereSurf[theta_] := ParametricPlot3D[
{xt[eta,theta,phi],yt[eta,theta,phi],zt[eta,theta,phi]},
{eta,0,4},{phi,0,4*Pi/3}]

TorusPlaneSurf[phi_] := ParametricPlot3D[
  {r*Cos[phi],r*Sin[phi],z},{r,0,10},{z,-2,8},
  PlotPoints->{20,20}]

ToroidalCoord = Show[
TorusSurf[1],TorusSphereSurf[Pi/6],TorusPlaneSurf[Pi/3],
DisplayFunction->$DisplayFunction];

(* Torii as functions of eta *)
c = 2;
TorusSurf[eta_] := ParametricPlot3D[
{xt[eta,theta,phi],yt[eta,theta,phi],zt[eta,theta,phi]},
{theta,-Pi,Pi},{phi,-Pi/4,5*Pi/4,DisplayFunction->
  $DisplayFunction]
TorusSurf[0.3];  TorusSurf[1];  TorusSurf[2];
```

By setting the options, avoid repetition below.

Define Cartesian coordinates in terms of toroidal coordinates.

Define torus surface as function of η.

Define sphere surface as function of θ.

Define half-plane surface as function of ϕ.

Show the three surfaces in one graphic.

Redefine torus with ϕ range changed.

Toroids in Figure 12.3.2.

The ConeCoords cell [12.4.1]

Execute this cell to see the parts of a spherical coordinate system that are relevant for using the conical functions in Section 12.4. You will get a 3D view of the surfaces of constant r (spheres), and constant θ (a cone), as shown in the adjacent motif.

Exploring with ConeCoords

The most interesting aspect of these spherical (*not* conical) coordinates is how the appearance of the cone varies as you run this cell with various values of the cone parameters θ. For example, the conical functions in Section 12.4.2 become problematic to compute efficiently and accurately (Table 12.4.1) as θ increases so that the cone includes points below the $x - y$ plane. Convince yourself that in most practical situations you could just reverse the z axis and avoid such problems.

```
(* ConeCoords cell *)

(* Cone Intercepting Spheres; Conical Functions *)

SetOptions[ParametricPlot3D,                          Set the options to
  PlotPoints->{20,20},Boxed->False,Axes->False,       avoid repetition
  ViewPoint->{0,-3,0},DisplayFunction->Identity];     below.


(* Using Spherical Coordinates (r,theta,phi) *)

xs[r_,theta_,phi_] := r*Sin[theta]*Cos[phi]           Define Cartesian
ys[r_,theta_,phi_] := r*Sin[theta]*Sin[phi]           coordinates in terms
zs[r_,theta_,phi_] := r*Cos[theta]                    of spherical polar
                                                      coordinates.

SphereSurf[r_] := ParametricPlot3D[                   Define sphere surface
  {xs[r,theta,phi],ys[r,theta,phi],zs[r,theta,phi]},  as function of r.
  {theta,Pi/4,Pi},{phi,-Pi/6,8*Pi/6}]

ConeSurf[theta_] := ParametricPlot3D[                 Define cone surface
  {xs[r,theta,phi],ys[r,theta,phi],zs[r,theta,phi]},  as function of θ.
  {r,1,6},{phi,-1.5*Pi/6,8.5*Pi/6}]

SphericalCoords = Show[                               Show two spheres
  SphereSurf[2],SphereSurf[4],ConeSurf[Pi/4],         and a cone in Figure
  DisplayFunction->$DisplayFunction];                 12.4.1.
```

Notebook Legendre

This notebook will draw for you many surfaces and curves that will help you understand the behavior of Legendre functions in the complex plane (Section 12.1), as well as the three types of Legendre functions that we consider; spherical (conventional, with integer degree n and order m, in Section 12.2), toroidal (half-integer degree and integer order, in Section 12.3), and conical (degree of $-1/2$ plus imaginary, zeroth order, in Section 12.4). Moreover, algebraic formulas for the spherical Legendre functions. plus numerical values of all the Legendre functions included in the *Atlas* can be generated.

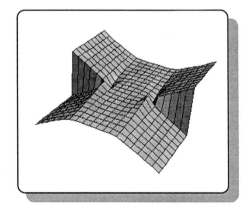

Each of the five cells in this notebook correspond to a major class of Legendre function; `LegendreP` and `LegendreQ` make the general Legendre function views and the spherical functions for the regular (*P*) and irregular (*Q*) functions, respectively; `ToroidP` and `ToroidQ` are for the toroidal functions; `Conical` is for the conical functions.

The `LegendreP` cell [12.1.1, 12.2.2]

By executing this cell you get many views of the Legendre functions of the first kind. Visualizations start with the general views of surfaces and curves that characterize the function for a range of values for the degree, *v*, and the order, *μ*, as in Section 12.1.1, where the branch-cut structure of the regular function is also shown (Figure 12.1.4).

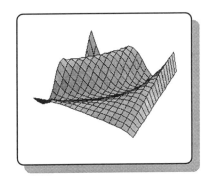

Cell `LegendreP` also generates graphics, algebraic expressions, and numerical values for the function of the first kind that are appropriate for spherical polar coordinate systems (*v* = *n* and *μ* = *n*, with *n* and *m* integers, in Section 12.2.2).

Exploring with `LegendreP`

Begin with the familiar Legendre functions for spherical polar coordinates, $P_n^m(x)$ with $|x| < 1$, as a preliminary to exploring the more exotic domains of the functions throughout the complex plane (as in Section 12.1.1) as well as the toroidal and conical functions (cells `ToroidP`, `ToroidQ`, and `Conical`).

```
(* LegendreP cell *)

(* Legendre Functions, P(n,m)(z) *)

(* Mathematica differs by phase (-1)^m from
   Abramowitz & Stegun and from Erdélyi *)

(* Legendre Functions, P: 3D & 2D views *)
(* 3D defined *)
LegP3D[n_,m_,ReOrIm_] := Plot3D[
  ReOrIm[(-1)^m*LegendreP[n,m,x+I*y]],
  {x,-1,1},{y,-0.5,0.5},
  PlotRange->All,PlotPoints->{30,30},
  Ticks->{{-1,0,1},{-0.5,0,0.5},None},
  AxesLabel->{"x","y",None}]

(* 2D defined *)
TS = ToString;
LegP2D[n_,m_] := Plot[
  (-1)^m*LegendreP[n,m,x],{x,-1,1},
  PlotRange->All,PlotPoints->20,
  Ticks->{{-1,1},{-1,1}},
  AxesLabel->
  {"x","P("<>TS[n]<>","<>TS[m]<>") (x)"}]
```

Define Legendre function 3D surface plots.

Define labelled curves of Legendre functions.

```
(* Views for    n = 3, m = 3,2,1,0 *)

LegP3D[3,3,Re]; LegP2D[3,3]; LegP3D[3,3,Im];
LegP3D[3,2,Re]; LegP2D[3,2]; LegP3D[3,2,Im];
LegP3D[3,1,Re]; LegP2D[3,1]; LegP3D[3,1,Im];
LegP3D[3,0,Re]; LegP2D[3,0]; LegP3D[3,0,Im];
```

Surface, curve, surface, for *m* = 3, 2, 1, 0.. as in Figure 12.1.1.

```
(* Legendre P; 3D views for x fixed *)

LegP3DXX[x_] := Plot3D[
  Re[(-1)^m*LegendreP[n,m,x]],{m,0,4},{n,0,4},
  PlotRange->All,PlotPoints->{20,20},
  Ticks->{{0,2,4},{0,2,4},None},
  AxesLabel->{"m","n",None}]
```

Define surface for *x* fixed.

```
LegP3DXX[-0.5]; LegP3DXX[0]; LegP3DXX[0.5];
```

Make Figure 12.1.2.

```
(* Legendre P; 3D views for y fixed *)

LegP3DYY[y_,ReOrIm_] := Plot3D[
  ReOrIm[(-1)^m*LegendreP[n,m,I*y]],
  {m,0,4},{n,0,4},
  PlotRange->All,PlotPoints->{20,20},
  Ticks->{{0,2,4},{0,2,4},None},
  AxesLabel->{"m","n",None}]
```

Define 3D views for *y* fixed.

```
LegP3DYY[-2,Re];   LegP3DYY[2,Re];
LegP3DYY[-2,Im];   LegP3DYY[2,Im];
```

Make Figure 12.1.3.

```
    (* Legendre Functions, P(n,m)(z) branch cuts *)
(* Mathematica phase conventions *)

(* 3D defined *)
LegPBC3D[n_,m_,ReOrIm_] := Plot3D[
  ReOrIm[LegendreP[n,m,x+I*y]],
  {x,-3,3},{y,-0.5,0.5},
  PlotRange->All,PlotPoints->{20,20},
  Ticks->{{-3,-2,-1,0,1,2,3},{-0.5,0,0.5},None},
  AxesLabel->{"x","y",None}]
```

Define surfaces of real or imaginary parts of regular function, *P*, for arbitrary *n* and *m*.

```
(* Views for n = 1.5, m = 1.5,0.5 *)

LegPBC3D[1.5,1.5,Re];   LegPBC3D[1.5,1.5,Im];
LegPBC3D[1.5,0.5,Re];   LegPBC3D[1.5,0.5,Im];
```

Show branch cut structure, Figure 12.1.4.

```
(*   Spherical Legendre Functions, P(n,m)(x)  *)


Off[Plot3D::gval,Plot3D::plnc];

(* Spherical Legendre Functions; P 3D views *)

LegPMN[n_,m_,x_] :=
  If[ n < m, 0, (-1)^m*LegendreP[n,m,x] ]
```

Define *P* with usual phase.

```
LegMP3D[m_] := Plot3D[
  LegPMN[n,m,x],{x,-1,1},{n,1,5},
  PlotRange->All,PlotPoints->{23,23},
  Ticks->{{-0.5,0,0.5},{1,2,3,4,5},None},
  AxesLabel->{"x","n",None}]

LegMP3D[0]; LegMP3D[1]; LegMP3D[2];
LegMP3D[3]; LegMP3D[4]; LegMP3D[5];

(* Spherical Legendre Functions, S(n,m)(x) *)

Off[Plot3D::gval,Plot3D::plnc];

(* Spherical Legendre Functions; (-1)^m S: 3D *)

SLegPMN[n_,m_,x_] :=
  If[ n < m, 0,
  (-1)^m*Sqrt[(n-m)!/(n+m)!]*LegendreP[n,m,x] ]

SLegMP3D[m_] := Plot3D[
  SLegPMN[n,m,x],{x,-1,1},{n,1,5},
  PlotRange->All,PlotPoints->{23,23},
  Ticks->{{-0.5,0,0.5},{1,2,3,4,5},None},
  AxesLabel->{"x","n",None}]

SLegMP3D[0]; SLegMP3D[1]; SLegMP3D[2];
SLegMP3D[3]; SLegMP3D[4]; SLegMP3D[5];

(* Legendre Functions: Formulas for
   P(n,m) and S(n,m) by Recurrence *)

P[n_,m_] :=
  If[ n < m, 0,
    If[ n == m, (2*m)!*(1-x^2)^(m/2)/(2^m*m!),
    ((2*n-1)*x*P[n-1,m]-(n+m-1)*P[n-2,m])/(n-m)]]

S[n_,m_] :=
  If[ n < m, 0,
    If[ n == m, Sqrt[(2*m)!]*(1-x^2)^(m/2)/(2^m*m!),
    ((2*n-1)*x*S[n-1,m]-Sqrt[(n-1)^2-m^2]*S[n-2,m])/
    Sqrt[n^2-m^2] ] ]

(* Tables of formulas from m = 0 to n *)
PPTable[n_] := Table[{m,Expand[P[n,m]]},{m,0,n}]
SSTable[n_] := Table[{m,Expand[S[n,m]]},{m,0,n}]

(* For example, n <= 3 *)
Do[Print[n,PPTable[n]],{n,0,3}]
Do[Print[n,SSTable[n]],{n,0,3}]

(* Spherical Legendre Functions: P Numerical *)

NumPNM[n_] := Table[{n,m,x,
  N[(-1)^m*LegendreP[n,m,x],10]},
  {m,0,n},{x,0,1,0.5}]

NumPNM[1]
NumPNM[2]
NumPNM[4]
```

LegendreP cell
continued

Define 3D plots for spherical Legendre P. Show surfaces for $m = 0$ to 5, as in Figure 12.2.2.

Define modified spherical Legendre function *S*, (12.2.4).

Define surface as function of *m*.

Show *S* surfaces, as in Figure 12.2.3.

Recurrence relation for P, first of (12.2.9).

Recurrence relation for S, second of (12.2.9).

Define tables of P and of S.

Make Table 12.2.1 then Table 12.2.2.

Define table of numerical P.

Tabulated values of P, as in Table 12.2.3.

The `LegendreQ` cell [12.1.2, 12.2.3]

Execute this cell to get many views of the Legendre functions of the second kind. Visualizations start with the general views of surfaces and curves characterizing the function for a range of values of the degree, v, and the order, μ, as in Section 12.1.2, where the branch-cut structure of the irregular function is also shown (Figure 12.1.9). Compared with the functions of the first kind (P), there is the additional complication that the Q functions are irregular at $x = \pm 1$.

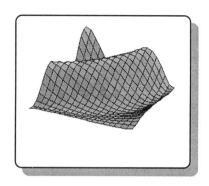

Cell `LegendreQ` also generates graphics, algebraic expressions, and numerical values for the function of the second kind that are appropriate for spherical polar coordinate systems ($v = n$ and $\mu = n$, with n and m integers, in Section 12.2.3).

Exploring with `LegendreQ`

Begin with the Legendre functions of the second kind that are appropriate for spherical polar coordinates, $Q_n^m(x)$ with $|x| < 1$, as a preliminary to exploring the more exotic domains of the functions throughout the complex plane (as in Section 12.1.2) as well as the irregular toroidal functions in cell `ToroidQ`.

```
(* LegendreQ cell *)


(*   Legendre Functions, Q(n,m)(z) *)


(* Mathematica differs by phase (-1)^m from
   Abramowitz & Stegun and from Erdélyi *)


(* Legendre Functions, Q: 3D views *)

(* 3D defined *)
LegQ3D[n_,m_,ReOrIm_] := Plot3D[
  ReOrIm[(-1)^m*LegendreQ[n,m,x+I*y]],
  {x,-0.8,0.8},{y,-0.5,0.5},
  PlotRange->All,PlotPoints->{30,30},
  Ticks->{{-0.8,0,0.8},{-0.5,0,0.5},None},
  AxesLabel->{"x","y",None}]


(* 2D defined *)
TS = ToString;
LegQ2D[n_,m_] := Plot[
  (-1)^m*LegendreQ[n,m,x],{x,-0.8,0.8},
  PlotRange->All,PlotPoints->20,
  Ticks->{{-0.8,0.8},{-1,1}},
  AxesLabel->
  {"x","Q("<>TS[n]<>","<>TS[m]<>") (x)"}]
```

Define Legendre function 3D surface plots.

Define labelled curves of Legendre functions of the second kind, Q.

```
(* Views for    n = 3, m = 3,2,1,0 *)

LegQ3D[3,3,Re]; LegQ2D[3,3]; LegQ3D[3,3,Im];
LegQ3D[3,2,Re]; LegQ2D[3,2]; LegQ3D[3,2,Im];
LegQ3D[3,1,Re]; LegQ2D[3,1]; LegQ3D[3,1,Im];
LegQ3D[3,0,Re]; LegQ2D[3,0]; LegQ3D[3,0,Im];

(* Legendre Q(m,n) (x): Removing singular behavior *)

(* Mathematica differs by phase (-1)^m from
   Abramowitz & Stegun and from Erdélyi *)

TS = ToString;
LegQSing[n_,m_] := Plot[
  (-1)^m*LegendreQ[n,m,x],{x,-0.7,0.7},
  PlotRange->All,PlotPoints->20,
  Ticks->{{-0.7,0.7},{-1,1}},
  AxesLabel->
   {"x","Q("<>TS[n]<>","<>TS[m]<>") (x)"},
  DisplayFunction->Identity]

Legqq[n_,m_] := Plot[
  (1-Abs[x])^(m/2)*(-1)^m*LegendreQ[n,m,x],
  {x,-0.999,0.999},PlotRange->All,PlotPoints->20,
  PlotStyle->{Dashing[{0.03,0.03}]},
  DisplayFunction->Identity]

ShowQq[m_,n_] := Show[LegQSing[m,n],Legqq[m,n],
  DisplayFunction->$DisplayFunction]

(* Views for    n = 3, m = 3,2,1 *)
ShowQq[3,3]; ShowQq[3,2]; ShowQq[3,1];


(* Legendre Q; 3D views for x fixed *)

LegQ3DXX[x_] := Plot3D[
  Re[(-1)^m*LegendreQ[n,m,x]],{m,0,4},{n,0,4},
  PlotRange->All,PlotPoints->{20,20},
  Ticks->{{0,2,4},{0,2,4},None},
  AxesLabel->{"m","n",None}]

LegQ3DXX[-0.5]; LegQ3DXX[0]; LegQ3DXX[0.5];

(* Legendre Q; 3D views for y fixed *)

LegQ3DYY[y_,ReOrIm_] := Plot3D[
  ReOrIm[(-1)^m*LegendreQ[n,m,I*y]],
  {m,0,4},{n,0,4},
  PlotRange->All,PlotPoints->{20,20},
  Ticks->{{0,2,4},{0,2,4},None},
  AxesLabel->{"m","n",None}]

LegQ3DYY[-2,Re];  LegQ3DYY[2,Re];
LegQ3DYY[-2,Im];  LegQ3DYY[2,Im];
```

`LegendreQ` cell
continued

Surface, curve, surface, for $m = 3, 2,$ 1, 0.. as in Figure 12.1.5.

Define curve of Q with x varying, n and m fixed.

Remove singularity factor near endpoints, (12.1.7).

Define Q and q together.

Make curves in Figure 12.1.6.

Define 3D views of Q for x fixed.

Make Figure 12.1.7.

Define surfaces of real or imaginary parts of regular function, Q, for y fixed.

Make Figure 12.1.8.

```
(* Legendre Functions, Q(n,m)(z) branch cuts *)

(* 3D views *)

LegQBC3D[n_,m_,ReOrIm_] := Plot3D[
 ReOrIm[LegendreQ[n,m,x+I*y]],
 {x,-3,3},{y,-0.5,0.5},
 PlotRange->All,PlotPoints->{20,20},
 Ticks->{{-3,-2,-1,0,1,2,3},{-0.5,0,0.5},None},
 AxesLabel->{"x","y",None}]

(* Views for    n = 1.5, m = 1.5,0.5 *)

LegQBC3D[1.5,1.5,Re];  LegQBC3D[1.5,1.5,Im];
LegQBC3D[1.5,0.5,Re];  LegQBC3D[1.5,0.5,Im];


(* Legendre Functions; Q(n,m) by Recurrence *)

Q[n_,m_] :=
 If[ n < m, 0,
  If[ n == m, (* Differentiate Qm m times *)
  Factor[(x^2-1)^(m/2)*D[
   (-1)^m*LegendreQ[m,x],{x,m}]],
  (* or use iteration *)
  Factor[((2*n-1)*x*Q[n-1,m]-(n+m-1)*Q[n-2,m])/
  (n-m)]]]

(* Tables of formulas from m = 0 to n *)
QQTable[n_] := Table[{m,Expand[Q[n,m]]},{m,0,n}]

(* For example, n <= 2 *)
Do[Print[n,QQTable[n]],{n,0,2}]


(*   Spherical Legendre Functions, Q(n,m)(x) *)

Off[Plot3D::gval,Plot3D::plnc];

(* Spherical Legendre Functions; Q 3D views *)

LegQMN[n_,m_,x_] :=
 If[ n < m, 0, (-1)^m*LegendreQ[n,m,x] ]


LegMQ3D[m_] := Plot3D[
 LegQMN[n,m,x],{x,-0.8,0.8},{n,1,5},
 PlotRange->All,PlotPoints->{20,20},
 Ticks->{{-0.4,0,0.4},{1,2,3,4,5},None},
 AxesLabel->{"x","n",None}]

LegMQ3D[0]; LegMQ3D[1]; LegMQ3D[2];
LegMQ3D[3]; LegMQ3D[4]; LegMQ3D[5];
```

Define 3D views of Q in the complex plane for n and m fixed.

Make Figure 12.1.9.

Recurrence relation for Q, (12.2.18).

Define tables of Q for n fixed.

Make Table 12.2.3.

Set to zero for $n < m$, else use different phase from *Mathematica*.

Define 3D plots for spherical Legendre Q.

Show surfaces for $m = 0$ to 5, as in Figure 12.2.5.

```
(* Legendre Function of Second Kind, Q
   degree n = 2, order m = 0,1,2, argument x *)

(* Define functions for each m and x range *)

(* m = 0; |x| < 1 then |x| > 1  *)
Q200[x_] := (0.75*x*x-0.25)*Log[(1+x)/(1-x)]-1.5*x
Q20P[x_] := (0.75*x*x-0.25)*Log[(x+1)/(x-1)]-1.5*x
Q20[x_] := Which[
 x<-1,-Q20P[-x], Abs[x]<1,Q200[x], x>1,Q20P[x] ]

(* m = 1 *)
Q210[x_] :=
 -1.5*x*Sqrt[1-x*x]*Log[(1+x)/(1-x)]-
 (3*x*x-2)/Sqrt[1-x*x]
Q21P[x_] := 1.5*x*Sqrt[x*x-1]*Log[(x+1)/(x-1)]-
 (3*x*x-2)/Sqrt[x*x-1]
Q21[x_] := Which[
 x<-1,-Q21P[-x], Abs[x]<1,Q210[x], x>1,Q21P[x] ]

(* m = 2 *)
Q220[x_] :=
 1.5*(1-x*x)*Log[(1+x)/(1-x)]+x*(5-3*x*x)/(1-x*x)
Q22P[x_] :=
 1.5*(x*x-1)*Log[(x+1)/(x-1)]+x*(5-3*x*x)/(x*x-1)
Q22[x_] := Which[
 x<-1,-Q22P[-x], Abs[x]<1,Q220[x], x>1,Q22P[x] ]

(* Finally, define the complete function *)
Q2M[m_,x_] := Switch[m,0,Q20[x],1,Q21[x],2,Q22[x]]

(* Legendre Functions, Q(2,m)(x)  2D views *)

LegQ2m[m_,DL_] := Plot[Q2M[m,x],
  {x,-1.5,1.5},PlotPoints->100,
  PlotRange->{-5,5},AxesLabel->{"x",None},
  Ticks->{{-1.5,-0.5,0,0.5,1.5},{-5,0,5}},
  PlotStyle->{Dashing[{DL,DL}]},
  DisplayFunction->Identity]

(* Views for    n = 2, m = 0,1,2 *)

Show[LegQ2m[0,0.01],LegQ2m[1,0.02],LegQ2m[2,0.03],
  DisplayFunction->$DisplayFunction];

(* Legendre Function of Second Kind: Numerical *)

Table[{n,m,x,N[LegendreQ[n,m,x],10]},
  {n,1,4},{m,0,2},{x,0.5,0.5,1}]
```

LegendreQ cell
continued

Explicit formulas for
Q:
First for m = 0,

then for *m* = 1,

then for *m* = 2.

Then assemble the
function.

Define curves of *Q*.

Composite curves of
Q, Figure 12.2.4.

Numerical Q values,
Table 12.2.5.

The `ToroidP` cell [12.3.2]

Use this cell to get many views of the toroidal functions of the first kind. The curves characterize the function for a range of values of the degree, $n - 1/2$, and the order, m, as in Section 12.3.2. Cell `ToroidP` also generates numerical values for the function of the first kind that are appropriate for the toroidal coordinate systems investigated in the `TorusCoords` cell and in Section 12.3.1.

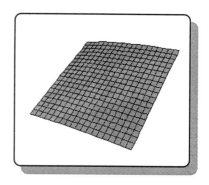

Exploring with `ToroidP`

Use cell `ToroidP` for the function of the first kind to convince yourself that the toroidal function of the first kind is a monotonous function of x, n, and m. This is shown in the motif above, which displays the logarithm of the function versus x (from 1.5 to 3.5 on the front axis) and n (from 2 to 5 on the side axis).

```
(* ToroidP cell *)

(* Toroidal Legendre Functions, P(n-1/2,m)(x) *)

(* Toroidal Functions; P 2D views *)

ToroidPNM[n_,m_,x_] :=
 Re[(-1)^(-m/2)*LegendreP[n-1/2,m,x]]
(* Log Scale: Values positive in ranges plotted *)
TS := ToString
ToroP2D[n_,m_] := Plot[
  Log[10,ToroidPNM[n,m,x]],{x,1.5,3.5},
  PlotRange->{-1,4},PlotPoints->30,
  Ticks->{{2,3},{-1,1,2,3,4}},AxesOrigin->{1.5,0},
  PlotLabel->"log P(n-1/2, "<>TS[m]<>", x)",
  AxesLabel->{"x",None},
  PlotStyle->{Dashing[{0.01*n,0.01*n}]},
  DisplayFunction->Identity]

ToroShowP[m_] := Show[ToroP2D[2,m],ToroP2D[3,m],
 ToroP2D[4,m],DisplayFunction->$DisplayFunction]

ToroShowP[0]; ToroShowP[1];
ToroShowP[2]; ToroShowP[3];

(* Toroidal Functions; P 3D views *)

Off[Plot3D::gval,Plot3D::plnc];

(* Log Scale; values positive in range plotted *)
ToroidP3D[m_] := Plot3D[
  Log[ToroidPNM[n,m,x]],{x,1.5,3.5},{n,2,5},
  PlotRange->All,PlotPoints->{20,20},
  Ticks->{{2,3},{2,3,4,5},None},
  AxesLabel->{"x","n","Log"}]

ToroidP3D[0]; ToroidP3D[1]; ToroidP3D[2];
```

Phase difference from *Mathematica*

Define regular (*P*) toroidal function 2D curves for *n* and *m* fixed.

Define curves for same *m* on same plot.

Show curves of *P*, Figure 12.3.3.

Define regular (*P*) toroidal function 3D surfaces for *m* fixed.

Show surfaces. (Not in text)

```
(* ToroidP cell continued *)

(* Toroidal Functions; P Numerical *)

TorPNM[n_,x_] := Table[{n,m,x,
  N[ToroidPNM[n,m,x],10]},{m,0,3}]

TorPNM[1,1.05]
TorPNM[1,1.55]
TorPNM[1,3.75]
TorPNM[2,1.05]
TorPNM[2,1.55]
TorPNM[2,3.75]
TorPNM[4,1.05]
TorPNM[4,1.55]
TorPNM[4,3.75]
```

Define table of
regular toroidal
function, *P*.

Make labelled tables,
as in Table 12.3.1.

The `ToroidQ` cell [12.3.3]

Run this cell to get views of the toroidal functions of the second kind, which are singular at
$x = 1$ ($\eta = 0$ in toroidal coordinates).   The
curves characterize the function for a range of
values of the degree, $n - 1/2$, and the order, m,
as in Section 12.3.3.   The cell `ToroidQ` also
generates numerical values for the function of
the second kind, as appropriate for toroidal coordinate systems generated by cell `Torus-`
`Coords` described in Section 12.3.1.

Exploring with `ToroidQ`

Use cell `ToroidQ` for the function of the second kind to convince yourself that the toroidal
function of the second kind is generally a monotonous function of x, n, and m (except for
an m-dependent phase factor).   This is shown in the motif above, which displays the logarithm of the function versus x (from 1.5 to 3.5) for $n = 2$ to 5 and $m = 2$.   Transections of
this surface are shown in Figure 12.3.4.

```
(* ToroidQ cell *)

(* Toroidal Legendre Functions, Q(n-1/2,m)(x) *)

(* Toroidal Functions; Q 3D views *)

ToroidQNM[n_,m_,x_] :=
  Re[(-1)^(-m/2)*LegendreQ[n-1/2,m,x]]

(* Log Scale; values positive in range plotted *)
ToroidQ3D[m_] := Plot3D[
  Log[ToroidQNM[n,m,x]],{x,1.5,3.5},{n,2,5},
  PlotRange->All,PlotPoints->{20,20},
  Axes->False,Boxed->False]
```

Phase difference
from *Mathematica*

Define regular (*P*)
toroidal function 3D
surfaces for *m* fixed.

```
(* m = 2 surface *)
ToroidQ3D[2];
```
Show surface for
$m = 2$. (Not in text)

```
(* Toroidal Functions; Q 2D views *)

(* Log Scale: Values positive in ranges plotted *)
TS := ToString
ToroQ2D[n_,m_] := Plot[
  Log[10,Abs[ToroidQNM[n,m,x]]],{x,1.5,3.5},
  PlotRange->{-3,1},PlotPoints->20,
  Ticks->{{2,3},{-3,-2,-1,0,1}},
  AxesOrigin->{1.5,0},
  PlotLabel->"log |Q(n-1/2, "<>TS[m]<>", x)|",
  AxesLabel->{"x",None},
  PlotStyle->{Dashing[{0.015*n,0.015*n}]},
  DisplayFunction->Identity]
```
Define curves for
same m on same plot.

```
ToroShow[m_] := Show[ToroQ2D[1,m],ToroQ2D[2,m],
  ToroQ2D[3,m],DisplayFunction->$DisplayFunction]
```
Merge curves for
$n = 1, 2, 3$.

```
ToroShow[0]; ToroShow[1]; ToroShow[2]; ToroShow[3];
```
Show curves of Q,
Figure 12.3.4.

```
(* Toroidal Functions; Q Numerical *)

TorQNM[n_,x_] := Table[{n,m,x,
  N[ToroidQNM[n,m,x],10]},{m,0,3}]
```
Define table of
values for m varying
and n and x fixed.

```
TorQNM[1,1.1]
TorQNM[1,1.55]
TorQNM[1,3.75]
TorQNM[2,1.1]
TorQNM[2,1.55]
TorQNM[2,3.75]
TorQNM[4,1.1]
TorQNM[4,1.55]
TorQNM[4,3.75]
```
Show numerical
values of toroidal
function of second
kind, Q, as in Table
12.3.2.

The `Conical` cell [12.4.2]

Run this cell to get views of the conical functions that are regular at $x = 1$ ($\theta = 0$ in spherical coordinates). The surfaces and curves characterize the function of zero order for a range of values of the degree, $-1/2 + it$ with t a real parameter, as in Section 12.4.2. The cell `Conical` also generates numerical values for the function of the second kind, as appropriate for coordinates generated by `ConeCoords` cell and described in Section 12.4.1.

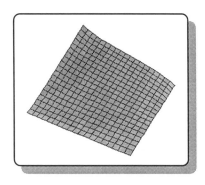

Exploring with `Conical`

Execute cell `Conical` to convince yourself that the conical function of the first kind is generally a monotonous function of x and t when the magnitude of x is less than unity.

This is shown in the motif above, which displays the logarithm of the function versus x (from −0.9 to 1.0 along the front) for $t = 0$ to 6 (from front to back). This surface and transections of it are shown in Figure 12.4.2, both as slices in the x direction and slices in the y direction.

```
(* Conical cell *)

(* Conical Functions, P(-1/2+It,0,x)  *)

ConicalP[t_,x_] := LegendreP[-1/2+I*t,x]
```

Define conical function.

```
(* Conical Functions: 3D & 2D views *)

Cone3D = Plot3D[Log[10,ConicalP[t,x]],
  {x,-0.9,1},{t,0,6},
  PlotRange->All,PlotPoints->{20,20},
  Ticks->{{-0.9,-0.5,0,0.5,1},{2,4,6},None},
  AxesLabel->{"x","t","log"}];
```

Make 3D surface as a function of x and t, as in Figure 12.4.2, left.

```
(* 2D views with  t  fixed *)
ConeTT2D[t_] := Plot[Log[10,ConicalP[t,x]],
  {x,-0.9,1},PlotRange->All,PlotPoints->20,
  Ticks->{{-0.9,-0.5,0,0.5,1},{0,1,2,3}},
  AxesLabel->{"x",None},
  DisplayFunction->Identity]
```

Define 2D curve for fixed t with x varying.

```
(* 2D views for  t = 1,2,3 *)

ShowConeXX = Show[ConeTT2D[1],ConeTT2D[2],
  ConeTT2D[3],DisplayFunction->$DisplayFunction];
```

Show curves, as in Figure 12.4.2, right top.

```
(* 2D views with  x  fixed *)

ConeXX2D[x_] := Plot[Log[10,ConicalP[t,x]],
  {t,-3,3},PlotRange->All,PlotPoints->20,
  Ticks->{{-3,-1,1,3},{0,1,2,3}},
  AxesLabel->{"t",None},
  DisplayFunction->Identity]
```

Define 2D curve for fixed x with t varying.

```
(* 2D views for  x = -0.5,  0,  0.5 *)

ShowConeTT = Show[ConeXX2D[-0.5],ConeXX2D[0],
  ConeXX2D[0.5],DisplayFunction->$DisplayFunction];
```

Show curves, as in Figure 12.4.2, right bottom.

```
(* Conical Functions: Numerical Values *)

ConeNum = Table[{t,x,N[Re[ConicalP[t,x]],10]},
  {t,0,10,5},{x,-0.8,0.8,0.8}]
```

Numerical values, as in Table 12.4.1.

20.13 SPHEROIDAL WAVE FUNCTIONS

Notebooks and Cells

Notebook SWFCoords; with cells ProlateCo-
ords, OblateCoords.   Notebook SWF;
cells SWFeigenvalue, SWFAngCoefficient, SWFAngWave, SWFRadCoefficient
SWFRadWave.

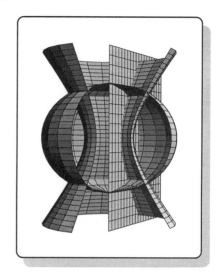

Overview

There are two *Mathematica* notebooks for
the spheroidal functions. The first notebook,
SWFCoords, generates graphics for two
coordinate systems, the spheroidal prolate
and spherical oblate systems, the second of
which is the adjacent motif.

The second *Mathematica* notebook is for
the spheroidal wave functions, including
their eigenvalues, their expansion coefficients for angular and radial functions, as
well as spheroidal angular and radial wave functions. This notebook is quite lengthy; most
of the formulas for the spheroidal wave functions are coded from first principles, since
Mathematica does not provide any spheroidal functions. To avoid cumbersome coding, the
cells in notebook SWF must be executed in a particular order, as explained in detail below.

Notebook SWFCoords [13.1]

Run this notebook to obtain surfaces that clarify the
coordinates used for spheroidal wave functions.

Check how spheroidal coordinates are generated
and understand their relation to spherical polar coordinates visualized in Section 12.2.1. For example,
vary parameter c for oblate coordinates and notice
how the deviation of the surfaces changes. A simplified example—showing only the ellipsoidal
surfaces of constant ξ—is in the motif at right; it has
($\xi = 1.0$, top) and for a surface that is more oblate
($\xi = 0.3$, bottom).

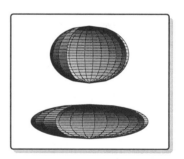

The ProlateCoords cell [13.1.1]

By executing this cell you get 3D views of the prolate
spheroidal surfaces of constant ξ ($\xi = 1.1$, left) and for
a surface that is more nearly spherical ($\xi = 2.0$, right).
Note that as $\xi \to 1$ the ellipsoid tends to a line
stretching along the z (vertical axis).

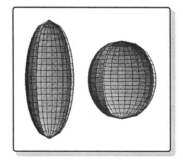

Exploring with Prolate Coords

Execute this cell to develop your insight into computing surfaces in spheroidal coordinates.

```
(* ProlateCoords cell *)

SetOptions[ParametricPlot3D,
 PlotPoints->{20,20},Boxed->False,Axes->False,
 ViewPoint->{0,-3,0},DisplayFunction->Identity];


(* Prolate Spheroidal Coordinates;
   (eta,chi,phi) in Flammer scheme *)

xs[eta_,chi_,phi_] :=
   Sqrt[(1-eta^2)*(chi^2-1)]*Cos[phi]

ys[eta_,chi_,phi_] :=
   Sqrt[(1-eta^2)*(chi^2-1)]*Sin[phi]

zs[eta_,chi_,phi_] := eta*chi

HyperboloidSurf[eta_] := ParametricPlot3D[
 {xs[eta,chi,phi],ys[eta,chi,phi],zs[eta,chi,phi]},
 {chi,1,2.5},{phi,-Pi/6,4*Pi/3}]

EllipsoidSurf[chi_] := ParametricPlot3D[
 {xs[eta,chi,phi],ys[eta,chi,phi],zs[eta,chi,phi]},
 {eta,-1,1},{phi,-Pi/6,4*Pi/3}]

PlaneSurf[phi_] := ParametricPlot3D[
 {r*Cos[phi],r*Sin[phi],z},{r,0,3},{z,-2.2,2.2}]

ProlateCoords = Show[
 HyperboloidSurf[0.8],HyperboloidSurf[-0.8],
 EllipsoidSurf[1.3],PlaneSurf[Pi/3],
 DisplayFunction->$DisplayFunction];
```

By setting the options, we avoid repetition below.

Define Cartesian coordinates in terms of prolate spheroidal coordinates.

Define hyperboloid surface as function of η.

Define ellipsoid surface as function of ξ.

Define half-plane surface as function of ϕ.

Show the three surfaces together.

The OblateCoords cell [13.1.1]

Execute this cell to get 3D views of the oblate spheroidal surfaces of constant ξ ($\xi = 0.2$, top) and for a surface that is more nearly spherical ($\xi = 1.0$, bottom). Note that as $\xi \to 0$ the ellipsoid tends to a line perpendicular to the z (vertical axis).

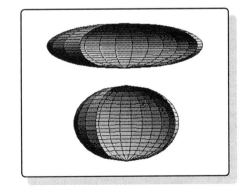

Exploring with OblateCoords

Execute this cell to develop your insight into visualizing surfaces in spheroidal coordinates. Change the values of eta, xi, and phi to see how the hyperboloid, ellipsoid, and half-plane surfaces depend upon these parameters.

```
(* OblateCoords cell *)

SetOptions[ParametricPlot3D,
 PlotPoints->{20,20},Boxed->False,Axes->False,
 ViewPoint->{0,-3,0},DisplayFunction->Identity];


(* Oblate Spheroidal Coordinates;
   (eta,chi,phi) in Flammer scheme *)

xs[eta_,chi_,phi_] :=
    Sqrt[(1-eta^2)*(chi^2+1)]*Cos[phi]
ys[eta_,chi_,phi_] :=
    Sqrt[(1-eta^2)*(chi^2+1)]*Sin[phi]
zs[eta_,chi_,phi_] := eta*chi

HyperboloidSurf[eta_] := ParametricPlot3D[
 {xs[eta,chi,phi],ys[eta,chi,phi],zs[eta,chi,phi]},
 {chi,0,2.5},{phi,-Pi/6,4*Pi/3}]

EllipsoidSurf[chi_] := ParametricPlot3D[
 {xs[eta,chi,phi],ys[eta,chi,phi],zs[eta,chi,phi]},
 {eta,-1,1},{phi,-Pi/6,4*Pi/3}]

PlaneSurf[phi_] := ParametricPlot3D[
 {r*Cos[phi],r*Sin[phi],z},{r,0,3},{z,-2.2,2.2}]

OblateCoords = Show[
 HyperboloidSurf[0.8],HyperboloidSurf[-0.8],
 EllipsoidSurf[1.3],PlaneSurf[Pi/3],
 DisplayFunction->$DisplayFunction];
```

By setting the options, we avoid repetition below.

Define Cartesian coordinates in terms of prolate spheroidal coordinates.

Define hyperboloid surface as function of η.

Define ellipsoid surface as function of ξ.

Define half-plane surface as function of ϕ.

Show the three surfaces together.

Notebook SWF

This is a very big notebook with a very short name; **S**pheroidal **W**ave **F**unctions. By executing the cells in this notebook you will be able to visualize and obtain to 10-digit accuracy numerical values of the following: spheroidal wave function eigenvalues and their anomalous roots as shown in the motif at right (the cell SWF-eigenvalue), coefficients for the angular functions (cell SWFAngCoefficient) then the angular functions themselves (cell SWFAngWave),

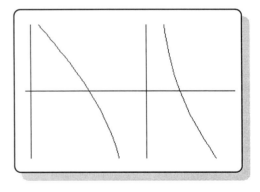

coefficients for the radial functions (cell SWFRadCoefficient) then the radial functions (cell SWFRadWave).

The output from running these cells provides all the figures and numerical values in Chapter 13. Since most of the *Mathematica* program is coded from scratch using elementary functions, these cells also provide the basic algorithms for the corresponding C functions also described in Chapter 13; the C driver program for the spheroidal wave functions is annotated in Section 21.13.

The SWFeigenvalue cell [13.1.3. 13.1.4]

Execute this cell to get the spheroidal wave function eigenvalues in both the λ_{mn} and t_{mn} forms (Sections 13.1.3 and 13.1.4, respectively). This cell also makes 3D views of the eigenvalues as functions of m and n with c fixed, for prolate and oblate coordinates, as shown in the motif at right for prolate coordinates with $c = 3$ (Figure 13.1.2, bottom left).

Exploring with SWFeigenvalue

First, change the accuracy parameter smax, which determines how many terms are used in the continued-fraction expansion for function U_2 in (13.1.16). Convince yourself that smax = 8 is sufficient for c up to about 5. (The case $m = 0$, $n = 0$, and oblate coordinates is most sensitive to smax.) Second, change the values of c used in this cell to see how the eigenvalues change with c. Note that the routine may not give correct eigenvalues for very large values of c; read the discussion above Figure 13.1.3 to clarify this.

```
(* SWFeigenvalue cell *)

(* Spheroidal Wave Function eigenvalues;
   Little and Corbato  tmn    *)

(* Modified Gamma[m,s] - tmn *)
GTT[m_,n_,s_,csq_,tmn_] := s*(2*n+s+1)+
 2*s*csq*(4*m^2-1)*(2*n+s+1)/
 ((2*n-1)*(2*n+3)*(2*n+2*s-1)*(2*n+2*s+3))-csq*tmn

 (* Beta[m,u] *)
B[m_,u_,csq_] :=  csq^2*u*(u-1)*(2*m+u)*(2*m+u-1)/
 ((2*(m+u)-1)^2*(2*(m+u)-3)*(2*(m+u)+1))

(* U1 series defined *)
t = x;
U1[m_,n_,csq_,tmn_] := GTT[m,n,0,csq,tmn] +
 (Do[t = -B[m,r,csq]/
  (GTT[m,n,r-2-(n-m),csq,tmn]+t),
  {r,If[EvenQ[n-m],0,1],n-m,2}]; t) /. x->0
```

Define modified gamma function, (13.1.21).

Define beta function, (13.1.14).

Finite continued fraction, (13.1.15).

```
(* U2 series defined *)
t = x;
U2[m_,n_,smax_,csq_,tmn_] :=
 (
 rBiggest = If[EvenQ[smax],smax,smax+1];
 Do[t = -B[m,n-m+2+rBiggest-r,csq]/
  (GTT[m,n,2+rBiggest-r,csq,tmn]+t),
  {r,0,rBiggest,2}]; t) /. x->0
```

Infinite continued fraction, (13.1.16), with about `smax` terms used.

```
(* Find the modified eigenvalue, tmn *)

SWFevalTMN[m_,n_,smax_,csq_] := N[FindRoot[
 U1[m,n,csq,tmn]+U2[m,n,smax,csq,tmn] == 0,
 {tmn,-0.01*csq}],10]
```

Define modified eigenvalue, as in Section 13.1.4.

```
(* to extract value from FindRoot object *)
NumRoot[Roots_] :=
 ( stringRoots = ToString[Roots];
  strXY = StringToStream[stringRoots];
  (* Locate arrow head position *)
  ListXY = Flatten[StringPosition[
      stringRoots,ToString[">"]]];
  SetStreamPosition[strXY, ListXY[[1]]+1];
  ExactZ = Read[strXY,Number] )
```

Root extraction function.

```
(* Spherical Wave Function Eigenvalues
   in  tmn  form; Numerical *)

smax = 8;
TmnTable[mBig_,nBig_,csq_] := N[Table[{m,n,If[ m>n,0,
  NumRoot[SWFevalTMN[m,n,smax,csq]]]},
  {m,0,mBig},{n,0,nBig}],10]
```

Define table of modified eigenvalues.

```
TmnTable[2,2, 2.0^2] (* Prolate with c = 2 *)
TmnTable[2,2,-2.0^2] (* Oblate with c = 2 *)
```

Tabulate, as in Tables 13.2.1 and 13.2.2.

```
(* Spherical Wave Function Eigenvalues
   in  Lmn  form; Numerical *)

(* Define eigenvalues in terms of  tmn  *)
Lmn[m_,n_,smax_,csq_] := n*(n+1)+
 (2*n*(n+1)-2*m^2-1)*csq/((2*n-1)*(2*n+3))+
 csq*NumRoot[SWFevalTMN[m,n,smax,csq]]
```

Define eigen-values.

```
LmnFormTable[mBig_,nBig_,csq_] :=
 N[Table[{m,n,If[ m>n, 0, Lmn[m,n,smax,csq]]},
  {m,0,mBig},{n,0,nBig}],10]
```

Define table form.

```
LmnFormTable[2,2, 2.0^2] (* Prolate with c = 2 *)
LmnFormTable[2,2,-2.0^2] (* Oblate with c = 2 *)
```

Tabulate; Tables 13.1.1 and 13.1.2.

```
(* Spherical Wave Function Eigenvalues;
   3D views of  [eigenvalues - n(n+1)]  *)

Lmn3DTable[mMax_,nMax_,csq_] := Table[If[m>n,  0,
 Lmn[m,n,smax,csq]-n*(n+1)],{n,1,nMax},{m,1,mMax}]

SWFShow[mMax_,nMax_,csq_] :=  ListPlot3D[
 Lmn3DTable[mMax,nMax,csq],PlotRange->{-3,6},
 Ticks->{Range[mMax],Range[nMax],{0,4}},
 AxesLabel->{"m","n",None}];

SWFShow[6,6,  2.0^2]; (* Prolate with c = 2 *)
SWFShow[6,6,-2.0^2]; (* Oblate with c = 2 *)
SWFShow[6,6,  3.0^2]; (* Prolate with c = 3 *)
SWFShow[6,6,-3.0^2]; (* Oblate with c = 3 *)

(* Anomalous roots *)
UU[m_,n_,smax_,csq_,LL_] :=
 (
  (* tmn  in terms of  LL  *)
  tmnLL = (LL - n*(n+1))/csq -
  (2*n*(n+1)-2*m^2-1)/((2*n-1)*(2*n+3));

 U1[m,n,csq,tmnLL]+U2[m,n,smax,csq,tmnLL]
 )
UUPlot[m_,n_,smax_,csq_,Lstart_,Lend_] :=
 Plot[UU[m,n,smax,csq,LL],{LL,Lstart,Lend},
 PlotPoints->41,PlotRange->{-10,10},
 AxesOrigin->{0,0},AxesLabel->{"L",None},
 Ticks->{{5,10,15,20,25,30,35,40},
   {-10,-5,0,5,10}}]

(* Multiple roots for oblate *)
m = 4;
n = 4;
smax = 8;
Lstart = 0;
Lend = 40;
Do[UUPlot[m,n,smax,Sign[c]*c^2,Lstart,Lend],
 {c,-8,-2,2}];
```

SWFeigenvalue
continued

Define table of
spheroidal eigen-
values, then
define 3D plotting
function.

Show surfaces of
eigenvalues, as in
Figure 13.1.2.

Define *U* function
whose roots are the
spheroidal eigen-
values.

Define 2D plots of
the *U* function.

Show the multiple
roots for oblate
coordinates, as in
Figure 13.1.3.

The SWFAngCoefficient cell [13.2.1]

Run this cell for the spheroidal angular function
coefficients $d_r^{mn}(c)$ in Section 13.2.1. This cell
also makes 3D views of the coefficients as functions of *r* and *c* with *m* and *n* fixed, for both prolate and oblate coordinates, as shown in the motif
at right for oblate coordinates with $m = 0$ and
$n = 4$ (Figure 13.2.2, bottom left).

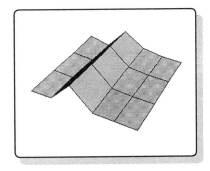

Exploring with SWFAngCoefficient

Try various combinations of *m* and *n* (both small,
with *n* not less than *m*), allowing *c* to vary over a

large range. At what values of c (especially for oblate coordinates) does the peak value of $d_r^{mn}(c)$ occur other than at $r = n - m$, the unique value for spherical coordinates?

The cell SWFAngCoefficient has considerable repetition of the SWFeigenvalue cell, since the eigenvalues are needed to calculate the $d_r^{mn}(c)$. We do this in order to allow flexibility in using the notebook. Note that cell SWFAngCoefficient must be run before the next cell, SWFAngWave, can be used.

```
(* SWFAngCoefficient cell *)

(* Spheroidal Angular Function Coefficients.
   Using  tmn  of Little & Corbato in [Stra56]
   with backward iteration *)

(* Modified Gamma[m,s] - tmn  *)
GTT[m_,n_,s_,csq_,tmn_] := s*(2*n+s+1)+
 2*s*csq*(4*m^2-1)*(2*n+s+1)/
 ((2*n-1)*(2*n+3)*(2*n+2*s-1)*(2*n+2*s+3))-csq*tmn

(* Beta[m,u] *)
B[m_,u_,csq_] := csq^2*u*(u-1)*(2*m+u)*(2*m+u-1)/
 ((2*(m+u)-1)^2*(2*(m+u)-3)*(2*(m+u)+1))

(* U1 series defined *)
t = x;
U1[m_,n_,csq_,tmn_] := GTT[m,n,0,csq,tmn]+
 (Do[t = -B[m,r,csq]/
   (GTT[m,n,r-2-(n-m),csq,tmn]+t),
   {r,If[EvenQ[n-m],0,1],n-m,2}]; t) /. x->0

(* U2 series defined *)
t = x;
U2[m_,n_,smax_,csq_,tmn_] :=
 (
 rrmax = If[EvenQ[smax],smax,smax+1];
 Do[t = -B[m,n-m+2+rrmax-r,csq]/
   (GTT[m,n,2+rrmax-r,csq,tmn]+t),
   {r,0,rrmax,2}]; t) /. x->0

(* Find the modified eigenvalue, tmn *)
SWFevalTMN[m_,n_,smax_,csq_] := N[FindRoot[
 U1[m,n,csq,tmn]+U2[m,n,smax,csq,tmn] == 0,
 {tmn,-0.01*csq}],10]

(* Extract eigenvalue from FindRoot object *)
NumRoot[Roots_] :=
 ( stringRoots = ToString[Roots];
   strXY = StringToStream[stringRoots];
   (* Locate arrow head position *)
   ListXY = Flatten[StringPosition[
            stringRoots,ToString[">"]]];
   SetStreamPosition[strXY, ListXY[[1]]+1];
   ExactZ = Read[strXY,Number] )
```

Using auxiliary function for eigenvalues, Section 13.1.4.

Define modified gamma function, (13.1.21).

Define beta function, (13.1.14).

Finite continued fraction, (13.1.15).

Infinite continued fraction, (13.1.16).

Extract root of U function as the modified eigenvalue.

Extract root from the character string to a numerical value.

```
(* Spherical Wave Function Eigenvalue as tmn *)

smax = 8;
tmn[m_,n_,smax_,csq_] := N[If[ m > n, 0,
 NumRoot[SWFevalTMN[m,n,smax,csq]]],10]

tmnForm[mBig_,nBig_,csq_] :=
 N[Table[{m,n,If[m>n,0,tmn[m,n,smax,csq]]},
   {m,0,mBig},{n,0,nBig}],10]

tmnForm[2,2, 4.0] (* Prolate with c = 2 *)
tmnForm[2,2,-4.0] (* Oblate with c = 2 *)

(* Angular Coefficients:
   Factors depending on r *)
aa[m_,r_,csq_] := (2*m+r+2)*(2*m+r+1)*csq/
 ((2*m+2*r+3)*(2*m+2*r+5))
bbtmn[m_,n_,s_,csq_] := s*(2*n+s+1)+
 csq*(2*s*(2*n+s+1)*(4*m^2-1)/
 ((2*n-1)*(2*n+3)*(2*n+2*s-1)*(2*n+2*s+3))-
 tmnEval)
cc[m_,r_,csq_] := r*(r-1)*csq/
 ((2*m+2*r-3)*(2*m+2*r-1))

(* Array of coefficients by backward iteration
   for r=rLarge down to r=rSt (0 or 1) by 2 *)
dmnArray[m_,n_,smax_,csq_,rExtra_] :=
 (  (* Auxiliary eigenvalue function *)
 tmnEval = tmn[m,n,smax,csq];

 If[ EvenQ[n-m], rSt = 0, rSt = 1 ];
 rLarge = n - m + rExtra;
 dmn[rLarge] = 1;
 Do[  (* downward iteration on  r  *)
  (
  If[csq == 0,
   If[r == n-m, dmn[r] = 1, dmn[r] = 0],
   If[r == rLarge,
    dmn[r-2] =
      -bbtmn[m,n,r-(n-m),csq]*dmn[r]/cc[m,r,csq],
    dmn[r-2] =
     -(bbtmn[m,n,r-(n-m),csq]*dmn[r]+
      aa[m,r,csq]*dmn[r+2])/cc[m,r,csq]]];
  ), {r,rLarge,rSt+2,-2}];
 dmnCoeffs = Table[dmn[r],{r,rSt,rLarge,2}];

(* Normalization in Flammer scheme *)
If[ rSt == 0,
 ( NormSumFla = Re[NSum[
    (-1)^(r/2)*(r+2*m)!*dmnCoeffs[[r/2+1]]/
    (2^r*(r/2)!*((r+2*m)/2)!),
    {r,rSt,rLarge,2}]];
```

SWFAng-
Coefficient
continued

Define table of
auxiliary functions,
(13.1.20).

Tables 13.2.1,
13.2.2 for auxiliary
functions.

Define recurrence
coefficients,
(13.2.4), then
(13.2.6),

then (13.2.5).

Spheroidal angular
coefficients.
Start with modified
eigenvalue.

Iterate down on *r*
by (13.2.3).

Spherical case is
special.

Starting iterate is
special.
Else, just use
(13.2.3).

Make a table.

Normalization
formula (13.2.7)
depends on
whether $n - m$ is
even or odd.

```
  NormFla = (-1)^((n-m)/2)*(n+m)!/
    (2^(n-m)*((n-m)/2)!*((n+m)/2)!)/NormSumFla;),

( NormSumFla = Re[NSum[
    (-1)^((r-1)/2)*(r+2*m+1)!*dmnCoeffs[[(r+1)/2]]/
    (2^r*((r-1)/2)!*((r+2*m+1)/2)!),
    {r,rSt,rLarge,2}]];

  NormFla = ((-1)^((n-m-1)/2)*(n+m+1)!/
    (2^(n-m)*((n-m-1)/2)!*((n+m+1)/2)!))/NormSumFla;)];
dmnFormTable = N[NormFla*dmnCoeffs,10];
)
```

Angular coeff-
icients in Flammer
normalized form.

```
(* SWF coefficients:
   Flammer normalization; 3D views *)

SWFCoeffShow[m_,n_,smax_,ProOb_,rExtra_] :=
( (* Tabulate normalized coefficients *)
  dmnArray[m,n,smax,ProOb*1,rExtra];
  dmnView1 = dmnFormTable;
  dmnArray[m,n,smax,ProOb*2^2,rExtra];
  dmnView2 = dmnFormTable;
  dmnArray[m,n,smax,ProOb*3^2,rExtra];
  dmnView3 = dmnFormTable;
  dmnArray[m,n,smax,ProOb*4^2,rExtra];
  dmnView4 = dmnFormTable;
```

Define collected
tables of coeff-
icients with c
varying, for prolate
or oblate coord-
inates.

```
  (* Make 3D view *)
  SWFCoeff3D = ListPlot3D[
   {dmnView1,dmnView2,dmnView3,dmnView4},
   PlotRange->{-1,1},Ticks->{None,None,{-1,0,1}},
   AxesLabel->{" "," ",None}];
)
```

Define surface of
spheroidal angular
coefficients.

```
rExtra = 6; (* accurate enough for clear plots *)
smax = 8; (* controls eigenvalue
              continued-fraction accuracy *)

ProOb = 1; (* Prolate coefficient plots *)
SWFCoeffShow[0,2,smax,ProOb,rExtra]
SWFCoeffShow[0,4,smax,ProOb,rExtra]
SWFCoeffShow[2,2,smax,ProOb,rExtra]
SWFCoeffShow[2,4,smax,ProOb,rExtra]
```

Make surface for
prolate angular
coefficients, as in
Figure 13.2.1.

```
ProOb = -1; (* Oblate coefficient plots *)
SWFCoeffShow[0,2,smax,ProOb,rExtra];
SWFCoeffShow[0,4,smax,ProOb,rExtra];
SWFCoeffShow[2,2,smax,ProOb,rExtra];
SWFCoeffShow[2,4,smax,ProOb,rExtra];
```

Make surface for
prolate angular
coefficients, as in
Figure 13.2.2.

```
(* Angular expansion coefficients: Numerical *)

rExtra = 14; (* ensures accurate normalization *)
dmnArray[0,0,8, 2.0^2,rExtra]; (* Prolate *)
dmnSavePro00 = dmnFormTable
dmnArray[0,1,8, 2.0^2,rExtra];
dmnSavePro01 = dmnFormTable
dmnArray[2,2,8, 2.0^2,rExtra];
dmnSavePro22 = dmnFormTable

dmnArray[0,0,8,-2.0^2,rExtra]; (* Oblate *)
dmnSaveOb00 = dmnFormTable
dmnArray[0,1,8,-2.0^2,rExtra];
dmnSaveOb01 = dmnFormTable
dmnArray[2,2,8,-2.0^2,rExtra];
dmnSaveOb22 = dmnFormTable
```

SWFAng-
Coefficient
continued

Make Table 13.2.3
for spheroidal
angular coeffic-
ients with prolate
coordinates.

Make Table 13.2.4
for spheroidal
angular coeffic-
ients with oblate
coordinates.

The `SWFAngWave` cell [13.2.2]

Use this cell to make 2D graphics and numerical values for the spheroidal angular functions $S_{mn}^{(1)}(c,\eta)$ and $S_{mn}^{(2)}(c,\eta)$ in Section 13.2.2. The graphics from this cell also compare the functions for prolate and oblate coordinates with the spherical functions, as shown in the motif at right for the regular function having $|c|=2$, $m=0$, and $n=1$ (Figure 13.2.3, bottom). The spherical function $P_1(\eta)=\eta$ is the solid line, the function for prolate coordinates is shown by long dashes, while the function for oblate coordinates has short dashes.

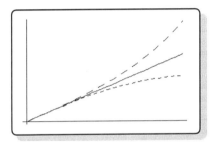

Note that the preceding cell, `SWFAngCoefficient`, must be run before cell `SWF-AngWave` can be used.

Exploring with `SWFAngWave`

Try various combinations of m and n (both small, with n not less than m), allowing c to also vary. For what values of c does the average of the spheroidal angular functions for corresponding prolate and oblate coordinates deviate noticeably from the spherical angular functions?

```
(* SWFAngWave cell *)

(* Spheroidal Angular Wave Functions
   in Flammer Normalization *)



(* You MUST execute cell SWFAngCoefficient first
   for angular function expansion coefficients *)
```

Spheroidal angular
functions are in the
Flammer normal-
ization.

```
(* Function of first kind expanded in P(n,m) *)
S1NMP[m_,n_,dmnSaved_,eta_] :=
 (
 If[ EvenQ[n-m], rSt = 0, rSt = 1 ];
 rLarge = n - m + rExtra;
 S1 = N[NSum[dmnSaved[[(r-rSt)/2+1]]*
  LegendreP[m+r,m,eta],{r,rSt,rLarge,2}],10]
 )
```

Define regular function by series (13.2.1).

```
(* Function of second kind expanded in Q(n,m) *)
S2NMQ[m_,n_,dmnSaved_,eta_] :=
 (
 If[ EvenQ[n-m], rSt = 0, rSt = 1 ];
 rLarge = n - m + rExtra;
 S2 = N[NSum[dmnSaved[[(r-
rSt)/2+1]]*LegendreQ[m+r,m,eta],
 {r,rSt,rLarge,2}],10]
 )
```

Define irregular function by series (13.2.2).

```
(* Regular function; Prolate coordinates *)
S100Pro[eta_] := S1NMP[0,0,dmnSavePro00,eta]
S101Pro[eta_] := S1NMP[0,1,dmnSavePro01,eta]
S122Pro[eta_] := S1NMP[2,2,dmnSavePro22,eta]
```

Define functions for each *m* and *n*. Regular prolate,

```
(* Regular function; Oblate coordinates *)
S100Ob[eta_] := S1NMP[0,0,dmnSaveOb00,eta]
S101Ob[eta_] := S1NMP[0,1,dmnSaveOb01,eta]
S122Ob[eta_] := S1NMP[2,2,dmnSaveOb22,eta]
```

then regular oblate,

```
(* Irregular function; Prolate coordinates *)
S200Pro[eta_] := S2NMQ[0,0,dmnSavePro00,eta]
S201Pro[eta_] := S2NMQ[0,1,dmnSavePro01,eta]
S222Pro[eta_] := S2NMQ[2,2,dmnSavePro22,eta]
```

then irregular prolate,

```
(* Irregular function; Oblate coordinates *)
S200Ob[eta_] := S2NMQ[0,0,dmnSaveOb00,eta]
S201Ob[eta_] := S2NMQ[0,1,dmnSaveOb01,eta]
S222Ob[eta_] := S2NMQ[2,2,dmnSaveOb22,eta]
```

and irregular oblate.

```
(* Spheroidal Angular Functions; 2D views *)
Off[NSum::precw];

SetOptions[Plot,PlotRange->All,AxesOrigin->{0,0},
 Ticks->{None,{-1,0,0,1,2,3,4,5,6}},
 DisplayFunction->Identity];
```

Set plotting options for brevity.

```
SMNPlot[m_,n_,SWFPro_,SWFOb_,go_,end_] :=
 ( SMNPro = Plot[SWFPro[eta],{eta,go,end},
  PlotStyle->Dashing[{0.02,0.02}]];
 SMNOb = Plot[SWFOb[eta],{eta,go,end},
  PlotStyle->Dashing[{0.04,0.04}]];
 PlotOfPMN = Plot[LegendreP[n,m,eta],{eta,go,end}];
 SWFAngPlot = Show[SMNPro,SMNOb,PlotOfPMN,
  DisplayFunction->$DisplayFunction]; )
```

Define plotted curves for prolate, oblate, and spherical functions.

```
SMNPlot[0,0,S100Pro,S1000b,0,1];
SMNPlot[0,1,S101Pro,S1010b,0,1];
SMNPlot[2,2,S122Pro,S1220b,0,1];
SMNPlot[0,0,S200Pro,S2000b,0,0.7];
SMNPlot[0,1,S201Pro,S2010b,0,0.7];
SMNPlot[2,2,S222Pro,S2220b,0,0.7];
```

Figures 13.2.3 (regular function) and 13.2.4 (irregular function).

```
(* Spheroidal Angular Functions; numerical *)

Do[ (* Regular function; prolate coordinates *)
  Print[S100Pro[eta]];
  Print[S101Pro[eta]];
  Print[S122Pro[eta]],{eta,0,1,0.5}]

Do[ (* Regular function; oblate coordinates *)
  Print[S1000b[eta]];
  Print[S1010b[eta]];
  Print[S1220b[eta]],{eta,0,1,0.5}]
```

Print numerical values for regular functions; Table 13.2.5 for prolate coordinates and Table 13.2.6 for oblate coordinates.

```
Do[ (* Irregular function; prolate coordinates *)
  Print[S200Pro[eta]];
  Print[S201Pro[eta]];
  Print[S222Pro[eta]],{eta,0.2,0.8,0.3}]

Do[ (* Irregular function; oblate coordinates *)
  Print[S2000b[eta]];
  Print[S2010b[eta]];
  Print[S2220b[eta]],{eta,0.2,0.8,0.3}]
```

Print numerical values for irregular functions; Table 13.2.7 for prolate coordinates and Table 13.2.8 for oblate coordinates.

The SWFRadCoefficient cell [13.3.1]

Run this cell for the spheroidal radial function coefficients $a_r^{mn}(c)$ in Section 13.3.1. The cell also makes 3D views of the coefficients as functions of r and c with m and n fixed, for both prolate and oblate coordinates, as shown in the motif at right for the oblate case having $m = 0$ and $n = 4$ (Figure 13.3.2, bottom left).

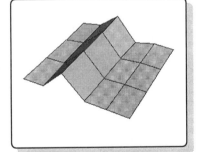

Exploring with SWFRadCoefficient

Try various combinations of m and n (both small, but with n not less than m), allowing c to vary over a large range. Discover at what values of c (especially for oblate coordinates) the peak value of $a_r^{mn}(c)$ occurs other than $r = n - m$, which is the unique value for spherical coordinates.

Cell SWFRadCoefficient has considerable repetition of the SWFeigenvalue and SWFAngCoefficient cells, since the eigenvalues are needed to calculate the angular coefficients, $d_r^{mn}(c)$, which are then used to compute the radial coefficients, $a_r^{mn}(c)$. This repetition, however, gives us flexibility in using the notebook. Note that cell SWFRad-Coefficient must be run before the next cell, SWFRadWave, can be used.

```
(* SWFRadCoefficient cell *)

(* Spheroidal Radial Coefficients.
   Using tmn of Little & Corbato in [Stra56]
   with backward iteration *)

(* Modified Gamma[m,s] - tmn  *)
GTT[m_,n_,s_,csq_,tmn_] := s*(2*n+s+1)+
 2*s*csq*(4*m^2-1)*(2*n+s+1)/
 ((2*n-1)*(2*n+3)*(2*n+2*s-1)*(2*n+2*s+3))-csq*tmn

(* Beta[m,u] *)
B[m_,u_,csq_] := csq^2*u*(u-1)*(2*m+u)*(2*m+u-1)/
 ((2*(m+u)-1)^2*(2*(m+u)-3)*(2*(m+u)+1))

(* U1 series defined *)
t = x;
U1[m_,n_,csq_,tmn_] := GTT[m,n,0,csq,tmn]+
 (Do[t = -B[m,r,csq]/
  (GTT[m,n,r-2-(n-m),csq,tmn]+t),
  {r,If[EvenQ[n-m],0,1],n-m,2}]; t) /. x->0

(* U2 series defined *)
t = x;
U2[m_,n_,smax_,csq_,tmn_] :=
 (
 rrmax = If[EvenQ[smax],smax,smax+1];
 Do[t = -B[m,n-m+2+rrmax-r,csq]/
  (GTT[m,n,2+rrmax-r,csq,tmn]+t),
  {r,0,rrmax,2}]; t) /. x->0

(* Find the modified eigenvalue, tmn *)
SWFevalTMN[m_,n_,smax_,csq_] := N[FindRoot[
  U1[m,n,csq,tmn]+U2[m,n,smax,csq,tmn] == 0,
  {tmn,-0.01*csq}],10]

(* Extract eigenvalue from FindRoot object *)
NumRoot[Roots_] :=
 ( stringRoots = ToString[Roots];
   strXY = StringToStream[stringRoots];
   (* Locate arrow head position *)
   ListXY = Flatten[StringPosition[
          stringRoots,ToString[">"]]];
   SetStreamPosition[strXY, ListXY[[1]]+1];
   ExactZ = Read[strXY,Number] )

(* Spherical Wave Function Eigenvalue as  tmn  *)
smax = 8;
tmn[m_,n_,smax_,csq_] := N[If[ m > n, 0,
 NumRoot[SWFevalTMN[m,n,smax,csq]]],10]
```

Using auxiliary function for eigen-values, Section 13.1.4.

Define modified gamma function, (13.1.21).

Define beta function, (13.1.14).

Finite continued fraction, (13.1.15).

Infinite continued fraction, (13.1.16).

Extract root of *U* function as the modified eigen-value.

Extract root from the character string to a numerical value.

Auxiliary eigen-function value, (13.1.20).

```
(* Factors depending on r in angular coefficients *)
aa[m_,r_,csq_] := (2*m+r+2)*(2*m+r+1)*csq/
  ((2*m+2*r+3)*(2*m+2*r+5))

bbtmn[m_,n_,s_,csq_] := s*(2*n+s+1)+
  csq*(2*s*(2*n+s+1)*(4*m^2-1)/
  ((2*n-1)*(2*n+3)*(2*n+2*s-1)*(2*n+2*s+3))-
  tmnEval)

cc[m_,r_,csq_] := r*(r-1)*csq/
  ((2*m+2*r-3)*(2*m+2*r-1))


(* Make unnormalized angular coefficients *)
amnRadArray[m_,n_,smax_,csq_,rExtra_] :=
 ( (* Auxiliary eigenvalue function *)
 tmnEval = tmn[m,n,smax,csq];

 If[ EvenQ[n-m], rSt = 0, rSt = 1 ];
 rLarge = n - m + rExtra;
 dmnAng[rLarge] = 1;


 Do[  (* downward iteration on  r  *)
  (
  If[csq == 0,
   If[r == n-m, dmnAng[r] = 1, dmnAng[r] = 0],
   If[r == rLarge,
    dmnAng[r-2] =
      -bbtmn[m,n,r-(n-m),csq]*dmnAng[r]/cc[m,r,csq],
    dmnAng[r-2] =
      -(bbtmn[m,n,r-(n-m),csq]*dmnAng[r]+
      aa[m,r,csq]*dmnAng[r+2])/cc[m,r,csq]]];
  ), {r,rLarge,rSt+2,-2}];
 dmnAngCoeffs = Table[dmnAng[r],{r,rSt,rLarge,2}];


(* Make unnormalized radial coefficients *)
 Do[ (* upward iteration on  r
        using precomputed angular coefficients *)
  amnRad[r] = (-1)^((r-n+m)/2)*(r+2*m)!*
              dmnAngCoeffs[[r/2+(2-rSt)/2]]/r!,
  {r,rSt,rLarge,2}];
 amnRadCoeffs = Table[amnRad[r],{r,rSt,rLarge,2}];


(* Normalization *)
 NormRadSum = Re[NSum[
  (-1)^((r-n+m)/2)*amnRadCoeffs[[r/2+(2-rSt)/2]],
  {r,rSt,rLarge,2}]];
 amnRadTable = N[amnRadCoeffs/NormRadSum,10];
 )

(* SWF Radial Coefficients: 3D views *)
```

SWFRad
Coefficient
continued

Define recurrence
coefficients,
(13.2.4), (13.2.6),
then (13.2.5).

Spheroidal angular
coefficients.
Start with modified
eigenvalue.

Iterate down on *r*
by (13.2.3).
Spherical case is
special.

Starting iterate is
special.
Else, just use
(13.2.3).

Make a table.

Get radial
coefficients
from angular
coefficients by
(13.3.2).

Normalization
condition (13.3.3).

Start 3D views of
radial coefficients.

```
SWFRadCoeffShow[m_,n_,smax_,ProOb_,rExtra_] :=
(  (* Tabulate normalized coefficients *)
 amnRadArray[m,n,smax,ProOb*1,rExtra];
 amnRadView1 = amnRadTable;
 amnRadArray[m,n,smax,ProOb*2^2,rExtra];
 amnRadView2 = amnRadTable;
 amnRadArray[m,n,smax,ProOb*3^2,rExtra];
 amnRadView3 = amnRadTable;
 amnRadArray[m,n,smax,ProOb*4^2,rExtra];
 amnRadView4 = amnRadTable;


 (* Make 3D view *)
 SWFRadCoeff3D = ListPlot3D[
   {amnRadView1,amnRadView2,amnRadView3,amnRadView4},
   PlotRange->{-1,1},Ticks->{None,None,{-1,0,1}},
   AxesLabel->{" "," ",None}];
)


rExtra = 6; (* accurate enough for clear plots *)
smax = 8; (* controls eigenvalue
              continued-fraction accuracy *)


ProOb = 1; (* Prolate coefficient plots *)
SWFRadCoeffShow[0,2,smax,ProOb,rExtra];
SWFRadCoeffShow[0,4,smax,ProOb,rExtra];
SWFRadCoeffShow[2,2,smax,ProOb,rExtra];
SWFRadCoeffShow[2,4,smax,ProOb,rExtra];
ProOb = -1; (* Oblate coefficient plots *)
SWFRadCoeffShow[0,2,smax,ProOb,rExtra];
SWFRadCoeffShow[0,4,smax,ProOb,rExtra];
SWFRadCoeffShow[2,2,smax,ProOb,rExtra];
SWFRadCoeffShow[2,4,smax,ProOb,rExtra];


(* Radial Expansion Coefficients: numerical *)


rExtra = 14; (* ensures accurate normalization *)
amnRadArray[0,0,8, 2.0^2,rExtra]; (* Prolate *)
amnRadPro00 = amnRadTable
amnRadArray[0,1,8, 2.0^2,rExtra];
amnRadPro01 = amnRadTable
amnRadArray[2,2,8, 2.0^2,rExtra];
amnRadPro22 = amnRadTable


amnRadArray[0,0,8,-2.0^2,rExtra]; (* Oblate *)
amnRadOb00 = amnRadTable
amnRadArray[0,1,8,-2.0^2,rExtra];
amnRadOb01 = amnRadTable
amnRadArray[2,2,8,-2.0^2,rExtra];
amnRadOb22 = amnRadTable
```

Define collected tables of coefficients with *c* varying, for prolate or oblate coordinates.

Define surface of spheroidal radial coefficients.

Make surface for prolate radial coefficients, as in Figure 13.3.1.

Make surface for prolate angular coefficients, as in Figure 13.3.2.

Numerical radial coefficients for prolate, as in Table 13.3.1.

Numerical radial coefficients for oblate, as in Table 13.3.2.

The `SWFRadWave` cell [13.3.2]

Use this cell to make 2D graphics and numerical values for the regular spheroidal radial functions $R_{mn}^{(1)}(c,\xi)$ shown in Figure 13.2.3, of which the motif at right shows $R_{01}^{(1)}(c,\xi)$ for $|c| = 2$ with ξ in the range 2 to 3. The prolate coordinate function is shown by long dashes, while the oblate coordinate function is indicated by short dashes.

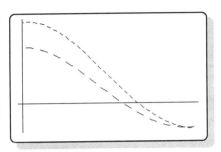

The preceding cell, `SWFRadCoeffic-ient`, must be run before cell `SWFRadWave` can be used.

Exploring with `SWFRadWave`

Try some combinations of m and n (both small for greatest sensitivity to c, but with n not less than m), allowing c to also vary. How small does c have to be for the spheroidal radial functions for prolate and oblate coordinates to be just noticeably different from each other?

```
(* SWFRadWave cell *)

(* Spheroidal Radial Wave Functions *)

(* You MUST execute cell SWFRadCoefficient
   first to get radial expansion coefficients *)

(* Function of first kind expanded in
   spherical Bessel function  j(m+r)   *)

(* Define spherical Bessel function *)
SpherBess[n_,x_] :=
 N[Sqrt[N[Pi]/(2*x)]*BesselJ[n+1/2,x],10]

(* Spheroidal radial function of first kind *)
R1NMP[m_,n_,sign_,c_,amnSaved_,chi_] :=
 (
 If[ EvenQ[n-m], rSt = 0, rSt = 1 ];
 rLarge = n - m + rExtra;
 R1 = N[(1-sign/chi^2)^(m/2)*
  NSum[amnSaved[[(r-rSt)/2+1]]*
  SpherBess[m+r,c*chi],{r,rSt,rLarge,2}],10]
 )

c = 2;
rExtra = 14;
(* Regular function; Prolate coordinates *)
R100Pro[chi_] := R1NMP[0,0,1,c,amnRadPro00,chi]
R101Pro[chi_] := R1NMP[0,1,1,c,amnRadPro01,chi]
R122Pro[chi_] := R1NMP[2,2,1,c,amnRadPro22,chi]
```

Spherical Bessel in terms of the general regular Bessel.

Define the regular spheroidal radial functions, (13.3.6) or (13.3.7).

Choose number of terms beyond peak coefficient as 14, then make functions for each m and n.

```
(* Regular function; Oblate coordinates *)
R100Ob[chi_] := R1NMP[0,0,-1,c,amnRadOb00,chi]
R101Ob[chi_] := R1NMP[0,1,-1,c,amnRadOb01,chi]
R122Ob[chi_] := R1NMP[2,2,-1,c,amnRadOb22,chi]

(* Spheroidal angular functions; 2D views *)
SetOptions[Plot,PlotRange->All,AxesOrigin->{1,0},
  Ticks->{None,{-0.4,-0.2,0,0.2,0.4,0.6}},
  DisplayFunction->Identity];

R1MNPlot[m_,n_,SWFPro_,SWFOb_,go_,end_] :=
  ( R1MNPro = Plot[SWFPro[chi],{chi,go,end},
    PlotStyle->Dashing[{0.02,0.02}]];
    R1MNOb = Plot[SWFOb[chi],{chi,go,end},
    PlotStyle->Dashing[{0.04,0.04}]];
    SWFRadPlot = Show[R1MNPro,R1MNOb,
    DisplayFunction->$DisplayFunction]; )

R1MNPlot[0,0,R100Pro,R100Ob,1.001,3];
R1MNPlot[0,1,R101Pro,R101Ob,1.001,3];
R1MNPlot[2,2,R122Pro,R122Ob,1.001,3];

(* Spheroidal Radial Functions; numerical *)
Do[ (* Prolate coordinates *)
 Print[R100Pro[chi]];
 Print[R101Pro[chi]];
 Print[R122Pro[chi]],{chi,2,3,0.5}]
Do[ (* Oblate coordinates *)
 Print[R100Ob[chi]];
 Print[R101Ob[chi]];
 Print[R122Ob[chi]],{chi,2,3,0.5}]
```

SWFRadWave
continued

Make functions for each *m* and *n*.

Plot options.

Define plotting function for regular spheroidal functions, prolate and oblate plotted together.

Plots for (*m*, *n*) pairs (0,0), (0,1), then (2,2).

Table 13.3.3 for prolate.

Table 13.3.4 for oblate.

20.14 BESSEL FUNCTIONS

Notebooks and Cells
Notebook BesselInteger; cells
BesselJY, BesselPQ, BesselYX,
BesselIK, Kelvin

Notebook BesselHalf; cells
Bessel_j, SphereBesselPQ,
Bessel_y, Bessel_i, BesselR,
Bessel_k

Notebook BesselAiry; cells
Airy, Airy.f&g, AiryAsymp-
totes, Airy.fp&gp, Airy-
PrimeAsymptotes

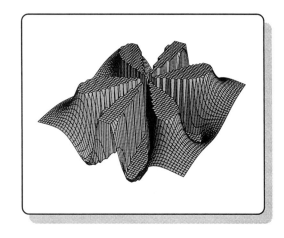

Overview

There are three *Mathematica* notebooks for Chapter 14 and each notebook has several cells. They provide 3D and 2D views of the 16 Bessel functions that we consider, as well as numerical values for the functions, which can be used with the C functions in the text and on the CD-ROM to test your C programs for correctness.

The notebooks correspond to the text as follows; `BesselInteger` covers the overview (Section 14.1) and the eight Bessel functions of integer order (Sections 14.2, 14.3), notebook `BesselHalf` includes the cells for the four functions of half-odd-integer order (Section 14.4), and notebook `BesselAiry` covers the two Airy functions (Bessel functions of order 1/3) and their first derivatives (Section 14.5).

Notebook `BesselInteger` [14.1, 14.2]

Run the cells in this notebook to get an overview of the general behavior of Bessel functions in terms of the complex variable z, to see them as functions of real variable x and order n, and to explore the singularities of the irregular functions.

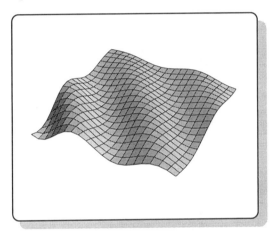

All the functions come as matched pairs of functions that are regular or irregular at $z = 0$; for example, the familiar regular cylindrical Bessel function $J_n(x)$ in Section 14.2.1 and the irregular function $Y_n(x)$ in Section 14.2.2. Even the regular functions have interesting behaviors, as for Kelvin function $\text{ber}_n x$ which is shown multiplied by an exponential damping factor (Figure 14.3.2) in the adjacent motif.

The `BesselJY` cell [14.1, 14.2]

Execute this cell to produce all of Figures 14.1.1–14.1.6 for the cylindrical Bessel functions shown in Sections 14.1 and 14.2. The first section—the overview of Bessel functions—which shows many 3D and 2D views of Bessel functions, also uses this cell.

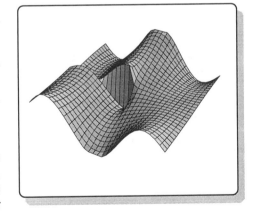

Exploring with `BesselJY`

Investigate the singularity structure of the irregular function $Y_n(z)$ in the complex plane, such as the branch cut along the negative x axis that is shown at right for $n = 0$. To understand our numerical algorithms for the C functions, extend the 2D views as in Figures 14.1.5 and 14.1.6 over a wider range of x and n.

```
(* BesselJY cell *)

(* Bessel Functions, Jn(z) & Yn(z)  *)

(* Bessel Functions, Jn(z) & Yn(z):
   3D & 2D views *)

(* 3D defined for n fixed, x & y vary *)
BessZ3D[n_,JorY_,ReOrIm_,Pnts_,BessRng_,
 ZTcks_] :=
Plot3D[
  ReOrIm[JorY[n,x+I*y]],{x,-5,5},{y,-2,2},
  PlotRange->{-BessRng,BessRng},
  PlotPoints->{Pnts,Pnts},
  Ticks->{{-4,-2,0,2,4},{-2,0,2},
  {-ZTcks,0,ZTcks}},AxesLabel->{"x","y",None}]

(* 3D defined; real z = x  & n varying *)
BessZXN3D[JorY_,ReOrIm_,BessRng_,ZTcks_] :=
 Plot3D[
  ReOrIm[JorY[n,x]],{x,-4,4},{n,0,5},
  PlotRange->{-BessRng,BessRng},
 PlotPoints->{30,30},
  Ticks->{{-4,-2,0,2,4},{0,2,4},
   {-ZTcks,0,ZTcks}},
 AxesLabel->{"x","n",None}]

(* 2D views defined; x varying & n fixed *)
BesselX2D[n_,JorY_,BessRng_,YTcks_,Dsh_] :=
 Plot[JorY[n,x],{x,0,10},
  PlotRange->{-BessRng,BessRng},PlotPoints->50,
  Ticks->{{0,2,4,6,8,10},{-YTcks,0,YTcks}},
  AxesLabel->{"x",None},
  PlotStyle->Dashing[{Dsh,Dsh}],
  DisplayFunction->Identity]

(* 2D views defined; n varying & x fixed *)
BesselNN2D[x_,JorY_,BessRng_,YTcks_,Dsh_] :=
 Plot[JorY[n,x],{n,0,10},
  PlotRange->{-BessRng,BessRng},PlotPoints->50,
  Ticks->{{0,2,4,6,8,10},{-YTcks,0,YTcks}},
  AxesLabel->{"n",None},
  PlotStyle->Dashing[{Dsh,Dsh}],
  DisplayFunction->Identity]

(* Complex plane 3D views for n = 0,3,6 *)

(* Regular Bessel; J  *)
BessZ3D[0,BesselJ,Re,20,2.5,1];
BessZ3D[0,BesselJ,Im,20,2.5,1];
BessZ3D[3,BesselJ,Re,20,1,0.5];
BessZ3D[3,BesselJ,Im,20,1,0.5];
BessZ3D[6,BesselJ,Re,20,0.1,0.05];
BessZ3D[6,BesselJ,Im,20,0.1,0.05];
```

Define 3D views of Bessel functions in the complex plane.

Define 3D views of Bessel functions against *x* and *n*.

Define 2D curves of Bessel functions against *x* with *n* fixed.

Define 2D curves of Bessel functions against *n* with *x* fixed.

Show surfaces of regular Bessel function in the complex plane, Figure 14.1.1.

```
(* Irregular Bessel; Y  *)
BessZ3D[0,BesselY,Re,30,2,1];
BessZ3D[0,BesselY,Im,30,2,1];
BessZ3D[3,BesselY,Re,30,10,5];
BessZ3D[3,BesselY,Im,30,10,5];
BessZ3D[6,BesselY,Re,30,10,5];
BessZ3D[6,BesselY,Im,30,10,5];


(*  x and n;  3D views  *)

(* Regular Bessel; J  *)
BessZXN3D[BesselJ,Re,1,0.5];
BessZXN3D[BesselJ,Im,1,0.5];

(* Irregular Bessel; Y  *)
BessZXN3D[BesselY,Re,10,5];
BessZXN3D[BesselY,Im,10,5];


(* x with n fixed; 2D views *)

(* Regular Bessel; J  *)
ShowRegX2D = Show[
 BesselX2D[0,BesselJ,1,0.5,0.06],
 BesselX2D[3,BesselJ,1,0.5,0.04],
 BesselX2D[6,BesselJ,1,0.5,0.02],
 DisplayFunction->$DisplayFunction];

(* Irregular Bessel; Y  *)
Off[Plot::plnr];(* blocks message for x=0 *)
ShowIrregX2D = Show[
 BesselX2D[0,BesselY,2,1,0.06],
 BesselX2D[3,BesselY,2,1,0.04],
 BesselX2D[6,BesselY,2,1,0.02],
 DisplayFunction->$DisplayFunction];


(* n with x fixed; 2D views *)

(* Regular Bessel; J  *)
ShowRegNN2D = Show[
 BesselNN2D[2,BesselJ,1,0.5,0.06],
 BesselNN2D[4,BesselJ,1,0.5,0.04],
 BesselNN2D[6,BesselJ,1,0.5,0.02],
 DisplayFunction->$DisplayFunction];

(* Irregular Bessel; Y  *)
ShowIrregNN2D = Show[
 BesselNN2D[2,BesselY,2,1,0.06],
 BesselNN2D[4,BesselY,2,1,0.04],
 BesselNN2D[6,BesselY,2,1,0.02],
 DisplayFunction->$DisplayFunction];
```

BesselJY *continued*

3D views of irregular cylindrical functions in complex plane, Figure 14.1.2.

Regular Bessel against x and n, Figure 14.1.3.

Irregular Bessel against x and n, Figure 14.1.4.

Show curves of regular then irregular Bessel against x, as in Figure 14.1.5.

Show curves of regular then irregular Bessel against n, as in Figure 14.1.6.

```
(* Regular Bessel functions: Numerical *)

BessJNTable = Table[{n,x,N[BesselJ[n,x],10]},
   {x,0,8,2},{n,0,6,3}]
BessJN20 = Table[{n,x,N[BesselJ[n,x],10]},
   {x,20,20,1},{n,0,6,3}]

(* Irregular Bessel functions: Numerical *)

BessYNTable = Table[{n,x,N[BesselY[n,x],10]},
   {x,2,10,2},{n,0,6,3}]
BessYN20 = Table[{n,x,N[BesselY[n,x],10]},
   {x,20,20,1},{n,0,6,3}]
```

Numerical values of regular cylindrical Bessel functions, Table 14.2.1.

Numerical values of irregular cylindrical Bessel functions, Table 14.2.3.

The `BesselPQ` cell [14.2.1]

This cell has the asymptotic series for the auxiliary functions P and Q in the asymptotic series for the regular cylindrical Bessel function $J_n(x)$ in Section 14.2.1, as calculated by (14.2.5), and for the irregular function $Y_n(x)$ in Section 14.2.2, as calculated by formula (14.2.15). Figures 14.2.1 shows the P and Q functions for $n = 0$ and $n = 1$.

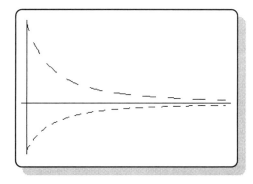

Exploring with `BesselPQ`

In the text we show and use only $n = 0$ and $n = 1$ for the auxiliary functions, since this sufficient for starting the upward iteration on n. Investigate the function for a wider range of x and n. Especially if you want the Bessel functions for only a single large n in the asymptotic region of x, investigate their convergence for that n value. Note that we are using asymptotic expansions (Section 3.4), which have no guarantee of convergence.

```
(* BesselPQ cell *)

(* Hankel's P series for asymptotic in x > 0 *)
(* Relative accuracy set to  eps = 10^(-12) *)

PP[n_,x_] :=
  ( mu = 4*n*n;
    atxs = 64.0*x*x;
    k = 0;
    term = 1.0;
    termold = 1.0;
    fk = -1.0;
    tk = 0.0;
    sumPP = 1.0;
    ratio = 10.0;
    eps = 10^(-12);
```

Define P function as in (14.2.6).

Mathematica code that is just like C; it works but it's not pretty

BesselPQ *continued*

```
While[ ratio > eps,
( k++;
  fk += 4.0;
  tk += 2.0;
  term *= -(mu-(fk-2.0)*(fk-2.0))*
   (mu-fk*fk)/(tk*(tk-1.0)*atxs);
  sumPP += term;
  ratio = Abs[term/sumPP];
  termnew = Abs[term];

  If[ termnew > termold,
   (
   Print
   ["\n\n!!In PP diverging after ",k," terms"];
   ratio = 0.0;
   ), termold = termnew; ]; ) ];

Return[sumPP]; )

(* Hankel's Q series for asymptotic in x > 0 *)
(* Relative accuracy set to  eps = 10^(-12) *)

QQ[n_,x_] :=
(
mu = 4*n*n;
atxs = 64.0*x*x;
k = 0;
term = 1.0;
termold = 1.0;
fk = 1.0;
tk = 1.0;
sumQQ = 1.0;
ratio = 10.0;
eps = 10^(-12);

While[ ratio > eps,
( k++;
  fk += 4.0;
  tk += 2.0;
  term *= -(mu-(fk-2.0)*(fk-2.0))*
   (mu-fk*fk)/(tk*(tk-1.0)*atxs);
  sumQQ += term;
  ratio = Abs[term/sumPP];
  termnew = Abs[term];

  If[ termnew > termold,
   (
   Print
   ["\n\n!!In QQ diverging after ",k," terms"];
   ratio = 0.0;
   ), termold = termnew; ]; ) ];

Return[(mu-1.0)*sumQQ/(8.0*x)]; )
```

Test convergence of
asymptotic expansion.

Define *Q* function as in
(14.2.7).

Test convergence of
asymptotic expansion.

```
(* 2D plot for P series *)

HankelPP[n_,Dsh_] := Plot[
  PP[n,x] - 1,{x,20,100},
  PlotRange->All,PlotPoints->100,
  Ticks->{Automatic,{-0.0001,0,0.0001}},
  AxesLabel->{"x","P(x) - 1"},
  PlotStyle->Dashing[{Dsh,Dsh}],
  DisplayFunction->Identity]
```

Define plots of *P* functions against *x* for fixed *n*.

```
(* View n = 0 and n = 1 together *)
PPShow = Show[HankelPP[0,0.03],HankelPP[1,0.06],
  DisplayFunction->$DisplayFunction];
```

Show *P* plots as in Figure 14.2.1 left.

```
(* 2D plot for Q series *)

HankelQQ[n_,Dsh_] := Plot[QQ[n,x],{x,20,100},
  PlotRange->All,PlotPoints->100,
  Ticks->{Automatic,{0,0.01,0.02}},
  AxesLabel->{"x","Q(x)"},
  PlotStyle->Dashing[{Dsh,Dsh}],
  DisplayFunction->Identity]
```

Define plots of *Q* functions against *x* for fixed *n*.

```
(* View n = 0 and n = 1 together *)

QQShow = Show[HankelQQ[0,0.03],HankelQQ[1,0.06],
  DisplayFunction->$DisplayFunction];
```

Show *Q* plots as in Figure 14.2.1 right.

The `BesselYX` cell [14.2.2]

This cell makes the graphics for the irregular cylindrical Bessel function $Y_n(x)$ in Section 14.2.2 and for the function defined by {14.2.13} that has the singularities removed (except when $n = 0$), as shown in the motif at right, where $X_n(x)$ is plotted against *x* and *n*.

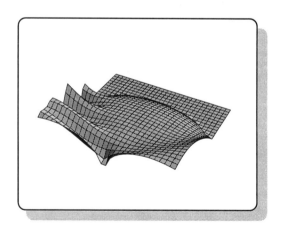

Views of $Y_n(z)$ and $X_n(z)$ (fixed *n*) in the complex plane, and of $Y_n(x)$ and $X_n(x)$ (*x* and *n* varying) that are more extensive are shown in Figures 14.2.2 and 14.2.3.

Exploring with `BesselYX`

Investigate the analytic structure of $Y_n(z)$—both the singularity near the origin and the branch cut along the negative *x* axis—by making views that are more detailed in the corresponding regions.

The annotated `BesselYX` cell is given on the following pages.

```
(* BesselYX cell *)

(* Irregular Bessel Yn *)

(* 3D defined for n fixed, x & y vary *)
BessYY3D[n_,ReOrIm_,Pnts_,BessRng_,ZTcks_] :=
 Plot3D[
  ReOrIm[BesselY[n,x+I*y]],{x,-3,3},{y,-2,2},
  PlotRange->{-BessRng,BessRng},
  PlotPoints->{Pnts,Pnts},
  Ticks->{{-4,-2,0,2,4},{-2,0,2},
  {-ZTcks,0,ZTcks}},AxesLabel->{"x","y",None}]
```

Define 3D views of irregular cylindrical Bessel function in the complex plane.

```
(* 3D defined; real z = x  & n varying *)
BessYYXN3D[ReOrIm_,BessRng_,ZTcks_] :=
 Plot3D[
  ReOrIm[BesselY[n,x]],{x,-3,3},{n,0,5},
  PlotRange->{-BessRng,BessRng},
  PlotPoints->{30,30},
  Ticks->{{-4,-2,0,2,4},{0,2,4},
  {-ZTcks,0,ZTcks}},AxesLabel->{"x","n",None}]
```

Define 3D views of irregular cylindrical Bessel function against x and n.

```
(* Modified Yn -> Xn *)

(* 3D defined for n fixed, x & y vary *)
BessXX3D[n_,ReOrIm_,Pnts_,BessRng_,
 ZTcks_] :=
 Plot3D[
  ReOrIm[((x+I*y)/2)^n*BesselY[n,x+I*y]],
  {x,-3,3},{y,-2,2},
  PlotRange->{-BessRng,BessRng},
  PlotPoints->{Pnts,Pnts},
  Ticks->{{-4,-2,0,2,4},{-2,0,2},
  {-ZTcks,0,ZTcks}},AxesLabel->{"x","y",None}]
```

Define 3D views of irregular cylindrical Bessel function in the complex plane with singularity removed for $n > 0$.

```
(* 3D defined; real z = x  & n varying *)
BessXXXN3D[ReOrIm_,BessRng_,ZTcks_] :=
 Plot3D[ReOrIm[(x/2)^n*BesselY[n,x]],
  {x,-3,3},{n,0,5},
  PlotRange->{-BessRng,BessRng},
  PlotPoints->{30,30},
  Ticks->{{-4,-2,0,2,4},{0,2,4},
  {-ZTcks,0,ZTcks}},AxesLabel->{"x","n",None}]
```

Define 3D views of irregular cylindrical Bessel function against x and n.with singularity removed for $n > 0$.

```
(* Irregular Bessel; Y & X  vs z  *)
BessYY3D[2,Re,30,10,5];
BessYY3D[2,Im,30,10,5];
BessXX3D[2,Re,30,2,1];
BessXX3D[2,Im,30,2,1];

BessYY3D[4,Re,30,10,5];
BessYY3D[4,Im,30,10,5];
BessXX3D[4,Re,30,2,1];
BessXX3D[4,Im,30,2,1];
```

Show irregular functions vs z for $n = 2$ in original and non-singular forms then the same for $n = 4$.

```
(*  Irregular Bessel; Y  & X  vs x & n *)

(* Irregular Bessel; Y *)
BessYYXN3D[Re,10,5];
BessYYXN3D[Im,10,5];
BessXXXN3D[Re,2,1];
BessXXXN3D[Im,2,1];
```

Show irregular functions vs *x* and *n* = 2 in original and non-singular forms.

The `BesselIK` cell [14.2.3]

With this cell you can make graphics of the hyperbolic Bessel functions, both for the regular function I_n in Section 14.2.3 and for the irregular function K_n in Section 14.2.4. Views of the functions in the complex plane for fixed *n*, and views with *x* and *n* varying, are shown in Figures 14.2.4–14.2.9. For example, you can see the very singular behavior of $K_n(x)$ for small *x*—especially as *n* increases—in the adjacent motif, in which the values have been truncated to create the mesa that expands in width as *n* increases.

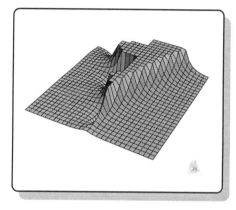

In order to help design the numerical algorithms that we use for the C functions `BessIN` and `BessKN`, it is helpful to see the functions as 2D curves, as in Figures 14.2.6 and 14.2.9; these are also produced when you run cell `BesselIK`.

Exploring with `BesselIK`

Explore the complicated analytic structure of $K_n(z)$—as seen in Figure 14.2.7 and the motif at the start of this section—by making views that are more detailed in the neighborhood of the origin. If your interests are more towards numerical results, look at the 2D curves in more detail.

```
(* BesselIK cell *)

(* Hyperbolic Bessel Functions, In(z) & Kn(z) *)

(* Bessel Functions, In(z) & Kn(z):
   3D & 2D views *)
(* 3D defined for n fixed, x & y vary *)
BessIK3D[n_,IorK_,ReOrIm_,Pnts_,
 BessRng_,ZTcks_] :=
 Plot3D[
  ReOrIm[IorK[n,x+I*y]],{x,-3,3},{y,-2,2},
  PlotRange->{-BessRng,BessRng},
  PlotPoints->{Pnts,Pnts},
  Ticks->{{-2,0,2},{-2,0,2},
  {-ZTcks,0,ZTcks}},AxesLabel->{"x","y",None}]
```

Define 3D views of *I* or *K* for order *n* fixed, choosing real or imaginary parts; number of plot points, range of plotted values, and position of vertical ticks are all parameters.

```
(* 3D defined; real z = x  & n varying *)
BessIKXN3D[IorK_,ReOrIm_,BessRng_,ZTcks_] :=
 Plot3D[
   ReOrIm[IorK[n,x]],{x,-3,3},{n,0,5},
   PlotRange->{-BessRng,BessRng},
  PlotPoints->{30,30},
   Ticks->{{-2,0,2},{0,2,4},
    {-ZTcks,0,ZTcks}},
   AxesLabel->{"x","n",None}]
```

BesselIK *continued*

Define 3D views of *I* or *K* with real argument *x* and order *n* fixed, choosing real or imaginary parts; range of plotted values, and position of vertical ticks are also parameters.

```
(* 2D views defined; x varying & n fixed *)
(* I and K on ln scales *)
BesselIK2D[n_,IorK_,BessRng_,YTcks_,Dsh_] :=
 Plot[Log[IorK[n,x]],{x,0,10},
   PlotRange->{-BessRng,BessRng},PlotPoints->50,
   Ticks->{{0,2,4,6,8,10},{-YTcks,0,YTcks}},
   AxesLabel->{"x",None},
   PlotStyle->Dashing[{Dsh,Dsh}],
   DisplayFunction->Identity]
```

Define 2D curves for *I* and *K* plotted on log scales with *n* fixed and *x* varying.

```
(* 2D views defined; n varying & x fixed *)
(* I and K on ln scales *)
BesselIKNN2D[x_,IorK_,BessRng_,YTcks_,Dsh_] :=
 Plot[Log[IorK[n,x]],{n,0,10},
   PlotRange->{-BessRng,BessRng},PlotPoints->50,
   Ticks->{{0,2,4,6,8,10},{-YTcks,0,YTcks}},
   AxesLabel->{"n",None},
   PlotStyle->Dashing[{Dsh,Dsh}],
   DisplayFunction->Identity]
```

Define 2D curves for *I* and *K* plotted on log scales with *x* fixed and *n* varying.

```
(* Complex plane 3D views for n = 0,3,6 *)

(* Regular hyperbolic Bessel; I *)
BessIK3D[0,BesselI,Re,20,6,3];
BessIK3D[0,BesselI,Im,20,6,3];
BessIK3D[3,BesselI,Re,20,1,0.5];
BessIK3D[3,BesselI,Im,20,1,0.5];
BessIK3D[6,BesselI,Re,20,0.1,0.05];
BessIK3D[6,BesselI,Im,20,0.1,0.05];
```

Show regular hyperbolic function in complex plane for *n* = 0, 3, 6, real and imaginary parts; Figure 14.2.4.

```
(* Irregular hyperbolic Bessel; K *)
BessIK3D[0,BesselK,Re,30,10,5];
BessIK3D[0,BesselK,Im,30,10,5];
BessIK3D[3,BesselK,Re,30,10,5];
BessIK3D[3,BesselK,Im,30,10,5];
BessIK3D[6,BesselK,Re,30,10,5];
BessIK3D[6,BesselK,Im,30,10,5];
```

Show irregular hyperbolic function in complex plane for *n* = 0, 3, 6, real and imaginary parts; Figure 14.2.7.

```
(*  x and n;  3D views  *)

(* Regular hyperbolic Bessel; I *)
BessIKXN3D[BesselI,Re,4,2];
BessIKXN3D[BesselI,Im,4,2];
```

Show regular function in *x–n* plane, real and imaginary parts; Figure 14.2.5.

```
(* Irregular hyperbolic Bessel; K *)
BessIKXN3D[BesselK,Re,20,10];
BessIKXN3D[BesselK,Im,20,10];
```

Show irregular function in *x*–*n* plane, real and imaginary parts; Figure 14.2.8.

```
(* x with n fixed; 2D views *)

(* Regular hyperbolic Bessel; ln(I)  *)
ShowRegIK2D = Show[
 BesselIK2D[0,BesselI,10,5,0.06],
 BesselIK2D[3,BesselI,10,5,0.04],
 BesselIK2D[6,BesselI,10,5,0.02],
 DisplayFunction->$DisplayFunction];
```

Curves of regular function against *x* for *n* = 0, 3, 6; Figure 14.2.6 left.

```
(* Irregular hyperbolic Bessel; ln(K)  *)
Off[Plot::plnr];(* blocks message for x=0 *)
ShowIrregIK2D = Show[
 BesselIK2D[0,BesselK,10,5,0.06],
 BesselIK2D[3,BesselK,10,5,0.04],
 BesselIK2D[6,BesselK,10,5,0.02],
 DisplayFunction->$DisplayFunction];
```

Curves of irregular function against *x* for *n* = 0, 3, 6; Figure 14.2.9 left.

```
(* n with x fixed; 2D views *)

(* Regular hyperbolic Bessel; ln(I)  *)
ShowRegNN2D = Show[
 BesselIKNN2D[2,BesselI,10,5,0.06],
 BesselIKNN2D[4,BesselI,10,5,0.04],
 BesselIKNN2D[6,BesselI,10,5,0.02],
 DisplayFunction->$DisplayFunction];
```

Curves of regular function against *n* for *x* = 2, 4, 6; Figure 14.2.6 right.

```
(* Irregular hyperbolic Bessel; ln(K)  *)
ShowIrregNN2D = Show[
 BesselIKNN2D[2,BesselK,10,5,0.06],
 BesselIKNN2D[4,BesselK,10,5,0.04],
 BesselIKNN2D[6,BesselK,10,5,0.02],
 DisplayFunction->$DisplayFunction];
```

Curves of irregular function against *n* for *x* = 2, 4, 6; Figure 14.2.9 right.

```
(* Regular hyperbolic Bessels: Numerical *)

BessIINTable = Table[{n,x,N[BesselI[n,x],10]},
 {x,0,8,2},{n,0,6,3}]
BessIIN20 = Table[{n,x,N[BesselI[n,x],10]},
 {x,20,20,1},{n,0,6,3}]
```

Test values of regular function; Table 14.2.4.

```
(* Irregular hyperbolic Bessels: Numerical *)

BessKKNTable = Table[{n,x,N[BesselK[n,x],10]},
 {x,2,10,2},{n,0,6,3}]
BessKKN40 = Table[{n,x,N[BesselK[n,x],10]},
 {x,40,40,1},{n,0,6,3}]
```

Test values of irregular function; Table 14.2.6.

The Kelvin cell [14.3]

Physicists and engineers will enjoy making pictures of the Kelvin functions—the invention of William Thomson (Lord Kelvin), a leader of science and technology in the nineteenth century. The regular functions, ber and bei in Section 14.3.1, as well as the irregular functions ker and kei in Section 14.3.2, can be computed and displayed. Views of the functions in the x–n plane are shown in Figures 14.3.1 and 14.3.6. For example, you can see the singular behavior of $\ker_n x$ for small x as n increases in the adjacent motif, where values are truncated to form table-top mountains that expand in x width as n increases.

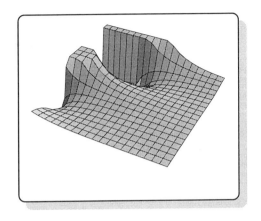

The Kelvin cell can be used to see how including an exponential factor of x smooths out the x and n dependences of Kelvin functions, as depicted in Figures 14.3.2 and 14.3.7. You can also generate test values of the four Kelvin functions.

To help design the numerical algorithms for our C functions BerBei and KerKei, it is helpful to show these functions by 2D curves, as in Figures 14.3.3, 14.3.4 (for ber and bei) and in Figures 14.3.8 and 14.3.9 (for ker and kei); these are also produced when you run cell Kelvin. The auxiliary functions for asymptotic expansions of Kelvin functions—f_n and g_n in (14.3.10)—are computed, displayed, and used to compute ker and kei for large x values.

Exploring with Kelvin

Kelvin functions were devised for a practical use—the skin effect in high-frequency electrical conducticity—and your interests are probably also toward numerical results. Therefore, look at the 2D curves in more detail and modify the relevant parts of the Kelvin cell so you can see the functions over a range of x and n values relevant to your needs.

```
(* Kelvin cell *)

(* Kelvin Functions;
   ber(n,x), bei(n,x); ker(n,x), kei(n,x) *)

(* Definitions in terms of Jn and Kn *)

Off[General::spell1];
ber[n_,x_] := Re[Exp[N[n*Pi*I]]*
 BesselJ[n,N[(1-I)*x/Sqrt[2]]]]
bei[n_,x_] := Im[Exp[N[n*Pi*I]]*
 BesselJ[n,N[(1-I)*x/Sqrt[2]]]]

ker[n_,x_] := Re[Exp[N[-n*Pi*I/2]]*
 BesselK[n,N[(1+I)*x/Sqrt[2]]]]
kei[n_,x_] := Im[Exp[N[-n*Pi*I/2]]*
 BesselK[n,N[(1+I)*x/Sqrt[2]]]]
```

Define ber and bei in terms of regular cylindrical Bessel, (14.3.2).

Define ker and kei in terms of irregular cylindrical Bessel, (14.3.13).

```
(* Kelvin Functions: 3D views defined;
   x & n varying *)

KelvinXN3D[KType_,xMin_,KRng_,ZTcks_] :=
 Plot3D[KType[n,x],{x,xMin,4},{n,0,5},
  PlotRange->{-KRng,KRng},
  PlotPoints->{20,20},
   Ticks->{{1,2,3},{0,2,4},{-ZTcks,0,ZTcks}},
  AxesLabel->{"x","n",None}]

(* Modified Kelvin Functions: 3D views defined;
   x & n varying *)

KelvinExp3D[KType_,Sign_,xMin_,KRng_,ZTcks_] :=
 Plot3D[Exp[Sign*x/Sqrt[2]]*KType[n,x],
  {x,xMin,20},{n,0,5},PlotRange->{-KRng,KRng},
  PlotPoints->{20,20},Ticks->{{4,8,12,16,20},
  {0,2,4},{-ZTcks,0,ZTcks}},
  AxesLabel->{"x","n",None}]

(* 2D views defined; x varying & n fixed *)

KelvinXX2D[KType_,n_,xStart_,KRng_,YTcks_,Dsh_]
 := Plot[KType[n,x],{x,xStart,4},
    PlotRange->{-KRng,KRng},PlotPoints->50,
    Ticks->{{0,2,4,6,8,10},{-YTcks,0,YTcks}},
    AxesLabel->{"x",None},
    PlotStyle->Dashing[{Dsh,Dsh}],
    DisplayFunction->Identity]

(* Define composite 2D view; x varying *)

ShowXX[KType_,xStart_] := Show[
 KelvinXX2D[KType,0,xStart,2,1,0.06],
 KelvinXX2D[KType,1,xStart,2,1,0.04],
 KelvinXX2D[KType,2,xStart,2,1,0.02],
 DisplayFunction->$DisplayFunction];

(* 2D views defined; n varying & x fixed *)

KelvinNN2D[KType_,x_,KRng_,YTcks_,Dsh_] :=
 Plot[KType[n,x],{n,0,4},
 PlotRange->{-KRng,KRng},PlotPoints->50,
 Ticks->{{0,2,4},{-YTcks,0,YTcks}},
 AxesLabel->{"n",None},
 PlotStyle->Dashing[{Dsh,Dsh}],
 DisplayFunction->Identity]

(* Define composite 2D views; n varying *)

ShowBBnn[BerBei_] := Show[
 KelvinNN2D[BerBei,1,2,1,0.06],
 KelvinNN2D[BerBei,2,2,1,0.04],
 KelvinNN2D[BerBei,3,2,1,0.02],
 DisplayFunction->$DisplayFunction];
```

Define 3D surfaces for Kelvin function `KType` with *x* and *n* varying.

Define 3D surfaces for Kelvin function `KType` with *x* and *n* varying and exponential factor (depending on `Sign`) included.

Define 2D curves for function `KType` with *n* fixed.

Define 2D views of function `KType` for $n = 0, 1, 2$.

Define 2D curves for function `KType` with *x* fixed.

Define 2D views of functions `ber and bei` for $x = 0, 1, 2, 3$.

```
ShowKKnn[KerKei_] := Show[
 KelvinNN2D[KerKei,3,0.4,0.2,0.06],
 KelvinNN2D[KerKei,4,0.4,0.2,0.03],
 DisplayFunction->$DisplayFunction];

(* Show regular functions: ber & bei *)

(* Display Kelvin functions in 3D *)
KelvinXN3D[ber,0,3,2];  KelvinXN3D[bei,0,3,2];

(* Display modified functions in 3D *)
KelvinExp3D[ber,-1,4,0.4,0.2];
KelvinExp3D[bei,-1,4,0.4,0.2];

(* Display composite 2D views; x varying *)
ShowXX[ber,0.0];   ShowXX[bei,0.0];

(* Display composite 2D views; n varying *)
ShowBBnn[ber];    ShowBBnn[bei];

(* Show irregular functions: ker & kei *)

(* Display Kelvin functions in 3D *)
KelvinXN3D[ker,0.5,4,2];
KelvinXN3D[kei,0.5,4,2];

(* Display modified functions in 3D *)
(**!! THIS GOES INTO INFINITE LOOP
KelvinExp3D[ker, 1,4,0.4,0.2];
KelvinExp3D[kei, 1,4,0.4,0.2];   !!*)

(* Display composite 2D views; x varying *)
ShowXX[ker,0.5];   ShowXX[kei,0.5];

(* Display composite 2D views; n varying *)
ShowKKnn[ker];    ShowKKnn[kei];

(* Regular Kelvin functions: Numerical *)

berTable = Table[{n,x,N[ber[n,x],10]},
 {x,2,8,2},{n,0,6,3}]
ber40 = Table[{n,x,N[ber[n,x],10]},
 {x,40,40,1},{n,0,6,3}]

beiTable = Table[{n,x,N[bei[n,x],10]},
 {x,2,8,2},{n,0,6,3}]
bei40 = Table[{n,x,N[bei[n,x],10]},
 {x,40,40,1},{n,0,6,3}]

(* Irregular Kelvin functions: Numerical *)
(*! INFINITE LOOP FOR x > 4  !*)
kerTable = Table[{n,x,N[ker[n,x],10]},
 {x,2,8,2},{n,0,6,3}]
ker20 = Table[{n,x,N[ker[n,x],10]},
 {x,20,20,1},{n,0,6,3}]
```

Kelvin *continued*

Define 2D views of functions ker and kei for x= 3,4.

Show ber and bei in 3D, Figure 14.3.1.

Show ber and bei in 3D with exponential factor, Figure 14.3.2.

Figure 14.3.3.

Figure 14.3.4.

Show ker and kei in 3D, Figure 14.3.6.

Show ker and kei in 3D with exponential factor, Figure 14.3.7; see last piece of cell.

Figure 14.3.8.

Figure 14.3.9.

Numerical ber values, Table 14.3.1.

Numerical bei values, Table 14.3.2.

Numerical ker values, Table 14.3.4.

```
keiTable = Table[{n,x,N[kei[n,x],10]},
 {x,2,8,2},{n,0,6,3}]
kei20 = Table[{n,x,N[kei[n,x],10]},
 {x,20,20,1},{n,0,6,3}]

(* Auxiliary functions for asymptotic
   expansion of Kelvin functions to
   relative accuracy  eps  *)
Fn[n_,sgnx_,x_,eps_] :=
 (
 mu = 4*n*n;
 If[ sgnx > 0, phase = -1, phase = 1.0];
 atex = 8.0*x;
 k = 0;  term = 1.0;  termold = 1.0;
 fk = 0.0;  tk = -1.0;
 FFnSum = 1.0; ratio = 10.0;
 While[ ratio > eps,
   ( k++;  fk += 1.0;  tk += 2.0;
   term *=  phase*(mu-tk*tk)/(fk*atex);
   FFnSum += term*Cos[fk*N[Pi]/4];
   ratio = Abs[term/FFnSum];
   termnew = Abs[term];
   If[ termnew > termold,
     ( Print[
       "!!In Fn diverging after ",k," terms"];
     ratio = 0.0; ), termold = termnew];
   ) ];
 Return[FFnSum];
 )
Gn[n_,sgnx_,x_,eps_] :=
 (
 mu = 4*n*n;
 If[ sgnx > 0, phase = -1, phase = 1.0];
 atex = 8.0*x;
 k = 0;  term = 1.0;  termold = 1.0;
 fk = 0.0;  tk = -1.0;
 GGnSum = 0.0;  ratio = 10.0;
 While[ ratio > eps,
   ( k++;  fk += 1.0;  tk += 2.0;
   term *=  phase*(mu-tk*tk)/(fk*atex);
   GGnSum += term*Sin[fk*N[Pi]/4];
   ratio = Abs[term/GGnSum];
   termnew = Abs[term];
   If[ termnew > termold,
     ( Print[
       "!!In Gn diverging after ",k," terms"];
     ratio = 0.0; ), termold = termnew];
   ) ];
 Return[GGnSum];
 )
```

Numerical kei values,
Table 14.3.5.

Define asymptotif series *f*
function in (14.3.10).

Check convergence.

Define asymptotif series *g*
function in (14.3.10).

Check convergence.

```
(* 2D plot for Fn series *)

KelvinFFn[n_,sgnx_,Dsh_] := Plot[
  Fn[n,sgnx,x,10^(-4)]-1,{x,20,100},
  PlotRange->All,PlotPoints->50,
  Ticks->{Automatic,{-0.010,-0.005,0.005}},
  AxesLabel->{"x",None},
  PlotStyle->Dashing[{Dsh,Dsh}],
  DisplayFunction->Identity]

(* View n = 0 and n = 1 together *)
FFnShow[sgnx_] := Show[
  KelvinFFn[0,sgnx,0.06],KelvinFFn[1,sgnx,0.03],
  DisplayFunction->$DisplayFunction]


(* 2D plot for Gn series *)

KelvinGGn[n_,sgnx_,Dsh_] := Plot[
  Gn[n,sgnx,x,10^(-4)],{x,20,100},
  PlotRange->All,PlotPoints->50,
  Ticks->{Automatic,{-0.010,-0.005,
  0,0.005,0.01}},
  AxesLabel->{"x",None},
  PlotStyle->Dashing[{Dsh,Dsh}],
  DisplayFunction->Identity]

(* View n = 0 and n = 1 together *)
GGnShow[sgnx_] := Show[
  KelvinGGn[0,sgnx,0.06],KelvinGGn[1,sgnx,0.03],
  DisplayFunction->$DisplayFunction]

FFnShow[ 1];  GGnShow[ 1];
FFnShow[-1];  GGnShow[-1];

(* (** Modified **) Asymptotic expansion
   for ker & kei *)

KerrAsym[n_,x_,eps_] :=
 ( betta = x/N[Sqrt[2]]+(4*n+1)*N[Pi]/8;
   Return[
     Sqrt[N[Pi]/(2*x)]*
     (** *Exp[-x/Sqrt[2]] **)
     (Fn[n,-1,x,eps]*Cos[betta]-
      Gn[n,-1,x,eps]*Sin[betta])]; )


KeiiAsym[n_,x_,eps_] :=
 ( betta = x/N[Sqrt[2]]+(4*n+1)*N[Pi]/8;
   Return[
     Sqrt[N[Pi]/(2*x)]*
     (** *Exp[-x/N[Sqrt[2]]] **)
     (-Fn[n,-1,x,eps]*Sin[betta]-
       Gn[n,-1,x,eps]*Cos[betta])]; )
```

Kelvin *continued*

Define 2D plot of *f* for given *n*.

Define combined display for given choice of sign of *x*.

Define 2D plot of *g* for given *n*.

Define combined display for given choice of sign of *x*.

Figure 14.3.5.
Figure 14.3.10.

Define asymptotic ker as in (14.3.20).

Include (** **) to get modified function.

Define asymptotic kei as in (14.3.20).

Include (** **) to get modified function.

```
(*  Modified Kelvin Functions:
    by Asymptotic Series:
    3D views defined;  x & n varying *)

KelvinExp3D[KType_,KRng_,ZTcks_,eps_] :=
  Plot3D[KType[n,x,eps],{x,8,20},{n,0,4},
    PlotRange->{-KRng,KRng},
    PlotPoints->{20,20},Ticks>{{8,12,16,20},
    {0,2,4},{-ZTcks,0,ZTcks}},
    AxesLabel->{"x","n",None}]

KelvinExp3D[KerrAsym,1,0.5,10^(-4)];
KelvinExp3D[KeiiAsym,1,0.5,10^(-4)];
```

Define 3D view of modified irregular Kelvin function.

Show modified irregular function for large x, Figure 14.3.7.

Notebook `BesselHalf` [14.4]

By running the cells in this notebook you will get views of the behavior of the spherical Bessel functions, that is, Bessel functions whose order is half an odd integer. There are four kinds of such functions—j and y (the regular and irregular analogues of J and Y), plus i and k (the analogues of I and K). For example, the irregular function $y_n(x)$ is shown in terms of x and n as the motif on the right.

Notebook `BesselHalf` also generates 2D curves of these four spherical Bessel functions, and it provides numerical values for them.

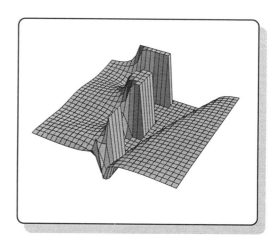

The `Bessel_j` cell [14.4.1]

Execute this cell—whose annotated listing is on the following pages—to produce Figures 14.4.1 and 14.4.2 for the regular spherical Bessel functions $j_n(x)$ in Section 14.4.1. For example, when n is allowed to be nonintegral $j_n(x)$ has a complicated dependence on x and n, as seen in the motif at right for the real part of the function over x from –4 to 4 and n from 0 to 5.

This cell also generates by recurrence the formulas for $j_n(x)$ from $n = 0$ to $n = 9$ in Table 14.4.1; it is easy to change this upper limit. Numerical values are also generated;

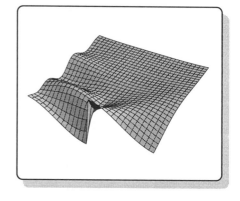

they can be used to test your implementation of the C function `SphereBessJ`.

Exploring with `Bessel_j`

Since spherical Bessel functions are usually needed for numerical applications, it is interesting to understand the dependence of $j_n(x)$ on x for fixed n or vice versa, as shown in Figure 14.4.2, which is generated by this cell. Here is the annotated *Mathematica* cell.

```
(* Bessel-j cell *)

(* Regular Spherical Bessel Functions, jn(x) *)

(* Recurrence relations for algebraic formulas;
   regular functions, j  *)

Print["n            j(n)"];
j[0] = r*s;          Print[0,"    ",j[0]];
j[1] = -r*c + r^2*s;  Print[1,"    ",j[1]];
For[ n = 1, n < 9, n++,
  ( jnp1 = Expand[(2*n+1)*r*j[n] - j[n-1]];
    j[n+1] = c*Factor[Coefficient[jnp1,c]]+
             s*Factor[Coefficient[jnp1,s]];
    Print[n+1,"   ",j[n+1]]; ) ]

(* Define regular spherical Bessel functions *)

SphereBessJJ[n_,x_] :=
  Sqrt[Pi/(2*x)]*BesselJ[n+1/2,x]
(* 3D & 2D views *)

(* 3D defined; real z = x  & n varying *)
BessJJXN3D[ReOrIm_,BessRng_,ZTcks_] :=
  Plot3D[
    ReOrIm[SphereBessJJ[n,x]],{x,-4,4},{n,0,5},
    PlotRange->{-BessRng,BessRng},
    PlotPoints->{30,30},
    Ticks->{{-4,-2,0,2,4},{0,2,4},
     {-ZTcks,0,ZTcks}},
    AxesLabel->{"x","n",None}]

(* 2D views defined; x varying & n fixed *)
BesselJJX2D[n_,BessRng_,YTcks_,Dsh_] :=
  Plot[SphereBessJJ[n,x],{x,0.0001,10},
    PlotRange->{-BessRng,BessRng},PlotPoints->50,
    Ticks->{{0,2,4,6,8,10},{-YTcks,0,YTcks}},
    AxesLabel->{"x",None},
    PlotStyle->Dashing[{Dsh,Dsh}],
    DisplayFunction->Identity]

(* 2D views defined; n varying & x fixed *)
BesselJJNN2D[x_,BessRng_,YTcks_,Dsh_] :=
  Plot[SphereBessJJ[n,x],{n,0,10},
    PlotRange->{-BessRng,BessRng},PlotPoints->50,
    Ticks->{{0,2,4,6,8,10},{-YTcks,0,YTcks}},
    AxesLabel->{"n",None},
    PlotStyle->Dashing[{Dsh,Dsh}],
    DisplayFunction->Identity]
```

Generate Table 14.4.1 for spherical Bessel function algebraic formulas from $n = 0$ to $n = 9$.

Define spherical Bessel functions in terms of general Bessel functions, (14.4.2).

Define 3D views of Bessel functions against x and n. Choose real or imaginary part, vertical scale, and position of vertical ticks.

Define 2D curves of spherical Bessel functions against x with n fixed.

Define 2D curves of spherical Bessel functions against n with x fixed.

```
(*  x and n;  3D views  *)

BessJJXN3D[Re,1,0.5];
BessJJXN3D[Im,1,0.5];

(* x with n fixed; 2D views *)

ShowRegX2D = Show[
 BesselJJX2D[0,1,0.5,0.06],
 BesselJJX2D[3,1,0.5,0.04],
 BesselJJX2D[6,1,0.5,0.02],
 DisplayFunction->$DisplayFunction];

(* n with x fixed; 2D views *)

(* Regular spherical Bessel; j  *)
ShowRegNN2D = Show[
 BesselJJNN2D[2,1,0.5,0.06],
 BesselJJNN2D[4,1,0.5,0.04],
 BesselJJNN2D[6,1,0.5,0.02],
 DisplayFunction->$DisplayFunction];

(* Regular spherical Bessels: Numerical *)

BessJJJNTable =
Table[{n,x,N[SphereBessJJ[n,x],10]},
 {x,2,8,2},{n,0,6,3}]
BessJJJN20 =
Table[{n,x,N[SphereBessJJ[n,x],10]},
 {x,20,20,1},{n,0,6,3}]
```

Show surfaces of spherical Bessel functions in the *x*–*n* plane, Figure 14.4.1.

Show 2D views of Bessel functions against *x* with $n = 0, 3, 6$, as in Figure 14.4.2 left.

Show 2D views of Bessel functions against *n* with $x = 2, 4, 6$, as in Figure 14.4.2 right.

Numerical spherical Bessel functions, Table 14.4.2.

The SphereBesselPQ cell [14.4.1]

By executing this cell you can produce Figure 14.4.3 for the auxiliary functions $p(n,x)$ and $q(n,x)$ that can used to compute the regular function $j_n(x)$ in Section 14.4.1 by using (14.4.6) and the irregular function $y_n(x)$ in Section 14.4.2 by using (14.4.13). The adjacent graphic shows $q(n,x)$ for $n = 1, 2, 3$, increasing in magnitude as *n* increases; it is shown with $p(n,x)$ in more detail in Figure 14.4.3.

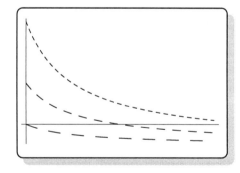

Exploring with SphereBesselPQ

Since $j_n(x)$ and $q(n,x)$ for a single *n* and *x* can be computed—algebraically or numerically—directly in terms of $p(n,x)$ and $q(n,x)$, it is interesting to modify the cell to check the algebraic expressions in Tables 14.4.1 and 14.4.3.

If you frequently need the spherical Bessel functions (regular or irregular) for a single *n* value and one or more *x* values, then it would be worthwhile for you to modify our C functions SphereBessJ or SphereBessY (Section 14.4.1 or 14.4.2, respectively) to use only the functions pp and qq, which are the C versions of pp[n,x] and qq[n,x] in the following *Mathematica* cell. This cell can be used to generate test values for the auxiliary functions.

```
(* SphereBesselPQ cell *)

(* Spherical Bessel Functions *)

(* Auxiliary functions  p  and  q  *)

Fact[n_,k_] := (n+k)!/(k!*(n-k)!)
```
Define factorial combination.

```
pp[n_,x_] :=
  ( sumpp = 0;
    For[ k = 0, k <= Floor[n/2], k++,
    sumpp += (-1)^k*Fact[n,2*k]/((2*x)^(2*k)) ];
    Return[sumpp]; )
```
Define *p* function as in (14.4.7).

```
qq[n_,x_] :=
  ( sumqq = 0;
    If[ n == 0, Return[0],
    For[ k = 0, k <= Floor[(n-1)/2], k++,
    sumqq += (-1)^k*Fact[n,2*k+1]/((2*x)^(2*k+1))
      ];
    Return[sumqq]; ]; )
```
Define *q* function as in (14.4.7).

```
(* 2D plot for P series *)

Spherepp[n_,Dsh_] := Plot[pp[n,x]-1,{x,20,100},
  PlotRange->All,PlotPoints->100,
  Ticks->{Automatic,{-0.03,-0.02,-0.01,0}},
  AxesLabel->{"x","p(x) - 1"},
  PlotStyle->Dashing[{Dsh,Dsh}],
  DisplayFunction->Identity]
```
Define 2D views of *p* functions against *x* for fixed *n*. Choose dash length for each *n*.

```
(* View n = 2,3 together *)
ppShow = Show[
  Spherepp[2,0.04],Spherepp[3,0.02],
  DisplayFunction->$DisplayFunction];
```
Show 2D curves of *p* functions for *n* = 2 and 3, Figure 14.4.3 left.

```
(* 2D plot for Q series *)

Sphereqq[n_,Dsh_] := Plot[qq[n,x],{x,20,100},
  PlotRange->All,PlotPoints->100,
  Ticks->{Automatic,{0,0.1,0.2,0.3}},
  AxesLabel->{"x","q(x)"},
  PlotStyle->Dashing[{Dsh,Dsh}],
  DisplayFunction->Identity]
```
Define 2D views of *q* functions against *x* for fixed *n*. Choose dash length for each *n*.

```
(* View n = 1,2,3 together *)
qqShow = Show[Sphereqq[1,0.06],
  Sphereqq[2,0.04],Sphereqq[3,0.02],
  DisplayFunction->$DisplayFunction];
```
Show 2D curves of *q* functions for *n* = 1, 2 and 3, Figure 14.4.3 right.

The `Bessel_y` cell [14.4.2]

Execute this cell to produce Figures 14.4.4 and 14.4.5 for the irregular functions $y_n(x)$ in Section 14.4.2. For example, when n is nonintegral $y_n(x)$ has a complicated dependence on x and n, as seen in the motif at right for the real part over x from –4 to 4 and n from 0 to 5. This cell also generates by recurrence the algebraic formulas for $y_n(x)$ from $n = 0$ to $n = 9$ in Table 14.4.3; it is easy to change this upper limit. Numerical values are also generated; you can test your implementation of C function `SphereBessY`.

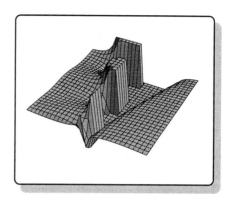

Exploring with `Bessel_y`

Irregular spherical Bessel functions are often needed for numerical applications, so it is interesting to understand the dependence of $y_n(x)$ on x for fixed n or vice versa, as shown in Figure 14.4.4, which is generated by the following annotated *Mathematica* cell.

```
(* Bessel-y cell *)

(* Irregular Spherical Bessel Functions,
   yn(x) *)

(* Recurrence relations for algebraic formulas;
   irregular functions, y  *)
Print["n            y(n)"];
y[0] = - r*c;            Print[0,"   ",y[0]];
y[1] = - r^2*c - r*s;  Print[1,"   ",y[1]];
For[ n = 1, n < 9, n++,
  ( ynp1 = Expand[(2*n+1)*r*y[n] - y[n-1]];
    y[n+1] = c*Factor[Coefficient[ynp1,c]]+
            s*Factor[Coefficient[ynp1,s]];
    Print[n+1,"   ",y[n+1]]; ) ]

(* Define irregular spherical Bessel function *)
SphereBessYY[n_,x_] :=
 Sqrt[Pi/(2*x)]*BesselY[n+1/2,x]

(* 3D & 2D views *)
(* 3D defined; real z = x  & n varying *)
BessYYXN3D[ReOrIm_,BessRng_,ZTcks_] :=
 Plot3D[
   ReOrIm[SphereBessYY[n,x]],{x,-4,4},{n,0,5},
   PlotRange->{-BessRng,BessRng},
   PlotPoints->{30,30},
   Ticks->{{-4,-2,0,2,4},{0,2,4},
     {-ZTcks,0,ZTcks}},
   AxesLabel->{"x","n",None}]
```

Generate Table 14.4.3 for irregular spherical Bessel function algebraic formulas from $n = 0$ to $n = 9$.

Define spherical Bessel functions in terms of general Bessel functions, (14.4.10).

Define 3D views of Bessel functions against x and n. Choose real or imaginary part, vertical scale, and position of vertical ticks.

```
(* 2D views defined; x varying & n fixed *)
BesselYYX2D[n_,BessRng_,YTcks_,Dsh_] :=
 Plot[SphereBessYY[n,x],{x,0.0001,10},
   PlotRange->{-BessRng,BessRng},PlotPoints->50,
   Ticks->{{0,2,4,6,8,10},{-YTcks,0,YTcks}},
   AxesLabel->{"x",None},
   PlotStyle->Dashing[{Dsh,Dsh}],
   DisplayFunction->Identity]
```

Bessel_y *continued*

Define 2D views of
irregularBessel functions
against *x* with *n* fixed.

```
(* 2D views defined; n varying & x fixed *)
BesselYYNN2D[x_,BessRng_,YTcks_,Dsh_] :=
 Plot[SphereBessYY[n,x],{n,0,10},
   PlotRange->{-BessRng,BessRng},PlotPoints->50,
   Ticks->{{0,2,4,6,8,10},{-YTcks,0,YTcks}},
   AxesLabel->{"n",None},
   PlotStyle->Dashing[{Dsh,Dsh}],
   DisplayFunction->Identity]
```

Define 2D views of
irregularBessel functions
against *n* with *x* fixed.

```
(*  x and n;  3D views  *)

BessYYXN3D[Re,10,5];
BessYYXN3D[Im,10,5];
```

Show 3D views as in Figure
14.4.4.

```
(* x with n fixed; 2D views *)

Off[Plot::plnr];(* blocks message for x=0 *)
ShowIrregX2D = Show[
 BesselYYX2D[0,2,1,0.06],
 BesselYYX2D[3,2,1,0.04],
 BesselYYX2D[6,2,1,0.02],
 DisplayFunction->$DisplayFunction];
```

Show 2D views of Bessel
functions against *x* with
n = 0, 3, 6, as in Figure
14.4.5 left.

```
(* n with x fixed; 2D views *)

ShowIrregNN2D = Show[
 BesselYYNN2D[2,2,1,0.06],
 BesselYYNN2D[4,2,1,0.04],
 BesselYYNN2D[6,2,1,0.02],
 DisplayFunction->$DisplayFunction];
```

Show 2D views of Bessel
functions against *n* with
x = 2, 4, 6, as in Figure
14.4.5 right.

```
(* Irregular spherical Bessels: Numerical *)

BessYYYNTable =
Table[{n,x,N[SphereBessYY[n,x],10]},
 {x,2,10,2},{n,0,6,3}]
BessYYYN20 =
Table[{n,x,N[SphereBessYY[n,x],10]},
 {x,20,20,1},{n,0,6,3}]
```

Numerical irregular spherical
Bessel functions, Table
14.4.4.

The `Bessel_i` cell [14.4.3]

Execute this cell to produce Figures 14.4.6–14.4.9 for the regular modified spherical Bessel functions $i_n(x)$ in Section 14.4.3. This function always decreases as n increases when x is fixed, as seen in the motif at right in which x varies from 0 to 3 for $n = 1, 2, 3$ from top (long dashes) to bottom (short dashes); more detail is given in Figure 14.4.7 left.

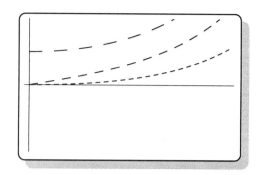

This cell also generates by recurrence the formulas for $i_n(x)$ from $n = 0$ to $n = 9$ in Table 14.4.5; it is easy to change the upper limit. Numerical values are also generated; they can be used to test implementation of the C function `SphereBessI`.

Exploring with `Bessel_i`

These modified spherical Bessel functions are usually needed for numerical applications, so it is interesting to understand the dependence of $i_n(x)$ on x for fixed n or vice versa, as shown in Figure 14.4.7, which is generated by this cell.

Noting that an exponential damping factor smooths out the x dependence of $i_n(x)$—as shown on the right sides of Figures 14.4.7 and 14.4.8—investigate in more detail whether this damped function is appropriate for your computations.

```
(* Bessel-i cell *)

(*   Regular Modified Spherical Bessel
     Functions, in(x) *)

(* Recurrence relations for algebraic formulas;
   regular functions, i  *)

Print["n            i(n)"];
i[0] = r*s;             Print[0," ",i[0]];
i[1] = r*c - r^2*s;  Print[1," ",i[1]];

For[ n = 1, n < 9, n++,
  ( inp1 = Expand[-(2*n+1)*r*i[n] + i[n-1]];
    i[n+1] = c*Factor[Coefficient[inp1,c]]+
             s*Factor[Coefficient[inp1,s]];
    Print[n+1," ",i[n+1]]; ) ]

(* Define regular modified spherical function *)

SphereBessII[n_,x_] :=
 Sqrt[Pi/(2*I*x)]*Exp[-n*Pi*I/2]*
 BesselJ[n+1/2,I*x]
```

Generate Table 14.4.5 for modified spherical Bessel function algebraic formulas from $n = 0$ to $n = 9$.

Define modified spherical functions in terms of general Bessel functions, (14.4.16).

```
(* 3D & 2D views *)

(* 3D defined; real z = x & n varying *)
BessIIXN3D[ReOrIm_,BessRng_,ZTcks_] :=
 Plot3D[
  ReOrIm[SphereBessII[n,x]],{x,-2,2},{n,0,5},
  PlotRange->{-BessRng,BessRng},
  PlotPoints->{30,30},
   Ticks->{{-2,-1,0,1,2},{0,2,4},
    {-ZTcks,0,ZTcks}},
  AxesLabel->{"x","n",None}]

(* 2D views defined; x varying & n fixed *)
BesselIIX2D[n_,BessRng_,YTcks_,Dsh_] :=
 Plot[Re[SphereBessII[n,x]],{x,0.0001,3},
   PlotRange->{-BessRng,BessRng},PlotPoints->50,
   Ticks->{{0,1,2,3},{-YTcks,0,YTcks}},
   AxesLabel->{"x",None},
   PlotStyle->Dashing[{Dsh,Dsh}],
   DisplayFunction->Identity]

(* Exponential removed;
   2D views defined; x varying & n fixed *)
BesselIIexpX2D[n_,BessRng_,YTcks_,Dsh_] :=
 Plot[Re[Exp[-
x]*SphereBessII[n,x]],{x,0.0001,3},
   PlotRange->{-BessRng,BessRng},PlotPoints->50,
   Ticks->{{0,1,2,3},{-YTcks,0,YTcks}},
   AxesLabel->{"x",None},
   PlotStyle->Dashing[{Dsh,Dsh}],
   DisplayFunction->Identity]

(* 2D views defined; n varying & x fixed *)
BesselIINN2D[x_,BessRng_,YTcks_,Dsh_] :=
 Plot[Re[SphereBessII[n,x]],{n,0,3},
   PlotRange->{-BessRng,BessRng},PlotPoints->50,
   Ticks->{{0,1,2,3},{-YTcks,0,YTcks}},
   AxesLabel->{"n",None},
   PlotStyle->Dashing[{Dsh,Dsh}],
   DisplayFunction->Identity]

(* Exponential removed;
   2D views defined; n varying & x fixed *)
BesselIIexpNN2D[x_,BessRng_,YTcks_,Dsh_] :=
 Plot[Re[Exp[-x]*SphereBessII[n,x]],{n,0,3},
   PlotRange->{-BessRng,BessRng},PlotPoints->50,
   Ticks->{{0,1,2,3},{-YTcks,0,YTcks}},
   AxesLabel->{"n",None},
   PlotStyle->Dashing[{Dsh,Dsh}],
   DisplayFunction->Identity]

(* x and n; 3D views *)
BessIIXN3D[Re,2,1];
BessIIXN3D[Im,2,1];
```

`Bessel_i` *continued*

Define 3D views of modified Bessel functions against x and n. Choose real or imaginary part, vertical range, and position of vertical ticks.

Define 2D curves of modified spherical Bessel functions against x with n fixed.

Define 2D curves of modified spherical Bessel functions with exponential factor removed; x varying with n fixed.

Define 2D curves of modified spherical Bessel functions against n with x fixed.

Define 2D curves of modified spherical Bessel functions against n with x fixed; exponential factor removed.

View modified regular spherical Bessel function, Figure 14.4.6.

```
(* x with n fixed; 2D views *)
ShowRegX2D = Show[
  BesselIIX2D[0,2,1,0.06],
  BesselIIX2D[1,2,1,0.04],
  BesselIIX2D[2,2,1,0.02],
  DisplayFunction->$DisplayFunction];
```

Curves of modified function
against x for $n = 1, 2, 3$,
Figure 14.4.7 left.

```
(* Exponential removed; x with n fixed;
   2D views *)
ShowRegExpX2D = Show[
  BesselIIexpX2D[0,1,0.5,0.06],
  BesselIIexpX2D[1,1,0.5,0.04],
  BesselIIexpX2D[2,1,0.5,0.02],
  DisplayFunction->$DisplayFunction];
```

Curves of modified function
against x with exponential
removed; for $n = 1, 2, 3$,
Figure 14.4.7 right.

```
(* n with x fixed; 2D views *)
ShowRegNN2D = Show[
  BesselIINN2D[1,4,2,0.06],
  BesselIINN2D[2,4,2,0.04],
  BesselIINN2D[3,4,2,0.02],
  DisplayFunction->$DisplayFunction];
```

Curves of modified function
against n for $x = 1, 2, 3$,
Figure 14.4.8 left.

```
(* Exponential removed; n with x fixed;
   2D views *)
ShowRegExpNN2D = Show[
  BesselIIexpNN2D[1,0.4,0.2,0.06],
  BesselIIexpNN2D[2,0.4,0.2,0.04],
  BesselIIexpNN2D[3,0.4,0.2,0.02],
  DisplayFunction->$DisplayFunction];
```

Curves of modified function
against n with exponential
removed; for $x = 1, 2, 3$,
Figure 14.4.8 right.

```
(* Regular spherical Bessels: Numerical *)
BessIINTable =
Table[{n,x,N[SphereBessII[n,x],10]},
  {x,1,3},{n,0,6,3}]
BessIIN10 = Table[{n,x,N[SphereBessII[n,x],10]},
  {x,10,10},{n,0,6,3}]
```

Numerical values, as in
Table 14.4.6.

The `BesselR` cell [14.4.3R]

Execute this cell to understand the x and n
dependence of the auxiliary function
$r(n,x)$ in (14.4.20) and (14.4.21), and to
produce Figure 14.4.9 which depicts the
function, similarly to the graphic at right.

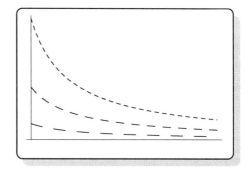

Exploring with `BesselR`

The auxiliary function is useful for calculating the regular modified function $i_n(x)$
for a single n value and several x values,
for which our recurrence scheme over n in

C function `SphereBessI` is not efficient. In such cases, it is interesting to understand the dependence of $r(n,x)$ on x and n, as in Figure 14.4.9.

```
(* Auxiliary function   r   for
   modified spherical Bessel functions *)

rr[n_,x_] :=
  ( sumrr = 0;
    For[ k = 0, k <= n, k++,
    sumrr += (n+k)!/(k!*(n-k)!*(2*x)^k) ];
    Return[sumrr]; )
```

Define auxiliary function as in (14.4.21).

```
(* 2D plot for r series *)

PosXrr[n_,Dsh_] :=
 Plot[rr[n,x]-1,{x,10,50},
   PlotRange->All,PlotPoints->40,
   Ticks->{Automatic,{0,0.2,0.4,0.6}},
   AxesLabel->{"x","r ( x) - 1"},
   PlotStyle->Dashing[{Dsh,Dsh}],
   DisplayFunction->Identity]
```

Define plot of r function for n fixed and x positive, with the dash length depending upon n.

```
(* View n = 1,2,3 together *)
rrPosShow = Show[PosXrr[1,0.06],
  PosXrr[2,0.04],PosXrr[3,0.02],
  DisplayFunction->$DisplayFunction];
```

Show curves for positive x together, Figure 14.4.9 left.

```
NegXrr[n_,Dsh_] :=
 Plot[rr[n,-x]-1,{x,10,50},
   PlotRange->All,PlotPoints->40,
   Ticks->{Automatic,{-0.4,-0.2,0}},
   AxesLabel->{"x","r (-x) - 1"},
   PlotStyle->Dashing[{Dsh,Dsh}],
   DisplayFunction->Identity]
```

Define plot of r function for n fixed and x negative, with the dash length depending upon n.

```
(* View n = 1,2,3 together *)
rrNegShow = Show[NegXrr[1,0.06],
  NegXrr[2,0.04],NegXrr[3,0.02],
  DisplayFunction->$DisplayFunction];
```

Show curves for negative x together, Figure 14.4.9 right.

The `Bessel_k` cell [14.4.4]

Run this cell for irregular modified spherical Bessel functions, $k_n(x)$, to get Figures 14.4.10–14.4.12, in addition to Table 14.4.7 for their algebraic expression. The singularity of order $n+1$ in the nth-order function is evident in the motif at right, which shows the real part over x from –2 to 2 and n from 0 to 5.

Numerical values of $k_n(x)$ are also generated by this cell; use them to test your version of C function `SphereBessK`.

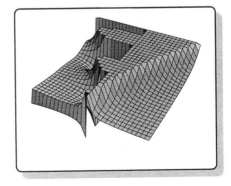

Exploring with `Bessel_k`

The irregular modified spherical Bessel functions, $k_n(x)$, are often used in numerical applications, so it is interesting to understand the dependence of $k_n(x)$ on x for fixed n or vice versa, as shown in Figure 14.4.11, generated by the following *Mathematica* cell.

```
(* Bessel k cell *)

(* Irregular Modified Spherical Bessel
   Functions, kn(x)  *)

(* Recurrence relations for algebraic formulas;
   irregular functions, k  *)
(* Symbol  p = Pi*r*Exp[-x]/2  *)

Print["n              k(n)"];
k[0] = p;           Print[0,"    ",k[0]];
k[1] = p*(1+r); Print[1,"    ",k[1]];
For[ n = 1, n < 9, n++,
 (
 k[n+1] = p*Coefficient[
  Expand[(2*n+1)*r*k[n] + k[n-1]],p];
 Print[n+1,"    ",k[n+1]];
 )]
```

Generate Table 14.4.7 for irregular modified spherical Bessel function algebraic formulas from $n = 0$ to $n = 9$.

```
(* Define irregular modified
   spherical function *)

SphereBessKK[n_,x_] :=
 Sqrt[Pi/(2*x)]*BesselK[n+1/2,x]
```

Define modified function in terms of general Bessel functions, (14.4.24).

```
(* 3D & 2D views *)

(* 3D defined; real z = x  & n varying *)
BessKKXN3D[ReOrIm_,BessRng_,ZTcks_] :=
 Plot3D[
  ReOrIm[SphereBessKK[n,x]],{x,-2,2},{n,0,5},
  PlotRange->{-BessRng,BessRng},
  PlotPoints->{30,30},
  Ticks->{{-2,-1,0,1,2},{0,2,4},
   {-ZTcks,0,ZTcks}},
  AxesLabel->{"x","n",None}]
```

Define 3D views of irregular functions against x and n. Choose real or imaginary part, vertical range, and position of vertical ticks.

```
(* 2D views defined; x varying & n fixed *)
BesselKKX2D[n_,BessRng_,YTcks_,Dsh_] :=
 Plot[Re[SphereBessKK[n,x]],{x,0.5,2},
  PlotRange->{0,BessRng},PlotPoints->50,
  Ticks->{{1,2},{0,YTcks}},
  AxesLabel->{"x",None},
  AxesOrigin->{0.5,0},
  PlotStyle->Dashing[{Dsh,Dsh}],
  DisplayFunction->Identity]
```

Define 2D views of irregular functions against x with n fixed. Choose vertical range, position of vertical ticks, and size of dashes.

```
(* Exponential & x inserted;
   2D views defined; x varying & n fixed *)
BesselKKexpX2D[n_,BessRng_,YTcks_,Dsh_] :=
 Plot[Re[x*Exp[x]*SphereBessKK[n,x]],{x,0.5,2},
   PlotRange->{0,BessRng},PlotPoints->50,
   Ticks->{{1,2},{0,YTcks}},
   AxesLabel->{"x",None},
   AxesOrigin->{0.5,0},
   PlotStyle->Dashing[{Dsh,Dsh}],
   DisplayFunction->Identity]
```

Bessel_k *continued*

Define 2D views of irregular modified spherical Bessel functions against *x* with *n* fixed; factors of *x* and an exponential are included.

```
(* 2D views defined; n varying & x fixed *)
BesselKKNN2D[x_,BessRng_,YTcks_,Dsh_] :=
 Plot[Re[SphereBessKK[n,x]],{n,0,3},
   PlotRange->{0,BessRng},PlotPoints->50,
   Ticks->{{0,1,2,3},{-YTcks,0,YTcks}},
   AxesLabel->{"n",None},
   PlotStyle->Dashing[{Dsh,Dsh}],
   DisplayFunction->Identity]
```

Define 2D views of irregular functions against *n* with *x* fixed.

```
(* Exponential & x inserted;
   2D views defined; n varying & x fixed *)
BesselKKexpNN2D[x_,BessRng_,YTcks_,Dsh_] :=
 Plot[Re[x*Exp[x]*SphereBessKK[n,x]],{n,0,3},
   PlotRange->{0,BessRng},PlotPoints->50,
   Ticks->{{0,1,2,3},{-YTcks,0,YTcks}},
   AxesLabel->{"n",None},
   PlotStyle->Dashing[{Dsh,Dsh}],
   DisplayFunction->Identity]
```

Define 2D views of irregular functions against *n* with *x* fixed; factors of *x* and an exponential are included.

```
(*  x and n;  3D views  *)

BessKKXN3D[Re,20,10];
BessKKXN3D[Im,20,10];  .
```

Show 3D views as in Figure 14.4.10.

```
(* x with n fixed; 2D views *)
ShowRegX2D = Show[
 BesselKKX2D[0,20,10,0.06],
 BesselKKX2D[1,20,10,0.04],
 BesselKKX2D[2,20,10,0.02],
 DisplayFunction->$DisplayFunction];
```

Show 2D views against *x* with *n* = 0, 1, 2, as in Figure 14.4.11 left.

```
(* Exponential inserted; x with n fixed;
   2D views *)
ShowRegExpX2D = Show[
 BesselKKexpX2D[0,10,5,0.06],
 BesselKKexpX2D[1,10,5,0.04],
 BesselKKexpX2D[2,10,5,0.02],
 DisplayFunction->$DisplayFunction];
```

Show 2D views against *x* with *n* = 0, 1, 2; *x* and exponential factor are included, as in Figure 14.4.11 right.

```
(* n with x fixed; 2D views *)

ShowRegNN2D = Show[
 BesselKKNN2D[1,20,10,0.06],
 BesselKKNN2D[2,20,10,0.04],
 BesselKKNN2D[3,20,10,0.02],
 DisplayFunction->$DisplayFunction];
```

Show 2D views against n with $x = 1, 2, 3$, as in Figure 14.4.12 left.

```
(* Exponential inserted; n with x fixed;
   2D views *)

ShowRegExpNN2D = Show[
 BesselKKexpNN2D[1,10,5,0.06],
 BesselKKexpNN2D[2,10,5,0.04],
 BesselKKexpNN2D[3,10,5,0.02],
 DisplayFunction->$DisplayFunction];
```

Show 2D views against n with $x = 1, 2, 3$, and exponential factor included. as in Figure 14.4.12 right.

```
(* Regular spherical Bessels: Numerical *)

BessKKNTable =
Table[{n,x,N[SphereBessKK[n,x],10]},
 {x,1,3},{n,0,6,3}]
BessKKN10 = Table[{n,x,N[SphereBessKK[n,x],10]},
 {x,10,10},{n,0,6,3}]
```

Numerical irregular modified spherical Bessel functions, Table 14.4.8.

Notebook `BesselAiry` [14.5]

Use the cells in this notebook to visualize the behavior of the Airy functions in Section 14.5, that is, of linear combinations of Bessel functions whose order is ±1/3. There are two functions—Ai and Bi. We also consider the first derivatives of these functions— Ai′ and Bi′. The real part of Bi(z) is shown in the complex plane in the motif on the right; the exponential increase with z when x is large and positive is truncated. More detail is provided in Figure 14.5.2.

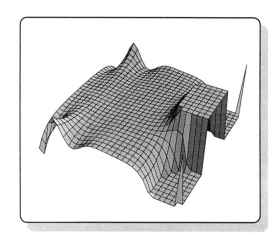

Notebook `BesselAiry` also generates auxiliary functions for computing Airy functions and their derivatives and it provides numerical values for them. Further, rational approximations needed for accurate numerical values when x is large in magnitude are developed in two of the cells—`AiryAsymptotes` and `AiryPrimeAsymptotes`. These illustrate the power of combining the high-accuracy and generality of *Mathematica* to produce very efficient and accurate algorithms for the corresponding C functions.

The `Airy` cell [14.5.1, 14.5.2]

Execute this cell to produce Figures 14.5.1–14.5.3 for the Airy functions Ai and Bi in Section 14.5.1, such as the curves for Ai(x) and Bi(x) seen in the motif at right for x from –4 to 4.

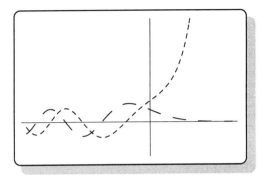

This cell also generates Figures 14.5.8–14.5.10 for the derivative functions, Ai'(x) and Bi'(x). Numerical values of these four functions are also generated; they can be used to test your implementation of the C functions `AiryAiBi` and `AiryPrimeAiBi`.

Exploring with `Airy`

Airy functions are applied in a wide variety of contexts, so choose a range of argument values (z or x) appropriate to your use of them, then explore this region graphically in more detail by modifying the argument range for the 3D or 2D plot functions in the following annotated *Mathematica* cell.

```
(* Airy cell *)

(* Airy Functions and their Derivatives  *)

(* Airy Functions, Ai(z), Bi(z),
   and their Derivatives *)

(*   Airy Functions: 3D & 2D views *)

(* 3D defined for x & y varying *)
Airy3D[ABorAPBP_,ReOrIm_,Pnts_,Rng_,ZTcks_] :=
 Plot3D[
   ReOrIm[ABorAPBP[x+I*y]],{x,-6,6},{y,-2,2},
   PlotRange->{-Rng,Rng},
   PlotPoints->{Pnts,Pnts},
   Ticks->{{-4,-2,0,2,4},{-2,0,2},
   {-ZTcks,0,ZTcks}},AxesLabel->{"x","y",None}]
```

Define 3D views of Airy functions or their derivatives in the complex plane. Choose real or imaginary part, number of points in each direction, vertical scale, and position of vertical ticks.

```
(* 2D views defined *)

AiryPlot[ABorAPBP_,Dsh_] :=
 Plot[ABorAPBP[x],{x,-6,4},
   PlotRange->{-1,3},PlotPoints->100,
   Ticks->{{-6,-4,-2,0,2,4},{-1,0,1,2,3}},
   AxesLabel->{"x",None},
   PlotStyle->Dashing[{Dsh,Dsh}],
   DisplayFunction->Identity]
```

Define 2D views of Airy functions or their derivatives against x, choosing dash length depending on function choice.

```
(* Complex plane 3D views *)

Airy3D[AiryAi,Re,30,4,2];
Airy3D[AiryAi,Im,30,4,2];
Airy3D[AiryBi,Re,30,20,10];
Airy3D[AiryBi,Im,30,20,10];
Airy3D[AiryAiPrime,Re,30,4,2];
Airy3D[AiryAiPrime,Im,30,4,2];
Airy3D[AiryBiPrime,Re,30,20,10];
Airy3D[AiryBiPrime,Im,30,20,10];
```

Show Ai real and imaginary parts in 3D.
Show Bi real and imaginary parts in 3D.
Show Ai′ real and imaginary parts in 3D.
Show Bi′ real and imaginary parts in 3D.

```
(* 2D views *)

Show2DAB = Show[
 AiryPlot[AiryAi,0.06],
 AiryPlot[AiryBi,0.02],
 DisplayFunction->$DisplayFunction];
```

Show Ai and Bi as long and short dashes in 2D, Figure 14.5.3.

```
Show2DAPBP = Show[
 AiryPlot[AiryAiPrime,0.06],
 AiryPlot[AiryBiPrime,0.02],
 DisplayFunction->$DisplayFunction];
```

Show Ai′ and Bi′ as long and short dashes in 2D, Figure 14.5.10.

```
(* Airy functions: Numerical *)

AiryTable[ABorAPBP_] :=
 Table[{x,N[ABorAPBP[x],10]},{x,-6,6,2}]
```

Define numerical table.

```
AiryTable[AiryAi]
AiryTable[AiryBi]
AiryTable[AiryAiPrime]
AiryTable[AiryBiPrime]
```

Show tables for Ai, Bi, and Ai′, and Bi′, as in Tables 14.5.2 and 14.5.4.

The `Airy.f&g` cell [14.5.1]

Use this cell to produce Figures 14.5.4 and 14.5.5 for the auxiliary functions $c_1 f(x)$ and $c_2 g(x)$ used to produce the power series expansions (14.5.5) for the Airy functions in Section 14.5.1. These two auxiliary functions are shown in the graphic motif at right; Figure 14.5.4 has more detail.

This cell also generates functions $F(\zeta)$ and $G(\zeta)$ used in (14.5.10) for the asymptotic expansion of the Airy functions Ai and Bi.

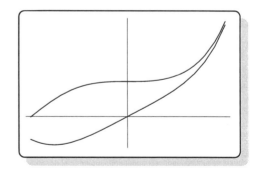

Exploring with `Airy.f&g`

In Section 14.5.1 we state that convergence of the power series method is slow when $|x| > 2$ and that subtractive cancellation between the two series leads to inaccurate values of Ai when $x > 2$, as is clear from the convergence of the two curves in the motif above. For the accuracy that you require in Ai, explore how large x can be. To do this, you need to modify the accuracy of the arithmetic in the following *Mathematica* cell and to change the summation range of k in the power series. Similarly, how low can you go before the asymptotic expansions (14.5.13) do not converge to satisfactory accuracy?

```
(* Airy.f&g cell *)

(* Auxiliary functions: c1f and c2g *)

c1f[x_] := NSum[
  x^(3*k)/(3^(2*k+2/3)*k!*Gamma[k+2/3]),
  {k,0,Infinity}]
```
Define *f* auxiliary function by (14.5.7).

```
c2g[x_] := NSum[
  x^(3*k+1)/(3^(2*k+4/3)*k!*Gamma[k+4/3]),
  {k,0,Infinity}]
```
Define *g* auxiliary function by (14.5.7).

```
Ai[x_] := c1f[x] - c2g[x]
```
Ai and Bi in terms of *f* and *g* by (14.5.5).

```
Bi[x_] := Sqrt[3]*(c1f[x] + c2g[x])
```

```
PlotForG[ForG_] := Plot[ForG[x],{x,-2,2},
  PlotRange->All,PlotPoints->50,
  Ticks->{{-2,-1,0,1,2},
   {-0.25,0,0.25,0.50,0.75,1.00}},
  AxesLabel->{"x",None},
  DisplayFunction->Identity]
```
Define 2D plot for *f* or *g*.

```
ShowFandG = Show[
  PlotForG[c1f],PlotForG[c2g],
  DisplayFunction->$DisplayFunction];
```
Show the two functions together, as in Figure 14.5.4.

```
(* Auxiliary functions  F  &  G  *)

FFAiryAux[zeta_] :=
  Sin[zeta+Pi/4]*Sqrt[N[Pi]*((1.5*zeta)^(1/3))]*
  AiryAi[-(1.5*zeta)^(2/3)]+
  Cos[zeta+Pi/4]*Sqrt[N[Pi]*((1.5*zeta)^(1/3))]*
  AiryBi[-(1.5*zeta)^(2/3)]
```
Define *F* in terms of Ai and Bi for any ζ by using (14.5.13).

```
GGAiryAux[zeta_] :=
  Sin[zeta+Pi/4]*Sqrt[N[Pi]*((1.5*zeta)^(1/3))]*
  AiryBi[-(1.5*zeta)^(2/3)]-
  Cos[zeta+Pi/4]*Sqrt[N[Pi]*((1.5*zeta)^(1/3))]*
  AiryAi[-(1.5*zeta)^(2/3)]
```
Define *G* in terms of Ai and Bi for any ζ by using (14.5.13).

```
(* Show  F - 1  vs  zeta  *)
FFAuxPlot = Plot[FFAiryAux[zeta]-1,{zeta,2,20},
  PlotRange->All,AxesOrigin->{2,0},
  Ticks->{{5,10,15,20},{-0.005,-0.0025,0}}];


(* Show  G  vs  zeta  *)
GGAuxPlot = Plot[GGAiryAux[zeta],{zeta,2,20},
  PlotRange->{0,0.035},AxesOrigin->{2,0},
  Ticks->{{5,10,15,20},{0,0.01,0.02,0.03}}];
```

Show $F-1$ against ζ, as in Figure 14.5.5 left.

Show G against ζ, as in Figure 14.5.5 right.

The `AiryAsymptote` cell [14.5.1]

Run this cell to generate the rational approximations to the series D, E, F, and G in Section 14.5.1 and in the C function `AiryAiBi`. Figures 14.5.6 and 14.5.7 —showing the accuracy of the approximations—are also made. For example, the fractional accuracy of the approximation to Ai for ζ in [2, 20] is always better than 12 digits and it varies with ζ as shown in the motif at right.

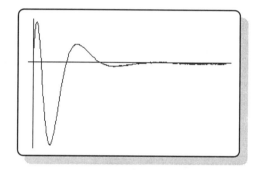

Exploring with `AiryAsymptote`

In Table 14.5.1 we list the orders of the polynomials in the numerator (p) and denominator (q) of the rational approximations of type (14.5.14) and the maximum error that results in each fitted function; these orders produce better than 10-digit accuracy in the Airy functions Ai and Bi, well within the accuracy goals of functions in the *Atlas*.

If your accuracy goals are more modest or if you need the functions over a smaller range, then p and q can be smaller, so the C function `AiryAiBi` will execute somewhat faster. If you need more than 10-digit accuracy, you will have to increase p and q and improve the accuracy of the arithmetic in *Mathematica* in order to produce improved rational-fraction approximations.

```
(* AiryAsymptote cell *)

<<NumericalMath`Approximations`

(* Fitting Ai for 2 <= zeta <= 20  *)

ClearAll[zeta];
DDAiry[zeta_] :=
  2*Sqrt[N[Pi]*((1.5*zeta)^(1/3))]*
  Exp[zeta]*AiryAi[(1.5*zeta)^(2/3)]

DDRatInt = RationalInterpolation[
  DDAiry[zeta],{zeta,5,6},{zeta,2,20},
  Bias->-0.8]
```

Define D in terms of Ai by first equation in (14.5.8).

Fit D by rational interpolation; 5th-order numerator, 6th-order denominator.

```
DDError = Plot[
 10^12*(DDAiry[zeta] - DDRatInt),{zeta,2,20},
 PlotRange->All,AxesOrigin->{2,0},
 AxesLabel->{"x",None},
 Ticks->{{2,5,10,15,20},
 {-0.75,-0.50,-0.25,0,0.25}}];
```

Plot error between D and its rational approximation.

```
(* Fitting Bi for 2 <= zeta <= 20  *)

ClearAll[zeta];

EEAiry[zeta_] :=
 Sqrt[N[Pi]*((1.5*zeta)^(1/3))]*
 Exp[-zeta]*AiryBi[(1.5*zeta)^(2/3)]
```

Define E in terms of Bi by third equation in (14.5.8).

```
EERatInt = RationalInterpolation[
 EEAiry[zeta],{zeta,9,10},{zeta,2,20},
 Bias->-0.5]
```

Fit E by rational interpolation; 9th-order numerator, 10th-order denominator.

```
EEError = Plot[
 10^11*(EEAiry[zeta] - EERatInt),{zeta,2,20},
 PlotRange->All,AxesOrigin->{2,0},
 AxesLabel->{"x",None},
 Ticks->{{2,5,10,15,20},{-0.5,0,0.5,1}}];
```

Plot error between E and its rational approximation.

```
(* Check Ai(x) and Bi(x) by interpolants
   in exponential asymptotic region *)

AAiExpError = Plot[10^13*
 (1 - DDRatInt/(2*Sqrt[Pi*((1.5*zeta)^(1/3))]*
 Exp[zeta])/AiryAi[(1.5*zeta)^(2/3)]),
 {zeta,2,20},PlotRange->All,AxesOrigin->{2,0},
 AxesLabel->{"x",None},
 Ticks->{{2,5,10,15,20},
 {-7.5,-5.0,-2.5,0,2.5}}];
```

Plot fractional error in rational approximation to Ai for $\zeta > 2$, Figure 14.5.6 left.

```
BBiExpError = Plot[10^12*
 (1 -EERatInt/(Sqrt[Pi*((1.5*zeta)^(1/3))]*
 Exp[-zeta])/AiryBi[(1.5*zeta)^(2/3)]),
 {zeta,2,20},PlotRange->All,AxesOrigin->{2,0},
 AxesLabel->{"x",None},
 Ticks->{{2,5,10,15,20},{-5.0,0,5.0}}];
```

Plot fractional error in rational approximation to Bi for $\zeta > 2$, Figure 14.5.6 right.

```
(* Fitting F for 2 <= zeta <= 20  *)

ClearAll[zeta];

FFAiry[zeta_] :=
 Sin[zeta+Pi/4]*Sqrt[N[Pi]*((1.5*zeta)^(1/3))]*
 AiryAi[-(1.5*zeta)^(2/3)]+
 Cos[zeta+Pi/4]*Sqrt[N[Pi]*((1.5*zeta)^(1/3))]*
 AiryBi[-(1.5*zeta)^(2/3)]
```

AiryAsymptote
continued

Define F in terms of Ai and Bi by first equation in (14.5.13).

```
FFRatInt = RationalInterpolation[
 FFAiry[zeta],{zeta,5,6},{zeta,2,20},
 Bias->-0.5]
```

Fit *F* by rational interpolation; 5th-order numerator, 6th-order denominator.

```
FFError = Plot[
10^12*(FFAiry[zeta] - FFRatInt),{zeta,2,20},
PlotRange->All,AxesOrigin->{2,0},
AxesLabel->{"x",None},
Ticks->{{2,5,10,15,20},{-3,-2,-1,0,1,2}}];
```

Plot error between *F* and its rational approximation.

```
(* Fitting G for 2 <= zeta <= 20  *)

ClearAll[zeta];
GGAiry[zeta_] :=
 Sin[zeta+Pi/4]*Sqrt[N[Pi]*((1.5*zeta)^(1/3))]*
 AiryBi[-(1.5*zeta)^(2/3)]-
 Cos[zeta+Pi/4]*Sqrt[N[Pi]*((1.5*zeta)^(1/3))]*
 AiryAi[-(1.5*zeta)^(2/3)]
```

Define *G* in terms of Ai and Bi by second equation in (14.5.13).

```
GGRatInt = RationalInterpolation[
 GGAiry[zeta],{zeta,5,6},{zeta,2,20},
 Bias->-0.5]
```

Fit *G* by rational interpolation; 5th-order numerator, 6th-order denominator.

```
GGError = Plot[
 10^14*(GGAiry[zeta] - GGRatInt),{zeta,2,20},
 PlotRange->All,AxesOrigin->{2,0},
 AxesLabel->{"x",None},
 Ticks->{{2,5,10,15,20},{-10,-5,0,5,10}}];
```

Plot error between *G* and its rational approximation.

```
(* Check Ai(-x) and Bi(-x) by interpolants
   in oscillatory asymptotic region *)

AAiError = Plot[10^13*
 (AiryAi[-(1.5*zeta)^(2/3)] -
 (Sin[zeta+Pi/4]*FFRatInt -
 Cos[zeta+Pi/4]*GGRatInt)/
 Sqrt[N[Pi]*((1.5*zeta)^(1/3))]),{zeta,2,20},
 PlotRange->All,AxesOrigin->{2,0},
 AxesLabel->{"x",None},
 Ticks->{{2,5,10,15,20},{-10,-5,0,5,10}}];
```

Plot fractional error in rational approximation to Ai for $\zeta < -2$, Figure 14.5.7 left.

```
BBiError = Plot[10^13*
 (AiryBi[-(1.5*zeta)^(2/3)] -
 (Cos[zeta+Pi/4]*FFRatInt +
 Sin[zeta+Pi/4]*GGRatInt)/
 Sqrt[N[Pi]*((1.5*zeta)^(1/3))]),{zeta,2,20},
 PlotRange->All,AxesOrigin->{2,0},
 AxesLabel->{"x",None},
 Ticks->{{2,5,10,15,20},{-10,-5,0,5,10}}];
```

Plot fractional error in rational approximation to Bi for $\zeta < -2$, Figure 14.5.7 right.

The `Airy.fp&gp` cell [14.5.2]

Run this cell to produce Figure 14.5.11 for the auxiliary functions $c_1 f'(x)$ and $c_2 g'(x)$; these produce the power series expansions (14.5.19) for the derivatives of the Airy functions, $Ai'(x)$ and $Bi'(x)$ in Section 14.5.2. The two auxiliary functions form the motif at right; Figure 14.5.11 has more detail.

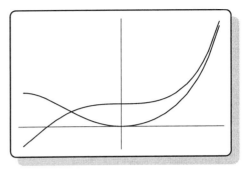

Cell `Airy.fp&gp` also makes functions $F_1(\zeta)$ and $G_1(\zeta)$ shown in Figure 14.5.12; they are used in (14.5.21) for the asymptotic expansion of the Airy functions $Ai'(x)$ and $Bi'(x)$.

Exploring with `Airy.fp&gp`

In Section 14.5.2 we state that convergence of the power series method for the derivatives is slow when $|x| > 2$ and that subtractive cancellation between the two series leads to inaccurate values of $Ai'(x)$ when $x > 2$, as is clear from the convergence of the two curves in the motif above. For the accuracy that you require in $Ai'(x)$, explore how large x can be. To do this, modify the accuracy of the arithmetic in the following *Mathematica* cell and change the summation range of k in the power series. Similarly, how low can you go in x before the asymptotic expansions do not converge to satisfactory accuracy?

```
(* Airy.fp&gp cell *)

(* Auxiliary functions for derivatives:
   c1fp and c2gp *)

c1fP[x_] := NSum[
  x^(3*k-1)/(3^(2*k-1/3)*(k-1)!*Gamma[k+2/3]),
  {k,1,Infinity}]
```
Define auxiliary f derivative function by (14.5.22).

```
c2gP[x_] := NSum[
  x^(3*k)/(3^(2*k+1/3)*k!*Gamma[k+1/3]),
  {k,0,Infinity}]
```
Define auxiliary g derivative function by (14.5.22).

```
AiP[x_] := c1fP[x] - c2gP[x]

BiP[x_] := Sqrt[3]*(c1fP[x] + c2gP[x])
```
Ai and Bi derivatives in terms of auxiliary derivative functions by (14.5.17).

```
PlotAiryP[FPorGP_] := Plot[FPorGP[x],{x,-2,2},
  PlotRange->All,PlotPoints->40,
  Ticks->{{-2,-1,0,1,2},
   {-0.25,0,0.50,0.75,1.00}},
  AxesLabel->{"x",None},
  DisplayFunction->Identity]
```
Define 2D plot for f or g derivatives.

```
ShowFPandGP = Show[
  PlotAiryP[c1fP],PlotAiryP[c2gP],
  DisplayFunction->$DisplayFunction];
```
Show both functions together, as in Figure 14.5.11.

```
(* Auxiliary derivative functions  F1 & G1  *)

FF1AiryPrime[zeta_] :=
  ((1.5*zeta)^(-1/6))*Sqrt[N[Pi]]*
  (Sin[zeta+Pi/4]*AiryBiPrime[-(1.5*zeta)^(2/3)]-
   Cos[zeta+Pi/4]*AiryAiPrime[-(1.5*zeta)^(2/3)])

GG1AiryPrime[zeta_] :=
  -((1.5*zeta)^(-1/6))*Sqrt[N[Pi]]*
  (Cos[zeta+Pi/4]*AiryBiPrime[-(1.5*zeta)^(2/3)]+
   Sin[zeta+Pi/4]*AiryAiPrime[-(1.5*zeta)^(2/3)])

(* Show  F1 - 1  vs  zeta  *)
FF1PrimePlot = Plot[FF1AiryPrime[zeta]-
1,{zeta,2,20},PlotRange->All,AxesOrigin->{2,0},
  Ticks->{{5,10,15,20},
    {0,0.0025,0.0050,0.0075}}];

(* Show  G1  vs  zeta  *)
GG1PrimePlot =
Plot[GG1AiryPrime[zeta],{zeta,2,20},
  PlotRange->{-0.05,0},AxesOrigin->{2,0},
  Ticks->{{5,10,15,20},{-0.04,-0.02,0}}];
```

Define F_1 in terms of Ai′ and Bi′ for any ζ by using (14.5.23).

Define G_1 in terms of Ai′ and Bi′ for any ζ by using (14.5.23).

Show $F_1-.1$ against ζ, as in Figure 14.5.12 left.

Show G_1 against ζ, as in Figure 14.5.12 right.

The `AiryPrimeAsymptote` cell [14.5.2]

By running this cell you will generate the rational approximations to the series D_1, E_1, F_1, and G_1 in Section 14.5.2 and in the C function `AiryPrimeAiBi`. In addition to graphics of these auxiliary functions, the graphics for Figures 14.5.13 and 14.5.14—showing the accuracy of the approximations to Ai′ and Bi′—are also made. For example, the fractional accuracy of the approximation to Bi′ for $-\zeta$ in [2, 20], which is the oscillatory region of

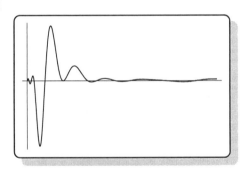

the derivative, is always better than 12 digits and it varies with ζ as shown in the motif at right.

Exploring with `AiryPrimeAsymptote`

In Table 14.5.3 we list the orders of the polynomials in the numerator (p) and denominator (q) of the rational approximations of type (14.5.24), along with the maximum error that results in each fitted function; these orders produce better than 10-digit accuracy in the Airy functions Ai′ and Bi′, well within the accuracy goals of functions in the *Atlas*.

If your accuracy goals are more modest or if you need the functions over a smaller range, then p and q can be smaller, so the C function `AiryPrimeAiBi` will execute slightly faster. To ensure more than 10-digit accuracy, you will have to increase p and q

and improve the accuracy of the arithmetic in *Mathematica* in order to produce improved rational-fraction approximations.

```
(* AiryPrimeAsymptote cell *)

(* Rational Approximations to Airy
   Derivatives Series in 2 <= |zeta| <= 20 *)


<<NumericalMath`Approximations`

(* Fitting Ai' for 2 <= zeta <= 20   *)

ClearAll[zeta];
DD1Airy[zeta_] :=
 -2*Sqrt[N[Pi]*((1.5*zeta)^(-1/3))]*
 Exp[zeta]*AiryAiPrime[(1.5*zeta)^(2/3)]

DD1RatInt = RationalInterpolation[
 DD1Airy[zeta],{zeta,5,6},{zeta,2,20},
 Bias->-0.7]

DD1Error = Plot[
 10^13*(DD1Airy[zeta] - DD1RatInt),{zeta,2,20},
 PlotRange->All,AxesOrigin->{2,0},
 Ticks->{{2,5,10,15,20},{-2,-1,0,1,2}}];


(* Fitting Bi' for 2 <= zeta <= 20   *)

ClearAll[zeta];
EE1Airy[zeta_] :=
 Sqrt[N[Pi]*((1.5*zeta)^(-1/3))]*
 Exp[-zeta]*AiryBiPrime[(1.5*zeta)^(2/3)]

EE1RatInt = RationalInterpolation[
 EE1Airy[zeta],{zeta,9,10},{zeta,2,20},
 Bias->-0.6]

EE1Error = Plot[
 10^11*(EE1Airy[zeta] - EE1RatInt),{zeta,2,20},
 PlotRange->All,AxesOrigin->{2,0},
 Ticks->{{2,5,10,15,20},{-0.5,0,0.5,1}}];

(* Check Ai'(x) and Bi'(x) by interpolants
   in exponential asymptotic region *)


AAiPrimeExpError = Plot[10^13*
 (1 - (DD1RatInt/(-2*Sqrt[Pi*
 ((1.5*zeta)^(-1/3))]*Exp[zeta]))/
 AiryAiPrime[(1.5*zeta)^(2/3)]),{zeta,2,20},
 PlotRange->All,AxesOrigin->{2,0},
 Ticks->{{2,5,10,15,20},
 {-1.5,-1.0,-0.5,0,0.5,1.0}}];
```

Define D_1 in terms of Ai$'$ by first equation in (14.5.20).

Fit D_1 by rational interpolation; 5th-order numerator, 6th-order denominator.

Plot error between D_1 and its rational approximation.

Define E_1 in terms of Bi$'$ by second equation in (14.5.20).

Fit E_1 by rational interpolation; 9th-order numerator, 10th-order denominator.

Plot error between E_1 and its rational approximation.

Plot fractional error in rational approximation to Ai$'$ for $\zeta > 2$, Figure 14.5.13 left.

```
BBiPrimeExpError = Plot[10^11*
 (1 - EE1RatInt/(Sqrt[Pi*((1.5*zeta)^(-1/3))]*
 Exp[-zeta])/
 AiryBiPrime[(1.5*zeta)^(2/3)]),{zeta,2,20},
 PlotRange->All,AxesOrigin->{2,0},
 Ticks->{{2,5,10,15,20},{-1.5,-1,-
0.5,0,0.5,1.0,1.5}}];
```

Plot fractional error in rational approximation to Bi′ for $\zeta > 2$, Figure 14.5.13 right.

```
(* Fitting F' for 2 <= zeta <= 20  *)
ClearAll[zeta];
FF1Airy[zeta_] :=
 Sin[zeta+Pi/4]*Sqrt[N[Pi]*((1.5*zeta)^(-1/3))]*
 AiryBiPrime[-(1.5*zeta)^(2/3)]-
 Cos[zeta+Pi/4]*Sqrt[N[Pi]*((1.5*zeta)^(-1/3))]*
 AiryAiPrime[-(1.5*zeta)^(2/3)]
```

Define F_1 in terms of Ai′ and Bi′ by first equation in (14.5.23).

```
FF1RatInt = RationalInterpolation[
 FF1Airy[zeta],{zeta,5,6},{zeta,2,20},
 Bias->-0.5]
```

Fit F_1 by rational interpolation; 5th-order numerator, 6th-order denominator.

```
FF1Error = Plot[
 10^12*(FF1Airy[zeta] - FF1RatInt),{zeta,2,20},
 PlotRange->All,AxesOrigin->{2,0},
 Ticks->{{2,5,10,15,20},{-3,-2,-1,0,1,2}}];
```

Plot error between F_1 and its rational approximation.

```
(* Fitting G' for 2 <= zeta <= 20  *)
ClearAll[zeta];
GG1Airy[zeta_] :=
-Sin[zeta+Pi/4]*Sqrt[N[Pi]*((1.5*zeta)^(-1/3))]*
 AiryAiPrime[-(1.5*zeta)^(2/3)]-
 Cos[zeta+Pi/4]*Sqrt[N[Pi]*((1.5*zeta)^(-1/3))]*
 AiryBiPrime[-(1.5*zeta)^(2/3)]
```

Define G_1 in terms of Ai′ and Bi′ by second equation in (14.5.23).

```
GG1RatInt = RationalInterpolation[
 GG1Airy[zeta],{zeta,5,6},{zeta,2,20},
 Bias->-0.6]
```

Fit G_1 by rational interpolation; 5th-order numerator, 6th-order denominator.

```
GG1Error = Plot[
 10^14*(GG1Airy[zeta] - GG1RatInt),{zeta,2,20},
 PlotRange->All,AxesOrigin->{2,0},
 Ticks->{{2,5,10,15,20},{-10,-5,0,5,10}}];
```

Plot error between G_1 and its rational approximation.

```
(* Check Ai'(-x) and Bi'(-x) by interpolants
   in oscillatory asymptotic region *)
AAiPrimeError = Plot[10^13*
 (AiryAiPrime[-(1.5*zeta)^(2/3)] -
 (-Cos[zeta+Pi/4]*FF1RatInt -
Sin[zeta+Pi/4]*GG1RatInt)/
 Sqrt[N[Pi]*((1.5*zeta)^(-1/3))]),{zeta,2,20},
 PlotRange->All,AxesOrigin->{2,0},
 Ticks->{{2,5,10,15,20},{-10,-5,0,5,10}}];
```

Plot fractional error in rational approximation to Ai′ for $\zeta < -2$, Figure 14.5.7 left.

```
BBiPrimeError = Plot[10^13*
  (AiryBiPrime[-(1.5*zeta)^(2/3)] -
  (Sin[zeta+Pi/4]*FF1RatInt -
Cos[zeta+Pi/4]*GG1RatInt)/
  Sqrt[N[Pi]*((1.5*zeta)^(-1/3))]),{zeta,2,20},
  PlotRange->All,AxesOrigin->{2,0},
  Ticks->{{2,5,10,15,20},{-10,-5,0,5,10}}];
```

AiryPrimeAsymptote
continued

Plot fractional error in
rational approximation to Bi′
for $\zeta < -2$, Figure 14.5.7
right.

20.15   STRUVE, ANGER, AND WEBER FUNCTIONS

Notebook and Cells
Notebook StruveAngerWeber;
cells StruveH, StruveL, AngerJ, and WeberE.

   Run the cells in this notebook
to visualize the behavior of the
Struve, Anger, and Weber functions described in Chapter 15.
These cells also generate numerical values that are useful when
checking your implementation of
the functions in C. For this purpose, run the *Mathematica* cells in
conjunction with the C driver programs whose annotated listings
are given in Section 21.15.

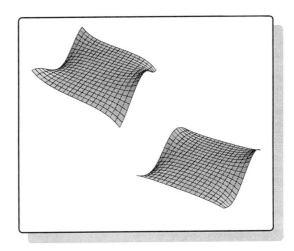

The StruveH cell [15.1.1]
Execute this cell to see Figures 15.1.1–
15.1.3 for the Struve functions $\mathbf{H}_v(z)$, such
as the surfaces for $\mathbf{H}_n(x)$ seen in the motif
at right, where x varies from –6 to 6 on the
front axis while on the side axis v varies
from –1 to 2.

Exploring with StruveH
Choose a range of argument values (z or x)
appropriate to your use of $\mathbf{H}_v(z)$, then explore this region graphically in more detail
by modifying the argument range for the
3D or 2D plot functions in the following

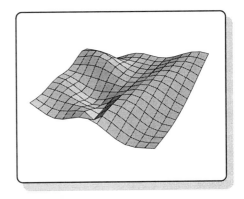

annotated *Mathematica* cell. In many practical situations—such as the vibrations of a dia-
phragm in a fluid—the order $v = n$, an integer that indicates the eigenmode of the vibration;
thus, n is usually small.

```
(* StruveH cell *)

(* Struve Functions *)

(* Power series definition *)
SetOptions[NSum,AccuracyGoal->12,
 PrecisionGoal->12,WorkingPrecision->19,
 NSumExtraTerms->16];
Off[NSum::precw];

HH[n_,z_] := If[ n == -1,
 If[ z == 0, 1/(Gamma[3/2]*Gamma[1/2]),
 1/(Gamma[3/2]*Gamma[1/2])+
  NSum[(-1)^k*(z/2)^(2*k)/
  (Gamma[k+3/2]*Gamma[k+1/2]),
  {k,1,Infinity},
  NSumTerms->Max[20,Floor[1.5*z-10]]]],
 If[ z == 0, 0, (z/2)^(n+1)*
 (1/(Gamma[3/2]*Gamma[n+3/2])+
 NSum[(-1)^k*(z/2)^(2*k)/
 (Gamma[k+3/2]*Gamma[k+n+3/2]),
 {k,1,Infinity},
 NSumTerms->Max[20,Floor[1.5*z-10]]])]]
```

This accuracy control is needed to get 10-digit accuracy in Struve functions.

Define the Struve functions in terms of their power series expansion (15.1.2). Several special cases must be checked, since *Mathematica* thinks that $0^0 \neq 1$.

```
(* 3D & 2D views *)

(* 3D defined for n fixed, x & y vary *)
HHZ3D[n_,ReOrIm_,Pnts_,HHRng_,ZTcks_] :=
 Plot3D[
  ReOrIm[HH[n,x+I*y]],{x,-8,8},{y,-3,3},
  PlotRange->{-HHRng,HHRng},
  PlotPoints->{Pnts,Pnts},
  Ticks->{{-6,-3,0,3,6},{-2,0,2},
  {-ZTcks,0,ZTcks}},AxesLabel->{"x","y",None}]
```

Define 3D views of Struve functions in the complex plane. Choose real or imaginary part, number of points in each direction, vertical scale, and position of vertical ticks.

```
(* 3D defined; real x & n varying *)
HHXN3D[ReOrIm_,HHRng_,ZTcks_] :=
 Plot3D[
  ReOrIm[HH[n,x]],{x,-8,8},{n,-1,2},
  PlotRange->{-HHRng,HHRng},
  PlotPoints->{20,20},
  Ticks->{{-6,-3,0,3,6},{-1,0,1,2},
   {-ZTcks,ZTcks}},
  AxesLabel->{"x","n",None}]
```

Define 3D views of Struve functions in the *x*–*n* plane. Choose real or imaginary part, vertical scale, and position of vertical ticks.

```
(* 2D views defined; x varying & n fixed *)
HHX2D[n_,HHRng_,YTcks_,Dsh_] :=
 Plot[HH[n,x],{x,0,10},
  PlotRange->{-HHRng,HHRng},PlotPoints->30,
  Ticks->{{0,2,4,6,8,10},{-YTcks,0,YTcks}},
  AxesLabel->{"x",None},
  PlotStyle->Dashing[{Dsh,Dsh}],
  DisplayFunction->Identity]
```

Define 2D views with *x* varying and *n* fixed. Plot range, position of vertical ticks, and size of dashes are parameters.

```
(* 2D views defined; n varying & x fixed *)

HHNN2D[x_,HHRng_,YTcks_,Dsh_] :=
  Plot[HH[n,x],{n,-1,6},
    PlotRange->{-HHRng,HHRng},PlotPoints->30,
    Ticks->{{-1,1,3,5},{-YTcks,0,YTcks}},
    AxesLabel->{"n",None},AxesOrigin->{-1,0},
    PlotStyle->Dashing[{Dsh,Dsh}],
    DisplayFunction->Identity]


(* Complex plane 3D views for  n = 0,1,2 *)
HHZ3D[0,Re,20,3,2]; HHZ3D[0,Im,20,3,2];
HHZ3D[1,Re,20,3,2]; HHZ3D[1,Im,20,3,2];
HHZ3D[2,Re,20,3,2]; HHZ3D[2,Im,20,3,2];


(*  x and n;  3D views  *)

HHXN3D[Re,2,1]; HHXN3D[Im,2,1];


(* x with n fixed; 2D views *)

ShowRegX2D = Show[
  HHX2D[-1,2,1,0.01],HHX2D[ 0,2,1,0.02],
  HHX2D[ 1,2,1,0.04],HHX2D[ 2,2,1,0.06],
  DisplayFunction->$DisplayFunction];


(* n with x fixed; 2D views *)

ShowRegNN2D = Show[HHNN2D[1,2,1,0.06],
  HHNN2D[3,2,1,0.04],HHNN2D[5,2,1,0.02],
  DisplayFunction->$DisplayFunction];


(* Struve functions: Numerical *)

HHJNTable = Table[{n,x,N[HH[n,x],10]},
  {x,0,6,2},{n,0,2}]
HHJN20 = Table[{n,x,N[HH[n,x],10]},
  {x,20,20,1},{n,0,2}]
```

StruveH cell
continued

Define 2D views with *n* varying and *x* fixed. Plot range, position of vertical ticks, and size of dashes are parameters.

Show Struve **H** function in the complex plane, as in Figure 15.1.1.

Show Struve **H** against *x* and *n*, real and imaginary parts, as in Figure 15.1.2.

Show curves of **H** against *x*, as in Figure 15.1.3 left.

Show curves of **H** against *n*, as in Figure 15.1.3 right.

Numerical test values, Table 15.1.1.

The StruveL cell [15.1.2]

Run this cell to see Figures 15.1.4–15.1.6 for the modified Struve functions $\mathbf{L}_\nu(z)$, such as the surface for the real part of $\mathbf{L}_0(z)$ in the motif at right, where *x* varies from –4 to 4 on the front axis while on the side axis *y* varies from –3 to 3; more detail is in Figure 15.1.4.

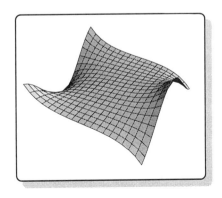

Exploring with StruveL

Using a range of argument values (*z* or *x*) appropriate to your application of $\mathbf{L}_\nu(z)$, explore this region graphically in more detail by modifying the argument range for the 3D or 2D plot functions in the annotated *Mathematica* cell on the next page. In most uses of $\mathbf{L}_n(x)$, the order n indicates the eigenmode of a vibration, so it is usually small.

```
(* StruveL cell *)

(* Modified Struve Functions *)

(* Power series definition *)

SetOptions[NSum,AccuracyGoal->6,
 PrecisionGoal->6,WorkingPrecision->10,
 NSumExtraTerms->6,NSumTerms->6];
Off[NSum::precw];

LL[n_,z_] := If[ n == -1,
 If[ z == 0, 1/(Gamma[3/2]*Gamma[1/2]),
 1/(Gamma[3/2]*Gamma[1/2])+
  NSum[(z/2)^(2*k)/
  (Gamma[k+3/2]*Gamma[k+1/2]),
  {k,1,Infinity}]],
 If[ z == 0, 0, (z/2)^(n+1)*
 (1/(Gamma[3/2]*Gamma[n+3/2])+
  NSum[(z/2)^(2*k)/
  (Gamma[k+3/2]*Gamma[k+n+3/2]),
  {k,1,Infinity}])]]


(* 3D & 2D views *)

(* 3D defined for n fixed, x & y vary *)
LLZ3D[n_,ReOrIm_,Pnts_,LLRng_,ZTcks_] :=
 Plot3D[
  ReOrIm[LL[n,x+I*y]],{x,-4,4},{y,-3,3},
  PlotRange->{-LLRng,LLRng},
  PlotPoints->{Pnts,Pnts},
  Ticks->{{-2,0,2},{-2,0,2},
  {-ZTcks,0,ZTcks}},AxesLabel->{"x","y",None}]


(* 3D defined; real x  & n varying *)
LLXN3D[ReOrIm_,LLRng_,ZTcks_] :=
 Plot3D[
  ReOrIm[LL[n,x]],{x,-4,4},{n,-1,2},
  PlotRange->{-LLRng,LLRng},
 PlotPoints->{20,20},
  Ticks->{{-2,0,2},{-1,0,1,2},
   {-ZTcks,ZTcks}},
  AxesLabel->{"x","n",None}]


(* 2D views defined; x varying & n fixed *)
LLX2D[n_,LLRng_,YTcks_,Dsh_] :=
 Plot[LL[n,x],{x,0,4},
  PlotRange->{-LLRng,LLRng},PlotPoints->20,
```

This accuracy is sufficient for graphs of the modified Struve functions.

Define the modified Struve functions in terms of their power series expansion (15.1.8). Several special cases must be checked, since *Mathematica* thinks that $0^0 \neq 1$.

Define 3D views of modified Struve functions in the complex plane. Choose real or imaginary part, number of points in each direction, vertical scale, and position of vertical ticks.

Define 3D views of modified Struve functions in the x–n plane. Choose real or imaginary part, vertical scale, and position of vertical ticks.

Define 2D views; x varying, n fixed. Plot range, position of vertical ticks, and size of dashes are parameters.

```
    Ticks->{{0,2,4},{-YTcks,0,YTcks}},
    AxesLabel->{"x",None},
    PlotStyle->Dashing[{Dsh,Dsh}],
    DisplayFunction->Identity]

(* 2D views defined; n varying & x fixed *)

LLNN2D[x_,LLRng_,YTcks_,Dsh_] :=
 Plot[LL[n,x],{n,-1,3},
   PlotRange->{-LLRng,LLRng},PlotPoints->20,
   Ticks->{{-1,0,1,2,3},{-YTcks,0,YTcks}},
   AxesLabel->{"n",None},AxesOrigin->{-1,0},
   PlotStyle->Dashing[{Dsh,Dsh}],
   DisplayFunction->Identity]

(* Complex plane 3D views for    n = 0,1 *)

LLZ3D[0,Re,20,10,5]; LLZ3D[0,Im,20,10,5];
LLZ3D[1,Re,20,10,5]; LLZ3D[1,Im,20,10,5];


(* x and n;  3D views  *)

LLXN3D[Re,10,5]; LLXN3D[Im,10,5];


(* x with n fixed; 2D views *)

ShowRegX2D = Show[
 LLX2D[-1,10,5,0.01],LLX2D[ 1,10,5,0.02],
 LLX2D[ 3,10,5,0.06],
 DisplayFunction->$DisplayFunction];


(* n with x fixed; 2D views *)

ShowRegNN2D = Show[LLNN2D[1,6,3,0.06],
  LLNN2D[2,6,3,0.04],LLNN2D[3,6,3,0.02],
  DisplayFunction->$DisplayFunction];


(* Struve functions: Numerical *)

(* Re-set options for more terms
   for large  x  *)
SetOptions[NSum,AccuracyGoal->12,
  PrecisionGoal->12,WorkingPrecision->16,
  NSumExtraTerms->10,NSumTerms->40];
Off[NSum::precw];

LLJNTable = Table[{n,x,N[LL[n,x],10]},
  {x,1,3,2},{n,0,6,3}]

LLJN50 = Table[{n,x,N[LL[n,x],10]},
  {x,50,50,1},{n,0,6,3}]
```

StruveL cell
continued

Define 2D views with *n*
varying and *x* fixed. Plot
range, position of vertical
ticks, and size of dashes are
parameters.

Show Struve **L** functions in
the complex plane, as in
Figure 15.1.4.

Show Struve **L** against *x* and
n, real and imaginary parts,
as in Figure 15.1.5.

Show curves of **L** against *x*,
as in Figure 15.1.6 left.

Show curves of **L** against *n*,
as in Figure 15.1.6 right.

Change accuracy controls in
NSum to get at least 10-digit
accuracy in numerical
values.

Numerical test values to 10
digits, as in Table 15.1.2.

The `AngerJ` cell [15.2.2]

Execute this cell to produce Figures 15.2.1 and 15.2.2 for the Anger functions $J_\nu(z)$, such as the surface for $J_\nu(x)$ vs x and ν in the motif at right, where x varies from –4 to 4 on the front axis while on the side axis n varies from –2 to 2; more detail is given in Figure 15.2.2.

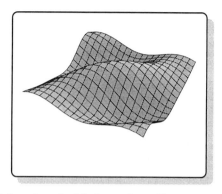

Exploring with `AngerJ`

Use a range of argument values appropriate to your application of $J_\nu(z)$, then explore this region graphically in more detail by modifying the argument range for the 3D or 2D plot functions in *Mathematica* cell `AngerJ`. Verify that when $\nu = n$, an integer, then $J_\nu(z)$ collapses to the regular cylindrical Bessel function, $J_n(z)$.

```
(* AngerJ cell *)


(* Anger Functions *)

(* Integral definition *)

Off[NIntegrate::precw];
SetOptions[NIntegrate,AccuracyGoal->6,
  PrecisionGoal->6,WorkingPrecision->16];

Anger[nu_,z_] := NIntegrate[
  Cos[nu*th - z*Sin[th]],
  {th,0,Pi}]/Pi


(* 3D & 2D views *)

(* 3D defined for nu fixed, x & y vary *)
AngerZ3D[nu_,ReOrIm_,Pnts_,Rng_,ZTcks_] :=
  Plot3D[
    ReOrIm[Anger[nu,x+I*y]],{x,-4,4},{y,-2,2},
    PlotRange->{-Rng,Rng},
    PlotPoints->{Pnts,Pnts},
    Ticks->{{-2,0,2},{-2,0,2},
    {-ZTcks,0,ZTcks}},AxesLabel->{"x","y",None}]


(* 3D defined; real x & nu varying *)
AngerXN3D[Rng_,ZTcks_] :=
  Plot3D[Anger[nu,x],{x,-4,4},{nu,-2,2},
    PlotRange->{-Rng,Rng},
    PlotPoints->{20,20},
    Ticks->{{-2,0,2},{-2,-1,0,1,2},
      {-ZTcks,ZTcks}},
    AxesLabel->{"x","nu",None}]
```

This accuracy is sufficient for graphs of the Anger functions.

Define Anger functions, **J**, in terms of integral (15.2.1).

Define 3D views of Anger functions in the complex plane. Choose real or imaginary part, number of points in each direction, vertical scale, and position of vertical ticks.

Define 3D views of Anger functions in the x–ν plane. Choose vertical scale and position of vertical ticks.

```
(* 2D views defined; x varying & nu fixed *)
AngerX2D[nu_,Rng_,YTcks_,Dsh_] :=
 Plot[Anger[nu,x],{x,0,4},
   PlotRange->{-Rng,Rng},PlotPoints->20,
   Ticks->{{0,2,4},{-YTcks,0,YTcks}},
   AxesLabel->{"x",None},
   PlotStyle->Dashing[{Dsh,Dsh}],
   DisplayFunction->Identity]

(* 2D views defined; nu varying & x fixed *)
AngerNN2D[x_,Rng_,YTcks_,Dsh_] :=
 Plot[Anger[nu,x],{nu,-2,2},
   PlotRange->{-Rng,Rng},PlotPoints->20,
   Ticks->{{-2,-1,0,1,2},{-YTcks,0,YTcks}},
   AxesLabel->{"nu",None},AxesOrigin->{0,0},
   PlotStyle->Dashing[{Dsh,Dsh}],
   DisplayFunction->Identity]

(* Complex plane 3D views:
    nu = -1/2, 1/2, 3/2  *)
AngerZ3D[-1/2,Re,20,2,1];
AngerZ3D[-1/2,Im,20,2,1];
AngerZ3D[ 1/2,Re,20,2,1];
AngerZ3D[ 1/2,Im,20,2,1];
AngerZ3D[ 3/2,Re,20,2,1];
AngerZ3D[ 3/2,Im,20,2,1];

(* x and nu;  3D views  *)
AngerXN3D[1,0.5];

(* x with nu fixed; 2D views *)
ShowRegX2D = Show[
 AngerX2D[-1/2,1,0.5,0.02],
 AngerX2D[ 1/2,1,0.5,0.04],
 AngerX2D[ 3/2,1,0.5,0.06],
 DisplayFunction->$DisplayFunction];

(* nu with x fixed; 2D views *)
ShowRegNN2D = Show[
 AngerNN2D[1,1,0.5,0.06],
 AngerNN2D[2,1,0.5,0.04],
 AngerNN2D[3,1,0.5,0.02],
 DisplayFunction->$DisplayFunction];

(* Anger functions: Numerical *)
(* Re-set options for more terms
    for large  x *)
SetOptions[NIntegrate,AccuracyGoal->10,
 PrecisionGoal->10,WorkingPrecision->16];
```

AngerJ cell
continued

Define 2D views; x varying, v fixed. Plot range, position of vertical ticks, and size of dashes are parameters.

Define 2D views with v varying and x fixed. Plot range, position of vertical ticks, and size of dashes are parameters.

Show Anger **J** functions in the complex plane, as in Figure 15.2.1.

Show Anger against x and v, as in Figure 15.2.2 left.

Show curves of **J** against x, as in Figure 15.2.2 right top.

Show curves of **J** against v, as in Figure 15.2.2 right bottom.

Change accuracy controls to get at least 10-digit accuracy in numerical values.

```
AngerJTable = Table[{nu,x,N[Anger[nu,x],10]},
  {x,1,3,2},{nu,0.3,2.3,1}]
AngerJ30 = Table[{nu,x,
 N[Anger[nu,x] - BesselJ[nu,x],10]},
 {x,30,30,1},{nu,0.3,2.3,1}]
```

Numerical test values to 10 digits, as in Table 15.2.1.

The WeberE cell [15.2.3]

By running this cell you will produce Figures 15.2.3 and 15.2.4 for the Weber functions $E_\nu(z)$. You can also visualize a surface such as $E_{5/2}(z)$ vs $z = x + iy$ in the motif at right, where x varies from –5 to 5 on the front axis while on the side axis y varies from –3 to 3.

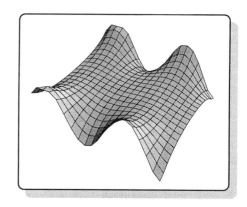

Exploring with WeberE

Choose a range of argument values appropriate to your use of $E_\nu(z)$, then explore this region graphically by modifying the argument range for the 3D or 2D plot functions in the *Mathematica* cell WeberE. For example, as x increases, how rapid is the approach to the negative of the irregular spherical Bessel function $Y_\nu(z)$, as predicted by (15.2.9)?

```
(* WeberE cell *)

(* Weber Functions *)

(* Integral definition *)

Off[NIntegrate::precw];
SetOptions[NIntegrate,AccuracyGoal->6,
 PrecisionGoal->6,WorkingPrecision->16];

Weber[nu_,z_] := NIntegrate[
 Sin[nu*th - z*Sin[th]],
 {th,0,Pi}]/Pi

(* 3D & 2D views *)

(* 3D defined for nu fixed, x & y vary *)
WeberZ3D[nu_,ReOrIm_,Pnts_,Rng_,ZTcks_] :=
 Plot3D[
   ReOrIm[Weber[nu,x+I*y]],{x,-4,4},{y,-2,2},
   PlotRange->{-Rng,Rng},
   PlotPoints->{Pnts,Pnts},
   Ticks->{{-2,0,2},{-2,0,2},
   {-ZTcks,0,ZTcks}},AxesLabel->{"x","y",None}]
```

This accuracy is sufficient for graphs of the Weber functions.

Define Weber functions, **E**, in terms of integral (15.2.2).

Define 3D views of Weber functions in the complex plane. Choose real or imaginary part, number of points in each direction, vertical scale, and position of vertical ticks.

```
(* 3D defined; real x  & nu varying *)

WeberXN3D[Rng_,ZTcks_] :=
 Plot3D[Weber[nu,x],{x,-4,4},{nu,-2,2},
  PlotRange->{-Rng,Rng},
  PlotPoints->{20,20},
  Ticks->{{-2,0,2},{-2,-1,0,1,2},
   {-ZTcks,ZTcks}},
  AxesLabel->{"x","nu",None}]
```

Define 3D views of Weber functions in the *x*–*v* plane. Choose vertical scale and position of vertical ticks.

```
(* 2D views defined; x varying & nu fixed *)

WeberX2D[nu_,Rng_,YTcks_,Dsh_] :=
 Plot[Weber[nu,x],{x,0,4},
  PlotRange->{-Rng,Rng},PlotPoints->20,
  Ticks->{{0,2,4},{-YTcks,0,YTcks}},
  AxesLabel->{"x",None},
  PlotStyle->Dashing[{Dsh,Dsh}],
  DisplayFunction->Identity]
```

Define 2D views; *x* varying, *v* fixed. Plot range, position of vertical ticks, and size of dashes are parameters.

```
(* 2D views defined; nu varying & x fixed *)

WeberNN2D[x_,Rng_,YTcks_,Dsh_] :=
 Plot[Weber[nu,x],{nu,-2,2},
  PlotRange->{-Rng,Rng},PlotPoints->20,
  Ticks->{{-2,-1,0,1,2},{-YTcks,0,YTcks}},
  AxesLabel->{"nu",None},AxesOrigin->{0,0},
  PlotStyle->Dashing[{Dsh,Dsh}],
  DisplayFunction->Identity]
```

Define 2D views with *v* varying and *x* fixed. Plot range, position of vertical ticks, and size of dashes are parameters.

```
(* Complex plane 3D views:
   nu = -1/2, 1/2, 3/2  *)

WeberZ3D[-1/2,Re,20,2,1];
WeberZ3D[-1/2,Im,20,2,1];
WeberZ3D[ 1/2,Re,20,2,1];
WeberZ3D[ 1/2,Im,20,2,1];
WeberZ3D[ 3/2,Re,20,2,1];
WeberZ3D[ 3/2,Im,20,2,1];
```

Show Weber **E** functions in the complex plane, as in Figure 15.2.3.

```
(* x and nu;  3D views  *)

WeberXN3D[1,0.5];
```

Show Weber against *x* and *v*, as in Figure 15.2.3 left.

```
(* x with nu fixed; 2D views *)

ShowRegX2D = Show[
 WeberX2D[-1/2,1,0.5,0.02],
 WeberX2D[ 1/2,1,0.5,0.04],
 WeberX2D[ 3/2,1,0.5,0.06],
 DisplayFunction->$DisplayFunction];
```

Show curves of **E** against *x*, as in Figure 15.2.3 right top.

```
(* nu with x fixed; 2D views *)

ShowRegNN2D = Show[
  WeberNN2D[1,1,0.5,0.06],
  WeberNN2D[2,1,0.5,0.04],
  WeberNN2D[3,1,0.5,0.02],
  DisplayFunction->$DisplayFunction];

(* Weber functions: Numerical *)

(* Re-set options for more terms
    for large  x  *)
SetOptions[NIntegrate,AccuracyGoal->10,
  PrecisionGoal->10,WorkingPrecision->16];

WeberETable = Table[{nu,x,N[Weber[nu,x],10]},
  {x,1,3,2},{nu,0.3,2.3,1}]
WeberE30 = Table[{nu,x,
  N[Weber[nu,x] + BesselY[nu,x],10]},
  {x,30,30,1},{nu,0.3,2.3,1}]
```

Show curves of **E** against *v*, as in Figure 15.2.3 right bottom.

Change accuracy controls to get at least 10-digit accuracy in numerical values.

Numerical test values to 10 digits, as in Table 15.2.2.

20.16 HYPERGEOMETRIC FUNCTIONS AND COULOMB WAVE FUNCTIONS

Notebook and Cells
Notebook `Hyprgmtrc`; cells `Hyper-geometric2F1`, `Confluent1F1`, `ConfluentU`, `CoulombF`, and `Cou-lombG`.

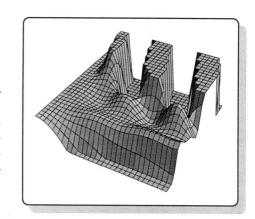

Run the cells in this notebook to visualize the hypergeometric and Coulomb wave functions in Chapter 16. These cells also generate numerical values for checking your C implementation of the functions. For this purpose, run the *Mathematica* cells in conjunction with the C driver programs whose annotated listings are in Section 21.16.

The `Hypergeometric2F1` cell [16.1]
Execute this cell to see Figures 16.1.1–16.1.3 for the hypergeometric functions, $_2F_1$, such as the surface for $b = 0.5$ and $c = 3.5$ with x and a varying seen in the motif at right, where x varies from −0.95 to 0.95 on the front axis while on the side axis a varies from −2 to 2.

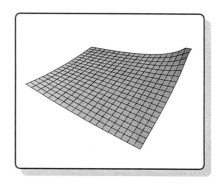

Exploring with `Hypergeometric2F1`
Choose a range of argument values with $|x| < 1$ and parameters a, b, and c appropriate to your use of $_2F_1$, then explore this region graphically

in more detail by modifying the argument range for the 3D or 2D plot functions in the following annotated *Mathematica* cell.

```
(* Hypergeometric2F1 cell *)

(* Hypergeometric function, 2F1 *)


(* 3D views defined;
   x & a vary, b & c fixed *)
Hyp21BBCC3D[b_,c_,RngLow_,RngHi_] :=
 Plot3D[Hypergeometric2F1[a,b,c,x],
 {x,-0.95,0.95},{a,-2,2},
 PlotRange->{RngLow,RngHi},PlotPoints->{20,20},
 Ticks->{{-0.5,0.5},{-2,0,2},
 {0,1,2,3}},AxesLabel->{"x","a",None}];

(* 3D views for b = 0.5, 1.5, c = 3.5, 4.5 *)
Hyp21BBCC3D[0.5,3.5,0.5,1.5];
Hyp21BBCC3D[0.5,4.5,0.5,1.5];
Hyp21BBCC3D[1.5,3.5,0.0,3.5];
Hyp21BBCC3D[1.5,4.5,0.0,3.5];


(* 2D views defined; x varying *)
Hyp21in2D[a_,b_,c_,Dsh_] :=
 Plot[Hypergeometric2F1[a,b,c,x],
  {x,-0.95,0.95},
  PlotRange->{0,2},PlotPoints->30,
  Ticks->{{-0.5,0,0.5},{0,2}},
  AxesOrigin->{0,0},AxesLabel->{"x",None},
  PlotStyle->Dashing[{Dsh,Dsh}],
  DisplayFunction->Identity]

ShowHyp21in2D[b_,c_] := Show[
 Hyp21in2D[-1.5,b,c,0.01],
 Hyp21in2D[-0.5,b,c,0.02],
 Hyp21in2D[ 0.5,b,c,0.04],
 Hyp21in2D[ 1.5,b,c,0.08],
 DisplayFunction->$DisplayFunction];

(* 2D for b = 0.5,1.5, c = 3.5,4.5 *)
ShowHyp21in2D[0.5,3.5];
ShowHyp21in2D[0.5,4.5];
ShowHyp21in2D[1.5,3.5];
ShowHyp21in2D[1.5,4.5];


(* 2F1 functions: Numerical *)
Hyp21Table[a_,b_] :=
 Table[{c,x,N[Hypergeometric2F1[a,b,c,x],
  10]},{c,0.5,1.5,0.5},{x,0.3,0.9,0.3}]

Hyp21Table[-2.0,-1.0]
Hyp21Table[ 1.0, 1.0]
```

Define 3D views of hypergeometric functions in *x*–*a* plane with *b* and *c* fixed. Choose lower and upper range for plot.

Show 3D views as in Figure 16.1.1.

Define 2D views with *x* varying and *a*, *b*, *c* fixed. Size of dashes is a parameter.

Define composite views with *b* and *c* fixed.

Show 2D views as in Figure 16.1.2.

Define table of values for *a* and *b* fixed.

Numerical values as in Tables 16.1.1, 16.1.2.

The `Confluent1F1` cell [16.2.1]

Run this cell to see Figures 16.2.1–16.2.3 for the confluent hypergeometric functions, $_1F_1$, such as the surface for $b = -1.5$ with x and a varying that is shown in the motif at right, where x varies from 0 to 8 on the front axis while on the side axis a varies from -3 to 1. The function range shown on the vertical scale is $[-40, 40]$, just twice that of the graphic in the top left corner of Figure 16.2.1.

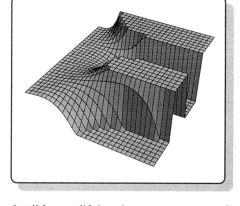

Exploring with `Confluent1F1`

Choose a range of argument values and parameters a and b appropriate to your use of $_1F_1$, then explore this region graphically in detail by modifying the argument range for the 3D and 2D plot functions in the following *Mathematica* cell.

```
(* Confluent1F1 cell *)

(* Confluent hypergeometric function, 1F1 *)

(* 3D views defined; x & a vary, b fixed *)
Hyp11BB3D[b_,RngLow_,RngHi_] :=
 Plot3D[Hypergeometric1F1[a,b,x],
 {x,0,8},{a,-3,1},
 PlotRange->{RngLow,RngHi},PlotPoints->{25,25},
 Ticks->{{0,2,4,6},{-3,-1,1},
 {-20,-10,0,10,20}},AxesLabel->{"x","a",None}];
```
Define 3D views of the confluent hypergeometric functions in *x–a* plane with *b* fixed. Choose lower and upper range for plot.

```
(* 3D views defined; 2Ix & a vary, b fixed *)
Hyp11ZZ3D[b_,ReOrIm_,RngLow_,RngHi_] :=
 Plot3D[ReOrIm[Hypergeometric1F1[a,b,2*I*x]],
 {x,0,8},{a,-2,2},
 PlotRange->{RngLow,RngHi},PlotPoints->{25,25},
 Ticks->{{0,2,4,6},{-2,0,2},
 {-20,0,20}},AxesLabel->{"x","a",None}];
```
Define 3D views of the confluent hypergeometric functions in *x–a* plane with *b* fixed. Argument is *2ix*. Choose lower and upper range for surface.

```
(*  3D views for b = -1.5, -0.5, 0.5, 1.5  *)
Hyp11BB3D[-1.5,-20,20];
Hyp11BB3D[-0.5,-20,20];
Hyp11BB3D[ 0.5,-20,20];
Hyp11BB3D[ 1.5,-20,20];
```
Show 3D views as in Figure 16.2.1.

```
(*  3D views for b = -1.5, -0.5, 0.5  *)
Hyp11ZZ3D[-1.5,Re,-30,30];
Hyp11ZZ3D[-1.5,Im,-30,30];
Hyp11ZZ3D[-0.5,Re,-30,30];
Hyp11ZZ3D[-0.5,Im,-30,30];
Hyp11ZZ3D[ 0.5,Re,-30,30];
Hyp11ZZ3D[ 0.5,Im,-30,30];
```
Show 3D views as in Figure 16.2.2.

```
(* 2D views defined; x varying *)

Hyp11Conf2D[a_,b_,Dsh_] :=
 Plot[Hypergeometric1F1[a,b,x],
  {x,-6.0,6.0},
  PlotRange->{-20,20},PlotPoints->30,
  Ticks->{{-6,-4,-2,0,2,4,6},{-10,10}},
  AxesOrigin->{0,0},AxesLabel->{"x",None},
  PlotStyle->Dashing[{Dsh,Dsh}],
  DisplayFunction->Identity]

ShowHyp11Conf2D[b_] := Show[
 Hyp11Conf2D[-1.5,b,0.01],
 Hyp11Conf2D[-0.5,b,0.02],
 Hyp11Conf2D[ 0.5,b,0.04],
 Hyp11Conf2D[ 1.5,b,0.08],
 DisplayFunction->$DisplayFunction];

(* 2D for b = -1.5,-0.5,0.5,1.5 *)
ShowHyp11Conf2D[-1.5];
ShowHyp11Conf2D[-0.5];
ShowHyp11Conf2D[ 0.5];
ShowHyp11Conf2D[ 1.5];

(* 1F1 functions: Numerical *)

Hyp11Table[a_] :=
 Table[{b,x,N[Hypergeometric1F1[a,b,x],10]},
  {b,0.5,1.5,0.5},{x,1,5,2}]

Hyp11Table[2.0]

Hyp11AsymTable[a_] :=
 Table[{b,x,N[Hypergeometric1F1[a,b,x],10]},
  {b,0.5,1.5,0.5},{x,50,100,50}]

Hyp11AsymTable[2.0]
```

Confluent1F1 cell
continued

Define 2D views with *x*
varying, *a* and *b* fixed. Size
of dashes is a parameter.

Define composite views with
b fixed.

Show 2D views as in Figure
16.2.3.

Define table of values for *a*
and *b* fixed.

Numerical values as in Table
16.2.1 for small *x* range then
for large *x* values.

The ConfluentU cell [16.2.2]

Run this cell to see Figures 16.2.4–16.2.6
for the irregular confluent hypergeometric
functions, *U*, such as the surface for *x* = 4.0
with *a* and *b* varying that is shown in the
motif at right, where *a* varies from –5 to –2
on the front axis while on the side axis *b*
varies from –2 to 3. The function range on
the vertical scale is [–200, 200], twice that
of the graphic in the bottom right corner of
Figure 16.2.5.

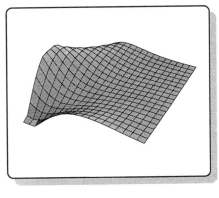

Exploring with ConfluentU

Choose a range of arguments and para-
meters *a* and *b* appropriate to your use of *U*, then explore this region graphically in detail

by modifying the argument range for the 3D and 2D plot functions in the following annotated *Mathematica* cell.

```
(* ConfluentU cell *)

(* Confluent hypergeometric function, U *)

(* 3D views defined; x & a vary, b fixed *)

HypUUBB3D[b_,RngLow_,RngHi_] :=
  Plot3D[HypergeometricU[a,b,x],
  {x,0.1,6},{a,-3,0},
  PlotRange->{RngLow,RngHi},PlotPoints->{20,20},
  Ticks->{{0,2,4,6},{-2,0},
  {-20,0,20}},AxesLabel->{"x","a",None}];

(*  3D views for b = -1.5, -0.5, 0.5, 1.5  *)
HypUUBB3D[-1.5,-20,40];
HypUUBB3D[-0.5,-20,40];
HypUUBB3D[ 0.5,-20,30];
HypUUBB3D[ 1.5,-20,30];

(* 3D views defined; a & b vary, x fixed *)

HypUUXX3D[x_,RngLow_,RngHi_] :=
  Plot3D[HypergeometricU[a,b,x],
  {a,-5,-2},{b,-2,3},
  PlotRange->{RngLow,RngHi},PlotPoints->{20,20},
  Ticks->{{-5,-3},{-2,0,2},{0,50,100}},
  AxesLabel->{"a","b",None}];

(*  3D views for x = 1,2,3,4 *)
HypUUXX3D[1,-50,150];
HypUUXX3D[2,-50,150];
HypUUXX3D[3,-50,150];
HypUUXX3D[4,-50,150];

(* 2D views defined; x varying *)

HypUUConf2D[a_,b_,Dsh_] :=
  Plot[HypergeometricU[a,b,x],
    {x,0.1,6.0},
    PlotRange->{-15,15},PlotPoints->20,
    Ticks->{{0,2,4,6},{-10,10}},
    AxesOrigin->{0,0},AxesLabel->{"x",None},
    PlotStyle->Dashing[{Dsh,Dsh}],
    DisplayFunction->Identity]

ShowHypUUConf2D[b_] := Show[
  HypUUConf2D[-3.0,b,0.01],
  HypUUConf2D[-2.0,b,0.02],
  HypUUConf2D[-1.0,b,0.04],
  DisplayFunction->$DisplayFunction];
```

Define 3D views of the U confluent hypergeometric functions in x–a plane with b fixed. Choose lower and upper range for plot.

Show 3D views as in Figure 16.2.4.

Define 3D views of the U confluent hypergeometric functions in a–b plane with x fixed. Choose lower and upper range for surface.

Show 3D views as in Figure 16.2.5.

Define 2D views with x varying, a and b fixed. Size of dashes is a parameter.

Define composite views with b fixed.

```
(* 2D for b = -1.5,-0.5,0.5,1.5 *)
ShowHypUUConf2D[-1.5];
ShowHypUUConf2D[-0.5];
ShowHypUUConf2D[ 0.5];
ShowHypUUConf2D[ 1.5];

(* U functions: Numerical *)

HypUUTable[a_,x_] :=
  Table[{a,x,b,N[HypergeometricU[a,b,x],10]},
  {b,3.0,5.0,1.0}]

HypUUTable[2.0,  0.1]
HypUUTable[2.0,  5.0]
HypUUTable[2.0,60.0]
```

ConfluentU cell
continued

Show 2D views as in Figure 16.2.6.

Define table of values for *a* fixed.

Numerical values as in Table 16.2.3.

The CoulombF cell [16.3.1]

Run this cell to make Figures 16.3.1–16.3.3 for the regular Coulomb wave functions, F_L, and their derivatives with respect to ρ, F'_L, such as the surface for $L = 4$ with ρ and η varying that is shown in the motif at right. Compare this with the graphic in the top right corner of Figure 16.3.1, which has the same variation of ρ on the front axis—0 to 15—and of η on the side axis—0 to 2—but it has angular momentum $L = 2$. Notice how for $L = 4$ the wave function is much smaller at the same ρ value, which is interpreted in wave mechanics as an increasing centripetal barrier.

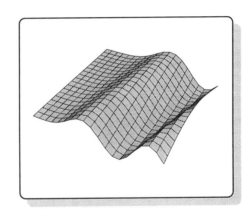

Exploring with CoulombF

Choose a range of L, η, and ρ appropriate to the quantum mechanics of the Coulomb interaction that you are investigating, then explore this region graphically in detail by modifying the argument range for the 3D plot functions in the following *Mathematica* cell. Note that you can vary L, η, and ρ in any combinations, and that you also see views of the derivative function F'_L.

```
(* CoulombF cell *)

(* Regular Coulomb wave function, F,
   and its derivative, F'  *)

(* Definitions in terms of Gamma & 1F1 *)
(* Coulomb factor *)
Clmb[L_,eta_] := 2^L*Exp[-Pi*eta/2]*
  Abs[Gamma[L+1+I*eta]]/Gamma[2*L+2]
```

Define Coulomb factor by (16.3.3).

```
(* F function *)

ClmbFF[L_,eta_,rho_] := Clmb[L,eta]*
  rho^(L+1)*Exp[-I*rho]*
  Hypergeometric1F1[L+1-I*eta,2*L+2,2*I*rho]
```

Define regular Coulomb function by (16.3.2).

```
(* F derivative function *)

ClmbFFPrime[L_,eta_,rho_] :=
  ((L+1)/rho+eta/(L+1))*ClmbFF[L,eta,rho]-
  Sqrt[1+(eta/(L+1))^2]*ClmbFF[L+1,eta,rho]
```

Define derivative of regular Coulomb function by (16.3.5).

```
(* 3D views defined; eta & rho vary, L fixed *)

ClmbFFLL3D[ForFP_,L_,RngLow_,RngHi_] :=
  Plot3D[Re[ForFP[L,eta,rho]],
  {rho,0.1,15},{eta,0,2},
  PlotRange->{RngLow,RngHi},PlotPoints->{20,20},
  Ticks->{{3,6,9,12},{0.5,1.5},{-1,0,1}},
  AxesLabel->{"rho","eta",None}];
```

Define 3D views of the function or its derivative with *L* fixed. Choose lower and upper range for plot.

```
(* 3D views; rho & L vary, eta fixed *)

ClmbFFEta3D[ForFP_,eta_,RngLow_,RngHi_] :=
  Plot3D[Re[ForFP[L,eta,rho]],
  {rho,0.1,15},{L,0,3},
  PlotRange->{RngLow,RngHi},PlotPoints->{20,20},
  Ticks->{{3,6,9,12},{0,1,2,3},{-1,0,1}},
  AxesLabel->{"rho","L",None}];
```

Define 3D views of the function or its derivative with η fixed. Choose lower and upper range for plot.

```
(* 3D views; eta & L vary, rho fixed *)

ClmbFFRho3D[ForFP_,rho_,RngLow_,RngHi_] :=
  Plot3D[Re[ForFP[L,eta,rho]],
  {eta,0,2},{L,0,3},
  PlotRange->{RngLow,RngHi},PlotPoints->{20,20},
  Ticks->{{0.5,1.5},{0,1,2,3},{-1,0,1}},
  AxesLabel->{"eta","L",None}];
```

Define 3D views of the function or its derivative with ρ fixed. Choose lower and upper range for plot.

```
(*  3D views for L = 0,2; F then F' *)

ClmbFFLL3D[ClmbFF,0,-2,2];
ClmbFFLL3D[ClmbFF,2,-2,2];
ClmbFFLL3D[ClmbFFPrime,0,-2,2];
ClmbFFLL3D[ClmbFFPrime,2,-2,2];
```

Show 3D views as in Figure 16.3.1 top then as in Figure 16.3.1 bottom.

```
(*  3D views for eta = 0.5,1.5; F then F' *)

ClmbFFEta3D[ClmbFF,0.5,-2,2];
ClmbFFEta3D[ClmbFF,1.5,-2,2];
ClmbFFEta3D[ClmbFFPrime,0.5,-2,2];
ClmbFFEta3D[ClmbFFPrime,1.5,-2,2];
```

Show 3D views as in Figure 16.3.2 top then as in Figure 16.3.2 bottom.

```
(*  3D views for rho = 5.0,10.0; F then F' *)

ClmbFFRho3D[ClmbFF,5.0,-2,2];
ClmbFFRho3D[ClmbFF,10.0,-2,2];
ClmbFFRho3D[ClmbFFPrime,5.0,-2,2];
ClmbFFRho3D[ClmbFFPrime,10.0,-2,2];
```

Show 3D views as in Figure 16.3.3 top then as in Figure 16.3.3 bottom.

```
(* Coulomb F & F' functions: Numerical *)
ClmbFFTable[ForFP_,eta_] :=
  Table[{L,rho,N[Re[ForFP[L,eta,rho]],7]},
  {L,0,8,4},{rho,2,16,8}]
ClmbFFTable[ClmbFF,0.0]
ClmbFFTable[ClmbFF,2.5]
ClmbFFTable[ClmbFF,5.0]
ClmbFFTable[ClmbFFPrime,0.0]
ClmbFFTable[ClmbFFPrime,2.5]
ClmbFFTable[ClmbFFPrime,5.0]
```

`CoulombF` cell
continued

Define table of values for η fixed.

Numerical values as in Tables 16.3.1–16.3.3.

The `CoulombG` cell [16.3.2]

Run this cell to make Figures 16.3.4–16.3.6 for irregular Coulomb wave functions, G_L, and their derivatives with respect to ρ, G_L', like the surface for $L = 4$ with ρ and η varying. shown in the motif. Compare this with the graphic in the top right of Figure 16.3.4, which has the same variation of ρ on the front axis and of η on the side axis but has $L = 2$. For $L = 4$ the wave function is larger for the same ρ and η because the singular behavior near $\rho = 0$ increases as L increases.

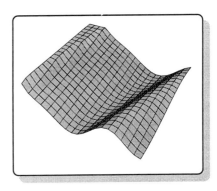

Exploring with `CoulombG`

Select L, η, and ρ appropriate to the Coulomb interaction you are investigating, then explore this region in detail by modifying the arguments for the 3D plot functions in the *Mathematica* cell. You can vary L, η, and ρ in any combination, and you also get views of the derivative function G_L'.

```
(* CoulombG cell *)
(* Irregular Coulomb wave function, G,
   and its derivative, G' *)
(* Definition in terms of integral *)
(* Coulomb factor *)
Clmb[L_,eta_] := 2^L*Exp[-Pi*eta/2]*
  Abs[Gamma[L+1+I*eta]]/Gamma[2*L+2]
(* G function *)
Off[NIntegrate::precw];
ClmbGG[L_,eta_,rho_,Prcsn_] :=
  (rho^(L+1)/
  (Exp[Pi*eta]*(2*L+1)!*Clmb[L,eta]))*
  NIntegrate[
  -(4/((Exp[t]+Exp[-t])^2))^(L+1)*
  Sin[rho*Tanh[t]-2*eta*t]
  +((1+t^2)^L)*Exp[-rho*t+2*eta*ArcTan[t]],
  {t,0,40},MaxRecursion->8,PrecisionGoal->Prcsn]
```

Define the Coulomb factor by (16.3.3).

Define the irregular Coulomb function by the integral (16.3.11); precision is a parameter (4 for graphics, 10 for numerics).

```
(* G derivative function *)
ClmbGGPrime[L_,eta_,rho_,Prcsn_] :=
  ((L+1)/rho+eta/(L+1))*ClmbGG[L,eta,rho,Prcsn]-
  Sqrt[1+(eta/(L+1))^2]*ClmbGG[L+1,eta,rho,Prcsn]
```

Define derivative of regular Coulomb function by (16.3.14).

```
(* 3D views defined; eta & rho vary, L fixed *)
(* 4-digit precision for graphics *)
ClmbGGLL3D[GorGP_,L_,RngLow_,RngHi_] :=
  Plot3D[Re[GorGP[L,eta,rho,4]],{rho,3,15},{eta,0,2},
  PlotRange->{RngLow,RngHi},PlotPoints->{20,20},
  Ticks->{{3,6,9,12},{0.5,1.5},{-1,0,1}},
  AxesLabel->{"rho","eta",None}];
```

Define 3D views of the function or its derivative with L fixed. Choose lower and upper plot range.

```
(* 3D views; rho & L vary, eta fixed *)
ClmbGGEta3D[GorGP_,eta_,RngLow_,RngHi_] :=
  Plot3D[Re[GorGP[L,eta,rho,4]],{rho,3,15},{L,0,3},
  PlotRange->{RngLow,RngHi},PlotPoints->{20,20},
  Ticks->{{3,6,9,12},{0,1,2,3},{-1,0,1}},
  AxesLabel->{"rho","L",None}];
```

Define 3D views of the function or its derivative with η fixed. Choose lower and upper plot range.

```
(* 3D views; eta & L vary, rho fixed *)
ClmbGGRho3D[GorGP_,rho_,RngLow_,RngHi_] :=
  Plot3D[Re[GorGP[L,eta,rho,4]],{eta,0,2},{L,0,3},
  PlotRange->{RngLow,RngHi},PlotPoints->{20,20},
  Ticks->{{0.5,1.5},{0,1,2,3},{-1,0,1}},
  AxesLabel->{"eta","L",None}];
```

Define 3D views of the function or its derivative with ρ fixed. Choose lower and upper plot range.

```
(* 3D views for L = 0, 2; G then G' *)
ClmbGGLL3D[ClmbGG,0,-2,3];
ClmbGGLL3D[ClmbGG,2,-2,3];
ClmbGGLL3D[ClmbGGPrime,0,-2,3];
ClmbGGLL3D[ClmbGGPrime,2,-2,3];
```

Show 3D views as in Figure 16.3.4 top then as in Figure 16.3.4 bottom.

```
(* 3D views for eta = 0.5, 1.5; G then G' *)
ClmbGGEta3D[ClmbGG,0.5,-2,3];
ClmbGGEta3D[ClmbGG,1.5,-2,3];
ClmbGGEta3D[ClmbGGPrime,0.5,-2,3];
ClmbGGEta3D[ClmbGGPrime,1.5,-2,3];
```

Show 3D views as in Figure 16.3.5 top then as in Figure 16.3.5 bottom.

```
(* 3D views for rho = 5.0, 10.0; G then G' *)
ClmbGGRho3D[ClmbGG,5.0,-2,3];
ClmbGGRho3D[ClmbGG,10.0,-2,3];
ClmbGGRho3D[ClmbGGPrime,5.0,-2,3];
ClmbGGRho3D[ClmbGGPrime,10.0,-2,3];
```

Show 3D views as in Figure 16.3.6 top then as in Figure 16.3.6 bottom.

```
(* Coulomb G & G' functions: Numerical *)
ClmbGGTable[GorGP_,eta_] :=
  Table[{L,rho,N[Re[GorGP[L,eta,rho,10]],7]},
  {L,0,8,4},{rho,2,18,8}]
ClmbGGTable[ClmbGG,0.0]
ClmbGGTable[ClmbGG,2.5]
ClmbGGTable[ClmbGG,5.0]
ClmbGGTable[ClmbGGPrime,0.0]
ClmbGGTable[ClmbGGPrime,2.5]
ClmbGGTable[ClmbGGPrime,5.0]
```

Define table of values for η fixed.

Numerical values as in Tables 16.3.7–16.3.12.

20.17   ELLIPTIC INTEGRALS AND ELLIPTIC FUNCTIONS

Notebook and Cells

Notebook `Elliptics`; cells `Ellip-ticF`, `Landen`, `EllipticE`, `Jaco-biZeta`, `HeumanLambda`, `Ellip-ticPi`, `JacobiSN`, `JacobiCN`, `JacobiDN`, `Theta`, and `LogDer-Theta`.

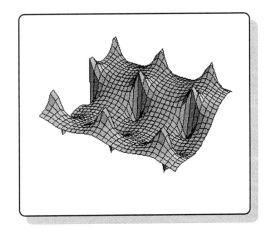

Run the cells in this notebook to visualize the behavior of the elliptic integrals, Jacobi elliptic functions, and auxiliary functions (Jacobi zeta, Heuman lambda, theta, and theta logarithmic derivative) described in Chapter 17. These cells also generate numerical values that are useful when checking your implementation of the functions in C. For this purpose, run the *Mathematica* cells in conjunction with the C driver programs whose annotated listings are given in Section 21.17.

The `EllipticF` cell [17.2.1]
Execute this cell to see Figures 17.2.1 and 17.2.2 for the elliptic integrals of the first kind, *F* and *K*, whose C function uses the Landen transform (AGM) scheme illustrated in the motif at right.

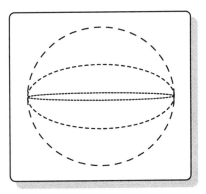

Exploring with `EllipticF`
If you need values of the elliptic integral for ϕ near $\pi/2$ and m close to unity, where the integral diverges, explore this region graphically in detail by modifying the argument range for the 3D and 2D plot functions in the following annotated *Mathematica* cell. For such arguments, formulas 17.4.13 and 17.4.14 in Abramowitz and Stegun may be useful.

```
(* EllipticF cell *)

(* Incomplete elliptic integral of the first
   kind *)

(* 3D views defined; complex  phi  varying *)

EllptcFFzz3D[m_,ReOrIm_,RngLow_,RngHi_] :=
  Plot3D[ReOrIm[EllipticF[phiX+I*phiY,m]],
  {phiX,0,Pi},{phiY,-2,2},
  PlotRange->{RngLow,RngHi},PlotPoints->{20,20},
```

Define 3D surface plot in terms of elliptic integral of the first kind.

```
Ticks->{{0.5,1.5,2.5},{-1,0,1},
 {-6,-3,0,3,6,9}},
 AxesLabel->{"phiX","phiY",None}];
(* 2D views defined; phi varying & m fixed *)

EllptcFphi2D[m_,Dsh_] :=
 Plot[EllipticF[phi,m],{phi,0,Pi/2},
   PlotRange->All,PlotPoints->30,
   Ticks->{{0,0.5,1.0,1.5},{0,1,2,3}},
   AxesLabel->{"phi",None},
   PlotStyle->Dashing[{Dsh,Dsh}],
   DisplayFunction->Identity]


(* 2D views defined; m varying & phi fixed *)
Off[Plot::plnr];
EllptcFmm2D[phi_,Dsh_] :=
 Plot[EllipticF[phi,m],{m,0,1},
   PlotRange->{0,3},PlotPoints->30,
   Ticks->{{0,0.2,0.4,0.6,0.8,1},{0,1,2,3}},
   AxesLabel->{"m",None},AxesOrigin->{0,0},
   PlotStyle->Dashing[{Dsh,Dsh}],
   DisplayFunction->Identity]

(* 3D views for m = 0.99. 1.01 *)

EllptcFFzz3D[0.99,Re,0,9];
EllptcFFzz3D[0.99,Im,-6,3];
EllptcFFzz3D[1.01,Re,0,9];
EllptcFFzz3D[1.01,Im,-6,3];


(* 3D views; real phi  & m varying *)

EllptcFF3D =
 Plot3D[EllipticF[phi,m],{phi,0,Pi/2},{m,0,1},
   PlotRange->{0,2},PlotPoints->{20,20},
   Ticks->{{0.5,1.0,1.5},{0,0.25,0.75,1},
   {0,1,2}},AxesLabel->{"phi","m",None}];


(* phi with m fixed; 2D views *)

ShowEllptcFphi2D = Show[
 EllptcFphi2D[0,0.01],EllptcFphi2D[0.5,0.02],
 EllptcFphi2D[0.75,0.04],EllptcFphi2D[0.9,0.08],
 DisplayFunction->$DisplayFunction];


(* m with phi fixed; 2D views *)

ShowEllptcFmm2D = Show[
 EllptcFmm2D[Pi/8,0.01],EllptcFmm2D[Pi/4,0.02],
 EllptcFmm2D[3*Pi/8,0.04],
 EllptcFmm2D[Pi/2,0.08],
 DisplayFunction->$DisplayFunction];
```

Define 2D views with ϕ varying and m fixed. Dash size is a parameter.

Define 2D views with m varying and ϕ fixed. Dash size is a parameter.

Show surfaces as real and imaginary parts, as in Figure 17.2.1.

Show surface as in Figure 17.2.2 left.

Show curves with m fixed, Figure 17.2.2 right top.

Show curves with ϕ fixed, Figure 17.2.2 right bottom.

```
(* Elliptic F & K  functions: Numerical *)


EllptcFTable =
  Table[{phi,m,N[EllipticF[phi,m],10]},
  {phi,0.5,2.0,0.5},{m,0.25,0.75,0.25}]


EllptcKKTable = Table[{m,N[EllipticK[m],10]},
  {m,0.25,0.75,0.25}]
```

EllipticF cell
continued

Test values to 10 digits, as in
Table 17.2.2.

The Landen cell [17.2.1]
Use this cell to make ellipses like those
in Figure 17.2.3 that show successive
stages of the Landen transform (Arithmetic Geometric Mean) scheme illustrated in the motif at right for the elliptic
integral parameter $m = 0.9975$.

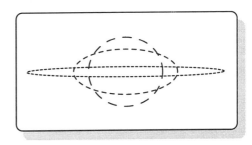

Exploring with Landen
In the following annotated *Mathematica*
cell use different initial choices of the ellipticity parameter m, then see how quickly successive Landen transforms convert the ellipses into shapes that are indistinguishable from
circles. For this, you will need to generalize the program, which is written specifically for
the case shown in Figure 17.2.3, namely $m = 0.9975$.

Since humans are very used to seeing ellipses and interpreting them as projections of
circles, it would be interesting for you to devise a graphic that displays on a suitable scale
the fractional difference between the transformed ellipse and a circle.

```
(* Landen Transform cell *)


Ellipse[an_,bn_,Dsh_] :=
  Show[Graphics[{AbsoluteThickness[1],
  Dashing[{Dsh,Dsh}],Circle[{0,0},{an,bn}]}],
  AspectRatio->Automatic,
  DisplayFunction->Identity]


LanEllipse = Show[Ellipse[1,0.05,0.01],
  Ellipse[0.525,0.2236,0.02],
  Ellipse[0.3743,0.3426,0.04],
  DisplayFunction->$DisplayFunction];


LanEllipseScale = Show[Ellipse[1,0.05,0.01],
  Ellipse[1,0.4259,0.02],
  Ellipse[1,0.9153,0.04],
  DisplayFunction->$DisplayFunction];
```

Define ellipse in terms of
major and minor axes. Dash
size is a parameter.

Show ellipse with $a = 1$,
$b = 0.05$, then succesive a
and b for descending transform.

Show scaled ellipse, Figure
17.2.3 right.

The EllipticE cell [17.2.2]
Run this cell to make Figures 17.2.4 and
17.2.5 for the elliptic integrals of the second
kind, *E*, whose real part is shown for *m* = 1.5
and complex ϕ in the motif at right.

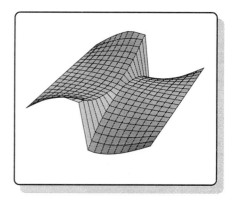

Exploring with EllipticE
If you need values of the elliptic integral of
the second kind for ϕ greater than $\pi/2$, increase the angular range for the 3D and 2D
plot functions in the following annotated
Mathematica cell so that you can verify the
periodicity of the function. In practice, the
incomplete elliptic integral for the part of ϕ
that is in excess of an integral multiple of $\pi/2$ can be added to the appropriate multiple of
the complete elliptic integral.

```
(* EllipticE cell *)

(* Incomplete elliptic integral of the
   second kind *)

Off[General::spell1];

(* 3D views defined; complex phi  & m varying *)

EllptcEEzz3D[m_,ReOrIm_,RngLow_,RngHi_] :=
 Plot3D[ReOrIm[EllipticE[phiX+I*phiY,m]],
 {phiX,0,Pi},{phiY,-2,2},
 PlotRange->{RngLow,RngHi},PlotPoints->{20,20},
 Ticks->{{0.5,1.5,2.5},{-1,0,1},{-4,-2,0,2,4}},
 AxesLabel->{"phiX","phiY",None}];


(* 2D views defined; phi varying & m fixed *)
EllptcEphi2D[m_,Dsh_] :=
 Plot[EllipticE[phi,m],{phi,0,Pi/2},
   PlotRange->All,PlotPoints->30,
   Ticks->{{0,0.5,1.0,1.5},{0,0.5,1,1.5}},
   AxesLabel->{"phi",None},
   PlotStyle->Dashing[{Dsh,Dsh}],
   DisplayFunction->Identity]


(* 2D views defined; m varying & phi fixed *)
Off[Plot::plnr];
EllptcEmm2D[phi_,Dsh_] :=
 Plot[EllipticE[phi,m],{m,0,1},
   PlotRange->All,PlotPoints->30,
   AxesOrigin->{0,0}, AxesLabel->{"m",None},
   Ticks->{{0,0.2,0.4,0.6,0.8,1},{0,1,2}},
   PlotStyle->Dashing[{Dsh,Dsh}],
   DisplayFunction->Identity]
```

Define 3D views with
complex ϕ varying and *m*
fixed. Upper range and
lower range are para-
meters.

Define 2D views with ϕ
varying and *m* fixed.
Dash size is a parameter.

Define 2D views with *m*
varying and ϕ fixed.
Dash size is a parameter.

```
(* 3D views for m = 0.5. 1.5 *)

EllptcEEzz3D[0.5,Re,-2,3];
EllptcEEzz3D[0.5,Im,-6,6];
EllptcEEzz3D[1.5,Re,-3,4];
EllptcEEzz3D[1.5,Im,-4,4];


(* 3D views; real phi  & m varying *)
EllptcEE3D =
  Plot3D[EllipticE[phi,m],{phi,0,Pi/2},{m,0,1},
  PlotRange->All,PlotPoints->{20,20},
  Ticks->{{0.5,1.0,1.5},{0,0.25,0.75},{0,1}},
  AxesLabel->{"phi","m",None}];


(* phi with m fixed; 2D views *)
ShowEllptcEphi2D = Show[
  EllptcEphi2D[0,0.02],EllptcEphi2D[0.5,0.04],
  EllptcEphi2D[1.0,0.08],
  DisplayFunction->$DisplayFunction];


(* m with phi fixed; 2D views *)
ShowEllptcEmm2D = Show[
  EllptcEmm2D[Pi/8,0.01],EllptcEmm2D[Pi/4,0.02],
  EllptcEmm2D[Pi/2,0.04],EllptcEmm2D[3*Pi/4,0.06],
  DisplayFunction->$DisplayFunction];


(* Elliptic E functions: Numerical *)

EllptcETable =
  Table[{phi,m,N[EllipticE[phi,m],10]},
  {phi,0.5,1.5,0.5},{m,0.25,0.75,0.25}]
EllptcEETable = Table[{m,N[EllipticE[m],10]},
  {m,0.25,0.75,0.25}]
```

`EllipticE` cell
continued

Show surfaces as real and imaginary parts, as in Figure 17.2.4.

Show surface as in Figure 17.2.5 left.

Show curves with m fixed, Figure 17.2.5 right top.

Show curves with ϕ fixed, Figure 17.2.5 right bottom.

Test values of elliptic integrals of the second kind, as in Table 17.2.3.

The `JacobiZeta` cell [17.2.3]
Run this cell to make Figures 17.2.6 and 17.2.7 for the Jacobi zeta function in Section 17.2.3, for example, the surface of the function shown at right for real ϕ and m.

Exploring with `JacobiZeta`
If you need values of the Jacobi zeta function for ϕ greater than $\pi/2$, increase the angular range for the 3D and 2D plot functions in the following annotated *Mathematica* cell so that you can verify the periodicity of the function. In practice, only the part of ϕ that is in excess of an integral multiple of $\pi/2$ needs to be known, since the zeta function vanishes for $\phi = \pi/2$.

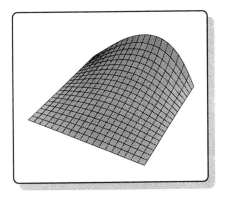

```
(* JacobiZeta cell *)

(* Jacobi Zeta function *)
Off[General::spell1];
(* 3D views defined; complex phi  & m varying *)

JacobiZeta3D[m_,ReOrIm_,RngLow_,RngHi_] :=
 Plot3D[ReOrIm[JacobiZeta[phiX+I*phiY,m]],
  {phiX,0,Pi},{phiY,-2,2},
  PlotRange->{RngLow,RngHi},PlotPoints->{20,20},
  Ticks->{{0.5,1.5,2.5},{-1,0,1},{-4,-2,0,2,4}},
  AxesLabel->{"phiX","phiY",None}];
```
Define 3D views with complex ϕ varying and *m* fixed. Upper range and lower range are parameters.

```
(* 2D views defined; phi varying & m fixed *)
JacobiZeta2D[m_,Dsh_] :=
 Plot[JacobiZeta[phi,m],{phi,0,Pi/2},
   PlotRange->All,PlotPoints->30,
   Ticks->{{0,0.5,1.0,1.5},{0,0.2,0.4}},
   AxesLabel->{"phi",None},
   PlotStyle->Dashing[{Dsh,Dsh}],
   DisplayFunction->Identity]
```
Define 2D views with ϕ varying and *m* fixed. Dash size is a parameter.

```
(* 2D views defined; m varying & phi fixed *)
Off[Plot::plnr];
JacobiZetaMM2D[phi_,Dsh_] :=
 Plot[JacobiZeta[phi,m],{m,0,0.95},
   PlotRange->All,PlotPoints->30,
   AxesOrigin->{0,0}, AxesLabel->{"m",None},
   Ticks->{{0,0.3,0.6,0.9},{0,0.2,0.4}},
   PlotStyle->Dashing[{Dsh,Dsh}],
   DisplayFunction->Identity]
```
Define 2D views with *m* varying and ϕ fixed. Dash size is a parameter.

```
(* 3D views for m = 0.5. 1.5 *)

JacobiZeta3D[0.5,Re,-2,3];
JacobiZeta3D[0.5,Im,-4,2];
JacobiZeta3D[1.5,Re,-4,4];
JacobiZeta3D[1.5,Im,-4,4];
```
Show surfaces as real and imaginary parts, as in Figure 17.2.6.

```
(* 3D views; real phi  & m varying *)
JacobiZetaPhiM3D =
 Plot3D[JacobiZeta[phi,m],{phi,0,Pi/2},{m,0,0.95},
  PlotRange->All,PlotPoints->{20,20},
  Ticks->{{0.5,1.0,1.5},{0,0.3,0.6},{0,0.2,0.4}},
  AxesLabel->{"phi","m",None}];
```
Show surface as in Figure 17.2.7 left.

```
(* phi with m fixed; 2D views *)
ShowJacobiZetaPhi2D = Show[
 JacobiZeta2D[0.3,0.02],JacobiZeta2D[0.6,0.04],
 JacobiZeta2D[0.9,0.08],
 DisplayFunction->$DisplayFunction];
```
Show curves with *m* fixed, Figure 17.2.7 right top.

```
(* m with phi fixed; 2D views *)
ShowJacobiZetaMM2D = Show[
 JacobiZetaMM2D[Pi/8,0.01],
 JacobiZetaMM2D[Pi/4,0.02],
 JacobiZetaMM2D[3*Pi/4,0.06],
 DisplayFunction->$DisplayFunction];
```

JacobiZeta cell
continued

Show curves with ϕ fixed,
Figure 17.2.7 right
bottom.

```
(* Jacobi Zeta functions: Numerical *)

JacobiZetaTable = Table[{phi,m,
 N[JacobiZeta[phi,m],10]},
 {phi,0.5,1.5,0.5},{m,0.25,0.75,0.25}]
```

Test values of zeta
functions, as in Table
17.2.4.

The `HeumanLambda` cell [17.2.4]

Execute this cell to make Figures 17.2.8 and 17.2.9 for the Heuman lambda function in Section 17.2.4, for example, the curves of the function shown at right for ϕ varying from 0 to $\pi/2$ and $m = 0$, 0.5, and 1 from top to bottom.

Exploring with `HeumanLambda`

If you need values of the Heuman lambda function for ϕ greater than $\pi/2$, increase the angular range for the 3D and 2D plot functions in the following annotated *Math-*

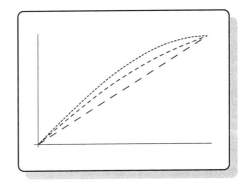

ematica cell and verify the periodicity of the function. In practice, only the part of ϕ that is in excess of an integral multiple of $\pi/2$ needs to be known, since the lambda function vanishes for $\phi = 0$ and is unity for $\phi = \pi/2$; this result is independent of m, as indicated in the curves above. Indeed, the small variation of the function within the analytic bounds given in (17.2.30) suggests that it can be accurately expanded in rapidly convergent series in terms of ϕ and m.

```
(* HeumanLambda cell *)

(* Heuman Lambda function *)

Off[General::spell1];

(* Define Heuman Lambda in terms of
   elliptic integrals *)

HeumanL[phi_,m_] :=
 If[ m == 1, 2*phi/Pi,
   If[ m == 0, Sin[phi],
     EllipticF[phi,1-m]/EllipticK[1-m]+
     2*EllipticK[m]*JacobiZeta[phi,1-m]/Pi ]]
```

Definition of function as
in (17.2.28).

```
(* 3D views defined; complex phi & m varying *)

HeumanL3D[m_,ReOrIm_,RngLow_,RngHi_] :=
 Plot3D[ReOrIm[HeumanL[phiX+I*phiY,m]],
 {phiX,0,Pi},{phiY,-2,2},
 PlotRange->{RngLow,RngHi},PlotPoints->{25,25},
 Ticks->{{0.5,1.5,2.5},{-1,0,1},{-4,-2,0,2,4}},
 AxesLabel->{"phiX","phiY",None}];
```

Define 3D views with complex ϕ varying and m fixed. Upper range and lower range are parameters.

```
(* 2D views defined; phi varying & m fixed *)

HeumanL2D[m_,Dsh_] :=
 Plot[HeumanL[phi,m],{phi,0,Pi/2},
  PlotRange->All,PlotPoints->30,
  Ticks->{{0,0.5,1.0,1.5},{0,0.5,1.0}},
  AxesLabel->{"phi",None},
  PlotStyle->Dashing[{Dsh,Dsh}],
  DisplayFunction->Identity]
```

Define 2D views with ϕ varying and m fixed. Dash size is a parameter.

```
(* 2D views defined; m varying & phi fixed *)

Off[Plot::plnr];

HeumanLMM2D[phi_,Dsh_] :=
 Plot[HeumanL[phi,m],{m,0,1},
  PlotRange->All,PlotPoints->30,
  AxesOrigin->{0,0}, AxesLabel->{"m",None},
  Ticks->{{0,0.5,1},{0,0.5,1.0,1.5}},
  PlotStyle->Dashing[{Dsh,Dsh}],
  DisplayFunction->Identity]
```

Define 2D views with m varying and ϕ fixed. Dash size is a parameter.

```
(* 3D views for m = 0.5. 1.5 *)

HeumanL3D[0.5,Re,-2,3];
HeumanL3D[0.5,Im,-4,3];

HeumanL3D[1.5,Re,-4,4];
HeumanL3D[1.5,Im,-4,4];
```

Show surfaces as real and imaginary parts, as in Figure 17.2.8.

```
(* 3D views; real phi  & m varying *)

HeumanLPhiM3D =
 Plot3D[HeumanL[phi,m],{phi,0,Pi/2},{m,0,1},
  PlotRange->All,PlotPoints->{20,20},
  Ticks->{{0.5,1.0,1.5},{0,0.25,0.75},{0,0.5,1}},
  AxesLabel->{"phi","m",None}];
```

Show surface as in Figure 17.2.9 left.

```
(* phi with m fixed; 2D views *)

ShowHeumanLPhi2D = Show[
 HeumanL2D[0,0.01],HeumanL2D[0.5,0.02],
 HeumanL2D[1.0,0.04],
 DisplayFunction->$DisplayFunction];
```

Show curves with m fixed, Figure 17.2.9 right top.

```
(* m with phi fixed; 2D views *)
ShowHeumanLMM2D = Show[
  HeumanLMM2D[0,0.001],
  HeumanLMM2D[Pi/8,0.01],
  HeumanLMM2D[Pi/4,0.02],
  HeumanLMM2D[3*Pi/4,0.04],
  DisplayFunction->$DisplayFunction];


(* Heuman Lambda functions: Numerical *)

HeumanLTable = Table[{phi,m,
  N[HeumanL[phi,m],10]},
  {phi,0.5,1.5,0.5},{m,0.25,0.75,0.25}]
```

HeumanLambda cell
continued

Show curves with ϕ fixed,
Figure 17.2.9 right
bottom.

Test values of lambda
functions, as in Table
17.2.5.

The `EllipticPi` cell [17.2.5]

Run this cell to make Figures 17.2.10 and 17.2.11 for the elliptic integrals of the third kind, Π, whose imaginary part is shown for $n = 1.5$ and $m = 1.5$ and complex ϕ in the motif at right. Travelers using the *Atlas* should beware the treacherous cliffs in the complex ϕ plane, which arise from the branch cuts in the square root that defines the integrand of Π in (17.2.31).

This cell also produces numerical values for testing the C function described in Section 17.2.5.

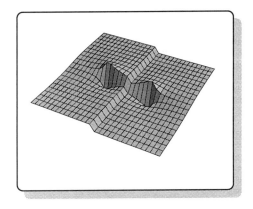

Exploring with `EllipticPi`

Our algorithm for the C function `EllptcPi` uses direct numerical integration in the definition (17.2.31). It thus avoids the many special cases that must be considered if the formulas in Section 17.7 of Abramowitz and Stegun are used. If, however, you need many values of this elliptic integral for n and m in a fixed relation then it is probably more efficient to code a C function in terms of the other elliptic integrals. To check out your coding, the numerical output from cell `EllipticPi` is very useful.

```
(* EllipticPi cell *)

(* Incomplete elliptic integral of the third kind *)

Off[General::spell1];

(* 3D views defined;
   complex phi with n & m fixed *)
```

```
EllptcPizz3D[n_,m_,ReOrIm_,RngLow_,RngHi_] :=
  Plot3D[ReOrIm[EllipticPi[n,phiX+I*phiY,m]],
  {phiX,0,Pi},{phiY,-2,2},
  PlotRange->{RngLow,RngHi},PlotPoints->{30,20},
  Ticks->{{0.5,1.5,2.5},{-1,1},
  {-6,-3,0,3,6,9}},
  AxesLabel->{"phiX","phiY",None}];
```

Define 3D views with complex ϕ varying and m fixed. Choice of real or imaginary, upper range, and lower range are parameters.

```
(* 3D views; n fixed, real  phi & m  varying *)
EllptcPi3D[n_] :=
  Plot3D[EllipticPi[n,phi,m],
  {phi,0,1.5},{m,0,0.9},
  PlotRange->{0,4},PlotPoints->{20,20},
  Ticks->{{0.5,1.0,1.5},{0.3,0.6,0.9},{0,2,4}},
  AxesLabel->{"phi","m",None}];
```

Define 3D surfaces for ϕ and m varying.

```
(* 3D views for  m values *)
EllptcPizz3D[0.5,0.5,Re,0,8];
EllptcPizz3D[0.5,0.5,Im,-4,4];
EllptcPizz3D[0.5,1.5,Re,0,6];
EllptcPizz3D[0.5,1.5,Im,-8,3];
EllptcPizz3D[1.5,0.5,Re,-3,3];
EllptcPizz3D[1.5,0.5,Im,-8,3];
EllptcPizz3D[1.5,1.5,Re,-5,3];
EllptcPizz3D[1.5,1.5,Im,-5,3];
```

Show surfaces as real and imaginary parts, as in Figure 17.2.10.

```
(* 3D views; n fixed, real phi & m varying *)
EllptcPi3D[0.0];  EllptcPi3D[0.3];
EllptcPi3D[0.6];  EllptcPi3D[0.9];
```

Show surfaces as in Figure 17.2.11.

```
(* EllipticPi functions: Numerical *)
EllptcETable[n_] := Table[{n,m,phi,
  N[EllipticPi[n,phi,m],10]},
  {phi,0.5,1.5,0.5},{m,0.5,1.5,0.25}]
EllptcETable[0.5]
EllptcETable[1.5]
```

Test values of elliptic integrals of the third kind, as in Table 17.2.6.

The JacobiSN, JacobiCN, and JacobiDN cells [17.3.1]

Run these three cells to make Figures 17.3.1–17.3.6 for the Jacobi elliptic functions sn, cn, and dn. The seven different views of each function should give you plenty of insight into their dependence on their parameters, u and m. For example, in the motif at right we see that the front curve of cn against real u and m is just $\cos u$, since m ranges from zero to 0.99 from the front to the back of this surface.

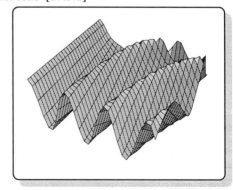

These cells also produce numerical values in Tables 17.3.1–17.3.3; these are useful to test C function `JacobiSnCnDn`

described in Section 17.3.1. Note that if your particular interest is any other of the twelve Jacobi elliptic functions, they can be formed easily from the copolar trio—sn, cn, and dn— by division, following Section 16.3 in Abramowitz and Stegun.

Exploring with `JacobiSN`, `JacobiCN`, and `JacobiDN`

Since of the three basic Jacobi functions has two parameters, u and m, there is plenty of territory to explore. For example, examine the functions in the complex u plane (Figures 17.3.1–17.3.3) in more detail to understand the regions of rapid variation.

The three cells are basically the same, so we list them sequentially.

```
(* JacobiSN cell *)

(* Jacobi  sn  function *)

Off[General::spell1];

(* 3D views defined;
   complex  u  with  m  fixed *)

JacobiSN3D[m_,ReOrIm_,RngLow_,RngHi_] :=
  Plot3D[ReOrIm[JacobiSN[uX+I*uY,m]],
  {uX,0,8},{uY,0,7},
  PlotRange->{RngLow,RngHi},PlotPoints->{30,30},
  Ticks->{{2,4,6},{0,2,4,6},{-4,-2,0,2,4}},
  AxesLabel->{"uX","uY",None}];


(* 2D views;  real u  with  m  fixed *)

JacobiSNUU2D[m_,Dsh_] := Plot[JacobiSN[u,m],
  {u,0,20},PlotRange->{-1,1},PlotPoints->30,
  Ticks->{{5,10,15},{-1,0,1}},
  AxesLabel->{"u",None},
  PlotStyle->Dashing[{Dsh,Dsh}],
  DisplayFunction->Identity];


(* 2D views;  m  varying with  u  fixed *)

JacobiSNMM2D[u_,Dsh_] := Plot[JacobiSN[u,m],
  {m,0,0.95},PlotRange->{-1,1},PlotPoints->30,
  Ticks->{{0,0.45,0.9},{-1,0,1}},
  AxesLabel->{"m",None},
  PlotStyle->Dashing[{Dsh,Dsh}],
  DisplayFunction->Identity]


(* 3D views for  m  pairs *)

JacobiSN3D[0.5,Re,-2,2];
JacobiSN3D[0.5,Im,-2,2];
JacobiSN3D[1.5,Re,-2,2];
JacobiSN3D[1.5,Im,-2,2];
```

JacobiSN cell

Define 3D views with complex u varying and m fixed. Choice of real or imaginary, upper range, and lower range are parameters.

Define 2D curves for u real with m fixed.

Define 2D curves for m real with u fixed.

Show surfaces as real and imaginary parts, as in Figure 17.3.1.

```
(* 3D views;  real u & m  vary *)

JacobiSNUM3D = Plot3D[JacobiSN[u,m],
  {u,0,20},{m,0,0.99},
  PlotRange->{-1,1},PlotPoints->{30,30},
  Ticks->{{5,10,15},{0.3,0.6,0.9},{-1,0,1}},
  AxesLabel->{"u","m",None}];
```

Show surface as in Figure 17.3.4 left.

```
(* 2D views; real u  with  m  fixed *)

ShowJacobiSNUU = Show[JacobiSNUU2D[0,0.01],
  JacobiSNUU2D[0.45,0.03],JacobiSNUU2D[0.90,0.06],
  DisplayFunction->$DisplayFunction];
```

Show curves with u varying and m fixed, as in Figure 17.3.4 right top.

```
(* 2D views;  m  varying with  u  fixed *)

ShowJacobiSNMM = Show[JacobiSNMM2D[5,0.01],
  JacobiSNMM2D[10,0.03],JacobiSNMM2D[20,0.06],
  DisplayFunction->$DisplayFunction];
```

Show curves with m varying and u fixed, as in Figure 17.3.4 right bottom.

```
(* Jacobi  sn  functions: Numerical *)

JacobiSNTable = Table[{m,u,
  N[JacobiSN[u,m],10]},
  {u,1,5,2},{m,0.3,0.9,0.3}]
```

Test values of Jacobi sn functions, as in Table 17.3.1.

```
(* JacobiCN cell *)
```

JacobiCN cell

```
(* Jacobi  cn  function *)

Off[General::spell1];
```

```
(* 3D views defined;
   complex  u  with  m  fixed *)

JacobiCN3D[m_,ReOrIm_,RngLow_,RngHi_] :=
  Plot3D[ReOrIm[JacobiCN[uX+I*uY,m]],
  {uX,0,8},{uY,0,7},
  PlotRange->{RngLow,RngHi},PlotPoints->{30,30},
  Ticks->{{2,4,6},{0,2,4,6},{-4,-2,0,2,4}},
  AxesLabel->{"uX","uY",None}];
```

Define 3D views with complex u varying and m fixed. Choice of real or imaginary, upper range, and lower range are parameters.

```
(* 2D views;  real u  with  m  fixed *)

JacobiCNUU2D[m_,Dsh_] := Plot[JacobiCN[u,m],
  {u,0,20},PlotRange->{-1,1},PlotPoints->30,
  Ticks->{{5,10,15},{-1,0,1}},
  AxesLabel->{"u",None},
  PlotStyle->Dashing[{Dsh,Dsh}],
  DisplayFunction->Identity];
```

Define 2D curves for u real with m fixed.

```
(* 2D views; m varying with u fixed *)
JacobiCNMM2D[u_,Dsh_] := Plot[JacobiCN[u,m],
  {m,0,0.95},PlotRange->{-1,1},PlotPoints->30,
  Ticks->{{0,0.45,0.9},{-1,0,1}},
  AxesLabel->{"m",None},
  PlotStyle->Dashing[{Dsh,Dsh}],
  DisplayFunction->Identity]

(* 3D views for m pairs *)
JacobiCN3D[0.5,Re,-2,2];
JacobiCN3D[0.5,Im,-2,2];
JacobiCN3D[1.5,Re,-2,2];
JacobiCN3D[1.5,Im,-2,2];

(* 3D views; real u & m vary *)
JacobiCNUM3D = Plot3D[JacobiCN[u,m],
  {u,0,20},{m,0,0.99},
  PlotRange->{-1,1},PlotPoints->{30,30},
  Ticks->{{5,10,15},{0.3,0.6,0.9},{-1,0,1}},
  AxesLabel->{"u","m",None}];

(* 2D views; real u with m fixed *)
ShowJacobiCNUU = Show[JacobiCNUU2D[0,0.01],
  JacobiCNUU2D[0.45,0.03],JacobiCNUU2D[0.90,0.06],
  DisplayFunction->$DisplayFunction];

(* 2D views; m varying with u fixed *)
ShowJacobiCNMM = Show[JacobiCNMM2D[5,0.01],
  JacobiCNMM2D[10,0.03],JacobiCNMM2D[20,0.06],
  DisplayFunction->$DisplayFunction];

(* Jacobi cn functions: Numerical *)
JacobiCNTable = Table[{m,u,
  N[JacobiCN[u,m],10]},
  {u,1,5,2},{m,0.3,0.9,0.3}]
```

JacobiCN cell
continued

Define 2D curves for *m*
real with *u* fixed.

Show surfaces as real and
imaginary parts, as in
Figure 17.3.2.

Show surface as in Figure
17.3.5 left.

Show curves with *u*
varying and *m* fixed, as in
Figure 17.3.5 right top.

Show curves with *m*
varying and *u* fixed, as in
Figure 17.3.5 right
bottom.

Test values of Jacobi cn
functions, as in Table
17.3.2.

```
(* JacobiDN cell *)
(* Jacobi dn function *)
Off[General::spell1];
(* 3D views defined;
   complex u with m fixed *)
JacobiDN3D[m_,ReOrIm_,RngLow_,RngHi_] :=
  Plot3D[ReOrIm[JacobiDN[uX+I*uY,m]],
  {uX,0,8},{uY,0,7},
  PlotRange->{RngLow,RngHi},PlotPoints->{30,30},
  Ticks->{{2,4,6},{0,2,4,6},{-4,-2,0,2,4}},
  AxesLabel->{"uX","uY",None}];

(* 2D views; real u with m fixed *)
JacobiDNUU2D[m_,Dsh_] := Plot[JacobiDN[u,m],
  {u,0,20},PlotRange->{0,1},PlotPoints->30,
```

JacobiDN cell

Define 3D views with
complex *u* varying and *m*
fixed. Choice of real or
imaginary, upper range,
and lower range are para-
meters.

Define 2D curves for *u*
real with *m* fixed.

```
Ticks->{{5,10,15},{0,1}},AxesLabel->{"u",None},
 PlotStyle->Dashing[{Dsh,Dsh}],
 DisplayFunction->Identity];

(* 2D views; m varying with u fixed *)
JacobiDNMM2D[u_,Dsh_] := Plot[JacobiDN[u,m],
 {m,0,0.95},PlotRange->{0,1},PlotPoints->30,
 Ticks->{{0,0.3,0.6,0.9},{0,1}},
 AxesLabel->{"m",None},
 PlotStyle->Dashing[{Dsh,Dsh}],
 DisplayFunction->Identity]

(* 3D views for m pairs *)
JacobiDN3D[0.5,Re,-2,2];
JacobiDN3D[0.5,Im,-2,2];
JacobiDN3D[1.5,Re,-2,2];
JacobiDN3D[1.5,Im,-2,2];

(* 3D views; real u & m vary *)
JacobiDNUM3D = Plot3D[JacobiDN[u,m],
 {u,0,20},{m,0,0.99},
 PlotRange->{0,1},PlotPoints->{30,30},
 Ticks->{{5,10,15,20},{0.3,0.6,0.9},{0,1}},
 AxesLabel->{"u","m",None}];

(* 2D views; real u with m fixed *)
ShowJacobiDNUU = Show[JacobiDNUU2D[0.3,0.01],
 JacobiDNUU2D[0.6,0.03],JacobiDNUU2D[0.9,0.06],
 DisplayFunction->$DisplayFunction];

(* 2D views; m varying with u fixed *)
ShowJacobiDNMM = Show[JacobiDNMM2D[5,0.01],
 JacobiDNMM2D[10,0.03],JacobiDNMM2D[20,0.06],
 DisplayFunction->$DisplayFunction];

(* Jacobi dn functions: Numerical *)
JacobiDNTable = Table[{m,u,
 N[JacobiDN[u,m],10]},
 {u,1,5,2},{m,0.3,0.9,0.3}]
```

Define 2D curves for *m* real with *u* fixed.

Show surfaces as real and imaginary parts, as in Figure 17.3.3.

Show surface as in Figure 17.3.6 left.

Show curves with *u* varying and *m* fixed, as in Figure 17.3.6 right top.

Show curves with *m* varying and *u* fixed, as in Figure 17.3.6 right bottom.

Test values of Jacobi dn functions, as in Table 17.3.3.

The `Theta` cell [17.3.2]

This cell makes all the graphics in Figures 17.3.7–17.3.10 for the four Jacobi theta functions. The many different views of each function—a surface and several transections at constant *q* and at constant *u*—give you plenty of insight into how they depend on these parameters. For example, in the motif at right we see how θ_4 (also called θ_0 or ϑ) varies with *u* for *q* = 0.25, 0.50, 0.75 indicated by short to long dashes.

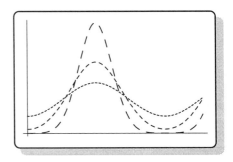

This cell also produces numerical values of the Jacobi theta functions in Tables 17.3.4–17.3.7. These are useful to test C function `JacobiTheta` described in Section 17.3.2.

Exploring with `Theta`

The Jacobi theta functions vary rapidly for u near multiples of $\pi/2$, as evident in the views of the surfaces in Figures 17.3.7—17.3.10. Away from these u values the surfaces appear to be featureless, but this is not so on closer inspection, as the matching curves in these figures show. It is therefore interesting to modify cell `Theta` so that you can examine regions of the $u - q$ plane in more detail.

```
(* Theta cell *)

(* Theta functions *)

Off[General::spell1];

(* 3D views defined; u  & q vary, a fixed *)

Theta3D[a_,RngLow_,RngHi_] :=
 Plot3D[EllipticTheta[a,u,q],
 {u,0,4},{q,0,0.99},
 PlotRange->{RngLow,RngHi},PlotPoints->{25,25},
 Ticks->{{1.5,3},{0,0.3,0.6,0.9},{-4,0,4}},
 AxesLabel->{"u","q",None}]
```

Define 3D views with u varying and q fixed. Upper range and lower range are parameters.

```
(* 2D views defined; u varying & q fixed *)
ThetaUU2D[a_,q_,Dsh_] :=
 Plot[EllipticTheta[a,u,q],{u,0,4},
 PlotRange->All,PlotPoints->30,
 Ticks->{{0,1,2,3,4},{-2,0,2,4}},
 AxesLabel->{"u",None},
 PlotStyle->Dashing[{Dsh,Dsh}],
 DisplayFunction->Identity]
```

Define 2D curves for u real with q fixed.

```
(* Composite 2D views; u with q fixed *)
ShowThetaUU2D[a_] := Show[
 ThetaUU2D[a,0.25,0.01],
 ThetaUU2D[a,0.50,0.02],
 ThetaUU2D[a,0.75,0.04],
 DisplayFunction->$DisplayFunction];
```

Composite curves.

```
(* 2D views defined; q varying & u fixed *)
Off[Plot::plnr];
ThetaQQ2D[a_,u_,Dsh_] :=
 Plot[EllipticTheta[a,u,q],{q,0,0.9},
 PlotRange->All,PlotPoints->30,
 AxesOrigin->{0,0}, AxesLabel->{"q",None},
 Ticks->{{0,0.25,0.5,0.75},
  {-4,-3,-2,-1,0,1,2,3,4}},
 PlotStyle->Dashing[{Dsh,Dsh}],
 DisplayFunction->Identity]
```

Define 2D curves for q real with u fixed.

```
(* Composite 2D views; q with u fixed *)
ShowThetaQQ2D[a_] := Show[
  ThetaQQ2D[a,1,0.01],
  ThetaQQ2D[a,2,0.02],
  ThetaQQ2D[a,3,0.04],
  DisplayFunction->$DisplayFunction];

For[ a = 1,  a <= 4,  a++,
  ( Theta3D[a,-4,4];
    ShowThetaUU2D[a];
    ShowThetaQQ2D[a];) ]

(* Theta functions: Numerical *)
ThetaTable[a_] := Table[{u,q,
  N[EllipticTheta[a,u,q],10]},
  {u,1,3,1},{q,0.25,0.75,0.25}]
ThetaTable[1]
ThetaTable[2]
ThetaTable[3]
ThetaTable[4]
```

Define composite curves.

Show surfaces and curves
for a = 1–4, as in Figures
17.3.7–17.3.10.

Test values of Jacobi theta
functions, as in Tables
17.3.4–17.3.7.

The `LogDerTheta` cell [17.3.3]

Run this cell to make the graphics in Figures 17.3.11–17.3.14 for logarithmic derivatives of the four Jacobi theta functions. Many views of each functio, a surface and several transections at constant q or constant u, show how the functions depend on these parameters. For example, in the motif at right we see how the logarithmic derivative of θ_4 (also called θ_0 or ϑ) varies with u for $q = 0.25, 0.50, 0.75$ indicated by short to long dashes.

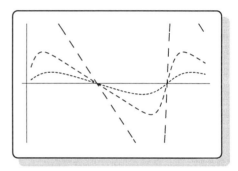

This cell also produces numerical values of the logarithmic derivatives, as in Tables 17.3.8–17.3.11. These are useful to test C function `LogDerTheta` in Section 17.3.3.

Exploring with `LogDerTheta`

The logarithmic derivatives of the four Jacobi theta functions vary rapidly for u near some multiples of $\pi/2$, as seen in the views of the surfaces in Figures 17.3.11—17.3.14. Away from these u values the surfaces appear to be featureless, but on closer inspection—as the matching curves in the figures show—there is much more detail. It is thus interesting to modify cell `LogDerTheta` and to examine regions of the $u - q$ plane in more detail.

```
(* LogDerTheta cell *)

(* Logarithmic derivatives of theta functions *)

Off[General::spell1];
```

```
(* Five-point formula for first derivative *)
FirstD[f_,a_,u_,q_,h_] :=
 (8*(f[a,u+h,q]-f[a,u-h,q])-
  (f[a,u+2*h,q]-f[a,u-2*h,q]))/(12*h)
```

First derivative of *f* at *u* and *q* using 5 points with equal steps *h*.

```
(* Numerical log derivative of a theta function *)
LDTheta[a_,u_,q_,eps_] :=
 FirstD[EllipticTheta,a,u,q,eps]/
 EllipticTheta[a,u,q]

eps = 10^(-4);
```

Specify *f* as a Jacobi elliptic theta function.

Step size for derivative.

```
(* 3D views defined; u  & q vary, a fixed *)

LDTheta3D[a_,RngLow_,RngHi_] :=
 Plot3D[LDTheta[a,u,q,eps],
 {u,0.1,4},{q,0.1,0.9},
 PlotRange->{RngLow,RngHi},PlotPoints->{25,25},
 Ticks->{{1.5,3},{0.3,0.6,0.9},{-6,-3,0,3,6}},
 AxesLabel->{"u","q",None}]
```

Define 3D views with *u* varying and *q* fixed. Upper range, and lower range are parameters.

```
(* 2D views defined; u varying & q fixed *)

LDThetaUU2D[a_,q_,Dsh_] :=
 Plot[LDTheta[a,u,q,eps],{u,0.1,4},
 PlotRange->{-6,6},PlotPoints->30,
 Ticks->{{1,2,3,4},{-6,-3,0,3,6}},
 AxesLabel->{"u",None},
 PlotStyle->Dashing[{Dsh,Dsh}],
 DisplayFunction->Identity]
```

Define 2D curves for *u* real with *q* fixed.

```
(* Composite 2D views; u with q fixed *)
ShowLDThetaUU2D[a_] := Show[
 LDThetaUU2D[a,0.25,0.01],
 LDThetaUU2D[a,0.50,0.02],
 LDThetaUU2D[a,0.75,0.04],
 DisplayFunction->$DisplayFunction];
```

Composite curves.

```
(* 2D views defined; q varying & u fixed *)

Off[Plot::plnr];
LDThetaQQ2D[a_,u_,Dsh_] :=
 Plot[LDTheta[a,u,q,eps],{q,0.1,0.9},
 PlotRange->{-12,12},PlotPoints->30,
 AxesOrigin->{0.1,0}, AxesLabel->{"q",None},
 Ticks->{{0.25,0.5,0.75},{-12,-6,0,6,12}},
 PlotStyle->Dashing[{Dsh,Dsh}],
 DisplayFunction->Identity]
```

Define 2D curves for *q* real with *u* fixed.

```
(* Composite 2D views; q with u fixed *)
ShowLDThetaQQ2D[a_] := Show[
 LDThetaQQ2D[a,0.5,0.01],
 LDThetaQQ2D[a,1.0,0.02],
 LDThetaQQ2D[a,2.0,0.04],
 DisplayFunction->$DisplayFunction];
```

Define composite curves.

```
For[ aa = 1, aa <= 4, aa++,
  ( LDTheta3D[aa,-6,6];
    ShowLDThetaUU2D[aa];
    ShowLDThetaQQ2D[aa];) ]

(* LDTheta functions: Numerical *)
LDThetaTable[aJ_] := Table[{u,q,
  N[LDTheta[aJ,u,q,eps],10]},
  {u,1,3,1},{q,0.25,0.75,0.25}]

LDThetaTable[1]
LDThetaTable[2]
LDThetaTable[3]
LDThetaTable[4]
```

`LogDerTheta` cell
continued

Show surfaces and curves
for *a* = 1–4, as in Figures
17.3.11–17.3.14.

Test values of Jacobi theta
functions, as in Tables
17.3.8–17.3.11.

20.18 PARABOLIC CYLINDER FUNCTIONS

Notebook and Cells
Notebook `ParCyl`; cells `Parabolic-Coords`, `ParabolicU`, `ParabolicV`

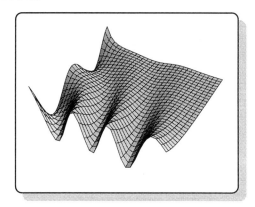

Use the cells in this notebook to understand the behavior of the parabolic cylinder functions *U* and *V*, both described in Chapter 18. These cells also generate numerical values that are useful to check your implementation of the functions in C. For this purpose, run the *Mathematica* cells in conjunction with the C driver programs whose annotated listings are in Section 21.18.

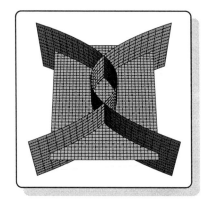

The `ParabolicCoords` cell [18.1.1]
Execute this cell to make Figure 18.1.1, showing the surfaces in the parabolic cylinder coordinate system described in Section 18.1.1.

By varying α and β when generating the graphics functions `ParaSurfPos` and `Para-SurfNeg`, respectively, you can explore how the parabolic cylinders depend upon these two coordinates.

Notice that the surfaces always intersect at right angles, since this is an orthogonal coordinate system.

```
(*  ParabolicCoords cell *)

(* Exploring Parabolic Coordinate Systems *)

SetOptions[ParametricPlot3D,Boxed->False,
  Axes->False,ViewPoint->{0,2,3.5},
  DisplayFunction->Identity];


(* Parabolic Coordinates;  (alpha,beta,z)  *)

xpc[alpha_,beta_,z_] := (alpha^2-beta^2)/2
ypc[alpha_,beta_,z_] := alpha*beta
zpc[alpha_,beta_,z_] := z

ParaSurfPos[alpha_] := ParametricPlot3D[
  {xpc[alpha,beta,z],ypc[alpha,beta,z],z},
  {beta,-5,5},{z,-6,6},PlotPoints->{40,15}]

ParaSurfNeg[beta_] := ParametricPlot3D[
  {xpc[alpha,beta,z],ypc[alpha,beta,z],z},
  {alpha,-5,5},{z,-6,6},PlotPoints->{40,15}]

PlaneSurf[z_] := ParametricPlot3D[{x,y,z},
  {x,-8,8},{y,-10,10},PlotPoints->{32,40}]

ParabolicCoords = Show[ParaSurfPos[2],
  ParaSurfNeg[2],PlaneSurf[0],
  DisplayFunction->$DisplayFunction];
```

Plot options; ViewPoint is important.

Define Cartesian coordinates in terms of parabolic cylinder coordinates.

Define parabolic cylinder for α constant.

Define parabolic cylinder for β constant.

Define planar surface for z constant.

Show the three surfaces together; here $\alpha = 2$, $\beta = 2$, $z = 0$.

The `ParabolicU` cell [18.2.1]

Execute this cell to produce Figures 18.2.2 and 18.2.3 for surfaces and integrands, respectively, of the parabolic cylinder function U.

For example, the motif at right shows for $a = -4.0$ the integrand as a function of the variable of integration w in (18.2.5) in [0, 20] along the front, and argument x in [5, 8] along the side. Integrating this to an accuracy of 10^{-10} would clearly be a challenge. Similar views for the oscillatory region are shown in the top row of Figure 18.2.3.

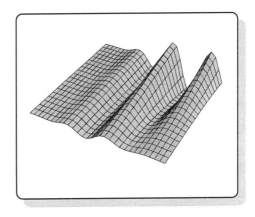

Exploring with `ParabolicU`

Two methods are used in Section 18.2.2 to compute the function $U(a,x)$—by confluent hypergeometric functions in (18.2.2) and by integration in (18.2.3). It is therefore interesting to explore how moving the switchover values that we have chosen ($x > 3.0$ and $a > 1.0$) affects the accuracy and efficiency of computing this parabolic cylinder function for the range of x and a values relevant to your applications. For example, if only very large x values (typically $x > 20$ for moderate a values) are needed, it might be worthwhile to program the asymptotic expansions in Section 19.8 of Abramowitz and Stegun [Abr64].

```
(* ParabolicCylinderU cell *)

(* Parabolic Cylinder Functions, U(a,x)  *)

(* Definition from 1F1 or integrals  *)
Hyp1F1[a_,b_,z_] := Hypergeometric1F1[a,b,z]
Off[NIntegrate::precw]

UU[a_,x_,Prcsn_] :=
If[ x <= 1,
(Sqrt[Pi]*Exp[-x*x/4]/(2^((a+1/2)/2)))*
 (Hyp1F1[a/2+1/4,1/2,x*x/2]/Gamma[3/4+a/2]-
 Sqrt[2]*x*Hyp1F1[a/2+3/4,3/2,x*x/2]/
 Gamma[1/4+a/2]),

If[ a < 0,   (* for  x > 1  *)
Sqrt[2/Pi]*Exp[x*x/4]*x^(a-1/2)*
 NIntegrate[Exp[-v*v/(2*x*x)]*v^(-(a+1/2))*
 Cos[v+(a+1/2)*Pi/2],{v,0,50},MaxRecursion->8,
 PrecisionGoal->Prcsn],

 Exp[-x*x/4]/((x^(a+1/2))*Gamma[a+1/2])*
 NIntegrate[Exp[-v-v*v/(2*x*x)]*v^(a-1/2),
 {v,0,40},
 MaxRecursion->6,PrecisionGoal->Prcsn] ] ]


(* 3D views; x & a vary *)
(* 4-digit precision *)

ParCylUU3D =
 Plot3D[UU[a,x,4],{x,-5,5},{a,-6,0},
 PlotRange->{-15,15},PlotPoints->{30,30},
 Ticks->{{-3,-1,1,3},{-6,-4,-2,0},{-10,0,10}},
 AxesLabel->{"x","a",None}];


(* 3D views; x & a vary, exponential factor *)

ParCylUUexp3D =
 Plot3D[Exp[x*x/4]*UU[a,x,4],
 {x,-3,3},{a,-6,0},
 PlotRange->All,PlotPoints->{30,30},
 Ticks->{{-1,1},{-6,-4,-2,0},{-50,0,50}},
 AxesLabel->{"x","a",None}];


(* 2D views defined; x varies, a fixed *)

ParCylUUxx2D[a_,Dsh_] :=
 Plot[UU[a,x,4],{x,-5.0,5.0},
  PlotRange->{-13,13},PlotPoints->30,
  Ticks->{{-4,-2,0,2,4},{-10,0,10}},
  AxesOrigin->{0,0},AxesLabel->{"x",None},
  PlotStyle->Dashing[{Dsh,Dsh}],
  DisplayFunction->Identity]
```

Abbreviate confluent hypergeometric and turn off message on integration precision.

Define U by hypergeometric functions,

or by integral (18.2.5).

Surface of U vs x and a, as in Figure 18.2.2 top left.

Surface of U with exponential factor removed vs x and a, as in Figure 18.2.2 bottom left.

Define curves of U vs x for fixed a, with dash length a parameter.

```
ShowParCylUUxx2D = Show[
 ParCylUUxx2D[-6.0,0.01],
 ParCylUUxx2D[-3.0,0.02],
 ParCylUUxx2D[ 0.0,0.04],
 DisplayFunction->$DisplayFunction];


(* 2D views defined; a varies, x fixed *)

ParCylUUaa2D[x_,Dsh_] :=
 Plot[UU[a,x,4],{a,-6.0,2.0},
  PlotRange->{-13,13},PlotPoints->30,
  Ticks->{{-5,-3,-1,1},{-10,0,10}},
  AxesOrigin->{0,0},AxesLabel->{"a",None},
  PlotStyle->Dashing[{Dsh,Dsh}],
  DisplayFunction->Identity]

ShowParCylUUaa2D = Show[
 ParCylUUaa2D[-4.0,0.01],
 ParCylUUaa2D[-2.0,0.02],
 ParCylUUaa2D[ 0.0,0.04],
 DisplayFunction->$DisplayFunction];


(* Integrand for  U(a, x > 0); w form  *)

UUIntgrnd[a_,x_,w_] :=
If[ a > -1/2, Exp[-w*w*w/(2*x*x)]*w^(a-1/2),
 Exp[-w*w/(2*x*x)]*w^(-a-1/2)*
 Cos[w+(a+1/2)*Pi/2] ]


(* 3D views defined; w & x vary, a fixed  *)

UUIntgrnd3D[a_,wMax_] :=
 Plot3D[UUIntgrnd[a,x,w],{w,0,wMax},{x,1,5},
 PlotRange->All,PlotPoints->{30,20},
 Ticks->{{0,5,10,15},{1,3,5},None},
 AxesLabel->{"w","x",None}];

(* Integrands in 3D *)
UUIntgrnd3D[-6,20];
UUIntgrnd3D[-2,20];
UUIntgrnd3D[ 2,20];
UUIntgrnd3D[ 6,20];


(* U functions: Numerical *)
(* 10-digit precision *)

ParCylUUTable[a_] :=
 Table[{a,x,N[UU[a,x,10],10]},
 {x,-5.0,5.0,5.0}]

ParCylUUTable[-5.0]
ParCylUUTable[ 0.0]
ParCylUUTable[ 5.0]
```

Show curves of U vs x for fixed a, as in Figure 18.2.2 top right.

Define curves of U vs a for fixed x, with dash length a parameter.

Show curves of U vs a for fixed x, as in Figure 18.2.2 bottom right.

Define integrand in (18.2.5).

Define surface of integrand for fixed a, with maximum w as parameter.

Show surfaces of integrands as in Figure 18.2.3.

Define table of numerical values of U.

Make test values table, as in Table 18.2.1.

The `ParabolicV` cell [18.2.2]

Run this cell to produce Figure 18.2.4 for surfaces and curves of $V(a,x)$, plus Figure 18.2.5 for detail of the function when a is near an integer. For example, the motif at right shows the surface of $e^{-x^2/4} V(a,x)$ for x in [–4, 2] along the front with a in the range [–4, 2] along the side. A similar view is in Figure 18.2.4 bottom left.

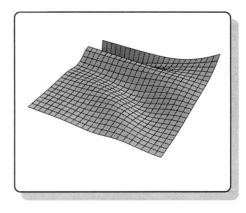

Exploring with `ParabolicV`

Above Figure 18.2.5 we discuss why $V(a,x)$ is sensitive to a when a is near an integer and $x < 0$. Explore this in more detail; it will depend upon the accuracy of your computer and of the sine function program.

```
(* ParabolicCylinderV cell *)

(* Parabolic Cylinder Functions, V(a,x)  *)

(* Definition of U(a,x) from 1F1 or integrals *)
Hyp1F1[a_,b_,z_] := Hypergeometric1F1[a,b,z]

UU[a_,x_,Prcsn_] :=
If[ x <= 1,
(Sqrt[Pi]*Exp[-x*x/4]/(2^((a+1/2)/2)))*
 (Hyp1F1[a/2+1/4,1/2,x*x/2]/Gamma[3/4+a/2]-
 Sqrt[2]*x*Hyp1F1[a/2+3/4,3/2,x*x/2]/
 Gamma[1/4+a/2]),

If[ a < 0,  (* for  x > 1  *)
Sqrt[2/Pi]*Exp[x*x/4]*x^(a-1/2)*
 NIntegrate[Exp[-v*v/(2*x*x)]*v^(-(a+1/2))*
 Cos[v+(a+1/2)*Pi/2],{v,0,50},MaxRecursion->8,
 PrecisionGoal->Prcsn],

 Exp[-x*x/4]/((x^(a+1/2))*Gamma[a+1/2])*
 NIntegrate[Exp[-v-v*v/(2*x*x)]*v^(a-1/2),
 {v,0,40},MaxRecursion->6,
 PrecisionGoal->Prcsn] ] ]

(* Definition of V(a,x) in terms of U(a,x) *)
VV[a_,x_,Prcsn_] := Gamma[a+1/2]*
 (Sin[Pi*a]*UU[a,x,Prcsn]+UU[a,-x,Prcsn])/Pi

(* 3D views; x & a vary *)
(* 4-digit precision *)
ParCylVV3D =
 Plot3D[VV[a,x,4],{x,-3,3},{a,-6,0},
 PlotRange->All,PlotPoints->{24,24},
 Ticks->{{-1,1},{-6,-4,-2,0},{-2,0,2}},
 AxesLabel->{"x","a",None}];
```

Abbreviate confluent hypergeometric and turn off message on integration precision.

Define U by hypergeometric functions,

or by integral (18.2.5).

Define V by (18.2.10).

Surface of V vs x and a, as in Figure 18.2.4 top left.

```
(* 3D views; x & a vary, exponential factor *)
ParCylVVexp3D = Plot3D[Exp[-x*x/4]*VV[a,x,4],
  {x,-3,3},{a,-6,0},
  PlotRange->All,PlotPoints->{30,30},
  Ticks->{{-1,1},{-6,-4,-2,0},{-0.5,0,0.5}},
  AxesLabel->{"x","a",None}];
```

Surface of *V* vs *x* and *a* with exponential factor removed, as in Figure 18.2.4 bottom left.

```
(* 2D views defined; x varies, a fixed *)
(* 4-digit precision *)
ParCylVVxx2D[a_,Dsh_] :=
 Plot[VV[a,x,4],{x,-5.0,5.0},
   PlotRange->{-2,2},PlotPoints->30,
   Ticks->{{-4,-2,0,2,4},{-1,0,1}},
   AxesOrigin->{0,0},AxesLabel->{"x",None},
   PlotStyle->Dashing[{Dsh,Dsh}],
   DisplayFunction->Identity]
```

Define curves of *V* vs *x* for fixed *a*; dash length is a parameter.

```
ShowParCylVVxx2D = Show[
 ParCylVVxx2D[-4.0,0.01],
 ParCylVVxx2D[-2.0,0.02],
 ParCylVVxx2D[ 0.0,0.04],
 DisplayFunction->$DisplayFunction];
```

Show curves of *V* vs *x* for fixed *a*, as in Figure 18.2.4 top right.

```
(* 2D views defined; a varies, x fixed *)
(* 4 digit precision *)
Off[Infinity::indet,Plot::plnr];
```

Turn off error message.

```
ParCylVVaa2D[x_,Dsh_] :=
 Plot[VV[a,x,4],{a,-6.0,2.0},
   PlotRange->{-2.5,2.5},PlotPoints->30,
   Ticks->{{-5,-3,-1,1},{-1,0,1}},
   AxesOrigin->{0,0},AxesLabel->{"a",None},
   PlotStyle->Dashing[{Dsh,Dsh}],
   DisplayFunction->Identity]
```

Define curves of *V* vs *a* for fixed *x*, with dash length a parameter.

```
ShowParCylVVaa2D = Show[
 ParCylVVaa2D[-1.5,0.01],
 ParCylVVaa2D[ 0.0,0.02],
 ParCylVVaa2D[ 1.5,0.04],
 DisplayFunction->$DisplayFunction];
```

Show curves of *V* vs *a* for fixed *x*, as in Figure 18.2.4 bottom right.

```
(* 2D views defined; a varies, x fixed,
   with different  a  range  *)
(* 4 digit precision *)
Off[Infinity::indet,Plot::plnr];
ParCylVVaa2D[x_,Dsh_] :=
 Plot[VV[a,x,4],{a,0,4},
   PlotRange->All,PlotPoints->29,
   Ticks->{{0,2,4},{-30,-20,-10,0,10}},
   AxesOrigin->{0,0},AxesLabel->{"a",None},
   PlotStyle->Dashing[{Dsh,Dsh}],
   DisplayFunction->Identity]
```

Define curve of *V* vs *a* for fixed *x*, with dash length as parameter.

```
ShowParCylVVaa2D = Show[
  ParCylVVaa2D[-2.0,0.01],
  ParCylVVaa2D[-1.5,0.02],
  ParCylVVaa2D[-1.0,0.04],
  DisplayFunction->$DisplayFunction];
```

ParabolicCylinderV
cell *continued*

Show curves of *V* vs *a*, as in
Figure 18.2.5 left.

```
(* Detail near  a = 2 *)
ParCylVVaNear2[x_,Dsh_] :=
  Plot[VV[a,x,4],{a,1.96,2.04},
    PlotRange->All,PlotPoints->29,
    Ticks->{{1.96,2.0,2.04},{-1,0,1}},
    AxesOrigin->{2,0},AxesLabel->{"a",None},
    PlotStyle->Dashing[{Dsh,Dsh}],
    DisplayFunction->Identity]
```

Define curve of *V* vs *a* for
fixed *x*, with dash length as
parameter. Small range of *a*.

```
ShowParCylVVaa2D = Show[
  ParCylVVaNear2[-2.0,0.01],
  ParCylVVaNear2[-1.5,0.02],
  ParCylVVaNear2[-1.0,0.04],
  DisplayFunction->$DisplayFunction];
```

Show curves of *V* near *a* = 2,
as in Figure 18.2.5 right.

```
(* V functions: Numerical *)
(* 10-digit precision *)
ParCylVVTable[a_] :=
  Table[{a,x,N[VV[a,x,10],10]},
  {x,-5.0,5.0,5}]
```

Define table of numerical
values of *V*.

```
ParCylVVTable[-5.0]
ParCylVVTable[-2.4]
ParCylVVTable[ 0.1]
ParCylVVTable[ 9.5]
```

Make test values, as in Table
18.2.2.

20.19 MISCELLANEOUS FUNCTIONS FOR SCIENCE AND ENGINEERING

Notebook and Cells

Notebook Miscellaneous; notebook cells
Debye, Sievert, Abramowitz, Spence,
Clausen, Voigt

Overview

The *Mathematica* notebook for Sections 19.1—19.7
is simple to use. It provides 3D and 2D views of
the functions, as well as numerical values for the
functions. These values may be used with the C
functions in the text and in machine-readable form
to test the C programs.

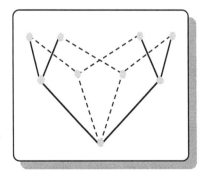

The basic definitions of all the functions in this chapter, except the angular momentum
coupling coefficients in Section 19.7, are in terms of integrals. Some of these integrals

expressions in terms of series expansions are therefore used for some of the functions. We also provide views of the integrands in these badlands so that you can avoid such treacherous territory.

The Debye cell [19.1]

Execution of this cell produces Figure 19.1.1 for values of the parameter n in the defining integral (19.1.1) between $n = 1$ and $n = 4$.

Exploring with Debye

Adventuresome users will push on to larger values of n and x. Since the integrand becomes singular as x approaches zero, small x values challenge the robustness of the *Mathematica* functions. In the C function Debye (Section 19.1) the first point in the integrand is computed specially, as described in Programming Notes.

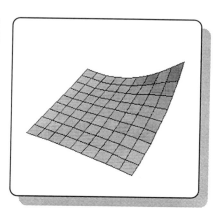

```
(* Debye cell *)

(* Debye functions by numerical integration *)
Debye[n_,x_] := NIntegrate[t^n/(Exp[t]-1),
  {t,10^-12,x},AccuracyGoal->12,PrecisionGoal
  ->12,WorkingPrecision->12];
```

Define Debye function by (19.1.1).

```
(* Modified Debye function *)
DebyeMod[n_,x_] := Debye[n,x]*n/(x^n);
```

Define modified Debye function (19.1.3).

```
(* Debye functions: 3D views *)
Plot3D[Debye[n,x],{x,0.1,4},{n,1,4},
  Ticks->{Automatic,Automatic,None},
  PlotRange->All,AxesLabel->{"x","n",""}];
```

Surface of Debye function, Figure 19.1.1 left.

```
(* Modified Debye functions: 3D views *)
Plot3D[DebyeMod[n,x],{x,0.1,4},{n,1,4},
  Ticks->{Automatic,Automatic,None},
  AxesLabel->{"x","n",""}];
```

Surface of modified Debye function, Figure 19.1.1 right.

```
(* Modified Debye functions: Numerical *)

DebyeModTable[n_] := N[Table[
  {n,x,DebyeMod[n,x]},{x,1,4}],10]
```

Define numerical table for fixed n.

```
DebyeModTable[1]
DebyeModTable[2]
DebyeModTable[3]
DebyeModTable[4]
```

Debye functions of order $n = 1, 2, 3, 4$, as in Table 19.1.1.

The `Sievert` cell [19.2]

Execute this cell to visualize the Sievert radiation integral (19.2.1) as a function of *x* and *θ*, as in Figure 19.2.2. The modified Sievert integral—more uniform as a function of *θ* than the original integral—is also shown in Figure 19.2.2.

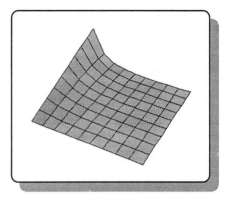

Exploring with `Sievert`

The integrand in the Sievert integral changes rapidly as *φ* approaches *π*/2. Check robustness of the *Mathematica* functions by increasing `AccuracyGoal`, `PrecisionGoal`, and the `WorkingPrecision` if values of *φ* near *π*/2 are important for your applications. The C function `Sievert` should then also be made more accurate by increasing `MAX` and `Nint` in the program. Note that—as a practical matter rather than merely a numerical curiosity—for such *φ* values the integral will depend very sensitively on *x*, as seen from the geometry in Figure 19.2.1. The utility of the Sievert integral is then more likely to depend upon the accuracy of determining *x* than on the accuracy of computing the function.

```
(* Sievert cell *)

(* Sievert integrals by numerical integration *)
Sievert[th_,x_] := NIntegrate[Exp[-x*Sec[t]],
  {t,0,th},AccuracyGoal->4,PrecisionGoal->4,
  WorkingPrecision->10];
```
Define Sievert integral by (19.2.1).

```
SievertMod[th_,x_] := Sievert[th,x]/th;
```
Modified Sievert integral (19.2.2).

```
(* Sievert integral surfaces: 3D views *)
Plot3D[Sievert[th,x],{x,0.01,5},{th,0.01,N[Pi/2]},
  Ticks->{Automatic,Automatic,None},
  PlotRange->All,AxesLabel->{"x","theta",""}];
```
Surface of Sievert integral as function of *x* and *θ*, Figure 19.2.2 left.

```
(* Modified Sievert integral surfaces: 3D views *)
Plot3D[SievertMod[th,x],{x,0.01,5},{th,0.01,N[Pi/2]},
  PlotRange->All,Ticks->{Automatic,Automatic,None},
  AxesLabel->{"x","theta",""}];
```
Surface of modified Sievert integral as function of *x* and *θ*, Figure 19.2.2 right.

```
(* Sievert integral: Numerical *)

thStep = N[Pi/9,10];
SievertTable = N[Table[
  {th,x,Sievert[th,x]},{th,thStep,3*thStep,thStep},
  {x,0,4,2}],10]
```
Numerical checks as in Table 19.2.1.

The `Abramowitz` cell [19.3]

By executing this cell you will get *Mathematica* to produce Figures 19.3.1–19.3.3 for the Abramowitz function (19.3.1) over a wide range of its parameters, m and x in the function, and over the integration variable t in (19.3.1).

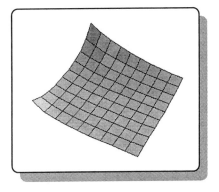

Exploring with `Abramowitz`

Adventuresome users will push on into the desert to larger values of m and x. The convergence of the series (19.3.4)–(19.3.6) used for the C functions can be checked by using this *Mathematica* cell to generate numerical values of the Abramowitz function with high accuracy.

```
(* Abramowitz cell *)

(* Abramowitz functions by numerical integration *)
Abrmwtz[m_,x_] := NIntegrate[t^m*Exp[-t^2-x/t],
  {t,0.01,20},AccuracyGoal->12,PrecisionGoal->12,
  WorkingPrecision->12]
```
Define Abramowitz function by integration, (19.3.1).

```
(* Abramowitz functions: 3D views *)
Plot3D[Abrmwtz[m,x],{x,0.1,3},{m,0,6},
  Ticks->{Automatic,Automatic,None},
  PlotRange->All,AxesLabel->{"x","m",""}];
```
Surface of Abramowitz function, Figure 19.3.1 left.

```
(* Abramowitz functions: 2D views *)
AbrmwtzPlot[m_] := Plot[Abrmwtz[m,x],{x,0.1,2},
  PlotRange->All,Ticks->{{0,0.5,1,1.5,2},None},
  Axes->{True,False},AxesOrigin->{0.6,0.0},
  AxesLabel->{"x",""},DisplayFunction->Identity]
```
Define curve of function for fixed m.

```
Show[AbrmwtzPlot[1],AbrmwtzPlot[5],
  DisplayFunction->$DisplayFunction];
```
Show curves for $m = 1$ and 5, Figure 19.3.1 right.

```
(* Integrand of Abramowitz function *)
AbrmwtzInt[m_,x_,t_] := t^m*Exp[-t^2-x/t]
```
Integrand of function.

```
(* Integrand of Abramowitz function: 3D views *)
PlotAbrmInt[m_] := Plot3D[AbrmwtzInt[m,x,t],
  {t,0.01,4},{x,0.1,3},PlotPoints->25,
  Ticks->{Automatic,Automatic,None},
  PlotRange->All,AxesLabel->{"t","x",""}];
```
Define integrand surface for fixed m.

```
PlotAbrmInt[0];  PlotAbrmInt[4];
```
Show integrand surface for $m = 0$ and 4, Figure 19.3.2.

```
(* Abramowitz function: Numerical *)

AbrmwtzTable = N[Table[
  {m,x,Abrmwtz[m,x]},{m,0,3},{x,0,1.5,0.5}],10]
```
Numerical values to 10 digits, as in Table 19.3.1.

The Spence cell [19.4]

Execute this cell to produce for the Spence integral (19.4.1) Figure 19.4.1 and numerical values, some of which appear in Table 19.4.1.

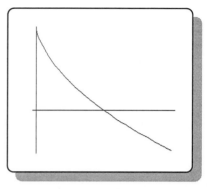

Exploring with Spence

Although the Spence integral—sketched in the adjacent motif—is well-behaved, the integrand in the *t* form in (19.4.1) becomes troublesome near *t* = 1 when *x* is near zero or unity. In Section 19.4 under Algorithm and Program, and also under Programming Notes, we explain our strategy for handling these trouble spots. Use *Mathematica* and modify the Spence cell to provide accurate check values for the C function.

To explore the behavior of the trouble zones, make a function then a graphics object to show the surface of the integrand of the Spence integral. A model for doing this is given in the Abramowitz cell immediately above.

```
(* Spence cell *)

(* Spence integral *)                        Define Spence integral as a
Spence[x_] := PolyLog[2,1-x];                dilogarithm by (19.4.2).

(* Spence integral: 2D views *)
Plot[Spence[x],{x,0,2},                       Curve of the Spence integral,
  PlotRange->All,AxesLabel->{"x","S(x) "}];  Figure 19.4.1.

(* Spence integral: Numerical *)
N[Table[{x,Spence[x]},{x,0,0.001,0.001}],10]  Numerical values of Spence
N[Table[{x,Spence[x]},{x,0.99,1.01,0.01}],10] integral to 10 digits for Table
N[Spence[2],10]                               19.4.1.
```

The Clausen cell [19.5]

If you execute this cell you will see Figures 19.5.1 and 19.5.2. You will also get numerical values, accurate to 10 digits, in Table 19.5.1.

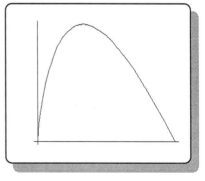

Exploring with Clausen

The integrand of Clausen's integral (19.5.2) becomes singular as *t* approaches zero, as shown in Figure 19.5.1. Therefore, if you want high accuracy for the integral as computed by the C function Clausen—which uses the infinite series in (19.5.2)—you will need a large number of terms. The *Mathematica* functions ClausenSum and ClausenInt in the Clausen cell use summation and integration methods, respectively. They can therefore be used to explore which method is better for a particular range of θ values.

```
(* Clausen cell *)

(* Clausen's integral *)

<<NumericalMath`NLimit`
```
Needs package for Euler limit of sum.

```
(* Clausen integral by summation *)
ClausenSum[theta_] := EulerSum[Sin[k*theta]/k^2,
  {k,1,Infinity},WorkingPrecision->40,
  Terms->300,ExtraTerms->50]
```
Defining Clausen integral by summation in (19.5.2).

```
(* Clausen integral by integration *)
ClausenInt[theta_] := - NIntegrate[Log[2*Sin[t/2]],
  {t,0,theta}, AccuracyGoal->12,PrecisionGoal->12,
  WorkingPrecision->12]
```
Defining Clausen integral directly in (19.5.2).

```
(* Integrand of Clausen integral: 2D view *)
Plot[-Log[2*Sin[t/2]],{t,10^-6,1.6},
  PlotLabel->"Integrand of C(theta)",
  Ticks->{{0,0.5,1,1.5},None},PlotPoints->40,
  PlotRange->All,AxesLabel->{" t","Log[2Sin[t/2] "}];
```
Curve of integrand of Clausen integral, Figure 19.5.1.

```
(* Clausen integral: 2D view *)
Plot[ClausenInt[theta],{theta,0,N[Pi]},
  PlotLabel->"Clausen's integral",
  Ticks->{{0,0.5,1.0,1.5,2.0,2.5,3.0},None},
  PlotRange->All,AxesLabel->{" theta","C(theta) "}];
```
Curve of Clausen integral as function of θ, Figure 19.5.2.

```
(* Clausen integral: Numerical by Integration *)
N[Table[ClausenInt[theta],
  {theta,N[Pi]/3,N[Pi],N[Pi]/3}],10]
```
Numerical values to 10 digits for Table 19.5.1.

The `Voigt` cell [19.6]

If you execute this cell you will visualize the Voigt profile and Hjerting functions through Figures 19.6.1–19.6.3. Numerical values for Table 19.6.1 are also generated.

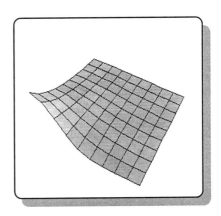

Exploring with `Voigt`

If you require high accuracy or speed of execution, or both, you should follow up the references in Section 19.6, which describe a wide variety of techniques. The *Mathematica* functions given here are typically about three orders of magnitude slower in execution than a compiled C program for the function.

```
(* Voigt cell *)

(* Voigt profile function *)

(* Voigt function from
    complementary error function *)
Voigt[u_,a_] :=
  Re[Exp[(a + I u)^2]*Erfc[a + I u]]

(* Voigt function: 3D views *)
VgtPlt3D = Plot3D[Voigt[u,a],{u,0,3},{a,0,2},
  Ticks->{Automatic,Automatic,None},
  PlotRange->All,AxesLabel->{"u","a",""}];

   (* Voigt function: 2D views *)
(* Define plot for fixed  a  *)
VoigtPlot[a_] := Plot[Voigt[u,a],{u,-3,3},
  PlotRange->{{-3,3},{0,1}},
  Axes->{True,False}, Ticks->None,
  DisplayFunction->Identity]

(* Layout of overlapping plots *)
RectPlot[Nummer_] := Rectangle[
  {0.4*Nummer,-0.05*Nummer},
  {1+0.4*Nummer,1-0.05*Nummer},VgtPlt[Nummer]]

(* Make 5 plots *)
For[ Nummer=0, Nummer <= 4, Nummer++,
     VgtPlt[Nummer] =
     VoigtPlot[0.4163*Nummer]]

(* and overlay them *)
Waldemar = Show[Graphics[
  {RectPlot[0],RectPlot[1],RectPlot[2],
  RectPlot[3],RectPlot[4]}],
  DisplayFunction->$DisplayFunction];

(* Compare Gaussian & Lorentzian of same FWHM
    and peak value *)
Gauss[nu_] := Exp[-nu^2]

GamSqdD4 = Log[2];

Lorentz[nu_] := GamSqdD4/(nu^2+GamSqdD4)

GaussPlot = Plot[Gauss[nu],{nu,-3,3},
  PlotRange->{{-3,3},{0,1}},
  Axes->{True,False}, Ticks->None,
  DisplayFunction->Identity];

LorentzPlot = Plot[Lorentz[nu],{nu,-3,3},
  PlotRange->{{-3,3},{0,1}},
  Axes->{True,False}, Ticks->None,
  PlotStyle->Dashing[{0.02,0.02}],
  DisplayFunction->Identity];
```

Define Voigt function in terms of complex error function by (19.6.9).

Surface of Voigt function in terms of u and a, Figure 19.6.1.

Define curve of Voigt function with a fixed and u varying.

Define layout of overlaying plots slightly displaced from each other.

Make the 5 plots into elements of array VgtPlt[].

Show the overlaid plots, Figure 19.6.2 left.

Gaussian curve.

Lorentzian curve with same FWHM and peak value as the Gaussian.

```
GaussLorentz = Show[GaussPlot,LorentzPlot,
  DisplayFunction->$DisplayFunction];
```

Show together, Figure 19.6.2 right.

```
(* Voigt function: Numerical *)
a= 0.0; N[Table[{u,Voigt[u,a]},{u,0,1,0.5}],10]
a= 0.1; N[Table[{u,Voigt[u,a]},{u,0,1,0.5}],10]
a= 0.2; N[Table[{u,Voigt[u,a]},{u,0,1,0.5}],7]
```

Numerical values to appropriate accuracy, as in Table 19.6.1.

```
(* Hjerting functions *)
FDaw[u_] := Re[I*Sqrt[Pi]*Erf[-
              I*u]/(2*Exp[u^2])]
```

Dawson integral.

```
HZero[u_] := Exp[-u^2]
HOne[u_] := 2*(u*FDaw[u]-1)/Sqrt[Pi]
HTwo[u_] := (1-2*u^2)*Exp[-u^2]
HThree[u_] := 4*((3-2*u^2)*u*FDaw[u]+u^2-1)/
              (3*Sqrt[Pi])
```

Define Hjerting functions by (19.6.12), (19.6.13).

```
SetOptions[Plot,PlotRange->{{0,2},{-1.5,1.5}},
  Ticks->{{0,1,2},{-1,0,1}},
  AxesLabel->{"u",None},
  DisplayFunction->Identity];
```

Plotting options.

```
FDawPlot = Plot[FDaw[u],{u,0,2}];
HZeroPlot = Plot[HZero[u],{u,0,2},
  PlotStyle->Dashing[{0.01,0.01}]];
HOnePlot = Plot[HOne[u],{u,0,2},
  PlotStyle->Dashing[{0.02,0.02}]];
HTwoPlot = Plot[HTwo[u],{u,0,2},
  PlotStyle->Dashing[{0.04,0.04}]];
HThreePlot = Plot[HThree[u],{u,0,2},
  PlotStyle->Dashing[{0.06,0.06}]];
```

Make graphic objects for Dawson integral and for Hjerting functions $n = 0$ to $n = 3$.

```
Hjerting = Show[FDawPlot,HZeroPlot,HOnePlot,
  HTwoPlot,HThreePlot,
  DisplayFunction->$DisplayFunction];
```

Show curves together, as in Figure 19.6.3.

Angular Momentum Coupling Coefficients [19.7]

We do not provide a *Mathematica* cell for angular momentum coupling coefficients—3-*j*, 6-*j*, and 9-*j* coefficients—described in Section 19.7. The 3-*j* and 6-*j* coefficients are built-in functions in *Mathematica*, being the `ThreeJSymbol` and `SixJSymbol` functions, respectively. The results are in terms of rational fractions, which is an advantage if exact values of the coefficients are needed. The C functions in Section 19.7 provide decimal approximations to these coefficients. Ways of testing these functions are also described in this section.

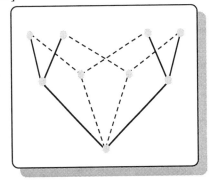

The references cited in Section 19.7 describe some of the ways these functions and the 9-*j* function can be visualized. Generally, however, the large number of parameters of each function (six for 3-*j* and 6-*j*, nine for 9-*j*) makes it difficult to visualize them, except for schematic graphical representations such as that for the 9-*j* coefficient in the motif on the previous page.

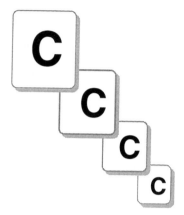

Chapter 21

THE C DRIVER PROGRAMS

This chapter of the *Atlas for Computing Mathematical Functions* contains all the C driver programs. These can be used to test the C functions that are described and listed in the *Atlas* and are also provided on the CD-ROM.

21.1   INTRODUCTION TO THE C DRIVER PROGRAMS

The C driver programs in this chapter are to be used with the C functions in Part I of the *Atlas*, where the functions are described from mathematical, pictorial, and programming viewpoints. The C functions and numerical test values for them are in Chapters 4 to 19. In the machine-readable forms of the programs the driver program immediately precedes the functions that it tests. In the *Atlas* we have put most of the drivers at the end, since they are often relatively long and consist mostly of control statements, input, and output, which would distract your attention from the C functions—the objects of primary interest.

This chapter brings together the driver (main) programs that test the C functions by allowing input (by scanf), function execution, and output (by using printf). A loop structure—which is very similar for all the drivers—allows you to use as many test sets of input values as you wish. Loops are terminated by testing for input values that are usually inappropriate for the functions to compute, such as arguments that would produce singularities of the functions or an accuracy control that requests exact computations (eps = 0). You are reminded of the termination values each time through the loop.

Test values for the functions are referenced in this chapter by the corresponding tables in Part I. Sometimes the main text also mentions tables in other sources, which may be useful if you wish to extend the range of parameters over which you use the function.

21.2   HOW THE C DRIVERS ARE ORGANIZED

The C driver programs in this chapter are ordered exactly as in Part I, where the functions are described. Also, section and subsection numbers are the same as the chapter and section numbers in Part I. The matching cell for *Mathematica* is in the section of Chapter 20 that matches the chapter in Part I. For example, if you know that the mathematics, graphics, and numerics of the incomplete gamma and beta functions are described in Section 6.3, then the C driver program will be in Section 21.6.3, and the corresponding *Mathematica* cell will be in the notebook in Section 20.6.

Each subsection of this chapter has a heading preceding the driver program listing. Below this heading we give the names of the C functions and of the driver program that calls them. We also point to the tables in Part I that contain test values for the functions.

21.3   ANNOTATIONS TO THE C DRIVER PROGRAMS

We provide annotations in the right margins of the listings of the C driver programs of this chapter. Such annotations do not appear in the source programs in machine-readable form, and should not be entered if you key the programs into a computer.

The main purpose of the annotations is to clarify the purpose, limitations, and possible peculiarities of the programs. The danger sign is used to alert you to hazards when using the C drivers. You should also refer to the main text for other warning signs, especially those referring to confusing function definitions and to limitations of our algorithms.

21.4   ELEMENTARY TRANSCENDENTAL FUNCTIONS

In this section we list and annotate the C driver programs for testing the elementary transcendental functions described in Chapter 4 of the *Atlas*.

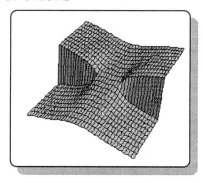

21.4.1   Exponential and Logarithmic Functions

C Functions and Driver; Test Values
ExpZ, LogZ, in driver Exp&Log.c; Tables
4.1.1, 4.1.2

```
#include <stdio.h>
#include <math.h>
main()
{
/* Function Names: ExpZ & LogZ */
/* Exponential and Logarithm (principal value)
   of complex argument  z = x + i y */
void ExpZ(double x, double y,
          double *ExpZRe, double *ExpZIm),
   LogZ(double x, double y,
          double *LogZRe, double *LogZIm);
double x,y,ExpZRe,ExpZIm,LogZRe,LogZIm;
printf("ExpZ & LogZ Tests");
x = 1;
while ( !(x == 0 && y == 0) )
   {
   printf("\n\nInput x,y ( x=0 & y=0  to end): ");
   scanf("%lf%lf",&x,&y);
```

Logarithm principal values are returned for complex arguments.

```
if ( x == 0 && y == 0 )
  { printf("\nEnd ExpZ & LogZ Tests");
   exit(0); }
else
  {
  ExpZ(x,y,&ExpZRe,&ExpZIm);
  printf
  ("\n ExpZ(%lf+i%lf) = %1.9E+i(%.9E)",
    x,y,ExpZRe,ExpZIm);
  LogZ(x,y,&LogZRe,&LogZIm);
  printf
  ("\n LogZ(%lf+i%lf) = %.9E+i(%.9E)",
    x,y,LogZRe,LogZIm);
  } } }
```

Driver loop ends at the singularity of the logarithm, $z = 0$ ($x = 0$, $y = 0$).

Compute and print to 10 digits the exponential then logarithm with same input x and y, Tables 4.1.1 and 4.1.2.

21.4.2   Circular and Inverse Circular Functions

For clarity we have separate driver programs for circular and inverse-circular functions.

Circular Functions: C Functions and Driver; Test Values

CosZ, SinZ, TanZ, in driver CosSinTan.c; Table 4.2.1

```
#include <stdio.h>
#include <math.h>
main()
{
/* Function Names: CosZ, SinZ, & TanZ */
/* Cosine, Sine, & Tangent
   of complex argument  z = x + i y  */
void CosZ(double x, double y,
          double *CosZRe, double *CosZIm),
    SinZ(double x, double y,
          double *SinZRe, double *SinZIm),
    TanZ(double x, double y,
          double *TanZRe, double *TanZIm);
double x,y,CosZRe,CosZIm,
  SinZRe,SinZIm,TanZRe,TanZIm;
printf("CosZ, SinZ, & TanZ Tests");
x = 1;
while ( x != 0  || y != 0 )
  {
  printf("\n\nInput x,y ( x=0 & y=0 to end): ");
  scanf("%lf%lf",&x,&y);
  if ( x == 0 && y == 0 )
    { printf("\nEnd CosZ, SinZ, & TanZ Tests");
     exit(0); }
  else
    {
    CosZ(x,y,&CosZRe,&CosZIm);
    printf("\n CosZ(%lf+i%lf) = %.9E+i(%.9E)",
      x,y,CosZRe,CosZIm);
    SinZ(x,y,&SinZRe,&SinZIm);
    printf("\n SinZ(%lf+i%lf) = %.9E+i(%.9E)",
      x,y,SinZRe,SinZIm);
    TanZ(x,y,&TanZRe,&TanZIm);
    printf("\n TanZ(%lf+i%lf) = %.9E+i(%.9E)",
      x,y,TanZRe,TanZIm);
    } } }
```

Cosine, sine, and tangent are computed for complex arguments input.

Terminate if trivial input argument $z = 0$.

Cosine, sine, and tangent to 10 digits, Table 4.2.1.

Inverse Circular Functions: C Functions and Driver; Test Values

ArcCosZ, ArcSinZ, ArcTanZ, in driver InvCosSinTan.c; Table 4.2.3

```
#include <stdio.h>
#include <math.h>

main()
{
/* Function Names: InvCosZ, InvSinZ, & InvTanZ */
/* Inverse Cosine, Sine, & Tangent
   of complex argument  z = x + i y  */
void InvCosZ(double x, double y,
           double *InvCosZRe, double *InvCosZIm),
   InvSinZ(double x, double y,
           double *InvSinZRe, double *InvSinZIm),
   InvTanZ(double x, double y,
           double *InvTanZRe, double *InvTanZIm);

double x,y,InvCosZRe,InvCosZIm,InvSinZRe,InvSinZIm,
  InvTanZRe,InvTanZIm;

printf("InvCosZ, InvSinZ, & InvTanZ Tests");
x = 1;
while (  !( x == 0 && y*y == 1)  )
  {
  printf("\n\nInput x,y ( x=0 & y*y=1 to end): ");
  scanf("%lf%lf",&x,&y);
  if (  ( x == 0 && y*y == 1)  )
    {
    printf(
      "\nEnd InvCosZ, InvSinZ, & InvTanZ Tests");
    exit(0);
    }
  else
    {
    InvCosZ(x,y,&InvCosZRe,&InvCosZIm);
    printf
    ("\n InvCosZ(%lf+i%lf) = %.9E+i(%.9E)",
      x,y,InvCosZRe,InvCosZIm);
    InvSinZ(x,y,&InvSinZRe,&InvSinZIm);
    printf
    ("\n InvSinZ(%lf+i%lf) = %.9E+i(%.9E)",
      x,y,InvSinZRe,InvSinZIm);
    InvTanZ(x,y,&InvTanZRe,&InvTanZIm);
    printf
    ("\n InvTanZ(%lf+i%lf) = %.9E+i(%.9E)",
      x,y,InvTanZRe,InvTanZIm);
    } } }
```

Inverse cosine, sine, and tangent are computed for complex arguments input. Principal values use branch cuts in Figure 4.2.3 and Table 4.2.2.

Terminate if (4.2.7) would fail for arctan$z$.

Principal value of inverse cosine,

inverse sine,

and inverse tangent, all to 10 digits, in Table 4.2.3.

21.4.3   Hyperbolic and Inverse Hyperbolic Functions

We have separate driver programs for hyperbolic and inverse-hyperbolic functions.

Hyperbolic Functions: C Functions and Driver; Test Values

CoshZ, SinhZ, TanhZ, in driver HypCosSinTan.c; Table 4.3.1

```
#include <stdio.h>
#include <math.h>
main()
{
/* Function Names: CoshZ, SinhZ, & TanhZ */
/* Hyperbolic Cosine, Sine, & Tangent
   of complex argument  z = x + i y */
void CoshZ(double x, double y,
          double *CoshZRe, double *CoshZIm),
   SinhZ(double x, double y,
          double *SinhZRe, double *SinhZIm),
   TanhZ(double x, double y,
          double *TanhZRe, double *TanhZIm);
double x,y,CoshZRe,CoshZIm,SinhZRe,
   SinhZIm,TanhZRe,TanhZIm;
printf("CoshZ, SinhZ, & TanhZ Tests");
x = 1;
while ( x != 0  || y != 0 )
  {
  printf("\n\nInput x,y ( x=0 & y=0 to end): ");
  scanf("%lf%lf",&x,&y);
  if ( x == 0 && y == 0 )
    { printf("\nEnd CoshZ, SinhZ, & TanhZ Tests");
    exit(0); }
  else
    { CoshZ(x,y,&CoshZRe,&CoshZIm);
    printf("\n CoshZ(%lf+i%lf) = %.9E+i(%.9E)",x,y,
      CoshZRe,CoshZIm);
    SinhZ(x,y,&SinhZRe,&SinhZIm);
    printf("\n SinhZ(%lf+i%lf) = %.9E+i(%.9E)",x,y,
      SinhZRe,SinhZIm);
    TanhZ(x,y,&TanhZRe,&TanhZIm);
    printf("\n TanhZ(%lf+i%lf) = %.9E+i(%.9E)",x,y,
      TanhZRe,TanhZIm);
    } } }
```

Hyperbolic cosine, sine, and tangent computed for complex arguments input.

Terminate if trivial input argument $z = 0$.

Cosh,

sinh,

and tanh, all to 10 digits, Table 4.3.1.

Inverse Hyperbolic Functions: C Functions and Driver; Test Values
`ArcCoshZ`, `ArcSinhZ`, `ArcTanhZ`, in driver `InvCSTHyp.c`; Table 4.3.3

```
#include <stdio.h>
#include <math.h>
main()
{
/* Function Names: InvCoshZ, InvSinhZ, & InvTanhZ */
/* Inverse Hyperbolic Cosine, Sine, & Tangent
   of complex argument  z = x + i y */
void InvCoshZ(double x, double y,
     double *InvCoshZRe, double *InvCoshZIm),
   InvSinhZ(double x, double y,
     double *InvSinhZRe, double *InvSinhZIm),
   InvTanhZ(double x, double y,
     double *InvTanhZRe, double *InvTanhZIm);
double x,y,InvCoshZRe,InvCoshZIm,InvSinhZRe,
   InvSinhZIm,InvTanhZRe,InvTanhZIm;
```

Inverse hyperbolic cosine, sine, and tangent computed for complex arguments input. Principal values use branch cuts in Figure 4.3.3 and Table 4.3.2.

```
printf("InvCoshZ, InvSinhZ, & InvTanhZ Tests");
y = 1;
while (  !( y == 0 && x*x == 1)  )
  {
  printf("\n\nInput x,y ( x*x=1 & y=0 to end): ");
  scanf("%lf%lf",&x,&y);
  if (  ( y == 0 && x*x == 1)  )
    { printf(
      "\nEnd InvCoshZ, InvSinhZ, & InvTanhZ Tests");
    exit(0); }
  else
    {
    InvCoshZ(x,y,&InvCoshZRe,&InvCoshZIm);
    printf
    ("\n InvCoshZ(%lf+i%lf) = %.9E+i(%.9E)",
        x,y,InvCoshZRe,InvCoshZIm);
    InvSinhZ(x,y,&InvSinhZRe,&InvSinhZIm);
    printf
    ("\n InvSinhZ(%lf+i%lf) = %.9E+i(%.9E)",
        x,y,InvSinhZRe,InvSinhZIm);
    InvTanhZ(x,y,&InvTanhZRe,&InvTanhZIm);
    printf
    ("\n InvTanhZ(%lf+i%lf) = %.9E+i(%.9E)",
        x,y,InvTanhZRe,InvTanhZIm);
    } } }
```

Driver `InvCSTHyp.c`
continued

Terminate if (4.3.8)
would fail for arctanh z.

Principal value of
inverse hyperbolic
cosine,

inverse hyperbolic sine,

and inverse hyperbolic
tangent, all to 10 digits,
Table 4.3.3.

21.5   EXPONENTIAL INTEGRALS AND RELATED FUNCTIONS

This section has annotated listings of the C driver
programs for testing the C functions for exponential,
cosine, and sine integrals, which are described in
Chapter 5 of the *Atlas*.

21.5.1   Exponential and Logarithmic Integrals

Exponential Integrals:
C Functions and Driver; Test Values
`ExpIntFirst`, `ExpIntSecond`, in driver `Exp-`
`Ints.c`; Tables 5.1.1, 5.1.2

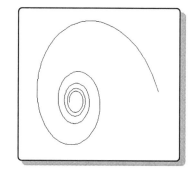

```
#include <stdio.h>
#include <math.h>

/* nearly smallest floating point number */
#define MINFP 1.0e-30

main()
{
/* Function Names: ExpIntFirst, ExpIntSecond */
/* Exponential Integrals of First & Second Kind */
double ExpIntFirst(int n, double x, double eps),
   x,eps,ExpIntSecond(double x, double eps);
int n;
```

Control for underflow
in continued fraction
expansion (5.1.7) of
exponential integral of
first kind.

```
printf("Exponential Integral Tests");
x = 1.0;
while ( x != 0 )
  {
  printf("\n\nInput n,x,eps ( x = 0 to end): ");
  scanf("%i%lf%lf",&n,&x,&eps);
  if ( x == 0 )
    { printf
    ("\nEnd Exponential Integral Tests");
    exit(0); }
  else
    {
    printf
    ("\n E(%i,%lf) = %.9E",n,x,ExpIntFirst(n,x,eps));
    printf
    ("\n Ei(%lf) = %.9E",x,ExpIntSecond(x,eps));
    } } }
```

Accuracy control eps .

Terminate if constraint
in (5.1.2) is violated.

Exponential integrals of
first and second kinds,
Table 5.1.1, 5.1.2.

Logarithmic Integral:
C Function and Driver; Test Values
LogIntegral, in driver LogInt.c; Table 5.1.3

```
#include <stdio.h>
#include <math.h>
/* nearly smallest floating point number */
#define MINFP 1.0e-30
main()
{
/* Function Name: LogIntegral */
/* Logarithmic Integral  */
double LogIntegral(double x, double eps),x,eps;
printf("Logarithmic Integral Tests");
x = 1.1;
while ( x > 1.0 )
  {
  printf("\n\nInput x,eps ( x = 1 to end): ");
  scanf("%lf%lf",&x,&eps);
  if ( x == 1.0 )
    { printf
    ("\nEnd Logarithmic Integral Tests");
    exit(0); }
  else
    { printf
    ("\n li(%lf) = %.9E",x,LogIntegral(x,eps));
    } } }
```

Control for underflow
in continued fraction
expansion (5.1.7) of
exponential integral of
first kind, which is
underlying function for
the logarithmic integral.
See Section 5.1.3.

Accuracy control eps .

Terminate if constraint
in (5.1.13) is violated.

Logarithmic integral to
10 digits, Table 5.1.3.

21.5.2   Cosine and Sine Integrals

C Functions and Driver; Test Values
CosIntegral, SinIntegral, in driver Cos&SinIntegral.c; Tables 5.2.1, 5.2.2

The driver program is listed on the following page.

```
#include <stdio.h>
#include <math.h>
main()
{
/* Function Names: CosIntegral & SinIntegral */
/* Cosine & Sine Integrals of positive argument x */
double CosIntegral(double x, double eps),
       SinIntegral(double x, double eps),x,eps;
printf("Cosine & Sine Integral Tests");
x = 1.0;
while ( x > 0 )
  {
  printf("\n\nInput x,eps ( x = 0 to end): ");
  scanf("%lf%lf",&x,&eps);
  if ( x == 0 )
    {
    printf("\nEnd Cosine & Sine Integral Tests");
    exit(0);
    }
  else
    {
    printf("\n CosIntegral(%lf) = %.9E",x,
     CosIntegral(x,eps));
    printf("\n SinIntegral(%lf) = %.9E",x,
     SinIntegral(x,eps));
    } } }
```

Driver program for cosine and sine integrals.

Accuracy control `eps`.

Terminate if $x = 0$, since cosine integral has a logarithmic singularity there. See (5.2.5).

Cosine and sine integrals to 10 digits, Tables 5.2.1, 5.2.2.

21.6   GAMMA AND BETA FUNCTIONS

In this section we give annotated listings of the C driver programs for testing C functions for the gamma and beta, psi and polygamma, and incomplete gamma and beta functions that are described in Chapter 6 of the *Atlas*.

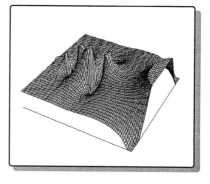

21.6.1   Gamma Function and Beta Function

Gamma Function:
C Function and Driver; Test Values
LogGamma in the driver LogGamma.c;
Table 6.1.1

```
#include <stdio.h>
#include <math.h>
main()
{
/* Function Name: LogGamma */
/* Logarithm of Gamma Function */
double LogGamma(double x),x;
```

Logarithm of gamma function from asymptotic expansion.

```
printf("Gamma Function Tests");
x = 1.0;
while ( x != 0 )
  {
  printf
  ("\n\nInput x ( x = 0 to end): ");
  scanf("%lf",&x);
  if ( x == 0 )
    { printf
    ("\nEnd Gamma Function Tests");
    exit(0); }
  else
    {
    printf
    ("\nLog[Gamma(%lf)] = %.9E",x,LogGamma(x));
    } } }
```

Terminate if *x* is zero,
violating use of (6.1.2).

Log of gamma function to 10
digits, Table 6.1.1.

Beta Function:
C Function and Driver; Test Values
BetaFunc in the driver BetaFunc.c; Table 6.1.2

```
#include <stdio.h>
#include <math.h>
main()
{
/* Function Name: BetaFunc */
/* Beta Function */
double BetaFunc(double x, double w),x,w;
printf("Beta Function Tests");
x = 1.0;
while ( x > 0 )
  {
  printf
  ("\n\nInput x,w ( x = 0 to end): ");
  scanf("%lf%lf",&x,&w);
  if ( x <= 0 )
    { printf
    ("\nEnd Beta Function Tests");
    exit(0); }
  else
    {
    printf
    ("\nBeta(%lf,%lf)] = %.9E",x,w,BetaFunc(x,w));
    } } }
```

Beta function is computed
in terms of log of gamma
function (included with
this driver).

Terminate if *x* would
violate use of (6.1.2) for
log gamma function.

Beta function to 10 digits,
Table 6.1.2.

21.6.2   Psi (Digamma) and Polygamma Functions

Psi Function:
C Function and Driver; Test Values
PsiFunc in the driver Psi.c; Table 6.2.3

The driver program is listed on the following page.

```
#include <stdio.h>
#include <math.h>
main()
{
/* Function Name: PsiFunc */
/* Psi (Digamma) Function */
double PsiFunc(double x),x;
printf("Psi Function Tests");
x = 1.0;
while ( x > 0 )
  {
  printf
  ("\n\nInput x ( x = 0 to end): ");
  scanf("%lf",&x);
  if (  x == 0 )
    { printf
    ("\nEnd Psi Function Tests");
    exit(0); }
  else
    {
    printf
    ("\nPsi(%lf) = %.9E",x,PsiFunc(x));
    } } }
```

Psi (digamma) function continued from previous page. Driver `Psi.c`.

Terminate if singularity of psi function at $x = 0$ (Figure 6.2.3) is attempted.

Print psi to 10 digits.

Polygamma Functions:
C Functions and Driver; Test Values
`Polygamma1`, `Polygamma2`, in driver `PolyGam.c`; Table 6.2.5

```
#include <stdio.h>
#include <math.h>
main()
{
/* Function Names: PolyGamma1, PolyGamma2 */
/* PolyGamma Functions for  n=1,2  */
double PolyGamma1(double x),PolyGamma2(double
x),x;
int n;
printf("Polygamma Function Tests");
n = 1;
while ( n > 0 )
  {
  printf
  ("\n\nInput x ( x = 0 to end): ");
  scanf("%lf",&x);
  if (  x <= 0 )
    { printf
    ("\nEnd Polygamma Function Tests");
    exit(0); }
  else
    {
    printf
      ("\nPsi(1,%lf) = %.9E",x,PolyGamma1(x));
    printf
      ("\nPsi(2,%lf) = %.9E",x,PolyGamma2(x));
    } } }
```

Polygamma functions for $n = 1, 2$. For $n > 2$ see Programming Notes in Section 6.2.2.

Terminate if singularity of polygamma function at $x = 0$ is input.

Trigamma function ($n = 1$), then the tetragamma function ($n = 2$). For nomenclature see Table 6.2.1.

21.6.3   Incomplete Gamma and Beta Functions

Incomplete Gamma Function:
C Function and Driver; Test Values
IncGamma in driver IncGamma.c; Table 6.3.1

```
#include <stdio.h>
#include <math.h>

#define SMALL 1.0e-30

main()
{
/* Function Name: IncGamma */
/* Regularized Incomplete Gamma Function */
double IncGamma(double a, double x, double eps),
 a,b,x,eps;
printf
 ("Regularized Incomplete Gamma Function Tests");
eps = 0.1;
while ( eps > 0 )
  {
  printf
  ("\n\nInput a,x,eps ( eps = 0 to end): ");
  scanf("%lf%lf%lf",&a,&x,&eps);
  if ( eps <= 0 )
    { printf
    ("\nEnd  Incomplete Gamma Function  Tests");
    exit(0); }
  else
    {
    printf
    ("\nP(%lf,%lf) = %.9E",
    a,x,IncGamma(a,x,eps));
    } } }
```

Control for underflow in continued fraction expansion (6.3.5).

Accuracy control eps .

Terminate if exact convergence of power series (6.3.3) is attempted.

Regularized incomplete gamma to accuracy eps, eps = 10^{-10} in Table 6.3.1.

Incomplete Beta Function:
C Function and Driver; Test Values
IncBeta in driver IncBeta.c; Table 6.3.2

```
#include <stdio.h>
#include <math.h>
main()
{
/* Function Name: IncBeta */
/* Incomplete Beta Function */
double IncBeta
  (double x, double a, double b, double eps),
  a,b,x,eps;
printf("Incomplete Beta Function Tests");
```

Input eps controls the accuracy of continued-fraction expansion (6.3.9).

```
eps = 0.1;
while ( eps > 0 )
  {
  printf
  ("\n\nInput x,a,b,eps ( eps = 0 to end): ");
  scanf("%lf%lf%lf%lf",&x,&a,&b,&eps);
  if ( eps <= 0 )
    { printf
    ("\nEnd  Incomplete Beta Function  Tests");
    exit(0); }
  else
    {
    printf
    ("\nI(%lf,%lf,%lf) = %.9E",
    a,b,x,IncBeta(x,a,b,eps));
    } } }
```

Incomplete beta function continued from previous page.

Accuracy control eps.

Terminate if exact convergence of power series (6.3.3) is attempted.

Incomplet beta function to accuracy eps. In Table 6.3.2 eps $= 10^{-10}$.

21.7   COMBINATORIAL FUNCTIONS

This section has annotated listings of the C driver programs for testing C functions for many combinatorial functions: factorials and rising factorials, binomials and mutinomials, Stirling numbers, Fibonacci and Lucas polynomials, Bernoulli and Euler numbers and polynomials. These functions are described and shown in Chapter 7 of the *Atlas*.

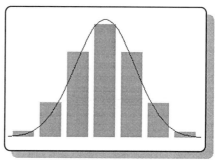

Most of the combinatorial functions in Chapter 7 are programmed in long-integer arithmetic. Although using long integers delays overflow, it will usually not prevent it if the arguments of the functions are large. The C functions provided can be adapted easily to use double-floating arithmetic, thus greatly extending the range of arguments that can be handled, but at the expense of approximate results whose errors depend upon the number of floating-point digits used in your computer system. Read carefully the material indicated by the warning signs!

21.7.1   Factorials and Rising Factorials

Factorial Function:
C Function and Driver; Test Values
Factorial in the driver Factorial.c; Table 7.1.1

```
#include <stdio.h>
#include <math.h>
main()
{
/* Function Name: Factorial */
int n;
long int i,Factorial();
```

```
printf("Factorial Tests");
n = 1;
while ( n >= 0 )
  {
  printf("\nInput n >= 0 ( n < 0 to end): ");
  scanf("%i",&n);
  if ( n < 0 )
    { printf("\nEnd Factorial Tests");
    exit(0); }
  else
    {
    printf("\n %i! = %li",n,Factorial(n));
    } } }
```

Largest *n* depends on word size. See Figure 7.1.2.

Terminate if factorial argument is inappropriate.

Long-integer factorial.

Rising Factorial Function:
C Function and Driver; Test Values
`RisingFactorial` in the driver `RsngFctrl.c`; Table 7.1.2

```
#include <stdio.h>
#include <math.h>
main()
{
/* Function Name: RisingFactorial */
int n,m;
long int RisingFactorial();
printf("Rising Factorial Tests");
n = 1;
while ( n >= 0 )
  {
  printf("\nInput n,m >= 0 (n<0 to end): ");
  scanf("%i%i",&n,&m);
  if ( n < 0 )
    { printf("\nEnd RisingFactorial Tests");
    exit(0); }
  else
    {
    printf("\n (%i)sub_%i = %li", n,m,
           RisingFactorial(n,m));
    } } }
```

Input nonnegative *n* and *m*.

Terminate if *n* < 0.

Long-integer rising factorial.

21.7.2   Binomial and Multinomial Coefficients

Binomial Coefficients:
C Function and Driver; Test Values
`Binomial` in the driver `Binomial.c`; Table 7.2.1

```
#include <stdio.h>
#include <math.h>
main()
{
/* Function Name: Binomial */
/* Needs RisingFactorial, Factorial */
int n,m;
long int Binomial(int n, int m),
  Factorial(),RisingFactorial(int n, int m);
```

Binomials are computed by (7.2.5) in terms of rising factorials and factorials, included in this file.

```
printf("Binomial Tests");
n = 1;
while ( n >= 0 )
  {
  printf("\n\nInput n,m ( n<0 to end): ");
  scanf("%i%i",&n,&m);
  if ( n < 0 )
    { printf("\nEnd Binomial Tests");
    exit(0); }
  else
    {
    printf("(%i %i) = %li", n, m,
      Binomial(n,m));
    } } }
```

Binomial *continued*

Terminate for illegal input.
Binomial will overflow for
large *n*; see Figure 7.1.2 and
Section 7.2.1.

Long-integer binomial
coefficient.

Multinomial Coefficients:
C Function and Driver; Test Values
Multinomial in the driver Multinomial.c; Table 7.2.2

```
#include <stdio.h>
#include <math.h>

#define MAX 11

main()
{
/* Function Name: Multinomial */
/* Needs Factorial, RisingFactorial */

int n,m,sumn,i,bin[MAX];
long int Multinomial(),
  Factorial(int n),RisingFactorial(int n, int m);
printf("Multinomial Tests");
n = 1;
while ( n >= 0 )
  {
  printf("\n\nInput n,m ( n<0 to end): ");
  scanf("%i%i",&n,&m);
  if ( n < 0 )
    { printf("\nEnd  Multinomial Tests");
    exit(0); }
  else
    {
    printf("\n Input %i values of n: ",m-1);
    /* last value is calculated from others */
    sumn = 0;
    for ( i = 1; i <= m-1; i++ )
      {
      scanf("%i",&bin[i]);
      sumn += bin[i];
      }
    bin[m] = n - sumn;
    if ( bin[m] < 0 )
      { printf("\n!! Sum exceeds n. Try again"); }
```

Maximum value of $m-1$
for multinomial. Recom-
pile if larger *m* needed.

Terminate if improper *n* is
used.

Input the first $m-1$ values
of n_i.

Calculate the last n_i from
sum rule in (7.2.6) and
check that it's non-
negative.

```
else
  {
  printf("\n (%i; ", n);
  for ( i = 1; i <= m; i++ )
        { printf("%i ",bin[i]); }
  printf(") = %li",
        Multinomial(n,m,bin));
  } } } }
```

Compute and print the multinomial in the same format as (7.2.6).

Multinomial will over-flow for large *n*; see Section 7.2.2. ⚠

21.7.3   Stirling Numbers of the First and Second Kinds

C Functions and Driver; Test Values
Stirl_1, Stirl_2, in driver Stirling.c; Tables 7.3.1, 7.3.2

```
#include <stdio.h>
#include <math.h>

main()

{
/* Function Names: Stirl_1, Stirl_2 */
int n,m;
long int Stirl_1(),Stirl_2();
printf("Stirling Number Tests");
n = 1;

while ( n >= 0 )
  {
  printf("\n\nInput n,m ( n<0 to end): ");
  scanf("%i%i",&n,&m);
  if ( n < 0 )
    { printf("\nEnd Stirling Number Tests");
    exit(0); }

  else
    {
    printf("Stirling1 = %li, Stirling2 = %li",
          Stirl_1(n,m),Stirl_2(n,m));
  } } }
```

Stirling numbers of both kinds are computed by this driver.

Terminate if improper *n* is input.

Stirling numbers may overflow for large *n*. See Sections 7.3.1, 7.3.2. ⚠

Compute and print Stirling numbers of both kinds.

21.7.4   Fibonacci and Lucas Polynomials

Fibonacci Polynomials:
C Function and Driver; Test Values
Fibonacci in the driver Fibonacci.c; Table 7.4.1

```
#include <stdio.h>
#include <math.h>

main()
{
/* Function Name: Fibonacci */
int n;
double x,Fibonacci(int n, double x);
printf("Fibonacci polynomial Tests");
n = 1;

while ( n > 0 )
  {
  printf("\n\nInput n, x ( n<0 to end): ");
  scanf("%i%lf",&n,&x);
  if ( n <= 0 )
    { printf("\nEnd Fibonacci polynomial Tests");
    exit(0); }

  else
    {
    printf("Fibonacci(%i,%g) = %.9E",
            n,x,Fibonacci(n,x) );
  } } }
```

Driver `Fibonacci`

Input *n* and *x*. Note that meaning of *n* varies. See Section 7.4.1.
  Terminate if improper *n* is input.

Fibonacci polynomials to 10 digits, as in Table 7.4.1.

Lucas Polynomials:
C Function and Driver; Test Values
`Lucas` in the driver `Lucas.c`; Table 7.4.2

```
#include <stdio.h>
#include <math.h>
main()

{
/* Function Name: Lucas */
int n;
double x,Lucas(int n, double x);
printf("Lucas polynomial Tests");
n = 1;

while ( n >= 0 )
  {
  printf("\n\nInput n, x ( n<0 to end): ");
  scanf("%i%lf",&n,&x);
  if ( n < 0 )
    { printf("\nEnd Lucas polynomial Tests");
    exit(0); }

  else
    {
    printf("Lucas(%i,%g) = %.9E",
            n,x,Lucas(n,x) );
  } } }
```

Input *n* and *x*. Note that meaning of *n* varies. See Section 7.4.2.
  Terminate if improper *n* is input.

Lucas polynomials to 10 digits, as in Table 7.4.2.

21.8   NUMBER THEORY FUNCTIONS

In this section we have annotated listings of the C driver programs for testing the number theory functions described in Chapter 8 of the *Atlas*. Note that many of the functions in Chapter 7 (Combinatorials), whose drivers are in Section 21.7, are also used in number theory.

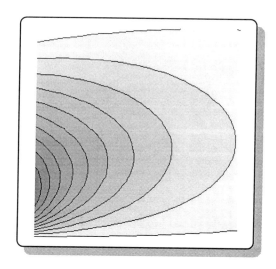

The functions included are for Bernoulli and Euler (both the numbers and the polynomials in Sections 8.1 and 8.2), the Riemann zeta function (Section 8.3), other sums of reciprocal powers (Section 8.4), and for polylogarithms (Section 8.5).

21.8.1   Bernoulli Numbers and Bernoulli Polynomials

C Functions and Driver; Test Values

BernoulliNum, BernoulliPoly in driver Bernoulli.c; Tables 8.1.1, 8.1.2

```
#include <stdio.h>
#include <math.h>

#define MAX 20

main()
{
/* Function Names: BernoulliNum, BernoulliPoly */
double x,B[MAX],BernOut,
  BernoulliPoly(int n, double x, double B[]);
int n;
void BernoulliNum(int n, double B[]);
printf("Bernoulli number & polynomial Tests");
n = 1;
while ( n >= 0 )
  {
  printf("\n\nInput n,x ( n<0 to end): ");
  scanf("%i%lf",&n,&x);
  if ( n < 0 )
  {printf("\nEnd Bernoulli number & polynomial Tests");
    exit(0); }
  else
    {
    BernoulliNum(n,B);
    BernOut = B[n];
    printf
    ("BernoulliNum[n] = %.9E, BernoulliPoly = %.9E",
      BernOut, BernoulliPoly(n,x,B));
  } } }
```

Maximum number of Bernoulli number entries.

Driver makes table of Bernoulli numbers into array B[] then makes the Bernoulli polynomial at *x*.

Terminate if *n* improper.

Make table of Bernoulli numbers and show the *n*th. Compute and print the *n*th Bernoulli polynomial. Both values to 10 digits.

21.8.2   Euler Numbers and Euler Polynomials

C Functions and Driver; Test Values
EulerNum, EulerPoly in driver Euler.c; Tables 8.2.1, 8.2.2

```
#include <stdio.h>
#include <math.h>
#define MAX 20
main()
{
/* Function Names: EulerNum, EulerPoly */
double x,EulerPoly(int n, double x, long int E[]);
int n;
long int E[MAX],EulOut;
void EulerNum(int n, long int E[]);
printf("Euler number & polynomial Tests");
n = 1;

while ( n >= 0 )
  {
  printf("\n\nInput n,x ( n<0 to end): ");
  scanf("%i%lf",&n,&x);
  if ( n < 0 )
    {printf("\nEnd Euler number & polynomial Tests");
    exit(0); }

  else
    {
    EulerNum(n,E);
    EulOut = E[n];
    printf("EulerNum[n] = %li, EulerPoly = %.9E",
           EulOut, EulerPoly(n,x,E));
    } } }
```

Maximum number of Euler number entries.

Terminate if *n* improper.

Make table into integer E[]. Note that E_n overflows quickly. Make and print Euler polynomial. at *x*.

21.8.3   Riemann Zeta Function

C Function and Driver; Test Values
RiemannZeta in driver RiemannZeta.c; Table 8.3.1

```
#include <stdio.h>
#include <math.h>
main()
{
/* Function Name: RiemannZeta */
double s,eps,RiemannZeta(double s, double eps);
printf("RiemannZeta function Tests");
s = 2;
```

```
while ( s > 1 )
  {
  printf("\n\nInput s,eps ( s<=1 to end): ");
  scanf("%lf%lf",&s,&eps);

  if ( s <= 1 )
    {
    printf("\nEnd RiemannZeta function Tests");
    exit(0);
    }
  else
    {
    /* Zeta(s) - 1 to fractional accuracy eps */
    printf("Zeta[s] - 1 = %.9E",RiemannZeta(s,eps));
    }
  }
}
```

Input *s* and the fractional accuracy eps.

Terminate if *s* is improper.

Make and print Riemann zeta minus unity to 10 digits.

21.8.4   Other Sums of Reciprocal Powers

C Functions and Driver; Test Values
EtaNum, LambdaNum, BetaNum in driver SumRecPwrs.c; Tables 8.4.1, 8.4.2

```
#include <stdio.h>
#include <math.h>

main()
{
/* Function Names: EtaNum,LambdaNum,BetaNum */

double s,eps,EtaNum(double s, double eps),
  LambdaNum(double s, double eps),
  BetaNum(double s, double eps);

printf("Sum of Reciprocal Powers Tests");

s = 2;
while ( s >= 1 )
  {
  printf("\n\nInput s,eps ( s<1 to end): ");
  scanf("%lf%lf",&s,&eps);
  if ( s < 1 )
    {
    printf("\nEnd Sum of Reciprocal Powers Tests");
    exit(0);
    }

  else
    {
    printf("\n\nEta[s] - 1 = %.9E",EtaNum(s,eps));
    printf("\n\nLambda[s] - 1 = %.9E",LambdaNum(s,eps));
    printf("\n\nBeta[s] - 1 = %.9E",BetaNum(s,eps));
    } } }
```

Input *s* and the fractional accuracy eps.

Terminate if *s* is improper.

Make and print eta, lambda, and beta minus unity to 10 digits.

21.8.5   Polylogarithms

C Function and Driver; Test Values
`PolyLog` in driver `PolyLog.c`; Table 8.5.2

```
#include <stdio.h>
#include <math.h>

main()
{
/* Function Name: PolyLog */
double x,eps,PolyLog(int n, double x, double eps);
int n;
printf("Polylogarithm Tests");
n = 2;
while ( n > 1 )
  {
  printf("\n\nInput n,x,eps ( n<2 to end): ");
  scanf("%i%lf%lf",&n,&x,&eps);
  if ( n < 2 )
    {
    printf("\nEnd Polylogarithm Tests");
    exit(0);
    }
  else
    {
    printf("\nLi(%i,%lf) = %.9E",n,x,PolyLog(n,x,eps));
    } } }
```

Input n, x, and the fractional accuracy eps.

Terminate if n is not at least for dilogarithm.

Make and print the polylogarithm to 10 digits.

21.9   PROBABILITY DISTRIBUTIONS

Here are the annotated listings of the C driver programs for testing the probability distribution functions (PDFs), for both the density and cumulative distributions, as described in Chapter 9 of the *Atlas*.

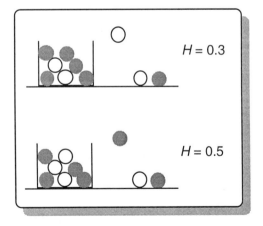

21.9.1   Organization of the PDFs

The functions in the main text are grouped into discrete PDFs (Section 9.2), normal distributions (in Section 9.3), and other probability distributions (Section 9.4).

21.9.2   Discrete Probability Distributions

Six discrete distributions are included in this section; binomial, negative binomial (Pascal), geometric, hypergeometric, logarithmic series, and Poisson PDFs, corresponding to the six subsections in Section 9.2 of the main text.

Binomial Distribution:
C Function and Driver; Test Values
BinomDist in driver BnmDist.c; Table 9.2.1

```
#include <stdio.h>
#include <math.h>

main()
{
/* Function Name: BinomDist */
/* Binomial Distribution Function */

double BinomDist(int N, double p, int k),p;
int N,k;
long int Combinatorial(),
  Factorial(),RisingFactorial();

printf("Binomial Distribution Tests");
N = 1;

while ( N > 0 )
  {
  printf("\nInput N,p,k ( N=0 to end): ");
  scanf("%i%lf%i",&N,&p,&k);
  if ( N <= 0 )
    {
    printf("\nEnd  Binomial Distribution Tests");
    exit(0);
    }

  else
    {
    printf
    ("\nP(%i,%lf,%i) = %.9E",N,p,k,BinomDist(N,p,k));
    } } }
```

Input three argu-
ments for binomial
distribution den-
sity.

Quit if *N* not
positive.

Binomial density
values to 10 digits.

Negative Binomial (Pascal) Distribution:
C Function and Driver; Test Values
NegBinomDist in driver N e g BnmDist.c; Table 9.2.2

```
#include <stdio.h>
#include <math.h>

main()
{
/* Function Name: NegBinomDist */
/* Negative Binomial Distribution Function */

double NegBinomDist(int N, double p, int k),p;
int N,k;
long int Combinatorial(),
  Factorial(),RisingFactorial();
printf("Negative Binomial Distribution Tests");
N = 1;
```

N e g BnmDst
continued

```
while ( N >= 0 )
  {
  printf("\n\nInput N,p,k ( N<0 to end): ");
  scanf("%i%lf%i",&N,&p,&k);
  if ( N < 0 )
    {
    printf
    ("\nEnd  Negative Binomial Distribution Tests");
    exit(0);
    }
  else
    {
    printf
    ("\nN(%i,%lf,%i) = %.9E",
      N,p,k,NegBinomDist(N,p,k));
    } } }
```

Input three arguments for negative binomial distribution density.

Quit if $N < 0$.

Negative binomial density values to 10 digits.

Geometric Distribution:
C Function and Driver; Test Values
GeometricDist in driver Gmtrc Dist.c; Table 9.2.2

```
#include <stdio.h>
#include <math.h>

main()
{
/* Function Name: GeometricDist */
/* Geometric Distribution Function */
double GeometricDist(double p, int k),p;
int k;
printf("Geometric Distribution Tests");
p = 0.1;
while ( p > 0 )
  {
  printf("\n\nInput p,k ( p<=0 to end): ");
  scanf("%lf%i",&p,&k);
  if ( p <= 0 )
    {
    printf
    ("\nEnd  Geometric Distribution Tests");
    exit(0);
    }
  else
    {
    printf
    ("\nG(%lf,%i) = %.9E",p,k,GeometricDist(p,k));
    } } }
```

Input two arguments for geometric distribution density.

Quit if p not positive.

Geometric density values to 10 digits.

Hypergeometric Distribution:
C Function and Driver; Test Values
HyprGmtrcDist in driver HyprGmtrcDist.c; Table 9.2.4

```
#include <stdio.h>
#include <math.h>

main()
{
/* Function Name: HyprGmtrcDist */
/* Hypergeometric Distribution Function */

double HyprGmtrcDist(int N, int N1, int n, int k);
int N,N1,n,k;
long int Combinatorial(),
  Factorial(),RisingFactorial();

printf("Hypergeometric Distribution Tests");
N = 1;
while ( N > 0 )
  {
  printf("\nInput N,N1,n,k ( N=0 to end): ");
  scanf("%i%i%i%i",&N,&N1,&n,&k);
  if ( N <= 0 )
    {
    printf
    ("\nEnd  Hypergeometric Distribution Tests");
    exit(0);
    }
  else
    {
    printf
    ("\nH(%i,%i,%i,%i) = %.9E",
      N,N1,n,k,HyprGmtrcDist(N,N1,n,k));
    } } }
```

Input four arguments for hypergeometric distribution density.

Quit if *N* not positive.

Hypergeometric density values to 10 digits.

Logarithmic Series Distribution:
C Function and Driver; Test Values
LogSeriesDist in driver LogSeriesDist.c; Table 9.2.5

```
#include <stdio.h>
#include <math.h>
main()
{
/* Function Name: LogSeriesDist */
/* Logarithmic Series Distribution Function */
double LogSeriesDist(double theta, int k),theta;
int k;
printf("Logarithmic Series Distribution Tests");
theta = 0.1;
while ( theta > 0 )
  {
  printf("\n\nInput theta,k ( theta<=0 to end): ");
  scanf("%lf%i",&theta,&k);
  if ( theta <= 0 )
    {
    printf
    ("\nEnd  Logarithmic Series Distribution Tests");
    exit(0);
    }
```

θ and *k* for logarithmic series distribution density.

Quit if θ not positive.

```
  else
   {
   printf
   ("\nL(%lf,%i) = %.9E",
     theta,k,LogSeriesDist(theta,k));
   } } }
```

LogSeriesDist
continued

Logarithmic series
density values to
10 digits, Table
9.2.5.

Poisson Distribution:
C Function and Driver; Test Values
PoissonDist in driver PoissonDist.c; Table 9.2.6

```
#include <stdio.h>
#include <math.h>

main()
{
/* Function Name: PoissonDist */
/* Poisson Distribution Function */

double PoissonDist(double m, int k),m;
int k;

printf("Poisson Distribution Tests");
m = 1;
while ( m > 0 )
   {
   printf("\n\nInput m,k ( m<=0 to end): ");
   scanf("%lf%i",&m,&k);
   if ( m <= 0 )
     {
     printf
     ("\nEnd  Poisson Distribution Tests");
     exit(0);
     }
   else
     {
     printf
     ("\nP(%lf,%i) = %.9E",
       m,k,PoissonDist(m,k));
     } } }
```

Input *m* and *k* for
Poisson distribu-
tion density.

Quit if *m* not
positive.

Print Poisson
density to 10
digits, Table 9.2.6.

21.9.3   Normal Probability Distributions

Six probability distributions closely related to the normal distributions are included in Section 9.3 of the main text of the *Atlas* and in the following C driver programs; Gauss (normal), bivariate normal, chi-square, *F*-(variance-ratio), Student's *t*, and lognormal PDFs.

Gauss (Normal) Distribution:
C Function and Driver; Test Values
GaussDist in driver GaussDist.c; Table 9.3.2

```
#include <stdio.h>
#include <math.h>
main()
{
/* Function Name: GaussDist */
/* Gauss Distribution Function;
  in standardized form */
double GaussDist(double x);
double x;

printf("Gauss Distribution Tests");
x = 1;
while ( fabs(x) <= 10 )
  {
  printf("\n\nInput x ( |x| > 10 to end): ");
  scanf("%lf",&x);
  if ( fabs(x) > 10 )
    {
    printf("\nEnd  Gauss Distribution Tests");
    exit(0);
    }
  else
    {
    printf
    ("\nG(%lf) = %.9E",x,GaussDist(x));
    } } }
```

Input *x* for Gauss distribution density.

Quit if *x* too large.

Print Gauss density to 10 digits, Table 9.3.2.

Bivariate Normal Distribution:
C Function and Driver; Test Values
BivNormDist in driver BivNormDist.c; Table 9.3.3

```
#include <stdio.h>
#include <math.h>

main()
{
/* Function Name: BivNormDist */
/* Bivariate Normal Distribution Function */

double BivNormDist(double x, double y, double r),
  x,y,r;

printf("Bivariate Normal Distribution Tests");
r = 0.5;
while ( fabs(r) < 1 )
  {
  printf("\n\nInput x,y,r ( |r|>1 to end): ");
  scanf("%lf%lf%lf",&x,&y,&r);
  if (  fabs(r) >= 1 )
    {
    printf
    ("\nEnd  Bivariate Normal Distribution Tests");
    exit(0);
    }
```

Input *x*, *y*, and correlation coefficient *r* for bivariate distribution density.

Quit if *r* too large in magnitude.

```
  else
    {
    printf
    ("\nB(%lf,%lf,%lf) = %.9E",
      x,y,r,BivNormDist(x,y,r));
    } } }
```

`BivNormDist`
continued

Print bivariate density to 10 digits, Table 9.3.3.

Chi-Square Distribution:
C Functions and Driver; Test Values

`ChiSqDnsty`, `ChiSqCum` in driver `ChiSqDist.c`; Tables 9.3.4, 9.3.5

```
#include <stdio.h>
#include <math.h>

main()
{
/* Function Names: ChiSqDnsty, ChiSqCum */
/* Chi-square Probability Function:
   Density & Cumulative Distribution Tests */

double ChiSqDnsty(double chisq,int nu),
     ChiSqCum(double chisq,int nu, double eps),
     chisq,eps;
int nu;

printf("Chi-square Probability Tests");
chisq = 1;
while ( chisq > 0 )
  {
  printf
  ("\n\nInput chisq,nu,eps ( chisq = 0 to end): ");
  scanf("%lf%i%lf",&chisq,&nu,&eps);
  if ( chisq <= 0 )
    {
    printf
    ("\nEnd  Chi-square Probability Tests");
    exit(0);
    }

  else
    {
    printf
    ("\nC(%lf,%i) = %.9E, P(%lf,%i,%lg) = %.9E",
    chisq,nu,ChiSqDnsty(chisq,nu),
    chisq,nu,eps,ChiSqCum(chisq,nu,eps));
    } } }
```

Input chi-square, v, and accuracy `eps`.

Quit if chi-square not positive.

Print density and cumulative distribution values to 10 digits, Tables 9.3.4, 9.3.5.

F-(Variance-Ratio) Distribution:
C Functions and Driver; Test Values

`FDnsty`, `FCum` in driver `FDist.c`; Tables 9.3.6, 9.3.7

```
#include <stdio.h>
#include <math.h>

main()
{
/* Function Names: FDnsty, FCum  */
/* F (Variance-Ratio) Distribution Function:
   Density & Cumulative Distribution Tests */

double FDnsty(int nu1, int nu2, double f),
 FCum(int nu1, int nu2, double f, double eps),
 IncBeta
 (double x, double a, double b, double eps),
 f,eps;
int nu1,nu2;

printf("F-Distribution Tests");
eps = 0.1;

while ( eps > 0 )
  {
  printf
  ("\n\nInput nu1,nu2,f,eps ( eps = 0 to end): ");
  scanf("%i%i%lf%lf",&nu1,&nu2,&f,&eps);
  if (  eps <= 0 )
    {
    printf("\nEnd F-Distribution Tests");
    exit(0);
    }

  else
    {
    printf("\nF(%i,%i,%lf) = %.9E",
      nu1,nu2,f,FDnsty(nu1,nu2,f));
    printf("\nQ(%i,%i,%lf) = %.9E",
      nu1,nu2,f,FCum(nu1,nu2,f,eps));
    } } }
```

Input two values of degrees of freedom, f, and the accuracy eps.

Quit if exact value requested.

Print density and cumulative distribution values to 10 digits, Tables 9.3.6, 9.3.7.

Student's *t* Distribution:
C Functions and Driver; Test Values
StudTDnsty, StudTCum in driver StudTDist.c; Tables 9.3.8, 9.3.9

```
#include <stdio.h>
#include <math.h>

main()
{
/* Function Names: StudTDnsty, StudTACum  */
/* Student's t Distribution Function:
   Density & Cumulative Distribution Tests */

double  StudTDnsty(int nu, double t),
 StudTACum(int nu, double t, double eps),t,eps;
int nu;
```

StudTDist
continued

```
printf(" Student's t-Distribution Tests");
eps = 0.1;

while ( eps > 0 )
  {
  printf
  ("\n\nInput nu,t,eps ( eps = 0 to end): ");
  scanf("%i%lf%lf",&nu,&t,&eps);
  if ( eps <= 0 )
    {
    printf("\nEnd Student's t-Distribution Tests");
    exit(0);
    }

  else
    {
    printf("\S(%i,%lf) = %.9E",
      nu,t, StudTDnsty(nu,t));
    printf("\nA(%i,%lf) = %.9E",
      nu,t,StudTACum(nu,t,eps));
    } } }
```

Input degrees of freedom (ν), t, and the accuracy eps.

Quit if exact value requested.

Print density and cumulative distribution values to 10 digits, Tables 9.3.8, 9.3.9.

Lognormal Distribution:
C Functions and Driver; Test Values

LogNDnsty, LogNCum in driver LogNDist.c; Tables 9.3.10, 9.3.11

```
#include <stdio.h>
#include <math.h>

main()
{
/* Function Names: LogNDnsty, LogNCum */
/* Lognormal Distribution Functions: */

double
  LogNDnsty(double zeta, double sigma, double x),
  LogNCum(double zeta, double sigma, double x,
  double eps),zeta,sigma,x,eps;

printf("Lognormal Distribution Tests");
eps = 10;

while ( eps > 0 )
  {
  printf
    ("\n\nInput zeta,sigma,x,eps (eps=0 to end): ");
  scanf("%lf%lf%lf%lf",&zeta,&sigma,&x,&eps);
  if ( eps == 0 )
    {
    printf("\nEnd Lognormal Distribution Tests");
    exit(0);
    }
```

Input ζ, σ, x, and the accuracy eps.

Quit if exact value requested.

```
else
   {
   printf
   ("\nL(%lf,%lf,%lf) = %.9E",
   zeta,sigma,x,LogNDnsty(zeta,sigma,x));
   printf
   ("\nP(%lf,%lf,%lf) = %.9E",
   zeta,sigma,x,LogNCum(zeta,sigma,x,eps));
   } } }
```

Print density and cumulative distribution values to 10 digits, Tables 9.3.10, 9.3.11.

21.9.4   Other Continuous Probability Distributions

Eight probability distributions that are not directly related to the normal distribution are included in Section 9.4 of the main text of the *Atlas* and in the following C driver programs; Cauchy (Lorentz), exponential, Pareto, Weibull, logistic, Laplace, Kolmogorov-Smirnov, and beta distributions. Both density- and cumulative-distribution functions are invoked by the driver programs.

Cauchy (Lorentz) Distribution:
C Functions and Driver; Test Values
CauchyDnsty, CauchyCum in driver CauchyDist.c; Table 9.4.2

```
#include <stdio.h>
#include <math.h>

main()
{
/* Function Names: CauchyDnsty,CauchyCum */
/* Cauchy Distribution Function;
  in standard form */

double CauchyDnsty(double x),
      CauchyCum(double x);
double x;

printf("Cauchy Distribution Tests");
x = 1;
while ( fabs(x) <= 10 )
   {
   printf("\n\nInput x ( |x| > 10 to end): ");
   scanf("%lf",&x);
   if ( fabs(x) > 10 )
      {
      printf("\nEnd  Cauchy Distribution Tests");
      exit(0);
      }
   else
      {
      printf
         ("\nC(%lf) = %.9E",x,CauchyDnsty(x));
      printf
         ("\nP(%lf) = %.9E",x,CauchyCum(x));
      } } }
```

Input *x* for Cauchy distribution.

Quit if *x* too large.

Print Cauchy density and cumulative to 10 digits, Table 9.4.2.

Exponential Distribution:
C Functions and Driver; Test Values
ExpDnsty, ExpCum in driver ExpDist.c; Table 9.4.3

```
#include <stdio.h>
#include <math.h>
main()
{
/* Function Names: ExpDnsty,ExpCum */
/* Exponential Distribution Function;
   in standard form */

double ExpDnsty(double x),
       ExpCum(double x);
double x;

printf(" Exponential Distribution Tests");
x = 1;
while ( x >= 0 )
  {
  printf("\n\nInput x ( x < 0 to end): ");
  scanf("%lf",&x);
  if ( x < 0 )
    {
    printf("\nEnd Exponential Distribution Tests");
    exit(0);
    }
  else
    {
    printf
      ("\nE(%lf) = %.9E",x,ExpDnsty(x));
    printf
      ("\nC(%lf) = %.9E",x,ExpCum(x));
    } } }
```

Input *x* for exponential distribution.

Quit if *x* too large.

Print exponential density and cumulative to 10 digits, Table 9.4.3.

Pareto Distribution:
C Functions and Driver; Test Values
ParetoDnsty, ParetoCum in driver ParetoDist.c; Table 9.4.4

```
#include <stdio.h>
#include <math.h>
main()
{
/* Function Names: ParetoDnsty,ParetoCum */
/* Pareto Distribution Function;
   in standard form */

double ParetoDnsty(double a, double k, double x),
       ParetoCum(double a, double k, double x);
double a,k,x;

printf(" Pareto Distribution Tests");
x = 1;
```

```
while ( x > 0 )
  {
  printf("\n\nInput a,k,x ( x=0 to end): ");
  scanf("%lf%lf%lf",&a,&k,&x);
  if ( x <= 0 )
     {
     printf("\nEnd Pareto Distribution Tests");
     exit(0);
     }
  else
     {
     printf
       ("\nP(%lf,%lf,%lf) = %.9E",
            a,k,x,ParetoDnsty(a,k,x));
     printf
       ("\nF(%lf%lf,%lf) = %.9E",
            a,k,x,ParetoCum(a,k,x));
  } } }
```

Input *x* for Pareto distribution.

Quit if *x* not positive.

Print Pareto density and cumulative to 10 digits, Table 9.4.4.

Weibull Distribution:
C Functions and Driver; Test Values
WeibullDnsty, WeibullCum in driver WbllDist.c; Tables 9.4.5, 9.4.6

```
#include <stdio.h>
#include <math.h>
main()
{
/* Function Names: WeibullDnsty,WeibullCum */
/* Weibull Distribution Function;
   in standard form */

double WeibullDnsty(double c, double x),
     WeibullCum(double c, double x);
double c,x;

printf(" Weibull Distribution Tests");
x = 1;
while ( x > 0 )
  {
  printf("\n\nInput c,x ( x=0 to end): ");
  scanf("%lf%lf",&c,&x);
  if ( x <= 0 )
     {
     printf("\nEnd Weibull Distribution Tests");
     exit(0);
     }
  else
     {
     printf
       ("\nW(%lf,%lf) = %.9E",
            c,x,WeibullDnsty(c,x));
     printf
       ("\nC(%lf,%lf) = %.9E",
            c,x,WeibullCum(c,x));
  } } }
```

Input *x* for Weibull distribution.

Quit if *x* not positive.

Print Weibull density and cumulative to 10 digits, Tables 9.4.5, 9.4.6.

Logistic Distribution:
C Functions and Driver; Test Values
LgstcDnsty, LgstcCum in driver LgstcDist.c; Table 9.4.7

```
#include <stdio.h>
#include <math.h>

main()
{
/* Function Names: LgstcDnsty,LgstcCum */
/* Logistic Distribution Function;
   in standard form */

double LgstcDnsty(double x),
       LgstcCum(double x);
double x;

printf(" Logistic Distribution Tests");
x = 1;
while ( fabs(x) <= 10 )
  {
  printf("\n\nInput x ( |x| > 10 to end): ");
  scanf("%lf",&x);
  if ( fabs(x) > 10 )
    {
    printf("\nEnd Logistic Distribution Tests");
    exit(0);
    }
  else
    {
    printf
      ("\nL(%lf) = %.9E",x,LgstcDnsty(x));
    printf
      ("\nC(%lf) = %.9E",x,LgstcCum(x));
    } } }
```

Input *x* for logistic distribution.

Quit if *x* too large.

Print logistic density and cumulative to 10 digits, Table 9.4.7.

Laplace Distribution:
C Functions and Driver; Test Values
LaplaceDnsty, LaplaceCum in driver LaplaceDist.c; Table 9.4.8

```
#include <stdio,h>
#include <math.h>

main()
{
/* Function Names: LaplaceDnsty,LaplaceCum */
/* Laplace Distribution Function;
   in standard form */

double LaplaceDnsty(double x),
       LaplaceCum(double x);
double x;

printf(" Laplace Distribution Tests");
x = 1;
```

```
while ( fabs(x) <= 10 )
 {
 printf("\n\nInput x ( |x| > 10 to end): ");
 scanf("%lf",&x);
 if ( fabs(x) > 10 )
   {
   printf("\nEnd Laplace Distribution Tests");
   exit(0);
   }

 else
   {
   printf
     ("\nL(%lf) = %.9E",x,LaplaceDnsty(x));
   printf
     ("\nC(%lf) = %.9E",x,LaplaceCum(x));
 } } }
```

Input *x* for Laplace distribution.

Quit if *x* too large.

Print Laplace density and cumulative to 10 digits, Table 9.4.8.

Kolmogorov-Smirnov Distribution:
C Functions and Driver; Test Values
KSDnsty, KSQCum in driver KSDist.c; Table 9.4.9

```
#include <stdio.h>
#include <math.h>

main()
{
/* Function Names: KSDnsty,KSQCum */
/* Kolmogorov-Smirnov Distribution Functions */

double KSDnsty(double x, double eps),
      KSQCum(double x, double eps);
double x,eps;

printf(" Kolmogorov-Smirnov Distribution Tests");
eps = 10;
while ( eps > 0 )
  {
  printf("\n\nInput x,eps ( eps=0 to end): ");
  scanf("%lf%lf",&x,&eps);
  if ( eps <= 0 )
    {
    printf
    ("\nEnd Kolmogorov-Smirnov Distribution Tests");
    exit(0);
    }

  else
    {
    printf
      ("\nK(%lf,%lf) = %.9E",x,eps,KSDnsty(x,eps));
    printf
      ("\nQ(%lf,%lf) = %.9E",x,eps,KSQCum(x,eps));
  } } }
```

Input *x* for K-S distribution, plus accuracy eps.

Quit if exact value attempted.

Print K-S density and Q-cumulative to 10 digits, Table 9.4.9.

Beta Distribution:
C Functions and Driver; Test Values
BetaDnsty, BetaCum in driver BetaDist.c; Tables 9.4.10, 9.4.11

```
#include <stdio.h>
#include <math.h>

main()
{
/* Function Names: BetaDnsty, BetaCum */
/* Beta Distribution Function:
   Density & Cumulative Distribution Tests */
double BetaDnsty(double p, double q, double x),
  BetaCum
  (double p, double q, double x, double eps),
  IncBeta
  (double a, double b, double x, double eps),
  p,q,x,eps;

printf("Beta Distribution Tests");

eps = 0.1;
while ( eps > 0 )
  {
  printf
  ("\n\nInput p,q,x,eps ( eps = 0 to end): ");
  scanf("%lf%lf%lf%lf",&p,&q,&x,&eps);
  if ( eps <= 0 )
    {
    printf("\nEnd Beta Distribution Tests");
    exit(0);
    }
  else
    {
    printf("\nB(%lf,%lf,%lf) = %.9E",
      p,q,x,BetaDnsty(p,q,x));
    printf("\nP(%lf,%lf,%lf) = %.9E",
      p,q,x,BetaCum(p,q,x,eps));
    } } }
```

Input *p, q, x* for beta distribution, plus accuracy eps.

Quit if exact value attempted.

Print beta density and cumulative to 10 digits, Tables 9.4.10, 9.4.11.

21.10   ERROR FUNCTION, FRESNEL AND DAWSON INTEGRALS

This section has the annotated listings of the C driver programs for testing the error function of real argument, the Fresnel cosine and sine integrals, and Dawson's integral. These functions are described in Chapter 10 of the *Atlas*.

21.10.1   Error Function
The C function computes the error function of real argument.

C Function and Driver; Test Values
ErfReal in the driver ErfReal.c; Table 10.1.1

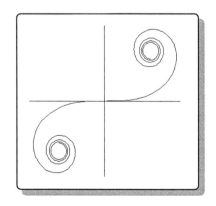

```
#include <stdio.h>
#include <math.h>

main()
{
/* Function Name: ErfReal */
/* Error Function of real argument  x
   to accuracy  eps  */

double ErfReal(double x, double eps),x,eps;

printf("Error Function Tests");
x = 1.0;
while ( x != 0 )
  {
  printf("\n\nInput x,eps ( x = 0 to end): ");
  scanf("%lf%lf",&x,&eps);
  if ( x == 0 )
    {
    printf("\nEnd Error Function Tests");
    exit(0);
    }
  else
    {
    printf
      ("\n erf(%lf) = %.9E",x,ErfReal(x,eps));
    } } }
```

Input argument *x* and the fractional accuracy eps.

Quit if *x* = 0.

Error function values to 10 digits, Table 10.1.1.

21.10.2   Fresnel Integrals

C Function and Driver; Test Values

FresnelCS in the driver Fresnel.c; Table 10.2.1

```
#include <stdio.h>
#include <math.h>

main()
{
/* Function Name: FresnelCS */
/* Fresnel Cosine & Sine Integrals of real argument
   by power series (to accuracy  eps ),
   rational approximation, and asymptotic expansion */

void FresnelCS(double x, double eps,
  double *FresnelC, double *FresnelS);
  double x,eps,FresnelC,FresnelS;

printf("Fresnel Integral Tests");
x = 1.0;
while ( x != 0 )
  {
  printf("\n\nInput x,eps ( x = 0 to end): ");
  scanf("%lf%lf",&x,&eps);
```

Function returns *two* values, FresnelC and FresnelS.

Input argument *x* and the fractional accuracy eps.

```
if ( x == 0 )
    {
    printf("\nEnd Fresnel Integral Tests");
    exit(0);
    }

  else
    {
    FresnelCS(x,eps,&FresnelC,&FresnelS);
    printf
      ("\n FresnelC(%lf) = %.9E",x,FresnelC);
    printf
      ("\n FresnelS(%lf) = %.9E",x,FresnelS);
    } } }
```

FresnelCS
continued

Quit if $x = 0$.

Fresnel cosine and
sine integrals to 10
digits, as in Table
10.2.1.

21.10.3   Dawson Integral

C Function and Driver; Test Values

FDawson in the driver Dawson.c; Table 10.3.1

```
#include <stdio.h>
#include <math.h>

main()
{
/* Function Name: FDawson */
/* Dawson Integral of real argument  x
   to fractional accuracy  eps  */

double FDawson(double x, double eps),x,eps;

printf("Dawson Integral Tests");

x = 1.0;

while ( x != 0 )
  {
  printf("\n\nInput x,eps ( eps = 0 to end): ");
  scanf("%lf%lf",&x,&eps);

  if ( eps <= 0 )
    {
    printf("\nEnd Dawson Integral Tests");
    exit(0);
    }

  else
    {
    printf
      ("\n F(%lf) = %.9E",x,FDawson(x,eps));
    } } }
```

Input argument x
and the fractional
accuracy eps.

Quit if an exact
value (eps = 0) is
requested.

Dawson integral to
10 digits, as in
Table 10.3.1.

21.11 ORTHOGONAL POLYNOMIALS

This section gives the annotated listings of the C driver programs that test the functions for the seven classical orthogonal polynomials—Chebyshev of the first and second kinds, Gegenbauer (ultraspherical), Hermite, Laguerre, Legendre, and Jacobi. The polynomials and functions are described in Chapter 11 of the *Atlas*.

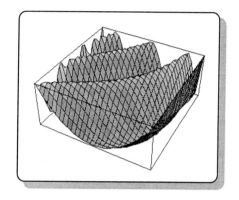

21.11.1 Orthogonal Polynomial Functions

The C functions for the orthogonal polynomials produce (with the exception of the Chebyshev polynomial of the second kind) results that are limited in accuracy only by computer roundoff errors, since the formulas and algorithms used are exact. For conformity with the rest of the *Atlas*, however, these C functions produce output with 10 digits of accuracy.

21.11.2 Chebyshev Polynomials

Chebyshev Polynomials of the First Kind:
C Function and Driver; Test Values
ChebyT in the driver ChebyT.c; Table 11.2.2

```c
#include <stdio.h>
#include <math.h>

main()
{
/* Function Name: ChebyT */
/* Chebyshev Polynomial of first kind, order n */

double ChebyT(int n, double x),x;
int n;

printf(" Chebyshev Polynomial (T) Tests");
x = 0;
while ( fabs(x) <= 1 )
  {
  printf("\n\nInput n,x  ( |x|>1 to end): ");
  scanf("%i%lf",&n,&x);
  if ( fabs(x) > 1 )
    {
    printf("\nEnd Chebyshev Polynomial (T) Tests");
    exit(0);
    }
  else
    {
    printf
      ("\nT(%i,%lf) = %lg",n,x,ChebyT(n,x));
    } } }
```

Input order n and argument x.

Quit if $x > 1$.

Chebyshev T to 10 digits, as in Table 11.2.2.

Chebyshev Polynomials of the Second Kind:
C Function and Driver; Test Values

ChebyU in the driver ChebyU.c; Table 11.2.4

```
#include <stdio.h>
#include <math.h>

main()
{
/* Function Name: ChebyU */
/* Chebyshev Polynomial of second kind, order n */

double ChebyU(int n, double x, double eps),x,eps;
int n;

printf(" Chebyshev Polynomial (U) Tests");
x = 0;
while ( fabs(x) <= 1 )
  {
  printf("\n\nInput n,x,eps ( |x|>1 to end): ");
  scanf("%i%lf%lf",&n,&x,&eps);
  if ( fabs(x) > 1 )
    {
    printf("\nEnd Chebyshev Polynomial (U) Tests");
    exit(0);
    }
  else
    {
    printf
      ("\nU(%i,%lf) = %.9E",n,x,ChebyU(n,x,eps));
    } } }
```

Input order n, argument x and accuracy control eps for x values close to ± 1.

Quit if $x > 1$.

Chebyshev U to 10 digits, as in Table 11.2.4.

21.11.3 Gegenbauer (Ultraspherical) Polynomials

C Function and Driver; Test Values

Gegenbauer in the driver Gegenbauer.c; Table 11.3.2

```
#include <stdio.h>
#include <math.h>

main()
{
/* Function Name: Gegenbauer */
/* Gegenbauer Polynomial, parameter a, order n */

double Gegenbauer(double a, int n, double x),a,x;
int n;

printf(" Gegenbauer Polynomial Tests");
a = 1;
while ( a > -0.5 )
  {
  printf("\n\nInput a,n,x ( a <= -0.5  to end): ");
  scanf("%lf%i%lf",&a,&n,&x);
```

Input a (α), order n, and argument x.

```
if ( a <= -0.5 )
  {
  printf("\nEnd Gegenbauer Polynomial Tests");
  exit(0);
  }
else
  {
  printf
    ("\nC((%lf),%i,%lf) = %.9E",
    a,n,x,Gegenbauer(a,n,x));
  } } }
```

Quit if a <= –0.5, for which Gegenbauer polynomials are not defined.

Gegenbauer to 10 digits, as in Table 11.3.2.

21.11.4 Hermite Polynomials

C Function and Driver; Test Values

Hermite in the driver Hermite.c; Table 11.4.2

```
#include <stdio.h>
#include <math.h>

main()
{
/* Function Name: Hermite */
/* Hermite Polynomial of order n */
double Hermite(int n, double x),x;
int n;

printf(" Hermite Polynomial Tests");
n = 1;
while ( n >= 0 )
  {
  printf("\n\nInput n,x ( n < 0  to end): ");
  scanf("%i%lf",&n,&x);
  if ( n < 0 )
    {
    printf("\nEnd Hermite Polynomial Tests");
    exit(0);
    }
  else
    {
    printf
      ("\nH(%i,%lf) = %.9E",n,x,Hermite(n,x));
    } } }
```

Input order n and argument x.

Quit for negative n.

Hermite polynomial to 10 digits, as in Table 11.4.2.

21.11.5 Laguerre Polynomials

C Function and Driver; Test Values

Laguerre in the driver Laguerre.c; Table 11.5.2

```
#include <stdio.h>
#include <math.h>

main()
{
/* Function Name: Laguerre */
/* Laguerre Polynomial, parameter a, order n */
double Laguerre(double a, int n, double x),a,x;
```

Generalized Laguerre polynomials if a $\pm$ 0.

```
int n;
printf(" Laguerre Polynomial Tests");
n = 1;

while ( n >= 0 )
  {
  printf("\n\nInput a,n,x ( n < 0  to end): ");
  scanf("%lf%i%lf",&a,&n,&x);
  if ( n < 0 )
    {
    printf("\nEnd Laguerre Polynomial Tests");
    exit(0);
    }

  else
    {
    printf
      ("\nL((%lf),%i,%lf) = %.9E",
      a,n,x,Laguerre(a,n,x));
    } } }
```

Laguerre
continued

Input a (α), order
n, and argument *x*.

Quit for negative
n.

Generalized La-
guerre polynomial
to 10 digits, as in
Table 11.5.2.

21.11.6 Legendre Polynomials

C Function and Driver; Test Values

Legendre in the driver Legendre.c; Table 11.6.2

```
#include <stdio.h>
#include <math.h>

main()
{
/* Function Name:.Legendre */
/* Legendre Polynomial of order n */

double Legendre(int n, double x),x;
int n;

printf(" Legendre Polynomial Tests");
n = 1;
while ( n >= 0 )
  {
  printf("\n\nInput n,x ( n < 0  to end): ");
  scanf("%i%lf",&n,&x);
  if ( n < 0 )
    {
    printf("\nEnd Legendre Polynomial Tests");
    exit(0);
    }

  else
    {
    printf
      ("\nL(%i,%lf) = %.9E",n,x,Legendre(n,x));
    } } }
```

Input order *n* and
argument *x*.

Quit if *n* < 0.

Print Legendre
polynomial to 10
digits.

21.11.7 Jacobi Polynomials

C Function and Driver; Test Values

Jacobi in the driver Jacobi.c; Tables 11.7.2, 11.7.3

```
#include <stdio.h>
#include <math.h>

main()
{
/* Function Name: Jacobi */
/* Jacobi Polynomial;
    parameters a & b, order n */

double Jacobi
   (double a, double b, int n, double x),a,b,x;
int n;

printf(" Jacobi Polynomial Tests");
n = 1;
while ( n >= 0 )
  {
  printf("\n\nInput a,b,n,x ( n < 0  to end): ");
  scanf("%lf%lf%i%lf",&a,&b,&n,&x);
  if ( n < 0 )
    {
    printf("\nEnd Jacobi Polynomial Tests");
    exit(0);
    }
  else
    {
    printf
      ("\nP((%lf,%lf),%i,%lf) = %.9E",
       a,b,n,x,Jacobi(a,b,n,x));
    } } }
```

Input order *n* and argument *x*.

Quit if *n* < 0.

Print Legendre polynomial to 10 digits.

21.12 LEGENDRE FUNCTIONS

This section gives the annotated listings of the C driver programs that test the six Legendre functions explored in Chapter 12 of the *Atlas*. These include the functions of integer degree, $v = n$, and order, $\mu = m$, often associated with spherical polar coordinates (Section 12.2.1) and described in Section 12.2.

Also included are the toroidal functions (Section 12.3), associated with toroidal coordinates (Section 12.3.1), as well as conical functions (Section 12.4.2) that are often encountered when solving the Laplace equation on a cone surface bounded by sections of spheres, as depicted in Section 12.4.1.

21.12.1 Overview of Legendre Functions

Section 12.1 has no C functions, but it is esential reading for understanding how to compute Legendre functions.

21.12.2 Spherical Legendre Functions

Legendre Functions of the First Kind for Integer *m* and *n*
C Function and Driver; Test Values

PNM in the driver PNM.c; Table 12.2.3

```
#include <stdio.h>
#include <math.h>

main()
{
/* Function Name: PNM */
/* Legendre Function of First Kind
   degree n, order m, argument x  */

double PNM(int n, int m, double x),x;
int n,m;
printf(" Legendre Function of First Kind Tests");
x = 0.5;
while ( fabs(x) <= 1.0 )
  {
  printf("\n\nInput n,m,x ( |x| > 1 to end): ");
  scanf("%i%i%lf",&n,&m,&x);
  if ( fabs(x) > 1.0 )
    {
    printf("\nEnd Legendre Function of First Kind Tests");
    exit(0);
    }
  else
    {
    printf
      ("\nP(%i,%i,%lf) = %.9E",n,m,x,PNM(n,m,x));
    } } }
```

(right margin notes:) Input degree *n*, order *m* (as integers), and argument *x*.

Print Legendre function P to 10 digits.

Legendre Functions of the Second Kind for Integer *m* and *n*
C Function and Driver; Test Values

QN in the driver QN.c (zero-order irregular function), QNM in the driver QNM.c (positive-order irregular function); Table 12.2.5

Here is the C driver for the zero-order function, $Q_n(x)$:

```
#include <stdio.h>
#include <math.h>

main()
{
/* Function Name: QN */
/* Legendre Function of Second Kind
   degree n (order zero), argument x in (-1,1)  */

double QN(int n, double x),x;
int n;
```

```
printf(" Legendre Function of Second Kind & Order Zero
  Tests");

x = 0.5;
while ( fabs(x) <= 1.0 )
  {
  printf("\n\nInput n,x ( |x| > 1 to end): ");
  scanf("%i%lf",&n,&x);

  if ( fabs(x) > 1.0 )
    {
    printf
    ("\nEnd Legendre Function of Second Kind & Order Zero
        Tests");
    exit(0);
    }

  else
    {
    printf
      ("\nQ(%i,%lf) = %.9E",n,x,QN(n,x));
    } } }
```

Input degree *n*,
(an integer), and
argument *x*.

Print Legendre
function Q to 10
digits.

The following C driver, in QNM.c, tests the positive-order ($m > 0$) irregular function, $Q_n^m(x)$:

```
#include <stdio.h>
#include <math.h>

main()
{
/* Function Name: QN */
/* Legendre Function of Second Kind
   degree n (order zero), argument x in (-1,1)   */

double QN(int n, double x),x;
int n;

printf(" Legendre Function of Second Kind & Order Zero
Tests");
x = 0.5;
while ( fabs(x) <= 1.0 )
  {
  printf("\n\nInput n,x ( |x| > 1 to end): ");
  scanf("%i%lf",&n,&x);
  if ( fabs(x) > 1.0 )
    {
    printf
    ("\nEnd Legendre Function of Second Kind & Order Zero
Tests");
    exit(0);
    }
  else
    {
    printf
      ("\nQ(%i,%lf) = %.9E",n,x,QN(n,x));
    } } }
```

Input degree *n*,
order *m* (as
integers), and
argument *x*.

Print Legendre
function Q to 10
digits.

21.12.3   Toroidal Legendre Functions

Toroidal Functions of the First Kind
C Function and Driver; Test Values
ToroidalP in the driver ToroidalP.c; Table 12.3.1

```
#include <stdio.h>
#include <math.h>

main()
{
/* Function Name: ToroidalP */
/* Toroidal Function of First Kind
   degree n, order m, argument  x > 1
   to fractional accuracy  eps  */

double ToroidalP(int n, int m, double x, double
eps),x,eps;
int n,m;

printf
 (" Toroidal Function of First Kind Tests");
x = 1.5;
while ( x > 1.0 )
   {
   printf("\n\nInput n,m,x,eps ( x <= 1 to end): ");
   scanf("%i%i%lf%lf",&n,&m,&x,&eps);
   if ( x <= 1.0 )
     {
     printf("\nEnd Toroidal Function of First Kind Tests");
     exit(0);
     }
   else
     {
     printf
     ("\nP(%i-1/2,%i,%lf) =
%.9E",n,m,x,ToroidalP(n,m,x,eps));
     } } }
```

Toroidal
functions have
$|x| > 1$.

Input degree n,
order m, and
argument x.

Regular toroidal
function to 10
digits, as in
Table 12.3.1.

Toroidal Functions of the Second Kind
C Function and Driver; Test Values
ToroidalQ in the driver ToroidalQ.c; Table 12.3.2

```
#include <stdio.h>
#include <math.h>

main()
{
/* Function Name: ToroidalQ */
/* Toroidal Function of Second Kind
   degree n, order m, argument  x > 1
   to fractional accuracy  eps  */

double ToroidalQ(int n, int m, double x, double
eps),x,eps;
int n,m;
```

```
printf
  (" Toroidal Function of Second Kind Tests");
x = 1.5;
while ( x > 1.0 )
  {
  printf("\n\nInput n,m,x,eps ( x <= 1 to end): ");
  scanf("%i%i%lf%lf",&n,&m,&x,&eps);
  if ( x <= 1.0 )
    {
    printf
    ("\nEnd Toroidal Function of Second Kind Tests");
    exit(0);
    }
  else
    {
    printf
    ("\nQ(%i-1/2,%i,%lf) = %.9E",
        n,m,x,ToroidalQ(n,m,x,eps));
    } } }
```

> Toroidal functions have $|x| > 1$.
>
> Input degree n, order m, and argument x.
>
> Irregular toroidal function to 10 digits, as in Table 12.3.2.

21.12.4 Conical Legendre Functions

C Function and Driver; Test Values

ConicalP in the driver ConicalP.c; Table 12.4.1

```
#include <stdio.h>
#include <math.h>

main()
{
/* Function Name: ConicalP */
/* ConicalP Function of First Kind
   degree -1/2+it, order 0, argument  x  ( |x| < 1 )
   to fractional accuracy  eps  */

double ConicalP(double t, double x, double eps),
 t,x,eps;

printf
  (" Conical Function Tests");
x = 0.5;
while ( fabs(x) < 1.0 )
  {
  printf("\n\nInput t,x,eps ( |x| >= 1 to end): ");
  scanf("%lf%lf%lf",&t,&x,&eps);
  if ( fabs(x) >= 1.0 )
    {
    printf("\nEnd Conical Function Tests");
    exit(0);
    }
  else
    {
    printf
    ("\nP(-1/2+i(%f),%lf) = %.9E",t,x,ConicalP(t,x,eps));
    } } }
```

> Conical functions have $|x| < 1$.
>
> Input parameter t and the argument x.
>
> Regular conical function to 10 digits, as in Table 12.4.1.

21.13 SPHEROIDAL WAVE FUNCTIONS

This section gives the annotated listings of the C driver programs for testing the seven spheroidal wave function modules explored in Chapter 13 of the *Atlas*. The C functions include computation of the wave-function eigenvalues λ_{mn} (Section 13.1.3) and the modified eigenvalue functions t_{mn} (Sections 13.1.4 and 13.2.1), the angular-function coefficients (Section 13.2.1) and the regular and irregular angular wave functions (Section 13.2.2). Finally, we include the expansion coefficients for the radial function (Section 13.3.1) and for the regular radial wave functions (Section 13.3.2).

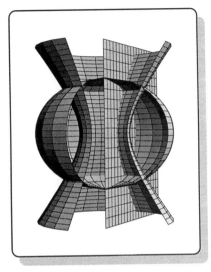

The various functions summarized above are strongly interdependent. Our first C driver program (Section 21.13.1) controls general computation of the eigenvalues only. The second driver program—listed at the end of Section 21.13.3—has options for computing in succession the eigenvalues, the angular then radial expansion coefficients, followed by the angular then the radial wave functions.

21.13.1 Overview of Spheroidal Wave Functions

This section has no C functions for the coordinate systems that are described and visualized in Section 13.1.1. Note, however, that the first *Mathematica* notebook described in Section 20.13, SWFCoords, can be used to explore the nature of the coordinates used for spheroidal wave functions.

Eigenvalues of spheroidal wave function—as λ_{mn} (Section 13.1.3)—are computed by the C function SWFevalue in Section 13.1.3. Test values for prolate and oblate coordinates are given in Tables 13.1.1 and 13.1.2, respectively. Here is the driver program for SWFevalue; it is in SWFevalue.c.

```
#include <stdio.h>
#include <math.h>
  /* maximum depth of continued fraction & arrays */
#define MAX 10
  /* maximum number of tries to bracket root */
#define NTRY 6
  /* maximum number of tries to find root */
#define KMAX 40

main()
{
/* Function Name: SWFevalue */
/* Spheroidal Wave Function Eigenvalue for
   degree n, order m, arguments c & sign;
   sign > 0 for prolate, sign < 0 for oblate   */
double SWFevalue(int m, int n, double sign,
  double c, int smax, double eps),sign,c,eps;
int m,n,smax;
```

Set control variables; these values will handle most values of *c*.

```
printf("Spheroidal Wave Function Eigenvalues");
m = 0;
n = 0;

while ( m <= n )
 {
 printf
   ("\n\nInput m,n,sign,c,smax,eps ( m>n to end): ");
 scanf("%i%i%lf%lf%i%lf",&m,&n,&sign,&c,&smax,&eps);
 if ( m > n )
  {
  printf
  ("\nEnd Spheroidal Wave Function Eigenvalue Tests");
  exit(0);
  }

 else
  {
  if ( (smax > MAX) || ((n-m+1)/2 > MAX) )
   {
   printf
   ("\n!!smax=%i or n-m=%i too large; increase MAX=%i",
     smax,n-m,MAX);
   }

 else
  { printf("\nLambda(%i,%i,%lf,%lf) = %.9E",
    m,n,sign,c,SWFevalue(m,n,sign,c,smax,eps)); }
 } } }
```

Input order m, degree n (as integers), sign (+1 for prolate, −1 for oblate), smax (number of terms in continued fraction) and accuracy of root, eps.

Print eigenvalue to 10 digits.

21.13.2 Spheroidal Angular Functions

Expansion Coefficients for Angular Functions
C Functions and Driver; Test Values
SWFevalTMN and SWFAngCoeff in the driver SMN.c; Tables 13.2.2–13.2.4

The driver program for the auxiliary functions for eigenvalues—the t_{mn} in Section 13.1.4—and for the expansion coefficients for angular functions—the $d_r^{mn}(c)$ in Section 13.2.1—is given at the end of Section 21.13.3, since all the functions from SWFeval-TMN onward are interdependent.

Spheroidal Angular Functions
C Functions and Driver; Test Values
S1MN and S2MN in the driver SMN.c; Tables 13.2.5–13.2.8

For the regular and irregular spheroidal angular functions—the $S_{mn}^{(1)}(c,\eta)$ and $S_{mn}^{(2)}(c,\eta)$ in Section 13.2.2—the driver program is given at the end of Section 21.13.3 because of the interdependence of the C functions for the spheroidal wave functions.

Note that the basis functions for the spheroidal angular functions, the Legendre functions $P_{m+r}^m(\eta)$ and $Q_{m+r}^m(\eta)$, are supplied with the functions and driver program, in concordance with our philosophy in the *Atlas* of "no assembly required."

21.13.3   Spheroidal Radial Functions

Expansion Coefficients for Radial Functions
C Functions and Driver; Test Values
SWFRadCoeff in the driver SMN.c; Tables 13.3.1, 13.3.2

The C driver program for the expansion coefficients for radial functions—the $a_r^{mn}(c)$ in Section 13.3.1—is given at the end of this section, since all the functions from SWFeval-TMN onward are interdependent.

Spheroidal Radial Functions
C Function and Driver; Test Values
R1MN in the driver SMN.c; Tables 13.3.3, 13.3.4

Our driver program for the regular spheroidal radial functions—the $R_{mn}^{(1)}(c, \xi)$ in Section 13.3.2—requires preliminary computation of all the preceding functions in this section because of the interdependence of the C functions for spheroidal wave functions. However, the driver program will guide you through the steps. If in doubt, read carefully the annotations to the following code.

The basis functions for the regular spheroidal radial functions, the spherical Bessel functions $j_{m+r}(c\xi)$, are computed as a table of values just before the radial functions are computed and printed.

```
#include <stdio.h>
#include <math.h>
  /* maximum depth of continued fraction & arrays */
#define MAX 10
  /* maximum number of tries to bracket root */
#define NTRY 10
  /* maximum number of tries to find root */
#define KMAX 40

  /* maximum number of coefficient entries */
#define CMAX 20
  /* maximum number of Bessel function entries,
     allowing for 6 inaccurate starting values */
#define NMAX 26


main()
{
/* Function Names: SWFAngCoeff, SWFevalTMN, SWFRadCoeff,
   R1MN, S1MN, S2MN  */
/* Spheroidal Angular and Radial Functions:
   Coefficients and Functions of First and Second Kinds;
   order m (m=0,1,2 for second kind), degree n,
   arguments  sign, c, eta.
   Accuracy controls for coefficients  smax, TMNeps.
   Accuracy control for wave function  Seps.
   Normalize angular coefficients in Flammer scheme. */

  /* Use functions  LogGamma, PMN, QMN, SpherBess  */
```

Control parameters for accuracy of eigenvalues.

More accuracy controls.

```
double S1MN(int m,int n,double eta,double dmn[],double
Seps),
 S2MN(int m,int n,double eta,double dmn[],double Seps),
 R1MN(int m,int n,double sign,double c,double chi,
       double amn[],double jn[],double Seps),
sign,c,eta,dmn[CMAX],TMNeps,tmn,Seps,amn[CMAX],chi,
jn[NMAX];
int m,n,smax,rStart,r,rLarge,nMax;
void SWFAngCoeff(int m, int n, double sign, double c,
 int smax, double TMNeps, double dmn[], double *tmn),
 SWFRadCoeff(int m, int n, double dmn[], double amn[]),
 SpherBess(int nMax, double x, double jn[]);
printf("Spheroidal Wave Function Tests");
m = 0;
n = 0;
while ( m <= n )
  {
  printf
  ("\n\nInput m,n,sign,c,smax,TMNepsr (m>n to end): ");
  scanf("%i%i%lf%lf%i%lf",
   &m,&n,&sign,&c,&smax,&TMNeps);
  if ( m > n )
    {
    printf("\nEnd Spheroidal Wave Function Tests");
    exit(0);
    }
  if ( (smax > MAX) || ((n-m+1)/2 > MAX) )
    {
    printf
    ("\n!!smax=%i or n-m=%i too large; increase MAX=%i",
      smax,n-m,MAX);
    }
  else
    {
    /* Make and print tables of expansion coefficients */
    rStart = ( (n-m) % 2 == 0 ) ? 0: 1;
    rLarge = n - m +14;/* coefficient range about 10^14 */

    /* Angular coefficients */
    SWFAngCoeff(m,n,sign,c,smax,TMNeps,dmn,&tmn);
    printf("\n\ntmn = %.9E",tmn);
    printf("\n\n r       dmn(r)\n");
    for ( r = rStart; r <= rLarge; r += 2 )
      { printf("\n%i   %.9E",r,dmn[r]); }

    /* Radial coefficients */
    SWFRadCoeff(m,n,dmn,amn);
    printf("\n\n r       amn(r)\n");
    for ( r = rStart; r <= rLarge; r += 2 )
      { printf("\n%i   %.9E",r,amn[r]); }

    /* Generate spheroidal angular functions */
    eta = 0.1;
    while ( fabs(eta) <= 1.0 )
      {
      printf("\n\nInput eta,Seps ( |eta| > 1 to end): ");
      scanf("%lf%lf",&eta,&Seps);
      if ( fabs(eta) > 1.0 )
        { printf("\nEnd Spheroidal Angular Functions"); }
```

Input order *m*, degree *n* (as integers), sign (+1 for prolate, −1 for oblate), smax (number of terms in continued fraction) and accuracy of root, eps.

Angular coefficient function calls the eigenvalue function. Flammer normalization.

Radial coefficient function uses dmn computed above.

Loop through choices of angular-function argument η.

```
    else                                                        SMN.c
                                                                continued
    {
    printf("\nS1(%i,%i,%lf,%lf,%lf) = %.9E",                     Print regular
                m,n,sign,c,eta,S1MN(m,n,eta,dmn,Seps));         angular
    if ( (m == 0) || (m  == 1) || (m == 2) )                    function.
    { printf("\nS2(%i,%i,%lf,%lf,%lf) = %.9E",
                m,n,sign,c,eta,S2MN(m,n,eta,dmn,Seps)); }        Print irregular
    } }                                                         angular
                                                                function.

    /* Generate spheroidal radial functions */
    chi = 1.0;
    while ( chi > 0.0 )
    {
    printf("\n\nInput chi,Seps ( chi <= 0 to end): ");          Loop through
    scanf("%lf%lf",&chi,&Seps);                                 choices of
    if ( chi <= 0.0 )                                           radial-function
    { printf("\nEnd Spheroidal Radial Functions"); }            argument ξ
    else
    {
    /* Table of spherical Bessel functions into  jn[]  */
    SpherBess(NMAX-8,c*chi,jn);
    printf("\nR1(%i,%i,%lf,%lf,%lf) = %.9E",                     Tabulate Bessel
    m,n,sign,c,chi,R1MN(m,n,sign,c,chi,amn,jn,Seps));           functions.
    } } } } }                                                   Print regular
                                                                radial function.
```

21.14 BESSEL FUNCTIONS

This section gives annotated listings of the C driver programs for testing the sixteen Bessel functions in Chapter 14 of the *Atlas*.

The C functions include the Bessel functions of integer order (J, Y, I, and K in Section 14.2), the Kelvin functions (ber, bei, ker, and kei in Section 14.3), the Bessel functions of half-integer order (j, y, i, and k in Section 14.4), and the Airy functions and their derivatives (Ai, Bi, Ai', and Bi' in Section 14.5).

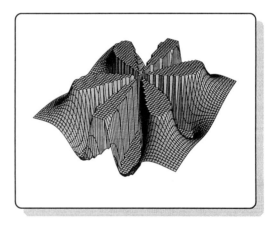

21.14.1 Overview of Bessel Functions

This section has no C functions for the overview of Bessel functions in Section 14.1; to understand how to compute Bessel functions, however, it is essential reading.

21.14.2 Bessel Functions of Integer Order

These driver programs are used for the following Bessel functions: $J_n(x)$, $Y_n(x)$, $I_n(x)$, and $K_n(x)$, with the n being nonnegative integers and x a real variable. Each function has its own driver program that generates an array of the functions from $n = 0$ up to a maximum input value.

Regular Cylindrical Bessel Function:
C Function and Driver; Test Values
BessJN in driver BessJN.c; Table 14.2.1

```
#include <stdio.h>
#include <math.h>
/* maximum number of table entries
   allowing for 6 inaccurate starting values  */
#define NMAX 46

main()
{
/* Function Name: BessJN */
/* Makes a table of regular Bessel Functions Jn(x)
   from order  n = 0  to  nMax <= NMAX-1  */

double x,JN[NMAX];
void BessJN(int nJMax, double x, double JN[]);
int nJMax,n;

printf(" Regular Bessel Function Tests");
nJMax = 10;

while ( nJMax >= 0 )
  {
  printf("\n\nInput nJMax,x ( nJMax < 0  to end): ");
  scanf("%i%lf",&nJMax,&x);
  if ( nJMax < 0 )
    {
    printf("\nEnd Regular Bessel Function Tests");
    exit(0);
    }
  else

    {
    if ( nJMax >= NMAX-6 )
      {
      printf("\n!!nJMax=%i too large; increase NMAX=%i",
             nJMax,NMAX);
      }
    else
      {
      BessJN(nJMax,x,JN);
      printf("\n\n n      J(n)\n");
      for ( n = 0; n <= nJMax; n++ )
        { printf("\n%i  %.9E",n,JN[n]); }
      } } } }
```

Set array size; this value will handle most needs.

Input maximum order and the argument x.

Quit if the maximum order is negative.

Print array of the regular cylindrical Bessel functions.

Irregular Cylindrical Bessel Function:
C Function and Driver; Test Values
BessYN in driver BessYN.c on the following page; Table 14.2.3

```
#include <stdio.h>
#include <math.h>
/* maximum number of table entries
   allowing for 6 inaccurate JN starting values  */
#define NMAX 46

main()
{
/* Function Name: BessYN */
/* Makes a table of irregular Bessel Functions Yn(x)
   from order  n = 0  to  nYMax < NMAX-6  */
double x,YN[NMAX];
void BessYN(int nYMax, double x, double YN[]);
int nYMax,n;
printf(" Irregular Bessel Function Tests");
x = 1.0;
while ( x > 0.0 )
  {
  printf
    ("\n\nInput nYMax,x; ( x <= 0  to end): ");
  scanf("%i%lf",&nYMax,&x);
  if ( x <= 0.0 )
    { printf("\nEnd Irregular Bessel Function Tests");
    exit(0); }
  else
    {
    if ( nYMax >= NMAX-6 )
      { printf("\n!!nYMax=%i too large; increase NMAX=%i",
            nYMax,NMAX); }
    else
      {
      BessYN(nYMax,x,YN);
      printf("\n\n n      Y(n)\n");
      for ( n = 0; n <= nYMax; n++ )
        { printf("\n%i  %.9E",n,YN[n]); }
      } } } }
```

Set array size; this value will handle most needs.

Irregular cylindrical Bessel functions

Input maximum order and the argument x.

Quit if x is nonpositive.

Print array of irregular cylindrical Bessel functions.

Regular Hyperbolic Bessel Function:
C Function and Driver; Test Values
BessIN in driver BessIN.c; Table 14.2.4

```
#include <stdio.h>
#include <math.h>
/* maximum number of table entries
   allowing for 6 inaccurate starting values  */
#define NMAX 46

main()
{
/* Function Name: BessIN */
/* Makes table of regular hyperbolic Bessel Functions
   In(x) from order  n = 0  to  nMax <= NMAX-1  */
double x,IN[NMAX];
void BessIN(int nIMax, double x, double IN[]);
int nIMax,n;
```

Regular hyperbolic Bessel functions

```
printf(" Regular Hyperbolic Bessel Function Tests");
nIMax = 10;
while ( nIMax >= 0 )
 {
 printf("\n\nInput nIMax,x ( nIMax < 0  to end): ");
 scanf("%i%lf",&nIMax,&x);
 if ( nIMax < 0 )
  {
  printf
  ("\nEnd Regular Hyperbolic Bessel Function Tests");
  exit(0);
  }

 else
  {
  if ( nIMax >= NMAX-6 )
   {
   printf("\n!!nIMax=%i too large; increase NMAX=%i",
          nIMax,NMAX);
   }
  else
   {
   BessIN(nIMax,x,IN);
   printf("\n\n n      I(n)\n");
   for ( n = 0; n <= nIMax; n++ )
    { printf("\n%i  %.9E",n,IN[n]); }
   } } } }
```

Input maximum order and the argument x.

Quit if maximum order is negative.

Print array of the regular hyperbolic Bessel functions, Table 14.2.4.

Irregular Hyperbolic Bessel Function:
C Function and Driver; Test Values
BessKN in driver BessKN.c; Table 14.2.6

```
#include <stdio.h>
#include <math.h>
/* maximum number of table entries
   allowing for 6 inaccurate starting values  */
#define NMAX 47

main()
{
/* Function Name: BessKN */
/* Makes table of irregular hyperbolic Bessel Functions
   Kn(x) from order  n = 0  to  nKMax <= NMAX-1  */

double x,KN[NMAX];
void BessKN(int nKMax, double x, double KN[]);
int nKMax,n;

printf(" Irregular Hyperbolic Bessel Function Tests");
x = 1.0;
while ( x > 0.0 )
 {
 printf("\n\nInput nKMax,x ( x <= 0  to end): ");
 scanf("%i%lf",&nKMax,&x);
```

Irregular hyperbolic Bessel functions

Input maximum order and the argument x.

```
 if ( x <= 0.0 )
   {
   printf
   ("\nEnd Irregular Hyperbolic Bessel Function Tests");
   exit(0);
   }
 else
   {
   if ( nKMax >= NMAX-6 )
     {
     printf("\n!!nKMax=%i too large; increase NMAX=%i",
           nKMax,NMAX);
     }
   else
     {
     BessKN(nKMax,x,KN);
     printf("\n\n n      K(n)\n");
     for ( n = 0; n <= nKMax; n++ )
       { printf("\n%i  %.9E",n,KN[n]); }
   } } } }
```

BessKN
continued

Quit if *x* is
nonpositive.

Print array of
the irregular
hyperbolic
Bessel
functions, Table
14.2.6.

21.14.3   Kelvin Functions

The annotated driver programs listed here are used for the following Bessel-type functions: ber, bei, ker, and kei—collectively known as the Kelvin (Thomson) functions. Each function has a nonnegative integer, n, as its order, and a real variable, x, as its argument. There are two driver programs; the first is for the regular Kelvin functions, ber and bei in Section 14.3.1, while the second is for the irregular functions, ker and kei in Section 14.3.2.

Regular Kelvin Functions ber, bei:
C Function and Driver; Test Values
BerBei in driver BerBei.c; Tables 14.3.1, 14.3.2

```
#include <stdio.h>
#include <math.h>
/* maximum number of table entries
   allowing for 6 inaccurate starting values   */
#define NMAX 50

main()
{
/* Function Name: BerBei */
/* Makes a table of regular Kelvin Functions
   ber n x  and  bei n x
   from order  n = 0  to  nBMax <= NMAX-1   */

double x,ber[NMAX],bei[NMAX];
void BerBei(int nBMax, double x,
 double ber[], double bei[]);
int nBMax,n;

printf(" Regular Kelvin Function Tests");
nBMax = 10;
```

Set array size;
this value will
handle most
needs.

```
while ( nBMax >= 0 )
 {
 printf("\n\nInput nBMax,x ( nBMax < 0  to end): ");
 scanf("%i%lf",&nBMax,&x);
 if ( nBMax < 0 )
  {
  printf("\nEnd Regular Kelvin Function Tests");
  exit(0);
  }

 else
  {
  if ( nBMax >= NMAX-6 )
   {
   printf("\n!!nBMax=%i too large; increase NMAX=%i",
        nBMax,NMAX);
   }

  else
   {
   BerBei(nBMax,x,ber,bei);
   printf("\n\n n      ber(n)            bei(n)\n");
   for ( n = 0; n <= nBMax; n++ )
    { printf("\n%i  %.9E    %.9E",n,ber[n],bei[n]); }
  } } } }
```

Input maximum order and the argument *x*.

Quit if the maximum order is negative.

Print array of the regular Kelvin functions, Tables 14.3.1, 14.3.2.

Irregular Kelvin Functions ker, kei:
C Function and Driver; Test Values

KerKei in driver KerKei.c; Tables 14.3.4, 14.3.5

```
#include <stdio.h>
#include <math.h>
/* maximum number of table entries
   allowing for 6 inaccurate starting values  */
#define NMAX 50

main()
{
/* Function Name: KerKei */
/* Makes a table of irregular Kelvin Functions
   ker n x  and  kei n x
   from order  n = 0  to  nKMax <= NMAX-1  */

double x,ker[NMAX],kei[NMAX];
void KerKei(int nKMax, double x,
 double ker[], double kei[]);
int nKMax,n;

printf(" Irregular Kelvin Function Tests");
x = 1.0;
while ( x > 0 )
 {
 printf("\n\nInput nKMax,x ( x <= 0  to end): ");
 scanf("%i%lf",&nKMax,&x);
```

Set array size; this value will handle most needs.

Input maximum order and the argument *x*.

```
if ( x <= 0.0 )
 {
  printf("\nEnd Irregular Kelvin Function Tests");
  exit(0);
 }
else
 {
 if ( nKMax >= NMAX-6 )
  {
   printf("\n!!nKMax=%i too large; increase NMAX=%i",
        nKMax,NMAX);
  }
 else
  {
  KerKei(nKMax,x,ker,kei);
  printf("\n\n n    ker(n)            kei(n)\n");
  for ( n = 0; n <= nKMax; n++ )
   { printf("\n%i  %.9E    %.9E",n,ker[n],kei[n]); }
 } } } }
```

KerKei
continued

Quit if the maximum order is negative.

Print array of the irregular Kelvin functions, Tables 14.3.4, 14.3.5.

21.14.4   Bessel Functions of Half-Integer Order

The four driver programs given here can be used to test our C functions that compute four spherical Bessel functions described in Section 14.4; the regular spherical function $j_n(x)$, the irregular spherical function $y_n(x)$, the regular modified spherical function $i_n(x)$, and the irregular modified spherical function $k_n(x)$.

Although we list and annotate here just the driver program, note that the CD-ROM that comes with the *Atlas* contains the complete program—driver, function, and any ancillary functions.

Regular Spherical Bessel Functions j :
C Function and Driver; Test Values
SphereBessJ in driver SphereBessJ.c; Table 14.4.2

```
#include <stdio.h>
#include <math.h>
 /* maximum number of table entries,
    allowing for 10 inaccurate starting values  */
#define NMAX 41

main()
{
/* Function Name: SphereBessJ */
/* Makes a table of regular spherical Bessel functions
    from order  n = 0  to  nJMax <= NMAX-1   */

double x,jn[NMAX];
void SphereBessJ(int nJMax, double x, double jn[]);
int nJMax,n;

printf(" Regular Spherical Bessel Function Tests");
nJMax = 12;
```

Set array size; this value will handle most needs.

```
while ( nJMax >= 0 )
  {
  printf("\n\nInput nJMax,x ( nJMax < 0  to end): ");
  scanf("%i%lf",&nJMax,&x);
  if ( nJMax < 0 )
    {
    printf
    ("\nEnd Regular Spherical Bessel Function Tests");
    exit(0);
    }
  else
    {
    if ( nJMax >= NMAX-10 )
      {
      printf("\n!!nJMax=%i too large; increase NMAX=%i",
            nJMax,NMAX);
      }
    else
      {
      SphereBessJ(nJMax,x,jn);
      printf("\n\n n        jn(x)\n");
      for ( n = 0; n <= nJMax; n++ )
        { printf("\n%i  %.9E",n,jn[n]); }
      } } } }
```

Input maximum order and the argument *x*.

Quit if the maximum order is negative.

Print array of the regular spherical Bessel functions, Table 14.4.2.

Irregular Spherical Bessel Functions *y* :
C Function and Driver; Test Values

SphereBessY in driver SphereBessY.c; Table 14.4.4

```
#include <stdio.h>
#include <math.h>
 /* maximum number of table entries  */
#define NMAX 41

main()
{
/* Function Name: SphereBessY */
/* Table of irregular spherical Bessel functions
   from order  n = 0  to  nYMax <= NMAX-1  */

double x,yn[NMAX];
void SphereBessY(int nYMax, double x, double yn[]);
int nYMax,n;

printf(" Irregular Spherical Bessel Function Tests");
x = 1.0;
while ( x > 0 )
  {
  printf("\n\nInput nYMax,x ( x <= 0  to end): ");
  scanf("%i%lf",&nYMax,&x);
  if ( x <= 0 )
    {
    printf
    ("\nEnd Irregular Spherical Bessel Function Tests");
    exit(0);
    }
```

Set array size; this value will handle most needs.

Input maximum order and the argument *x*.

Quit if the maximum order is negative.

```
else
{
if ( nYMax >= NMAX )
  {
  printf("\n!!nYMax=%i too large; increase NMAX=%i",
        nYMax,NMAX);
  }

else
  {
  SphereBessY(nYMax,x,yn);
  printf("\n\n n      yn(x)\n");
  for ( n = 0; n <= nYMax; n++ )
    { printf("\n%i  %.9E",n,yn[n]); }
} } } }
```

SphereBessY
continued

Check that array
large enough.

Print array of
the irregular
spherical Bessel
functions, Table
14.4.4.

Regular Modified Spherical Bessel Functions i :
C Function and Driver; Test Values
SphereBessI in driver SphereBessI.c; Table 14.4.6

```
#include <stdio.h>
#include <math.h>
 /* maximum number of table entries,
    allowing for 6 inaccurate starting values  */
#define NMAX 37

main()
{
/* Function Name: SphereBessI */
/* Table of regular modified spherical Bessel
    functions from order  n = 0  to  nIMax <= NMAX-1  */

double x,in[NMAX];
void SphereBessI(int nIMax, double x, double in[]);
int nIMax,n;

printf
 (" Regular Modified Spherical Bessel Function Tests");
nIMax = 12;
while ( nIMax >= 0 )
  {
  printf("\n\nInput nIMax,x ( nIMax < 0  to end): ");
  scanf("%i%lf",&nIMax,&x);
  if ( nIMax < 0 )
    {
    printf
    ("\nEnd Regular Modified Spherical Bessel Function
      Tests");
    exit(0);
    }
```

Set array size;
this value will
handle most
needs.

Input maximum
order and the
argument x.

Quit if the maxi-
mum order is
negative.

```
  else
  {
  if ( nIMax >= NMAX-6 )
    {
    printf("\n!!nIMax=%i too large; increase NMAX=%i",
          nIMax,NMAX);
    }
  else
    {
    SphereBessI(nIMax,x,in);
    printf("\n\n n      in(x)\n");
    for ( n = 0; n <= nIMax; n++ )
    { printf("\n%i   %.9E",n,in[n]); }
  } } } }
```

Print array of regular modified spherical Bessel functions, Table 14.4.6.

Irregular Modified Spherical Bessel Functions
C Function and Driver; Test Values

SphereBessK in driver SphereBessK.c; Table 14.4.8

```
#include <stdio.h>
#include <math.h>
 /* maximum number of table entries  */
#define NMAX 41

main()
{
/* Function Name: SphereBessK */
/* Table of irregular modified spherical Bessel
   functions from order  n = 0  to  nKMax <= NMAX-1   */

double x,kn[NMAX];
void SphereBessK(int nKMax, double x, double kn[]);
int nKMax,n;

printf
(" Irregular Modified Spherical Bessel Function Tests");
x = 1.0;
while ( x > 0 )
  {
  printf("\n\nInput nKMax,x ( x <= 0  to end): ");
  scanf("%i%lf",&nKMax,&x);
  if ( x <= 0 )
    {
    printf
    ("\nEnd Irregular Modified Spherical Bessel Function
Tests");
    exit(0);
    }
  else
    {
    if ( nKMax >= NMAX )
      { printf("\n!!nKMax=%i too large; increase NMAX=%i",
              nKMax,NMAX); }
```

Set array size; this value will handle most needs.

Input maximum order and the argument x.

Quit if the argument is negative.

Check that array is large enough.

```
  exit(0);
  }
else
  { if ( nKMax >= NMAX )
    {
    printf("\n!!nKMax=%i too large; increase NMAX=%i",
           nKMax,NMAX); }
  else
    {
    SphereBessK(nKMax,x,kn);
    printf("\n\n n      kn(x)\n");
    for ( n = 0; n <= nKMax; n++ )
      { printf("\n%i  %.9E",n,kn[n]); }
  } } } }
```

SphereBessK
continued

Check that array
large enough.

Print array of
irregular modi-
fied spherical
Bessel
functions, Table
14.4.8.

21.14.5   Airy Functions

The two driver programs given here can be used to test our C functions that compute the Airy functions, Ai and Bi, described in Section 14.5.1, and also their first derivatives, Ai′ and Bi′, in Section 14.5.2. Although we list and annotate here just the driver program, the CD-ROM that comes with the *Atlas* contains the complete program—driver, function, and the auxiliary functions.

Airy Functions Ai, Bi :
C Function and Driver; Test Values
AiryAiBi in driver AiryAiBi.c; Table 14.5.2

```
#include <stdio.h>
#include <math.h>

main()
{
/* Function Name: AiryAiBi */
/* Airy Functions Ai(x)  and  Bi(x)  to
   fractional accuracy  eps  */

double x,eps,Ai,Bi;
void AiryAiBi(double x, double eps, double *Ai,
  double *Bi);

printf("Airy Function Tests");
eps = 1.0;
while ( eps > 0.0 )
  {
  printf("\n\nInput x,eps ( eps <= 0 to end): ");
  scanf("%lf%lf",&x,&eps);
  if ( eps <= 0.0 )
    { printf("\nEnd Airy Function Tests");
    exit(0); }
  else
    {
    AiryAiBi(x,eps,&Ai,&Bi);
    printf
      ("\nAi(%lf) = %.9E   Bi(%lf) = %.9E",x,Ai,x,Bi);
  } } }
```

Airy Ai and Bi

Input argument
x and fractional
accuracy eps

Print Ai and Bi
to 10 digits,
Table 14.5.2.

Derivatives of Airy Functions Ai′, Bi′ :
C Function and Driver; Test Values
AiryPrimeAiBi in driver AiryPrimeAiBi.c; Table 14.5.4

```
#include <stdio.h>
#include <math.h>

main()
{
/* Function Name: AiryPrimeAiBi */
/* Derivatives of Airy Functions, Ai'(x) and Bi'(x)
   to fractional accuracy  eps  */

double x,eps,AiPrime,BiPrime;
void AiryPrimeAiBi(double x, double eps,
 double *AiPrime, double *BiPrime);

printf("Airy Function Derivatives Tests");
eps = 1.0;
while ( eps > 0.0 )
  {
 printf("\n\nInput x,eps ( eps <= 0 to end): ");
 scanf("%lf%lf",&x,&eps);
 if ( eps <= 0.0 )
   {
  printf("\nEnd Airy Function Derivatives Tests");
  exit(0);
  }
 else
  {
  AiryPrimeAiBi(x,eps,&AiPrime,&BiPrime);
  printf
   ("\nAi'(%lf) = %.9E   Bi'(%lf) = %.9E",
    x,AiPrime,x,BiPrime);
  } } }
```

Airy Ai′ and Bi′

Input argument
x and fractional
accuracy eps

Print Ai′ and Bi′
to 10 digits,
Table 14.5.4.

21.15 STRUVE, ANGER, AND WEBER FUNCTIONS

This section gives the annotated listings of the C driver programs for checking the four functions in Chapter 15 of the *Atlas*.

The C functions are the Struve functions **H** and **L** in Section 15.1, the Anger function **J** that is described in Section 15.2.2, and the Weber function **E** in Section 15.2.3.

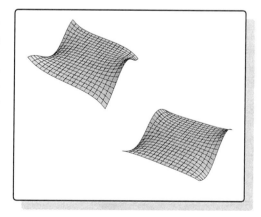

21.15.1 Struve Functions

The two driver programs for testing **H** and **L** compute arrays of these functions for order $n = 0$ up to some maximum specified by the user. The function listings in Section 15.1 show only functions that are not listed elsewhere in the *Atlas*, but the version on CD-ROM comes completely assembled.

Struve Function H:
C Function and Driver; Test Values
StruveH in driver StruveH.c; Table 15.1.1

```
#include <stdio.h>
#include <math.h>

/* maximum number of table entries */
#define NMAX 37

main()
{
/* Function Name: StruveH */
/* Makes a table of Struve functions Hn(x)
   from order  n = 0 to nHMax < NMAX for x >= 0
   to fractional accuracy   eps  */

double x,eps,HN[NMAX];
void StruveH(int nHMax,double x,double eps,double HN[]);
int nHMax,n;

printf(" StruveH Function Tests");
x = 1.0;
while ( x >= 0.0 )
  {
  printf
    ("\n\nInput nHMax,x,eps; ( eps = 0  to end): ");
  scanf("%i%lf%lf",&nHMax,&x,&eps);
  if ( eps == 0.0 )
    {
    printf("\nEnd StruveH Function Tests");
    exit(0);
    }

  else
    {
    if ( nHMax >= NMAX-6 )
     {
     printf
       ("\n!!nHMax=%i too large; increase NMAX=%i",
        nHMax,NMAX);
     }

   else
     {
     StruveH(nHMax,x,eps,HN);
     printf("\n\n n      H(n)\n");
     for ( n = 0; n <= nHMax; n++ )
       { printf("\n%i  %.9E",n,HN[n]); }
     }
    }
  }
}
```

Set array size; this value will handle up to $n = 30$.

Input maximum order, argument x, and fractional accuracy eps.

Quit if accuracy limit is zero.

Check that array is large enough.

Make and print array of Struve **H** functions.

Struve Function L:
C Function and Driver; Test Values
StruveL in driver StruveL.c; Table 15.2.1

```
#include <stdio.h>
#include <math.h>
/* maximum number of table entries */
#define NMAX 30

main()
{
/* Function Name: StruveL */
/* Makes a table of modified Struve functions Ln(x)
   from order  n = 0 to nLMax < NMAX for x >= 0
   to fractional accuracy  eps  */

double x,eps,LN[NMAX];
void StruveL(int nLMax,double x,double eps,double LN[]);
int nLMax,n;

printf(" StruveL Function Tests");
x = 1.0;

while ( x >= 0.0 )
 {
 printf
   ("\n\nInput nLMax,x,eps; ( eps = 0  to end): ");
 scanf("%i%lf%lf",&nLMax,&x,&eps);
 if ( eps == 0.0 )
   {
   printf("\nEnd StruveL Function Tests");
   exit(0);
   }

 else
   {
   if ( nLMax >= NMAX-1 )
     { printf
       ("\n!!nLMax=%i too large; increase NMAX=%i",
       nLMax,NMAX); }

   else
     {
     StruveL(nLMax,x,eps,LN);
     printf("\n\n n       L(n)\n");
     for ( n = 0; n <= nLMax; n++ )
       { printf("\n%i  %.9E",n,LN[n]); }
     } } } }
```

Set array size; this value will handle up to *n* = 29.

Input maximum order, argument *x*, and fractional accuracy eps.

Quit if accuracy limit is zero.

Check that array is large enough.

Make and print array of Struve **L** functions, as in Table 15.1.2.

21.15.2 Anger and Weber Functions

There is a single driver program for testing the Anger and Weber functions—**J** and **E** respectively—because they are generated by using the same two Lommel functions as auxiliary functions. The driver computes arrays of **J** and **E** for order $n = 0$ up to some maximum specified by the user. The function listings in Section 15.1 show only functions that are not listed elsewhere in the *Atlas*, but the version on CD-ROM comes completely assembled.

C Function and Driver; Test Values

AngerWeber in driver AngerWeber.c; Table 15.2.2

```
#include <math.h>
/* maximum number of table entries  */
#define NMAX 10

main()
{
/* Function Name: AngerWeber */
/* Makes a table of Anger and Weber Functions
   from noninteger order  numMin > 0  for
   nAWMax < NMAX-1  values; fixed x >= 0.
   Fractional accuracy is   eps  */

double nuMin,x,eps,Anger[NMAX],Weber[NMAX];
void AngerWeber(int nAWMax, double nuMin, double x,
 double eps, double Anger[], double Weber[]);
int nAWMax,n;

printf(" Anger and Weber Function Tests");
nuMin = 0.5;
while ( nuMin > 0.0 )
 {
 printf
  ("\n\nInput nAWMax,nuMin,x,eps (nuMin=0 to end): ");
 scanf("%i%lf%lf%lf",&nAWMax,&nuMin,&x,&eps);
 if ( nuMin <= 0 )
  { printf("\nEnd Anger and Weber Function Tests");
    exit(0); }

 else
  {
  if ( nAWMax >= NMAX )
   { printf("\n!!nAWMax=%i too large; increase NMAX=%i",
     nAWMax,NMAX); }
  else
   { /* Make array of Anger & Weber functions */
   AngerWeber(nAWMax,nuMin,x,eps,Anger,Weber);
   if ( x < 30.0 )
    {
    printf
    ("\n\n   nu        Anger(nu)        Weber(nu)\n");
    }

   else
    {
    printf
    ("\n\n   nu    Anger(nu)-J(nu)  Weber(nu)+Y(nu)\n");
    }
   for ( n = 1; n <= nAWMax; n++ )
    { printf("\n%.3E   %.9E   %.9E",
      n-1+nuMin,Anger[n],Weber[n]); }
 } } } }
```

Set array size; this value will handle up to $n = 30$.

Input maximum order, argument x, and fractional accuracy eps.

Quit if accuracy limit is zero.

Check that array is large enough.

Make arrays of the **J** and **E** functions.

For $x < 30$, **J** and **E** are printed.

For larger x, **J**–*J* and **E**+*Y* are printed.

21.16 HYPERGEOMETRIC FUNCTIONS AND COULOMB WAVE FUNCTIONS

This section gives annotated listings of the four C driver programs for checking the seven functions in Chapter 16 of the *Atlas*. There are C drivers for the $_2F_1$, $_1F_1$, and U hypergeometric functions (Sections 16.1 and 16.2), as well as a driver for the Coulomb wave functions—F_L, F'_L, G_L, and G'_L in Section 16.3. The *Mathematica* versions of the functions are in the annotated notebook in Section 20.16.

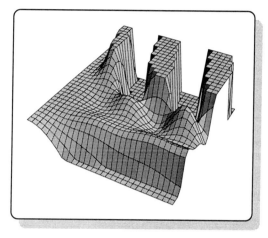

21.16.1 Hypergeometric Functions

C Function and Driver; Test Values

Hyper2F1 in driver `Hyper2F1.c`; Tables 16.1.1, 16.1.2

The driver program for this section tests the $_2F_1$ hypergeometric function that is described in Section 16.1.

```
#include <stdio.h>
#include <math.h>

main()
{
/* Function Name: Hyper2F1 */
/* Hypergeometric Function 2F1 with parameters
    a, b, and c, argument  x, with |x| < 1,
    to fractional accuracy  eps  */

double Hyper2F1(double a, double b, double c,
  double x, double eps),a,b,c,x,eps;

printf(" Hypergeometric Function 2F1 Tests");
x = 0.5;

while ( fabs(x) < 1.0 )
  {
  printf("\n\nInput a,b,c,x,eps ( eps <= 0 to end): ");
  scanf("%lf%lf%lf%lf%lf",&a,&b,&c,&x,&eps);
  if ( eps <= 0.0 )
  { printf("\nEnd Hypergeometric Function 2F1 Tests");
  exit(0); }
  else
    {
    printf("\n2F1(%lf,%lf,%lf,%lf) = %.9E",
      a,b,c,x,Hyper2F1(a,b,c,x,eps));
    } } }
```

Input parameters *a*, *b*, and *c*, argument *x*, and fractional accuracy eps.

Make and print the hypergeometric function.

21.16.2 Confluent Hypergeometric Functions

Regular Function $_1F_1$
C Function and Driver; Test Values
Hyper1F1 in driver Hyper1F1.c; Table 16.2.1

The driver program for this section tests the $_1F_1$ confluent hypergeometric function that is described in Section 16.2.1.

```
#include <stdio.h>
#include <math.h>

main()
{
/* Function Name: Hyper1F1 */
/* Confluent Hypergeometric Function 1F1
   with parameters a and b, argument  x,
   to fractional accuracy  eps  */

double Hyper1F1(double a, double b,
 double x, double eps),a,b,c,x,eps;

printf
 (" Confluent Hypergeometric Function Tests");
eps = 1.0;
while ( eps > 0.0 )
 { printf("\n\nInput a,b,x,eps ( eps <= 0 to end): ");
 scanf("%lf%lf%lf%lf",&a,&b,&x,&eps);
 if ( eps <= 0.0 )
 { printf
 ("\nEnd Confluent Hypergeometric Function Tests");
 exit(0); }
 else
 { printf("\n1F1(%lf,%lf,%lf) = %.9E",
 a,b,x,Hyper1F1(a,b,x,eps)); } } }
```

Input parameters a, and b, argument x, and fractional accuracy eps.

Make and print the confluent hypergeometric function.

Irregular Function U
C Function and Driver; Test Values
HyperU in driver HyperU.c; Table 16.2.3

The driver program for this section tests the U confluent hypergeometric function in Section 16.2.1.

```
#include <stdio.h>
#include <math.h>

main()
{
/* Function Name: HyperU */
/* Confluent Hypergeometric Function U
   with parameters a and b, argument  x,
   to fractional accuracy  eps  */
double HyperU(double a, double b,
 double x, double eps),a,b,c,x,eps;
printf(" Confluent Hypergeometric Function U Tests");
x = 1.0;
while ( x > 0.0 )
 { printf("\n\nInput a,b,x,eps ( x <= 0 to end): ");
```

```
    scanf("%lf%lf%lf%lf",&a,&b,&x,&eps);
    if ( x <= 0.0 )
    { printf
    ("\nEnd Confluent Hypergeometric Function U Tests");
     exit(0); }
    else
    { printf("\nU(%lf,%lf,%lf) = %.9E",
      a,b,x,HyperU(a,b,x,eps)); } } }
```

Input parameters a, b, and c, argument x, and fractional accuracy eps.

Make and print the hypergeometric function.

21.16.3 Coulomb Wave Functions

C Function and Driver; Test Values
CoulombFG in driver CoulombFG.c; Tables 16.3.1–16.3.12

The driver program for this section tests the Coulomb wave functions that is described in Section 16.2.1.

```
#include <stdio.h>
#include <math.h>
#define MAX 51 /* for maximum  L value 30  */

main()
{
/* Function Name: CoulombFG */
/* Regular Coulomb Wave Functions F & G plus their
   derivatives F' & G' and the Coulomb phase shifts.
   Parameters L, eta, rho */
void CoulombFG(int LMAX, double eta, double rho,
 double F[], double FP[], double G[], double GP[],
 double Sigma[]);
double F[MAX], FP[MAX], G[MAX], GP[MAX], Sigma[MAX];
double eta,rho;
int LMAX,L;

printf(" Coulomb Wave Function Tests");
rho = 1.0;
while ( rho >= 0.0 )
 {
 printf("\n\nInput LMAX,eta,rho ( rho <= 0 to end): ");
 scanf("%i%lf%lf",&LMAX,&eta,&rho);
 if ( rho <= 0.0 )
 { printf("\nEnd Coulomb Wave Function Tests");
   exit(0); }
 else
   { if ( LMAX > MAX - 21 )
     { printf("\n!!LMAX=%i too large for array size"); }
     else
     { /* F(L), FP(L), G(L), GP(L),  L = 0 to LMAX;
         Ignore 10 highest L values  */
     CoulombFG(LMAX+20,eta,rho,F,FP,G,GP,Sigma);
     printf("\n L,  F,  FP,  G,  GP,  Sigma\n");
     for ( L = 0; L <= LMAX; L++ )
       { printf("\n %i %.6E %.6E %.6E %.6E %.6E",
                L,F[L],FP[L],G[L],GP[L],Sigma[L]); }
 } } } }
```

Input parameters L, η, and argument ρ.

Make and print array of Coulomb functions and their derivatives.

21.17 ELLIPTIC INTEGRALS AND ELLIPTIC FUNCTIONS

This section gives annotated listings of the C driver programs for checking the fourteen functions in Chapter 17 of the *Atlas*. There are C functions for the three kinds of elliptic integrals, the Jacobi zeta function, and the Heuman lambda function, all of which are described in Section 17.2 and in the annotated *Mathematica* notebook in Section 20.17.

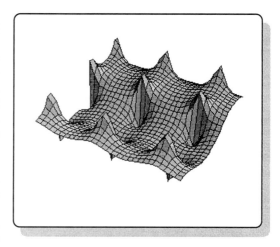

We also describe the C drivers for the three basic Jacobi elliptic functions (sn, cn, and dn), for the four theta functions, as well as for their logarithmic derivatives; these functions are described in Section 17.3 of the *Atlas* and their notebooks are also described in Section 20.17.

21.17.1   Overview of Elliptic Integrals and Elliptic Functions

Section 17.1 in the *Atlas* has no C driver programs; it should, however, be read in order to understand the notations that we use.

21.17.2   Elliptic Integrals

There are two driver programs for this section; one tests the complete and incomplete elliptic integrals of the first kind (*F* and *K*) and the other tests elliptic integrals of the second kind (*E*). The function listings in Section 17.2 show only functions that are not listed elsewhere in the *Atlas*, but the version on CD-ROM comes completely assembled.

Elliptic Integrals of the First Kind, *F* and *K*:
C Function and Driver; Test Values
EllptcFK in driver EllptcFK.c; Table 17.2.2

```
#include <stdio.h>
#include <math.h>

main()
{
/* Function Name: EllptcFK */
/* Elliptic integrals of first kind
    by AGM method; F(phi|m)  and  K(m)  */

double phi,m,eps,Km;
double EllptcFK(double phi, double m,
  double eps, double *Km);
printf
(" Elliptic Integral of First Kind (F) Tests");
```

```
m = 0.5;
while ( m < 1 )
  {
  printf
  ("\n\nInput phi,m, eps ( m >= 1 to end): ");
  scanf("%lf%lf%lf",&phi,&m,&eps);
  if ( m >= 1 )
    {
    printf("\nEnd Elliptic Integral (F) Tests");
    exit(0);
    }

  printf
   ("F(%.9E,%.9E) =
   %.9E\n",phi,m,EllptcFK(phi,m,eps,&Km));
  printf("K(%.9E) =  %.9E\n",m,Km);
  } }
```

Input argument
ϕ, parameter m,
and fractional
accuracy eps.

Quit if m makes
the integral
complex.

Make and print
incomplete (F)
then complete
(K) integrals of
first kind.

Elliptic Integrals of the Second Kind, E:
C Function and Driver; Test Values
EllptcEE in driver EllptcEE.c; Table 17.2.3

```
#include <stdio.h>
#include <math.h>

main()
{
/* Function Name: EllptcEE */
/* Elliptic integrals of second kind by AGM method;
   E(phi|m)  for phi <= Pi/2  and  E(m)  */

double phi,m,eps,Em;
double EllptcEE(double phi, double m,
 double eps, double *Em);

printf
  (" Elliptic Integral  of Second Kind (E) Tests");

m = 0.5;
while ( m <= 1 )
  {
  printf
  ("\n\nInput phi ( <= Pi/2), m, eps ( m > 1 to end): ");
  scanf("%lf%lf%lf",&phi,&m,&eps);

  if ( m > 1 )
    {
    printf("\nEnd Elliptic Integral (E) Tests");
    exit(0);
    }

  printf
   ("E(%.9E,%.9E) =
%.9E\n",phi,m,EllptcEE(phi,m,eps,&Em));
   printf("E(%.9E) =  %.9E\n",m,Em);
  } }
```

Input argument
ϕ, parameter m,
and fractional
accuracy eps.

Quit if m makes
integral com-
plex.

Make and print
incomplete (E)
then complete
(Em) integrals of
second kind.

Jacobi Zeta Function, Z :
C Function and Driver; Test Values

JacobiZeta in driver JacobiZeta.c; Table 17.2.4

```
#include <stdio.h>
#include <math.h>

main()
{
/* Function Name: JacobiZeta */
/* Jacobi Zeta function  by combining AGM methods
   for elliptic integrals of first and second kinds */

double phi,m,eps;
double JacobiZeta(double phi, double m, double eps);

printf
  (" Jacobi Zeta function Tests");

m = 0.5;
while ( m < 1 )
  {
  printf
   ("\n\nInput phi,m, eps ( m >= 1 to end): ");
  scanf("%lf%lf%lf",&phi,&m,&eps);
  if ( m >= 1 )
   {
   printf("\nEnd Jacobi Zeta function Tests");
   exit(0);
   }
  printf
   ("Z(%.9E,%.9E) =
%.9E\n",phi,m,JacobiZeta(phi,m,eps));
  } }
```

Input argument ϕ, parameter m, and fractional accuracy eps. Quit if m makes integral complex.

Make and print Jacobi zeta function.

Heuman Lambda Function, Λ_0 :
C Function and Driver; Test Values

HeumanLambda in driver HeumanLambda.c; Table 17.2.5

```
#include <stdio.h>
#include <math.h>

main()
{
/* Function Name: HeumanLambda */
/* Heuman Lambda function  by combining AGM methods
   for elliptic integrals of first and second kinds */

double phi,m,eps;
double HeumanLambda(double phi, double m, double eps);

printf
  (" Heuman Lambda function Tests");
```

```
m = 0.5;
while ( ( m <= 1.0 ) && ( m >= 0.0) )
  {
  printf
  ("\n\nInput phi,m, eps ( m>1 or m<0 to end): ");
  scanf("%lf%lf%lf",&phi,&m,&eps);
  if ( ( m > 1.0 )  || ( m < 0.0) )
    {
    printf("\nEnd Heuman Lambda function Tests");
    exit(0);
    }
  printf
  ("Lambda0(%.9E,%.9E) =  %.9E\n",
    phi,m,HeumanLambda(phi,m,eps));
  } }
```

Input argument ϕ, parameter m, and fractional accuracy eps. Quit if m makes integral complex.

Make and print Heuman lambda function.

Elliptic Integrals of the Third Kind, Π :
C Function and Driver; Test Values
EllptcPi in driver EllptcPi.c; Table 17.2.6

```
#include <stdio.h>
#include <math.h>

main()
{
/* Function Name: EllptcPi */
/* Elliptic integral of third kind by direct integration.
   Accuracy is at least 10^(-10) for
   n <= 0.94, m <= 0.94, phi = Pi/2   */

double n,phi,m;
double EllptcPi(double n, double phi, double m);

printf
  (" Elliptic Integral of Third Kind (Pi) Tests");
m = 0.5;
while ( m < 1.0 )
  {
  printf
  ("\n\nInput n, phi ( <= Pi/2), m ( m >= 1 to end): ");
  scanf("%lf%lf%lf",&n,&phi,&m);
  if ( m >= 1.0 )
    {
    printf("\nEnd Elliptic Integral (Pi) Tests");
    exit(0);
    }
  printf("Pi(%.9E,%.9E,%.9E) =  %.9E\n",
    n,phi,m,EllptcPi(n,phi,m));
  } }
```

Input argument ϕ, parameter m, and fractional accuracy eps. Quit if m makes integral complex.

Make and print elliptic integral of third kind.

21.17.3   Jacobi Elliptic Functions and Theta Functions

There are three driver programs for this section; one tests the Jacobi elliptic functions (Section 17.3.1), one tests the Jacobi theta functions (Section 17.3.2), and the third tests the logarithmic derivatives of the theta functions (Section 17.3.3).

Jacobi Elliptic Functions, sn, cn, dn
C Function and Driver; Test Values

JacobiSnCnDn in driver JacobiSnCnDn.c; Tables 17.3.1–17.3.3

```
#include <stdio.h>
#include <math.h>
/* Maximum number of array elements */
#define MAX 10


main()
{
/* Function Name: JacobiSnCnDn */
/* Jacobi elliptic functions by AGM method;
   sn, cn, and  dn  */

double u,m,eps,sn,cn,dn;
void JacobiSnCnDn(double u, double m,
 double eps, double *sn, double *cn, double *dn);

printf
 (" Jacobi elliptic function (sn,cn,dn) Tests");
m = 0.5;
while ( m <= 1 )
 {
 printf
 ("\n\nInput u,m, eps ( m > 1 to end): ");
 scanf("%lf%lf%lf",&u,&m,&eps);
 if ( m > 1 )
  {
  printf("\nEnd Jacobi elliptic function Tests");
  exit(0);
  }
 JacobiSnCnDn(u,m,eps,&sn,&cn,&dn);
 printf("\nsn(%.9E,%.9E)  =  %.9E",u,m,sn);
 printf("\ncn(%.9E,%.9E)  =  %.9E",u,m,cn);
 printf("\ndn(%.9E,%.9E)  =  %.9E",u,m,dn);
 } }
```

Arrays store amplitudes in AGM scheme.

Three Jacobi elliptic function values are returned in argument list.

Input argument u, parameter m, and fractional accuracy eps.

Quit if m makes the function complex.

Make and print three Jacobi elliptic functions.

Theta Functions, θ_α
C Function and Driver; Test Values

JacobiTheta in driver JacobiTheta.c; Tables 17.3.4–17.3.7

```
#include <stdio.h>
#include <math.h>

main()
{
/* Function Name: JacobiTheta */
/* Jacobi theta functions by series */

double u,q,eps,theta1,theta2,theta3,theta4;
void JacobiTheta(double u, double q, double eps,
 double *theta1, double *theta2,
 double *theta3, double *theta4);
```

Four Jacobi theta function values are returned through the argument list.

```
printf
 (" Jacobi theta functions tests");
q = 0.5;
while ( fabs(q) < 1 )
 {
 printf
 ("\n\nInput u,q, eps ( |q| >= 1 to end): ");
 scanf("%lf%lf%lf",&u,&q,&eps);
 if ( fabs(q) >= 1 )
  {
  printf("\nEnd Jacobi theta functions tests");
  exit(0);
  }

 JacobiTheta(u,q,eps,&theta1,&theta2,
  &theta3,&theta4);
 printf("\ntheta1(%.9E,%.9E) =  %.9E",u,q,theta1);
 printf("\ntheta2(%.9E,%.9E) =  %.9E",u,q,theta2);
 printf("\ntheta3(%.9E,%.9E) =  %.9E",u,q,theta3);
 printf("\ntheta4(%.9E,%.9E) =  %.9E",u,q,theta4);
 } }
```

Input argument u, nome q, and absolute accuracy eps.

Quit if q makes the series diverge.

Make then print the four Jacobi theta functions.

Logarithmic Derivatives of Theta Functions, $\theta'_\alpha / \theta_\alpha$
C Function and Driver; Test Values

LogDerTheta in driver LogDerTheta.c; Tables 17.3.8–17.3.11

```
#include <stdio.h>
#include <math.h>

main()
{
/* Function Name: LogDerTheta */
/* Log derivatives of Jacobi theta functions by
series */

double u,q,eps,LDtheta1,LDtheta2,LDtheta3,LDtheta4;
void LogDerTheta(double u, double q, double eps,
 double *LDtheta1, double *LDtheta2,
 double *LDtheta3, double *LDtheta4);

printf
 (" Log Derivatives of Theta functions tests");
q = 0.5;
while ( fabs(q) < 1 )
 {
 printf
 ("\n\nInput u,q, eps ( |q| >= 1 to end): ");
 scanf("%lf%lf%lf",&u,&q,&eps);

 if ( fabs(q) >= 1 )
  {
  printf("\nEnd Log Derivatives of Theta functions
    tests");
  exit(0);
  }
```

Four Jacobi theta function values are returned through the argument list.

```
  LogDerTheta(u,q,eps,&LDtheta1,&LDtheta2,
    &LDtheta3,&LDtheta4);

  printf("\nLDtheta1(%.9E,%.9E) =
    %.9E",u,q,LDtheta1);
  printf("\nLDtheta2(%.9E,%.9E) =
    %.9E",u,q,LDtheta2);
  printf("\nLDtheta3(%.9E,%.9E) =
    %.9E",u,q,LDtheta3);
  printf("\nLDtheta4(%.9E,%.9E) =
    %.9E",u,q,LDtheta4);
  } }
```

`LogDerTheta` driver
continued

Make then print the four
logarithmic derivatives
of the Jacobi theta
functions.

21.18 PARABOLIC CYLINDER FUNCTIONS

This section gives annotated listings of the C driver programs for checking the functions in Chapter 18 of the *Atlas*. The corresponding annotated *Mathematica* notebooks are given in Section 20.18.

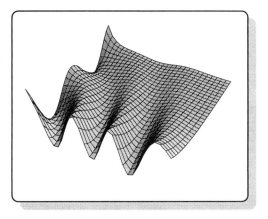

21.18.1 Parabolic Cylinder Functions
Section 18.1 in the *Atlas* has two C driver programs for the parabolic cylinder functions. The first is for `ParCylU` that computes $U(a,x)$ in Section 18.1.2, and the second is for `ParCylV` that computes $V(a,x)$ in Section 18.1.3. Function listings in these sections show only functions not listed elsewhere in the *Atlas*; the version on the CD-ROM comes completely assembled.

Parabolic Cylinder Functions *U*
C Function and Driver; Test Values
`ParCylU` in driver `ParCylU.c`; Table 18.2.1

```
#include <stdio.h>
#include <math.h>

main()
{
/* Function Name: ParCylU */
/* Parabolic Cylinder Function U(a,x)
   with parameter a, argument  x.
   Fractional accuracy is  eps  for x <= 2
   and is at least 10 digits for  x > 2   */
double ParCylU(double a, double x, double eps),
  a,x,eps;
```

```
printf
 (" Parabolic Cylinder Function (U) Tests");
eps = 1.0;
while ( eps > 0.0 )
 {
 printf("\n\nInput a,x,eps ( eps <= 0 to end): ");
 scanf("%lf%lf%lf",&a,&x,&eps);

 if ( eps <= 0.0 )
 {
 printf
 ("\nEnd Parabolic Cylinder (U) Function Tests");
 exit(0);
 }

 else
   { printf("\nU(%lf,%lf) = %.9E",
      a,x,ParCylU(a,x,eps)); } } }
```

Input arguments *a* and *x* then fractional accuracy eps.

Quit if eps = 0.

Make and print parabolic cylinder function *U*.

Parabolic Cylinder Functions *V*
C Function and Driver; Test Values
ParCylV in driver ParCylV.c; Table 18.2.2

```
#include <stdio.h>
#include <math.h>

main()
{
/* Function Name: ParCylV */
/* Parabolic Cylinder Function V(a,x)
   with parameter a, argument  x.
   Fractional accuracy goal is  eps  for series */

double ParCylV(double a, double x, double eps),
 a,x,eps;

printf
 (" Parabolic Cylinder Function (V) Tests");
eps = 1.0;
while ( eps > 0.0 )
 {
 printf("\n\nInput a,x,eps ( eps <= 0 to end): ");
 scanf("%lf%lf%lf",&a,&x,&eps);
 if ( eps <= 0.0 )
 {
 printf
 ("\nEnd Parabolic Cylinder (V) Function Tests");
 exit(0);
 }

 else
   { printf("\nV(%lf,%lf) = %.9E",
      a,x,ParCylV(a,x,eps)); } } }
```

Input arguments *a* and *x* then fractional accuracy eps.

Quit if eps = 0.

Make and print parabolic cylinder function *V*.

21.19 MISCELLANEOUS FUNCTIONS FOR SCIENCE AND ENGINEERING

This section gives annotated listings of the C driver programs for checking the functions in Sections 19.1–19.6 of the *Atlas*. The corresponding annotated *Mathematica* notebooks are in Section 20.19. Drivers for the angular momentum coupling coefficients in Section 19.7 are in that section.

21.19.1 Debye Functions

The C driver program for Debye functions. in Section 19.1 is very simple, controlling the choice of order, n, and argument, x, and the number of integration points, `Nint`.

C Function and Driver; Test Values

`Debye` in driver `Debye.c`; Table 19.1.1

```
#include <stdio.h>
#include <math.h>
#define MAX 31

main()
{
/* Driver program for Debye
   -- Debye functions by Simpson integration
      using Nint points */

double x,Debye(int n, double x, int Nint);
int n,Nint;

printf("Debye Functions\n");

n = 1;
while ( n > 0 )
  {
  printf
  ("\nInput n, x, Nint (n=0 to end):\n");
  scanf("%i%lf%i",&n,&x,&Nint);

  if ( n > 0 )
    {
    printf("%i %6.2lf %i",n,x,Nint);
    if ( Nint < MAX )
    printf("%.9E\n",Debye(n,x,Nint));
    else
    printf("\n!!Nint=%i !< MAX=%i\n",Nint,MAX);
    }
  } /* end  n & x  loop */
printf("\nEnd  Debye Functions");
}
```

Maximum number of integration points.

Input order, n, argument, x, and number of integration points, `Nint`.

Check that number of points not too large. Make and print the Debye function.

21.19.2 Sievert Integral

The C driver program for the Sievert integral in Section 19.2 is very simple, controlling the angle argument, θ, and the distance argument, x.

C Function and Driver; Test Values

`Sievert` in driver `Sievert.c`; Table 19.2.1

```
#include <stdio.h>
#include <math.h>
#define MAX 31

main()
{
/* Driver program for Sievert
   -- Sievert integrals by Simpson integration
      using Nint points */

double theta,x,
  Sievert(double theta, double x, int Nint);
int Nint;

printf("Sievert Functions\n");

theta = 1;
while ( theta > 0 )
  {
  printf("\nInput theta (rad), x, Nint");
  printf("(theta=0 to end):\n");
  scanf("%lf%lf%i",&theta,&x,&Nint);
  if ( theta > 0 )
    {
    printf("%6.4lf %6.3f %i",theta,x,Nint);
    if ( Nint < MAX )
    printf("%.9E\n",Sievert(theta,x,Nint));
    else
    printf("\n!!Nint=%i !< MAX=%i\n",Nint,MAX);
    }
  } /* end  theta & x  loop */
printf("\nEnd  Sievert  Functions");
}
```

Maximum number of integration points.

Input order, n, argument, x, and number of integration points, `Nint`.

Check that number of points not too large. Make and print the Sievert integral..

21.19.3 Abramowitz Function

The C driver program for the Abramowitz function in Section 19.3 controls the order $m = 0, 1, \ldots$ and the argument, x.

C Function and Driver; Test Values

`Abrmwtz` in driver `Abrmwtz.c`; Table 19.3.1

```
#include <stdio.h>
#include <math.h>


#define KMAX 21
#define MMAX 7
```

Maximum number of summation points, `KMAX`, and maximum order, `MMAX`.

```
main()
{
/* Driver program for Abrmwtz
   -- Abramowitz functions by series */

double x,eps,
Abrmwtz(int m, double x, double eps, int first);
int m,first;

printf("Abramowitz Functions\n");
m = 1;
first = 1;
while ( m >= 0 )
  {
  printf("\nInput m, x, eps");
  printf(" (m<0 to end):\n");
  scanf("%i%lf%lf",&m,&x,&eps);
  if ( m >= 0 )
    {

    if ( m > MMAX-1 )
    { printf
    ("\n!!m=%i larger than MMAX-1=%i\n",m,MMAX-1);
    }

    printf("%i %6.4lf %6.4le",m,x,eps);
    printf("%.9E\n",Abrmwtz(m,x,eps,first));
    first = 0;
    }
  } /* end  m & x  loop */
printf("\nEnd  Abramowitz  Functions");
}
```

Abrmwtz
continued

Input order, *m*, and argument, *x*.

Check that order is not too large.

Make and print Abramowitz function.

21.19.4 Spence Integral

The C driver program for the Spence integral in Section 19.4 is simple, since only the argument, *x*, and the number of integration points, `Nint`, are to be input.

C Function and Driver; Test Values

`Spence` in driver `Spence.c`; Table 19.4.1

```
#include <stdio.h>
#include <math.h>
#define MAX 141

main()
{
/* Driver program for Spence;
   Spence integral by series & integration */

double x, Spence(double x, int Nint);
int Nint;

printf("Spence Integral");

x = 1.0;
```

Maximum number of integration points is `MAX-1`.

```
while ( x > 0 )
 {
 printf("\n\nInput  x, Nint ( x<=0 to end): ");
 scanf("%lf%i",&x,&Nint);

 if ( x > 0 )
  {
  if ( Nint < MAX )
   {
   printf
    ("\nSpence(%.9E) = %.9E",x,Spence(x,Nint));
   }

  else
   {
   printf
    ("\n!!Integration points=%i >= MAX=%i",Nint,MAX);
   }

  }
 } /* end   theta  loop */
printf("\nEnd  Spence Integral");
}
```

Input argument, *x*, and number of integration points `Nint`.

Check that number is not too large.

Make and print Spence integral.

Increase array size, `MAX`, then recompile and run.

21.19.5 Clausen Integral

The C driver program for the Clausen integral in Section 19.5 requires the argument, θ, and the maximum number of coefficient values, KMAX, to be input.

C Function and Driver; Test Values

Clausen in driver Clausen.c; Table 19.5.1

```
#include <stdio.h>
#include <math.h>

#define KMAX 21

main()
{
/* Driver program for Clausen
   -- Clausen integral by
   Bernoulli series to accuracy  eps */

double theta,eps,
Clausen(double theta, double eps, int first);
int first;

printf("Clausen Integral");
first = 1;
theta = 1.0;

while ( theta >= 0.0 )
  {
  printf("\n\nInput theta,eps (theta<0 to end): ");
  scanf("%lf%lf",&theta,&eps);
```

Maximum number of coefficient points is KMAX–1.

Set flag for first time through `Clausen`.

Input argument θ and accuracy ε.

```
    if ( theta >= 0.0 )
      {
      printf("\nClausen(%10.8lf) = %.9E",
        theta,Clausen(theta,eps,first));
      first = 0;
      }
    } /* end   theta  loop */
  printf("\nEnd Clausen Integral");
  }
```

Clausen continued

Make and print the
Clausen integral to
10 digits.

21.19.6 Voigt (Plasma Dispersion) Function

The C driver program for the Voigt function in Section 19.6 needs the arguments u and a.

C Function and Driver; Test Values

Voigt in driver Voigt.c; Table 19.6.1

```
#include <stdio.h>
#include <math.h>

main()
{
/* Driver program for Voigt
   -- Voigt function by series expansion*/

double u,a,Voigt(double u,double a);

printf("Voigt Functions\n");
a = 1.0;
while ( a >= 0 )
  {
  printf("\n\nInput u,a (a<0 to end): ");
  scanf("%lf%lf",&u,&a);
  if ( a >= 0 )
    {
    printf
      ("\nVoigt(%lf,%lf) = %.9E",u,a,Voigt(u,a));
    }
  } /* end  a  loop */
printf("\nEnd Voigt Function");
}
```

Input arguments u
and a.

Voigt function with
accuracy as in Table
19.6.1.

21.19.7 Angular Momentum Coupling Coefficients

The C driver programs for the 3-*j*, 6-*j*, and 9-*j* coefficients in Section 19.7 are listed in that section with the C functions, since the drivers illustrate how the three, six, or nine arguments can be checked in the driver program before the coefficient function is invoked.

3-*j* Coefficients
C Function and Driver; Test Values

_3j in driver _3j.c; Table 19.7.1

6-*j* Coefficients
C Function and Driver; Test Values
_6j in driver _6j.c; Tables 19.7.2, 19.7.3

9-*j* Coefficients
C Function and Driver; Test Values
_9j in driver _9j.c; Test Values in Section 19.7.3

APPENDIX

FILE NAMES FOR PC-BASED SYSTEMS

In order to maintain compatibility with file names for those computer systems that restrict the length of components of file names to be at most 8 characters, folders ACMFwin.m and ACMFwin.c are provided on the CD-ROM that comes with the *Atlas*.

The correspondence between the long file names given in the main text and the short file names for each *Mathematica* notebook and C function is given in the following two tables.

TABLE A.1 Long and short file names for *Mathematica*.

Long file name	Short file name	Long file name	Short file name
BesselAiry.m	BslAry.m	LegendreCoords.m	LgndrCrd.m
BesselHalf.m	BslHlf.m	Miscellaneous.m	Msclns.m
BesselInteger.m	BslIntgr.m	NPDF.m	NPDF.m
Combinatorial.m	Cmbntrl.m	NumberTheory.m	NmbrThry.m
DPDF.m	DPDF.m	ocPDF.m	ocPDF.m
ElemTF.m	ElemTF.m	OrthoP.m	OrthP.m
ExpInt.m	ExpInt.m	OrthoPIter.m	OrthPItr.m
ErfFD.m	ErfFD.m	ParCyl.m	ParCyl.m
Elliptics.m	Ellptcs.m	StruveAngerWeber.m	StrAngWb.m
GmBt.m	GmBt.m	SWF.m	SWF.m
Hyprgmtrc.m	Hprgmtrc.m	SWFCoords.m	SWFCrds.m
Legendre.m	Lgndr.m		

TABLE A.2   Long and short file names for C.

Long file name	Short file name	Long file name	Short file name
_3j.c	_3j.c	Exp&Log.c	Exp&Log.c
_6j.c	_6j.c	ExpDist.c	ExpDst.c
_9j.c	_9j.c	ExpInts.c	ExpInts.c
Abrmwtz.c	Abrmwtz.c	Factorial.c	Fctorial.c
AiryAiBi.c	AiryAB.c	FDist.c	FDst.c
AiryPrimeAiBi.c	AiryPrAB.c	Fibonacci.c	Fbonacci.c
AngerWeber.c	AngrWbr.c	Fresnel.c	Fresnel.c
Bernoulli.c	Brnoulli.c	GaussDist.c	GaussDst.c
BerBei.c	BerBei.c	Gegenbauer.c	Ggenbr.c
BessIN.c	BessIN.c	GmtrcDist.c	GmtrcDst.c
BessJN.c	BessJN.c	Hermite.c	Hermite.c
BessKN.c	BessKN.c	HeumanLambda.c	HmnLmbda.c
BessYN.c	BessYN.c	HypCosSinTan.c	HypCST.c
BetaDist.c	BetaDist.c	Hyper1F1.c	Hyper1F1.c
BetaFunc.c	BetaFunc.c	Hyper2F1.c	Hyper2F1.c
Binomial.c	Binmial.c	HyperU.c	HyperU.c
BivNormDist.c	BvNrmDst.c	HypGmtrcDist.c	HypGmDst.c
BnmDist.c	BnmDst.c	IncBeta,c	IncBeta,c
CauchyDist.c	CchyDst.c	IncGamma.c	IncGamma.c
ChebyT.c	ChebyT.c	InvCosSinTan.c	InCST.c
ChebyU.c	ChebyU.c	InvCSTHyp.c	InCSTHyp.c
ChiSqDist.c	ChiSqDst.c	Jacobi.c	Jacobi.c
Clausen.c	Clausen.c	JacobiSnCnDn.c	JcobiSCD.c
ConicalP.c	ConicalP.c	JacobiTheta.c	JcobiTht.c
Cos&SinIntegral.c	CSIntgrl.c	JacobiZeta.c	JcobiZt.c
CosSinTan.c	CsSnTn.c	KerKei.c	KerKei.c
CoulombFG.c	ClmbFG.c	KSDist.c	KSDst.c
Dawson.c	Dawson.c	Laguerre.c	Laguerre.c
Debye.c	Debye.c	LaplaceDist.c	LplacDst.c
EllptcEE.c	EllptcEE.c	Legendre.c	Legendre.c
EllptcFK.c	EllptcFK.c	LgstcDist.c	LgstcDst.c
EllptcPi.c	EllptcPi.c	LogDerTheta.c	LgDerTht.c
ErfReal.c	ErfReal.c	LogGamma.c	LgGamma.c
Euler.c	Euler.c	LogInt.c	LgInt.c

Long file name	Short file name	Long file name	Short file name
LogNDist.c	LgNDst.c	Sievert.c	Sievert.c
LogSeriesDist.c	LgSrsDst.c	SMN.c	SMN.c
Lucas.c	Lucas.c	Spence.c	Spence.c
Multinomial.c	Multnoml.c	SphereBessI.c	SphrBssI.c
NegBnmDist.c	NgBnmDst.c	SphereBessJ.c	SphrBssJ.c
ParCylU.c	ParCylU.c	SphereBessK.c	SphrBssK.c
ParCylV.c	ParCylV.c	SphereBessY.c	SphrBssY.c
ParetoDist.c	ParetDst.c	Stirling.c	Stirling.c
PNM.c	PNM.c	StruveH.c	StruveH.c
PoissonDist.c	PssnDst.c	StruveL.c	StruveL.c
PolyGam.c	PolyGam.c	StudTDist.c	StudTDst.c
PolyLog.c	PolyLog.c	SumRecPwrs.c	SmRcPwrs.c
Psi.c	Psi.c	SWFevalue.c	SWFeval.c
QN.c	QN.c	ToroidalP.c	TroidalP.c
QNM.c	QNM.c	ToroidalQ.c	TroidalQ.c
RiemannZeta.c	RmannZt.c	WbllDist.c	WbllDst.c
RsngFctrl.c	RsngFctr.c	Voigt.c	Voigt.c

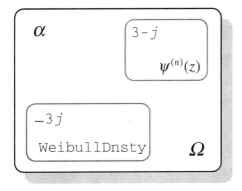

INDEXES

INDEX OF FUNCTION NOTATIONS

Roman Letters

Greek Letters

$\boxed{\alpha}$
$\boxed{\Omega}$

INDEX OF PROGRAMS AND DEPENDENCIES

The following index lists alphabetically all functions in the *Atlas*, plus any other functions they may use. The following functions are not listed; the *Atlas* driver programs (Chapter 21), the ANSI/ISO standard mathematical function library `math.h`, and the ANSI/ISO standard input/output library `stdio.h`.

   Note that this index serves only as a checklist, since the source programs for the driver and all its functions come fully assembled on the CD-ROM accompanying the *Atlas*. See the discussion in The C Functions: No Assembly Required in the Computer Interface section of the Introduction.

   The nomenclature for this index can be understood from the following example:

   `StruveH (BessY01 (BessJN (PP, QQ)), HHSeries)`   15.1.1

This notation indicates that the `StruveH` function that computes Struve's **H** function in Section 15.1.1 uses functions `BessY01` and `HHSeries`. In turn, `BessY01` uses the function `BessJN`, which uses `PP` and `QQ`.

INDEX OF SUBJECTS AND AUTHORS

Several authors of mathematical handbooks and other aids to computation are cited very frequently in the *Atlas*. They are therefore listed here, but not in the following index.

Abramowitz, M., and I. A. Stegun, *Handbook of Mathematical Functions,* Dover, New York, 1964 [Abr64]

Erdélyi, A., et al, *Higher Transcendental Functions*, Vol. 1, McGraw-Hill, New York, 1953; reprint edition, Krieger, Malabar, Florida, 1981 [Erd53a]

Erdélyi, A., et al, *Higher Transcendental Functions*, Vol. 2, McGraw-Hill, New York, 1953; reprint edition, Krieger, Malabar, Florida, 1981 [Erd53b]

Feller, W., *An Introduction to Probability Theory and Its Applications*, Wiley, New York, third edition, 1968 [Fel68]

Flammer, C., *Spheroidal Wave Functions*, Stanford University Press, Stanford, 1957 [Fla57]

Gradshteyn, I. S., and I. M. Ryzhik, *Table of Integrals, Series, and Products*, Academic Press, Orlando, fifth edition, 1994 [Gra94]

Jahnke, E., and F. Emde, *Tables of Functions*, Dover, New York, fourth edition, 1945 [Jah45]

Johnson, N. L., and S. Kotz, *Distributions in Statistics: Continuous Univariate Distributions-2*, Houghton Mifflin, Boston, 1970 [Joh70]

Johnson, N. L., S. Kotz, and A. W. Kemp, *Univariate Discrete Distributions*, Wiley, New York, second edition, 1992 [Joh92]

Johnson, N. L., S. Kotz, and N. Balakrishnan, *Distributions in Statistics: Continuous Univariate Distributions-1*, Wiley, New York, second edition, 1994 [Joh94a]

Johnson, N. L., S. Kotz, and N. Balakrishnan, *Distributions in Statistics: Continuous Univariate Distributions-2*, Wiley, New York, second edition, 1994 [Joh94b]

Press, W. H., S. A. Teukolsky, W. T. Vetterling, and B. P. Flannery, *Numerical Recipes*, Cambridge University Press, New York, second edition, 1992 [Pre92]

Spanier, J., and K. B. Oldham, *An Atlas of Functions*, Hemisphere Publishing, Washington, D.C., 1987 [Spa87]

Thompson, W. J., *Computing for Scientists and Engineers*, Wiley, New York, 1992 [Tho92]

Watson, G. N., *A Treatise on the Theory of Bessel Functions*, Cambridge University Press, New York, second edition, 1944 [Wat44]

Whittaker, E. T., and G. N. Watson, *A Course of Modern Analysis*, Cambridge University Press, Cambridge, England, fourth edition, 1927 [Whi27]

Wolfram, S., *Mathematica: A System for Doing Mathematics by Computer*, Addison-Wesley, Redwood City, California, second edition, 1991 [Wol91]